The Enzymes

VOLUME XVII

CONTROL BY PHOSPHORYLATION
Part A

General Features
Specific Enzymes (I)

Third Edition

THE ENZYMES

Edited by

Paul D. Boyer

Department of Chemistry and Biochemistry
and Molecular Biology Institute
University of California
Los Angeles, California

Edwin G. Krebs

Howard Hughes Medical Institute
and Department of Pharmacology
University of Washington
Seattle, Washington

Volume XVII

CONTROL BY PHOSPHORYLATION
Part A

General Features
Specific Enzymes (I)

THIRD EDITION

1986

ACADEMIC PRESS, INC.

Harcourt Brace Jovanovich, Publishers
Orlando   San Diego   New York   Austin
Boston   London   Sydney   Tokyo   Toronto

ACADEMIC PRESS, INC.
Orlando, Florida 32887

United Kingdom Edition published by
ACADEMIC PRESS INC. (LONDON) LTD.
24–28 Oval Road, London NW1 7DX

Library of Congress Cataloging in Publication Data
(Revised for vol.: 17, pt. A)

The Enzymes.

   Includes bibliographical references.
   1. Enzymes—Collected works.   I. Boyer, Paul D., ed.
[DNLM: 1. Enzymes. QU 135 B791e]
QP601.E523   574.19'25   75-117107
ISBN 0—12—122717—0

PRINTED IN THE UNITED STATES OF AMERICA

86 87 88 89   9 8 7 6 5 4 3 2 1

Contents

8. Phosphoprotein Phosphatases

LISA M. BALLOU AND EDMOND H. FISCHER

Section II. Control of Specific Enzymes

9. Glycogen Phosphorylase

NEIL B. MADSEN

10. Phosphorylase Kinase

CHERYL A. PICKETT-GIES AND DONAL A. WALSH

11. Muscle Glycogen Synthase

PHILIP COHEN

12. Liver Glycogen Synthase

PETER J. ROACH

Preface

Over the past two decades there has been a remarkable increase in the recognition of the salient importance and the wondrous complexity of the control of enzyme catalysis. The modulation of enzymic and other protein-dependent processes by protein phosphorylation or dephosphorylation has emerged as the most widespread and important control achieved by covalent modification. So much information has emerged that adequate coverage in two volumes (XVII and XVIII) was a challenging task. The editors are gratified that the contributing authors have commendably met this challenge.

The first portion of Volumes XVII and XVIII concerns the "machinery" of control by protein phosphorylation and dephosphorylation and includes coverage of the major types of protein kinases and of phosphoprotein phosphatases. The central core of the volumes presents chapters on the control of specific enzymes. This is followed by a substantial final section on the control of biological processes.

The selection of authors for various chapters was a rewarding experience, but made somewhat difficult because for most topics there was more than one well-qualified potential author. The quality of the volumes was assured by the welcome acceptance of the invitation to participate by nearly all of the invited authors.

The reversible covalent modification of enzymes and of proteins with other functions is now known to occur in all types of cells and in virtually all cellular compartments and organelles. Enzymes as a group constitute those proteins whose function and control are best understood in molecular terms. The treatment of enzymes gains additional importance because their regulation provides prototypic examples to guide investigators studying less well defined and often less abundant proteins. The versatility of protein control by phosphorylation finds expression in ion channels, hormone receptors, protein synthesis, contractile processes, and brain function. Chapters in these areas point the way for future exciting developments.

Although the breadth of coverage is in general regarded as satisfying, there are other topics or areas that may have warranted inclusion. These include the developing knowledge of the control by phosphorylation of histones of the nucleus and the messenger-independent casein kinases, whose role is not as clear as that of the major protein kinases that respond to regulatory agents.

The quality of the volumes has been crucially dependent on the editorial assistance of Lyda Boyer and the fine cooperation provided by the staff of Academic Press. We record our thanks here.

As readers of this Preface have likely discerned, it is a pleasure for the editors to have volumes of high quality to present to the profession.

Paul D. Boyer
Edwin G. Krebs

Section I

Enzymology of Control by Phosphorylation

1

The Enzymology of Control by Phosphorylation

EDWIN G. KREBS

Howard Hughes Medical Institute and
Department of Pharmacology
University of Washington
Seattle, Washington 98195

I. Historical Aspects of Protein Phosphorylation

Of the many types of posttranslational modification of proteins that occur in cells relatively few are readily reversible. Those that are include acetylation,

THE ENZYMES, Vol. XVII

methylation, adenylylation, uridylylation, phosphorylation, and possibly one or two others such as ADP ribosylation. By all odds the most common type of reversible protein modification is phosphorylation, the process with which this volume is concerned. As has become abundantly clear nature has chosen phosphorylation–dephosphorylation as an almost universal mechanism for regulating the function of proteins, not only those that display enzymic activity but also proteins involved in many other biological processes.

General recognition that protein phosphorylation has a major role in regulating protein function developed over a protracted period of time rather than being appreciated immediately as happened for allosteric regulation after the revelations by Monod *et al.* (*1*) in the mid-1960s. The first dynamic protein phosphorylation–dephosphorylation system to be elucidated was that involving glycogen phosphorylase, an enzyme that had been known to exist in two interconvertible forms, phosphorylases *b* and *a* (*2, 3*). In the mid-1950s these forms were shown to be nonphosphorylated and phosphorylated species of the eyzyme, which could be interconverted through the action of a protein kinase and a phosphoprotein phosphatase (*4, 5*). In 1959 evidence was obtained that the kinase involved in this process, phosphorylase kinase, was itself regulated by phosphorylation–dephosphorylation (*6*). A few years later it was determined that glycogen synthase also exists in interconvertible phosphorylated and nonphosphorylated forms (*7*).

The fact that the first three phosphorylatable enzymes to be discovered all involved glycogen metabolism suggested to some that the process might be restricted to this area, but this idea was soon abandoned as further reports began to appear implicating enzymes that act in other pathways (*8, 9*). In particular, the studies by Lester Reed and his associates (*10, 11*) showing that pyruvate dehydrogenase is regulated by phosphorylation–dephosphorylation served to broaden people's perspective regarding the scope of the process. At about this same time evidence was obtained that the enzyme that catalyzed the phosphorylation and activation of phosphorylase kinase was a cyclic AMP-dependent protein kinase that could catalyze the phosphorylation of other proteins, thus making it a probable mediator of the diverse actions of cyclic AMP (*12, 13*). These events served to stimulate general interest in the process of protein phosphorylation, and during the next fifteen years numerous enzymes were found to be regulated in this manner. In addition, many important functional proteins other than enzymes were shown to undergo reversible phosphorylation and to be controlled by this process. More detailed accounts of early work on protein phosphorylation have been reviewed elsewhere (*14, 15*).

II. Protein Phosphorylation–Dephosphorylation Reactions

Phosphoproteins are formed in cells through the action of protein kinases, and they also occur as transient intermediates formed as part of the mechanism of action of certain enzymes and transport proteins [reviewed in Ref. (16)]. It is with the former type of protein phosphorylation that the various chapters in this volume are concerned.

A. GENERAL PROPERTIES

The reactions of protein phosphorylation–dephosphorylation involving protein kinases and phosphoprotein phosphatases are shown in Eqs. (1) and (2).

$$\text{Protein} + n\text{NTP} \xrightarrow{\text{Protein kinase}} \text{Protein-P}_n + n\text{NDP} \tag{1}$$

$$\text{Protein-P}_n + n\text{H}_2\text{O} \xrightarrow{\text{Phosphoprotein phosphatase}} \text{Protein} + n\text{P}_i \tag{2}$$

As is implied in Eq. (1), a given protein kinase often catalyzes the transfer of phosphate to more than a single site in its protein substrate. In this sense, glycogen phosphorylase, the first enzyme shown to undergo phosphorylation–dephosphorylation and the prototype for many studies in this field, is unusual inasmuch as it is phosphorylated at a single site.

In protein phosphorylation reactions it is common for a single protein to serve as the substrate for more than one kinase. In some instances different kinases catalyze the phosphorylation of identical sites in the protein substrate, but more commonly each kinase has its own site "preference" since, as discussed in Section II,C,2, protein kinases exhibit a moderately high degree of specificity for particular amino acid sequences surrounding phosphorylatable amino acid residues. The phosphorylation of an identical site by two or more protein kinases probably occurs due to a high degree of exposure of that site combined with the fact that protein kinases do not exhibit absolute specificities. The most intensively studied set of protein kinase reactions involving a single substrate are those that occur with glycogen synthase, which can be phosphorylated in vitro by no less than ten different protein kinases. It should be noted, however, that all of the kinases capable of phosphorylating glycogen synthase in vitro may not function in this manner physiologically (see Chapters 11 and 12 in this volume).

As with essentially all phosphotransferase reactions, protein kinase reactions require divalent metal ions, Mg^{2+} probably serving as the physiologically significant cation in all instances, although manganous ions are nearly always effective in vitro. The actual substrate for the reaction depicted in Eq. (1) is the $NTP–Me^{2+}$ complex. In some instances, however, a role for divalent ions in addition to forming the metal–nucleotide complex has been noted. This is seen, for example, with the cyclic AMP-dependent protein kinase, which binds free metal ions at a separate site (17, 18).

The protein kinases constitute a very diverse set of enzymes, the total number of which is only now beginning to be appreciated. These enzymes are commonly regulated through their interaction with "second messengers" generated within cells in response to hormones and other extracellular agents (see below). The reactions catalyzed by the kinases are also regulated through the interaction of metabolites with their substrates, that is, through "substrate level" control (see, for example, Chapter 9, this volume and Chapter 2, Volume XVIII).

The phosphoprotein phosphatases appear to be smaller in total number than the protein kinases, and they probably exhibit broader specificities. This in itself implies that these enzymes may not be as actively involved in regulating phosphorylation–dephosphorylation cycles as are the kinases, since the broader the specificity the greater the number of pathways that would be affected simultaneously by the regulation of a given enzyme. A more passive regulatory role for the phosphatases as compared to the kinases is also suggested by the fact that investigators have not discovered as many diverse forms of regulation for this set of enzymes. With respect to the last point, however, it is possible that the phosphatases have simply been more difficult to characterize than the kinases, and additional regulatory mechanisms may be found in the future. These matters are considered more fully in Chapter 8.

B.   SPECIFICITY FOR PHOSPHORYL DONORS IN PROTEIN KINASE REACTIONS

The protein kinase reaction (Eq. 1) is written as being dependent on a nucleoside triphosphate. The physiologically significant donor in nearly all instances is probably ATP (*19*), although several protein kinases can also use GTP effectively *in vitro*. The last category includes casein kinase II [reviewed in Ref. (*20*)], a histone H1 kinase from tumor cells (*21*) and pp60^{src} (*22*). An unusual protein kinase from rabbit skeletal muscle, which utilizes phosphoenolpyruvate as a phosphate donor was reported by Khandelwal *et al.* (*23*). The kinase studied by this group was found to be *activated* by CTP (*24*). Phosphoenolpyruvate has long been known to serve as the phosphoryl donor in the phosphorylation of the enzyme involved in the first step of sugar transport in bacteria [reviewed in Ref. (*25*)], but this process belongs in a category set apart from typical protein kinase reactions as previously indicated.

C.   PROTEIN SUBSTRATE SPECIFICITY OF PROTEIN KINASES AND PHOSPHOPROTEIN PHOSPHATASES

1.   *Amino Acid Residues That Serve as Acceptors of Phosphoryl Groups*

Most of the acid-stable protein-bound phosphate found in cells is present as phosphoserine and phosphothreonine and is formed as a result of the action of protein serine and threonine kinases. A much smaller fraction, usually less than 0.2% of the total, is present as phosphotyrosine and arises as a result of the action of protein tyrosine kinases [reviewed in Ref. (*26*)]. The possible existence of a hydroxylysine kinase has also been indicated (27). In addition to protein kinases that catalyze the previously mentioned phosphorylations, there are indications that protein kinases exist that can catalyze the formation of acid-labile phosphohistidine and phospholysine in proteins (28–30). No work has been done on phosphatases that reverse the action of this last set of kinases, so it is not known whether such phosphorylations represent reversible protein modifications.

2. Structural Determinants of Specificity

The phosphorylation site sequences in various protein substrates for a number of different protein kinases have been determined and investigators using synthetic peptides as substrates have also elucidated a number of the structural requirements for substrate specificity of protein kinases. It has become clear that although the primary structure of a phosphorylation site sequence does not tell the whole story, it is usually possible to distinguish differences between the site specificities of the various kinases. For the cAMP-dependent protein kinase, for example, it can be predicted that if a given protein has an exposed -Arg-Arg-X-Ser-X- sequence, it will be phosphorylated by this enzyme. Similarly, casein kinase II phosphorylates exposed serine (or threonine) residues followed by one or more glutamic acids residues one position removed from the serine (e.g., -X-X-Ser-X-Glu-). The entire specificity pattern of a protein kinase is not revealed, however, simply by knowledge of its preferred primary amino acid sequence. Higher orders of protein structure also play a part in determining whether a given protein can be phosphorylated at an appreciable rate by a given kinase. These considerations are discussed in chapters that refer to specific kinases.

The specificity of the phosphoprotein phosphatases is not as well understood as that of the kinases. One of the problems in studying this set of enzymes has been that earlier investigators were never certain whether their preparations contained more than one enzyme, and conclusions regarding specificity were obviously difficult to reach. Later studies, however, and particularly in the laboratory of Dr. Philip Cohen, made considerable headway in defining and classifying these enzymes. The relative activities of different phosphatases toward a number of different phosphoprotein substrates have been determined. Several general observations can be made: First, there is probably more overlapping of specificity among the phosphatases than is seen for the kinases. Second, it is evident that the pattern of specificity is *not* one in which a given phosphatase is "designed" to dephosphorylate a set of proteins phosphorylated by a given kinase. Thus, although the kinases recognize specific amino acid sequences in their substrates, this is not so readily apparent for the phosphatases. This suggests that higher orders of protein structure may be more important in governing substrate specificity for the phosphatases than for the kinases. If this is so, then one might anticipate that regulation of phosphatase activity might commonly occur at the "substrate level" (i.e., through the interaction of metabolites with the phosphoproteins involved). Through such interactions and the ensuing conformational changes phosphatase activities could be regulated in a highly specific manner. Finally, it is apparent that some phosphatases exist that may attack low-molecular-weight substrates as well as proteins under physiological conditions. Synthetic phosphopeptides have been employed only rarely in studies on phosphoprotein phosphatases. A pioneering effort in this regard was the work of Titanji *et al.* (*31*).

D. THE REVERSIBILITY OF PROTEIN KINASE REACTIONS

The phosphorylation reactions that result in the formation of phosphoserine and phosphotyrosine in proteins, and presumably those that result in phosphothreonine, can all be demonstrated to be reversible albeit not under physiological conditions. Rabinowitz and Lipmann (*32*) were the first to describe the reversibility of protein phosphorylation. Using phosphorylated phosvitin as a phosphate donor, they demonstrated the formation of ATP from ADP in the presence of phosvitin kinase (probably casein kinase II) from yeast or brain and postulated that runs of adjacent phosphoserines in phosvitin may have accounted for the apparent high free energy of hydrolysis of the bound phosphate in this protein. That such structural relationships are not essential for reversibility of protein kinase reactions was brought out by the work of Shizuta *et al.* (*33*), using the cyclic AMP-dependent protein kinase as enzyme and substrates in which no adjacent phosphoserines are present. These workers determined the equilibrium position in a well-defined protein kinase reaction and calculated that the free energy of hydrolysis of serine phosphate in this substrate (^{32}P-labeled reduced carboxymethylated maleylated lysozyme) was -6.5 kcal mol^{-1}. The reversibility of a number of other protein serine kinase reactions has been reported (*34–37*). Fukami and Lipmann (*38*) described reversibility of Rous sarcoma-specific immunoglobulin phosphorylation catalyzed by the *src* gene kinase and reported a free energy of hydrolysis of -9.48 kcal mol^{-1} for protein-bound tyrosine phosphate, appreciably higher than the value of -6.5 kcal reported for protein serine phosphate by Shizuta *et al.* (*33*). However, the latter workers assumed a different $\Delta G^{\circ\prime}$ for hydrolysis of ATP (-8.4 kcal mol^{-1}) than that used by Fukami and Lipmann (-10 kcal mol^{-1}). When the same value is used, a free energy of hydrolysis of -8.1 kcal mol^{-1} is obtained for protein serine phosphate (i.e., only slightly lower than that for tyrosine phosphate). A standard free energy of hydrolysis for serine phosphate in pyruvate kinase was found to be -6.6 kcal mol^{-1} by El-Maghrabi *et al.* (*37*). These workers used -8.4 kcal mol^{-1} for the free energy of hydrolysis of ATP.

Very little has been made of the possible physiological significance of the reversal of protein kinase reactions, but it is possible that under some circumstances the process could be important. The fact that protein-bound serine phosphate and tyrosine phosphate is relatively "energy rich" would indicate that a significant rate of turnover of this phosphate could occur in the absence of protein phosphatase activity. Potentially, the phosphate could be transferred to receptors other than proteins. In connection with the possible physiological significance of the reversal of protein kinase reactions, it is of interest that reversal of the phosphorylase kinase reaction required the presence of glucose or glycogen demonstrating that the reaction might be subject to regulation (36).

E. AUTOPHOSPHORYLATION REACTIONS

Almost all protein kinases catalyze autophosphorylation reactions (i.e., reactions in which the kinase serves as its own substrate). Flockhart and Corbin (19) published a comprehensive list of the kinases that had been shown by the time of their review (1982) to exhibit this property. Their list could now be extended to include glycogen synthase kinase 3 (39), protein kinase C (40), the insulin receptor [reviewed in Ref. (41)] and a number of other oncogene-encoded protein tyrosine kinases (26).

Autophosphorylation reactions can be intramolecular or intermolecular. This aspect was first examined with respect to the autophosphorylation of phosphorylase kinase, in which case the latter mechanism was shown to prevail (42). The autophosphorylation reaction involving type II cyclic AMP-dependent protein kinase, however, was found to occur by an intramolecular process (43). An intramolecular reaction was also found to be involved in autophosphorylation of the cyclic GMP-dependent protein kinase (44) and the insulin receptor (see Chapter 11).

The significance of autophosphorylation reactions of protein kinases is readily apparent in some instances but not in all. With phosphorylase kinase autophosphorylation causes marked enhancement of activity, although there is no evidence that this is the physiologically significant process (see Chapter 10). Autophosphorylation of type II cyclic AMP-dependent protein kinase affects cyclic AMP binding and the dissociation–reassociation reactions of the enzyme in the presence of cyclic AMP [Ref. (45) and Chapter 3], an effect that may be of physiological significance, although again this has not been demonstrated. Of great interest is the fact that autophosphorylation of the insulin receptor renders this protein kinase independent of insulin (46). In some instances, particularly with those protein kinases in which autophosphorylation occurs by an intermolecular process and in which no functional change occurs in the activity of the kinase as a result of autophosphorylation, the reaction may simply be an expression of the lack of absolute specificity of protein kinases. In looking for autophosphorylation investigators often employ high concentrations of a kinase, and under these conditions autophosphorylation is observed even though the enzyme is actually an extremely poor substrate for itself.

A possible unifying concept with respect to the significance of the autophosphorylation reactions of protein kinase is that many of these enzymes may be kept in an inactive or inhibited state as a result of interaction between their autophosphorylation sites and their protein substrate binding sites. This phenomenon was first observed with type II cyclic AMP-dependent protein kinase. Here the autophosphorylation site on the regulation subunit (R_{II}), which possesses an amino acid sequence closely resembling a typical substrate sequence, is believed to interact with the active center of the catalytic subunit (see Chapter 3). Autophosphorylation of R_{II} renders it less potent as an inhibitor of R_{II} and facilitates activation by cyclic AMP. The latter activator, however, is capable of fully activating the enzyme with or without autophosphorylation. A similar, albeit not identical, situation appears to exist with the insulin receptor in which activation occurs as a result of an autophosphorylation reaction involving a site whose binding to the active center may well render the enzyme inactive in its dephospho form. The role of autophosphorylation in other protein tyrosine kinases is discussed in Chapters 6 and 7. For some protein kinases a functional autophosphorylation site may not exist (e.g., type I cyclic AMP-dependent protein kinase); instead, pseudoautophosphorylation sites are present. In these instances sites homologous to protein substrate phosphorylation sites are present but the sites lack phosphorylatable amino acid residues. For such enzymes activation is dependent on activators that alter conformation of the kinases and cause displacement of the pseudosubstrate sites from the active centers of these enzymes. Skeletal muscle myosin light chain kinase may be a second example of a protein kinase that is kept in its inactive form as a result of interaction of a pseudosubstrate site with the active center. This enzyme possesses what appears to be a site that shows strong homology to the phosphorylation site in myosin light chains. Interestingly, this site is close to or a part of the calmodulin binding site in this enzyme *(47, 48)*.

III. Classification of Protein Kinases and Phosphoprotein Phosphatases

A. PROTEIN KINASES

How many different protein kinases are there and how are they classified and named? The latter presents a difficult problem for a number of reasons. Classically, enzymes are grouped in accordance with the kind of chemical reaction that they catalyze and individual enzymes are then delineated and named on the basis of the specific substrate on which they act. This system breaks down for the protein kinases for several reasons. First, as previously indicated, a given protein kinase usually catalyzes the phosphorylation of a number of different proteins, so the selection of one particular substrate loses its meaning. Second, more often than not one protein can serve as a substrate for more than one kinase. At times, even the same phosphorylation site is involved with two or more kinases. Because of these complications, no systematic approach for naming this set of enzymes has been developed.

In practice, it appears that protein kinases can be divided initially into broad classes based on whether the amino acid acceptor of the phosphoryl residue is (1) a protein alcohol group (i.e., the protein serine and protein threonine kinases), (2) a protein phenolic group (i.e., the protein tyrosine kinases), or (3) a protein nitrogen-containing group (i.e., the protein, histidine and lysine kinases). The existence of this last class of protein kinases is strongly suggested by the work of Roberts Smith and his collaborators (28, 29) and the report by Huebner and Mathews (30). Within each of the main classes of kinases individual groups of enzymes or subclasses exist, which more often than not can be delineated on the basis of the regulation of their activities. Table I presents a possible classification and nomenclature scheme for the protein serine and threonine kinases and Table II a scheme for the protein tyrosine kinases. Some of the groups or subclasses of protein serine and threonine kinases have only one known member at this time, but this situation may change. With respect to the grouping of the protein serine and threonine kinases shown in Table I, Group 6 consists of the so-called "independent" protein kinases (i.e., those that are not definitely known to be regulated through the interaction of a messenger(s) directly with the kinase). Some of the kinases in Group 6 are regulated as a result of the interaction of metabolites with their substrates.

The individual protein serine and threonine kinases of Table I are not discussed here. The most intensively studied groups are the cyclic nucleotide-dependent and calcium–calmodulin-dependent enzymes. Each of these groups is the subject of a separate chapter in this volume (see Chapters 3 and 4). The protein kinase that is regulated by diacylglycerol in the presence of calcium ions and phospholipid, protein kinase C, is also discussed in a separate chapter (see Chapter 5). One of the calcium–calmodulin (CaM)-dependent protein kinases, phosphorylase kinase, is itself an enzyme subject to regulation and is treated in the set of specific enzymes regulated by phosphorylation–dephosphorylation (see Chapter 10). Many of the protein serine and threonine kinases of Table I are discussed in chapters concerned with specific regulatory systems in which they function. For example, the double-stranded RNA-dependent and the hemin-inhibited protein kinases are handled in the chapter on the regulation of protein synthesis. Pyruvate dehydrogenase kinase and branched-chain ketoacid dehydrogenase kinase are discussed in Volume XVIII, Chapters 3 and 4 on the regulation of pyruvate dehydrogenase and branched chain keto acid dehydrogenase, respectively. The last set of enzymes (i.e., the two mitochondrial dehydrogenases) are arbitrarily listed in Table I as independent kinases. They are both subject to regulation by metabolites that most logically would be thought to act by combin-

TABLE I

Classification and Nomenclature of Protein Kinases That Catalyze the Transfer of
Phosphate to Protein Alcohol Groups (Protein Serine and Threonine Kinases)

Group	Regulatory agent(s)	Specific enzyme
1	Cyclic nucleotides	Type I cyclic AMP-dependent protein kinases
		Type II (H) cyclic AMP-dependent protein kinase[a]
		Type II (B) cyclic AMP-dependent protein kinase[a]
		Cyclic GMP-dependent protein kinase
2	Ca^{2+} and calmodulin	Phosphorylase kinase (GSK 2)[b]
		Skeletal muscle myosin light chain kinase
		Smooth muscle myosin light chain kinase
		Multifunctional Ca^{2+}-calmodulin-dependent protein kinase
3	Diacylglycerol (Ca^{2+} and phospholipid)	Protein kinase C
4	Double-stranded RNA	Double-stranded RNA-dependent protein kinase
5	Hemin	Hemin-inhibited protein kinase
6	Independent of regulation except at substrate level	Pyruvate dehydrogenase kinase
		Branched-chain ketoacid dehydrogenase kinase
		Casein kinase I
		Casein kinase II
		GSK 3 (Factor F_A)[c]
		GSK 4
		Rhodopsin kinase
		HMG-CoA reductase kinase
		HMG-CoA reductase kinase kinase
		Histone HI kinase
		Histone kinase II
		H4-specific protease-activated protein kinase,
		Protease-activated kinase II

[a] "H" and "B" signify heart and brain isozymes.

[b] GSK stands for glycogen synthase kinase (see Chapter 11).

[c] Factor F_A, name applied to a phosphoprotein phosphatase-activating factor that was later found to be a glycogen synthase kinase (49).

ing with their respective protein substrates. It should be noted, however, that acetyl-CoA, NADH, ADP, and pyruvate have all been reported to affect the activity of pyruvate dehydrogenase kinase when it is acting on a small synthetic peptide substrate, suggesting that the kinase itself interacts with these metabolites (see Volume XVIII, Chapter 3). Thus, it is possible that pyruvate dehydrogenase should be handled differently than shown in Table I (49).

Some protein serine and threonine kinases have not been included in Table I because they have not been studied sufficiently to be certain of their status and relationship to other kinases. In this category the double-stranded DNA-depen-

TABLE II

CLASSIFICATION AND NOMENCLATURE OF PROTEIN KINASES
THAT CATALYZE THE TRANSFER OF PHOSPHATE
TO PROTEIN PHENOLIC GROUPS (PROTEIN TYROSINE KINASES)

Group	Regulatory agent(s)	Specific enzymes
1	Polypeptide hormones and growth factors	EGF receptor
		PDGF receptor
		Insulin receptor
		IGF-I receptor
2	Unknown	v-*src* and c-*src*-Encoded protein kinases
		v-*yes* and c-*yes*-Encoded protein kinases
		v-*abl* and c-*abl*-Encoded protein kinases
		v-*fgr* and c-*fgr*-Encoded protein kinases
		v-*fps* and c-*fps*-Encoded protein kinases
		v-*fes* and c-*fes*-Encoded protein kinases
		v-*ros* and c-*ros*-Encoded protein kinases

dent protein kinase (50) and the polypeptide-dependent protein kinase (51) could be mentioned as examples. These and many new independent kinases will probably be implicated as having specific regulatory roles in the future.

The protein tyrosine kinases have arbitrarily been divided into the two groups shown in Table II (i.e., the hormone or growth factor-dependent enzymes (see Chapter 7) and the independent kinases for which no regulatory agents are known (see Chapter 6). The latter set of kinases are encoded by certain retroviral oncogenes and their closely related cellular counterparts. Other independent protein tyrosine kinases not listed in Table II consist of the LSTRA kinase (52), a possibly related kinase from spleen (53), and TPK-75 from liver (54). The intriguing possibility that protein tyrosine kinases are regulated by intracellular second messengers is raised by reports suggesting that calcium–CaM and polyamines can stimulate tyrosine phosphorylation (55).

The protein serine and tyrosine kinases all contain homologous catalytic domains indicating that they are members of a single family. This homology is shown in Fig. 1 in which the sequence of a portion (residues 49–253) of the catalytic subunit of beef cyclic AMP-dependent protein kinase (56) is used as a basis for calculating alignment scores (57) for comparing this sequence with the catalytic domains of other protein kinases. Comparisons are made with beef lung cyclic GMP-dependent protein kinase (58), the γ-subunit of rabbit skeletal muscle phosphorylase kinase (59), rabbit skeletal muscle myosin light chain kinase (60), P90$^{gag-raf}$ (61), pp60^{src} (62), the EGF receptor (63), and the insulin receptor (64). As can be seen there is a highly significant degree of homology for the catalytic domains in all of these protein kinases. As might be expected the

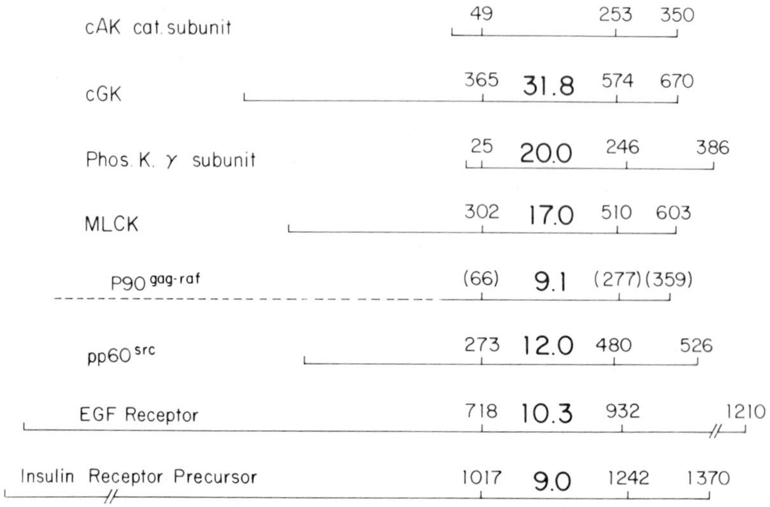

FIG. 1. Amino acid sequence homology between the catalytic domains of representative protein kinases. Each protein is represented by a horizontal bar, the length of which is proportional to the molecular weight of the protein as determined by its amino acid sequence (except for a portion of P90$^{gag-raf}$). The small numbers indicate residue numbers. The large numbers indicate Barker–Dayhoff alignment scores (57) of the catalytic domains to the various protein kinases compared to the segment between residues 49 and 253 in the catalytic subunit of the cyclic AMP-dependent protein kinase (cAK cat. subunit). cGK = cyclic GMP-dependent protein kinase; Phos. K = phosphorylase kinase; MLCK = myosin light chain kinase; P90$^{gag-raf}$ = chimeric protein containing raf oncogene-encoded sequence and the gag sequence; pp60^{src} = src-encoded protein tyrosine kinase.

degree of homology between the catalytic subunit of the cyclic AMP-dependent protein kinase and other protein serine and threonine kinases is higher than it is with the protein tyrosine kinase, except that this is not seen with respect to the oncogene-encoded protein P90$^{gag-raf}$, which is thought to be a protein serine kinase (see Chapter 6). Phosphotransferases other than protein kinases have not been found to contain catalytic domains homologous to the catalytic subunit of the cyclic AMP-dependent protein kinase.

B. PHOSPHOPROTEIN PHOSPHATASES

Early work on the phosphoprotein phosphatases was insufficient to make it feasible to classify or name them in a systematic manner. Impure preparations were often characterized and treated as single entities, when in fact more than one enzyme was present. What were thought of as different phosphatases often turned out to be dissociation products of more complex holoenzymes. In some instances "new" phosphatases were later recognized to be proteolytic artifacts.

Fortunately, these enzymes that have been so recalcitrant to experimentation in the past are beginning to reveal their secrets. They are treated as a group in Chapter 8 and no attempt is made to classify them here.

IV. Protein Phosphorylation and the Regulation of Biological Processes

A. PHOSPHORYLATION–DEPHOSPHORYLATION VERSUS OTHER CONTROL MECHANISMS

Given the many different mechanisms that are available for regulating biological functions what is special about protein phosphorylation? Does this process have a unique role that cannot be carried out by other mechanisms? The last question is particularly pertinent when one considers the enormous potential for regulation through allosteric control involving the reversible binding of ligands to proteins. In fact, examination of the mechanisms for regulation of many of the specific enzymes described in this volume reveals that the kinetic parameters that are altered by ligand binding are usually the same as those changed by phosphorylation. Why attach a ligand (phosphate) covalently, when the reversible binding of metabolite can apparently accomplish the same purpose? (See Chapter 2, however, for a discussion of signal amplification by means of allosteric control versus cyclic covalent modification casades.) Reference to the role of protein phosphorylation in the regulation of glycogen phosphorylase may help to provide an answer to this question.

In Chapter 9 Neil Madsen notes that conversion of phosphorylase b to phosphorylase a allows phosphorylase "to *escape* from the allosteric controls" to which it is subject when present in its b form; that is, after phosphorylation phosphorylase becomes fully active in the absence of $5'$-AMP and is no longer subject to inhibition by ATP or glucose 6-P. The implication is that there may be an advantage in having a system in which under certain circumstances an enzyme can be "frozen" in its active (or inactive) configuration regardless of what happens to metabolite levels. The same line of thinking may also apply to the purpose of a third form of regulation in which the *amount* of a given enzyme is altered through changes in its rate of biosynthesis or destruction. Dramatic changes in the concentration of an enzyme provides a means of superceding control by phosphorylation–dephosphorylation as well as allosteric control. Implicit in this form of reasoning is the often-voiced idea that allosteric control provides an immediate or almost instantaneous response to a given situation, control by protein phosphorylation provides an intermediate form of response on a time scale, and control through changes in the levels of particular proteins provides a slow but enduring response.

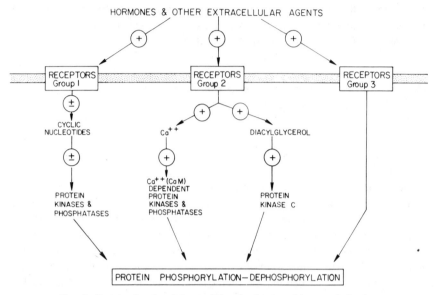

Fig. 2. Protein phosphorylation and the transduction of hormonal signals.

Regardless of these considerations, although not entirely unrelated, is the fact that protein phosphorylation as a process is geared to the handling of signals that have an extracellular origin (hormones, growth factors, etc.), whereas allosteric control is utilized primarily for the handling of signals that arise intracellularly. The major pathways involving protein phosphorylation as a mechanism for relaying hormonal signals are shown in Fig. 2. One group of hormones, arbitrarily designated as Group 1, act through their receptors to raise or lower cyclic nucleotide levels in cells and thus affect the state of phosphorylation of cyclic nucleotide-dependent protein kinase substrates. A second large group of hormones and other extracellular agents (Group 2) trigger the breakdown of phosphatidylinositol diphosphate generating diacylglycerol and inositol triphosphate, the latter causing elevated intracellular calcium ion levels. The increase in calcium ions causes activation of a set of Ca^{2+}–CaM-dependent protein kinases and one phosphoprotein phosphatase, leading to the modulation in the state of phosphorylation of protein substrates for these enzymes. The diacylglycerol formed as a result of the action of this group of hormones causes activation of protein kinase C, which again leads to protein phosphorylation. Finally, the third set of hormones (Group 3) act through receptors that possess protein kinase activity (i.e., protein tyrosine kinase activity). Although the nature of the protein substrates for this set of enzymes have not been fully elucidated (see Chapters 6 and 7) there appears to be little doubt that they constitute important phosphorylatable proteins involved in the action of these hormones. The same proteins affected by the transmembrane signaling mechanism portrayed in Fig. 2, are also often targets for allosteric control through their interactions with metabolites generated within the cell.

The concept that protein phosphorylation functions primarily to handle extracellular signals whereas allosteric control is utilized to handle intracellular signals is in keeping with the finding that eukaryotic cells contain much more phosphoprotein than is found in the prokaryotes (65, 66). This is not to say that protein phosphorylation is never used as a regulatory device in bacteria (see Volume XVIII, Chapter 14) but the extent to which it is employed is much less than in higher organisms in which appropriate response to extracellular signals is so important. On the other hand, allosteric control as a means of response to intracellular signals is highly developed in bacteria. It is of interest that whereas many protein kinases are regulated by second messengers generated in response to extracellular signals (Tables I and II and Fig. 2), relatively few of them are regulated by ordinary metabolites. The (at least partial) immunity of protein kinases to metabolite control may help to provide a mechanism whereby regulation by phosphorylation–dephosphorylation is segregated from allosteric control.

B. DIVERSITY OF PHOSPHORYLATION–DEPHOSPHORYLATION
   CONTROL MECHANISMS

The existence of phosphoproteins first became known as a result of their nutritional significance. Thus, the early work in this field was concerned with the caseins of milk and the various phosphoproteins of egg yolk which serve as a source of phosphorus and amino acids for developing organisms. The second set of proteins to receive attention as targets for phosphorylation–dephosphorylation were the enzymes, many of which are reviewed in this volume. Concomitantly with work on enzyme phosphorylation–dephosphorylation, however, investigators extended their studies of this process to include a host of nonenzymic proteins involved in many different cellular functions, some of which (e.g., ion movements, receptor activity, and muscle contraction) are also treated here. Protein phosphorylation reactions are important in modulating nuclear events within the cell as is evidenced by the extensive work on histone and protamine phosphorylation as well as studies on the phosphorylation of nonhistone proteins. The exact nature of the role of protein phosphorylation reactions within the nucleus remains elusive, however, inasmuch as the precise function of the proteins involved is not always clear. An unusual role for phosphoproteins is that seen for a highly phosphorylated dentine protein that appears to function in the calcification process in teeth (67).

Even under circumstances in which investigators are well aware of the physiological function of a particular protein known to be phosphorylated, they are not always able to find an effect of phosphorylation on the activity of these proteins.

Reactions in this category are often referred to euphemistically as "silent phosphorylations." A possible explanation for such phosphorylations is that they may be targeting particular proteins for destruction. This possibility is discussed in Volume XVIII Chapters 2 and 7 in relation to pyruvate kinase and HMG-CoA reductase respectively; for each of these proteins phosphorylation causes the formation of a configuration that leads to enhanced susceptibility to proteases. A related role for protein phosphorylatin would be the part that it may play in the processing of certain proteins by proteases (68, 69). Other possible explanations for silent phosphorylations should also be considered. Protein–protein interactions within the cell may be regulated by phosphorylation–dephosphorylation. The intracellular localization of proteins may be influenced by phosphorylation. These and many other protein functions are difficult to measure outside of the natural intracellular environment. Finally, with respect to silent phosphorylations, it is possible that some protein phosphorylation–dephosphorylations serve no particular purpose other than generating heat and simply constitute futile cycles.

REFERENCES

1. Monod, J., Changeux, J.-P., and Jacob, F. (1963). *JMB* **6**, 306–329.
2. Cori, G. T., and Green, A. A. (1943). *JBC* **151**, 31–38.
3. Cori, G. T., and Cori, C. F. (1945). *JBC* **158**, 321–332.
4. Fischer, E. H., and Krebs, E. G. (1955). *JBC* **216**, 121–132.
5. Sutherland, E. W., and Wosilait, W. D. (1955). *Nature (London)* **175**, 169–171.
6. Krebs, E. G., Graves, D. J., and Fischer, E. H. (1959). *JBC* **234**, 2867–2873.
7. Friedman, D. L., and Larner, J. (1963). *Biochemistry* **2**, 669–675.
8. Rizak, M. A. (1964). *JBC* **239**, 392–395.
9. Mendicino, J., Beaudreau, C., and Bhattachryya, R. N. (1966). *ABB* **116**, 436–445.
10. Linn, T. C., Pettit, F. H., Hucho, F., and Reed, L. J. (1969). *PNAS* **64**, 227–234.
11. Linn, T. C., Pettit, F. H., and Reed, L. J. (1969). *PNAS* **62**, 234–241.
12. Walsh, D. A., Perkins, J. P., and Krebs, E. G. (1968). *JBC* **243**, 3763–3765.
13. Kuo, J. F., and Greengard, P. (1969). *JBC* **244**, 3417–3419.
14. Krebs, E. G. (1983). *Philos. Trans. R. Soc. London, Ser. B* **302**, 3–11.
15. Krebs, E. G. (1985). *Trans. Biochem. Soc.* **13**, 813–820.
16. Krebs, E. G. (1972). *Curr. Top. Cell. Regul.* **5**, 99–133.
17. Armstrong, R. N., Kando, H., Geanot, J., Kaiser, E. T., and Mildvan, A. S. (1979). *Biochemistry* **18**, 1230–1238.
18. Hixson, C. S., and Krebs, E. G. (1979). *JBC* **254**, 7509–7514.
19. Flockhart, D. A., and Corbin, J. D. (1982). *CRC Crit. Rev. Biochem.* **13**, 133–186.
20. Hathaway, G. M., and Traugh, J. A. (1982). *Curr. Top. Cell. Regul.* **21**, 101–127.
21. Quirin-Steicker, C., and Schmitt, M. (1981). *EJB* **118**, 165–172.
22. Graziani, Y., Erikson, E., and Erikson, R. L. (1983). *JBC* **258**, 6344–6351.
23. Khandelwal, R. L., Mattoo, R. L., and Waygood, E. B. (1983). *FEBS Lett.* **162**, 127–132.
24. Mattoo, R. L., Waygood, E. B., and Khandelwal, R. L. (1984). *FEBS Lett.* **165**, 117–120.
25. Simoni, R. D., and Postma, P. W. (1975). *Annu. Rev. Biochem.* **44**, 523–554.
26. Hunter, T., and Cooper, J. A. (1985). *Annu. Rev. Biochem.* **54**, 897–930.

27. Urishizaki, Y., and Seifter, S. (1985). *Biophys. J.* **47,** 233a.
28. Smith, D. L., Chen, C. C., Bruegger, B. B., Holtz, S. L., Halpern, R. M., and Smith, R. A. (1974). *Biochemistry* **13,** 3780–3785.
29. Smith, R. A., Halpern, R. M., Bruegger, B. B., Dunlop, A. K., and Fricke, O. (1978). *Methods Cell Biol.* **14,** 153–159.
30. Huebner, V. D., and Mathews, H. R. (1985). *FP* **44,** Abstr. 3882.
31. Titanji, V. P. K., Ragnarsson, U., Humble, E., and Zetterquist, O. (1980). *JBC* **255,** 11339–11343.
32. Rabinowitz, M., and Lipmann, F. (1960). *JBC* **235,** 1043–1056.
33. Shizuta, Y., Beavo, J. A., Bechtel, P. J., Hofmann, F., and Krebs, E. G. (1975). *JBC* **250,** 6891–6896.
34. Lerch, K., Muir, L. W., and Fischer, E. H. (1975). *Biochemistry* **14,** 2015–2023.
35. Rosen, O. M., and Erlichman, J. (1975). *JBC* **250,** 7788–8894.
36. Shizuta, Y., Khandelwal, R. L., Maller, J. L., Vandenheede, J. R., and Krebs, E. G. (1977). *JBC* **252,** 3408–3413.
37. El-Maghrabi, M. R., Haston, W. S., Flockhart, D. A., Claus, T. H., and Pilkis, S. J. (1980). *JBC* **255,** 668–675.
38. Fukami, Y., and Lipmann, F. (1983). *PNAS* **80,** 1872–1876.
39. Hemmings, B. A., Yellowlees, D., Kernohan, J. C., and Cohen, P. (1981). *EJB* **119,** 443–451.
40. Kikkawa, U., Takai, Y., Minakuchi, R., Inohara, S., and Nishizuka, Y. (1982). *JBC* **257,** 13341–13348.
41. Kahn, C. R., White, M. F., Grigorescu, F., Takayama, S., Haring, H. U., and Crettaz, M. (1985). *In* "Molecular Basis of Insulin Action" (M. P. Czech, ed.), pp. 67–93. Plenum, New York.
42. DeLange, R. J., Kemp, R. G., Riley, W. D., Cooper, R. A., and Krebs, E. G. (1968). *JBC* **243,** 2200–2208.
43. Rangel-Aldao, R., and Rosen, O. M. (1976). *JBC* **251,** 7526–7529.
44. Lincoln, T. M., Flockhart, D. A., and Corbin, C. D. (1978). *JBC* **253,** 6002–6009.
45. Erlichman, J., Rosenfeld, R., and Rosen, O. M. (1974). *JBC* **249,** 5000–5003.
46. Rosen, O. M., Herrara, R., Olowe, Y., Petruzzeli, L. M., and Cobb, M. (1983). *PNAS* **80,** 3237–3240.
47. Blumenthal, D. K., Takio, K., Edelman, A. M., Charbonneau, H., Titani, K., Walsh, K. A., and Krebs, E. G. (1985). *PNAS* **82,** 3187–3191.
48. Edelman, A. M., Takio, K., Blumenthal, D. K., Hansen, R. S., Walsh, K. A., Titani, K., and Krebs, E. G. (1985). *JBC* **260,** 11275–11285.
49. Vandenheede, J. R., Yang, S. D., Goris, J., and Merlevede, W. (1980). *JBC* **255,** 11768–11774.
50. Walker, A. I., Hunt, T., Jackson, R. H., and Anderson, C. W. (1985). *EMBO J.* **4,** 139–145.
51. Racker, E., Abdel-Ghany, M., Sherril, K., Riegler, C., and Blair, E. A. (1984). *PNAS* **81,** 4250–4254.
52. Casnellie, J. E., Gentry, L. E., Rhorschneider, L. R., and Krebs, E. G. (1984). *PNAS* **81,** 6676–6680.
53. Swarup, G., Dasgupta, J. D., and Garbers, D. L. (1983). *JBC* **258,** 10341–10347.
54. Goldberg, A. R., and Wong, T. W. (1984). *Adv. Enzyme Regul.* **22,** 289–308.
55. Migliaccio, A., Rotondi, A., and Auricchio, F. (1984). *PNAS* **81,** 5921–5925.
56. Shozo, S., Parmelee, D. C., Wade, R.D., Kumar, S., Ericsson, L. H., Walsh, K. A., Neurath, H., Long, G. L., Demaille, J. G., Fischer, E. H., and Titani, K. (1981). *PNAS* **78,** 848–851.
57. Barker, W. C., and Dayhoff, M. O. (1982). *PNAS* **79,** 2836–2839.
58. Takio, K., Wade, R. D., Smith, S. B., Krebs, E. G., Walsh, K. A., and Titani, K. (1984). *Biochemistry* **23,** 4207–4218.

59. Reimann, E. M., Titani, K., Ericsson, L. H., Wade, R. D., Fischer, E. H., and Walsh, K. A. (1984). *Biochemistry* **23**, 4185–4192.
60. Takio, K., Blumenthal, D. K., Edelman, A. M., Walsh, K. A., Krebs, E. G., and Titani, K. (1985). *Biochemistry* **24** (in press).
61. Mark, G. E., and Rapp, U. R. (1984). *Science* **224**, 285–289.
62. Czernilofsky, A. P., Levinson, A. D., Varmus, H. E., Bishop, J. M., Tischer, E., and Goodman, H. (1980). *Nature (London)* **287**, 198–203.
63. Ullrich, A., Cousens, L., Hayflick, J. S., Dull, T. J., Gray, A., Tam, A. W., Lee, J., Yarden, Y., Libermann, T. A., Schlessinger, J., Downward, J., Mayes, E. L. V., Whittle, N., Waterfield, M. D., and Seeburg, P. H. (1984). *Nature (London)* **309**, 418–425.
64. Ullrich, A., Bell, J. R., Chen, E. Y., Herrera, R., Petruzzelli, L. M., Dull, T. J., Gray, A., Coussens, L., Liao, Y.-C., Tsubokawa, M., Mason, A., Seeburg, P. H., Grunfeld, C., Rosen, O. M., and Ramachandran, J. (1985). *Nature (London)* **313**, 756–761.
65. Sedmak, J., and Ramaley, R. (1968). *BBA* **170**, 440–442.
66. Forsberg, H., Zetterqvist, O., and Engström, L. (1969). *BBA* **181**, 171–175.
67. Pal, B. K., and Roy-Berman, P. (1978). *BBRC* **81**, 344–350.
68. Kaufman, J. F., and Strominger, J. L. (1979). *JBC* **76**, 6304–6308.
69. Veis, A., and Perry, A. (1967). *Biochemistry* **6**, 2409–2416.

2

Cyclic Cascades
and Metabolic Regulation

EMILY SHACTER* • P. BOON CHOCK •
SUE GOO RHEE • EARL R. STADTMAN

Laboratory of Biochemistry
National Heart, Lung, and Blood Institute
National Institutes of Health
Bethesda, Maryland 20892

I. Perspectives

Reversible covalent modification of an interconvertible protein involves the enzymic transfer and removal of a modifying group from a donor molecule to a specific amino acid residue. The different forms of reversible covalent modification are listed in Table I (*1–14*). Each of these modifications and demodifications

---

*Present address: Laboratory of Genetics, National Cancer Institute, National Institutes of Health, Bethesda, Maryland 20892

THE ENZYMES, Vol. XVII

TABLE I

Reversible Covalent Modifications of Proteins

Modification	Donor molecule	Amino residue(s)	References
Phosphorylation	ATP, GTP	Serine	(1,2)
		Threonine	(1,2)
		Tyrosine	(3)
		Hydroxylysine	(4)
ADP-ribosylation	NAD⁺	Arginine	(5)
		Glutamate	(6)
		Lysine (terminal COOH)	(6)
		Diphthamide	(7)
Nucleotidylylation	ATP, UTP	Tyrosine	(8)
(adenylylation		Serine	(9)
and uridylylation)			
Methylation	S-Adenosyl-methionine	Aspartate	(10)
		Glutamate	(10)
		Lysine	(11)
		Histidine	(10)
		Glutamine	(10)
Acetylation	Acetyl-CoA	Lysine	(12)
Tyrosylation	Tyrosine	Carboxyl terminus	(13)
Sulfation	3-Phosphoadenosine 5-phosphosulfate	Tyrosine	(14)

is catalyzed by specific converter enzymes such as protein kinases and phosphoprotein phosphatases. The unification of two opposing converter enzymes and an interconvertible enzyme forms a cyclic cascade system. Thus, any enzyme or protein that undergoes reversible covalent modification is a member of a cyclic cascade.

The number of enzymes and pathways regulated by cyclic cascade systems has increased dramatically since the discovery in 1955 by Sutherland and Wosilait (15) and by Fischer and Krebs (16) that glycogen phosphorylase exists in two forms: an active phosphorylated form and a relatively inactive unmodified form. A comprehensive list of interconvertible enzymes and proteins is given in Table II (17–159). Proteins that undergo covalent modification but that have not been functionally defined, such as miscellaneous lens proteins (160), are not included. Nonetheless, Table II is almost five times as long as a similar list published in 1978 (161) and its volume continues to increase. It should be pointed out that tyrosine sulfation is not known to be cyclic nor has it been shown to regulate any enzyme activity. However, in light of evidence that the membranal receptor for interleukin-2 contains tyrosine sulfate (88) and other results showing that the level of tyrosine sulfate decreases in cells that are transformed by retroviruses

TABLE II

INTERCONVERTIBLE ENZYMES (PROTEINS)

Enzyme	References
Phosphorylation	
Acetyl-CoA carboxylase	*(17,18)*
Acetylcholine receptor (nicotinic)	*(19,20)*
Actin	*(21,22)*
Acyl coenzyme A: cholesterol acyltransferase	*(23)*
Alpha-ketoacid dehydrogenase complex	*(24)*
Aminoacyl-tRNA synthetase	*(25)*
Angiotensin	*(26)*
Asialoglycoprotein receptor	*(27)*
Asparaginase (*L. michotii*)	*(28)*
ATP citrate lyase	*(29,30)*
Beta-adrenergic receptor	*(31)*
Beta-glucuronidase	*(32)*
Branched-chain 2-oxoacid dehydrogenase complex	*(33)*
Ca^{2+}-ATPase	*(34)*
C-protein (cardiac muscle)	*(35)*
Casein	*(36)*
cAMP-dependent protein kinase (regulatory subunit)	*(37)*
cAMP-dependent protein kinase (catalytic subunit)	*(38)*
cGMP-dependent protein kinase	*(39)*
Cholesterol esterase	*(40,41)*
Cholesterol 7α-hydroxylase	*(42)*
Cyclic nucleotide phosphodiesterase	*(43,44)*
Cytochrome *P*-450 (adrenocortical)	*(45)*
DARPP-32	*(46)*
DNA topoisomerase I	*(47)*
DNA topoisomerase II	*(48)*
Enolase	*(49)*
Epidermal growth factor receptor	*(50)*
Erythrocyte membrane band 3 protein	*(51)*
Eukaryotic intiation factor-2 (eIF-2)	*(52–56)*
Eukaryotic initiation factor-3 (eIF-3)	*(57)*
Eukaryotic peptide elongation factor-1	*(58)*
Fibrinogen	*(59)*
Fodrin (nonerythroid spectrin)	*(60)*
Fructose-1,6-bisphosphatase	*(61)*
Fructose-2,6-bisphosphatase–fructose-6-P-2-kinase	*(62)*
γ-Aminobutyric acid (GABA) receptor complex	*(63)*
Glucocorticoid receptor (liver)	*(64)*
Glutamate dehydrogenase, NAD-dependent (yeast)	*(65)*
Glycerol phospate acyltransferase	*(66)*
Glycogen phosphorylase	*(67)*
Glycogen synthase	*(8,68–70)*
G-substrate (brain)	*(71)*

(continued)

TABLE II   (*Continued*)

Enzyme	References
Guanylate cyclase	(72)
High-mobility group (HMG) proteins	(73,74)
Histones	(75)
HLA antigens (Class I)	(76)
Hormone-sensitive lipase–diglyceride lipase	(77,78)
3-Hydroxy-3-methylglutaryl coenzyme A reductase	(79,80)
Hydroxymethylglutaryl coenzyme A reductase kinase	(81,82)
Immunoglobulin G	(83)
Insulin receptor	(84–86)
Interleukin-2 (IL-2, T-cell growth factor) receptor	(87,88)
Isocitrate dehydrogenase	(89)
Keratin (Type I)	(90)
Lactate dehydrogenase	(49)
Microtubule-associated protein-2 (MAP-2)	(91)
Myelin basic protein	(92)
Myeloperoxidase	(93)
Myosin heavy chains	(94)
Myosin light chains	(95–97)
Myosin light chain kinase	(98)
Na^+ channel, α subunit	(99)
Na^+–K^+-ATPase	(100)
Ornithine decarboxylase	(101)
3-Oxo-5α-steroid Δ^4-dehydrogenase	(102)
Phenylalanine hydroxylase	(103,104)
Phosphofructokinase	(105)
Phosphoglycerate mutase	(49)
Phospholamban	(106)
Phospholipid methyltransferase	(107)
Phosphoprotein phosphatase inhibitor-1	(108)
Phosphoprotein phosphatase inhibitor-2	(109–111)
Phosphorylase kinase	(108,112)
Phosvitin	(113)
Platelet-derived growth factor receptor	(83,114)
Poly(A) polymerase	(115,116)
Polyoma T antigen	(117)
Prolactin	(118)
Protamine	(119)
Pyruvate dehydrogenase	(120,121)
Pyruvate kinase	(122)
Pyruvate, Pi, dikinase (leaf)	(123)
Retroviral oncogenes	(83)
Reverse transcriptase	(124)
Rhodopsin	(125)
Ribosomal protein S6	(126,127)
RNA Polymerase (DNA-dependent)	(128)
Synapsin I (brain)	(129)

TABLE II   (*Continued*)

Enzyme	References
Tau factor	(*130*)
Troponins I and T	(*131*)
Tubulin	(*132*)
Tyrosine hydroxylase	(*133,134*)
Vinculin	(*135*)
ADP-ribosylation	
Actin	(*136*)
Adenylate cyclase	(*137*)
Ca^{2+},Mg^{2+}-dependent endonuclease	(*138*)
DNA ligase II	(*139*)
DNA polymerases α and β	(*140*)
DNA topoisomerase I	(*141*)
Elongation factor II (*E. coli*)	(*142*)
Glutamine synthetase (mamm.)	(*143*)
High-mobility group proteins	(*144*)
Histones	(*145*)
Micrococcal nuclease	(*146*)
Nitrogenase (bacterial)	(*147*)
Phosphorylase kinase (skeletal muscle)	(*148*)
Poly(ADP-ribose) synthetase	(*145*)
Protamines	(*149*)
RNA polymerase (*E. coli*)	(*150*)
RNase (bovine)	(*146*)
SV40 T antigen (large)	(*151*)
Terminal deoxynucleotidyl transferase	(*140*)
Transducin	(*152*)
Nucleotidylylation	
Aspartokinase (*E. coli*)	(*153*)
Glutamine synthetase (*E. coli*)	(*8*)
Regulatory protein, P_{II} (*E. coli*)	(*8*)
SV40 T antigen (large)	(*9*)
Methylation	
ACTH	(*154*)
Actin	(*10*)
Calmodulin	(*154*)
Citrate synthase	(*10*)
Cytochrome *c*	(*11*)
EF-Tu	(*10*)
γ-Globulin	(*154*)
High-mobility group (HMG) proteins -1 and -2	(*10*)
Histones	(*11*)
Methyl-accepting chemotaxis proteins (MCPs)	(*155*)
Myosin	(*155*)
Opsin	(*10*)
Ovalbumin	(*154*)
Ribonuclease	(*154*)

(*continued*)

TABLE II   (*Continued*)

Enzyme	References
Acetylation	
High-mobility group (HMG) proteins	(*12*)
Histones	(*12*)
Tyrosinolation	
Tubulin α-subunit	(*13*)
Sulfation	
Complement C4	(*157*)
Fibrinogens, fibrins	(*14*)
Gastrin II	(*14*)
Phyllokinin	(*14*)
Caerulein	(*14*)
Cholecystokinin	(*14*)
Leu-enkephalin	(*14*)
Fibronectin	(*158*)
Immunoglobulins	(*159*)
Interleukin-2 (IL-2, T-cell growth factor) receptor	(*88*)

(*158*), it seems possible that sulfation may have a regulatory function and, therefore, it has been included in the tables.

Perusal of Table II reveals that a broad array of cellular pathways, such as protein, carbohydrate, lipid, and nucleic acid metabolism; interferon action (*162*); DNA repair (*163*); viral oncogenesis (*164–166*); muscle contraction (*98*); and membranal signal transduction are regulated by cyclic covalent modification. In addition, a number of regulatory enzymes undergo multiple covalent modifications. For example, glycogen synthase is phosphorylated on seven different sites by five different protein kinases (*108*), each of which is regulated by different stimuli and effectors. Furthermore, some proteins, such as phosphorylase kinase, histones, and the interleukin-2 receptor undergo more than one form of modification. Such multisite modification not only expands the potential of regulatory networks numerically, but also creates the possibility for competition between different converter enzymes and, hence, between different cyclic cascade systems. This has been suggested specifically in the case of ADP-ribosylation and phosphorylation of phosphorylase kinase and miscellaneous nuclear proteins (*148, 167*).

Cyclic cascade systems respond to metabolic requirements by shifting an interconvertible protein between different extents of modification. In the case of an interconvertible enzyme, this produces a change in its specific activity either by increasing or decreasing the K_m or V_{max} of the enzyme or by modulating its response to allosteric effectors. For example, phosphorylation of glycogen phosphorylase results in activation of the enzyme by increasing the V_{max}, whereas phosphorylation of myosin light chain kinase by cAMP-dependent protein kinase inhibits the enzyme by decreasing its affinity for the essential activation complex, Ca^{2+}–calmodulin. Analogously, in the case of a binding protein (e.g., eIF-2, histones), covalent modification can either enhance or diminish binding efficiency.

Key to their pivotal role in cellular regulation is the fact that cyclic cascade systems are regulated by allosteric interactions of metabolic effectors with either the converter enzymes or the interconvertible enzymes. Moreover, as previously mentioned, covalent modification can modulate the allosteric interactions between metabolic effectors and an interconvertible enzyme. Thus, these two major mechanisms of enzyme regulation are inextricably intertwined [see Ref. (*168*) for a review of the multimodulation of enzyme activity].

Covalent modification has also been implicated in regulating specific protein degradation in the cell. That is, phosphorylation of several regulatory enzymes has been shown to increase their susceptibility to proteolysis both *in vitro* and *in vivo;* these include HMG-CoA reductase (*169*), pyruvate kinase (*170*), glutamate dehydrogenase from yeast, (*171*), fructose-1,6-bisphosphatase (*172*), and cardiac troponin (*173*). Similarly, ubiquitination of proteins (which is not known to be cyclic) has been proposed to be involved in marking proteins for ATP-dependent proteolysis (*174*). Thus, covalent modification may be a mediator for the third major mechanism of metabolic regulation—that of controlling intracellular enzyme levels. Clearly, a thorough understanding of cyclic cascades is requisite to our ability to understand and manipulate normal and abnormal cell function.

The following sections present a brief update and summary of the characteristics of cyclic cascade systems that make them so well suited for their central position in cellular regulation. More detailed reviews can be found in Refs. (*8*) and (*161*).

II. Features of Cyclic Cascade Systems

A. Unidirectional versus Interconvertible Cascades

Covalent modifications of enzymes (proteins) is catalyzed by converter enzymes such as kinases and phosphatases. They involve the action of one enzyme upon another and are therefore referred to as cascade systems (*8, 161, 175*). These systems can be divided into two classes—unidirectional cascades and cyclic cascades. Unidirectional cascades are irreversible and are usually involved in proteolytic cleavage of a specific peptide bond, as occurs in the activation of zymogens (*176*). Well-recognized unidirectional cascades include the blood-clotting cascades (*177*) and the cascade involved in complement fixation (*178*). They are designed as amplifiers that, in response to certain alarm signals, generate an avalanche of product required to meet specific biological challenges. When the need subsides, the cascades are terminated. Therefore, unidirectional cascades are contingency systems serving as biological switches that can be turned ON to meet occasional emergency situations.

In contrast, cyclic cascades involve the derivatization of one or more specific amino acid residues within the protein as occurs in the ATP-dependent phosphorylation of the hydroxyl groups of serines, threonines, or tyrosines (1), in the ATP-dependent adenylylation of the hydroxyl group of a specific tyrosine in E. coli glutamine synthetase (179, 180), and others described in Table I. These enzyme derivatizations are cyclic processes resulting from the coupling of two opposing cascades—one concerned with the covalent modification and the other with the demodification of an interconvertible enzyme. The properties of cyclic cascades elucidated by the theoretical analysis discussed in the following section show that they are endowed with unique characteristics that provide the cell with the capacity to regulate and coordinate a multitude of metabolic pathways.

B. THEORETICAL ANALYSIS

In view of the vast number of interconvertible proteins listed in Table II, many of which are key enzymes in metabolic control, we have carried out a theoretical analysis designed to reveal the advantages of such a complex mechanism for cellular regulation (161, 181). In this analysis, it has been ascertained that the covalent modification of enzymes does not function simply as an ON–OFF switch for various metabolic pathways, but rather that it is part of a dynamic process in which the fractional activities of the interconvertible enzymes can be varied progressively over a wide range. This concept derived from the experiments of Brown et al. (182) demonstrating that the adenylylation of glutamine synthetase is not an all-or-none process; instead, a steady state is established and its level is modulated by the concentrations of effectors involved. Similar observations have been reported by Pettit et al. (183) for the mammalian pyruvate–dehydrogenase complex.

The cyclic cascade model, as depicted in Fig. 1, derives from the coupling of a forward cascade and a reverse cascade. The forward cascade involves the activation of the inactive converter enzyme, PK, by an allosteric effector, e_1. The activated converter enzyme, PK_a, then catalyzes the modification (i.e., phosphorylation) of the interconvertible enzyme from its unmodified form, S, to its modified form, S-P. In the reverse cascade, the inactive converter enzyme, PT, is activated by an allosteric effector, e_2, and the active converter enzyme catalyzes the demodification (i.e., dephosphorylation) of the modified interconvertible enzyme, S-P, to its unmodified form. The coupling of these two cascades results from the fact that the substrate of the converter enzyme in one cascade is the product of the opposing cascade. Note also that for each complete cycle, a molecule of ATP is hydrolyzed to ADP and P_i. When the ATP concentration is maintained in excess relative to the enzymes involved and at a fairly constant

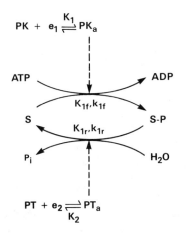

Fig. 1. Schematic representation of a monocyclic cascade. K_1, K_2, K_{1f}, K_{1r} are dissociation constants for PK_a, PT_a, $S-PK_a$, and $S-P \cdot PT_a$, respectively; k_{1f} and k_{1r} are specific rate constants for the reactions designated.

level, a steady state is established in which the rate of S-P formation is equal to the rate of S regeneration. Using the parameters shown in Fig. 1, it is possible to derive a relatively simple equation to quantitate the fractional phosphorylation of the interconvertible enzyme (*161, 181*).

When the modified interconvertible enzyme in one cycle catalyzes the modification of an interconvertible enzyme in another cycle, the two cycles are coupled such that the fractional modification of the second interconvertible enzyme is a function of all the quantitative parameters that define both cycles. Similarly, when the cascade is composed of n interconvertible enzymes and the modified interconvertible enzyme in one cycle functions as the converter enzyme for the next interconvertible enzyme, a multicyclic cascade composed of n cycles is obtained (*161, 184*).

Quantitative analysis of cyclic cascades reveals (*8, 161, 181, 184*) that (*a*) they are endowed with an enormous capacity for *signal amplification*. As a result, they can respond to primary effector (e$_1$ in Fig. 1) concentrations well below the dissociation constant of the effector–enzyme complex; (*b*) they can modulate the *amplitude* of the maximal response that an interconvertible enzyme can accomplish at saturating concentrations of allosteric effectors; (*c*) they can enhance the *sensitivity* of modification of the interconvertible enzyme to changes in the concentrations of allosteric effectors (i.e., they are capable of eliciting apparent positive and negative cooperativity in response to increasing concentrations of allosteric effectors); (*d*) they serve as *biological integration systems* that can sense simultaneous fluctuations in the intracellular concentrations of numerous metabolites and adjust the specific activity of the interconvertible enzymes accordingly; (e) they are highly *flexible* with respect to allosteric regulation and are capable of exhibiting a variety of responses to primary allosteric stimuli; and (f) they serve as *rate amplifiers* and therefore are capable of responding extremely rapidly to changes in metabolite levels (*185, 186*).

1. Signal Amplification

This is a time-independent parameter defined (*161, 181*) as the ratio of the concentration of the primary allosteric effector (e_1 in Fig. 1) required to attain 50% activation of the converter enzyme PK to the concentration required to produce 50% modification of the interconvertible enzyme S. This exceptional property derives from the fact that enzymes (catalysts) act as intermediaries between the metabolic effectors (signals) and the target interconvertible enzymes. Signal amplification can be quantified using the steady-state expression for the fractional modification of the interconvertible enzyme. It is a multiplicative function of the variables shown in Fig. 1; each of these parameters is susceptible to modulation by allosteric effectors. Because of the multiplicative nature of the fractional modification expression, small changes in several of the parameters can lead to enormous changes in fractional modification of the intercovertible enzyme in response to effector. Moreover, signal amplification increases exponentially as a function of the number of cycles in the cascade. Due to signal amplification, interconvertible enzymes can respond to effector concentrations that are well below the dissociation constant of the effector–converter enzyme complex. In other words, only a relatively small fraction of converter enzyme need be activated in order to obtain a significant fractional modification of the interconvertible enzyme.

It should be pointed out that the signal amplification described here is distinctly different from *catalytic amplification,* which is solely a function of the expansion of the relative concentrations and catalytic efficiencies of the converter and interconvertible enzymes. For the unique case in which the maximal specific catalytic activities of the converter enzyme, PK, and the interconvertible enzyme, S, are the same, the catalytic amplification potential is equal to the concentration ratio [S]/[PK]. In fact, in most cascades that have been studied, there exists a pyramidal increase in the concentration of the cascade enzymes; that is, the concentration of converter enzymes is significantly lower than that of the interconvertible enzyme substrate. Therefore, they possess a high catalytic amplification potential. This enhances the signal amplification of the cascade.

2. Amplitude

Amplitude is defined (*161*) as the maximal value of fractional modification of the interconvertible enzyme attainable with saturating concentrations of an effector. By changing the magnatude of the cascade parameters, the amplitude can change smoothly from nearly 100% to nearly 0%. Therefore, even at saturating levels of an effector (e.g., e_1 in Fig. 1) interconvertible enzymes do not function as ON–OFF switches. Gresser has pointed out (*186a*) that all modification reactions would proceed to their individual equilibrium. For physiological cascades, e.g., phosphorylaton/dephosphorylation, adenylylation/deadenylylation, etc., almost all interconvertible proteins would be present in only one form when one considers the individual equilibria for either the modification or demodification reaction. Thus, under certain extreme conditions, such as the temporary loss of either the regeneration cascade or forward cascade, enzyme cascades can effectively function as ON–OFF switches for enzymic activity.

3. *Sensitivity*

Cyclic cascades can generate either apparent positive or negative cooperative responses of fractional modification (i.e., enzymic activity) of the interconvertible enzyme to increasing concentrations of an allosteric effector (*161*). These apparent cooperativities can derive from the synergistic or antagonistic effects that a single allosteric effector exerts on two or more steps in the cascade. Thus, a sigmoidal response need not reflect positive cooperativity in the binding of an effector to multiple binding sites on the converter enzyme. Instead, it can be accomplished when an effector activates the forward converter enzyme and inactivates the reverse converter enzyme, or vice versa. Consequently, an effective way for obtaining high sensitivity is to have both forward and reverse converter enzyme activities combined in a single polypeptide such that binding of one effector can activate one activity while inactivating the other activity. Four such bifunctional enzymes have been isolated and characterized. They are the uridylyltransferase–uridylyl-removing enzyme activities that catalyze the uridylylation–deuridylylation of the P_{II} regulatory protein in the *E. coli* glutamine synthetase cascade (*187*), the adenylyltransferase that catalyzes the adenylylation–deadenylylation of glutamine synthetase (*188, 189*), a protein kinase-phosphatase that catalyzes the phosphorylation and dephosphorylation of isocitrate dehydrogenase in *E. coli* (*190*), and the 6-phosphofructose-2-kinase-fructose-2, 6-bisphosphatase that catalyzes the synthesis and breakdown of fructose 2, 6-bisphosphate (*191, 192*). All four of these bifunctional enzymes are involved in cyclic cascade systems.

A sigmoidal response can also be achieved by the regulatory mechanism reported by Jurgensen (193) in which the type II regulatory subunit of cAMP-dependent protein kinase inhibits the MgATP-dependent protein phosphatase. In addition, apparent cooperativity can be obtained when the active converter enzyme forms a tight complex with the interconvertible enzyme as shown by Shacter *et al.* (*194, 195*) and independently by Goldbetter and Koshland (*196*). The latter authors showed that when the converter enzyme is saturated by its interconvertible substrate, the cyclic cascade can exhibit a sigmoidal response to effector concentration. They call this effect "zero-order ultrasensitivity."

4. *Flexibility and Biological Integration*

There are two aspects of flexibility that need to be considered in discussing the properties of cyclic cascades; namely, the flexibility for generating various allosteric control patterns and the flexibility in regulation by multiple metabolites. The control pattern depicted in Fig. 1 illustrates just one of many variations that can be derived by changing the nature of the interactions between the allosteric effectors, e_1 and e_2, and the converter enzymes, PK and PT. For the case in which the converter enzyme PK is activated by effector e_1, four different regulatory mechanisms may result depending upon whether e_1 and e_2 activate or inhibit the activity of converter enzyme PT (*181*). Numerical analysis of these four mechanistic schemes shows that they can yield an array of patterns of fractional modification of the interconvertible enzyme in response to increasing concentrations of e_1 (*161, 181*). These patterns differ with respect to their amplitude, signal amplification, and sensitivity to changes in e_1 concentration. In fact, three of these four regulatory patterns have been observed in regulation of the mammalian pyruvate dehydrogenase cascade (*183, 197*). Furthermore, because the number of converter enzymes in multicyclic cascades is greater than in monocyclic cascades, it is possible to obtain an even greater number of unique regulatory patterns in response to positive and negative allosteric interactions.

The fact that a minimum of two converter enzymes and one interconvertible enzyme is required to form a single interconversion cycle, and each enzyme can be a separate target for one or more allosteric effectors, cyclic cascades provide a high degree of flexibility for metabolic input. Through allosteric interactions with the cascade enzymes, fluctuations in the concentrations of numerous metabolites lead to automatic adjustments in the activities of the converter enzymes that determine the steady-state levels of fractional modification (specific activity) of the interconvertible enzymes. In essence, cyclic cascades serve as biological integrators that can sense changes in the concentrations of innumerable metabolites and modulate the activities of pertinent enzymes accordingly.

5. *Rate Amplification*

Kinetic analysis of multicyclic cascades reveals that the rate of covalent modification of the last interconvertible enzyme in the cascade is a multiplicative function of the rate constants of all the reactions that lead to the formation of the modified enzyme (*8, 186*). Therefore, following an initial lag period, cyclic cascades can function as rate amplifiers to generate an almost explosive increase in catalytic activity of the target interconvertible enzyme in response to stimuli. This rate amplification potential increases with the number of cycles in the cascade. The magnitude of the rate amplification is further enhanced if the multicyclic cascade involved possesses a pyramidal relationship with respect to the concentrations of its interconvertible enzymes. Employing reasonable estimates for the rate constants, it has been demonstrated that multicyclic cascades are capable of generating large biochemical responses to primary stimuli in the millisecond time range. Moreover, if the converter and interconvertible enzymes are topographically positioned close to each other, an even greater rate of response is possible. Such topographic positioning does occur in the case of mammalian pyruvate dehydrogenase (198) and the enzymes of the glycogen cascade, which have been shown to be adsorbed to glycogen particles (199). Experimentally, Cori and his associates reported (200) that following electrical stimulation of frog sartorius muscle at 30°C, phosphorylase b is converted to phosphorylase a with a half-time of 700 msec. The fact that cyclic cascades can respond to stimuli in the millisecond time range, together with their capacity for signal amplification, indicates that they can be involved in the regulation of neurochemical processes (186).

6. ATP Flux

As shown in Fig. 1, for each complete cycle of a phosphorylation–dephosphorylation cascade, one equivalent of ATP is consumed and one equivalent each of ADP and P_i is generated. Similarly, the energy-rich donor molecules for all other forms of covalent modification (see Table I) also are degraded continuously when the cyclic cascade is in operation. The rate of ATP hydrolysis is regulated by the parameters that control the fractional modification of the intercovertible enzyme. In the theoretical analysis (161, 181), the ATP concentration is assumed to be constant because, in vivo, the concentration of ATP is metabolically maintained at fairly constant levels which are several orders of magnitude greater than the concentrations of the enzymes involved in the cascades. This ATP flux is an essential feature of the cyclic cascade regulatory mechanism because it provides the free energy required to maintain the steady-state distribution between the modified forms of the interconvertible enzyme at metabolite-specified levels which are different from those specified by thermodynamic considerations (see Section IV).

III. Experimental Verification of the Cyclic Cascade Model

To verify the properties of cyclic cascades described in the previous sections, the following systems have been investigated: (a) a simple in vitro phosphorylation–dephosphorylation cyclic cascade which was developed to study the validity of the theoretical predictions; (b) the bicyclic cascade of glutamine synthetase employing both purified proteins and permeabilized cells.

The in vitro system consists of type II cAMP-dependent protein kinase and a

38-kDa, type 2A protein phosphatase (*201*) as converter enzymes, and a nano-peptide as interconvertible substrate (*195*). Both converter enzymes were pu-rified to near homogeneity from bovine heart. The nanopeptide, Leu-Arg-Arg-Ala-Ser-Val-Ala-Gln-Leu, is homologous to the phosphorylation site of rat liver pyruvate kinase (*202*). The allosteric effectors used are cAMP, an activator of the kinase, and P_i, an inhibitor of the phosphatase. This model cascade is useful both as a tool for studying the mechanism whereby enzymes regulate each other and as a relatively simple system to aid our conceptualization of metabolic regulation through cyclic cascades. The results show that when the ATP con-centration is maintained at a relatively constant level, a steady state is established for the fractional phosphorylation of the nanopeptide. Under this condition, ATP hydrolysis continues at a constant rate, which is a measure of the interconversion rate of the nanopeptide between its phosphorylated and dephosphorylated forms. In addition, as predicted by the cyclic cascade analysis, this monocyclic cascade exhibits both the capacity for signal amplification and the capacity to generate a cooperative response to increasing effector concentrations. Due to signal ampli-fication, only one-tenth of the cAMP concentration required to half-activate the kinase is required to obtain a 50% phosphorylation of the nanopeptide. Because of an increase in sensitivity in this system, phosphorylation of the nanopeptide responds more sharply to increasing concentrations of cAMP than does the activation of the cAMP-dependent protein kinase. This apparent cooperativity derives from the fact that the catalytic subunit of the protein kinase forms a tight complex with the nanopeptide. Furthermore, in the presence of P_i, an inhibitor of the phosphatase, both the sensitivity and signal amplification were enhanced considerably (*195*).

Experiments have been carried out also using purified proteins of the bicyclic glutamine synthetase cascade of *E. coli* (*189, 203*) and permeabilized *E. coli* cells (*204*). Despite complications resulting from the dual roles of the effectors glutamine and α-ketoglutarate in the adenylylation and deadenylylation reac-tions, and the fact that six of the converter enzyme–effector complexes are catalytically active, the data confirmed the predictions of the theoretical analysis. They illustrated that cyclic cascades can serve as a signal amplifier and can elicit a cooperative response to increasing effector concentrations. Similar results were obtained both *in vitro* and with permeabilized cells. In addition, comparative studies (*204*) show that cells in which both cycles of the bicyclic cascade are functioning possess a higher signal amplification and are more sensitive to changes in metabolite concentrations than those with only one cycle functioning. In essence, most of the predicted properties of cyclic cascades have been con-firmed experimentally *in vitro* and *in vivo* through studies of the phosphoryla-tion–dephosphorylation monocycle and the glutamine synthetase bicyclic cas-cade.

IV. Energy Consumption

The steady states that develop in cyclic cascades are dynamic, energetic conditions in which the interconvertible proteins are being cycled constantly between modified and unmodified states. The capacity of a cyclic cascade system to maintain a steady state is dependent upon a constant supply of metabolic energy to drive the modification reactions, for in the absence of adequate donor molecules, the interconvertible protein would be converted completely to the unmodified form. Thus, the constant flux of metabolic energy through the cyclic cascade is the fuel required to maintain such an exquisite mechanism of cellular regulation.

The question arises, then, as to how much energy is consumed by reversible covalent modification systems. Detailed quantitation of the ATP flux through a model monocyclic phosphorylation–dephosphorylation cascade (205) demonstrated that in the presence of a relatively constant amount of ATP, a number of general characteristics are observed, which can be summarized as follows:

1. The ATP turnover is directly proportional to the concentration of both converter enzymes in the system; i.e., the higher the concentration of both protein kinase and phosphatase, the higher the cycling rate and, hence, the higher the ATP flux.

2. Attainment of a specific steady-state level of phosphorylation is dependent upon the ratio of concentrations of protein kinase and phosphatase and is independent of their absolute concentrations.

3. The time required to reach a given steady state is inversely proportional to the concentrations of converter enzymes.

4. For a given concentration of protein kinase, phosphatase, and phosphorylatable protein, the rate of ATP consumption is directly proportional to the steady-state level of phosphorylation (determined by the effector concentrations for the converter enzymes); in other words, to maintain a protein in a highly phosphorylated, thermodynamically unstable state, proportionately more energy must be expended than to maintain it at a lower steady-state level of modification.

Note that these characteristics are relevant for all forms of reversible covalent modification.

It seems likely that the cell must maintain a delicate balance between a requirement to reach a new steady state within the metabolic time frame while not expending an excessive amount of energy. In fact, the levels of converter enzymes in the cell are relatively high (in the μM range), and, not unexpectedly, steady-state levels of phosphorylation of specific proteins *in vivo* are reached within seconds after an extracellular stimulus (205). It was of interest, therefore, to estimate what the actual ATP turnover in an *in vivo* cyclic cascade might be (*205*). To this end, all relevant experimental parameters were compiled for two well-characterized cellular cyclic cascade systems: phosphorylation of hepatic pyruvate kinase by cAMP-dependent protein kinase and phosphorylation of skeletal muscle glycogen phosphorylase by phosphorylase kinase following neural stimulation. The values were used to quantitate the ATP flux through each monocyclic cascade. It was estimated that each of these cyclic covalent modification systems consumes less than 0.02% of the total cellular energy flux. This analysis did not take into account the fact that multiple cascades are activated simultaneously following hormonal and neuronal stimuli, nor did it include the energy conservation mechanisms inherent in the cyclic cascade system (*205*). Nevertheless, even allowing for a large margin of error, this result suggests that cyclic cascade systems not only are exceptional in their regulatory potential, but they are highly energy efficient as well.

V. Covalent Interconversion versus Simple Allosteric Control

Although the cyclic cascade model was derived for enzyme systems undergoing covalent interconversion, the reaction scheme shown in Fig. 1 theoretically does not require covalent modification. Hence, it is reasonable to question whether allosteric interactions alone between metabolites and enzymes can exhibit the properties of cyclic cascade systems, such as signal amplification, flexibility, and sensitivity with respect to increasing effector concentrations. It should be remembered that the cyclic cascade model utilizes both covalent modification of enzymes and allosteric control.

To accomplish signal amplification by means of simple allosteric control, the following conditions would have to be met: (*a*) very tight binding between the allosteric effector and the target enzyme; and (*b*) a reaction that possesses catalytic properties such that one effector can activate more than one target enzyme molecule. Note that signal amplification in cyclic covalent modification cascades is achieved without a requirement for tight binding between effector and converter enzyme. Because the binding rate for the effector is limited by the diffusion rate, a slow off-rate for the enzyme-bound effector would be required to achieve tight binding. However, tight binding would reduce the temporal efficiency of the control process. Furthermore, in order to achieve a catalytic effect in a simple allosteric model, the effector would first have to bind to the target enzyme, induce an active conformation, and then dissociate from the active enzyme which would have to remain in the active conformation. Such a mechanism has been implicated in the past (*206*). However, to remain regulatable by the effector, the active enzyme would have to be able to relax back to its inactive form. This kind of mechanism is thermodynamically unfavorable (*207*). Finally, without the presence of converter enzymes, the capacity for allosteric interactions would be reduced considerably. Nevertheless, the apparent cooperativity that provides the sensitivity observed in cyclic cascade systems can be accomplished by allosteric interaction alone, particularly if the enzyme involved contains multiple subunits. So, some of the advantages derived from cyclic cascade regulation cannot be achieved without invoking reversible covalent modification, while others can be accomplished but with less regulatory efficiency.

VI. Concluding Remarks

The cyclic cascade model, derived mainly from data based upon detailed studies of glutamine synthetase, is applicable to all covalent interconvertible enzyme systems. It reveals many regulatory advantages such as signal amplification, rate amplification, sensitivity, and flexibility. This regulatory mechanism makes use of both covalent modification and allosteric interactions. By means of allosteric interactions with one or more enzymes, cyclic cascades can continuously monitor fluctuations in the concentrations of a multitude of metabolites and adjust the specific activities of the target enzymes according to biological requirements. Thus, they serve as biological integrators. Although a cyclic cascade modulates the specific activity of the interconvertible enzyme smoothly and continuously over a wide range of conditions, it can under extreme physiological situations serve as an ON–OFF switch to turn on or off an interconvertible enzyme. The energy for maintaining such an efficient regulatory mechanism is the consumption of ATP and other energy-rich donor molecules. However, the amount of ATP consumed is negligible compared to the total cellular ATP hydrolysis. In view of the unique properties of cyclic cascades, it is not surprising that a large number of key enzymes are regulated by this mechanism.

REFERENCES

1. Krebs, E. G., and Beavo, J. A. (1979). *Annu. Rev. Biochem.* **48**, 923–959.
2. Engström, L., Ekman, P., Humble, E., Ragnarsson, U., and Zetterqvist, O. (1984). "Methods in Enzymology," Vol. 107, pp. 130–153.
3. Cooper, J. A., Sefton, B. M., and Hunter, T. (1983). "Methods in Enzymology," Vol. 99, pp. 387–402.
4. Urushizaki, Y., and Seifter, S. (1985). *PNAS* **82**, 3091–3095.
5. Moss, J., and Vaughan, M. (1982). *In* "ADP-Ribosylation Reactions: Biology and Medicine" (O. Hayaishi and K. Ueda, eds.), pp. 637–645. Academic Press, New York.
6. Ueda, K., and Hayaishi, O. (1984). "Methods in Enzymology," Vol. 106, pp. 450–461.
7. Bodley, J. W., Van Ness, B. G., and Howard, J. B. (1980). *In* "Novel ADP-Ribosylations of Regulatory Enzymes and Proteins" (M. Smulson and T. Sugimura, eds.), pp. 413–422. Elsevier/North-Holland, New York.
8. Chock, P. B., Rhee, S. G., and Stadtman, E. R. (1980). *Annu. Rev. Biochem.* **49**, 813–843.

9. Bradley, M. K., Hudson, J., Villanueva, M. S., and Livingston, D. M. (1984). *PNAS* **81**, 6574–6578.
10. Paik, W. K. (1984). "Methods in Enzymology," Vol. 106, pp. 265–268.
11. Paik, W. K., and Di Maria, P. (1984). "Methods in Enzymology," Vol. 106, pp. 274–287.
12. Allfrey, V. G., Di Paola, E. A., and Sterner, R. (1984). "Methods in Enzymology," Vol. 107, pp. 224–240.
13. Flavin, M., and Murofushi, H. (1984). "Methods in Enzymology," Vol. 106, pp. 223–237.
14. Huttner, W. B. (1984). "Methods in Enzymology," Vol. 107, pp. 200–223.
15. Sutherland, E. W., Jr., and Wosilait, W. D. (1955). *Nature (London)* **175**, 169–170.
16. Fischer, E. H., and Krebs, E. G. (1958). *JBC* **231**, 65–71.
17. Holland, R., Hardie, D. G., Clegg, R. A., and Zammit, V. A. (1985). *BJ* **226**, 139–145.
18. Kim, K.-H. (1983). *Curr. Top. Cell. Regul.* **22**, 143–176.
19. Davis, C. G., Gordon, A. S., and Diamond, I. (1982). *PNAS* **79**, 3666–3670.
20. Huganir, R. L., Miles, K., and Greengard, P. (1984). *PNAS* **81**, 6968–6972.
21. Walsh, M. P., Hinkins, S., and Hartshorne, D. J. (1981). *BBRC* **102**, 149–157.
22. Steinberg, R. A. (1980). *PNAS* **77**, 910–914.
23. Field, F. J., Henning, B., and Mathur, S. N. (1984). *BBA* **802**, 9–16.
24. Paxton, R., and Harris, R. A. (1982). *JBC* **257**, 14433–14439.
25. Damuni, Z., Caudwell, B., and Cohen, P. (1982). *EJB* **129**, 57–65.
26. Wong, T. W., and Goldberg, A. R. (1984). *JBC* **259**, 3127–3131.
27. Takahashi, T., Nakada, H., Okumura, T., Sawamura, T., and Tashiro, Y. (1985). *BBRC* **126**, 1054–1060.
28. Jerebzoff, S., and Jerebzoff-Quintin, S. (1984). *FEBS Lett.* **171**, 67–71.
29. Ramakrishna, S., Pucci, D. L., and Benjamin, W. B. (1981). *JBC* **256**, 10213–10216.
30. Houston, B., and Nimmo. H. G. (1985). *BBA* **844**, 193–199.
31. Sibley, D. R., Peters, J. R., Nambi, P., Caron, M. G., and Lefkowitz, R. J. (1984). *JBC* **259**, 9742–9749.
32. Fujita, M., Taniguchi, N., Makita, A., Ono, M., and Oikawa, K. (1985). *BBRC* **126**, 818–824.
33. Cook, K. G., Bradford, A. P., Yeaman, S. J., Aitken, A., Fearnley, I. M., and Walker, J. E. (1984). *EJB* **145**, 587–591.
34. Michalak, M., Famulski, K., and Carafoli, G. (1984). *JBC* **259**, 15540–15547.
35. Hartzell, H. C., and Glass, D. B. (1984). *JBC* **259**, 15587–15596.
36. Hathaway, G. M., and Traugh, J. A. (1982). *Curr. Top. Cell. Regul.* **21**, 101–127.
37. Erlichman, J., Rangel-Aldao, R., and Rosen, O. M. (1983). "Methods in Enzymology," Vol. 99, pp. 176–186.
38. Shoji, S., Titani, K., Demaille, S. G., and Fischer, E. H. (1979). *JBC* **254**, 6211–6214.
39. Lincoln, T. M., and Corbin, J. D. (1983). *Adv. Cyclic Nucleotide Res.* **15**, 139–192.
40. Boyd, G. S., and Gorban, A. M. (1980). *Mol. Aspects Cell. Regul.* **1**, 95–134.
41. Khoo, J. C., Mahoney, E. M., and Steinberg, D. (1981). *JBC* **256**, 12659–12661.
42. Scallen, T. J., and Sanghvi, A. (1983). *PNAS* **80**, 2477–2480.
43. Sharma, R. K., Wang, T. H., Wirch, E., and Wang, J. H. (1980). *JBC* **255**, 5916–5923.
44. Sharma, R. K., and Wang, J. H. (1985). *PNAS* **82**, 2603–2607.
45. Vilgrain, I., Defaye, G., and Chambaz, E. M. (1984). *BBRC* **125**, 554–561.
46. Hemmings, H. C., Jr., Greengard, P., Tung, H. Y., and Cohen, P. (1984). *Nature (London)* **310**, 503–505.
47. Durban, E., Roll, D., Beckner, G., and Busch, H. (1981). *Cancer Res.* **41**, 537–545.
48. Ackerman, P., Glover, C. V. C., and Osheroff, N. (1985). *PNAS* **82**, 3164–3168.
49. Cooper, J. A., Reiss, N., Schwartz, R. J., and Hunter, T. (1983). *Nature (London)* **302**, 218–223.
50. Cohen, S. (1983). "Methods in Enzymology," Vol. 99, pp. 379–387.
51. Dekowski, S. A., Rybicki, A., and Drickamer, K. (1983). *JBC* **258**, 2750–2753.

52. Jagus, R., Crouch, D., Konieczny, A., and Safer, B. (1982). *Curr. Top. Cell. Regul.* **21,** 35–64.
53. Ochoa, S., de Haro, C., Siekierka, J., and Grosfeld, H. (1981). *Curr. Top. Cell. Regul.* **18,** 421–435.
54. Petryshyn, R., Levin, D. H., and London, I. M. (1983). "Methods in Enzymology," Vol. 99, pp. 346–362.
55. Kramer, G., and Hardesty, B. (1981). *Curr. Top. Cell. Regul.* **20,** 185–203.
56. Lengyel, P. (1982). *Annu. Rev. Biochem.* **51,** 251–282.
57. Traugh, J. A., and Lundak, T. S. (1978). *BBRC* **83,** 379–384.
58. Tuhackova, Z., Ullrichova, J., and Hradec, J. (1985). *EJB* **146,** 161–166.
59. Engström, L., Edlund, B., Ragnarsson, U., Dahlqvist-Edberg, U., and Humble, E. (1980). *BBRC* **96,** 1503–1507.
60. Kadowaki, T., Nishida, E., Kasuga, M., Akiyama, T., Takaku, F., Ishikawa, M., Sakai, H., Kathuria, S., and Fujita-Yamaguchi, Y. (1985). *BBRC* **127,** 493–500.
61. Ekman, P., and Dahlqvist-Edberg, U. (1981). *BBA* **662,** 265–270.
62. Claus, T. H., El-Maghrabi, M. R., Regen, D. M., Stewart, H. B., McGrane, M., Kountz, P. D., Nyfeler, F., Pilkis, J., and Pilkis, S. J. (1984). *Curr. Top. Cell. Regul.* **23,** 57–86.
63. Wise, B. C., Guidotti, A., and Costa, E. (1984). *Adv. Cyclic Nucleotide Protein Phosphorylation Res.* **17,** 511–519.
64. Singh, V. B., and Moudgil, V. K. (1985). *JBC* **260,** 3684–3690.
65. Hemmings, B. A. (1978). *JBC* **253,** 5255–5258.
66. Nimmo, G. A., and Nimmo, H. G. (1984). *BJ* **224,** 101–108.
67. Krebs, E. G. (1981). *Curr. Top. Cell. Regul.* **18,** 401–419.
68. Roach, P. J. (1981). *Curr. Top. Cell Regul.* **20,** 45–105.
69. Woodgett, J. R., and Cohen, P. (1984). *EJB* **143,** 267–272.
70. Akatsuka, A., Singh, T. J., Nakabayashi, H., Lin, M. C., and Huang, K. P. (1985). *JBC* **260,** 3239–3242.
71. Detre, J. A., Nairn, A. C., Aswad, D. W., and Greengard, P. (1984). *J. Neurosci.* **4,** 2843–2849.
72. Zwiller, J., Revel, M. O., and Malviya, A. N. (1985). *JBC* **260,** 1350–1353.
73. Taylor, S. S. (1982). *JBC* **257,** 6056–6063.
74. Walton, G. M., Spiess, J., and Gill, G. N. (1982). *JBC* **257,** 4661–4668.
75. Langan, T. A., Zeilig, C., and Leichtling, B. (1981). *In* "Protein Phosphorylation" (O. M. Rosen and E. G. Krebs, eds.), pp. 1039–1052. Cold Spring Harbor Lab., Cold Spring Harbor, New York.
76. Feurstein, N., Monos, D. S., and Cooper, H. L. (1985). *BBRC* **126,** 206–213.
77. Belfrage, P., Fredrikson, G., Olsson, H., and Stralfors, P. (1984). *Adv. Cyclic Nucleotide Protein Phosphorylation Res.* **17,** 351–359.
78. Steinberg, D. (1976). *Adv. Cyclic Nucleotide Res.* **7,** 157–198.
79. Ingebritsen, T. S., and Gibson, D. M. (1980). *Mol Aspects Cell. Regul.* **1,** 63–93.
80. Beg, Z. H., Stonik, J. A., and Brewer, H. B., Jr. (1985). *JBC* **260,** 1682–1687.
81. Ingebritsen, T. S. (1983). *Biochem. Soc. Trans.* **11,** 644–646.
82. Beg, Z. H., Stonik, J. A., and Brewer, H. B., Jr. (1984). *BBRC* **119,** 488–498.
83. Swarup, G., Dasgupta, J. D., and Garbers, D. L. (1984). *Adv. Enzyme Regul.* **22,** 267–288.
84. Cobb, M. H., and Rosen, O. M. (1984). *BBA* **738,** 1–8.
85. Kasuga, M., Karlsson, F. A., and Kahn, C. R. (1982). *Science* **215,** 185–187.
86. Zick, Y., Whittaker, J., and Roth, J. (1983). *JBC* **258,** 3431–3434.
87. Shackelford, D. A., and Trowbridge, I. S. (1984). *JBC* **259,** 11706–11712.
88. Leonard, W. J., Depper, J. M., Kronke, M., Robb, R. J., Waldmann, T. A., and Greene, W. C. (1985). *JBC* **260,** 1872–1880.
89. Garland, D., and Nimmo, H. G. (1984). *FEBS Lett.* **165,** 259–264.
90. Gibbs, P. E., Zouzias, D. C., and Freedberg, I. M. (1985). *BBA* **824,** 247–255.

91. Theurkauf, W. G., and Vallee, R. B. (1982). *JBC* **257**, 3284–3290.
92. Turner, R. S., Chou, C. H., Mazzei, G. J., Dembure, P., and Kue, J. F. (1984). *J. Neurochem.* **43**, 1257–1264.
93. Hasilik, A., Pohlmann, R., Olsen, R. L., and von Figura, K. (1984). *EMBO J.* **3**, 2671–2676.
94. Albanesi, J. P., Fujisaki, H., and Korn, E. D. (1984). *JBC* **259**, 14184–14189.
95. England, P. J. (1984). *J. Mol. Cell. Cardiol.* **16**, 591–595.
96. Walsh, M. P., Hinkins, S., Dabrowska, R., and Hartshorne, D. J. (1983). "Methods in Enzymology," Vol. 99, pp. 279–288.
97. Westwood, S. A., Hudlicka, O., and Perry, S. V. (1984). *BJ* **218**, 841–847.
98. Adelstein, R. S. (1983). *J. Clin. Invest.* **72**, 1863–1866.
99. Costa, M. R., Casnellie, J. E., and Catterall, W. A. (1982). *JBC* **257**, 7918–7921.
100. Mardh, S. (1983). *Curr. Top. Membr. Transp.* **19**, 999–1004.
101. Atmar, V. J., and Kuehn, G. D. (1983). "Methods in Enzymology," Vol. 99, pp. 366–372.
102. Golf, S. W., and Graef, V. (1984). *J. Clin. Chem. Clin. Biochem.* **22**, 705–709.
103. Donlon, J., and Kaufman, S. (1978). *JBC* **253**, 6657–6659.
104. Wretborn, M., Humble, E., Ragnarsson, U., and Engström, L. (1980). *BBRC* **93**, 403–408.
105. Söling, H.-D., and Brand, I. A. (1981). *Curr. Top. Cell. Regul.* **20**, 107–138.
106. Le Peuch, C. J., Haiech, J., and Demaille, J. G. (1979). *Biochemistry* **18**, 5150–5157.
107. Villalba, M., Varela, I., Mérida, I., Pajares, M. A., Martinez del Pozo, A., and Mato, J. M. (1985). *BBA* **847**, 273–279.
108. Cohen, P. (1982). *Nature (London)* **296**, 613–620.
109. Resink, T. S., Hemmings, B. A., Tung, H. Y. L., and Cohen, P. (1983). *EJB* **133**, 455–461.
110. Jurgenson, S., Shacter, E., Huang, C. Y., Chock, P. B., Yang, S.-D. Vandenheede, J. R., and Merlevede, W. (1984). *JBC* **259**, 5864–5870.
111. Villa-Moruzzi, E., Ballou, L. M., and Fischer, E. H. (1984). *JBC* **259**, 5857–5863.
112. Singh, T. J., Akatsuka, A., and Huang, K. P. (1984). *JBC* **259**, 12857–12864.
113. Katōh, N., and Kubo, S. (1983). *BBA* **760**, 61–68.
114. Stiles, C. D. (1983). *Cell* **33**, 653–655.
115. Jacob, S. T., and Rose, K. M. (1984). *Adv. Enzyme Regul.* **22**, 485–497.
116. Stetler, D. A., Seidel, B. L., and Jacob, S. T. (1984). *JBC* **259**, 14481–14485.
117. Schaffhausen, B. S., and Benjamin, T. L. (1979). *Cell* **18**, 935–946.
118. Oetting, W. S., Tuazon, P. T., Traugh, J. A., and Walker, A. M. (1986). *JBC* **261**, 1649–1652.
119. Qi, D. F., Turner, R. S., and Kuo, J. F. (1984). *J. Neurochem.* **42**, 458–465.
120. Reed, L. J. (1981). *Curr. Top. Cell. Regul.* **18**, 95–106.
121. Pettit, F. H., Yeaman, S. J., and Reed, L. J. (1983). "Methods in Enzymology," Vol. 99, pp. 331–336.
122. Engström, L. (1978). *Curr. Top. Cell. Regul.* **13**, 29–51.
123. Burnell, J. N., and Hatch, M. D. (1985). *Trends Biochem. Sci.* **10**, 288–291.
124. Lee, S. G., Miceli, M. V., Jungmann, R. A., and Hung, P. P. (1975). *PNAS* **72**, 2945–2949.
125. Shichi, H., Somers, R. L., and Yamamoto, K. (1983). "Methods in Enzymology," Vol. 99, pp. 362–366.
126. Cobb, M. H., and Rosen, O. M. (1983). *JBC* **258**, 12472–12481.
127. Trevillyan, J. M., Perisic, O., Traugh, J. A., and Byus, C. V. (1985). *JBC* **260**, 3041–3044.
128. Rose, K. M., Duceman, B. W., Stetler, D., and Jacob, S. T. (1983). *Adv. Enzyme Regul.* **21**, 307–320.
129. Nestler, E. J., Walaas, I., and Greengard, P. (1984). *Science* **225**, 1357–1364.
130. Yamamoto, H., Fukunaga, K., Goto, S., Tanaka, E., and Miyamoto, E. (1985). *J. Neurochem.* **44**, 759–768.
131. Perry, S. V. (1979). *Biochem. Soc. Trans.* **7**, 593–617.
132. Gard, D. L., and Kirschner, M. W. (1985). *J. Cell Biol.* **100**, 769–774.

133. Albert, K. A., Helmer-Matyjek, E., Nairn, A. C., Muller, T. H., Haycock, J. W., Greene, L. A., Goldstein, M., and Greengard, P. (1984). *PNAS* **81**, 7713–7717.
134. Cahill, A. L., and Perlman, R. L. (1984). *PNAS* **81**, 7243–7247.
135. Werth, D. K., Niedel, J. E., and Pastan, I. (1983). *JBC* **258**, 11423–11426.
136. Kun, E. Romaschin, A. D., Claisdell, R. J., and Jackowski, G. (1981). *In* "Metabolic Interconversion of Enzymes" (H. Holzer, ed.), pp. 280–293. Springer-Verlag, Berlin and New York.
137. Moss, J., and Vaughan, M. (1984). "Methods in Enzymology," Vol. 106, pp. 411–418.
138. Nomura, H., Tanigawa, Y., Kitamura, A., Kawakami, K., and Shimoyama, M. (1981). *BBRC* **98**, 806–814.
139. Creissen, D., and Shall, S. (1982). *Nature* **296**, 271–272.
140. Yoshihara, K., Itaya, A., Tanaka, Y., Ohashi, Y., Ito, K., Teraoka, H., Tsukada, K., Matsukage, A., and Kamiya, T. (1985). *BBRC* **128**, 61–67.
141. Ferro, A. M., and Olivera, B. M. (1984). *JBC* **259**, 547–554.
142. Van Ness, B. G., Howard, J. B., and Bodley, J. W. (1980). *JBC* **255**, 717–720.
143. Moss, J., Watkins, P. A., Stanley, S. S., Purnell, M. R., and Kidwell, W. R. (1984). *JBC* **259**, 5100–5104.
144. Poirier, G. G., Niedergang, C., Champagne, M., Mazen, A., and Mandel, P. (1982). *EJB* **127**, 437–442.
145. Ueda, K., Ogata, N., Kawaichi, M., Inada, S., and Hayaishi, O. (1982). *Curr. Top. Cell. Regul.* **21**, 175–187.
146. Tanaka, Y., Yoshihara, K., Ohashi, Y., Itaya, A., Nakano, T., Ito, K., and Kamiya, T. (1985). *Anal. Biochem.* **145**, 137–143.
147. Pope, M. R., Murrell, S. A., and Ludden, P. W. (1985). *PNAS* **82**, 3173–3177.
148. Tsuchiya, M., Tanigawa, Y., Ushiroyama, T., Matsunra, R., and Shimoyama, M. (1985). *EJB* **147**, 33–40.
149. Wong, N. C. W., Poirier, G. G., and Dixon, G. H. (1977). *EJB* **77**, 11–21.
150. Goff, C. G. (1984). "Methods in Enzymology," Vol. 106, pp. 418–429.
151. Goldman, N., Brown, M., and Khoury, G. (1981). *Cell* **24**, 567–572.
152. Abood, M. E., Hurley, J. B., Pappone, M. C., Bourne, H. R., and Stryer, L. (1982). *JBC* **257**, 10540–10543.
153. Niles, E. G., and Westhead, E. W. (1973). *Biochemistry* **12**, 1723–1729.
154. Kim, S. (1984). "Methods in Enzymology, Vol. 106, pp. 295–309.
155. Stock, J. B., Clarke, S., and Koshland, D. E., Jr. (1984). "Methods in Enzymology," Vol. 106, pp. 310–321.
156. Huszar, G. (1984). "Methods in Enzymology," Vol. 106, pp. 287–295.
157. Karp, D. R. (1983). *JBC* **258**, 12745–12748.
158. Liu, M.-C., and Lipmann, F. (1984). *PNAS* **81**, 3695–3698.
159. Baeuerle, P. A., and Huttner, W. B. (1984). *EMBO J.* **3**, 2209–2215.
160. Garland, D., and Russell, P. (1985). *PNAS* **82**, 653–657.
161. Stadtman, E. R., and Chock, P. B. (1978). *Curr. Top. Cell. Regul.* **13**, 53–95.
162. Johnston, M. I., and Torrence, P. F. (1984). *In* "Interferon, 3: Mechanisms of Production and Action" (R. M. Friedman, ed.), pp. 189–298. Elsevier, Amsterdam.
163. Sugimura, T., and Miwa, M. (1983). *Carcinogenesis (London)* **4**, 1503–1506.
164. Sefton, B. M., and Hunter, T. (1984). *Adv. Cyclic Nucleotide Protein Phosphorylation Res.* **18**, 195–226.
165. Collett, M. S., and Erikson, R. L. (1978). *PNAS* **75**, 2021–2024.
166. Levinson, A. D., Oppermann, H., Levintow, L., Varmus, H. E., and Bishop, J. M. (1978). *Cell* **15**, 561–572.
167. Tanigawa, Y., Tsuchiya, M., Imai, Y., and Shimoyama, M. (1983). *BBRC* **113**, 135–141.
168. Sols, A. (1981). *Curr. Top. Cell. Regul.* **19**, 77–101.

169. Parker, R. A., Miller, S. J., and Gibson, D. M. (1984). *BBRC* **125**, 629–635.
170. Bergström, G., Ekman, P., Humble, E., and Engström, L. (1978). *BBA* **532**, 259–267.
171. Hemmings, B. A. (1980). *FEBS Lett.* **122**, 297–302.
172. Muller, D., and Holzer, H. (1981). *BBRC* **103**, 926–933.
173. Toyo-oka, T. (1982). *BBRC* **107**, 44–50.
174. Hershko, A., and Ciechanover, A. (1982). *Annu. Rev. Biochem.* **51**, 335–364.
175. MacFarlane, R. G. (1964). *Nature (London)* **202**, 498–499.
176. Neurath, H., and Walsh, K. A. (1976). In "Proteolysis and Physiological Regulation" (E. W. Ribbons and K. Brew, eds.), pp. 29–40. Academic Press, New York.
177. Davie, E. W., and Fujikawa, K. (1975). *Annu. Rev. Biochem.* **44**, 798–829.
178. Müller-Eberhard, H. J. (1975). *Annu. Rev. Biochem.* **44**, 697–724.
179. Kingdon, H. S., Shapiro, B. M., and Stadtman, E. R. (1967). *PNAS* **58**, 1703–1710.
180. Wulff, K., Mecke, D., and Holzer, H. (1967). *BBRC* **28**, 740–745.
181. Stadtman, E. R., and Chock, P. B. (1977). *PNAS* **74**, 2761–2765.
182. Brown, M. S., Segal, A., and Stadtman, E. R. (1974). *ABB* **161**, 319–327.
183. Pettit, F. H., Pelley, J. W., and Reed, L. J. (1975). *BBRC* **65**, 575–582.
184. Chock, P. B., and Stadtman, E. R. (1977). *PNAS* **74**, 2766–2770.
185. Chock, P. B., and Stadtman, E. R. (1979). In "Modulation of Protein Function" (D. E. Atkinson and C. F. Fox, eds.), pp. 185–202. Academic Press, New York.
186. Stadtman, E. R., and Chock, P. B. (1979). In "The Neurosciences: Fourth Study Program" (F. O. Schmitt, ed.), pp. 801–817. MIT Press, Cambridge, Massachusetts.
187. Garcia, E., and Rhee, S. G. (1983). *JBC* **258**, 2246–2253.
188. Caban, C. E., and Ginsburg, A. (1976). *Biochemistry* **15**, 1569–1580.
189. Rhee, S. G., Park, R., Chock, P. B., and Stadtman, E. R. (1978). *PNAS* **75**, 3138–3142.
190. La Porte, D. C., and Koshland, D. E., Jr. (1982). *Nature (London)* **300**, 458–460.
191. El-Maghrabi, M. R., Claus, T. H., Pilkis, J., Fox, E., and Pilkis, S. J. (1982). *JBC* **257**, 7603–7607.
192. Van Schaftingen, E., Davies, D. R., and Hers, H. G. (1982). *EJB* **142**, 143–149.
193. Jurgensen, S. R., Chock, P. B., Taylor, S. S., Vandenheede, J. R., and Merlevede, W. (1985). *FP* **44**, 1052.
194. Shacter-Noiman, E., Chock, P. B., and Stadtman, E. R. (1983). *Philos. Trans. R. Soc. London* **302**, 157–166.
195. Shacter, E., Chock, P. B., and Stadtman, E. R. (1984). *JBC* **259**, 12252–12259.
196. Goldbetter, A., and Koshland, D. E., Jr. (1981). *PNAS* **78**, 6840–6844.
197. Hucho, F., Randell, D. D., Roche, T. E., Burgett, M. W., Pelley, J. W., and Reed, L. J. (1972). *ABB* **151**, 328–340.
198. Reed, L. J. (1969). *Curr. Top. Cell. Regul.* **1**, 233–251.
199. Meyer, F., Heilmeyer, L. M. G., Jr., Haschke, R. H., and Fischer, E. H. (1970). *JBC* **245**, 6642–6648.
200. Danforth, W. H., Helmreich, E., and Cori, C. F. (1962). *PNAS* **48**, 1191–1199.
201. Shacter-Noiman, E., and Chock, P. B. (1983). *JBC* **258**, 4214–4219.
202. Titanji, V. P. K., Ragnarsson, U., Humble, E., and Zetterqvist, O. (1980). *JBC* **255**, 11339–11343.
203. Rhee, S. G., Chock, P. B., and Stadtman, E. R. (1985). In "The Enzymology of Post-Translational Modification of Proteins" (R. Freedman, ed.), Vol. 2, pp. 273–297. Academic Press, New York.
204. Mura, U., Chock, P. B., and Stadtman, E. R. (1981). *JBC* **256**, 13022–13029.
205. Shacter, E., Chock, P. B., and Stadtman, E. R. (1984). *JBC* **259**, 12260–12264.
206. Hatfield, G. W., and Burns, R. O. (1970). *Science* **167**, 75–76.
207. Astumian, D., and Chock, P. B. (1985). Unpublished results.

3

Cyclic Nucleotide-Dependent Protein Kinases

STEPHEN J. BEEBE • JACKIE D. CORBIN

Howard Hughes Medical Institute
Department of Molecular Physiology
and Biophysics
Vanderbilt University
Nashville, Tennessee 37232

THE ENZYMES, Vol. XVII
Copyright © 1986 by Academic Press, Inc.

I. Introduction

Since the first report of a protein kinase in liver by Burnett and Kennedy (*1*) and the classical work on interconvertible forms of enzymes involved in glycogen metabolism, as well as investigations of protein kinases reported by Rabinowitz (*2*), phosphorylation and dephosphorylation reactions have been the most intensely studied mechanisms of posttranslational modification. Many different protein kinases have since been identified and studied, but the cyclic nucleotide-dependent protein kinases have probably received the most attention. The primary impetus for research in this area was the Nobel Prize-winning work of Sutherland and co-workers who developed the concept of cAMP as an intracellular second messenger of hormone action (*3, 4*), and the discovery of the cAMP-dependent protein kinase (*5*). It was eventually recognized that in mammalian cells the cAMP-dependent protein kinase is the major, if not the only, intracellular receptor for cAMP. The activation–inactivation reaction is indicated by the following equation (*6*):

$$R_2C_2 + 4 \text{ cAMP} \rightleftharpoons R_2(\text{cAMP})_4 + 2C \qquad (1)$$
$$\text{(inactive)} \qquad\qquad \text{(active)}$$

In the holoenzyme form (R_2C_2), the catalytic subunit (C) is inhibited by the regulatory subunit (R). The cAMP-dependent protein kinase is represented by two different major types of isozymes which are operationally defined by the salt gradient elution behavior from DEAE-cellulose. Type I elutes as NaCl concentrations less than $0.1 M$ and type II elutes at concentrations greater than $0.1 M$ (*7*). Both holoenzymes are tetramers composed of dimeric regulatory subunits with four cAMP-binding sites (two sites/monomeric chain) (*6–9*) and two monomeric catalytic subunits. When cAMP binds to the regulatory subunit of protein kinase the equilibrium shifts to the right and the catalytic subunit is released from the inhibition imposed by the regulatory subunit. It is the free, active catalytic subunit that mediates protein phosphorylation, which is established to be the primary, if not the only, mechanism of cAMP action in mammals. When cAMP is hydrolyzed by cAMP phosphodiesterases (see Chapter 2), the equilibrium shifts back to the left, the catalytic subunit reassociates with the regulatory subunit and phosphotransferase activity is terminated (*10–15*). In order for the unmodified form of the protein substrate to be regenerated, a phosphoprotein phosphatase catalyzes a dephosphorylation reaction (see Chapter 8). The steady-state, phosphorylation–dephosphorylation equilibrium between an active and inactive form of a protein is dynamically regulated by protein kinases, phosphoprotein phosphatases, and their respective effectors. This constitutes a cyclic cascade system and can provide the cell with an efficient and sensitive control mechanism that has potential for amplification and cooperativity of response (16–20, 20a). The ubiquity of the cyclic nucleotide protein kinases in nature and the central role of cyclic nucleotides in eukaryotic metabolic regulation makes the involvement of cyclic nucleotides as effectors of this system of particular interest and importance.

The cGMP-dependent protein kinase, which has received less attention, may also play a role in the phosphorylation–dephosphorylation control of cellular events. It has a different mechanism of activation–inactivation as indicated by the following equation (21):

$$E_2 + 4\ cGMP \rightleftharpoons E_2 \cdot cGMP_4 \tag{2}$$
$$\text{(inactive)} \qquad \text{(active)}$$

The cGMP kinase is a dimer with each monomeric chain containing both a regulatory, cGMP-binding component and a catalytic component (21–24). There are four cGMP-binding sites per enzyme dimer or two sites in each of the two regulatory components. When cGMP binds to the regulatory domain of the enzyme, the equilibrium shifts to the right and the active catalytic domain carries out the phosphorylation reactions. No separation of subunits occurs (see Ref. 29) as is the case for activation of the cAMP-dependent protein kinase, where the regulatory and catalytic subunits physically separate. When cGMP dissociates from the active form of the enzyme and is hydrolyzed by cGMP phosphodiesterase, the equilibrium shifts back to the left and the inactive conformation of the enzyme is reestablished.

The reader is referred to several reviews (5, 10–15, 25–40) and the monograph edited by Rosen and Krebs (41), which are related to cyclic nucleotide-dependent protein kinases and their role in phosphorylation–dephosphorylation mechanisms. The aim of this chapter is to review the characteristics and functions of cyclic nucleotide-dependent protein kinases. A brief review of methods for purifying the enzyme is presented. The characteristics of the enzymes are described as they relate to functional aspects of the isozymes, to evolutionary relationships and homologies among protein kinases and other cyclic nucleotide binding proteins, and to the mechanisms of catalytic and regulatory subunit action. The regulation of enzyme activity by cyclic nucleotides and other effectors is reviewed and the role the enzymes play in cellular function is assessed. The role of these kinases in the short-term control of cellular function through the regulation of enzyme activities and the long-term control of processes such as transcription, protein induction and cell growth, and differentiation are highlighted. In addition, several methods used for deducing the biological role of the kinases are compared and critiqued.

II. Purification

A. CYCLIC AMP-DEPENDENT PROTEIN KINASE

1. *Regulatory Subunit*

Before the development of affinity chromatography the regulatory subunit of the cAMP-dependent protein kinase could be obtained by purification of the respective holoenzymes followed by subunit dissociation. Since then the regulatory subunit can be more easily purified to homogeneity by affinity chromatography using immobilized cAMP analogs (*42–45*). Both type I and type II regulatory subunits can be purified using this approach, but it is often advantageous to use different affinity ligands for each (*46*). The regulatory subunits can be eluted from affinity columns using 8 M urea, followed by urea removal and renaturation. Alternatively, they can be specifically eluted by cAMP. The urea elution method allows the recovery of a relatively cAMP-free regulatory subunit but it usually has altered properties compared to the subunit prepared by cAMP elution (*47*). Even though elution with cAMP alleviates some anomalies, removal of bound cAMP from the regulatory subunit without deleterious effects is very difficult (*6*).

The type I regulatory subunit binds very effectively to N^6-(2-aminoethyl)amino-cAMP-Sepharose (*44*) and is readily purified to homogeneity by specific cAMP elution (*44, 46*). The type I subunit is not readily eluted from 8-(6-aminohexyl)amino-cAMP-Sepharose using cAMP but can be eluted using 8 M urea (*46*). This subunit has also been purified using a N^6-(6-aminohexyl)-cAMP derivative attached to Sepharose (*45*). The subunit is not eluted from this ligand using 2.5 M salt or 0.5 mM cAMP but is eluted with 0.5 mM cAMP in the presence of a low concentration of catalytic subunit.

The type II regulatory subunit has been prepared to homogeneity by Corbin *et al.* (*6, 46*) by using a cAMP elution from 8-(6-aminohexyl)amino-cAMP-Sepharose. The type II subunit has also been purified using a cAMP or cGMP (0.1 mM) elution from N^6-aminoethyl-cAMP-Sepharose (*48*). Since cGMP binds weakly and is thus more easily removed, it has the advantage of allowing the preparation of a relatively cyclic nucleotide-free regulatory subunit. In order to obtain a nondenatured, cAMP-free regulatory subunit, Seville and Holbrook (*49*) used a method where cGMP is exchanged for cAMP while the subunit is bound to DEAE-cellulose. The subunit is then eluted with salt and the weakly bound cGMP is removed by washing and dialysis. In general, the type II regulatory subunit is eluted from cyclic nucleotide affinity columns at lower cAMP concentrations than is the type I subunit. For example, the elution of type II subunit from 8-(6-aminohexyl)amino-cAMP-Sepharose is carried out with 10 mM cAMP (*6*) but elution of type I requires urea denaturation (*46*). While the type II subunit is eluted from N^6-aminoethyl-cAMP-Sepharose at 0.1 mM cAMP or 0.1 mM cGMP, the type I subunit is eluted with 30 mM cAMP (48).

2. Catalytic Subunit

The catalytic subunit of the cAMP-dependent protein kinase has been purified from a large number of tissues including bovine heart; rabbit, rat, and porcine skeletal muscle; bovine and rabbit liver [see Ref. (38) for review]; rabbit kidney and porcine stomach mucosa (50); rat adipose tissue (51); and porcine kidney (52). The most efficient purification procedures are based on the facts that the holoenzyme and regulatory subunit have different ion exchange properties from the catalytic subunit, and that the subunits specifically dissociate in the presence of cAMP or cAMP analogs (50). Two different approaches have been used to purify the catalytic subunit to homogeneity. A partially purified preparation of type I and/or type II holoenzyme is treated with cAMP before chromatography on carboxymethylcellulose which binds the catalytic subunit but not the regulatory subunit (53, 54). Alternatively the holoenzyme can be treated with cAMP while it is bound to DEAE-cellulose which does not retain the dissociated catalytic subunit (50). Subsequent purification or concentration of catalytic subunit may be necessary on hydroxylapatite (50, 55), carboxymethylcellulose (53), Sephadex G-100 (52, 54), or Blue Dextran (52).

3. Holoenzymes

The holoenzymes are usually obtained in pure form by first purifying the regulatory and catalytic subunits by the procedures previously described. These subunits are then combined. It is not clear, however, whether or not in all cases the holoenzyme produced by this recombination resembles in all respects the native holoenzymes. Although the methods are more difficult, it may be necessary for some studies to purify the holoenzymes as such. The partial purification of the cAMP-dependent protein kinase was first reported by Walsh et al. (56). Since then the holoenzymes have been purified to homogeneity in a number of laboratories. The type I isozyme has been purified from rabbit (53, 57–59) and porcine skeletal muscle (60, 61) and the type II holoenzyme from bovine heart (62–64). The rabbit skeletal muscle type I and the bovine heart type II are often taken as the prototype isozymes. These tissues are ideal for purifying the respective enzymes since they contain predominately one isozyme. Conventional purification steps have been used to purify the enzymes. The isozymes are first separated on DEAE-cellulose based on their different salt elution (7). C_6-Aminoalkyl-agarose chromatography has also been used to separate the isozymes (65). The procedures for most homogeneous preparations of either holoenzyme have successfully employed DEAE-cellulose chromatography, ammonium sulfate fractionation, alumina C_γ chromatography, and gel filtration techniques. Beavo et al. (53, 54) and Hofmann et al. (57) used negative chromatography by batch adsorption to remove some contaminating proteins on carboxymethylcellulose. Hydroxylapatite has also been used (*58, 59, 61–64*), but Rubin *et al.* (*63*) found that in their procedure, rechromatography on DEAE-cellulose could substitute for the hydroxylapatite step. Hofmann *et al.* (*57*) used histone IIA-Sepharose to purify the bovine heart type II kinase, and Taylor *et al.* (*60, 61*) used 8-(6-aminohexyl)amino-ATP-Sepharose and 3-aminopyridine-NAD$^+$-Sepharose affinity chromatography in addition to isoelectric focusing to purify porcine muscle type I holoenzyme to homogeneity. Hydrophobic chromatography on hexyl-Sepharose (*58*) or phenyl-Sepharose (*64*) has also been used as very effective purification steps for type I and type II isozymes, respectively. Cobb and Corbin (*64*) have successfully used Bio-Rad 1EX545-DEAE and Bio-sil TSK-250 high-performance liquid chromatography in addition to conventional procedures to purify the type II holoenzyme from bovine heart to homogeneity. These homogeneous preparations are generally purified 1000–3000-fold depending on the tissue and the isozyme.

B.   CYCLIC GMP-DEPENDENT PROTEIN KINASE

The cyclic GMP-dependent protein kinase has been purified to homogeneity from soluble fractions of bovine lung by Lincoln *et al.* (*66*), Gill *et al.* (*22*), Corbin and Døskeland (*21*), and MacKenzie (*24*) and from bovine heart by Flockerzi *et al.* (*67*). A unique form of the enzyme has also been highly purified from intestinal brushborder membranes by de Jonge (*68*). The purification procedures generally utilize DEAE-cellulose chromatography, ammonium sulfate precipitation, and cyclic nucleotide affinity chromatography.

All of the cyclic nucleotide-dependent protein kinases bind to DEAE-cellulose, and this is a convenient early step in the purification procedure. The cGMP-dependent protein kinase from bovine lung binds more tightly to DEAE-cellulose than does the type I cAMP-dependent protein kinase and less tightly than does the type II isozyme of the cAMP protein kinase (*22, 66*), and is separated from the latter enzymes using NaCl (*12, 60*) or ammonium sulfate (*22, 66, 67, 69, 70*). Corbin and Døskeland (*21*) also used a procedure through the DEAE-cellulose step designed for purification of guanylate cyclase (*71*). Triethanolamine (pH 7.5) elutes the cGMP-dependent protein kinase slightly before it elutes the cyclase on this column. At least three different immobilized cyclic nucleotide analogs have been used to obtain pure preparations of the cGMP kinase. Sepharose-bound 8-(2-aminoethyl)thio-cGMP was first used by Lincoln *et al.* (*66*) and 8-(2-aminoethyl)amino cAMP by Gill *et al.* (*22*) to purify the enzyme to homogeneity. Corbin and Døskeland (*21*) and MacKenzie (*24*) used 8-(6-aminohexyl)amino-cAMP-Sepharose to purify the cGMP kinase in their studies. When N^6-[(6-aminohexyl)carbamoylmethyl]-cAMP was used to purify the enzyme, it was not homogeneous (*24*). Since the cAMP-dependent protein kinases also bind to these affinity resins, preliminary purification on DEAE-cellulose is important. Additionally, specific elution of the enzyme from the affinity columns with cGMP allows for a high degree of purification. If trace contamination by regulatory subunit is still present at this stage, it can be removed by sucrose gradient centrifugation (*21*). The purification of the cGMP-dependent protein kinase from intestinal brushborder membranes requires a modified procedure (*68*). Following the extraction of the enzyme from the membranes with detergent and high salt concentration, it is further purified on 8-(2-aminoethyl)-amino-cAMP-Sepharose. Except for the presence of 0.1% Triton and high salt concentration during all steps of affinity chromatography, the procedure is similar to the ones for purification of the lung enzyme.

Several factors can be manipulated to maximize or minimize binding of these cyclic nucleotide protein kinases to affinity supports (*48, 72*). In addition to the relative affinities for binding, other factors such as steric hindrance, density of substitution, and availability of ligand can be of potential importance.

III. Characterization and Physical Properties

A. CYCLIC AMP-DEPENDENT PROTEIN KINASES

There are two major classes of cAMP-dependent protein kinase isozymes designated type I and type II (*58*). Each isozyme has a tetrameric structure consisting of two monomeric catalytic subunits and a dimeric regulatory subunit (*65–68*). The catalytic subunits from both isozymes are indistinguishable by a large number of criteria. They have similar chromatographic, chemical, physical, immunological, and catalytic properties, and can reassociate with both type I and type II regulatory subunits (*11, 12, 14, 35, 38, 55*). However, several investigators have found two or three different froms of catalytic subunit with different isoelectric points from both type I and type II isozymes (*55, 73–78*). Sugden *et al.* identified at least three forms of the bovine liver catalytic subunit with pI values of 6.72, 7.04, and 7.35 (*55*). Yamamura *et al.* (*74–76*) found that the rabbit skeletal muscle type I and the rat liver type II catalytic subunits each contained two forms with isoelectric points of 7.4 and 8.2. These forms have similar heat stability, K_m for ATP, and rate of phosphorylation of several proteins; they phosphorylate the same serine and threonine residues in histone, protamine, glycogen synthase, and phosphorylase (*74–76*). The significance of these different isoelectric forms of the catalytic subunit is not clearly understood.

It has been discovered that there are at least two bovine genes coding for the catalytic subunit. One of these gene codes for a protein of 351 amino acids that is 98% homologous with the bovine heart catalytic subunit (*78a*). The second gene codes for a protein closely related to the bovine heart catalytic subunit. The nucleotide sequences of the two genes are 85–93% homologous, with differences clustered in the sequences coding for the amino terminal portion of the molecules, believed to be the ATP binding site (79), in the carboxy terminal region of the protein, and in the 3'-untranslated portion of the gene (S. McKnight and R. Maurer, personal communication). It is presently unclear whether the products of these two genes are related to the different isoelectric forms of the catalytic subunit discussed above. However, this is possible since the two genes code for proteins which would be expected to have different isoelectric points and, consequently, different tryptic peptides. It is also possible that the second gene could be related to the "mute" catalytic subunit like the one isolated and characterized from rat skeletal muscle by Reed et al. (78b). The "mute" subunit is released from the regulatory subunit by cAMP but must be activated by a heat- and acid-stable modulator. Some evidence suggests that these two genes may have different tissue distributions. For example, the mRNA for the second gene appears to be more abundant in brain (S. McKnight and R. Mauer, personal communications). Although there is presently no evidence to correlate the association of either of these catalytic subunit gene products with one or the other regulatory subunits, it is known that the type IIB (85) or neural type II regulatory subunit (125, 126) is also abundant in the brain. These recent developments suggest a potential for greater specificity of protein kinase action, and firm conclusions regarding possible functional differences should be forthcoming. Future work will undoubtedly focus on the expression of these two genes and possibly others, so that their protein products can be characterized.

The amino acid sequence of the catalytic subunit from type II bovine heart has been determined to consist of 350 residues (79), giving it a molecular weight of 40,862, including a myristyl amino terminal blocking group and phosphates at threonine-197 and serine-338. This molecular weight is in fairly good agreement with a molecular weight of 39,000–42,000 determined by SDS-gel electrophoresis, sedimentation-equilibrium centrifugation, amino acid analysis, or calculated from the Stokes radius (2.7 nm) and sedimentation coefficient ($s_{20,w}$ = 3.6) (38). The protein has a frictional ratio of 1.2 and an axial ratio of 4.5, indicating a globular, symmetrical shape (38, 55).

At present it is known that there is at least one gene for the type I regulatory subunit and at least two genes for the type II regulatory subunit. The type I subunit has been cloned from bovine testes (78a), and one of the type II subunits has been cloned from rat ovary. The gene for the type I regulatory subunit codes for a protein with the same amino acid sequence as the rabbit skeletal muscle type I. The DNA for the rat ovary type II subunit codes for a M_r = 52,000 protein, which is distinct from the bovine heart type II subunit (79a). These two type II subunits are homologous, have similar amino-terminal and carboxyl-terminal amino acid sequences, and have two duplicated sequences which are presumably the cAMP binding domains. However, these two duplicated sequences and the sequences surrounding the autophosphorylation site are different between the two proteins. Sequences around the autophosphorylation site of the rat ovary type II regulatory subunit are also different from sequences in the same area of the rat heart regulatory subunit (J. S. Richards, personal communication). Further developments in this field will undoubtedly strengthen our understanding of these proteins and their functions.

The type I and type II isozymes are homologous proteins which have the same tetrameric structure, similar mechanisms of cAMP activation (see Section II,C), and two different intrasubunit cyclic nucleotide-binding sites on the regulatory subunit (6, 59). Although the sequence homologies of the regulatory subunits are strongest in the cAMP-binding domains, the sequences of other domains are less well conserved (see Section II,C) and several features of the regulatory subunits or their respective holoenzymes can be used to differentiate them. These include physical, immunological, and kinetic differences (10, 11, 14, 38, 57). The type II subunit has an axial ratio of 12 and a frictional rateo of 1.6 (82), indicating an oblong asymmetric structure. It consists of 43% α helices, 23–30% β-strands, and has 23 β-turns (81). The type I regulatory subunit has an axial ratio of 8.5 and a frictional ratio of 1.47 (86). The molecular weights have been determined from amino acid sequence to be 42,804 for the bovine skeletal muscle type I regulatory subunit (379 residues) (80) and 45,004 for the bovine heart type II subunit (400 residues) (81). These values are slightly lower than the molecular weights determined by sedimentation equilibrium centrifugation (15, 82) or non-denaturing gel electrophoresis (34, 62, 83) according to the method of Hedrick and Smith (84), but considerably lower than the apparent molecular weights determined from SDS-gel electrophoresis [M_r = 49,000 for the type I from porcine, bovine, and rabbit skeletal muscle and 56,000–58,000 for the bovine heart type II subunit Refs. (14, 57)]. Robinson-Steiner et al. (85) showed that the apparent molecular weight of the type II regulatory subunits determined by SDS-gel electrophoresis differ significantly among different species and tissues. In addition, the extent of migration differs for some forms of type II regulatory subunit depending on the presence of phosphate in the autophosphorylation site (33, 34, 85). The type I regulatory subunit does not undergo autophosphorylation and shows similar but slightly less pronounced discrepancies in molecular weights when determined by different methods (14, 59). The explanation for the error in apparent molecular weight for the type II regulatory subunit determined by SDS-gel electrophoresis is not certain. It is possible that the protein structure may not be completely denatured by SDS and therefore does not migrate strictly according to molecular weight. If this is correct, the presence or absence of phosphate on the autophosphorylation site affects the binding of SDS more to some than to other forms of type II regulatory subunit (85).

The physical properties of the holoenzymes have been reviewed and conveniently tabulated by Nimmo and Cohen (14) and by Carlson et al. (38). The molecular weights of the holoenzymes are now known from amino acid sequences to be 167,332 for type I (79, 80) and 171,732 for type II (79, 81). Estimates of molecular weights from sedimentation equilibrium centrifugation and nondenaturing gel electrophoresis (62, 83) are essentially the same as these values.

The isolation of the cAMP-dependent protein kinases by anion-exchange chromatography, using DEAE cellulose, has been used as a standard procedure to define, separate, and quantitate the type I and type II isozymes (7, 35). The different elution behaviors are predicted from their respective isoelectric points (pI) and differences in amino acid composition. The type I subunit has a higher pI (5.45–5.57) (87, 88) and elutes at lower salt concentrations than does the type II subunit, which has a pI of 5.34–5.40. In addition the type II isozyme has a higher acidic amino acid content (6). Using a standard, low-ionic-strength phosphate buffer, the type I and type II isozymes elute from DEAE cellulose at less than and greater than 100 mM NaCl, respectively.

The type I and type II holoenzymes have distinct antigenic determinants (89–92). When the holoenzymes or the regulatory subunits are used as antigens, antibodies are produced that recognize the individual regulatory subunits (89, 92). Specific antibodies to the catalytic subunit are generated only when the free subunit is used as the antigen, and the immunological properties appear to be similar among the subunits from distinct tissues and species (15, 54, 55, 93). Although antisera are isozyme specific, they lack absolute species specificity. For example, Fleischer et al. (89) generated antibodies against the bovine heart type II isozyme that cross-reacted identically with the type II isozymes from other bovine tissues, but reacted in a nonparallel manner with the type II isozymes from rat tissues and procine heart. Weldon et al. (94) prepared monoclonal antibodies against the type II bovine heart regulatory subunit that had an antigenic site localized in a region near the dimerization domain, the autophosphorylation site, and the cAMP-binding site 2. This was a conserved sequence in the bovine and porcine heart isozymes and was recognized with similar affinities in both species by one monoclonal antibody.

The type I and type II isozymes are also different in some of their kinetic properties and these have been used to distinguish the isozymes when mixtures are present. The type I holoenzyme is more readily dissociated than the type II holoenzymes in the presence of histone and high salt concentration (7, 35, 95). The presence of MgATP prevents the dissociation of the type I holoenzyme under these conditions (7, 57) probably by inhibition of cAMP binding (57). The type II holoenzyme reassociates rapidly in the absence of salt but slowly in the presence of salt, while the type I holoenzyme associates slowly in the presence or absence of salt (7, 35). The presence of phosphate in the autophosphorylation site of the type II regulatory subunit [serine-95 of the primary sequence (81, 96)] increases the dissociation constant for the regulatory and catalytic subunit complex and slows the rate of reassociation (*9, 34, 97, 98*). Although the physiological significance of the autophosphorylation reaction is not clear, it has been proposed that the type II holoenzyme exists *in vivo* primarily in the phosphorylated form (*99*), which is apparently favored for dissociation (*9*). Recently, Scott and Mumby (*100*) demonstrated that both phosphorylated and dephosphorylated forms exist in intact trachael smooth muscle and that the relative amounts are changed by intracellular cAMP. The type II regulatory subunit is also phosphorylated *in vitro* at serines-44 and -47 by glycogen synthase kinase 3 (*101*) and at serines-74 and -76 by casein kinase II (*101, 102*). These latter two serines are also phosphorylated *in vivo* (*102*). The physiological significance of these phosphorylations, if any, has not been established.

Other kinetic differences have been revealed through the use of cAMP analogs. Although the cAMP binding domains have been highly conserved (see Section II,C), differences in cyclic nucleotide binding to the type I and type II isozymes have been found. Although both the 3′ and 5′ oxygens in the ribose portion of cAMP are important for activation of both isozymes, use of analogs substituted with sulfur at these positions suggests that differences exist between the two isozymes in one or both of the binding sites that recognize the 4′ position of the ribose ring and the 3′,5′ cyclic phosphate (*103*). In addition, C8-aminoalkyl-cAMP analogs preferentially activate the type I isozyme and 2-phenyl-1, N^6-etheno-cAMP, the only significant one of over 100 analogs tested, preferentially activates the type II isozyme (*104*).

More striking differences between the binding sites of cyclic nucleotide-dependent isozymes are evident when both cAMP-binding sites are considered (Fig. 1). For the cAMP-dependent protein kinase isozymes as well as the cGMP-dependent protein kinase, site-1 (site B) has a slower cAMP dissociation rate (*59, 86, 105–107*) and a relative selectivity for cAMP analogs modified at the C8 carbon of the base (C8 analogs) (*86, 105, 107*), while site-2 (site A) has a faster dissociation rate and a relative selectivity for analogs modified at the C6 position (C6 analogs) with the cAMP-dependent protein kinase isozymes. This site has a relative selectivity for analogs modified at the C1 position (C1 analogs) for cGMP analogs with the cGMP-dependent protein kinase (*108*). [³H]cIMP (site-2 selective), 8-azido-[³²P]cAMP (site-1 selective), and [³H]cAMP-binding experiments establish that binding of cyclic nucleotides at either site stimulates binding at the other site for both type I and type II protein kinase isozymes (*107, 109, 110*). This positive cooperativity in cyclic nucleotide binding is reflected by a positive cooperativity of protein kinase activation (*111*). The cAMP and cGMP kinases have Hill coefficients of approximately 1.6 for their respective activators (*12, 85, 112*). In addition, a synergism of protein kinase activation for each isozyme occurs using a combination of a site 1 and site 2 selective analog (*111, 113*). An important point in this regard is that the analog combinations that cause synergism of protein kinase activation differ for the isozymes. The type I iso-

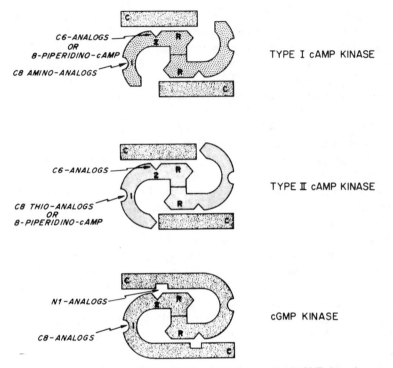

FIG. 1. Proposed structural homologies between cAMP-dependent and cGMP-dependent protein kinases. The cyclic nucleotide analog selectivities for the two intrasubunit binding sites are indicated.

zyme generally shows better relative selectivity at site 1 for C8 amino analogs and the type II isozyme generally shows a better relative selectivity at site 1 for C8 thio analogs, while both isozymes show selectivity at site 2 for C6 analogs. Consequently, a combination of a C6 and a C8 amino analog causes a synergism of type I but not type II activation while a combination of a C6 and a C8 thio analog causes a synergism of type II but not type I activation (*111, 113*).

An unusual analog, 8-piperidino-cAMP, has been characterized (*104*). This analog is selective for site 1 on the type II isozyme but selective for site 2 on the type I isozyme. Therefore, 8-piperidino-cAMP is used with C8 amino analogs as a type I directed-analog pair or used with C6 analogs as a type II directed-analog pair. [^{3}H]cGMP-binding to the cGMP-dependent protein kinase is stimulated by analogs that are selective for site 2, but not by those selective for site 1 (*108*). This enzyme is synergistically activated by a combination of a C1 and a C8 analog. Interestingly, cGMP itself also exhibits synergism with a C1 analog (*108*).

By using cAMP analogs that are highly selective for one site on the regulatory subunit of one isozyme, analog combinations can be chosen that maximize the synergism of the response of a single isozyme in a tissue that contains both isozymes. Ogreid *et al.* (*104*) have used a quantitative *in vitro* approach which should predict the best analog combination to use for *in vivo* experiments so that only a single isozyme is activated. The C8 amino analogs are most highly selective for site 1 on the type I isozyme, but analogs that are optimally selective for other sites on the respective isozymes are unavailable. Analog pairs that have been used in intact tissues (*113*) may not measure the optimal synergism potential for a given isozyme. However, when appropriate site-selective analogs are used in intact tissues at relatively low concentrations so that the protein kinase(s) is activated only slightly above its basal state (5–15% of maximum), a slight elevation of the physiological response occurs. The synergism observed with an appropriate type I- or type II-directed analog pair can then be attributed to a single isozyme with a minimum synergistic activation of the other isozyme (*114*). Regardless of whether optimal pairs of analogs are used or not, optimal synergism will be observed only when the protein kinase and the cell responses are slightly activated above the basal state with single analogs.

While the classification of isozymes into major type I and II classes is useful, the development of more advanced techniques to prove the isozyme structure indicates that at least the type II kinase is represented by a continuum of microheterogeneous forms. The DEAE-cellulose elution behavior of several type II isozymes is different (*7, 35, 112*). Malkinson *et al.* (*115*) found that an apparent murine adipose tissue type I elutes at a higher salt concentration and contaminates the type II isozyme. Toru-Delbauffe *et al.* (*116*) reported a similar result with the type II isozyme from rat thyroid. Robinson-Steiner *et al.* (*85*) found that a second fractionation on DEAE-cellulose is sometimes required to completely resolve the two isozymes. This is especially the case when the type I to type II ratio is high. However, several potential artifacts can result in an incorrect isozyme identification from DEAE-cellulose analysis alone. The free regulatory subunits of both isozymes elute at higher salt concentrations than do the holoenzymes (*35, 117*), and type I regulatory subunit can contaminate the type II holoenzyme. This can be a complication when the holoenzymes are artifactually activated by the homogenization procedure. It is also difficult to rule out that during chromatography free catalytic subunit reassociates with free regulatory subunit and modifies the protein kinase elution behavior. Limited proteolysis of the two regulatory subunit isozymes produces similar fragments with apparent molecular weights of 30,000 to 40,000 (*118*), and the DEAE-cellulose elution behavior of partially proteolyzed type I and type II isozymes may be less well resolved (*119*). It has been demonstrated that limited trypsinization of the type II bovine heart holoenzyme produces a cAMP-dependent dimeric enzyme, containing a proteolyzed regulatory subunit of $M_r = 45,000$–$48,000$ on denaturing gels and a single intact catalytic subunit (*120, 120a*) (see Fig. 4 and Section III,A). A similar molecule was found in aged preparations of type II protein kinase from rat liver (*121*) and adipose tissue (S. J. Beebe and J. D. Corbin, unpublished). Since both the dimeric enzyme from bovine heart (*120*) and the native adipose tissue holoenzyme (*35, 112*) elute at a relatively low salt concentration on DEAE-cellulose and each has a regulatory subunit with an apparent molecular weight by SDS-gel electrophoresis of $M_r = 48,000$ (*120*) and $M_r = 51,000$ (*85, 112*), respectively, isozyme classification can be erroneous. Since the type II, but not the type I, regulatory subunit is autophosphorylated (*57, 122, 123*), a simple method is available to aid the identification of the type II isozyme. The type I isozyme can be identified since MgATP inhibits cAMP binding to this holoenzyme but not to the type II holoenzyme (*57, 124*).

When the existence of the two main isozyme types of the cAMP-dependent protein kinase was first described (*35, 112*), it was pointed out that there is more than one class of type II in the same animal species. It was found that rat adipose tissue and rat heart type II elute at different NaCl concentrations from DEAE-cellulose and that mixtures of the enzymes from the two tissues can be separated by chromatography on these columns. In spite of these physical differences, they were found to exhibit similar kinetic properties. From immunological studies of type II regulatory subunit, Erlichman *et al.* (*125, 126*) described neural and nonneural subclasses. Corbin *et al.* (*85, 112*) suggested that this classification may not include all type II subforms. By comparing mobility upon gel electrophoresis, Stokes radii measurements, and the effect of autophosphorylation and proteolysis on the type II regulatory subunit from different tissues and species, several different forms could be resolved (*85*). An operational classification separates type II regulatory subunits into those that shift mobility on SDS gels after autophosphorylation (type IIA) and those that do not (type IIB) (*85*). These forms are further distinguished by apparent molecular weights. An examination of the distribution of types IIA and IIB in different species of heart tissue revealed a definite pattern (Fig. 2) (*85*). Type IIB is present in hearts of rodents, lagomorphs, and primates: while type IIA is present in hearts of carnivores and ungulates. These two broad groups of species diverged from each other about 70 million years ago. That type IIB is a more recent evolutionary development is suggested by the finding of type IIA in chicken heart. More than one form is also found in different tissues of the same species. Bovine lung contains equal amounts of $M_r = 56,000$ type IIA (like bovine heart) and $M_r = 52,000$ type IIB. Bovine brain contains a small amount ($\sim 15\%$) of this same type IIA and a predominant $M_r = 52,000$ type IIB. The same forms are also found in different species. Both rat brain and bovine brain contain type IIB. The rat adipose tissue ($M_r = 51,000$ type IIB) and the bovine heart ($M_r = 56,000$ type IIA) holoenzymes differ in several other properties, including Stokes radius, calculated molecular weight, and frictional ratio (*85, 112*). Even though the adipose tissue enzyme is quite similar to other type II forms in the kinetics of cAMP action, the

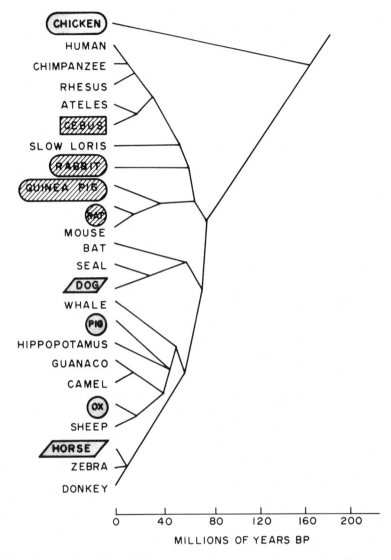

FIG. 2. Distribution of types IIA and B in hearts of various species. Shaded symbols represent type IIA and cross-hatched symbols represent type IIB. Different geometric symbols indicate molecular weight differences as determined by SDS-polyacrylamide gels: circle, $M_r = 56,000$ (dephospho-); parallelogram, $M_r = 54,000$ (dephospho-); ellipse, $M_r = 52,000$; rectangle, $M_r = 51,000$. The evolutionary tree is from Goodman (103a).

cAMP-binding sites are different from them based on the kinetics of activation by certain cAMP analogs. Several analogs modified at the N^6-position of the adenine ring have higher apparent K_a values for protein kinase activation for the adipose tissue enzyme than for the bovine heart and several other heart isozymes (112). The $M_r = 51,000$ type IIB enzymes from bovine brain and monkey heart, like that from adipose tissue, show similar but less striking differences compared to the bovine heart and other heart isozymes (S. J. Beebe and J. D. Corbin, unpublished).

Jahnsen et al. (127) purified and characterized three isoforms ($M_r = 54,000$, 52,000, and 51,000) of the type II regulatory subunits from rat ovarian granulosa cells using immunological, electrophoretic, photoaffinity labeling, and phosphorylation criteria. Antiserum against rat heart type II regulatory subunit recognizes the $M_r = 54,000$ form while antibovine heart type II regulatory subunit serum recognizes the other two forms. The $M_r = 51,000$ and 52,000 forms, unlike the $M_r = 54,000$ form, are regulated by hormone in preovulatory follicles and do not readily change mobility on SDS-gels following phosphorylation. In addition, the $M_r = 54,000$ form has a distinct peptide map. The authors suggest that the hormone-regulated forms are distinct gene products.

Other apparent type II forms have been described. Schwartz and Rubin (128) have identified two distinct forms of $M_r = 54,000$ and 52,000 from Friend erythroleukemic cells using monoclonal antibodies and peptide map analysis. Further distinctions between subclasses of the type II isozyme are indicated by differences in the amino acid composition, two dimensional tryptic peptide maps, and the peptide containing the autophosphorylation site between porcine brain and skeletal and cardiac muscles (129). Weldon et al. (130) have reported that the type II regulatory subunit from bovine brain binds only 2 mol of cAMP/ mol subunit, although the two different classes of binding sites are present. This is in contrast to bovine heart and other regulatory subunit isozymes which bind 4 mol cAMP/mol subunit. The explanation for the lower total cAMP-binding to the brain subunit requires further study. Furthermore, two different monoclonal antibodies with antigenic determinants at the NH_2-terminal third of the heart regulatory subunit class react very poorly with the brain subunit. This information, in conjunction with comparative amino acid sequence analysis of this region suggest that these two forms of type II regulatory subunit are unique gene products (130).

In summary, the demonstration of microheterogeneous forms of the type II isozyme tends to blur the distinction between isozyme forms. Consequently a careful examination using a series of tests is required to positively identify an isozyme. Although the number of isozymes studied in detail is small, the kinetic differences between the isozymes is most reliable. In addition, no type I isozymes have been reported to undergo autophosphorylation.

B. CYCLIC GMP-DEPENDENT PROTEIN KINASE

The soluble cGMP-dependent protein kinase from bovine lung is a dimer composed of two identical monomeric chains each containing 670 amino acid residues (131). The monomer has a molecular weight of 76,331 determined from the amino acid sequence. Each chain contains a regulatory domain with two homologous cGMP-binding domains and a catalytic domain carboxyl-terminal to the binding domains (131). The dimer has a molecular weight of 152,662. This is in fairly good agreement with the molecular weight of 150,000–162,000 determined from SDS–gel electrophoresis (22, 66) or calculated from the sedimentation coefficient (6.7S–7.8S) (66, 132), and the Stokes radius (5.0–5.2 nm) (66, 132). The molecular weight of the cGMP kinase is therefore only slightly less than that of the cAMP kinase and considerably higher than that of the catalytically active component of the cAMP kinase. The cGMP-dependent kinase has an isoelectric point of 5.4 and is probably somewhat less asymmetric than the cAMP kinase since both the axial ratio of 7.4 and the frictional ratio of 1.4–1.5 are slightly less (132). Like the cAMP kinase, the cGMP kinase is autophosphorylated (70). Although this occurs in the presence of either cGMP or cAMP (25, 133), autophosphorylation with cAMP is distinctly different from that with cGMP. First, 2–6 mol phosphate per mol subunit is incorporated, instead of 0.75 mol/mol with cGMP (134). Stimulation of phosphate incorporation into serines-50 and -72 and theonines-58 and -84 (134) occurs with the occupancy of only one of the two intrasubunit binding sites (135). The most rapidly phosphorylated residue, theonine-58, is the major site phosphorylated in the presence of cGMP (134). Second, autophosphorylation in the presence of cAMP, but not cGMP, causes a 10-fold reduction in the concentration of cAMP required for half-maximal activation of phosphotransferase activity (133). Furthermore, following maximal autophosphorylation (total of ~4 mol phosphate/mol subunit) the dissociation rate of cGMP from site 1, measured with an excess of cold cGMP, is decreased approximately 10-fold. Autophosphorylation therefore primarily affects the binding at site 1 and elimates cooperative binding at this site (136). Other properties of the soluble cGMP-dependent protein kinase have been more completely reviewed by others (25–29) and the effects of cyclic nucleotide analogs are discussed in conjunction with the cAMP-dependent protein kinase in Section II,A.

De Jonge (68) has characterized an intestinal, brush border membrane-specific cGMP-dependent protein kinase (type II) that is distinct from the enzyme characterized in lung, heart, and smooth muscle (type I). In contrast to the soluble enzyme, the brush border cGMP kinase has an apparent molecular weight of 86,000 and is anchored to the membrane or to the contractile core of the microvilli by a 15,000-dalton fragment containing a site which is preferentially phosphorylated *in situ* in a cGMP-dependent manner. An apparent $M_r = 71,000$ form, which contains the cGMP-binding and catalytic domains, is generated from the apparent $M_r = 86,000$ form by proteolysis. The brushborder cGMP kinase is further differentiated from the soluble enzyme by isoelectric point (7.5) and phosphopeptide pattern following limited proteolysis. However, the brushborder cGMP kinase is immunologically similar to the soluble form and also undergoes autophosphorylation in the absence of cGMP.

C. EVOLUTIONARY RELATIONSHIPS AMONG CYCLIC AMP-BINDING PROTEINS AND PROTEIN KINASES

The first proposal of homology between protein kinases was based on similarities in physical and kinetic properties, and in amino acid composition, between cAMP- and cGMP-dependent protein kinases (*23, 29*). A model for structural homology between the two enzymes was also proposed (*25*), and it was suggested that the two intrasubunit cyclic nucleotide-binding sites on each enzyme were evolved by contiguous gene duplication (*10, 21, 25, 86*). Structural models, slightly modified from the original, are shown in Fig. 3. Conclusive proof of a common progenitor for these kinases was the subsequent finding of a high degree of amino acid sequence homology in both the regulatory and in the catalytic domains (*131*). In the case of the regulatory component, the homology occurs not only between the two intrasubunit cyclic nucleotide-binding sites in the carboxy terminal two-thirds of this sequence, suggesting a tandem gene duplication, but also between the corresponding binding sites of the two different cyclic nucleotide-dependent protein kinases. There is much less apparent homology between these kinases in the amino terminal one-third of the primary sequence, which contains the dimerization domain. As expected, the regulatory subunits of the two main isozymic forms of cAMP-dependent protein kinase, types I and II, also exhibit strong homology. The finding of several microheterogeneous forms of type II regulatory subunit, which can generally be classified into types IIA and IIB subclasses (*85*), and the finding that the primary amino acid sequence of a short segment of at least one type IIB form is very different from that of type IIA (*85*), are additional lines of evidence that the entire primary sequence of the regulatory subunit has not been well conserved.

The regulatory and catalytic components of cyclic nucleotide-dependent protein kinases have been independently derived in the course of evolution since there is no apparent homology between them. Each of them is a member of one or the other of two distinct families, the cAMP-binding proteins and protein kinases, as illustrated in Fig. 3. All higher eucaryotes examined so far contain the regulatory components of cyclic nucleotide-dependent protein kinases. The enzymes in fungi (*137–145*), slime mold (*146*), and insects (*147–149*) appear to have the same basic structure as the mammalian enzymes. Vardanis has reported

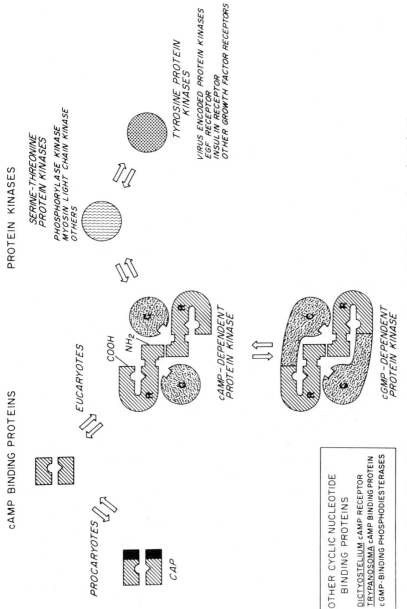

Fig. 3. Possible evolutionary relationships among cyclic nucleotide binding proteins and protein kinases.

the existence in insects of a protein kinase that lacks specificity with regard to cAMP and cGMP (*150*), and of a cAMP-dependent protein kinase in nematodes, which does not dissociate into subunits during activation (*151*). There is no consistent evidence of cyclic nucleotide-dependent protein kinases in procaryotes, but these organisms contain a cAMP-binding protein (*152, 153*) named catabolite gene activator protein (CAP). This protein serves a similar function in *E. coli* as the regulatory subunit serves in the mammalian liver: when sugar availability declines, cAMP is elevated, binding of cAMP to the respective protein occurs and this leads to increased sugar availability. In mammals sugar is made available by increased glycogenolysis and gluconeogenesis; in bacteria sugar is made available by the switching on of catabolic gene operons for proteins involved in various sugar transport and utilization. Thus, CAP contains, in addition to a cAMP-binding domain, a domain at the carboxy terminus (indicated in black in the model of Fig. 3) which binds to DNA and regulates gene expression. This domain is known to have structural and sequence homologies with the DNA-binding protein of bacterial and viral repressors (*154, 155*). The cAMP-binding domain of CAP is strongly homologous with the cAMP-binding domains of cyclic nucleotide-dependent protein kinases (*156*). From both X-ray crystallographic analysis of cAMP-bound CAP and from sequence homology between CAP and regulatory subunit, certain predictions of cAMP binding to cyclic nucleotide-dependent protein kinases have been made (*156*). It has been suggested that each cAMP-binding site contains a β-roll structure and perhaps an α-helix. If the regulatory components of the kinases are like CAP they could contain a deep pocket for cAMP binding between the β-roll and α-helix. Each site presumably also contains an essential arginine for binding the ribose phosphate moiety of cAMP.

The CAP protein is an asymmetric dimer, the monomeric components presumably containing identical amino acid sequences (*155*). Two molecules of cAMP bind to the dimer, and each molecule binds to both monomers. The binding of cAMP to CAP exhibits cooperativity (*157*). In several respects the two cAMP-binding sites of the dimeric CAP resemble the two intrasubunit cAMP-binding sites of the regulatory components of the protein kinases. This suggests that one possible evolutionary precursor of the cAMP-binding sites of each protein is a dimeric, asymmetric precursor as illustrated in Fig. 3. CAP could have evolved from it by acquiring a DNA-binding domain through gene fusion. The regulatory subunit could have evolved from it by acquiring at least a catalytic subunit-binding domain and a dimerization domain at the amino terminus. In addition, by gene duplication and fusion, two cAMP-binding sites on the same protein chain could have been produced. These two sites might then have similar relative emplacements as the two sites on the dimeric CAP. If this theory is correct, then one might predict that the two cooperative intrasubunit cyclic nucleotide-binding sites of the regulatory component of protein kinase could be close to each other, and that each cyclic nucleotide binds to components of both binding domains.

Other cyclic nucleotide-binding proteins which are not components of a protein kinase have been described. These should be examined to see if they are related in evolution to the CAP proteins and to the cyclic nucleotide-binding sites of cyclic nucleotide-dependent protein kinases. In the cellular slime mold *Dictyostelium discoideum*, cAMP acts as a chemotactic agent to induce cell aggregation during exhaustion of food supplies (*158*). This effect is mediated by a cell surface receptor which has physical and kinetic properties different from either CAP or the cyclic nucleotide-dependent protein kinases (*159, 160*). In species of *Trypanosoma*, a parasitic lower eucaryote, cAMP is believed to be involved in the regulation of cell growth and differentiation. These organisms are low or completely devoid in cyclic nucleotide-dependent protein kinase activity, but they contain a cAMP-binding protein of $M_r = 62,000$ which differs from the kinase or CAP protein in several respects (*161*). Another well-known cyclic nucleotide-binding protein is the family of phosphodiesterases which possess a cGMP-specific-binding site in addition to a catalytic site (*162–164*).

As seen in Fig. 3, the protein kinase family of proteins includes the catalytic component of cAMP- and cGMP-dependent protein kinases. This family can conveniently be divided into serine–threonine-specific protein kinases and tyrosine-specific protein kinases (see Chapter 6). As is the case for cyclic nucleotide-dependent protein kinases, two Ca^{2+}–calmodulin regulated protein kinases, phosphorylase kinase and myosin light chain kinase, catalyze phosphorylation of serine or threonine in protein substrates. These kinases are also known to be closely related to the cyclic nucleotide kinases by amino acid sequence homology (*165, 166*). More distantly related to the catalytic component of cyclic nucleotide-dependent protein kinases is the group of tyrosine-specific protein kinases (*167*). A lysyl residue that binds the ATP analogue affinity label [(fluorosulfonyl)-benzol]adenosine is apparently conserved in all of the protein kinases and could represent a part of the active site that interacts with the terminal phosphate of ATP (*162, 164*). It is not unreasonable to expect that other protein kinases, which have similar physical and kinetic properties but which have not yet been examined at the protein chemical level of those shown in Fig. 3, will be shown to be in this same family of proteins related to each other in evolution.

Speculation is in order for the mechanism by which "marriage" of ancestral protein kinase catalytic proteins and inhibitory proteins such as the regulatory subunit occurred. One possibility is that originally the regulatory protein was simply a protein kinase substrate. Through mutational alteration, this protein could have developed such a strong affinity that it prevented, in a competitive manner, the phosphorylation of other protein substrates. Support for this theory is derived from the fact that the regulatory components of the cyclic nucleotide-dependent kinases contain either a phosphorylation site or phosphorylation analog site for the catalytic component of the enzymes. Furthermore, the phosphorylation site is believed to be located in the catalytic component inhibitory region of the regulatory component (*6,* see Ref. *174*), even though the phosphorylation site is known not to be the only component of the inhibitory domain of the regulatory subunit (*4*). Finally, it is known that several different protein kinases catalyze autophosphorylation (*10*). Whether or not the phosphorylation site of any of these enzymes is in an inhibitory domain has not been established.

IV. Mechanism of Action

A. Mechanism of Regulatory Subunit Action

The precise mechanism by which the regulatory subunit inhibits the catalytic subunit is not certain, although some progress has been made in understanding this phenomenon.

In higher organisms all of the isozymes that have been characterized have a tetrameric structure (R_2C_2), but the functional unit of these forms appears to be a dimer (RC). This proposal is based on the observation that proteolytically modified type II holoenzyme (R'C) from either bovine liver (*121*), rabbit skeletal muscle (*120*), or bovine heart (*120, 120a*) behaves as a dimer (Fig. 4) on sucrose density gradients and gel filtration. The modified enzymes of muscle and heart have been studied in detail and it has been determined that the site of proteolysis, using either endogenous proteases or added trypsin, is in the amino terminal one-fourth of the molecule (*120, 120a*). It has been further established that the carboxy terminal monomeric fragment, which contains the two cAMP-binding sites and the autophosphorylation site, retains virtually unaltered ability to inhibit the catalytic subunit. Thus, even though the native dimeric structure of regulatory subunit has been highly conserved, it is apparently not required for the inhibitory action. Perhaps all forms of holoenzyme do not contain a dimeric regulatory subunit.

The inhibitory domain(s) of the type II regulatory subunit must be contained, at least in part, around the autophosphorylation site (residue 95) of the primary sequence (total residues = 400). This conclusion is based on the observation that modifications of this region abolish the inhibitory activity (*6, 10, 168,* see Ref. *174*), and supports the suggestion (*6, 10, 174*) that the regulatory subunit inhibits the catalytic subunit by acting as a "substrate analog" of high affinity which shields the protein or peptide substrate binding site of the catalytic subunit. The autophosphorylation site could be the substrate analog site in question. From the magnetic resonance studies of Granot and co-workers (*9*), the type II regulatory subunit might be considered to be a "dead-end" substrate since the respective holoenzyme binds MgADP with high affinity. Type I isozyme does not undergo autophosphorylation, but it does contain a homologous autophosphorylation analog site (*80*) that could serve a similar purpose. In addition to the autophosphor-

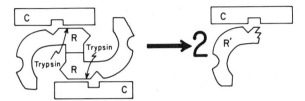

FIG. 4. Formation of the functionally competent dimer by trypsin treatment of the tetrameric holoenzyme.

ylation primary sequence of type II, the inhibitory domain must also include a structure(s) of higher order since heat denaturation destroys the regulatory subunit inhibitory activity, but not its ability to act as a substrate for autophosphorylation (*174*). With regard to this second postulated structure, since autophosphorylation can take place, the regulatory subunit does not block the MgATP-binding site of the catalytic subunit, but it could block the transfer of phosphate from ATP to substrate (*10*). It follows that the inhibitory action of the regulatory subunit could be exerted in two ways. First, it may act as a competitive inhibitor of high affinity; and second, it could block the ability of catalytic subunit to transfer phosphate from ATP to substrate.

Several isoforms of a heat-stable protein kinase inhibitor of high specificity for the cAMP-dependent protein kinase are present in mammalian tissues (*169–171*). Although the physiological role of this inhibitor has not been established conclusively, it has been quite useful in studies of mechanism and cellular role of the kinase. As is probably the case for the regulatory subunit, this inhibitor may act at least in part by serving as a competitive substrate analog of the cAMP-dependent protein kinase (*172*). Scott *et al.* have isolated a proteolytically derived fragment of the inhibitor that retains inhibitory activity (*173*). This fragment contains a substrate analog sequence as indicated by the presence of tandem arginines followed by alanine in the usual position of a phosphorylatable serine or threonine of natural protein kinase substrates. These investigators have synthesized a potent (K_i = 0.8 μM) peptide kinase inhibitor of an amino acid sequence corresponding to the first twenty amino acids of the proteolytically derived fragment. The amino acid sequence of this peptide resembles the sequence in the regulatory subunit which, as previously noted, is part of the domain that inhibits the catalytic subunit.

Several findings are indicative that a ternary complex is formed between cAMP and holoenzyme during activation of the kinase (*175–177*). Although difficulties have been encountered in isolating and studying a stable ternary complex, several reports suggest that this is possible. Cobb *et al.* (*64, 64a*) separated and characterized two distinct peaks of homogeneous Type II bovine heart tetrameric holoenzymes. Using HPLC-DEAE chromatography, both peaks had similar molecular weights as calculated from their Stoke's radii and $s_{20,w}$

values or determined by denaturing electrophoresis. They contained equimolar amounts of regulatory and catalytic subunits and were highly cAMP-dependent. One of these peaks was identified as a relatively stable ternary complex which appeared to be holoenzyme (R_2C_2) with approximately half-filled cAMP binding sites. This ternary complex was further distinguished from the cAMP-deficient holoenzyme by an enhanced positive cooperativity in nonequilibrium [^{3}H]cAMP binding and in kinase activation. This complex could represent an enzyme form that is primed for activation. At least two other reports have provided evidence for the presence of a ternary complex between cAMP and a trimeric form of cAMP-dependent protein kinase. Rangel-Aldao and Rosen (*122a*) and Connelly *et al.* (*83*) identified an intermediate assigned an R_2C structure. In contrast to the ternary complex reported by Cobb *et al.* (*64, 64a*), these trimer complexes were formed during reassociation of catalytic and regulatory subunits. The significance of these complexes remain to be fully determined.

It is assumed that cAMP binds in a step-wise manner to the four binding sites of the holoenzyme. The binding exhibits positive cooperativity with reported Hill constants of 1.6–1.8 for kinase activation (*10*). More direct demonstration of positive cooperativity is by the use of site-selective cAMP analogs. These investigations have shown that binding of cyclic nucleotide to either of the two intrasubunit sites (sites 1 and 2) has a strong stimulatory effect on binding to the other site (*109*). There may be several mechanisms for these stimulations. Although it cannot be ruled out that all four binding sites of the holoenzyme are accessible to the first mole of cAMP, the results of several experiments using type II suggest that binding to site 2 is blocked by the presence of catalytic subunit, but this site is made available by site 1 occupancy (*12, 109*). Thus, cAMP probably binds first to site 1, which then stimulates site 2 to bind. Using the free regulatory subunit of type I, it has been found that when cAMP occupies site 2, the dissociation of cAMP from site 1 is retarded (*179*). These characteristics could explain at least in part the observed stimulation of binding to both sites 1 and 2 noted above. The additional finding, using the free regulatory subunit of type I, that binding to site 2 of one subunit retards dissociation of cAMP from site 2 of the other subunit could also explain part of the stimulation (*15*). This latter intersubunit stimulatory effect is probably minor using type II since the cooperative binding is essentially unaltered in the dimeric (RC) holoenzyme as compared with the native tetramer (R_2C_2) (*2*). However, it has been calculated that an intersubunit positive cooperativity is necessary in order to explain the high Hill constant for cAMP activation of the protein kinase (*180*).

The precise mechanism by which binding of cAMP to the enzyme causes activation is not yet clear. At the time of the discovery of the two intrasubunit cAMP-binding sites, it was suggested that, because of the high affinity of each, both are involved in activating the kinase (*6*). Subsequent experimental evidence has verified this claim (*111, 181*). The exact involvement of site 1 or site 2, and of the binding of the four total cAMP molecules to this enzyme in the activation process is not certain, however. Binding of cAMP to one of the sites could simply serve the function of stimulating the other site to bind without being directly involved in kinase activation. From the results of fluorescence–polarization spectroscopy studies, Seville *et al.* have concluded that the binding of two cAMP molecules to the holoenzymes is sufficient to cause the dissociation of both catalytic subunits (*182*). Studies of cGMP-dependent protein kinase suggest that at least binding to site 1 of this enzyme is directly involved in partially activating the kinase (*21*). Binding to site 2 causes an additional increase in enzyme activity. Even though the cAMP-dependent protein kinase is a homologue of this enzyme, it cannot be assumed that it behaves identically with respect to kinase activation. Since the functional unit of the enzyme is a hetero-dimer (RC), this implies that one C at a time is replaced from the native tetramer (R_2C_2) during activation by cAMP. This suggests the existence of an intermediate form (R_2C) (*83, 122a*).

The process of cAMP-dependent protein kinase inactivation has also been investigated. It has been known for many years that the catalytic subunit induces release of cAMP from the R·cAMP complex (*183*). For type I enzyme, this process is stimulated by the presence of MgATP (*183*). Øgreid and Døskeland have examined the effect of the catalytic subunit on the release of cAMP from each of the binding sites (*184*). From the experimental results, it has been determined that cAMP is first released from site 2, and that site 2 vacancy retards the release of cAMP from site 1. It has been also determined that both of the cAMP molecules are released from one subunit of the dimer before cAMP is released from the other subunit.

B. Mechanism of Catalytic Subunit Action

The catalytic subunit action has been reviewed extensively (*23*) and will be addressed only briefly in this chapter. The available evidence indicates that the catalytic subunit has a single active site on each monomer. The monomer incorporates a single substituent when photoaffinity labeled either with the 2′,5′-dialdehyde derivative of ATP (*185*) or with *p*-fluorosulfonylbenzoyl adenosine (*186, 187*). In the latter case, the label is attached to lysine-71 of the primary amino acid sequence of the catalytic subunit. The sequence of this region of the molecule, which presumably represents at least part of the active site, is Leu-Val-Lys-His-Lys-Glu-Thr-Gly-Asn-His-Phe-Ala-Met-Lys*-Ile-Leu-Asp-Lys-Glu-Lys-Val-Val-Lys-Leu-Lys-Gln-Ile. The catalytic subunit can also be labeled in a single site per monomer with *o*-phthalaldehyde (*188*) or with peptide substrates containing reactive groups such as 3-nitro-2-pyridinesulfenyl (*189*).

It is well known that sulfhydryl modifying agents abolish activity of the catalytic subunit and that MgATP protects against these agents (*54, 55, 77, 79,*

190). Of two -SH groups in the catalytic subunit, cysteine-199 appears to be located in the active site (*189*). The amino acid sequence around this residue is Gly-Arg-Thr-Thr-Thr-Leu-Cys*-Gly-Thr-Pro-Glu-Tyr-Leu-Ala (*79*). From the observation that catalytic subunit modified at both lysine-71 and cysteine-199 does not form an isoindole derivative on treatment with *o*-phthalaldehyde, together with consideration of transition energy measurements on the *o*-phthalaldehyde modified enzyme, the lysine and cysteine residues in the active site are about 3 Å apart and are proposed to be in a hydrophobic environment (*188*).

From NMR (*191*) and ATP analog specificity studies (*192*), it is known that the ATP is bound in the *anti*-conformation in the active site. Data from several laboratories (*191*) indicate that the catalytic subunit utilizes β,γ-bidentate MgATP with the geometry shown below. The coordinates of the binding of this complex in the active site have been mapped (*191*).

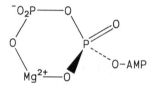

There have been numerous inconsistencies concerning the kinetic mechanism by which the catalytic subunit catalyzes the transfer of the γ-phosphate of ATP to substrate (*193–198*). Whitehouse *et al.* have carefully analyzed the reaction mechanism using the enzyme, MgATP, peptide substrate, substrate analogs, and protein kinase inhibitor (*198*). They have concluded that the steady-state kinetics follow an ordered bi bi mechanism in which ATP binds first. The terminal anhydride bond is then proposed to undergo a conformational change, induced either as a consequence of ATP binding *per se* or by the subsequent binding of protein substrate. Catalysis then ensues, leading to the sequential release of phospho-substrate and ADP.

The substrate specificities of cAMP- and cGMP-dependent protein kinases have been reviewed extensively (*13, 14, 31, 38, 39*). Although there are slight differences *in vitro* in specificity, in general, they both require a pair of basic amino acids on the amino terminal side of phosphorylated serines or threonines. Small peptides with amino acid sequences of this type are excellent substrates for the enzymes. Granot *et al.* determined by a process of elimination that if a particular conformation of protein or peptide substrate is required in the active site, then it is probably a coil (*199*). From results of induced circular dichroism, Reed and Kinzel have concluded that binding of the protein substrate results in a conformational change at the ATP-binding site (*200a*). This change apparently takes place in at least two steps, one dependent on the presence of a phosphorylatable serine or threonine in the substrate, and the other dependent on the pair of basic amino acids. The authors have also postulated from the results of competition experiments that the binding site closes over the substrate protein following the initial binding. The catalytic subunit from porcine heart has been crystallized, which should allow for more detailed studies of its structure and mechanism (*200*).

V. Biological Role of Protein Kinases

A. CRITERIA TO ESTABLISH BIOLOGICAL ROLE

Several approaches have been used to elucidate the roles of cAMP and the cAMP-dependent protein kinase in cell function. Sutherland (*4*) first proposed criteria that should be satisfied to justify that a given hormone acts through cAMP. These criteria include the demonstration that the hormone activates adenylate cyclase and elevates the intracellular level of cAMP. Additionally, the response should be potentiated by phosphodiesterase inhibitors and should be mimicked by the addition of exogenous cAMP or cAMP analogs. The discovery of the cAMP-dependent protein kinase (*5, 56*) led to the hypothesis that all of the effects of cAMP are mediated by phosphotransferase reactions catalyzed by this enzyme. Subsequently, Krebs proposed a set of criteria that should be satisfied before a cAMP-mediated response could be established to be carried out by the cAMP-dependent protein kinase (*201*). These criteria have also been reevaluated (*11, 13, 37*). It should be demonstrated that a protein substrate, which is shown to be involved in the response, can be stoichiometrically phosphorylated (and dephosphorylated) at an appropriate rate *in vitro* and *in vivo* and that this is correlated with an appropriate coordinant change in the function of the substrate. It is the *in vivo* aspects of these criteria that are the most difficult to satisfy.

In many instances the measurement of the cAMP-dependent protein kinase activity ratio (kinase activity in the absence of cAMP divided by activity in the presence of cAMP) in crude extracts from hormone-stimulated intact cells or tissues is a valid indication of the activation state of the enzyme. The cooperative binding of cAMP to the regulatory subunit causes a cooperative activation of protein kinase (*12, 85, 112*). It can therefore be expected that small changes in the level of cAMP will result in large changes in the activation of protein kinase. Consequently, since this enzyme is at a pivotal point in the overall metabolic regulation of the cell, an accurate determination of the protein kinase activity ratio is probably a more informative indicator of the potential of a tissue to respond to hormone stimulation than is determination of the level of cAMP. The difficulty in detecting small changes in the level of cAMP may be partially responsible for some controversies concerning the involvement of cAMP in mediating certain physiological responses. Although activity ratio measurements can be a useful and valid method for evaluating hormone action and protein kinase activation, it is critical that these studies are carried out carefully and that certain pitfalls are recognized and avoided. Since Flockhart and Corbin have reviewed the method and potential pitfalls (*10, 202*), only brief comments and later developments are included here.

The most fundamental requirement for valid activity ratio determination concerns maintaining the intracellular protein kinase activation state following homogenization. Postextraction activation or inactivation causes an overestimation or underestimation of the activity ratio, respectively. Both of these potential artifacts can be controlled (*10, 202*). The most important advance in this procedure is the use of a synthetic peptide as substrate (*203*). In general, use of this substrate instead of histone is recommended since it is more specific and sensitive, does not cause protein kinase dissociation, is not inhibited by NaCl under any recommended homogenizing technique (S. J. Beebe and J. D. Corbin, unpublished), and is a poor substrate for phosphoprotein phosphatases. The cGMP-dependent protein kinase also phosphorylates this substrate but this generally is not a problem since this kinase activity is a minor component of most tissues.

Another approach used to define the biological role of the cAMP-dependent protein kinase is genetic analysis using mutant cells that have a single lesion in the pathway of cAMP action, preferably an altered or absent protein kinase [see Ref. (*204*) for a review]. If the cAMP-dependent protein kinase were required for a given response, a mutant deficient in protein kinase would be unable to elicit the response. Such mutant cells provide advantages since, in contrast to the wild-type cells, their growth is not inhibited by cAMP or cAMP analogs. Several criteria have been used to determine that a phenotypic variation is due to a mutation (*205*). Cyclic AMP-resistant cells must arise spontaneously, be clonally inherited, and the phenotype must be stable. In addition, the mutation must arise at a frequency consistent with mutation in microbial systems and increase in frequency in the presence of known bacterial mutagens. Conclusions derived from these studies are valid only if the mutation in question is not a pleiotropic one affecting multiple cellular functions.

Several other approaches have been used to establish that the protein kinase mediates a given response. The direct introduction of the catalytic subunit of protein kinase or its specific heat-stable inhibitor into a cell can be achieved by microinjection. This approach was used in *Xenopus* oocytes (*206*). Although the technique is generally restricted to relatively large cells, cAMP and the subunits of the cAMP-dependent protein kinase were successfully injected into isolated guinea pig ventricular myocytes (*207, 208*). A similar method was used to fuse vesicles with isolated or cultured cells. Culpepper and Liu (*209*) fused cAMP-containing and 8-azido-cAMP-containing vesicles with H4 and H35 hepatoma cells, and Boney *et al.* (*210*) incorporated catalytic subunit and protein kinase inhibitor into H35 hepatoma cells using protein-loaded human erythrocyte ghosts. Bkaily and Sperelakis (*211*) fused cultured myocytes with phosphotidylcholine liposomes containing catalytic subunit and protein kinase inhibitor.

Another approach that has been used is incubation of intact cells or tissues in the presence and absence of hormone or other agent. The potential protein substrate is then isolated and phosphorylated *in vitro* by addition of [^{32}P]ATP and the catalytic subunit of protein kinase. The decrease in ^{32}P content of the protein substrate that occurred in the presence of the agent compared to the control can be assumed to have occurred *in vivo* (*212*).

A technique has been used that is based on a variation of the Sutherland criteria requiring that cAMP analogs mimic a given cAMP-mediated response. It is basically a shortcut to the Krebs criteria, which were designed to establish that a given response is mediated by the cAMP-dependent protein kinase. The procedure is modified to account for the presence of two different intrasubunit cyclic nucleotide-binding sites on the regulatory subunit of protein kinase. Intact cells are incubated with two cAMP analogs either alone or in combination (*113, 114, 114a*). The procedure takes advantage of two properties of cAMP analogs, (*a*) the passage of analogs across cell membranes and the direct activation of the protein kinase; and (*b*) the site selectivity of the two different analogs each of which binds to one or the other of two distinct intrachain binding sites (Fig. 1). This occurs in such a way as to facilitate the positively cooperative activation of protein kinase. The advantages are that it is highly specific to the protein kinase isozymes and it does not require the measurement of cAMP binding to or activation of protein kinase. Instead, measurements of the response are made due to protein kinase activation in the intact tissue. Furthermore, proper use of this procedure can potentially differentiate between type I and type II protein kinase-mediated metabolic events. Since the technique measures the metabolic response itself, it does not specifically require the identification of the phosphorylated protein which leads to the response. Although it does establish that the response is due to protein kinase activation, it does not specifically correlate the response to a phosphorylation reaction. For example, it does not rule out that the free regulatory subunit is involved in the response. In addition, the use of analogs cannot necessarily predict hormone responses, but it can modify them. For instance, cAMP analogs have been shown to block the endogenous cAMP elevation in glucagon-stimulated hepatocytes (*213*). Furthermore, the insulin blockade of adipocyte lipolysis or hepatocyte glycogenolysis depends on the cAMP analog used to elicit the response (*214*). Several other procedures have been used to correlate a metabolic response with the activation of one or the other of the protein kinase isozymes. These are discussed in Section V,C.

B. DISTRIBUTION OF ISOZYMES

An examination of the relative tissue and species distribution of the isozymes of the cAMP kinase does not provide proof for specific roles for the isozymes but does present some interesting considerations. While the ratios of the regulatory subunit to the catalytic subunit of the cAMP-dependent protein kinase are relatively constant among various tissues (90, 216), some tissues contain either predominantly type I or type II protein kinase while others contain an equal mixture of the two isozymes (7, 14). For instance the brain, stomach mucosa, and adipose tissue from many species contain predominantly the type II isozyme while rabbit psoas muscle, bovine corpus lutea, rat testes, and bovine neutrophils contain primarily the type I protein kinase. Rabbit soleus muscle, rat liver, rabbit reticulocytes and erythrocytes, and human neutrophils contain a mixture of both isozymes. Cardiac tissues are particularly interesting regarding isozymes. While bovine and guinea pig heart contain predominantly type II, rat and mouse heart contain type I and rabbit and human heart contain an equal mixture of the isozymes (215). In addition, cardiac tissue has been shown to contain a cAMP-dependent protein kinase in the particulate fraction (216), which has been reported to be primarily associated with sarcolemma (217–223) and sarcoplasmic reticulum (224–228). Although in most cells and tissues the enzyme is predominantly in the soluble fraction, membrane-bound cAMP-dependent kinase activity has also been found in brain (229), erythrocytes (230), corpus luteum (231), sperm (232, 233), and thyroid (116). The membrane-associated kinase is usually found to be the type II isozyme, but the type I isozyme has been reported in membranes from erythrocytes (230), sperm (232, 233), and thyroid (116). The holoenzyme is apparently attached to particulate material by its regulatory subunit (216). However, the catalytic subunit has been reported to nonspecifically bind to membranes under conditions of low ionic strength (234). Both the particulate-bound holoenzyme and the regulatory subunit are readily solubilized by a number of conditions, which, according to the criteria of Singer (235), classifies them as "peripheral," as opposed to "integral" proteins.

Immunochemical studies, using antibodies that are specific for the regulatory subunit of the cyclic nucleotide-dependent protein kinases, have allowed a more precise localization of the isozymes and subunits. The advantages and limitations of these techniques have been reviewed (236–238). An increase in nuclear protein kinase has been reported in regenerating liver (239), ACTH-stimulated adrenal medulla (240), and growing human breast cancer cells (241). Van Sande et al. (242) reported that the type II regulatory subunit is present in the nucleus of thyroid follicular cells and the type I regulatory subunit, catalytic subunit, and cGMP-dependent protein kinase are primarily in the cytoplasm and associated with the apical membrane. Jungmann and co-workers, using an indirect colloidal immunogold technique (243), found that only the catalytic subunit is present in the nucleus of glucagon- or dibutyryl-cAMP-stimulated hepatocytes (244), but the catalytic subunit and both type I and type II regulatory subunits are in the nuclei of regenerating liver cells (245). Fletcher and Byus (246, 247), using a fluorescein-conjugated heat-stable protein kinase inhibitor, determined that catalytic subunit appears rapidly (5–15 min) in the cytoplasm and nucleolus but slowly (1 h) in the nucleus of glucagon- or dibutyryl-cAMP-stimulated H35 hepatoma cells. In contrast, Murtaugh *et al.* (*248*) were unable to detect any apparent redistribution of catalytic subunit in 8-bromo-cAMP-stimulated CHO cells and observed a diffuse staining pattern in several other cultured cell types. It is presently not clear if these differences are due to variations in basic mechanisms among cell types or to technical differences. The type II isozyme has been reported to be bound to a number of proteins, including microtubule-associated protein 2, a brain cytoskeletal protein (*249–252*); the mitotic spindle and nuclei of human breast cancer cells during different phases of growth (*253*); calcineurin (*254*), a calcium-calmodulin-activated protein phosphatase (*255*); P75, an unidentified brain protein with $M_r = 75,000$ (*256*); and to some other brain proteins which are distinct from several type II regulatory subunit-bound proteins in heart (*252*). It is interesting that many of these examples are calcium-calmodulin-binding proteins which serve as substrates for the cAMP-dependent protein kinase.

The cGMP-dependent protein kinase was first discovered in arthropod tissues (*257*) and was subsequently demonstrated in mammalian tissues (*258*). In contrast to the cAMP kinase, the cGMP-dependent protein kinase represents a minor component in most cells and is more restricted in its distribution (*259–261*). However, the cGMP kinase is found in significant concentrations in heart, lung, intestine, adrenal cortex, cerebellum, and smooth muscle tissue (*25*). Immunocytochemical studies localize the cGMP kinase in the cerebellum and the smooth muscle cells of major and minor blood vessels, intestinal wall and respiratory tract (*262*). It is presently not clear whether there are specialized functions in various tissues which require a specific cyclic nucleotide-dependent protein kinase isozyme or if there is a special advantage for a given tissue to contain a specific isozyme or isozyme mixture. An alternative explanation, that either isozyme can serve the physiological function equally well, and that the presence of different isozyme ratios in the various tissues is only fortuitous, has not been proved. However, the cGMP and the cGMP-dependent protein kinase has been reported to mediate the effects of certain agents, such as nitroglycerin and atrial natriuretic factor, on smooth muscle relaxation and kidney functions (*262a, 262b*).

C. Selective Activation of Cyclic AMP-Dependent Protein Kinase Isozymes

Several factors regarding the physiological roles of the isozymes can be considered. One of these relates to the subunit structure of the kinases and the proposed mechanism of regulation of metabolism in the cell. As already described, both isozymes are composed of an inhibitory regulatory subunit dimer and two catalytic subunits. It is the difference between the regulatory subunits that determine the isozyme type. According to the prevailing view, in higher organisms all of the effects of cAMP are mediated by the cAMP-dependent protein kinase and all regulatory processes are believed to be brought about by phosphorylation reactions catalyzed by the catalytic subunit of the enzyme. Since the catalytic subunit is identical in both isozymes, it can be argued that either isozyme can mediate the responses initiated by elevations in intracellular concentrations of cAMP. This does not exclude the possibility that an isozyme may be compartmentalized so that a single isozyme is activated by hormone stimulation. One should interpret with caution the finding that the type I isozyme is activated *in vitro* at lower concentrations of cAMP than is the type II isozyme. Activation of type I isozyme is inhibited by physiological concentrations of MgATP and autophosphorylation of the type II regulatory subunit inhibits subunit reassociation. Hofmann presented *in vitro* evidence indicating that in the presence of MgATP and sodium chloride the concentration of cAMP required to dissociate each isozyme is similar (57). Other factors such as dissociation of the isozymes by basic protein substrates (263, 264) and differential salt activation of the isozymes may be important *in vivo*. The extent of activation will also depend upon the relative concentrations of the regulatory and catalytic subunit present (265).

As previously mentioned enzyme compartmentalization may be an important factor if selective activation of a single isozyme occurs. Buxton and Brunton (266) demonstrated selective activation of protein kinase based on the subcellular localization of the enzyme in a homogeneous population of cardiac myocytes. Specifically, isoproterenol and prostaglandin E_1 cause an elevation of cytosolic cAMP and an activation of cytosolic protein kinase. However, isoproterenol, but not prostaglandin E_1, causes an elevation of particulate cAMP, a translocation of protein kinase activity from the particulate to the cytosolic fraction, and an activation of glycogen phosphorylase. This isoproterenol-specific response is rapid and is temporally related to the phosphorylase activation. That differential calcium availability is responsible for the observed differences is largely ruled out (266). These results suggest that the β-receptor adenylate cyclase, protein kinase, and phosphorylase are spatially isolated from the prostaglandin E_1 system in cardiac myocytes. This agonist-specific protein kinase activation has been demonstrated in isolated, perfused hearts from several species (267–269).

In the following discussion, reports supporting selective isozyme activation are examined and then reports of simultaneous activation of the isozymes are reviewed. Selective activation of the cAMP-dependent protein kinase isozymes has been reported in intact organisms and in intact cells. Schwoch (270) evaluated the extent of activation of the type I and type II protein kinases in rat liver after the animals were injected with glucagon. This analysis was based on DEAE-cellulose separation of inactive holoenzyme and free catalytic subunit and the property of the type II isozyme to rapidly associate in low ionic strength.

Although the technique has a potential for artifactual enzyme activation (see Section II,A), it was one of the first studies designed to evaluate the physiological significance of two cAMP-dependent protein kinase isozymes. The data suggested a sequential activation of type I and then type II protein kinase. More specifically, 2 min after glucagon injection the type I isozyme was completely activated for at least 60 min while the type II isozyme was only partially activated. The type II kinase remained active for only about 30 min while the levels of cAMP were relatively high.

Byus *et al.* (*271*) also found that a sequential activation of type I and type II protein kinase occurs when normal hepatocytes are incubated with glucagon or dibutyryl-cAMP. They used a method to distinguish type I and type II kinase activation based on separation of free catalytic subunit, type I and type II protein kinase using C6-aminoalkyl agarose chromatography. Maximal glycogenolysis was correlated with a selective activation of type I protein kinase, which was predominantly activated by the lowest effective concentrations of either agonist.

Hunzicker-Dunn (*272*) also used a technique that separated subunits of cAMP-dependent protein kinase on DEAE-cellulose. Data from this study suggested a preferential activation of the type I isozyme in the corpora lutea obtained from ovaries of 4-day pseudopregnant rabbits treated with a single injection of human chorionic gonadotropin.

Livesey and co-workers (*273*) developed a rapid batch elution method for separating the two isozymes of protein kinase on DEAE-cellulose columns. In these studies rigorous attempts were made to exclude the possibility of postextraction activation of protein kinase. One study evaluated the effects of parathyroid hormones and prostaglandin E_2 on the activation of type I and type II protein kinase from normal and neoplastic osteocytes from rat calveria (*273*). These data suggested that the isozyme response not only is specific for a particular hormone effector but also depends upon the cell type. Parathyroid hormone predominantly activated the type I isozyme in the neoplastic osteocytes but activates both isozymes to the same extent in the normal cells. Prostaglandin E_2 also caused a predominantly type I isozyme activation in the malignant cells but specifically activated the type II isozyme in the normal calveria cells. This technique was modified and validated to evaluate the effects of conditions and prostaglandin E_2 on protein kinase activation in two human breast cancer cells lines, T47D and MCF7 (*274, 275*). In both of these cells, calcitonin selectively activated the type II isozyme, however, the duration of the response was different. While type II activation in the MCF7 cells was transient over 4–6 h, it was persistent in the T47D cells for at least 24 h.

Mizuno *et al.* (*276*) analyzed the cytosolic protein kinases from the submandibular glands of control and isoproterenol-treated rats. Using DEAE-cellulose, they concluded that 10 min after isoproterenol injection, the type II isozyme readily dissociated into regulatory and catalytic subunits which are separated on the column. The type I isozyme eluted from the column at 0.1 M KCl but had an activity ratio of about 1.0; they concluded that the type I isozyme was less dissociable than the type II isozyme, but was highly active in an undissociated form. Therefore, activation of both isozymes correlated with salivary secretion but the type I remains active for as long as an hour while the activity ratio of the type II isozyme decreases slightly. This latter result is similar to the time-dependent behavior of glucagon-stimulated rat liver protein kinase isozymes reported by Schwoch (270).

Chew (277) investigated histamine-stimulated acid secretion in parietal cells isolated from rabbit gastric mucosa. These cells contained the type I isozyme in the cytosol and the type II kinase in the cytosol and particulate-fraction. Histamine activated only the type I isozyme while forskolin activated both isozymes. This conclusion was obtained from results of DEAE-cellulose chromatography, differential isozyme reassociation after cAMP removal by Sephadex G-25 chromatography, and 8-azido-[^{32}P]cAMP photoaffinity labeling of crude extracts. The results suggest that the type I isozyme mediates histamine-stimulated acid secretion and that this isozyme is compartmentalized with the histamine H_2 receptor-coupled adenylate cyclase.

Other techniques have been used to evaluate the biological role and the physiological significance of the two isozymes of the cAMP-dependent protein kinase. Litvin and co-workers (278) used an approach based on specific antibody precipitation of type I or type II protein kinase. They demonstrated a parallel release of ACTH and a selective type I protein kinase activation when AtT20 mouse pituitary tumor cells were stimulated with corticotropin releasing factor. Maximal ACTH release was seen with a 2-fold increase in cAMP and only a slight activation of the type II isozyme. When 0.5 mM 3-methylisobutylxanthine was added, no further increase in ACTH release occurred and the type II isozyme is now activated by 50%. 3-Methylisobutylxanthine alone (0.5 mM) fully activates type I, while type II was activated by about 25%.

The earliest report of simultaneous activation of types I and II in a tissue was that of Corbin and Keely (215). This study was done using epinephrine-perfused hearts from several mammalian species which were known to possess different relative amounts of the two isozymes. From examination of the degree of reversal of the hormone effect by Sephadex G-25 chromatography, no apparent selectivity was observed and it was concluded that both isozymes are activated by epinephrine.

Ekanger et al. (279) developed an approach in which endogeneous cAMP bound to the regulatory subunit of type I or type II protein kinase was quantitated after specific adsorption to protein A-agarose coated with antibodies directed against the respective isozymes. While the work of Schwoch with rat liver (270) and Byus et al. with isolated hepatocytes (271) indicated a preferential activation of type I protein kinase, Ekanger and co-workers demonstrated that both type I and type II isozymes were equally activated in isolated hepatocytes as the concentration of glucagon is varied over a wide range.

As discussed in Section II,A, certain cAMP analogs are selective for one or the other of two intrasubunit cyclic nucleotide binding sites and cAMP analog combinations can distinguish between synergistic activation of type I and type II protein kinase (Fig. 1). These *in vitro* findings have been extended to demonstrate a synergism of cAMP-mediated responses in a number of isolated cell and tissue preparations (*113, 114, 114a*). It has been demonstrated that the analogs used in these studies have no apparent special properties other than site selectivity (*105, 107, 109, 111, 113*). Furthermore, when hepatocyte glycogenolysis is stimulated by a series of equipotent concentrations of glucagon or 8-thio-parachlorophenyl-cAMP [and the analog effectively removed (Ref. *213*)], the degree of phosphorylase activation was closely correlated with the cAMP-dependent protein kinase activity ratio for both agonists (T. W. Gettys and J. D. Corbin, unpublished). Therefore, data demonstrating that a combination of site-selective cAMP analogs generate a synergism of a physiological response can be used to prove that protein kinase mediates the response. This technique also suggests that the cooperativity of cAMP binding and protein kinase activation measured *in vitro* are important to the function of protein kinase *in vivo* and that the cooperativity may be a mechanism of sensitivity amplification at the protein kinase step in the intact tissue (*113*).

When cyclic nucleotide analogs are added to isolated cells, they cross the cell membrane, bypass the hormone receptor–adenylate cyclase system and directly activate the cAMP-dependent protein kinase isozymes (*113, 114, 114a, 214*). When a site-1- and a site-2-selective analog are added in combination to intact cells synergism of a physiological response occurs only if the response is mediated by the cAMP-dependent protein kinase. Furthermore, if the type I protein kinase mediates the response, synergism should occur only when two analogs selective for site 1 and site 2, respectively, for this isozyme are combined (type I directed-analog pair), and if the type II isozyme mediates the response synergism should only occur when two analogs selective for site 1 and 2 of this isozyme are combined (type II directed-analog pair) (see Section II,A, and Fig. 1). It is important to use the analogs at relatively low concentrations, which allows for a wider window for examining synergistic effects (*113, 114, 114a*). It is also important to use the analogs at concentrations that result in a linear dose–response relationship for the measured effect. Additionally, controls should be included to determine that a combination of two analogs that are selective for the same site do not cause a synergism of the response.

This method, which is fully described elsewhere, has now been tested in a number of different systems (*114a*) including adipocyte lipolysis (*113*), hepatocyte phosphorylase activation (*114*), increases in mRNA for phosphoenolpyruvate carboxykinase in continuous cultures of H4IIE hepatoma cells (*280*), induction of LH receptors and increased progesterone synthesis in primary cultures of porcine granulosa cells (281), and thyroid hormone release in dog thyroid slices (J. Van Sande and J. E. Dumont, personal communication).

This technique was first tested by measuring adipocyte lipolysis (113). Adipocytes contain predominately, if not exclusively, the type II isozyme (7). Synergism of protein kinase activation in vitro and glycerol release from intact cells was observed only with type II directed-analog pair. Synergism did not occur for either process with type I directed-analog pair, or with a combination of two site-1 or two site-2 selective analogs. These data demonstrate that the type II isozyme mediates the lipolytic response.

Since the first demonstration of synergism in intact cells using pairs of site-selective cAMP analogs, new analogs have been developed that, when combined with another analog selective for the opposite site, provide a more selective synergistic activation of only one isozyme. One of these analogs is 8-piperidino-cAMP (104). In addition to using N^6-benzoyl- and 8-aminohexylamino-cAMP as a type I directed-analog pair and N^6-benzoyl- and 8-thiomethyl-cAMP as a type II directed-analog pair, further experiments were carried out using 8-piperidino- and 8-piperidino-cAMP as a type I directed-pair and N^6-benzoyl- and 8-piperidino-cAMP as a type II directed-pair (see Fig. 1). In these experiments, phosphorylase activation was measured in various cell types which contain different protein kinase isozyme ratios. Bovine neutrophils (85% type I), rat adipocytes (>95% type II), and rat hepatocytes (type I $\cong$ type II) were used. For all isozymes tested in vitro, type I directed-analog pairs synergistically activate only type I isozymes while type II directed-analog pairs synergistically activate only type II isozymes. In close agreement with these results, bovine neutrophil phosphorylase was synergistically activated almost exclusively by type I directed-analog pairs, adipocyte phosphorylase was synergistically activated almost exclusively by type II directed-analog pairs, and rat hepatocytes were synergistically activated by both type I directed- and type II directed-analog pairs (114). These data indicate that either type I or type II protein kinase can regulate phosphorylase activation.

These techniques have been used to determine whether the cAMP-dependent protein kinase has a regulatory role at the nuclear level in H4IIE hepatoma cells. Since it is known that cAMP analogs increase the amount of mRNA for phosphoenolpyruvate carboxykinase (PEPCK) in these cells (283), it was of interest to see if cAMP analog combinations could act synergistically to produce this response. It was also possible to test if a synergism of PEPCK gene transcription could be demonstrated. To test these possibilities, various site-1- and site-2-selective cAMP analogs were added alone (in the linear dose-response range) and in combination to H4II cell type II cAMP-dependent protein kinase in vitro and to H4 cells in culture. Determinations were then made of the extent of synergism for the isolated activation, increase in mRNA[PEPCK], and PEPCK gene transcription. All analog pairs that resulted in a synergistic activation of protein kinase also resulted in a synergistic increase in mRNA^{PEPCK} and PEPCK gene transcription. Synergism, as quantitated by ratio of responses of an analog pair divided by sum of single analog responses, is approximately 2 for protein kinase activation, 3–4 for the increase in mRNA^{PEPCK}, and 10–12 for PEPCK gene transcription. For all responses tested, synergism does not occur when two site-1-selective or two site-2-selective analogs are combined. The occurrence of synergism only when a site-1- and a site-2-selective analog are combined is a protein kinase-specific characteristic. These results clearly indicate the involvement of protein kinase in the cAMP regulation of PEPCK gene transcription. It is presently not clear if this effect is due to a phosphorylation event or to some other mechanism such as the cAMP-bound regulatory subunit acting in a manner analogous to the *E. coli* catabolite gene activator protein (CAP).

Segaloff *et al.* (*281*) used these site-selective cAMP analog combinations in primary cultures of porcine granulosa cells to test whether or not induction of LH receptors and the increase in progesterone synthesis are cAMP-dependent protein kinase-mediated events. Photoaffinity 8-azido-[^{32}P]cAMP labeling of cell extracts suggested that both regulatory subunit isozymes are present. Two specifically labeled bands with apparent M_r = 56,000 and 49,000 comigrate with homogeneous rabbit skeletal muscle type I and type II regulatory subunit standards. The type II isozyme is induced 5- to 10-fold by cholera toxin and represents the predominant isozyme while type I represents only a minor band. Incubation of cells in primary culture with type I directed- and type II directed-analog pairs resulted in a synergism of LH receptor induction, as determined by ^{125}I human choriogonadotropin binding, and in a synergism in progesterone synthesis, as measured by radioimmunoassay. The synergism of both of these responses was much greater with a type II directed- than a type I directed-analog pair. No synergism of either response was observed with two site 1-selective analogs or with two site-2 selective analogs. These data indicate that both the induction of LH receptors and the increase in progesterone are mediated by the cAMP-dependent protein kinase and that the predominant type II isozyme is primarily responsible for both of these responses.

This method is presently being tested by J. Van Sande and J. E. Dumont (personal communication), using thyroid hormone secretion in dog thyroid slices. Secretion was measured by butanol extraction of ^{131}I from the medium as a percentage of total radioactivity present in the slices at the beginning of a 5 h incubation. Both type I and type II protein kinase were present in the dog thyroid (*242*). When tissue slices were incubated with type I directed- and type II directed-analog pairs, synergism of thyroid hormone secretion was present to approximately the same extent for both isozyme synergistic pairs. This result suggests that both type I and type II isozymes mediate thyroid hormone secretion.

Data from studies using type I directed- and type II directed-analog pairs suggest that, for the various parameters measured, either isozyme can mediate cAMP-dependent protein kinase generated responses. The extent of the contribution of each of the isozymes seems to correlate approximately with the amount of each of the isozymes present. These conclusions appear to be valid for the limited number of experimental models tested, and this technique appears to be generally applicable. It should be pointed out that through the use of cAMP analogs, the hormone–receptor–adenylate cyclase system is bypassed. It is indeed possible that the intact cell is compartmentalized to the extent that the hormone is coupled to only a single kinase isozyme. Furthermore, although for the type I and type II isozymes tested, the defined analog pairs appear to specifically cause a synergistic activation of only a single protein kinase isozyme, it cannot be ruled out that exceptions to these generalizations may exist. This is a concern given data describing microheterogeneous subforms of the type II isozyme from various tissues and species (see Section II,A). Similar microheterogeneous subforms of the type I isozyme may also exist. However, for the responses of type I and type II protein kinase activation tested *in vitro,* the defined cAMP analog combinations appear to be isozyme specific. This is based on experiments of type I and type II isozymes from rabbit skeletal muscle, rat heart, rat hepatocyte, and bovine neutrophil, as well as the type II isozyme from rat liver plasma membrane, rat adipocytes, cultured rat hepatoma (H4IIE) cells, dog thyroid, and bovine heart. It should be pointed out that in spite of the microheterogeneous differences between the type II isozymes from bovine heart, rat hepatocyte, and rat adipocyte, similar effects of the cAMP analog combinations are observed with each isozymic form.

D. Use of Cyclic Nucleotide Analogs in Intact Cells

As mentioned previously, cyclic nucleotide derivatives were originally designed to be used in studies of the structural requirements of cyclic nucleotide activation of protein kinase, to develop analogs which are more potent than cAMP, and to develop antagonists of cAMP (*281*). These studies have been expanded to accommodate the presence of two types of cAMP-dependent protein kinases, designated type I and type II (*7*), as well as the cGMP-dependent protein kinase. Furthermore, it is now clear that the regulatory components of these three cyclic nucleotide-dependent protein kinase isozymes have two different intrasubunit cyclic nucleotide binding sites, termed site 1 or site B and site 2 or site A, which exchange bound, labeled cyclic nucleotide slowly and rapidly, respectively (*6, 8, 21, 24, 59, 86, 105*). The development and uses of cyclic nucleotide analogs have closely paralleled and sometimes led the development of our understanding of the cyclic nucleotide-dependent protein kinases. In fact, derivatives of cyclic nucleotides which had been available for many years were used to define and characterize the two different binding sites on the respective kinases (*21, 105, 107, 108, 181*). Consequently, studies are being directed toward understanding structural and functional differences among the cyclic nucleotide-dependent protein kinase isozymes, finding analogs that are more potent activators of one or the other isozymes, and developing analogs that are selective for one or the other different intrasubunit binding sites. In addition to these *in vitro* studies, experiments applying the available knowledge concerning the mechanism of protein kinase isozyme activation, phosphodiesterase hydrolysis, and potential for analogs to be permeable to membranes have been and are being conducted using cyclic nucleotide analogs on a large number of *in vivo* preparations from whole animals to isolated cells.

Because of the hundreds of analogs that have been synthesize, it has proved important to develop a systematic and rational approach to using analogs. Miller and Jastorff and their collaborators have synthesized and tested a large number of analogs (*103, 285–292*) that have proved useful for *in vitro* and *in vivo* studies. The work of these and others have greatly advanced our knowledge concerning the cyclic nucleotide-dependent protein kinases and their role in metabolic regulation.

Both of the intrasubunit sites on the cAMP-dependent and cGMP-dependent protein kinases have high specificity for the ribose-phosphate moiety and lower specificity for the base moiety of the respective cyclic nucleotide (*108*). The cAMP- and the cGMP-dependent protein kinases distinguish between cAMP and cGMP by specific recognition of the respective bases of the molecules. The cAMP kinase isozymes are specific for a C6-amino group (*290*). Although the base is not required for activation, it is thought to interact with the regulatory subunit by hydrophobic and/or π electron bonding interactions rather than by hydrogen bond interactions (*290, 293, 294*). Substitution at C6 results in reduced specificity for cAMP kinase activation and if the C6-amino group is changed to a C6-imino group, the resulting analog shows an increased specificity for the cGMP kinase (*290, 291*). As expected, the cGMP-dependent protein kinase appears to prefer both a C6-oxygen and a C2-amino group for activation. The C6-oxygen apparently accepts a proton from and the C2-amino donates a proton to the enzyme (*290, 291*).

By systematically altering the cyclic nucleotide molecule, structural requirements for cyclic nucleotide binding and protein kinase activation have been partially defined. Evidence indicates that the primary determinant of cyclic nucleotide activation of the respective isozymes is the hydrophilic cyclic phosphate moiety (*103, 285*). Alterations in this portion of the molecule are generally not tolerated. Activation of both cAMP protein kinase isozymes requires a charged cyclic phosphate in addition to a 3' oxygen and a 2' hydroxyl group both in the *ribo* conformation. These are probably important for hydrogen bonding (*103, 289, 287*). Additionally, the *syn* conformation of the cyclic nucleotide is preferred (*287, 289*) and the orientation of the purine-cyclic phosphate is critical (*287, 289*). Although the cAMP- and cGMP-dependent protein kinases are believed to be homologous proteins with highly conserved cyclic nucleotide binding sites, it is clear that differences do exist. These considerations also extend to the two isozymes of the cAMP-dependent protein kinase. Recent reports indicate that the type II isozymes from various tissues and species appear to be microheterogeneous (see Section II,A), and may contain differences in the binding sites. It should be kept in mind that the established structure–activity relationships for cyclic nucleotide recognition by the protein kinases are from a limited number of isozyme types.

Cyclic AMP analogs have been used in a number of isolated cell and tissue preparations. The efficacy of cyclic nucleotide analogs as agonists of protein kinase-mediated responses in intact cells or in intact organisms is dependent upon the concentration of the analog at its site of action. The efficacy of all analogs tested for adipocyte lipolysis and hepatocyte phosphorylase activation are explained by consideration of at least three basic analog properties (*113, 114a, 214*) (favorable lipophilicity, low K_a for protein kinase activation, and resistance to low K_m phosphodiesterase hydrolysis). The lipid character and porosity of the cell membrane and other cellular constituents that could "trap the analog" should also be considered (*214*). Table I shows the concentration of analogs required for a half maximal activation of adipocyte lipolysis, hepatocyte phosphorylase *a*, and induction of mRNA for phosphoenolpyruvate carboxykinase in H4IIE hepatoma cells. There are only minor differences in the cyclic nucleotide specificities between the adipocyte and hepatocyte protein kinases and between the low K_m phosphodiesterases from the respective cells types. Therefore, lipid content of cellular constituents, membrane porosity, and phosphodiesterase activity appear to be potentially important when comparing the efficacy of analogs in adipocytes and hepatocytes (*114a, 214*). Adipocyte lipolysis is sensitive to analogs in the millimolar concentration range and hepatocyte phosphorylase is activated in the micromolar concentration range. Generally, hepatocytes are 100–10,000 times more sensitive than adipocytes to cAMP analog stimulation when comparing these two physiological processes. The nuclear response measured in H4IIE hepatoma cells is intermediate in sensitivity to cAMP analogs. It is possible that other physiological responses in these cell types may show a different sensitivity to cAMP analogs. The sensitivity of a cellular response to cAMP analogs may also depend upon enzyme compartmentalization, the basal activation state of protein kinase (thereby affecting sensitivity amplification at the protein kinase step), and the potential for magnitude amplification (determined to some extent by the number of steps between the initial activation event and the final response). Free *et al.* (*295*) demonstrated that cAMP analogs were generally more potent in stimulating adrenal cell steroidogenesis than in stimulating adipocyte lipolysis. These studies

TABLE I

Concentrations of Cyclic Nucleotide Analogs
for Phosphoenolpyruvate Carboxykinase Activation

Cyclic nucleotide analog	EC_{50} Adipocyte lipolysis (μM)	EC_{50} Hepatocyte phos a (μM)	EC_{50} mRNA[PEPCK] (μM)
8-Thioethyl	900	0.3	—
8-Aminomethyl	7,900	6.0	>5,000
N^6-Diethyl	2,000	—	—
8-Bromo	3,900	1.3	—
8-Thio-p-chlorophenyl	1,000	0.1	25
6-Thiomethyl	600	—	—
8-Thiomethyl	1,000	0.5	330
8-Amino	6,100	—	—
8-Thioisopropyl	500	2.0	—
8-Thiobenzyl	1,900	—	50
8-Thio-p-nitrobenzyl	1,200	0.1	—
N^6-Aminohexylcarbamoylmethyl	10,000	—	—
$N^6,O^{2'}$-Dibutyryl	1,100	—	—
N^6-Carbamoylpropyl	500	3.0	—
N^6-Benzoyl	700	0.5	580
N^6-Butyryl	1,900	4.8	2,250
8-Hydroxy	900	—	—
8-Aminohexylamino	8,600	60	1,000
8-Aminobenzyl	>15,000	—	780

illustrate a degree of cellular specificity and suggest the possibility of using cAMP analogs in whole organisms to selectivity stimulate some cell types and not others.

One of the originally defined objectives for the synthesis of cyclic nucleotide analogs was to find an antagonist of protein kinase (*284*). Some cAMP analogs have proven to be competitive antagonists of cAMP binding to the catabolite gene activator protein (CAP) and prevent specific DNA binding and stimulation of gene transcription in *E. coli* (*296, 297*). For example, cGMP, cIMP, and N^6-monobutyryl-cAMP apparently interfere with optimal hydrogen bonding while the bulky groups of 8-bromo- and 8-thio-cAMP cause steric interferences. Presumably, these competitive antagonists and cAMP both bind to CAP but the antagonists do not induce the necessary conformational change required for specific DNA binding. The competitive antagonists of cAMP *E. coli* gene transcription mentioned above are agonists of protein kinase activation. The structural requirements for antagonism of protein kinase activation are therefore considerably different than for the CAP protein (*298, 299*).

Several reports using diastereomers of cyclic monophosphorothioate and cyclic monophosphodimethylamidate derivatives of cAMP [cAMPS and cAMP-N(CH$_3$)$_2$, respectively], which contain chiral phosphorus centers (RRp and SSp) suggest that some of these analogs may fulfill the criteria as useful protein kinase antagonists (*300–306*). These criteria include (*a*) binding to but not activating protein kinase; (*b*) having a relatively good affinity for protein kinase so that competition with cAMP at the cyclic nucleotide binding site is effective; and (*c*) having structural features that allow membrane penetration so that antagonists can be used in intact tissues. This latter criterion may vary for cells with different lipid characters.

DeWit *et al.* (*294, 305*) used the diastereomeric forms of both of the above analogs to determine the characteristics of cyclic nucleotide binding to site 1 (stable site) on the type I isozyme from rabbit skeletal muscle. They observed that all the cAMP analogs tested, except Rp cAMPs and both enantiomers of cAMP-N(CH$_3$)$_2$, have lower affinity for site 1 of the holoenzyme than for the free regulatory subunit. This suggests that, unlike other analogs tested, these three analogs did not compete with cAMP for binding to the regulatory subunit. Furthermore, Rp cAMPS competed with cAMP for binding to the holoenzyme but was ineffective for protein kinase activation (*294, 300, 305*).

O'Brian *et al.* (*300*) used Rp and Sp cAMPS to study the stereochemistry of cyclic nucleotide binding to and activation of the type II protein kinase. They reported that both enantiomers bind to and activate the kinase but the Sp conformer is much more potent for both of these actions than the Rp conformer. They also determined that the Sp conformer binds preferentially to site 2 while the Rp conformer binds preferentially to site 1.

Botelho *et al.* have described a number of studies using Sp and Rp cAMPS in isolated hepatocytes (*301, 302, 304, 306*). Sp cAMPS is a full agonist that stimulates glucose production half-maximally and maximally at 0.8 and 10 μM, respectively. Rp cAMPS appears to have no significant potency as an agonist of glucose production (*301, 302*), phosphorylase activation, or glycogen synthase inactivation (*304*). Instead, Rp cAMPS antagonizes these agonist-stimulated reactions. When 5 μM Rp cAMPS is present, 10-fold and 6-fold higher concentrations of Sp cAMPS and glucagon, respectively, are required for half-maximal glucose production. Similar increases in agonist concentrations are required for the activation of the cAMP-dependent protein kinase and phosphorylase, and inactivation of glycogen synthase. The putative antagonist (3 μM – 100 μM) produced a maximal inhibition of 50–74% when glucose production was stimulated by 1 nM glucagon, depending on the time of preincubation with Rp. Rp cAMPS appears to be an *in vivo* competitive protein kinase antagonist. It has been estimated that for Sp cAMPS-stimulated glucose production, a Rp cAMPS to Sp cAMPS molar ratio of 3 and 30 are required for half-maximal and maximal inhibition, respectively. Further research using these protein kinase antagonists and the development of others can provide useful and important tools to more completely define the role of protein kinase in the regulation of cell function.

A potential complication of using cyclic nucleotide analogs in intact cells is a change in the intracellular cAMP concentration. For example, it is possible that analogs could serve as competitive substrates for phosphodiesterases and thereby increase the cAMP concentration. However, it has been shown that cAMP analog-stimulated hepatocytes have lower cAMP concentrations than unstimulated cells. The cAMP decrease appears to be mediated by a cAMP-dependent protein kinase mechanism (213). This suggests that through a feedback mechanism, such as phosphoidiesterase(s) stimulation and/or adenylate cyclase inhibition, the cAMP-dependent protein kinase can regulate its own activation state.

The effects of insulin on cAMP analog-stimulated adipocyte lipolysis, hepatocyte glycogenolysis, and cardiac myocyte glycogenolysis and glycogenesis have been investigated (214, 307). In several mammalian tissues insulin is known to block most of the metabolic effects of hormones that elevate cAMP levels. If lowering of cAMP by phosphodiesterase activation is a necessary step in some of the effects of insulin, then this hormone may not work if the phosphodiesterase is prevented from catalyzing hydrolysis of cyclic nucleotides. This would be the case when cellular cAMP-dependent protein kinase is activated by phosphodiesterase-resistant cAMP analogs, instead of by either hormonal elevation of cAMP or by phosphodiesterase-sensitive cAMP analogs. Some, but not all, cAMP analogs are quite resistant to hydrolysis by phosphodiesterase (214). In cardiomyocytes insulin does not block the effects of the cAMP analogs tested on phosphorylase and glycogen synthase activity (307). In adipocytes and hepatocyges insulin blocks the metabolic effects of hormones that elevate cAMP, but when cAMP analogs are used, the results are different from those obtained in cardiomyocytes. Although insulin blocks the effects of some cAMP analogs it does not block the effects of analogs that are extremely resistant to hydrolysis by the low K_m, hormone-sensitive phosphodiesterase (214). The results suggest that at least in adipocytes and hepatocytes insulin antagonism of the metabolic effects examined can be explained by phosphodiesterase activation.

E. THE ROLE OF CYCLIC AMP AND THE CYCLIC
AMP-DEPENDENT PROTEIN KINASE ISOZYMES IN THE CELL
CYCLE, PROLIFERATION, AND DIFFERENTIATION

The primary emphasis of this section is placed on cAMP and the cAMP-dependent protein kinase since far more literature is available on this system as compared to cGMP and the cGMP-dependent protein kinase. Three levels of cyclic nucleotide involvement in the regulation of these developmental processes are discussed. The role of cAMP in these processes is only briefly reviewed here since this subject has been intensely studied over the years and has been the target of several excellent reviews (308–318). The role of the cAMP-dependent protein kinase in these processes has been studied less frequently but has also been reviewed (239, 319, 320). The third level of involvement examines the possibility that specific cellular functions may be regulated by specific protein kinase isozymes.

The literature in this field is inconclusive and often contradictory. It is difficult to arrive at definitive conclusions because methodological limitations require conservative interpretations (237, 312, 315, 318). Much of the early work was conducted without an appreciation for some of the pitfalls inherent in the experimental techniques. The events that regulate the cell cycle and determine if a cell will continue to proliferate or undergo differentiation are extremely complex. Research is being conducted against a background of incomplete and insufficient information.

Several lines of evidence have accumulated to suggest a role for cAMP in the modulation of the cell cycle. One of the most fundamental observations supporting this is that the cAMP level in populations of synchronized cells oscillates during the cell cycle (311, 312, 314, 315, 317, 321–323). These cAMP fluctuations have a regular pattern and have been observed in a number of cell types, including normal and malignant cells, and using several different methods of cell synchronization. The most common finding is that the cAMP level is at a minimum during mitosis, gradually increases during G_1, reaches a peak near the G_1–S border and declines again during the S phase. The G_2 phase is also often associated with a transient cAMP elevation. It is often seen that during the G_1 phase two peaks of cAMP are observed (311, 314, 318, 324–326). When two peaks occur, it appears to be the second peak that is required for DNA synthesis (318, 325, 327, 328).

Although the cause of these cAMP surges is not known for certain, in general terms it is expected that a high ratio of adenylate cyclase to phosphodiesterase activities accounts for the elevation of cAMP and a low ratio of these enzyme activities accounts for the cAMP decrease (329). Although this seems reasonable, the hypothesis has not been rigorously tested. These relationships, however, have been observed in RPMI-8866 human lymphoid cells (330). In regenerating rat liver, an increase in the number of beta receptors on the hepatocyte membranes contributes to an increase in adenylate cyclase activity (331). In addition, a transient increase in the levels of calmodulin have been reported to coincide with the second cAMP increase seen in this tissue (318, 332). This could conceivably contribute to a stimulatory effect on adenylate cyclase and/or a stimulation of a phosphodiesterase (332, 333). Several investigators have demonstrated that cAMP elevation leads to an induction of phosphodiesterase (334–339). Liu reported in 3T3-L1 cells that treatment with dibutyryl-cAMP for 2–7 days leads to a 4- to 6-fold increase in the specificity activity of phosphodiesterase (337). Similar results were obtained in S-49 mouse lymphoma cells (338) and mouse neuroblastoma cells (339). The observation that indo-methacin, an inhibitor of prostaglandin synthesis, blocks the increase in cAMP and subsequent DNA synthesis in mouse lymphocytes suggests that prostaglan-din, a known stimulator of adenyate cyclase, may be involved in the cAMP elevation (340). Rozengurt et al. (341) reported that prostaglandin E_1 causes an increase in intracellular cAMP and stimulates DNA synthesis in Swiss 3T3 cells when insulin is present. These investigators (342) have also demonstrated that partially purified porcine platelet-derived growth factor, a potent mitogen for un-transformed fibroblastic cells, causes an increase in production of prostaglandins of the E series and a marked, indomethacin-sensitive elevation of cAMP in the presence of phosphodiesterase inhibitors.

The origin of the signals that regulate these cAMP fluctuations are presently not known. There has been some speculation that signals that regulate the cell cycle are internal or intrinsic to the cells. There are examples of this type in nature. The intrinsic cardiac rhythmicity and electroencephalographic wave rhythms are of internal origin. Generation of cAMP may be one of these inherent signals of cAMP regulation superimposed on other inherent regulatory signals (317). In the intact organism, cells of a developing tissue do not function in isolation, and external signals, which act at cell surface or internal receptors, may affect cAMP levels. Hormones and growth factors are required for prolifera-tion or differentiation. The cell cycle and these developmental processes may be regulated by both internal and external signals similar to circadian rhythms or other biological clock mechanisms. Rozengurt et al. have suggested (342), from their work and the work of others (343–345), that cells may secrete substances such as prostaglandins which then act on receptors of surrounding cells, as well as on their own.

The original second-messenger hypothesis (3) presented a unifying theory describing the elevation of cAMP as an intracellular response generated by an external hormone signal. Therefore, cAMP is described as a "Director of For-eign Affairs" (3). That cAMP and the cAMP-dependent protein kinase are involved in cell development, which could be "internal affairs," is not neces-sarily implicit in this theory (312). Although phosphorylation reactions are the only known mechanism of protein kinase action, it cannot necessarily be as-sumed that all the effects of cAMP are mediated by them. It should not be ruled out that the regulatory subunit or other unrecognized cAMP-binding proteins may play a role of their own (see Section IV,F.). Nevertheless, knowing the mechanism of activation of protein kinase by cAMP and the importance of this equilibrium reaction in the regulation of metabolism of cells in the quiescent, G_0 state, it is likely that protein kinase-mediated phosphorylation is involved in some aspects of regulation or modulation of cell cycle events.

If cAMP and/or the protein kinase are involved in the regulation of develop-mental processes, an important question is at what site or sites do they act. Using the role of cAMP in hormonal regulation as an analogy, it is likely that the cAMP-dependent protein kinase acts at transition phases; for example, it could act at sites between the end of one cell division and the DNA synthesis that precedes the next division. However, there is some controversy concerning the nature of events that comprise this transition (314, 317, 318). One hypothesis proposes that growth-arrested cells are in a distinct quiescent G_0 state which they must leave to reenter the G_1 phase (346–348). Friedman (317) has discussed the possible roles of cAMP at this point but whether or not protein kinase mediates these events, and if so, how, is not known. An alternative theory proposes that nonproliferating cells remain in the G_1 phase but have a very low probability of continuing into the S phase due to a number of factors related to requirements for DNA synthesis and mitosis (349). Pledger and co-workers (350) showed that quiescent BALB-c/3T3 cells must pass through at least two phases, which are separately regulated, before they can synthesize DNA. They suggested that these cells require a platelet-derived growth factor to become "competent," that is to leave the G_0 phase and enter the cell cycle. Other factors are required for competent cells to progress through the cell cycle. These factors are present in platelet-poor plasma and are effective in competent but not in incompetent cells. This progression factor appears to be somatomedin C or other members of this family of growth factors and epidermal growth factor. Interestingly, simian virus 40 provides both competence and progression activity. Boynton and Whitfield (318) have reviewed this theory in detail and have suggested that there are at least four different types of prereplicative phases in eucaryotic cells.

There has been considerable controversy concerning the role of cAMP in the cell cycle as a positive or negative regulator (314, 315, 317, 318). That cAMP is a negative regulator of proliferation is based on three general types of evidence. One is the observation that the levels of cAMP are often high in the quiescent state and then drop as the cells are stimulated to proliferate. The second is that the use of exogenous cyclic nucleotides or agents that elevate intracellular cAMP arrests many cells in the G_1 or G_2 phase of the cycle. A third type of evidence is derived from the use of cell varients that are deficient in the cAMP-dependent protein kinase (351–360). However, evidence is also available to support a stimulatory role for cAMP in the cell cycle. Although the addition of high concentrations of cAMP or cAMP analogs at critical points during the cell cycle inhibits cell proliferation in some cell types, at appropriate times cAMP or cAMP analogs stimulate proliferation in other cell types (314, 317, 318).

Studies conducted by Wang et al. (340) using conconavalin A-stimulated mouse lymphocytes indicate that a rise and a fall in cAMP is required for the progression of these cells into the S phase of the cell cycle (318, 340, 361). When conconavalin A is used to stimulate lymphocytes to proliferate, RNA synthesis precedes and DNA synthesis follows the increase in cAMP. When indomethacin, which stimulates adenylate cyclase, is used to block synthesis of prostaglandins, both the cAMP increase and DNA synthesis are blocked. This condition is reversed when indomethacin is removed. The addition of 8-bromo-cAMP or the phosphodiesterase inhibitor RO-20-1724 after the cAMP rise, but before DNA synthesis, blocks the progression of concanavalin A-stimulated cells into the S phase. When the first peak of cAMP is prevented from falling in serum-starved human fibroblasts, DNA synthesis is not blocked (362). However, if the subsequent late G_1 phase cAMP surge is prevented from subsiding, DNA synthesis is arrested. Likewise, prevention of the natural rise and fall of cAMP in the late G_2 phase of human lymphoid cells prevents mitosis (363). It should be kept in mind that it is difficult to generalize the effects of cAMP on cell development, and the effects of the cyclic nucleotide may be tissue- and/or cell stage-specific (314, 315, 317).

Other evidence not only supports a role in cell development for cAMP but also for the cAMP-dependent protein kinase isozymes (237, 317–320). The method generally used to quantitate the cAMP-dependent protein kinase isozymes is by the use of DEAE-cellulose chromatography. Although this technique can be used for reasonably accurate estimations of protein kinase isozymes, it is important to realize and avoid the potential pitfalls of the method. Specific antibodies for each isozyme have been used and provide a more accurate determination of isozyme levels (237).

The levels of the cAMP-dependent protein kinase isozymes have been measured during the cell cycle in several studies. In these studies the levels of both isozymes are reported to oscillate during cell cycle traverse. While the oscillation of isozymes suggests that each may have a separate and specific function, there are little, if any, data to indicate what these functions might be. Costa *et al.* (364) observed in Chinese hamster ovary cells that the specific activity of cAMP-dependent protein kinase first decreases about twofold and then increases 1.5 to 3.5-fold such that a maximum was reached at the G_1–S border. As the cells proceed through the G_1 phase, the ratio of type I to type II protein kinase changes reciprocally. The increase in specific protein kinase activity is due to a selective elevation of the type II isozyme while the type I isozyme decreases. Haddox *et al.* (365) observed that as the Chinese hamster ovary cells traverse G_1 and approach the S phase, the type II isozyme increases as a function of time and cAMP concentration, but the specific activity of type I does not change throughout the cycle. However, if dibutyryl- or 8-bromo-cAMP is included in the growth medium after detachment, cell growth is arrested and a dramatic, cycloheximide-sensitive increase in the type I isozyme occurs. The type II isozyme decreases by this treatment. Estimation of the half-lives of the isozymes indicates that the cyclic nucleotide treatment causes a selective turnover of the type II isozyme.

Friedman and co-workers correlated the levels of cAMP with the levels and the activation state of the cAMP-dependent protein kinase in HeLa cells (366). Cells in the log phase of growth have nearly equal amounts of type I and type II isozymes but the enzyme activity ratio changes are relatively insensitive to high or changing levels of cAMP. This may be due to a higher K_a for cAMP activation at high levels of protein kinase. However, during mitosis, the cAMP levels and the activity ratio decrease to their lowest levels. The protein kinase levels remain high during mitosis and then fall after mitosis.

Although no separate and specific roles for the cAMP-dependent protein kinase isozymes have been clearly established, it has been suggested that the type I isozyme is a positive effector of growth and the type II isozyme relates more to tissue differentiation (319, 320). A number of different approaches have been used to study these phenomena. One approach has been to study the relative proportions of the cAMP-dependent protein kinase isozymes during whole-organ development. Lee et al. (367) demonstrated a rapid increase in rat testicular protein kinase activity during the first postnatal week which is correlated primarily with an increase in the type I isozyme. An increase in the type II isozyme occurs coincidently with the onset of complete spermatogenesis. Eppenberger et al. (368), studying postnatal uterine development, indicated that the ratio of type I to type II protein kinase decreases from about 0.55 on day 1 to about 0.1 on day 20, then increases to the day 1 level by day 40. These ratio changes are due to a decrease and then an increase in the type I isozyme with no change in the type II isozyme. Wittmaack et al. (369) used immunoprecipatation techniques in developing rat liver and malignant hepatic tissues. In developing liver, the total inhibitor-sensitive protein kinase activity steadily increases during normal liver development until the maximum levels are reached in the adult liver. The type I–type II ratio is about 1.2 in the fetus (4 days prepartum), 0.7 at birth, 1.8 on day 17 after birth, and 1.2 in the adult.

Another approach to demonstrate a specific role for one of the isozymes in proliferation is to evaluate isozyme ratios in normal and malignant tissues from the same organ. Fossberg et al. (370), comparing extracts of normal and carcinoma-involved human renal cortex, and Handschin and Eppenberger, studying normal and malignant human mammary tissue (371), found that the type I to type II ratio is approximately two times higher in extracts from malignant tissue, although the total protein kinase activity is the same in both normal and malignant tissue.

Weber et al. (372) and Wittmaack et al. (369) suggest that a simple correlation of one isozyme with proliferation and the other with differentiation may not hold. They point out that in some of these systems changes in proliferation are also associated with changes in differentiation. Lymphocytes from normal patients and patients with chronic lymphocytic leukemia represent pure cell-types that show negligible proliferation and differ only in the state of differentiation (372). The tumor lymphocytes contain lower levels of cAMP, inhibitor-sensitive cAMP-dependent protein kinase activity and correspondingly lower levels of regulatory subunit. By using SDS-gel electrophoretic separation of ^{32}P-labeled

8-azido-cAMP regulatory subunit and type I- and type II-specific immunotitration, differences in the regulatory subunit pattern are seen between the normal and leukemic lymphocytes. Approximately 60% of the total cAMP-binding sites are associated with the type I isozyme in both normal and leukemic cells. About 40% of the cAMP-binding sites are associated with the type II isozyme in leukemic lymphocytes, but in normal cells 20% of the sites are associated with the type II isozyme and about 20% are represented as low immunoreactive type I binding sites. Wittmaack *et al.* (*369*) have also demonstrated that the type I to type II ratio in rapidly proliferating AH-130 hepatoma cells is higher than in the normal adult liver but does not change when the cells go into the stationary phase. In addition, the undifferentiated AH-130 hepatoma cells have a lower ratio than do the well-differentiated 9618-A cells. Consequently, these data do not support the type I-proliferation, type II-differentiation hypothesis but suggest that the ratio of type I to type II relates more to the terminal differentiation of the organ than to the proliferation rates.

Byus *et al.* (*373*) reported that proliferation of concanavalin A-stimulated human peripheral lymphocytes leads to a cAMP increase and a selective activation of the type I isozyme of the cAMP-dependent protein kinase, even though both isozymes are present. Incubation of lymphocytes with both concanavalin A and dibutyryl-cAMP causes an activation of both type I and type II isozymes and prevents both RNA and DNA synthesis. These data led them to conclude that the type I isozyme is a positive modulator of lymphocyte proliferation while activation of the type II isozyme, or activation of both isozymes, inhibits proliferation. The conclusion of Wang *et al.* (*340*), that there is a requirement for a rise and fall in cAMP, is not inconsistent with these data. The conconavalin A could allow the rise and fall in cAMP and subsequent cell division but the inclusion of dibutyryl-cAMP would not allow the fall in cyclic nucleotide, and proliferation could be prevented.

Several lines of evidence indicate that the cAMP-dependent protein kinase isozyme ratio changes following viral or chemical transformation. Gharrett *et al.* (*374*) and Wehner *et al.* (*375*) characterized and compared the cAMP-dependent protein kinase isozymes in normal 3T3 cells with SV40-transformed (*374, 375*), spontaneously transformed (*375*), and methylcholanthrene-transformed (*375*) cell counterparts. While normal cells contain only the type II isozyme, transformed cells contain an equivalent amount of kinase activity but these cells contain both type I and type II activities as determined by all three of these techniques. Ledinko and Chan (*376*) also observed that the type I isozyme is higher in rat 3Y1 cells transformed by human adenovirus type 12 compared to untransformed 3Y1 cells. Little or no change occurs in the type II isozyme following transformation.

However, Haddox *et al.* (*365*) determined that Rous sarcoma virus-transformed fibroblasts have a greater protein kinase specific activity and a lower ratio of type I to type II isozyme than do the normal cells (0.42 in the transformed cells and 0.85 in the normal cells). The higher amount of the type II isozyme in these transformed cells compared to normal cells is in contrast to the higher amount of the type I isozyme in the SV40-transformed 3T3 cells observed by Gharrett *et al.* (*374*) and Wehner *et al.* (*375*) and in the adenosine-transformed 3Y1 cells described by Ledinko and Chan (*376*). It is not clear if these differences are due to the cell type, the transforming virus, or both.

Several laboratories have reported changes in cAMP-dependent protein kinase isozymes during cellular differentiation. Conti *et al.* (*377*) separated germ cells from mouse testes and determined the quantity of the two isozymes of protein kinase in preparations enriched (70–90%) in middle-late pachytene spermatocytes, round spermatids, and elongated spermatids by measuring cAMP-dependent kinase activity using protamine as substrate and [³H]cAMP binding activity following DEAE-cellulose chromatography. Pachytene spermatocytes had a type I to type II isozyme ratio of about 2.0. Cells in later stages of spermatogenesis had lower amounts of the type I isozyme and increasing amounts of the type II isozyme. Round spermatids had a type I to type II isozyme ratio of about 1.0 while elongated spermatids contained almost exclusively the type I isozyme. Elongated spermatids appeared to contain lower total cAMP-dependent protein kinase activity than do pachytene spermatocytes or rounded spermatids.

Fakunding and Means (*378*) demonstrated similar isozyme changes during rat Sertoli cell maturation by measuring [³H]cAMP binding and inhibitor-sensitive, cAMP-dependent histone kinase activity after DEAE-cellulose chromatography. At 12-days of age, Sertoli cells contained 2.5–3.0 times more type I than type II isozyme. Further development was associated with a decrease in the type I to type II isozyme ratio such that after 20-days of age the ratio was approximately 0.8.

Schwartz and Rubin (*379*) have used cAMP-dependent histone phosphorylation and [³H]cAMP binding following DEAE-cellulose chromatography in combination with type I and type II specific immunoprecipitation to study changes in the amounts of regulatory and catalytic subunits in Friend erythroleukemic cells before and after stimulation of differentiation with dimethylsulfoxide. During differentiation, the concentration of the type II regulatory subunit increased threefold and the type I regulatory subunit decreases to one-third of the control level, resulting in a change of the type I to type II regulatory subunit from 0.8 in the undifferentiated cells to 0.1 after induction of differentiation. Changes in the catalytic activity were proportional to changes in binding activity, indicating that the ratio of regulatory to catalytic subunit did not change. When the cells were treated with 8-bromo-cAMP and 3-isobutyl-1-methylxanthine for two days, similar changes in the regulatory subunit occurred without proportional changes in the catalytic subunit, but differentiation did not appear.

Liu (*337*) studied changes in the cAMP-dependent isozymes during differentiation of 3T3-L1 cells from fibroblasts to adipocytes. The regulatory subunit was determined by 8-azido-cAMP photoaffinity labeling and catalytic activity was determined by cAMP-dependent histone phosphorylation. In the undifferentiated fibroblasts, the type II isozyme was the predominant cAMP-dependent protein kinase. After promotion of differentiation with 3-isobutyl-1-methylxanthine, dexamethasone, and insulin, an increase in the total cAMP-dependent protein kinase activity was attributed to a 3- to 6-fold increase in the type I isozyme. This was due to an increase in both the regulatory and the catalytic subunits. Similar results were obtained with spontaneously differentiating 3T3-L1 cells, demonstrating that the selective increase in the type I isozyme appeared to be directly related to adipocyte differentiation rather than to the drug and hormone treatment used to promote differentiation. Furthermore, 3T3-C2 cells, which had a lower intrinsic ability to undergo differentiation, did not show the selective type I increase when stimulated in the same manner.

How cAMP and protein kinases are involved in growth and development is poorly understood. It is likely that protein kinase is integrated with primary, intrinsic regulatory signals and other second messengers such as calcium and phosphoinositides. The consequences of this latter second messenger are only beginning to be unraveled but it is pertinent to ask how it may be involved in cell development (*380, 381*). Calcium is particularly interesting since it appears to play a central role in many regulatory processes (*380, 381*). For example, both phospholipase A_2, which generates prostaglandin-like precursors (*382*), and phospholipase C, which generates phosphoinositides and diacylglycerol (*380, 381*), are calcium-sensitive enzymes. Furthermore, calcium can also regulate cAMP metabolism via regulation of adenylate cyclase and phosphodiesterase (*383*). A circular and more complicated, but potentially tightly regulated picture, emerges when it is considered that prostaglandins activate adenylate cyclase, phosphoinositides release intracellular calcium, and diacylglycerol activates protein kinase C, a calcium-sensitive enzyme. Cyclic AMP, via protein kinase, is known to regulate calcium levels. It will be interesting to follow future developments involving the interactions of cAMP, Ca^{2+}, and phosphoinositides in the regulation of growth, development, and metabolism in general.

F.   VARIATIONS IN THE REGULATORY
     SUBUNIT–CATALYTIC SUBUNIT RATIO

The ratio of regulatory to catalytic subunit in terminally differentiated tissues has been reported to be approximately 1.0. This has been demonstrated in several species of heart (*216*), in several tissues from adult rabbits (*90*), and in the postnatal developing rat liver (*369*) and rat brain (*384*). This suggests that the two subunits are coordinately regulated. Uno *et al.* (*385*), analyzing yeast mutants altered in cAMP metabolism, have determined that the subunits of protein kinase are not coded by the same gene locus, suggesting the possibility that the levels of regulatory and catalytic subunits could be independently regulated.

There are several reports in the literature indicating that, at least under certain circumstances in some cells, there are greater amounts of regulatory subunit than catalytic subunit. Richards and Rolfes (386) studied ovarian granulosa cells from immature, hypophysectomized rats treated with estrogen and human follicle-stimulating hormones. Under the influence of these hormones, granulosa cells luteinize and produce steroids. These investigators reported that the type II regulatory subunit is induced 10- to 20-fold in treated animals, with little if any corresponding increase in catalytic activity. Darbon *et al.* (387) reported a similar finding. Segaloff *et al.* (281) have shown that porcine granulosa cells in primary culture respond to treatment with cholera toxin, follicle-stimulating hormone, and cAMP analogs by an increased specific activity of the type II, but not of the type I regulatory subunit. A 5- to 10-fold increase in the regulatory subunit determined [^{3}H]cAMP binding and densitometric scans of autoradiograms from specific 8-azido[^{32}H]cAMP photoaffinity labeling of the regulatory subunit, is correlated with a 2- to 4-fold increase in the specific catalytic subunit activity, measured by heptapeptide phosphorylation.

Walter *et al.* (388) and Lohmann *et al.* (389) studied the potential role of protein kinase in the dibutyryl-cAMP-stimulated differentiation of neuroblastoma-glioma hybrid cells. Treatment of hybrid cells with dibutyryl-cAMP or addition of agents that elevate intracellular cAMP results in a selective increase in the type I regulatory subunit with no change in the catalytic subunit or type II regulatory subunit. The type I regulatory subunit is detected by 8-azido[^{32}P]cAMP photoaffinity labeling (388) and by an enzyme-linked immunosorbent assay and radioimmunolabeling followed by transfer from SDS-gels to nitrocellulose (389). These experiments indicate that the hybrid cells separately regulate the levels of regulatory and catalytic subunit and also suggest that the free type I regulatory subunit could possibly be involved with the expression of some differentiating function(s). Furthermore, Morrison *et al.* (390) have demonstrated that there are increases in the levels of mRNA coding for the regulatory subunit. It has been suggested that this altered regulatory subunit, termed R′, may be elevated due to increased synthesis and decreased degradation. It is interesting that most of the known examples of tissues that appear to show an elevation of regulatory subunit, but not of catalytic subunit, are of nervous or reproductive origin or from a mutant cell line (386, 394). It will be interesting to see if this is a unique characteristic of these tissues or if other tissues also show this phenomenon.

Although it appears in all of these studies that the type II regulatory subunit is present in excess over the catalytic subunit, it has not been ruled out that some inhibitor(s) of protein kinase, other than the regulatory subunit, is also increased by the same treatment. The presence of a heat-stable protein kinase inhibitor in various tissues has been known for some time (32). Beale *et al.* (395) and Tash *et al.* (396) reported that a low-molecular-weight heat-stable protein kinase inhibitor is increased in the Sertoli cells of rat testes after follicle-stimulating hormone treatment. Because of potential errors in determining the levels of the protein kinase catalytic activity (389), specific antibody techniques or cDNA clones for the catalytic subunit may help resolve these uncertainties.

If the regulatory subunit is present in excess compared to the catalytic subunit, the mechanism that maintains these subunits in equal proportions in normal, differentiated tissues is apparently disrupted. Although the levels of many proteins are regulated at the level of transcription, regulation can also occur at the level of translation and/or degradation of messenger RNA, and at the level of protein degradation (see Section V,G). It is therefore possible that the two subunits are regulated coordinately at one level (e.g., transcription) but regulated differently at another level (e.g., translation or degradation). These two subunits may be generally maintained in stoichiometric amounts by proteolysis of either subunit that is present in excess regardless of the mechanism of regulation of individual subunit levels. For example, the free regulatory subunit is known to be much more sensitive to proteases than is the regulatory subunit which is bound to the catalytic subunit (10).

The nature of the events that determine the protein kinase subunit levels has not been determined. However, the experiments outlined in this section suggest that cAMP may be important in determining these levels. Recent results by Ratoosh *et al.* (397) indicate that FSH, via cAMP, may increase synthesis of rabbit granulosa cell type II regulatory subunit by altering the levels of mRNA for this protein. If it is assumed that the mechanism of cAMP action is through the cAMP-dependent protein kinase, it is possible that this enzyme regulates its own levels.

Although either the type I or type II regulatory subunit may be bound to the particulate fraction (see Section V,B.), there may be an additional function for this subunit. Evidence has appeared to suggest such a role. The regulatory subunit may participate in the transient inhibition of phosphoprotein phosphatase(s) that dephosphorylates various protein kinase substrates. In 1977, Gergely and Bot (398) reported that the cAMP-dependent protein kinase, in the presence of cAMP, inhibited the phosphoprotein phosphatase-catalyzed dephosphorylation of phosphorylase *a*, but not a ^{32}P-labeled tetradecapeptide, suggesting a substrate-mediated effect. Khatra *et al.* (399) have demonstrated that homogeneous preparations of type II regulatory subunits, or homogeneous type II cAMP-dependent protein kinase in the presence of cAMP, inhibits a high-molecular-weight phosphoprotein phosphatase from rabbit skeletal muscle. This inhibition of dephosphorylation occurs when phosphorylase, glycogen synthase (phosphorylated at sites 2 and 3) or histone is used as substrate, suggesting an enzyme-directed inhibition. This inhibition occurs *in vitro* at concentrations of regulatory subunit that occur *in vivo*. Through the decrease in dephosphorylation rate, the regulatory subunit of protein kinase could possibly magnify the increases in phosphate content of proteins brought about by activation of the catalytic subunit. Thus, when tissue cAMP is elevated, the phosphoprotein phosphatase is inhibited, and when cAMP declines, as occurs in some tissues by insulin action, this enzyme is activated. It remains to be proved if this mechanism has importance in the physiological regulation of phosphoprotein phosphatase, but, if so, it suggests the possibility of a multifunctional regulatory subunit. It has been reported by Constantinou *et al.* (*399a*) that in the presence of cAMP the phosphorylated form of the rat liver type II regulatory subunit possesses topoisomerase activity. There are indications to suggest that this regulatory subunit can form DNA-phospho-regulatory subunit-cAMP complexes and relax superhelixes of DNA. If this report can be confirmed and extended, a new concept in the transcriptional action of the cAMP-dependent protein kinase may emerge.

Further experimentation will be required to elucidate the mechanisms that regulate the levels of protein kinase and to determine if the free regulatory subunit has a role(s) other than regulating protein kinase catalytic activity. If the regulatory subunit does have another function, the nomenclature of this enzyme as a protein kinase, which specifies a phosphorylation reaction, will be appropriate only for the phosphotransferase function. Although it has not been conclusively established that the regulatory subunit has a catalytic function, it may serve as a regulatory protein in capacities other than for protein kinase activity. Since it has some sequence homologies with the catabolite activator protein (CAP) of *E. coli* (*156*), it is tempting to speculate that it may serve as a regulator of gene transcription in mammalian cells. However, there is presently no direct evidence to support this idea, and it is known that the regulatory subunit does not contain the corresponding DNA binding domain found in CAP.

G. REGULATION OF THE AMOUNT OF CERTAIN PROTEINS BY CYCLIC AMP-DEPENDENT PROTEIN KINASES

The postsynthetic modification of some preexisting enzymes by cAMP-dependent phosphorylation is known to alter the activities of these enzymes and thereby modify various short-term cellular functions. In recent years it has become clearer that cAMP is also involved in more long-term regulation of cellular functions through the regulation of the amounts of specific proteins.

Since protein phosphorylation is a well-established mechanism of cAMP-dependent protein kinase action, this mechanism may be involved in the phosphorylation of nuclear proteins, which could regulate gene transcription or other nuclear events. Although both histone and nonhistone nuclear proteins are known to be phosphorylated by the cAMP-dependent protein kinase *in vitro*, the most convincing evidence for *in vivo* cAMP-mediated nuclear protein phosphorylation is for histone H1. Langan demonstrated an *in vivo* increase in [^{32}P]phosphate incorporation of histone H1 in the liver of rats injected with glucagon or dibutyryl-cAMP (*31*). To rule out artifactual phosphorylation during the extraction procedure, Wicks *et al.* (*400*) extracted histone H1 with sulfuric acid and suggested a role for the cAMP-dependent protein kinase in the induction of tyrosine aminotransferase in Reuber H35 hepatoma cells. Harrison *et al.* (*401*) observed complex changes in the [^{32}P]phosphate content of histone H1 subspecies in rat C6 glioma cells. Isoproterenol elicited both increases and decreases in [^{32}P]phosphate levels at several phosphorylation sites. Although cAMP-independent phosphorylation apparently occurs, cAMP-dependent phosphorylation at serine-37 of the amino terminus of histone H1-1 and H1-2 *in vitro* correlated with increased ^{32}P content at these sites *in vivo* following treatment of these cells with isoproterenol. Dibutyryl-cAMP also caused complex changes in histone H1 phosphorylation but these are different from those for isoproterenol.

The physiological significance of histone phosphorylation is still not clear and has recently been reviewed by Johnson (*402*). Briefly, histone H1 is believed to be involved in the formation of higher-order chromatin structures and it has been suggested that phosphorylation of this protein alters the circular dichroism of histone–DNA complexes and the template activity of chromatin. Phosphorylation of serine-38 of histone H1 reduces histone binding to DNA and may function to unwind tightly coiled chromatin to expose regions for RNA polymerase binding and transcription. Langan (*31*) has determined that more liver histone is phosphorylated at serine-38 than would be required to actively transcribe specific sequences in response to hormones or cAMP. This and other data suggest that specific base sequence recognition due to histone H1 phosphorylation is unlikely. Alternatively, Harrison *et al.* (*401*) calculated that only a very small number of glioma cell histone H1 molecules are phosphorylated and suggest that these may occur at selected gene loci to alter chromatin function.

Several reports have appeared documenting the *in vitro* cAMP-dependent and cAMP-independent protein kinase phosphorylation of RNA polymerase II, the enzyme that catalyzes the transcription of genes coding for messenger RNA. Some investigators report that phosphorylation is correlated with increases in polymerase activity (*403–406*) while others are unable to support this finding (*407–409*). Kranias *et al.* (*405*) reported that nuclear cAMP-dependent protein kinase leads to incorporation of 0.5 mol [^{32}P]phosphate/mol enzyme into the 25,000-dalton subunit of calf thymus RNA polymerase II with a concomitant 3-fold increase in enzyme activity. Dephosphorylation by *E. coli* phosphatase results in a loss of ^{32}P label and a corresponding decrease in enzyme activity. The *in vivo*, isoproterenol-stimulated phosphorylation of rat C6 glioma cell RNA polymerase II has been reported by Lee *et al.* (*410*). Stimulation of cells results in ^{32}P incorporation into serine residues of all six polymerase subunits with a total incorporation of 0.5–2.0 mol phosphate/mol enzyme. Propranolol inhibits and dibutyryl-cAMP mimics the isoproterenol effects. The phosphorylation of RNA polymerase I, which catalyzes the transcription of ribosomal genes, and RNA polymerase III, which transcribes genes coding for 5 S ribosomal RNA, have also been reported [see Ref. (402)].

If the cAMP-stimulated induction of specific proteins requires phosphorylation event(s), the presence of the cAMP-dependent protein kinase or its catalytic subunit in the nucleus or a translocation to the nucleus would be important. Since cAMP is known to be synthesized at the plasma membrane, it seems probable that some mechanism may be required to translocate either cAMP and/or protein kinase to the nucleus. There is, however, some evidence to indicate that cAMP-dependent protein kinases are in the nucleus and translocation of protein kinase to the nucleus may also occur (see Section V,B.). Agents that elevate the intracellular cAMP concentration have been shown to result in an apparent translocation of protein kinase activity from the cytoplasm to the nucleus in several tissues [see Refs. (36, 237, 402, 404) for reviews]. It has been suggested that some of these earlier studies may be subject to artifacts due to enzyme redistribution and nonspecific ionic interaction of the catalytic subunit of protein kinase with subcellular binding sites when homogenization is carried out in low-ionic-strength buffers (234). A number of other potential pitfalls of translocation studies have been reviewed, including cytoplasmic contamination and cAMP-independent protein kinases other than the catalytic subunit of protein kinase (10). However, nonaqueous procedures are now generally used to minimize redistribution and nonspecific binding during nuclei isolation (411). Cho-Chung (412, 413) proposed that tumor regression is dependent on translocation of a cAMP-protein kinase type II holoenzyme ternary complex from the cytoplasm to the nucleus. Although these studies and others (36, 237, 402, 404) suggest a translocation of protein kinase activity to the nucleus, they are not in agreement on whether it is the holoenzyme or free catalytic subunit that is translocated.

Histochemical or immunocytochemical techniques have also been used to address the question of protein kinase translocation. This technique complements the purely biochemical approaches, and provides a sensitive technique to visually localize protein kinases which may be present at low concentrations in some cells or cell compartments. This approach avoids certain potential artifacts of in vitro biochemical methods and allows a more definite assignment of the presence of protein kinases in the nucleus (also see Section V,B.). Although it cannot be totally ruled out that the appearance of immunoreactive protein kinase in the nucleus is due to "uncovering" enzyme already present in the nucleus, the data are consistent with biochemical data suggesting translocation of protein kinase from the cytoplasm to the nucleus.

Early experiments concerning the cAMP-stimulated induction of tyrosine aminotransferase and phosphoenolpyruvate carboxykinase suggested that regulation occurs at a posttranscriptional level, most likely involving an enhancement in the rate of translation of preexisting mRNAs (414, 415). This conclusion was based on studies using inhibitors of protein synthesis and comparing the kinetics of cAMP- and glucocorticoid-stimulated protein induction. While these earlier studies determined protein induction by measuring enzyme activities, later studies quantitated specific mRNA levels by indirect *in vitro* translation assays and in some cases more directly by measuring mRNA levels and/or specific gene transcription, using specific cDNA probes. Results from these later studies suggested regulation at a pretranslational step and emphasized the need to reevaluate the cAMP mechanism for the specific induction of enzymes in eucaryotes.

Using indirect quantitation of specific mRNA levels by *in vitro* translation, it has been determined that cAMP regulates the functional levels of mRNA or mRNA translational efficiency for phosphoenolpyruvate carboxykinase (416–418), tyrosine aminotransferase (419, 420), lactate dehydrogenase A subunit (421), and alkaline phosphatase (422). The proportional changes in the specific mRNAs and the enzymes for which they code indicate that translation is not the primary site of regulation for these proteins. Since this approach measures translationally active mRNAs, it is not possible to determine if cAMP affects transcription, posttranscriptional modification of primary gene transcripts, increases in translatability of mRNAs, or decreased breakdown or stability of mRNAs. Experiments using specific cDNA probes indicate that cAMP regulates the abundance of mRNA for albumin in mouse hepatoma cells (423), phosphoenolpyruvate carboxykinase in rat liver (424–426), and lactate dehydrogenase A subunit in rat glioma cells (427). Evidence for transcriptional action by cAMP on induction of pituitary prolactin (428), phosphoenolpyruvate carboxykinase (429–433), lactate dehydrogenase A subunit (434), and tyrosine aminotransferase (435, 436) have also been reported.

The cAMP-stimulation of phosphoenolpyruvate carboxykinase gene transcription has been shown to be via the cAMP-dependent protein kinase (280). The use of cAMP analog combinations to specifically cause a synergistic increase in mRNA[PEPCK] and gene transcription clearly indicate that these effects occur following protein kinase activation. The type II isozyme is the predominant isozyme and it is primarily responsible for mediating the response (see Section V,C.). Since the catalytic subunit is common to both isozymes and determines the substrate specificity, and because all known cAMP-dependent protein kinase effects are due to phosphorylation, it is likely that PEPCK gene transcription is mediated by a phosphorylation event. Results obtained from fusion of hepatoma cells with red cell ghosts loaded with either catalytic subunit or protein kinase inhibitor indicate that the catalytic subunit and not the regulatory subunit is responsible for regulation of tyrosine aminotransferase gene transcription (210). However, the finding by Constantinou *et al.* (399a) that a complex of phospho-type II regulatory subunit-cAMP may be a topoisomerase implicates the regulatory subunit in the regulation of transcription.

In summary, these studies generally indicate that cAMP regulates the levels of some proteins by some mechanism that involves an enhanced transcription of specific genes coding for the regulated proteins. In the case of PEPCK and tyrosine aminotransferase, the cAMP-dependent protein kinase appears to be the mediator of the transcription of these genes but the putative phosphorylated proteins responsible have not been identified. It has been reported that chromosome-associated proteins such as histones and high-mobility group proteins are phosphorylated but demonstration of associated functional alterations of these proteins has been difficult. It is reasonable to consider that phosphorylation of these proteins may play some role in gene transcription, but since specificity is in question, this may serve a general function which may facilitate another mechanism mediating specific gene transcription. A specific mechanism could possibly involve phosphorylation and activation of mRNA polymerase II. However, to clearly demonstrate the physiological significance of this, a rigorous analysis of these reactions is required. It also remains to be clearly demonstrated whether or not the regulatory subunit plays a role in the regulation of gene transcription.

ACKNOWLEDGMENTS

The authors are grateful to Dr. D. Friedman, Dr. D. Granner, and Dr. R. Uhing for helpful discussions during the course of this work. We would also like to thank the many scientists who provided us with published and unpublished data during the preparation of this text. We are also very appreciative of Mrs. Penny Stelling, who spent many hours typing and editing this manuscript.

REFERENCES

1. Burnett, G., and Kennedy, E. P. (1954). *JBC* **211,** 969.
2. Rabinowitz, M. (1962). "The Enzymes," 2nd ed., Vol. 6, p. 119.
3. Robinson, G. A., Butcher, R. W., and Sutherland, E. W. (1971). "Cyclic AMP." Academic Press, New York.
4. Sutherland, E. W. (1972). *Science* **177,** 401.
5. Krebs, E. G. (1972). *Curr. Top. Cell. Regul.* **5,** 99.
6. Corbin, J. D., Sugden, P. H., West, L., Flockhart, D. A., Lincoln, T. M., and McCarthy, D. (1978). *JBC* **253,** 3997.
7. Corbin, J. D., Keely, S. L., and Park, C. R. (1975). *JBC* **250,** 218.
8. Weber, W., and Hilz, H. (1979). *BBRC* **90,** 1073.
9. Granot, J., Mildvan, A. S., Hiyama, K., Kondo, H., and Kaiser, E. T. (1980). *JBC* **255,** 4569.
10. Flockhart, D. A., and Corbin, J. D. (1982). *CRC Crit. Rev. Biochem.* **12,** 133.
11. Beavo, J. A., and Mumby, M. C. (1982). *Handb. Exp. Pharmakol.* **581,** Part 1, 363.
12. Døskeland, S. O., and Øgreid, D. (1981). *Int. J. Biochem.* **13,** 1.
13. Krebs, E. G., and Beavo, J. A. (1978). *Annu. Rev. Biochem.* **48,** 923.
14. Nimmo, H. G., and Cohen, P. (1977). *Adv. Cyclic Nucleotide Res.* **8,** 146.
15. Beavo, J. A., Bechtel, P. J., and Krebs, E. G. (1975). *Adv. Cyclic Nucleotide Res.* **5,** 241.
16. Chock, P. B., and Stadtman, E. R. (1977). *PNAS* **74,** 2766.

17. Goldbeter, A., and Koshland, D. E., Jr. (1981). *PNAS* **78**, 6840.
18. Laporte, D. C., and Koshland, D. E., Jr. (1983). *Nature (London)* **305**, 286.
19. Koshland, D. E., Goldbeter, A., and Stock, J. B. (1982). *Science* **217**, 220.
20. Shacter, E., Chock, P. B., and Stadtman, E. R. (1984). *JBC* **259**, 12252.
20a. Meinke, M. H., Bishop, J. S., and Edstrom, R. D. (1986). *PNAS* (in press).
21. Corbin, J. D., and Døskeland, S. O. (1983). *JBC* **258**, 11391.
22. Gill, G. N., Holdy, K. E., Walton, G. M., and Kanstein, C. B. (1976). *PNAS* **73**, 3918.
23. Lincoln, T. M., and Corbin, J. D. (1977). *PNAS* **8**, 3239.
24. MacKenzie, C. W., III (1982). *JBC* **257**, 5589.
25. Lincoln, T. M., and Corbin, J. D. (1983). *Adv. Cyclic Nucleotide Res.* **15**, 139.
26. Gill, G. N., and McCune, R. W. (1979). *Curr. Top. Regul.* **15**, 1.
27. Corbin, J. D., and Lincoln, T. M. (1978). *Adv. Cyclic Nucleotide Res.* **9**, 159.
28. Lincoln, T. M., and Corbin, J. D. *J. Cyclic Nucleotide Res.* **4**, 3.
29. Gill, G. N. (1977). *J. Cyclic Nucleotide Res.* **3**, 153.
30. Walsh, D. A., and Krebs, E. G. (1973). *Enzymes* **8**, 555.
31. Langan, T. A. (1973). *Adv. Cyclic Nucleotide Res.* **3**, 99.
32. Walsh, D. A., and Ashby, C. D. (1973). *Recent Prog. Horm. Res.* **29**, 329.
33. Rubin, C. S., and Rosen, O. M. (1975). *Annu. Rev. Biochem.* **44**, 831.
34. Rosen, O. W., Erlichman, J., and Rubin, C. S. (1975). *Adv. Cyclic Nucleotide Res.* **5**, 253.
35. Corbin, J. D., Keely, S. L., Soderling, T. R., and Park, C. R. (1975). *Adv. Cyclic Nucleotide Res.* **5**, 265.
36. Jungmann, R. A., and Russell, D. H. (1977). *Life Sci.* **20**, 1787.
37. Cohen, P. (1978). *Curr. Top. Cell. Regul.* **14**, 118.
38. Carlson, G. M., Bechtel, P. J., and Graves, D. J. (1979). *Adv. Enzymol.* **50**, 41.
39. Glass, D. B., and Krebs, E. G. (1980). *Annu. Rev. Pharmacol. Toxicol.* **20**, 363.
40. Robinson-Steiner, A., and Corbin, J. D. (1986). *In* "Handbook of Cardiology" (in press).
41. Rosen, O. M., and Krebs, E. G. (1981). *Cold Spring Harbor Conf. Cell Proliferation* **8**.
42. Ramsyer, J., Kaslow, H. R., and Gill, G. N. (1974). *BBRC* **59**, 813.
43. Dills, W. L., Beavo, J. A., Bechtel, P. J., and Krebs, E. G. (1975). *Adv. Cyclic Nucleotide Res.* **5**, 829.
44. Dills, W. L., Beavo, J. A., Bechtel, P. J., and Krebs, E. G. (1975). *BBRC* **62**, 70.
45. Rieke, E., Pauitz, N., Eigel, A., and Wagner, K. G. (1975). *Hoppe-Seyler's Z. Physiol. Chem.* **356**, 1177.
46. Rannels, S. R., and Corbin, J. D. (1983). "Methods in Enzymology," Vol. 99, p. 55.
47. Corbin, J. D., and Rannels, S. R. (1981). *JBC* **256**, 11671.
48. Goodwin, C. D., Beavo, J. A., Dills, W. L., and Krebs, E. G. (1977). *FP* **36**, 728.
49. Seville, M., and Holbrook, J. J. (1984). *Anal. Biochem.* **137**, 330.
50. Reimann, E. M., and Beham, R. A. (1983). "Methods in Enzymology," Vol. 99, p. 51.
51. Strålfors, P., and Belfrage, P. (1982). *BBA* **721**, 434.
52. Muniyappa, K., Leibach, F. H., and Mendicino, J. (1983). *Mol. Cell. Biochem.* **50**, 157.
53. Beavo, J. A., Bechtel, P. J., and Krebs, E. G. (1974). "Methods in Enzymology," Vol. 38C, p. 299.
54. Bechtel, P. J., Beavo, J. A., and Krebs, E. G. (1977). *JBC* **252**, 2691.
55. Sugden, P. H., Holladay, L. A., Reimann, E. M., and Corbin, J. D. (1976). *BJ* **159**, 409.
56. Walsh, D. A., Perkins, J. P., and Krebs, E. G. (1968). *JBC* **243**, 3763.
57. Hofmann, F., Beavo, J. A., Bechtel, P. J., and Krebs, E. G. (1975). *JBC* **250**, 7795.
58. Hoppe, F., and Wagner, K. G. (1977). *FEBS Lett.* **74**, 95.
59. Døskeland, S. O. (1978). *BBRC* **83**, 542.
60. Taylor, S. S., Lee, C., Swain, L., and Stafford, P. H. (1976). *Anal. Biochem.* **76**, 45.
61. Taylor, S. S., and Stafford, P. H. (1978). *JBC* **253**, 2284.

62. Rubin, C. S., Erlichman, J., and Rosen, O. M. (1972). *JBC* **247**, 36.
63. Rubin, C. S., Erlichman, J., and Rosen, O. M. (1974). "Methods in Enzymology," Vol. 38, p. 308.
64. Cobb, C., and Corbin, J. D. (1987). "Methods in Enzymology" (in press).
64a. Cobb, C. E., Beth, A. H., and Corbin, J. D. (1986). *F.P.* **45**, 1800.
65. Rosen, O. M., Rangel-Aldao, R., and Erlichman, J. (1977). *Curr. Top. Cell. Regul.* **12**, 39.
66. Lincoln, T. M., Dills, W. L., and Corbin, J. D. (1977). *JBC* **252**, 4269.
67. Flockerzi, V., Speichermann, N., and Hofmann, F. (1978). *JBC* **253**, 3395.
68. de Jonge, H. R. (1981). *Adv. Cyclic Nucleotide Res.* **14**, 315.
69. Nakazawa, K., and Sano, M. (1975). *JBC* **250**, 7415.
70. de Jonge, H. R., and Rosen, O. M. (1977). *JBC* **252**, 2780.
71. Gerzer, R., Hofmann, F., and Schultz, G. (1981). *EJB* **116**, 479.
72. Dills, W. L., Goodwin, C. D., Lincoln, T. M., Beavo, J. A., Bechtel, P. J., Corbin, J. D., and Krebs, E. G. (1979). *Adv. Cyclic Nucleotide Res.* **10**, 199.
73. Chen, L. J., and Walsh, D. A. (1971). *Biochemistry* **10**, 3614.
74. Yamamura, H., Inoue, Y., Shimomura, R., and Nishizuka, Y. (1972). *BBRC* **46**, 589.
75. Kumon, A., Nishiyama, K., Yamamura, H., and Nishizuka, Y. (1972). *JBC* **247**, 3726.
76. Yamamura, H., Nishiyama, K., Shimomura, R., and Nishizuka, Y. (1973). *Biochemistry* **12**, 856.
77. Peters, K. A., Demaille, J. G., and Fischer, E. H. (1977). *Biochemistry* **16**, 5691.
78. Kubler, D., Gagelman, M., Pyerin, W., and Kinzel, V. (1979). *Hoppe-Seyler's Z. Physiol. Chem.* **360**, 1421.
78a. Uhler, M. D., Carmichael, D. F., Lee, D. C. Chriva, J. C., Krebs, E. G., and McKnight, G. S. (1986). *PNAS* **83**, 1300.
78b. Reed, J., Gagelmann, M., and Kinzel, V. (1983). *ABB* **222**, 276.
79. Shoji, S., Ericsson, L. H., Walsh, K. A., Fischer, E. H., and Titani, K. (1983). *Biochemistry* **22**, 3702.
79a. Jahnsen, T., Hedin, L., Kidd, V. J., Schulz, T. Z., and Richards, J. S. (1987). "Methods in Enzymology" (in press).
80. Titani, K., Sasagawa, T., Ericsson, L. H., Kumar, S., Smith, S. B., Krebs, E. G., and Walsh, K. A. (1984). *Biochemistry* **23**, 4193.
81. Takio, K., Smith, S. B., Krebs, E. G., Walsh, K. A., and Titani, K. (1982). *PNAS* **79**, 2544.
82. Erlichman, J., Rubin, C. S., and Rosen, O. M. (1973). *JBC* **248**, 7607.
83. Connelly, P. A., Hastings, T. G., and Reimann, E. M. (1986). *JBC* **261**, 2325.
84. Hedrick, J. L., and Smith, A. J. (1968). *ABB* **126**, 155.
85. Robinson-Steiner, A. M. Beebe, S. J., Rannels, S. R., and Corbin, J. D. (1984). *JBC* **259**, 10596.
86. Rannels, S. R., and Corbin, J. D. (1980). *J. Cyclic Nucleotide Res.* **6**, 203.
87. Uno, I., Udea, T., and Greengard, P. (1977). *JBC* **252**, 5164.
88. Nesterova, M. V., Sashchenco, L. P., Vasiliev, V. Y., and Severin, E. S. (1975). *BBA* **377**, 271.
89. Fleischer, N., Rosen, O. M., and Reichlin, M. (1976). *PNAS* **73**, 54.
90. Hofmann, F., Bechtel, P. J., and Krebs, E. G. (1977). *JBC* **252**, 1441.
91. Rubin, C. S., Rangel-Aldao, A., Sarkar, D., Erlichman, J., and Fleischer, N. (1979). *JBC* **254**, 5797.
92. Rapoor, C. L., Beavo, J. A., and Steiner, A. L. (1979). *JBC* **254**, 12427.
93. Zoller, M. J., Kerlavage, A. R., and Taylor, S. S. (1979). *JBC* **254**, 2408.
94. Weldon, S. L., Mumby, M. C., Beavo, J. A., and Taylor, S. S. (1983). *JBC* **258**, 1129.
95. Corbin, J. D., Soderling, T. R., and Park, C. R. (1973). *JBC* **248**, 1813.
96. Huang, T. S., Feramisco, J. R., Glass, D. B., and Krebs, E. G. (1979). *Miami Winter Symp.* **16**, 449.

97. Rangel-Aldao, R., and Rosen, O. M. (1976). *JBC* **251**, 3375.
98. Rangel-Aldao, R., and Rosen, O. M. (1977). *JBC* **252**, 7140.
99. Rangel-Aldao, R., Kupiec, J. W., and Rosen, O. M. (1979). *JBC* **254**, 2499.
100. Scott, C. W., and Mumby, M. C. (1985). *JBC* **260**, 2274.
101. Hemmings, B. A., Aiten, A., Cohen, P., Rymond, M., and Hofmann, F. (1982). *EJB,* **127,** 473.
102. Carmichael, D. F., Geahlen, R. L., Allen, S. M., and Krebs, E. G. (1982). *JBC* **257**, 10440.
103. Yagura, T. S., and Miller, J. P. (1981). *Biochemistry* **20**, 879.
103a. Goodman, M. (1981). *Prog. Biophys. Mol. Biol.* **37**, 105.
104. Øgreid, D., Ekanger, R., Suva, R. H., Miller, J. P., Sturm, P., Corbin, J. D., and Døskeland, S. O. (1985). *EJB* **150**, 219.
105. Rannels, S. R., and Corbin, J. D. (1980). *JBC* **255**, 7085.
106. Øgreid, D., and Døskeland, S. O. (1980). *FEBS Lett.* **121**, 340.
107. Corbin, J. D., Rannels, S. R., Flockhart, D. A., Robinson-Steiner, A. M., Tigani, M. C., Døskeland, S. O., Suva, R., and Miller, J. P. (1982). *EJB* **125**, 259.
108. Corbin, J. D., Øgreid, D., Miller, J. P., Suva, R. H., Jastorff, B., and Døskeland, S. O. (1986). *JBC* **261**, 12081.
109. Rannels, S. R., and Corbin, J. D. (1981). *JBC* **256**, 7871.
110. Robinson-Steiner, A. M., and Corbin, J. D. (1982). *JBC* **257**, 5482.
111. Robinson-Steiner, A. M., and Corbin, J. D. (1983). *JBC* **258**, 1032.
112. Beebe, S. J., and Corbin, J. D. (1984). *Mol. Cell. Endocrinol.* **36**, 67.
113. Beebe, S. J., Holloway, R., Rannels, S. R., and Corbin, J. D. (1984). *JBC* **259**, 3539.
114. Beebe, S. J., Blackmore, P. F., Chrisman, T. D., and Corbin, J. D. (1987). "Methods in Enzymology" (in press).
114a. Beebe, S. J., Blackmore, P. F., Segaloff, D. L., Koch, S. R., Burks, D., Limbird, L. E., Granner, D. K., and Corbin, J. D. (1986). *Eur. Symp. Horm. Cell Reg. 10th,* 1986 (in press).
115. Malkinson, A. M., Beer, D. S., Wehner, J. M., and Sheppard, J. R. (1983). *BBRC* **12**, 214.
116. Toru-Delbauffe, D., Lognonne, J., Ohayon, R., Garaset, J., and Pavlovic-Hournae, M. (1982). *EJB* **125**, 267.
117. Corbin, J. D., Soderling, T. R., Sugden, P. H., Keely, S. L., and Park, C. R. (1976). *In* "Eukaryotic Cell Function and Growth" (J. E. Dumon, B. L. Brown, and N. J. Marshall, eds.), p. 231. Plenum, New York.
118. Potter, R. L., and Taylor, S. S. (1979). *JBC* **254**, 2413.
119. Fossberg, T. M., Døskeland, S. O., and Ueland, P. M. (1978). *ABB* **189**, 372.
120. Rannels, S. R., Cobb, C. E., Landiss, L. R., and Corbin, J. D. (1985). *JBC* **260**, 3423.
120a. Reimann, E. M. (1986). *Biochemistry* **25**, 119.
121. Sugden, P. H., and Corbin, J. D. (1976). *BJ* **159**, 423.
122. Erlichman, J., Rosenfeld, R., and Rosen, O. M. (1974). *JBC* **249**, 5000.
122a. Rangel-Aldao, R., and Rosen, O. M. (1977). *JBC* **252**, 7140.
123. Rosen, O. M., and Erlichman, J. (1975). *JBC* **250**, 7786.
124. Beavo, J. A., Bechtel, P. J., and Krebs, E. G. (1974). *PNAS* **71**, 3580.
125. Erlichman, J., Sarkar, D., Fleischer, N., and Rubin, C. S. (1980). *JBC* **255**, 8179.
126. Rubin, C. S., Rangel-Aldao, R., Sarkar, D., Erlichman, J., and Fleischer, N. (1979). *JBC* **254**, 3797.
127. Jahnsen, T., Lohmann, S. M., Walter, U., Hedin, L., and Richards, J. S. (1985). *JBC* **260**, 15980.
128. Schwartz, D. A., and Rubin, C. R., (1985). *JBC* **260**, 6296.
129. Hartl, F. T., and Roskoski, R. (1983). *JBC* **258**, 3950.
130. Weldon, S. L., Mumby, M. C., and Taylor, S. S. (1985). *JBC* **260**, 6440.
131. Takio, K., Wade, R. D., Smith, S. B., Krebs, E. G., Walsh, K. A., and Titani, K. (1984). *Biochemistry* **23**, 4207.

132. Gill, G. N., Walton, G. M., and Sperry, P. J. (1977). *JBC* **252**, 6443.
133. Hofmann, F., and Flockerzi, V. (1983). *EJB* **130**, 599.
134. Aitken, A., Hemmings, B. A., and Hofmann, F. (1984). *BBA* **790**, 219.
135. Hofmann, F., Gensheimer, H. P., and Gobel, C. (1983). *FEBS Lett.* **164**, 350.
136. Hofmann, F., Gensheimer, H. P., and Gobel, C. (1985). *EJB* **147**, 361.
137. Takai, Y., Yamamura, H., and Nishizuka, Y. (1974). *JBC* **249**, 530.
138. Grankowshi, N., Kudlicki, W., and Gasior, E. (1974). *FEBS Lett.* **47**, 103.
139. Becker-Ursic, D., and Davies, J. (1976). *Biochemistry* **15**, 2289.
140. Hixson, C. S., and Krebs, E. G. (1980). *JBC* **255**, 2137.
141. Sy, J., and Roselle, M. (1982). *PNAS* **79**, 2874.
142. Uno, I., Matsumoto, K., Adachi, K., and Ishikawa, T. (1984). *JBC* **259**, 12508.
143. Moreno, S., and Paseron, S. (1980). *ABB* **199**, 321.
144. Juliani, M. H., and Maia, J. C. C. (1979). *BBA* **567**, 347.
145. Trevillyan, J. M., and Pall, M. L. (1982). *JBC* **257**, 3978.
146. Majorfeld, I. H., Leichtling, B. H., Meligeni, J. A., Spitz, E., and Rickenberg, H. V. (1984). *JBC* **259**, 654.
147. Garcia, J. L., Haro, A., and Muncio, A. M. (1983). *ABB* **220**, 509.
148. Rutherford, C. L., Vaughan, R. L., Cloutier, M. J., Ferris, D. K., and Brickley, D. A. (1984). **23**, 4611.
149. Foster, J. L., Guttman, J. J., Hall, L. M., and Rosen, O. M. (1984). *JBC* **259**, 13049.
150. Vardanis, A. (1980). *JBC* **255**, 7238.
151. Vardanis, A. (1984). *BBRC* **125**, 947.
152. Riggs, A. D., Reiness, G., and Zubay G. (1971). *PNAS* **68**, 1222.
153. Anderson, W. B., Schneider, A. B., Perlman, R. L., and Pastan, I. (1971). *JBC* **246**, 5929.
154. Anderson, W. F., Ohlendorf, D. H., Takeda, Y., and Matthews, B. (1981). *Nature (London)* **290**, 754.
155. Pabo, C., and Lewis, M. (1982). *Nature (London)* **298**, 443.
156. Weber, I. T., Takio, K., Titani, K., and Steitz, T. A. (1982). *PNAS* **79**, 7679.
157. Takahashi, M., Blazy, B., and Bandras, A. (1980). *Biochemistry* **19**, 5124.
158. Bonner, J. T., Barkley, D. S., Hall, E. M., Konijn, T. M., Mason, J. W., O'Keefe, G., and Wolfe, P. B. (1969). *Dev. Biol.* **20**, 72.
159. Hutchins, B. L. M., and Frazier, W. A. (1984). *JBC* **259**, 4379.
160. Theibert, A., Klein, P., and Devreotes, P. N. (1984). *JBC* **259**, 12318.
161. Rangel-Aldao, R., Tovar, G., and de Ruiz, M. L. (1983). *JBC* **258**, 6979.
162. Francis, S. H., Lincoln, T. M., and Corbin, J. D. (1980). *JBC* **255**, 620.
163. Hamet, P., and Coquil, J. F. (1978). *J. Cyclic Nucleotide Res.* **4**, 281.
164. Hurwitz, R. L., Hansen, R. S., Harrison, S. A., Martins, J. T., Mumby, M. C., and Beavo, J. A. (1984). *Adv. Cyclic Nucleotide Protein Phosphorylation Res.* **16**, 89.
165. Reimann, E. M., Titani, K., Ericsson, L. H., Wade, R. D., Fischer, E. H., and Walsh, K. A. (1984). *Biochemistry* **23**, 4185.
166. Takio, K. T., Blumenthal, D. K., Edelman, A. M., Walsh, K. A., Krebs, E. G., and Titani, K. (1985). *Biochemistry* **24**, 6028.
167. Barker, W. C., and Dayhoff, M. O. (1982). *PNAS* **79**, 2836.
168. Weldon, S. L., and Taylor, S. S. (1985). *JBC* **260**, 4203.
169. Walsh, D. A., Ashby, C. D., Gonzalez, C., Calkins, D., Fischer, E. F., and Krebs, E. G. (1971). *JBC* **246**, 1977.
170. Ferraz, C., Demaille, J. G., and Fischer, E. H. (1979). *Biochimie* **61**, 645.
171. McPherson, J. M., Whitehouse, S., and Walsh, D. A. (1979). *Biochemistry* **18**, 4835.
172. Demaille, J. G., Peters, K. A., and Fischer, E. H. (1977). *Biochemistry* **16**, 3080.
173. Scott, J. D., Fischer, E. H., Demaille, J. F., and Krebs, E. G. (1985). *PNAS* **82**, 4379.

174. Flockhart, D. A., Watterson, D. M., and Corbin, J. D. (1980). *JBC* **255**, 4435.
175. Tsuzuki, J., and Kiger, J. A., Jr. (1978). *Biochemistry* **17**, 2961.
176. Chan, V., Huang, L. C., Romero, G., Giltonen, R. L., and Huang, C. (1980). *Biochemistry* **19**, 924.
177. Builder, S. E., Beavo, J. A., and Krebs, E. J. (1980). *JBC* **255**, 3514.
178. Armstrong, R. N., and Kaiser, E. T. (1978). *Biochemistry* **17**, 2840.
179. Døskeland, S. O., and Øgreid, D. (1984). *JBC* **259**, 2291.
180. Swillens, S. (1983). *EJB* **137**, 581.
181. Øgreid, D., Døskeland, S. O., and Miller, J. P. (1983). *JBC* **258**, 1041.
182. Seville, M., England, P. J., and Holbrook, J. J. (1984). *BJ* **217**, 633.
183. Broström, C. O., Corbin, J. D., King, C. A., and Krebs, E. G. (1983). *PNAS* **68**, 2444.
184. Øgreid, D., and Døskeland, S. O. (1983). *Biochemistry* **22**, 1686.
185. Hartl, F. T., Roskoski, R., Jr., Rosendahl, M. S., and Leonard, N. J. (1983). *Biochemistry* **22**, 2347.
186. Zoller, M. J., and Taylor, S. S. (1979). *JBC* **254**, 8363.
187. Zoller, M. J., Nelson, N. C., and Taylor, S. S. (1981). *JBC* **256**, 10837.
188. Rajinder, N. P., Bhatnager, D., and Roskoski, R., Jr. (1985). *FP* **44**, 702 (abstr.).
189. Bramson, H. N., Thornas, N., Matsueda, R., Nelson, N. C., Taylor, S. S., and Kaiser, E. T. (1982). *JBC* **257**, 10575.
190. Kupfer, A., Jimenez, J. S., and Shaltiel, S. (1980). *BBRC* **96**, 77.
191. Bramson, H. N., Kaiser, E. T., and Mildvan, A. S., (1984). *CRC Crit. Rev. Biochem.* **15**, 93.
192. Flockhart, D. A., Freist, W., Hoppe, J., Lincoln, T. M., and Corbin, J. D. (1984). *EJB* **140**, 289.
193. Moll, G. W., Jr., and Kaiser, E. T., (1976). *JBC* **251**, 3993.
194. Matsuo, M., Chang, L., Huang, C., and Villar-Palasi, C. (1978). *FEBS Lett.* **87**, 77.
195. Pomerantz, A. H., Allfrey, V. G., Merrfield, R. B., and Johnson, E. M. (1977). *PNAS* **74**, 4261.
196. Kochetkov, S. N., Bulargina, T. V., Sashchenko, L. P., and Severin, E. S. (1977). *EJB* **81**, 111.
197. Cook, P. F., Neville, M. E., Jr., Vrana, K., Hartl, F. T., and Roskoski, R., Jr. (1982). *Biochemistry* **21**, 5794.
198. Whitehouse, S., Feramisco, J. R., Casnellie, J. E., Krebs, E. G., and Walsh, D. A. (1983). *JBC* **258**, 3693.
199. Granot, J., Mildvan, A. S., Bramson, H. N., Thomas, N., and Kaiser, E. T. (1981). *Biochemistry* **20**, 602.
200. Reed, J., and Kinzel, V. (1984). *Biochemistry* **23**, 968.
200a. Sowadski, J. M., Xuong, N., Anderson, D., and Taylor, S. S. (1985). *J. Mol. Biol.* **182**, 617.
201. Krebs, E. G. (1973). *Int. Congr. Ser.—Excerpta Med.* **273**, 17.
202. Corbin, J. D. (1983). "Methods in Enzymology," Vol. 99, p. 227.
203. Kemp, B. E., and Clark, M. G. (1978). *JBC* **253**, 5147.
204. Steinberg, R. A. (1984). *Biochem. Actions Horm.* **11**, 25.
205. Coffino, P., Bourne, H. R., Friedrich, U., Hochman, J., Insel, P. A., Lamaire, I., Melmon, K., and Tomkins, G. M. (1975). *Recent Prog. Horm. Res.* **32**, 669.
206. Maller, J. L., and Krebs, E. G. (1977). *JBC* **252**, 1712.
207. Osterrieder, W., Brum, G., Herscheler, J., Trautwain, W., Flockerzi, V., and Hofmann, F. (1982). *Nature (London)* **298**, 576.
208. Brum, G., Flockerzi, V., Hofmann, F., Osterrieder, W., and Trautwein, W. (1983). *Pflueger's Arch.* **398**, 147.
209. Culpepper, J. A., and Liu, A. Y.-C. (1981). *J. Cell Biol.* **88**, 89.
210. Boney, C., Fink, D., Schlichter, D., Carr, K., and Wicks, W. D. (1983). *JBC* **258**, 4911.

211. Bkaily, G., and Sperelakis, S. (1984). *Am. J. Physiol.* **246,** H630.
212. Huttunen, J. K., Steinberg, D., and Mayer, S. E. (1970). *PNAS* **67,** 290.
213. Corbin, J. D., Beebe, S. J., and Blackmore, P. F. (1985). *JBC* **260,** 8731.
214. Beebe, S. J., Redmon, J. B., Blackmore, P. F., and Corbin, J. D. (1985). *JBC* **260,** 15781.
215. Corbin, J. D., and Keely, S. L. (1977). *JBC* **252,** 910.
216. Corbin, J. D., Sugden, P. H., Lincoln, T. M., and Keely, S. L. (1977). *JBC* **252,** 3854.
217. Krause, E. G., Will, H., Schirpke, B., and Wollenberger, A. (1975). *Adv. Cyclic Nucleotide Res.* **5,** 473.
218. Hui, C., Drummond, M., and Drummond, G. I. (1976). *ABB* **173,** 415.
219. Jones, L. R., Maddock, S. W., and Besch, H. R., Jr. (1980). *JBC* **255,** 9971.
220. Horwood, D. M., and Singhal, R. L. (1976). *J. Mol. Cell. Cardiol.* **8,** 29.
221. Jones, L. R., Maddock, S. W., and Hathaway, D. R. (1981). *BBA* **641,** 242.
222. Lamers, J. M. J., Stinis, H. T., and DeJunge, H. R. (1981). *FEBS Lett.* **127,** 139.
223. Manalan, A. S., and Jones, L. R. (1982). *JBC* **257,** 10052.
224. Kirschberger, M. A., Tada, M., and Katz, A. M. (1974). *JBC* **249,** 6166.
225. LaRaia, P. J., and Morkin, E. (1974). *Circ. Res.* **35,** 293.
226. Entman, M. L., Kaniike, K., Goldstein, M. A., Nelson, T. E., Bornet, E. P., Futch, T. W., and Schwartz, A. (1976). *JBC* **251,** 3140.
227. Kranias, E. G., Mandel, F., and Schwartz, A. (1980). *BBRC* **92,** 1370.
228. Kranias, E. G., Schwartz, A., and Jungmann, R. A. (1982). *BBA* **709,** 28.
229. Uno, I., Ueda, T., and Greengard, P. (1976). *JBC* **251,** 2192.
230. Rubin, C. S., Erlichman, J., and Rosen, O. M. (1972). *JBC* **247,** 6135.
231. Menon, K. M. J. (1973). *JBC* **248,** 494.
232. Schoff, P. K., Forrester, I. T., Haley, B. E., and Atherton, R. W. (1982). *J. Cell. Biochem.* **19,** 1.
233. Horowitz, J. A., Toeg, H., and Orr, G. A. (1984). *JBC* **259,** 832.
234. Keely, S. L., Corbin, J. D., and Park, C. R. (1975). *PNAS* **72,** 1501.
235. Singer, S. J. (1974). *Adv. Immunol.* **19,** 1.
236. Kapoor, C. L., and Steiner, A. L. (1982). *Handb. Exp. Pharmakol.* **58,** Part 1, 333.
237. Lohmann, S. M., and Walter, U. (1984). *Adv. Cyclic Nucleotide Protein Phosphorylation Res.* **18,** 63.
238. Walter, U., DeCamilli, P., Lohmann, S. M., Miller, P., and Greengard, P. (1981). *Cold Spring Harbor Conf. Cell Proliferation* **8,** 141.
239. Steiner, A. L., Koide, Y., Earp, H., Bechtel, P. J., and Beavo, J. (1978). *Adv. Cyclic Nucleotide Res.* **9,** 691.
240. Harper, J. F., Wallace, R. W., Cheung, C. Y., and Steiner, A. L. (1981). *Adv. Cyclic Nucleotide Res.* **14,** 581.
241. Kapoor, C. L., and Cho-Chung, Y. S. (1983). *Cancer Res.* **43,** 295.
242. Van Sande, J., Huang, H. L., Steiner, A., and Dumont, J. E. (1983). *Cell Biol. Int. Rep.* **7,** 981.
243. Jungman, R. A., Kuettel, M. R., Squinto, S. P., Kwast-Welfeld, J. (1987). "Methods in Enzymology" (in press).
244. Kuettel, M. R., Schwoch, G., and Jungmann, R. A. (1984). *Cell Biol. Int. Rep.* **8,** 949.
245. Squinto, S. P., Kelley-Geraghty, D. C., and Jungmann, R. A. (1985). *J. Cyclic Nucleotide Protein Phosphorylation Res.* **10,** 65.
246. Fletcher, W. H., and Byus, C. V. (1982). *J. Cell Biol.* **93,** 719.
247. Byus, C. V., and Fletcher, C. H. (1982). *J. Cell Biol.* **93,** 727.
248. Murtaugh, M. P., Steiner, A. L., and Davies, P. J. A. (1982). *J. Cell Biol.* **95,** 64.
249. Vallee, R. B., DiBartolomeis, M. J., and Theurkauf, W. E. (1981). *J. Cell Biol.* **90,** 568.
250. Theurkauf, W. E., and Vallee, R. B. (1982). *JBC* **257,** 3284.

251. Miller, P., Walter, U., Theurkauf, W. E., Vallee, R. B., and DeCamilli, P. (1982). *PNAS* **79**, 5562.

252. Lohmann, S. M., DeCamilli, P., Einig, I., and Walter, U. (1984). *PNAS* **81**, 6723.

253. Kapoor, C. L., and Cho-Chung, Y. S. (1983). *Cell Biol. Int. Rep.* **7**, 49.

254. Hathaway, D. R., Adelstein, R. S., and Klee, C. B. (1981). *JBC* **256**, 8183.

255. Stewart, A. A., Ingebritsen, T. S., Manallan, A., Klee, C. B., and Cohen, P. (1982). *FEBS Lett.* **137**, 80.

256. Sarkar, D., Erlichman, J., and Rubin, C. S. (1984). *JBC* **259**, 9840.

257. Kuo, J. K., and Greengard, P. (1970). *JBC* **245**, 2493.

258. Kuo, J. K. (1974). *PNAS* **71**, 4037.

259. Walter, U., Uno, I., Liu, A. Y.-C., and Greengard, P. (1977). *JBC* **252**, 6494.

260. Walter, U., DeCamilli, P., Lohmann, S. M., Miller, P., and Greengard, P. (1981). *Cold Spring Harbor Conf. Cell Proliferation* **8**, 141.

261. Lincoln, T. M., Hall, C. L., Park, C. R., and Corbin, J. D. (1976). *PNAS* **73**, 2559.

262. Lohmann, S. M., Walter, U., Miller, P. E., Greengard, P., and DeCamilli, P. (1981). *PNAS* **78**, 653.

262a. Tremblay, J., Gerzer, R., Vinay, P., Pang, S. C. Beliveau, R., and Hamet, P. (1985). *FEBS Lett.* **181**, 17.

262b. Fiscus, R. R., Robles, B. T., Rapoport, R. M., Waldman, S. A., and Murad, F. (1984). *Fed. Proc.* **44**, 698.

263. Miyamoto, E., Petzold, G. L., Harris, J. S., and Greengard, P. (1971). *BBRC* **44**, 305.

264. Tao, M. (1972). *BBRC* **46**, 56.

265. Hofmann, F. (1980). *JBC* **255**, 1559.

266. Buxton, I. L. O., and Brunton, L. L. (1983). *JBC* **258**, 10233.

267. Keely, S. L. (1977). *Res. Commun. Chem. Pathol. Pharmacol.* **18**, 283.

268. Hayes, J. S., Brunton, L. L., Brown, J. H., Reese, J. B., and Mayer, S. E. (1979). *PNAS* **76**, 1570.

269. Hayes, J. S., Brunton, L. L., and Mayer, S. E. (1980). *JBC* **255**, 5113.

270. Schwoch, G. (1978). *BJ* **170**, 469.

271. Byus, C. V., Hayes, J. S., Brendel, K., and Russell, D. H. (1979). *Mol. Pharmacol.* **19**, 941.

272. Hunzicker-Dunn, M. (1981). *JBC* **256**, 12185.

273. Livesey, S. A., Kemp, B. E., Re, C. A., Partridge, N. C., and Martin, T. J. (1982). *JBC* **257**, 14983.

274. Hg, K. W., Livesey, S. A., Larkins, R. G., and Martin, T. J. (1983). *Can. Res.* **43**, 794.

275. Livesey, S. A., Collier, G., Zajac, J. D., Kemp, B. E., and Martin, T. S. (1984). *BJ* **224**, 361.

276. Mizuno, T., Itzuka, K., Ikarashi, A., and Nohara, H. (1982). *Arch. Oral Biol.* **27**, 589.

277. Chew, C. S. (1985). *JBC* **260**, 7540.

278. Litvin, Y., PasMantier, R., Fleischer, N., and Erlichmann, J. (1984). *JBC* **259**, 10296.

279. Ekanger, R., Sand, T. E., Øgreid, D., Christoffersen, T., and Døskeland, S. O. (1985). *JBC* **260**, 3393.

280. Beebe, S. J., Koch, S. R., Granner, D. K., and Corbin, J. D. (1985). *Int. Congr. Biochem.*, *13th*, 649.

281. Segaloff, D. L., Beebe, S. J., Burks, D., Corbin, J. D., and Limbird, L. E. (1985). *Endocrinology (Baltimore)* **116**, Suppl, 79.

282. Cherrington, A. D., Assimacopoulos, F. D., Harper, S. C., Corbin, J. D., Park, C. R., and Exton, J. H. (1976). *JBC* **251**, 5209.

283. Wicks, W.. (1974). *Adv. Cyclic Nucleotide Res.* **4**, 335.

284. Posternak, T., Sutherland, E. W., and Henion, W. F. (1962). *BBA* **65**, 558.

285. Meyer, R. B., Jr., and Miller, J. P. (1974). *Life Sci.* **14**, 1019.

286. Meyer, R. B., Jr., Uno, H., Robins, R. K., Simon, L. N., and Miller, J. P. (1975). *Biochemistry* **14**, 3315.
287. Jastorff, B. (1976). *In* "Eucaryotic Cell Function and Growth" (J. E. Dumont, B. L. Brown, and M. J. Marshall, eds.), p. 379. Plenum, New York.
288. Yagura, T. S., Sigman, C. C., Sturm, P. A., Reiut, E. J., Johnson, H. L., and Miller, J. P. (1980). *BBRC* **92**, 463.
289. Miller, J. P. (1981). *Adv. Cyclic Nucleotide Res.* **14**, 335.
290. Miller, J. P., Uno, H., Christensen, L. F., Robins, R. K., and Meyers, R. B., Jr. (1981). *Biochem. Pharmacol.* **30**, 509.
291. Jastorff, B. (1982). *In* "Cell Regulation by Intracellular Signals" (S. Swillens and J. E. Dumont, eds.), p. 195. Plenum, New York.
292. Coulsen, R., Baraniak, J., Stec, W. J., and Jastorff, B. (1983). *Life Sci.* **32**, 1489.
293. Jastorff, B., Hoppe, J., and Morr, M. (1979). *EJB* **101**, 555.
294. DeWit, R. J. W., Hoppe, J., Stec, W. J., Baraniak, J., and Jastorff, B. (1982). *EJB* **122**, 95.
295. Free, C. A., Chasin, M., Paik, V. S., and Hers, S. M. (1971). *Biochemistry,* **10** 3785.
296. deChrombrugghe, B., Perlman, R., Vormus, H., and Pastan, J. (1969). *JBC* **244**, 5828.
297. Anderson, W. B., Perlman, R. L., and Pastan, L. (1972). *JBC* **247**, 2717.
298. McKay, D. B., Weber, J. T., and Steitz, T. A. (1982). *JBC* **257**, 9518.
299. Scholubbers, H. G., Van Knippenberg, P. H., Baraniak, J., Stec, W. J., Morr, M., and Jastorff, B. (1984). *EJB* **138**, 101.
300. O'Brian, C. A., Roczniak, S. O., Bramson, H. N., Baraniak, J., Stec, W. J., and Kaiser, E. T. (1982). *Biochemistry* **21**, 4371.
301. Rothermel, J. D., Stec, W. J., Baraniak, J., Jastorff, B., and Parker Botelho, L. H. (1983). *JBC* **258**, 12125.
302. Rothermel, J. D., Jastorff, B., and Parker Botelho, L. H. (1984). *JBC* **259**, 8151.
303. VanHaastert, P. J. M., Van Driel, R., Jastorff, B., Baraniak, J., Stec, W. J., and deWit, R. J. W. (1984). *JBC* **259**, 10020.
304. Rothermel, J. D., Perillo, N. L., Marks, J. S., and Parker Botelho, L. H. (1984). *JBC* **259**, 15294.
305. DeWit, R. J. W., Hekstra, D., Jastorff, B., Stec, W. J., Baraniak, J., Van Driel, R., and Van Haastert, P. J. M. (1984). *EJB* **142**, 225.
306. Dragland-Meserve, C. J., Rothermel, J. D., Houlihan, M. J., and Botelho, L. H. P. (1985). *J. Cyclic Nucleotide Protein Phosphoryl. Res.* **10**, 371.
307. Corbin, J. D., Beebe, S. J., Blackmore, P. F., Redmon, J. B., Sheorain, V. S., and Gettys, T. W. (1986). *In* "Fernstrom Symposium on Insulin Action" (in press).
308. Abell, C. W., and Monahan, T. W. (1973). *J. Cell Biol.* **59**, 549.
309. Pastan, I., and Johnson, G. S. (1974). *Adv. Cancer Res.* **19**, 303.
310. Ryan, W. L., and Heidrick, M. L. (1974). *Adv. Cyclic Nucleotide Res.* **4**, 81.
311. MacManus, J. P., Whitfield, J. F., Boynton, A. L., and Rixon, R. H. (1975). *Adv. Cyclic Nucleotide Res.* **5**, 719.
312. Chlapowski, F. S., Kelley, L. A., and Butcher, R. W. (1975). *Adv. Cyclic Nucleotide Res.* **6**, 245.
313. Pastan, I., Johnson, G. S., and Anderson, N. B. (1975). *Annu. Rev. Biochem.* **44**, 491.
314. Friedman, D. L. (1971). *Physiol. Rev.* **56**, 652.
315. Friedman, D. L., Johnson, R. A., and Zeilig, C. E. (1976). *Adv. Cyclic Nucleotide Res.* **7**, 69.
316. Rebhun, L. I. (1977). *Int. Rev. Cytol.* **49**, 1.
317. Friedman, D. L. (1982). *Handb. Exp. Pharmakol.* **58**, Part 2, 151.
318. Boynton, A. L., and Whitfield, J. F. (1983). *Adv. Cyclic Nucleotide Res.* **15**, 193.
319. Jungman, R. A., and Russell, D. H. (1977). *Life Sci.* **20**, 1787.
320. Russell, D. H. (1978). *Adv. Cyclic Nucleotide Res.* **9**, 493.

321. Burger, M. M., Bombik, B. M., Breckenridge, B., and Sheppard, J. R. (1972). *Nature (London) New Biol.* **239**, 161.
322. Seifert, W., and Rudland, P. S. (1974). *PNAS* **71**, 4920.
323. Russell, D. H., and Stambrook, P. J. (1975). *PNAS* **72**, 1482.
324. MacManus, J. P., Franks, D. J., Youdale, T., and Braceland, B. M. (1972). *BBRC* **49**, 1201.
325. Thrower, S., and Ord, M. G. (1974). *BJ* **144**, 361.
326. Short, J., Tsukada, K., Rudert, W. A., and Leiberman, I. (1975). *JBC* **250**, 3602.
327. Tsang, B. K., Rixon, R. H., and Whitfield, J. F. (1980). *J. Cell. Physiol.* **102**, 19.
328. MacManus, J. P., Braceland, B. M., Youdale, T., and Whitfield, J. F. (1973). *J. Cell. Physiol.* **82**, 15.
329. Ebina, Y., Iwai, H., Fukui, N., Ohtsuka, H., and Miura, Y. (1975). *J. Biochem. (Tokyo)* **77**, 641.
330. Millis, A. J. T., Forrest, G. A., and Pious, D. A. (1974). *Exp. Cell Res.* **83**, 335.
331. Bronstad, G., and Christoffersen, T. (1980). *FEBS Lett.* **120**, 89.
332. Boynton, A. L., and Whitfield, J. P. (1981). *Adv. Cyclic Nucleotide Res.* **14**, 411.
333. Klee, C. B., Crouch, T. H., and Richman, P. G. (1980). *Annu. Rev. Biochem.* **49**, 489.
334. Manganiello, V., and Vaughan, M. (1972). *PNAS* **69**, 269.
335. D'Armiento, M., Johnson, G. S., and Pastan, I. (1972). *PNAS* **69**, 459.
336. Russell, T. R., and Pastan, I. (1974). *JBC* **249**, 7764.
337. Liu, A. Y. C. (1982). *JBC* **257**, 298.
338. Bovine, H. R., Tomkins, G. M., and Dion, S. (1973). *Science* **181**, 952.
339. Prasad, K. N., Sahu, S. K., and Sinha, P. K. (1976). *J. Natl. Cancer Inst. (U.S.)* **57**, 619.
340. Wang, T., Sheppard, J. R., and Foker, J. E. (1978). *Science* **201**, 255.
341. Rozengurt, E., Collins, M. K. L. and Keechan, M. (1983). *J. Cell. Physiol.* **116**, 379.
342. Rozengurt, E., Stroobant, P., Waterfield, M. D., Deuel, T. F., and Kechan, M. (1983). *Cell* **34**, 265.
343. Hammarström, S. (1977). *EJB* **74**, 7.
344. Levine, L. (1981). *Adv. Cancer Res.* **35**, 49.
345. Hammarström, S. (1982). *ABB* **214**, 431.
346. Levine, E. M., Becker, Y., Boone, C. W., and Eagle, H. (1965). *PNAS* **53**, 350.
347. Salas, J., and Green, H. (1971). *Nature (London), New Biol.* **229**, 165.
348. Becker, H., and Stanners, C. P. (1972). *J. Cell. Physiol.* **80**, 51.
349. Smith, J. A., and Martin, L. (1973). *PNAS* **70**, 1263.
350. Pledger, W. J., Stiles, C. D., Antoniades, H. N., and Scher, C. D. (1979). *PNAS* **74**, 4481.
351. Daniel, V., Litwack, G., and Tomkins, G. (1973). *PNAS* **70**, 76.
352. Daniel, V., Bourne, H. R., and Tomkins, G. M. (1973). *Nature (London) New Biol.* **244**, 167.
353. Insel, P. A., Bourne, H. R., Coffino, P., and Tomkins, G. M. (1975). *Science* **190**, 896.
354. Bourne, H. R., Coffino, P., Melmon, K. L., Tomkins, G. M., and Weinstein, Y. (1975). *Adv. Cyclic Nucleotide Res.* **5**, 771.
355. Coffino, P., Bourne, H. R., Friedrich, U., Hochman, J., Insel, P. A., Lemaire, I., Melmon, K. L., and Tomkins, G. M. (1976). *Recent Prog. Horm. Res.* **32**, 669.
356. Steinberg, R. A., O'Farrell, P. H., Friedrich, U., and Coffino, P. (1977). *Cell* **10**, 381.
357. Gottesman, M. M. (1980). *Cell* **22**, 329.
358. Gottesman, M. (1983). "Methods in Enzymology," Vol. 97, p. 197.
359. Gottesman, M. M., Roth, C., Leutschuh, M., Richert, N., and Pastan, I. (1984). *J. Cell. Biochem.*
360. Steinberg, R. A. (1984). *Biochem. Actions Hormo.* **9**, 25.
361. Whitfield, J. F., MacManus, G. P., Rixon, R. H., Boynton, A. L., and Yondale, T. (1976). *In Vitro* **12**, 1.

362. Rechler, M. M., Bruni, C. B., Podskalny, J. M., Warner, W., and Carchman, R. A. (1977). *Exp. Cell Res.* **104,** 411.
363. Millis, A. J. T., Forrest, G. A., and Pious, D. A. (1974). *Exp. Cell Res.* **83,** 335.
364. Costa, M., Gerner, E. W., and Russell, D. H. (1976). *JBC* **251,** 3313.
365. Haddox, M. K., Magun, B. E., and Russell, D. H. (1980). *PNAS* **77,** 3445.
366. Gray, J. P., Johnson, R. A., and Friedman, D. L. (1980). *ABB* **202,** 259.
367. Lee, P. C., Radlo, H. D., Schweppe, J. S., and Jungman, R. A. (1976). *JBC* **251,** 914.
368. Eppenberger, U., Roos, W., Fabbro, D., Sury, A., Weber, J., Bechtel, E., Huber, P., and Jungmann, R. A. (1979). *EJB* **98,** 253.
369. Wittmaack, F. M., Weber, W., and Hilz, H. (1983). *EJB* **129,** 669.
370. Fossberg, T. M., Døskeland, S. O., and Ueland, P. M. (1978). *ABB* **189,** 372.
371. Handschin, J. C., and Eppenberger, U. (1979). *FEBS Lett.* **106,** 301.
372. Weber, W., Schwoch, G., Wideckens, K., Garteman, A., and Hilz, H. (1981). *EJB* **120,** 585.
373. Byus, C. V., Klimpel, G. R., Lucas, D. O., and Russell, D. H. (1977). *Nature (London)* **268,** 63.
374. Gharrett, A. J., Mackinson, A. M., and Sheppard, J. R. (1976). *Nature (London)* **264,** 673.
375. Wehner, J. M., Malkinson, A. M., Wiser, M. F., and Sheppard, J. R. (1981). *J. Cell. Physiol.* **108,** 175.
376. Ledinko, N., and Chan, D. (1984). *Cancer Res.* **44,** 2622.
377. Conti, M., Geremia, R., and Monesi, V. (1974). *Mol. Cell. Endocrinol.* **13,** 137.
378. Fakunding, J. L., and Means, A. R. (1977). *Endocrology (Baltimore)* **101,** 1358.
379. Schwartz, D. A., and Rubin, C. S. (1983). *JBC* **258,** 777.
380. Berridge, M. J. (1984). *Nature (London)* **312,** 315.
381. Berridge, M. J. (1984). *BJ* **220,** 345.
382. Larrue, J., Dorian, B., Daret, D., Demond-Henri, J., and Bricaud, H. (1984). *Adv. Cyclic Nucleotide Protein Phosphorylation Res.* **17,** 585.
383. Cheung, W. Y., and Storm, D. R. (1982). *Handb. Exp. Pharmakol.* **58,** Part 1, 301.
384. Lohmann, S. M., Walter, U., and Greengard, P., (1978). *J. Cyclic Nucleotide Res.* **4,** 445.
385. Uno, I., Matsumoto, K., Adachi, K., and Ishikawa, T. (1984). *JBC* **259,** 12508.
386. Richards, J. S., and Rolfes, A. I. (1980). *JBC* **255,** 5481.
387. Darbon, J. M., Knect, M., Ranta, T., Dufau, M. L., and Catt, K. J. (1984). *JBC* **259,** 14778.
388. Walter, U., Costa, M. R. C., Breakefield, X. O., and Greengard, (1979). *PNAS* **76,** 3251.
389. Lohmann, S. M., Schwoch, G., Reiser, G., Port, R., and Walter, U. (1983). *Embo J.* **2,** 153.
390. Morrison, M. R., Pardue, S., Prashad, N., Croall, D. E., and Brodeur, R. (1980). *EJB* **106,** 463.
391. Prashad, N., Rosenberg, R. N., Wischmeyer, G., Ulrich, C., and Sparkman, D. (1979). *Biochemistry* **18,** 2717.
392. Liu, A. Y. C., Chan, T., and Chen, K. Y. (1981). *Can. Res.* **41,** 4579.
393. Singh, T. J., Roth, C., Gottesman, M. M., and Pastan, I. H. (1981). *JBC* **256,** 926.
394. Steinberg, R. A., VanDaalen Wetters, T., and Coffino, P. (1978). *Cell* **15,** 1351.
395. Beale, E. G., Dedman, J. R., and Means, A. R. (1977). *JBC* **252,** 6322.
396. Tash, J. S., Dedman, J. R., and Means, A. R. (1979). *JBC* **254,** 1241.
397. Ratoosh, S. L., Jahnsen, T., and Richards, J. S. (1985). *Endocrinology (Baltimore)* **116,** Suppl., 79.
398. Gergely, P., and Bot, G. (1977). *FEBS Lett.* **82,** 269.
399. Khatra, B. S., Printz, R., Cobb, C. E., and Corbin, J. D. (1985). *BBRC* **130,** 567.
399a. Constantinou, A. I., Squinto, S. P., and Jungmann, R. A. (1985). *Cell* **42,** 429.
400. Wicks, W. D., Koontz, J., and Wagner, K. (1975). *J. Cyclic Nucleotide Res.* **1,** 49.
401. Harrison, J. J., Schwoch, G., Schweppe, J. S., and Jungmann, R. A. (1982). *JBC* **257,** 13602.
402. Johnson, E. M. (1982). *Handb. Exp. Pharmakol.* **58,** Part 1, 507.

403. Dahmas, M. E. (1976). *Biochemistry* **15,** 1821.
404. Jungmann, R. A., and Kranias, E. G. (1977). *Int. J. Biochem.* **8,** 819.
405. Kranias, E. G., Schweppe, J. S., and Jungmann, R. A. (1977). *JBC* **252,** 6750.
406. Duccman, B. W., Rose, K. M., and Jacob, S. T. (1981). *JBC* **256,** 10755.
407. Bell, G. I., Valenzuela, P., and Rutter, W. J. (1977). *JBC* **252,** 3082.
408. Dahmas, M. E. (1981). *JBC* **256,** 3332.
409. Breant, B., Buhler, J. M., Sentenac, A., and Fromageot, P. (1983). *EJB* **130,** 247.
410. Lee, S.-K., Schweppe, J. S., and Jungmann, R. A. (1984). *JBC* **259,** 14695.
411. Spielrogal, A. M., Mednicks, M. I., Eppenberger, U., and Jungmann, R. A. (1977). *EJB* **73,** 199.
412. Cho-Chung, Y. S. (1980). *Adv. Cyclic Nucleotide Res.* **12,** 111.
413. Cho-Chung, Y. S. (1980). *J. Cyclic Nucleotide Res.* **6,** 163.
414. Wicks, W. D. (1971). *JBC* **246,** 217.
415. Wicks, W. D., Barnett, C. A., and McKibben, J. B. (1974). *FB* **33,** 1105.
416. Iynedjian, P. B., and Hanson, R. W. (1977). *JBC* **252,** 655.
417. Nelson, K., Cimbala, M., and Hanson, R. (1980). *JBC* **255,** 8509.
418. Beale, E., Katzen, C., and Granner, D. (1981). *Biochemistry* **20,** 4878.
419. Ernest, M. J., and Feigelson, P. (1978). *JBC* **253,** 319.
420. Noguchi, T., Diesterhof, M., and Granner, D. (1978). *JBC* **253,** 1332.
421. Derda, D. F., Miles, M. F., Schweppe, J. S., and Johnson, R. A. (1980). *JBC* **255,** 11112.
422. Firestone, G. L., and Heath, E. C. (1981). *JBC* **256,** 1396.
423. Brown, P. C., and Papaconstantinou, J. (1979). *JBC* **254,** 9397.
424. Yoo-Warren, H., Cimbala, M., Felz, K., Monahan, J., Leis, J., and Hanson, R. (1981). *JBC* **256,** 10224.
425. Beale, E., Hartley, J., and Granner, D. (1982). *JBC* **257,** 2022.
426. Beale, E., Andreone, T., Koch, S., Granner, M., and Granner, D. (1984). *Diabetes* **33,** 328.
427. Miles, M. F., Hung, P., and Jungmann, R. A. (1981). *JBC* **256,** 12445.
428. Maurer, R. A. (1981). *Nature (London)* **294,** 94.
429. Cimbala, M. A., Lamers, W. H., Nelson, K., Monahan, J. E., Yoo-Warren, H., and Hanson, R. W. (1982). *JBC* **257,** 7629.
430. Lamers, W., Hanson, R., and Meisner, H. (1982). *PNAS* **79,** 5137.
431. Granner, D., Andreone, T., Sasaki, K., and Beale, E. (1983) *Nature (London)* **305,** 549.
432. Sasaki, K., Cripe, T. P., Koch, S. R., Andreone, T. L., Peterson, D. D., Beale, E. G., and Granner, D. K. (1984). *JBC* **259,** 15242.
433. Hod, Y., Morris, S. M., and Hanson, R. W. (1984). *JBC* **259,** 15603.
434. Jungmann, R. A., Kelly, D. C., Miles, M. F., and Milkowski, D. M. (1983). *JBC* **258,** 5312.
435. Culpepper, J. A., and Lee, A. Y. (1983). *JBC* **258,** 3812.
436. Hashimoto, S., Schmid, W., and Schutz, G. (1984). *PNAS* **81,** 6637.

4

Calmodulin-Dependent Protein Kinases

JAMES T. STULL • MARY H. NUNNALLY •
CAROLYN H. MICHNOFF

Department of Pharmacology and
Moss Heart Center
University of Texas Health Science Center
at Dallas
Dallas, Texas 75235

THE ENZYMES, Vol. XVII

I. Introduction

A. CA²⁺ AS A SECOND MESSENGER

In 1883, Ringer (*1–3*) found that extracellular Ca^{2+} was required to maintain contractile activity in isolated frog hearts. Locke subsequently reported that extracellular Ca^{2+} was required for transmission of impulses from nerve to muscle (*4*). These fundamental observations were the first to implicate Ca^{2+} as an important regulator of biological processes. During the 30–40 years following these pioneering experiments, a number of physiologists demonstrated that the absence of extracellular Ca^{2+} affected a wide variety of cellular processes (*5, 6*).

The general biological importance of Ca^{2+} was extended to a role as an intracellular regulator by the observation in 1942 that Ca^{2+} activated myosin ATPase activity (*7, 8*). In 1947, Heilbrunn and Wiercinski (*9*) showed that the injection of Ca^{2+} into muscle fibers caused contraction. These biochemical and physiological observations stimulated research on the mechanisms by which Ca^{2+} could stimulate muscle contraction. Ebashi and his colleagues (*10, 11*) found that native tropomyosin, consisting of troponin and tropomyosin, was the Ca^{2+}–receptor complex on thin filaments in striated muscle. They proposed that Ca^{2+} binding to troponin causes a conformational change in F-actin transmitted via tropomyosin that results in stimulation of actin and myosin interactions. About the same time, Meyer *et al.* (*12*) proposed that Ca^{2+} may be required for phosphorylation of phosphorylase by phosphorylase kinase. This suggestion was subsequently confirmed by Ozawa *et al.* (*13*) who showed that Ca^{2+} concentrations required for phosphorylase kinase activity were similar to the Ca^{2+} concentrations required for activation of contractile elements in skeletal muscle. Thus, Ca^{2+} was identified as a common link between glycogenolysis and contraction.

The identification of troponin and phosphorylase kinase as intracellular receptor proteins for Ca^{2+} provided important biochemical models for subsequent investigations on many Ca^{2+}-dependent processes such as cell motility, secretion, cell division, metabolism, and membrane permeability. The second messenger concept proposed by Sutherland and co-workers (*14*) to describe the effects of cAMP on a wide variety of cellular responses was extended to Ca^{2+} (*15*). It is now generally appreciated that Ca^{2+} exists in low concentrations in the cytoplasm of all resting, nonactivated cells. A variety of extracellular stimuli appropriate to a particular cell results in an increase in cytoplasmic Ca^{2+} concentrations and initiation of Ca^{2+}-dependent responses. The two second messengers, cAMP and Ca^{2+}, may both affect a cellular response (i.e., cell function may be simultaneously regulated by one or both of these second-messenger systems) (*15–17*).

The compelling evidence that Ca^{2+} was a second messenger led to investigations to identify mechanisms for Ca^{2+} selectivity to biological responses. In particular, proteins that detect and respond to a Ca^{2+} signal have been identified (5, 6). These Ca^{2+} binding proteins include calmodulin, troponin C, parvalbumin, intestinal Ca^{2+}-binding protein, S 100 protein, and regulatory light chain of myosins. Calmodulin is unique because it is found in most, if not all, eukaryotic cells and mediates Ca^{2+} control of a large number of enzymes. These two general properties confer to calmodulin the role of mediator of many Ca^{2+}-dependent cellular processes.

Calmodulin is known to activate distinct protein kinases, enzymes that catalyze the transfer of the terminal phosphate of ATP to serine or threonine residues in protein substrates. The introduction of a phosphate moiety into a protein may result in marked changes in its biochemical properties and thereby provide a means for regulating particular biological processes. In this chapter, we present information on the properties of two types of Ca^{2+} and calmodulin-dependent protein kinases, myosin light chain kinase and a multifunctional calmodulin-dependent protein kinase. We use the term calmodulin-dependent to mean that an enzyme or biological process is dependent upon both Ca^{2+} and calmodulin for activity unless noted otherwise. Phosphorylase kinase, another Ca^{2+}-dependent protein kinase, is discussed in Chapter 10. The information presented in this chapter relates, in particular, to the chapters entitled ''Regulation of Contractile Activity'' and ''Phosphorylation of Brain Proteins'' in Volume XVIII, Chapters 13 and 9.

The name myosin light chain kinase is used to identify the enzyme originally described as a protein kinase that catalyzes the calmodulin-dependent phosphorylation of a single serine on the regulatory or phosphorylatable light chains of myosins. Although this protein kinase has also been referred to as myosin kinase, we feel this terminology should be avoided because of the possible confusion with enzymes that phosphorylate myosin heavy chains (18–20).

Various names have also been used to identify other calmodulin-dependent protein kinases. A type of enzyme that phosphorylates many different proteins has been purified from many tissues. We refer to this enzyme as the multifunctional calmodulin-dependent protein kinase, but the reader should be aware that some other aliases include calmodulin-dependent glycogen synthase kinase, calmodulin-dependent protein kinase II, tubulin kinase, and calmodulin-dependent multiprotein kinase. Since we have not contributed to investigations on these protein kinases, we will not be presumptuous in recommending that multifunctional calmodulin-dependent protein kinase is necessarily the most appropriate name. However, for the purpose of this chapter, it is used to emphasize, in contrast to myosin light chain kinase, the broad specificity of this enzyme in regards to protein substrates. This chapter also includes pertinent information on calmodulin that is not historically comprehensive, and we have relied upon reviews by other authors for the citation of certain developments and perspectives.

B. CALMODULIN

1. *Physicochemical Properties*

Calmodulin has been purified and characterized from many types of cells, including plants and protozoa as well as vertebrate tissues (*21–24*). The distribution of calmodulin does not correspond to the distribution of any particular receptor protein such as calmodulin-stimulated phosphodiesterase and myosin light chain kinase. Furthermore, this ubiquitous distribution of calmodulin distinguishes it from other Ca^{2+}-binding proteins. These findings indicate calmodulin may serve in a general regulatory role in a wide variety of animal species and tissues.

The properties of vertebrate calmodulin have been studied in considerable detail. It is an acidic 148-residue protein that contains no tryptophan, cysteine, phosphate, or carbohydrate (*25*). Based upon a comparison of calmodulin's amino acid sequence to other Ca^{2+}-binding proteins such as parvalbumin and troponin C, four Ca^{2+}-binding domains have been postulated as well as putative Ca^{2+}-binding amino acid residues within those domains. These Ca^{2+}-binding domains are separated by regions of α-helical structure. There is homology among the four Ca^{2+}-binding domains in calmodulin, with greater homology between domains 1 and 3 than between 2 and 4. Sequence data also indicate that the primary structure of calmodulin is highly conserved among widely divergent animal and plant species. This apparent structural constraint may be related to the consequence of calmodulin regulating many intracellular functions.

Calmodulin binds 4 Ca^{2+} per mol of calmodulin with high affinity, and studies on the Ca^{2+}-binding properties of calmodulin indicate multiple classes of sites of negative cooperativity (*26–30*). However, positive cooperativity at low Ca^{2+} concentrations (*31*) or equivalent sites (*32*) has also been described. An extensive study of the binding of cations to calmodulin under a variety of conditions has been performed by Haiech *et al.* (*33*). Results from various studies have yielded four different macroscopic binding constants for the different Ca^{2+}-binding sites (Table I). However, Potter *et al.* (*34*) believe that calmodulin binds 4 mol Ca^{2+} per mol of calmodulin at essentially the same affinity for each of the four sites. These authors discuss the technical and theoretical limitations in determining Ca^{2+}-binding affinities for a protein that has multiple Ca^{2+}-binding sites.

Upon binding Ca^{2+}, calmodulin undergoes conformational changes as indicated by circular dichroism, optical rotary dispersion, nuclear magnetic resonance, and uv-difference spectroscopy (*21, 35*). Low- and high-affinity Ca^{2+}-

TABLE I

Macroscopic Ca^{2+} Dissociation Constants
for Calmodulin

K_1 (μM)	K_2 (μM)	K_3 (μM)	K_4 (μM)	Reference
2.0	1.9	7.3	61	(33)
5.3	3.6	22	83	(128)
5.0	4.5	17	50	(130)
7.5	2.7	31	31	(131)

binding sites of calmodulin have been detected by these procedures. Based upon differences in binding properties of the different sites for Ca^{2+}, it has been suggested that various calmodulin-related enzymes might be activated by distinct conformers of calmodulin containing different amounts of Ca^{2+} (21, 35).

2. Calmodulin-Dependent Enzymes

Cheung (36) and Kakiuchi et al. (37) first reported an activator for the Ca^{2+}-stimulated cyclic nucleotide phosphodiesterase. Teo and Wang (26) showed that the activator protein (calmodulin) bound Ca^{2+}; Teshima and Kakiuchi (38) demonstrated that calmodulin formed a Ca^{2+}-dependent complex with phosphodiesterase. From these and other investigations a general mechanism for the Ca^{2+}-dependent activation of various enzymes by calmodulin has been described (21–24). The two reactions involving the activation process are shown in Eqs. (1) and (2).

$$nCa^{2+} + CaM \rightleftharpoons Ca_n^{2+} \cdot CaM^* \tag{1}$$

$$Ca_n^{2+} \cdot CaM^* + E \rightleftharpoons Ca_n^{2+} \cdot CaM^* \cdot E^* \tag{2}$$

Calmodulin (CaM) binds Ca^{2+}, which is accompanied by specific conformational changes. The active conformer of calmodulin (CaM*) binds to a receptor enzyme (E) which may be inactive or partially active. The binding of calmodulin to the enzyme results in a conformational change associated with activation or stimulation of catalysis (E*). This general scheme ignores some of the important questions regarding stoichiometries of Ca^{2+}-binding to calmodulin that may be required for activation of a particular enzyme. Calmodulin has four Ca^{2+}-binding domains and it has been suggested that different classes of receptor proteins may interact with calmodulin at different extents of Ca^{2+} occupancy in the four different domains (21). It should also be noted that this general scheme is not meant to imply a particular stoichiometry for calmodulin binding to receptor proteins. These particular issues are discussed in detail in the following sections regarding the different protein kinases.

The ability of calmodulin to increase enzyme activities in a Ca^{2+}-dependent manner has been described for many enzymes from a variety of tissues and animal species including a cyclic nucleotide phosphodiesterase (36, 39), brain adenylate cyclase (40), phosphorylase kinase (41), NAD kinases (42, 43), guanylate cyclase (44), membrane ATPases (45–48), a phospholipase activity (49), myosin light chain kinases (50–52), and other Ca^{2+}-dependent protein kinases (53–55). Considering the ubiquitous nature of calmodulin and its ability to affect activation of numerous enzymes, it seems likely that the list of biological processes regulated by calmodulin will continue to grow. Many of these enzymes have not been purified to homogeneity so that quantitative information regarding mechanisms of enzyme activation are limited. As pointed out by others (23), calmodulin stimulation of enzyme activity may not necessarily reflect a physiological function of calmodulin. For example, calmodulin can substitute for another Ca^{2+}-binding protein that regulates enzyme activity (i.e., troponin C regulation of actomyosin ATPase activity) (56, 57).

II. Myosin Light Chain Kinases

A. INTRODUCTION

Eukaryotic cells contain actin, myosin, and related proteins that are of primary importance in muscle contractility and motility of nonmuscle cells (58–64). In general, muscle contraction is thought to consist of sliding of interdigitating thick myosin filaments and thin actin filaments past each other. The driving force for this process is the cyclic attachment and detachment of the globular head region of myosin to the actin filament with the energy for this process provided by ATP (see Volume XVIII, Chapter 13).

The release of Ca^{2+} into the cytoplasm is the primary event in excitation–contraction coupling, and the subsequent binding of Ca^{2+} to particular sites on regulatory proteins associated with or acting on the contractile apparatus ultimately results in contraction. Although there may be similarities in the properties of interaction between the contractile proteins actin and myosin from different kinds of muscle and nonmuscle cells, the biochemical mechanisms by which Ca^{2+} triggers activation of the contractile elements may be markedly different (63, 65). In nonmuscle cells there may be additional mechanisms for regulating cell motility (58, 59, 66).

Phosphorylation and dephosphorylation of myofibrillar proteins has been considered a potentially important biochemical mechanism for regulating contraction since the discovery that myosin (67) and troponin (68, 69) purified from rabbit skeletal muscle were phosphoproteins, and that purified subunits of these proteins could be phosphorylated by different types of protein kinases (70).

Myosins purified from all types of vertebrate tissues contain a single class of light chains that have similar biochemical properties. The masses of these myosin light chains range from 18 to 20 kDa, and they are capable of binding divalent cations. All of the light chain subunits from vertebrate tissues that fall into this class are also capable of being phosphorylated by myosin light chain kinase. These myosin light chains from vertebrate tissues have also been referred to as phosphorylatable or P-light chains (71). Early investigations demonstrated that P-light chain of myosin from white, fast-twitch skeletal (67), cardiac (71), red, slow-twitch skeletal (71), and smooth (72) muscles was reversibly phosphorylated and dephosphorylated. Myosin isolated from smooth muscles as well as from nonmuscle cells requires phosphorylation of P-light chain for actin activation of the Mg^{2+}-dependent myosin ATPase activity (73, 74). A more detailed description of the function of the phosphorylation of myosin in regulating actin and myosin interactions may be found in Volume XVIII, Chapter 13.

The original investigations by Perry and his colleagues demonstrated a protein kinase present in rabbit skeletal muscle that catalyzed the phosphorylation of P-light chain. This enzyme had biochemical properties that distinguished it from other types of protein kinases such as cyclic nucleotide-dependent protein kinases and phosphorylase kinase, and it was named myosin light chain kinase (75). The enzyme was highly specific for catalyzing the phosphorylation of P-light chains as compared to other protein substrates. It was also dependent upon Ca^{2+} for activity. It was subsequently demonstrated that myosin light chain kinase requires two protein components for activity. Dabrowska and Hartshorne (50) showed that calmodulin was required for the activation of the catalytic subunit of smooth-muscle myosin light chain kinase. At the same time, Yagi *et al.* (76) found that skeletal-muscle myosin light chain kinase also required calmodulin for activation. Because myosin light chain kinases were dependent upon Ca^{2+} and calmodulin for activity, the possibility was naturally considered that there may be a link between activation of this kinase and muscle contraction.

B. PHYSICOCHEMICAL PROPERTIES

Myosin light chain kinases have been purified to homogeneity from a number of vertebrate muscles. In addition, myosin light chain kinases have been identified and partially purified from several vertebrate nonmuscle tissues. One of the primary problems in establishing the physicochemical properties of the native forms of myosin light chain kinases has been the marked sensitivity of these kinases to limited proteolysis during purification. Daniel and Adelstein (77) found myosin light chain kinase purified from platelets was a 78-kDa enzyme that was not dependent upon calmodulin for activity. Dabrowska and Hartshorne (50) partially purified calmodulin-dependent myosin light chain kinases from both human platelet and bovine brain that were similar sizes, 105 kDa. Hathaway and Adelstein (51) subsequently reported purification of a human platelet myosin light chain kinase, with an apparent mass of 105 kDa. However, later work by Hathaway et al. (78) suggested that the bovine brain myosin light chain kinase was a 130-kDa species and, presumably, isolation of the smaller brain and platelet kinases was due to proteolysis during the enzyme purification.

An early report on the purification of the chicken gizzard smooth-muscle kinase indicated that this enzyme was 105 kDa (79). However, inclusion of protease inhibitors results in purification of larger myosin light chain kinases from gizzard smooth muscle [turkey gizzard, 130 kDa (80, 81); chicken gizzard, 130–136 kDa (82, 83)]. The molecular weight of the kinase determined by sedimentation equilibrium experiments is 124,000, indicating that the enzyme probably exists as a monomer in its native state (80). The turkey gizzard kinase is asymmetrical with a Stokes radius of 75 Å, a sedimentation coefficient of 4.45 S, and a frictional ratio of 1.85. It also binds 1 mol calmodulin per mol kinase.

Adachi et al. (84) claimed that myosin light chain kinase from chicken gizzard is 136 kDa. This form of the kinase was identified by immunoblots with a monoclonal antibody to the 130-kDa form. Subsequently, Ngai et al. (83) purified this 136-kDa form of the chicken gizzard myosin light chain kinase. The authors proposed that after extraction from the tissue the kinase undergoes proteolytic cleavage resulting in the size reduction from 136 to 130 kDa. They suggest that it is the partially degraded 130-kDa form of the gizzard kinase that has been routinely purified (80, 81). However, the small change in size prevents a definitive conclusion regarding the form of the kinase purified in different laboratories.

Myosin light chain kinases have also been purified from mammalian smooth muscle. The mammalian smooth-muscle kinases appear to be as large or larger than the purified chicken and turkey smooth-muscle myosin light chain kinases [steer stomach, 155 kDa (85); steer trachea, 150–160 kDa (86, 87); steer aorta, 142 kDa (88); hog myometrium, 130 kDa (89); hog carotid artery, 140 kDa (J. T. Stull and S. M. Moreland, unpublished observation)]. The reasons for these slight differences in reported mass for the mammalian myosin light chain kinases is not clear, but may reflect animal species differences (90) in addition to variations in determining sizes among different laboratories. However, it seems clear that mammalian myosin light chain kinases, in general, are of slightly larger size than the gizzard kinase. Whether this difference is due to a difference in primary structure or an undetermined posttranslational modification has yet to be determined. There are no published reports on the hydrodynamic properties of mammalian smooth-muscle myosin light chain kinases.

The myosin light chain kinases that have been purified from mammalian striated muscle are considerably smaller than the smooth-muscle kinases. The relative masses determined by polyacrylamide gel electrophoresis in the presence of sodium dodecyl sulfate for rabbit skeletal-muscle myosin light chain kinase range from 77 to 94 kDa (*75, 90–96*). The native molecular weight of rabbit skeletal-muscle kinase determined by gel filtration and sedimentation equilibrium is between 70,000 and 80,000 (*75, 91, 93, 95*). One report presented a value of 103,000 for the native molecular weight determined by sedimentation equilibrium (*96*). Thus, rabbit skeletal-muscle myosin light chain kinase, similar to the gizzard smooth-muscle enzyme, is a monomeric enzyme. In addition, the skeletal-muscle kinase is a highly asymmetrical molecule with a Stokes radius of 54 Å, sedimentation coefficient of 3.2 S and a frictional ratio of ~2.0 (*93, 95*).

Myosin light chain kinase has also been purified from steer cardiac muscle and is reported to be 85 (*97*) and 94 kDa (*98*). A Stokes radius of 44 Å (*97*) suggests that the cardiac kinase also is asymmetrical. It should be noted that the reported sizes of the cardiac enzyme are similar to values for the enzyme from rabbit skeletal muscle, which are smaller than the smooth-muscle kinases.

Sellers and Harvey (*99*) reported the purification of myosin light chain kinase from *Limulus* striated muscle. Polyacrylamide gel electrophoresis in the presence of sodium dodecyl sulfate demonstrated that the kinase was composed of a protein doublet of 39 and 37 kDA, respectively. Both polypeptides bound [^{125}I]-calmodulin. Electrophoresis under nondenaturing conditions resulted in separation of the two polypeptides and both forms had catalytic activity. As pointed out by the authors, it is not clear whether this calmodulin-dependent myosin light chain kinase is derived from a larger form by limited proteolysis or, more interestingly, whether it represents the native form in a more primitive animal species.

The large disparity in size between the smooth- and striated-muscle kinases, as well as their susceptibility to proteolysis, indicated that the smaller striated-muscle kinase (77 to 94 kDa) could be proteolytic fragments of larger native enzymes. Walsh and Guilleux (*100*) partially purified a 150-kDa myosin light chain kinase from dog skeletal and cardiac muscle. The specific enzyme activity was very low and was never unequivocally shown to be associated with the 150-kDa protein. In addition, they also proposed that striated-muscle myosin light chain kinases were bound to contractile proteins. Guerriero *et al.* (*101*) showed that myosin light chain kinase identified by immunoblots in extracts of chicken brain, heart, skeletal and smooth muscle has identical masses, 130 kDa. Therefore, they suggested that the same enzyme existed in these different tissues. However, the immunoblots were performed with antibodies raised to the smooth-muscle kinase and the amount of cross-reactive material in the skeletal-muscle extract was considerably less than the gizzard-muscle extract, although these two muscles contain similar amounts of kinase activity. The antibodies may have recognized smooth-muscle kinase present in vascular smooth muscle contaminating the skeletal-muscle tissue.

To establish that the purified 87-kDa rabbit skeletal-muscle myosin light chain kinase represented the native form of the enzyme, we (*90*) used an immunoblot analysis of rabbit skeletal-muscle to identify the form of the kinase *in situ*. Our results demonstrated that the size of the kinase in skeletal muscle was identical to that of the purified enzyme and there was no evidence of proteolysis. Essentially all of the kinase was solubilized by the low-ionic-strength extraction procedure; there was no detectable kinase associated with the myofibrillar pellet. Thus, there was no indication of a putative larger myofibrillar-associated myosin light chain kinase.

We extended our immunoblot analyses to several other mammalian species and determined that the masses of skeletal-muscle myosin light chain kinases varied greatly among animal species (Table II). The masses as determined by immunoblot analyses ranged from 68 (human) to 108 kDa (steer). The mass was constant within a mammalian species, regardless of skeletal-muscle fiber type. None of the mammalian skeletal-muscle kinases were as large as the purified smooth-muscle kinases. Unfortunately, although the antibodies used in these studies inhibited chicken skeletal-muscle myosin light chain kinase activity, they did not crossreact in the immunoblots with chicken kinase. The antibodies also did not crossreact (inhibition enzyme activity or immunoblots) with mammalian smooth-muscle myosin light chain kinases.

The report that the chicken skeletal- and gizzard smooth-muscle myosin light chain kinases might be similar, or identical, enzymes (*101*) prompted us to purify the chicken skeletal-muscle myosin light chain kinase. The purified chicken skeletal-muscle myosin light chain kinase is significantly larger (150 kDa; Table II) than any of the other skeletal-muscle kinases identified previously and larger than the 130-kDa gizzard enzyme (*102*). Affinity-purified antibodies to rabbit skeletal-muscle myosin light chain kinase crossreact with the chicken skeletal-muscle kinase, but not with the gizzard smooth-muscle kinase. The two skeletal-muscle kinases have very similar catalytic properties that are distinct from the gizzard smooth-muscle kinase.

TABLE II

RELATIVE MASSES OF ANIMAL SKELETAL MUSCLE
MYOSIN LIGHT CHAIN KINASES

Animal species	Mass (kDa)	Reference
Human	68	(*180*)
Mouse	75	(*90*)
Rat	82	(*90*)
Guinea pig	83	(*90*)
Rabbit	87	(*90*)
Dog	100	(*90*)
Steer	108	(*90*)
Chicken	150	(*102*)

An important question to consider is how are the rabbit and chicken skeletal-muscle kinases structurally related since they have similar catalytic properties and are dependent upon calmodulin for activity? Partial peptide maps generated by *Staphylococcus aureus* V8 protease following the method of Cleveland *et al.* (*103*), indicated that, although these two kinases had several peptides of similar molecular weight, there were also a number of peptides unique to each kinase (*102*). Analysis of complete tryptic digests of these two kinases by high-performance reverse-phase chromatography also indicated similar and different peptides for each kinase (*102*). Therefore, it does not appear that the large chicken skeletal-muscle kinase represents simply an extension of the rabbit skeletal-muscle myosin light chain kinase primary structure. The common peptides may represent portions of the enzymes from the catalytic and calmodulin-binding domains, but direct evidence involving sequence analyses are needed to establish these points.

Rabbit skeletal-muscle myosin light chain kinase can be proteolytically cleaved to generate specific domains. Mayr and Heilmeyer (*95*) obtained a calmodulin-dependent active fragment after partial digestion with trypsin. This 36-kDa active fragment was globular, high in α-helix content, and termed the "head" fragment. Another fragment (33 kDa) was catalytically inactive, rich in proline, and highly asymmetrical (axial ratio greater than 10). This fragment was termed the "tail" fragment. The two fragments accounted for the complete native enzyme based on amino acid composition. Mayr and Heilmeyer (*95*) constructed a model for the skeletal-muscle kinase based on the hydrodynamic properties of the fragments they purified. The model depicted an end-to-end arrangement of the head and tail fragments yielding an overall length for the intact kinase of approximately 380 Å. The tail fragment had a length of 270 Å and width of 18 Å. The head fragment was shorter and more compact, with a mean width of 40 Å.

Srivastava and Hartshorne (*104*) generated a calmodulin-independent fragment of rabbit skeletal-muscle myosin light chain kinase by limited cleavage with α-chymotrypsin. The specific activity of this 65-kDa fragment was similar to the activity of the native calmodulin-dependent enzyme. Thus, within the active head domain of the rabbit skeletal-muscle myosin light chain kinase, the calmodulin-binding domain appears to be distinct from the catalytic site and is located near the end of the molecule, which allows it to be removed proteolytically.

Blumenthal *et al.* (*105*) have purified and sequenced a fragment of rabbit skeletal-muscle myosin light chain kinase that represents the calmodulin-binding domain of the enzyme. The peptide is 27 residues in length and represents the carboxyl terminus of myosin light chain kinase. The peptide sequence shows that it contains a single tryptophan, no acidic or prolyl residues, and has a high probability of α-helix formation. The data of Blumenthal *et al.* (*105*) corroborate the findings of Mayr and Heilmeyer (*95*) and Srivastava and Hartshorne (*104*) which suggested that the skeletal-muscle kinase can be divided into at least two functional domains involving calmodulin-binding and catalytic activity. The functional domains do not overlap within the primary structure, and the calmodulin-binding domain appears to be distal to the catalytic site and represents the carboxyl terminus of the kinase (105).

Of particular interest is the finding of Blumenthal et al. (105) that the calmodulin-binding domain of the rabbit skeletal-muscle kinase does not have significant sequence homology to either troponin I or the γ-subunit of phosphorylase kinase. Both of these proteins bind calmodulin with high affinity, but the nature of that binding must be different enough to allow for completely different amino acid sequences. However, the peptide did have structural features similar to amphiphilic peptides that bind to calmodulin (106, 107). These features include clusters of basic residues, hydrophobic residues adjacent to these clusters, and an α-helical structure.

Structural studies have also been performed with the gizzard smooth-muscle myosin light chain kinase to define functional domains. Adelstein et al. (108) subjected turkey gizzard myosin light chain kinase to limited tryptic digestion, releasing a 22-kDa fragment that contained both cAMP-dependent protein kinase phosphorylation sites. These authors suggested that the two phosphorylation sites must be near one terminus of the 130-kDa kinase. Subsequent work by Walsh et al. (109) demonstrated that limited digestion of the turkey gizzard kinase with α-chymotrypsin resulted in the generation of a calmodulin-independent, catalytically active 80-kDa myosin light chain kinase. Generation of the calmodulin-independent kinase required that the proteolytic digestion be performed with calmodulin bound to the kinase (i.e., in the presence of Ca^{2+} and calmodulin). The calmodulin-independent myosin light chain kinase was incapable of binding to calmodulin-Sepharose, suggesting that it had lost its calmodulin-binding site. In addition, both sites of phosphorylation by cAMP-dependent protein kinase were removed by the α-chymotrypsin cleavage. If the digestion were performed in the absence of bound calmodulin, the generated 95-kDa fragment retained Ca^{2+}·calmodulin-dependent activity.

Foyt et al. (110) determined the structural domains of chicken gizzard myosin light chain kinase by limited and sequential digestions with different proteolytic enzymes. They concluded that the functional domains of the catalytic and calmodulin-binding sites, as well as the sites phosphorylated by cAMP-dependent protein kinase, did not overlap within the primary structure of the molecule. The authors presented a linear model with the calmodulin-binding domain between the catalytic site and the two phosphorylation sites. The two phosphorylation sites appear to be located within a 3-kDa peptide and the phosphorylation site modulated by calmodulin binding is closest to the calmodulin-binding region.

In summary, myosin light chain kinases are asymmetric, monomeric enzymes that vary considerably in mass depending upon tissue source and animal species.

They have a calmodulin-binding domain that is distinct from the catalytic domain. In addition, the smooth-muscle myosin light chain kinases contain two sites that are phosphorylated by cAMP-dependent protein kinase. These sites, one of which effects calmodulin activation of the enzyme, are not in the catalytic or calmodulin-binding domains.

C. Ca^{2+} AND CALMODULIN ACTIVATION

Although myosin light chain kinases, in general, require Ca^{2+} and calmodulin for activity, there are some exceptions. A Ca^{2+}-independent myosin light chain kinase was isolated from aortic smooth-muscle cells grown in culture whereas myosin light chain kinase purified under identical conditions from uncultured aortic smooth-muscle cells yielded a Ca^{2+}-dependent enzyme (51). In addition, a Ca^{2+}-independent myosin light chain kinase was found in cultured prefusion myoblasts of skeletal muscle, whereas a Ca^{2+}-dependent enzyme was found in the more differentiated cell form (111). Platelet myosin light chain kinase was originally described as a Ca^{2+}-independent enzyme (77), but it was later found that this form represented a proteolytic fragment of the native calmodulin-dependent enzyme (51). Bremel and Shaw (112) found Ca^{2+}-insensitive myosin light chain kinase activity in extracts of mammary glands. Addition of Ca^{2+}-sensitive myosin light chain kinase purified from gizzard smooth muscle to mammary actomyosin led to the rapid loss of Ca^{2+}-dependent kinase activity. Protease activity in cardiac myofibrils could also convert gizzard myosin light chain kinase to a Ca^{2+}-independent form (113). Incubation of purified myosin light chain kinases from gizzard smooth muscle (109, 114) and rabbit skeletal muscle (104, 114) with chymotrypsin or trypsin results in the conversion of enzyme to a lower-molecular-weight form and loss of calmodulin-dependent kinase activity. Thus, based upon these studies and the susceptibility of myosin light chain kinases to proteolysis as described in the preceding section, it is not clear whether calmodulin-independent forms of myosin light chain kinase reflect distinct forms or proteolyzed products. As emphasized previously, partial proteolysis does not necessarily lead to Ca^{2+}-independent myosin light chain kinase activity. For example, Mayr and Heilmayer (95) found that myosin light chain kinase purified from rabbit skeletal muscle can be converted to a smaller fragment, 36 kDa, that is dependent upon Ca^{2+} and calmodulin for activity. In this regard, it is interesting that myosin light chain kinase(s) purified from *Limulus* muscle have masses of 39 and 37 kDa and are dependent upon calmodulin for activity (99). Whether these low-molecular-mass forms of *Limulus* myosin light chain kinase represent native forms of a more primitive enzyme or a partially proteolyzed product has not been determined.

Evidence for the physical interaction of myosin light chain kinase with calmodulin was originally obtained by affinity chromatography techniques. For example, myosin light chain kinases from skeletal (91), smooth (80), and cardiac muscles (98), and platelets (51) bind to calmodulin-Sepharose in the presence of Ca^{2+}. After the free Ca^{2+} concentration is decreased to less than 10 nM with the addition of EGTA, kinase is eluted. Myosin light chain kinase is not retained on calmodulin-Sepharose if Mg^{2+} is substituted for Ca^{2+}. Thus, these data provide evidence that myosin light chain kinases and calmodulin form a Ca^{2+}-dependent complex.

A number of studies indicate that the stoichiometry for activation of myosin light chain kinase is 1 mol calmodulin per mol myosin light chain kinase including enzyme from rabbit skeletal muscle (92, 93, 96, 115, 116), steer cardiac muscle (98), and various smooth muscles (80, 85, 88, 89, 117). The one-to-one interaction of calmodulin with myosin light chain kinase has also been demonstrated by sedimentation equilibrium experiments (80), fluorescence titrations in the presence of 9-anthroylcholine (118), and myosin light chain kinase tryptophan fluorescence (119). In general, the concentration of calmodulin required for half-maximal activation of myosin light chain kinases under conditions where all four Ca^{2+}-binding sites in calmodulin are saturated with Ca^{2+} is approximately 1 nM (80, 85, 88, 89, 92, 96, 98, 115–118, 120). Formation of the holoenzyme complex follows simple hyperbolic kinetics which is consistent with the notion that calmodulin forms a stoichiometric 1 : 1 complex with the catalytic subunit of myosin light chain kinases. The reports on a lower affinity of calmodulin (K_d= 50 nM) for rabbit skeletal-muscle myosin light chain kinase (93, 119) may be explained by the high ionic strength in the enzyme assay. Activation of myosin light chain kinase activity by calmodulin is inhibited appreciably by increases in ionic strength (116).

An analysis of the stoichiometry and affinity of the Ca_n^{2+}·calmodulin species which is in equilibrium with activated myosin light chain kinase from rabbit skeletal muscle has been performed (92). The quantitative studies indicate that the activation is a sequential, fully reversible process, and the first step of the activation requires binding of Ca^{2+} to all four divalent metal-binding sites on calmodulin to form the complex, Ca_4^{2+}·calmodulin. This complex then binds to and activates the inactive catalytic subunit of rabbit skeletal-muscle myosin light chain kinase to form the active holoenzyme complex, Ca_4^{2+}·calmodulin·myosin light chain kinase (92). Similar conclusions regarding activation of calmodulin-stimulated cyclic nucleotide phosphodiesterase were subsequently obtained (121). Both groups of investigators emphasized that the activation of myosin light chain kinase and cyclic nucleotide phosphodiesterase, respectively, was positively cooperative in regard to Ca^{2+}. Thus, the enzymes may be activated by small changes in cytoplasmic Ca^{2+} concentrations, with the activation mechanism having the property of an on-and-off switch process. Furthermore, the actual Ca^{2+} concentrations required for activation depend not only upon the calmodulin concentrations, but also on the ratio of calmodulin to its receptor protein (92, 121, 122).

Cox and colleagues (*123–127*) have analyzed the activation mechanism by Ca^{2+} and calmodulin of the erythrocyte Ca^{2+}-dependent, Mg^{2+}-activated ATPase, cyclic nucleotide phosphodiesterase, phosphorylase kinase, and adenylate cyclase. The general conclusions are similar to the previous reports regarding activation, except that these investigators propose that Ca_3^{2+}·calmodulin *and* Ca_4^{2+}·calmodulin may activate these enzymes. The reasons for the discrepancy regarding the Ca_n^{2+}·calmodulin species required for activation in these studies is not apparent, but may be related to unique assumptions intrinsic in the modeling of these systems, such as Ca^{2+} binding properties of calmodulin. In any event, these investigators agree that most, if not all, of the Ca^{2+}-binding sites on calmodulin must be occupied by Ca^{2+} to activate these calmodulin-dependent enzymes.

Johnson *et al.* (*119*) measured the Ca^{2+} dependence of the calmodulin-induced increase in tryptophan flourescence in rabbit skeletal-muscle myosin light chain kinase. They proposed that the calmodulin and myosin light chain kinase interaction occurs before saturation of all four Ca^{2+}-binding sites on calmodulin and that activation could occur as a result of Ca^{2+} binding to one or perhaps two sites. This conclusion was based on the Ca^{2+}-dependence of the calmodulin-induced increase in tryptophan fluorescence in the enzyme as compared to the Ca^{2+}-binding properties of calmodulin alone. However, others have shown that the apparent dissociation constant for the four Ca^{2+}-binding sites is shifted from 14 μM to 1.7 μM in the presence of a calmodulin-binding protein, troponin I (*128*). Furthermore, the affinity of calmodulin for Ca^{2+} is increased up to 38-fold in the presence of myosin light chain kinase (*129*). The free-energy coupling for the interaction of Ca^{2+} and myosin light chain kinase with calmodulin is -8.44 kcal/4 mol Ca^{2+}. This value is similar to -8.00 kcal obtained with phosphorylase kinase (*127*), and both values are greater than -5.00 kcal/4 mol Ca^{2+} obtained with troponin I. Thus, a greater proportion of the Ca^{2+}-binding energy is transferred through calmodulin to myosin light chain kinase than to troponin I. These observations indicate that concentrations of Ca^{2+} required for Ca^{2+}-binding to calmodulin alone cannot be directly compared to the concentrations of Ca^{2+} required for calmodulin activation of myosin light chain kinase. Taking these factors into account, the results obtained by Johnson *et al.* (*119*) are in fact consistent with the notion that most, if not all, of the Ca^{2+}-binding sites on calmodulin must be saturated for activation of myosin light chain kinase.

The reaction scheme for association of Ca^{2+}, calmodulin, and myosin light chain kinase may be extended (Fig. 1). Although investigators have obtained significantly different K_d values for the individual reactions (including investigations on the same receptor enzyme), the general importance of different pathways for activation versus inactivation of calmodulin-activated enzymes in a biological system has been emphasized (*65, 121, 128–130*). Based upon these biochemical studies, activation associated with an increase in cytoplasmic Ca^{2+} concentrations *in vivo* is the result of Ca^{2+} binding first to calmodulin, with

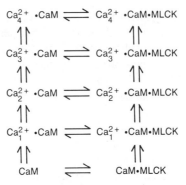

$$Ca_4^{2+} \cdot CaM \rightleftharpoons Ca_4^{2+} \cdot CaM \cdot MLCK$$
$$\Updownarrow \qquad\qquad \Updownarrow$$
$$Ca_3^{2+} \cdot CaM \rightleftharpoons Ca_3^{2+} \cdot CaM \cdot MLCK$$
$$\Updownarrow \qquad\qquad \Updownarrow$$
$$Ca_2^{2+} \cdot CaM \rightleftharpoons Ca_2^{2+} \cdot CaM \cdot MLCK$$
$$\Updownarrow \qquad\qquad \Updownarrow$$
$$Ca_1^{2+} \cdot CaM \rightleftharpoons Ca_1^{2+} \cdot CaM \cdot MLCK$$
$$\Updownarrow \qquad\qquad \Updownarrow$$
$$CaM \rightleftharpoons CaM \cdot MLCK$$

FIG. 1. This general scheme emphasizes the multiple steps involved in the association of Ca^{2+}, calmodulin (CaM), and myosin light chain kinase (MLCK).

subsequent binding to and activation of a calmodulin-dependent enzyme. Inactivation of enzyme activity due to a decrease in cytoplasmic Ca^{2+} concentrations, however, is not simply a reversal of these reactions and follows a different pathway. The rate of inactivation is much faster when Ca^{2+} first dissociates from the $Ca_4^{2+} \cdot$calmodulin·enzyme complex. For example, the rate of inactivation of cyclic nucleotide phosphodiesterase is about three orders of magnitude faster when Ca^{2+} first dissociates from the $Ca_{3\,or\,4}^{2+} \cdot$calmodulin·enzyme complex (130). The calculated Ca^{2+} off-rate constant (assuming a diffusion-controlled Ca^{2+} on-rate constant) is about 8 s^{-1}. However, the $Ca_3^{2+} \cdot$calmodulin off-rate constant is approximately 10^{-2} s^{-1}. These ideas are supported by experimental determination of dissociation rate constants and inactivation rates of myosin light chain kinase and phosphodiesterase (Table III). The rate constant determined by stopped flow techniques for Ca^{2+} dissociating from $Ca_4^{2+} \cdot$calmodulin·cyclic nucleotide phosphodiesterase was 4 s^{-1} (131). The rate of enzyme deactivation had a similar rate constant. Chau et al. (131) estimated that the rapid initial dissociation of Ca^{2+} from the $Ca_4^{2+} \cdot$calmodulin·phosphodiesterase complex was due to the dissociation of only one or, at the most, two Ca^{2+} from the protein complex. Johnson et al. (119) found that Ca^{2+} removal from $Ca_4^{2+} \cdot$calmodulin·myosin light chain kinase complex with EGTA resulted in a reversal of the tryptophan flourescence at a rate of 2 s^{-1} (Table III). The rate of deactivation of $Ca_4^{2+} \cdot$calmodulin·myosin light chain kinase by EGTA was similar [Ref. (132) and Table III]. It is noteworthy that these rates of decrease in tryptophan fluorescence and enzyme activity determined with purified myosin light chain kinase are similar to the apparent rate of inactivation of myosin light chain kinase in vivo (122, 133).

Although the rates of activation of myosin light chain kinase have not been measured experimentally, Johnson et al. (119) concluded that Ca^{2+} binds to calmodulin, activates calmodulin, which subsequently binds to myosin light

TABLE III

RATE CONSTANTS FOR CA²⁺ DISSOCIATION AND ENZYME DEACTIVATION

Complex	Measurement	Rate constant (s^{-1})	Comment	Reference
$Ca_4^{2+} \cdot CaM \cdot PDE^a$	Ca^{2+}-dissociation	4	Initial rapid release measured	(131)
$Ca_4^{2+} \cdot CaM \cdot PDE$	Enzyme deactivation	4.5	Single exponential	(131)
$Ca_4^{2+} \cdot CaM \cdot MLCK^a$	Decrease MLCK fluorescence	2	Single exponential	(119)
$Ca_4^{2+} \cdot CaM \cdot MLCK$	Enzyme deactivation	1	Single exponential	(132)

^{a} Abbreviations include CaM, calmodulin; PDE, calmodulin-stimulated cyclic nucleotide phosphodiesterase; MLCK, rabbit skeletal-muscle myosin light chain kinase.

chain kinase very rapidly (greater than 70 s⁻¹). The increase in tryptophan fluorescence associated with the binding of calmodulin to myosin light chain kinase occurs as a biphasic process with rates of 65 s⁻¹ and 6 s⁻¹, respectively. The relationship of these structural changes to myosin light chain kinase activity have not been determined.

The binding of Ca^{2+} to calmodulin results in the exposure of a hydrophobic domain which is thought to serve as the interface for Ca^{2+}-dependent interactions of calmodulin with receptor proteins (134, 135). Various hydrophobic ligands including the phenothiazine antipsychotics and naphthalene sulfonamides are also presumed to bind in this domain and thereby antagonize calmodulin–protein interactions (134–137). Hydrophobic interactions between calmodulin and rabbit skeletal-muscle myosin light kinase are necessary for the activation of the enzyme although other types of molecular interactions appear to also make significant and probably obligatory contributions to the activation process (116).

Investigations on rabbit skeletal-muscle myosin light chain kinase indicate that the interaction of Ca^{2+} with calmodulin is highly cooperative in the presence of kinase and substrates. In the presence of substrates, including the phosphorylatable skeletal-muscle myosin P-light chain and the ATP analog 5′-adenylylimidodiphosphate, the apparent dissociation constant for calmodulin decreases approximately 3- to 5-fold (129). The binding of calmodulin to gizzard smooth-muscle myosin light chain kinase enhances the binding of both ATP and 9-anthroylcholine, which appear to compete for a common binding site. The dissociation constants for ATP and 9-anthroylcholine are 120 and 20 μM, respectively, for myosin light chain kinase alone versus 43 and 6.4 μM for the $Ca_4^{2+} \cdot$calmodulin·myosin light chain kinase complex. Likewise, the apparent K_m value for myosin P-light chain decreases over 20-fold in the presence of $Ca_4^{2+} \cdot$calmodulin, while the V_{max} value increases only 1.5-fold (138). These data indicate that there are energy coupling contributions between bound ligands with myosin light chain kinase. The free energy coupling for binding of calmodulin and substrates of skeletal-muscle myosin light chain kinase is -0.95 kcal/mol calmodulin (129). This free energy coupling value is smaller than the contributions derived from Ca^{2+} and myosin light chain kinase coupling to calmodulin.

The calmodulin binding domain of skeletal-muscle myosin light chain kinase has been identified and sequenced (105). Following cyanogen bromide digestion, a peptide fragment of skeletal-muscle myosin light chain kinase inhibited activation of undigested myosin light chain kinase by calmodulin competitively. The peptide did not inhibit calmodulin-independent activity of myosin light chain kinase obtained by limited proteolysis. The primary structure of the peptide that represents the carboxyl terminus of rabbit skeletal-muscle myosin light chain kinase is K R R W K K N F I A V S A A N R F K K I S S S G A L M. The peptide contains a high percentage of basic and hydrophobic residues but no acidic residues. Secondary structural predictions indicate that the peptide has a high probability of forming an α-helix structure. The clusters of basic amino acid residues adjacent to hydrophobic residues with a predicted α-helical structure represent physical properties common to other amphiphilic peptides that have a high affinity for calmodulin (107, 139, 140). The synthetic peptide of identical sequence binds to calmodulin-Sepharose in a Ca^{2+}-dependent manner (although 2 M urea is required to elute the peptide in addition to Ca^{2+} chelation). The synthetic peptide undergoes a significant increase in intrinsic tryptophan fluorescence in the presence of Ca^{2+} and calmodulin, which is maximal at a peptide–calmodulin ratio of 1 : 1. This finding is similar to the increase in tryptophan fluorescence produced by calmodulin in native myosin light chain kinase (119). It will be important to determine whether other calmodulin-dependent enzymes have calmodulin-binding domains that are similar, if not identical, in physicochemical properties.

The work of Klee and colleagues (21, 35) suggests that many calmodulin-dependent enzymes have structurally different calmodulin-binding domains. A peptide fragment of calmodulin prevents activation of phosphodiesterase by calmodulin and by itself does not stimulate activity (141). The same fragment fully activates phosphorylase kinase. Cyclic AMP-dependent protein kinase and phosphodiesterase bind to the calmodulin fragment containing residues 1–77, whereas calcineurin does not (142). A covalent adduct of norchlorpromazine and calmodulin competitively inhibits activation of phosphodiesterase and gizzard myosin light chain kinase by calmodulin (143). However, the adduct partially activates the protein phosphatase, calcineurin.

D. CATALYTIC PROPERTIES

A feature of phosphotransferase reactions catalyzed by myosin light chain kinases is the specificity for myosin P-light chains (70, 75, 80, 97, 144). Myosin light chain kinases from skeletal muscle (75, 115), gizzard smooth muscle (80), cardiac muscle (97), platelets (51), and thyroid gland (145) phosphorylate casein, phosvitin, histone, and phosphorylase b at rates less than 3% of the rate with myosin P-light chain. In addition, skeletal-muscle myosin light chain kinase does not phosphorylate troponin I, phosphorylase kinase, myosin heavy chain, the 15- and 22-kDa myosin light chains, or the molluscan adductor myosin light chain at significant rates (75, 115). The narrow protein substrate specificity of myosin light chain kinases distinguishes them from the majority of other protein kinases including multifunctional calmodulin-dependent protein kinase and cyclic nucleotide-dependent protein kinases, which phosphorylate a broad spectrum of protein substrates (146).

Based upon differences in catalytic properties, myosin light chain kinases from skeletal and smooth muscles fall into two general classes. Interestingly, the effect of myosin P-light chain phosphorylation on actin-activated Mg^{2+} ATPase activity is also different for these two muscle types (see Volume XVIII, Chapter 13). Catalytic properties are characterized best for rabbit skeletal-muscle and gizzard smooth-muscle myosin light chain kinases and provide the primary basis for distinction of these two classes.

Gizzard smooth-muscle myosin light chain kinase phosphorylates gizzard myosin with apparent K_m and V_{max} values of approximately 15 μM and 15 $\mu mol/min/mg$, respectively (147). These values are comparable to the kinetic constants for isolated gizzard myosin P-light chains (Table IV). Various investigators have shown that filamentous gizzard smooth-muscle myosin is phosphorylated in vitro by an ordered process (147–149). First-order rate constants are greater for phosphorylation of P-light chain of the first head of myosin than for the second head (1.4 min^{-1} and 0.3 min^{-1}, respectively). Phosphorylation of both myosin P-light chains is required for significant increases in actin-activated myosin ATPase activity. In contrast to these conclusions, Sobieszek (150) reported a single population of myosin heads for chicken gizzard and pig stomach smooth-muscle myosins with respect to phosphorylation by myosin light chain kinase. The binding of myosin light chain kinase with equal affinity to nonphosphorylated and phosphorylated myosin accounted for significant product inhibition, and first-order kinetics at both low and high substrate concentrations.

Smooth muscle and nonmuscle myosin light chain kinases exhibit specificity for myosin light chains from smooth muscle or nonmuscle tissues (51, 111, 144, 151, 152). Kinetic constants are shown in Table IV for gizzard smooth-muscle myosin P-light chain with myosin light chain kinases from gizzard smooth-muscle and some nonmuscle tissues. As noted in Table IV, there is a decrease in V_{max} values for striated muscle P-light chains compared to gizzard P-light chain with gizzard myosin light chain kinase. The ratio of V_{max} to K_m for gizzard smooth-muscle P-light chain is about 2 whereas the ratio for cardiac and skeletal muscle P-light chains, respectively, is 10- and 100-fold less. These results are consistent with the lower phosphorylation rates obtained with cardiac and skel-

TABLE IV

KINETIC CONSTANTS FOR MYOSIN LIGHT CHAIN KINASES

Myosin light chain kinase	MgATP^{2-} K_m (μM)	Myosin P-light chain			Reference
		Source	K_m (μM)	V_{max}[a] (μmol/min/mg)	
Rabbit skeletal-muscle	200–400	Rabbit skeletal muscle	9–53	30–87	(95,96,102,115)
		Chicken skeletal muscle	8	53	(102)
		Cardiac muscle	9	54	(102)
		Gizzard smooth muscle	15	73	(102)
Chicken skeletal-muscle	163	Rabbit skeletal muscle	10	29	(102)
		Chicken skeletal muscle	8	42	(102)
		Cardiac muscle	20	56	(102)
		Gizzard smooth muscle	10	68	(102)
Cardiac muscle	175–220	Cardiac muscle	11–20	20–30	(97,98)
Gizzard smooth-muscle	50–68	Gizzard smooth muscle	5–14	15–36	(80,102,109,147,155)
		Cardiac muscle	63	12	(102)
		Rabbit or chicken skeletal muscle	95	4.0	(102)
Platelets	121	Gizzard smooth muscle	18	4.7[b]	(51)
Pancreatic islet cells	70	Gizzard smooth muscle	44	0.062[b]	(221)
Proliferative myoblasts		Platelets	2	0.077[b]	(151)
		Cardiac muscle	11	0.077[b]	(151)
		Skeletal muscle	35	0.077[b]	(151)

[a] Some rates corrected to a value at 30°C assuming a Q_{10} of 2 (116).
[b] Specific activities of partially purified myosin light chain kinase preparations.

etal muscle P-light chains by Adelstein and Klee (*80*). However, these investigators found no major differences in the K_m values for these three different types of P-light chains, whereas Nunnally *et al.* (*102*) found the K_m values were increased 6–8-fold for the striated-muscle P-light chains. Nonmuscle myosin light chain kinases from platelets (*51*), astrocytes (*151*), and lymphocytes (*152*) phosphorylate striated-muscle myosin P-light chains at rates lower than those obtained with myosin P-light chains from smooth-muscle or nonmuscle tissues. However, these catalytic studies on nonmuscle myosin light chain kinases have not been performed with purified enzyme. Owing to the promiscuous nature of gizzard P-light chain as a substrate for protein kinases (*153*), any conclusions regarding the specific catalytic properties of nonmuscle myosin light chain kinases should be regarded as tentative until the purified enzymes have been characterized.

In contrast to the properties of smooth-muscle myosin phosphorylation, skeletal-muscle myosin is phosphorylated *in vitro* by an apparently random-order mechanism (*154*). Kinetic constants for phosphorylation of rabbit skeletal-muscle myosin by rabbit skeletal-muscle myosin light chain kinase are 19 μ*M* for K_m and 47 μmol/min/mg for V_{max} (*96*). The properties of phosphorylation of free P-light chain and P-light chain bound to myosin are similar (Table IV). It should be noted that some of the earlier reported V_{max} values for rabbit skeletal-muscle myosin light chain kinase are significantly lower than the later reported values. Both Mayr and Heilmeyer (*95*) and Nagamoto and Yagi (*96*) have emphasized that shorter times for purification result in greater enzyme specific activity. The reasons for these differences are not clear at this time.

Myosin light chain kinases from skeletal muscle exhibit a broader protein substrate specificity than smooth-muscle and nonmuscle myosin light chain kinases. Rabbit and chicken skeletal-muscle myosin light chain kinases phosphorylate skeletal, cardiac, and smooth-muscle myosin P-light chains with similar K_m and V_{max} values (Table IV). Thus, in contrast to the smooth- and nonmuscle myosin light chain kinases, skeletal-muscle myosin light chain kinases show no apparent specificity in phosphorylating P-light chains from different tissues and cells. Furthermore, the rabbit and chicken skeletal-muscle enzymes, although differing substantially in size, have basically identical catalytic specifity.

There are differences in the primary structures surrounding the phosphorylatable serine residue in myosin P-light chains from smooth, skeletal, and cardiac muscle (Table V). The importance of the primary structure around the phosphorylatable serine has been examined with synthetic peptide substrates homologous with the primary structure of native gizzard smooth-muscle myosin P-light chain (*155, 156*). The apparent K_m and V_{max} values for a 23-residue synthetic peptide identical to the amino terminus of gizzard smooth-muscle myosin P-light chain (Table V) are 2.7 μ*M* and 3.0 μmol/min/mg, respectively (*156*). These values are less than values obtained for purified gizzard myosin P-light chain

TABLE V

AMINO ACID SEQUENCES OF MYOSIN P-LIGHT CHAINS[a]

Source	Region[a] III	II	I	Reference
Chicken gizzard smooth muscle	S S K R A K T	T K K R - P Q R	- A T S̲ N V F	(252)
Chicken skeletal muscle	P K K - A K R R	- A A E - G S S̲	N V F	(253)
Rabbit skeletal muscle	P K K - A K R R	A A A E G G S S̲	N V F	(254,255)
Chicken cardiac muscle A	P K K - A K K R	- - I E G A N S̲	N V F	(256)
Chicken cardiac muscle B	P K K - A K K K	- - V E - G G S̲	N V F	(256)

[a] The abbreviations used for amino acids include S, serine; K, lysine; R, arginine; A, alanine; T, threonine; P, proline; Q, glutamine; N, asparagine; V, valine; F, phenylalanine; E, glutamic acid; G, glycine; L, leucine; I, isoleucine; B, asparagine-aspartic acid; D, aspartic acid; Z, glutamine-glutamic acid; and S̲, serine that is phosphorylated.

(Table IV). Removal of the first seven residues from the synthetic peptide of gizzard P-light chain did not significantly affect apparent K_m and V_{max} values (156). The K_m/V_{max} ratio increased greater than 70-fold for the 12–23 residue peptide compared to the 11–23 residue peptide (156). The major phosphorylated amino acid in the longer peptides was phosphoserine; however, an increased fraction of phosphothreonine was found with decreased peptide length (156). The single substitution of alanine residues for basic residues at positions 11, 12, or 13 produced a 27- to 44-fold increase in K_m/V_{max} ratios (156). These results indicate that the six to eight basic residues towards the amino terminus from the phosphorylatable serine (Region II, Table V) influence catalysis of gizzard smooth-muscle myosin light chain kinase.

The arginine at position 16 in the sequence of smooth-muscle P-light chain (Region I, Table V) has a significant influence on the kinetic properties that is similar to the influence of basic residues in Region II (157). The location of arginine-16 in relation to the phosphorylatable serine-19 as well as the distance between arginine-16 (Region I) and arginine-13 (Region II) affect the kinetic properties and site of phosphorylation. Placing arginine at position 15 instead of 16 results in phosphorylation of threonine-18 instead of serine-19. Increasing or decreasing the number of alanine residues between arginine-13 and arginine-16 results in an increase in the apparent K_m values and phosphorylation of both threonine-18 and serine-19 (157).

Synthetic peptide substrates identical to skeletal- or smooth-muscle myosin P-light chains have been used with skeletal-muscle myosin light chain kinases. Rabbit and chicken skeletal-muscle myosin light chain kinases exhibit similar phosphorylation kinetics with the peptide substrates which indicates recognition of similar substrate determinants for these two enzymes (158). Both low K_m and V_{max} values are obtained with peptides identical to the amino terminus of chicken or rabbit skeletal-muscle myosin P-light chain (Table V). These values are about 2 μM (K_m) and 1.0 $\mu mol/min/mg$ (V_{max}) for both kinases. Substitution of basic residues, lysine-2 and lysine-3, with alanine residues in the chicken skeletal-muscle peptide (Table V) results in a 40-fold increase in apparent K_m values without a change in the V_{max} value. Thus, basic amino acids located 10–11 residues toward the amino terminus from the phosphorylatable serine in the chicken skeletal-muscle myosin light chain significantly influence K_m values (Region III, Table V). Basic residues in Region II of the chicken skeletal muscle peptide influence K_m values in a similar manner, substitution of alanine residues for arginine-6 and arginine-7 produces a 50-fold increase in K_m values. Substitution of an arginine residue for glutamic acid at position 10 in the chicken skeletal-muscle myosin light chain peptide results in a 3-fold decrease in apparent K_m, and a 7-fold increase in the V_{max} values. Thus, the presence of an acidic residue in Region I (Table V) seems to be a negative substrate determinant.

Skeletal-muscle myosin light chain kinases phosphorylate the synthetic peptide identical to smooth-muscle myosin P-light chain with apparent kinetic constants of about 1.5 μM for K_m and 26 μmol/min/mg for V_{max} (158). These values are similar to the kinetic constants for native smooth-muscle myosin P-light chain (Table IV). It is not clear why the synthetic peptides for the smooth-muscle P-light chain have significantly higher V_{max} values than the synthetic peptides for skeletal-muscle P-light chains. Removal of residues 1–5 from the synthetic peptide of smooth-muscle myosin P-light chain does not influence apparent K_m or V_{max} values; however, apparent K_m values are increased 8–10-fold with the removal of the first 10 residues (Region III, Table V). Substitution of alanine residues for basic residues at positions 12 or 13 (Region II, Table V) produced a 4–7-fold or 9–12-fold, respectively, increase in apparent K_m without influencing V_{max} values.

Skeletal- and smooth-muscle myosin light chain kinases share some general properties regarding the influence of basic residues on catalysis. The activity of myosin light chain kinases appears to be influenced by three regions in the primary structure of myosin P-light chains (Table V). In regards to gizzard smooth-muscle myosin light chain kinase, Regions I and II are particularly important, whereas the basic residues in Region III which are nearer the amino terminus are of lesser importance. For skeletal-muscle myosin light chain kinases, the basic residues in Regions II and III seem to be more important. The general importance of basic residues as substrate determinants for both classes of myosin light chain kinases is similar to their importance for other protein kinases such as phosphorylase kinase (159), cAMP-dependent protein kinase (160), and cGMP-dependent protein kinase (161) although the optimal positions of basic residues in relation to the phosphorylatable serine are unique.

Myosin light chain kinases utilize MgATP^{2-} for the metal ion–nucleotide substrate (70, 75, 80, 97, 144). Neither ATP^{4-}, MnATP^{2-}, nor CaATP^{2-} are substrates, and a 1.5 to 1.0 molar ratio of Mg^{2+} to ATP^{4-} results in maximal kinase activity (75, 144). The reported K_m values for MgATP^{2+} shown in Table IV are 1.7- to 4-fold greater with skeletal- and cardiac-muscle myosin light chain kinases than with gizzard smooth-muscle and nonmuscle myosin light chain kinases. The nucleotide triphosphates ITP, GTP, UTP, and CTP are utilized by skeletal-muscle myosin light chain kinase at rates less than 5% of the rates obtained with ATP (75). Catalytic activities of skeletal-muscle and gizzard smooth-muscle myosin light chain kinases were inhibited by ADP^{3-} and ATP^{4-}, but were not inhibited by AMP^{2-} or cyclic AMP. Furthermore, gizzard smooth-muscle myosin light chain kinase activity was not affected by ITP^{3-}, cyclic GMP, or adenosine (75, 144).

Catalytic activities of myosin light chain kinases are affected similarly by temperature, pH, and ionic strength. They have a Q_{10} of 2 and are less stable at 35–37° than at 30°C or lower (116, 162). Myosin light chain kinases from skeletal, smooth, and cardiac muscles have a broad pH optimum from pH 6.5 to 8.0 (97, 116, 162).

Ionic strength affects myosin light chain kinase activities. Salt concentrations greater than 0.22 M inhibit phosphorylation rates significantly (*97*, *116*, *162*). The K_m value for skeletal-muscle myosin P-light chain increases approximately 3-fold in 0.22 M KCl while V_{max} values decrease 1.5- and 2.5-fold in 0.12 and 0.22 M KCl, respectively (*116*). Phosphorylation of smooth-muscle myosin P-light chain decreases approximately 50 and 80% in 0.1 and 0.2 M KCl, respectively, with the proteolytically produced calcium-independent smooth-muscle myosin light chain kinase (*109*).

Limited information is available on the catalytic mechanism of myosin light chain kinases or characterization of specific chemical and physical interactions between enzyme and substrates. Analyses of the catalytic properties indicate a sequential kinetic mechanism for skeletal-muscle myosin light chain kinase (*91*). Evidence was presented in kinetic studies with rabbit skeletal-muscle myosin light chain kinase for a rapid equilibrium bi-bi kinetic mechanism (*159a*). In addition it appears that the kinase can form a dead-end complex with ADP and unphosphorylated myosin light chain. The catalytic activity of skeletal-muscle myosin light chain kinase is inhibited by 86% after chemical modification of approximately 30% of the tyrosine residues in the enzyme, but modification of cysteine residues has no effect (*96*). No information is available on the catalytic mechanisms of myosin light chain kinases from heart, smooth muscle, or non-muscle tissues.

E. PHOSPHORYLATION

A common feature of many protein kinases is the ability to undergo auto-phosphorylation or to serve as substrates for other protein kinases (*146*). Phosphorylation and dephosphorylation of protein kinases have been characterized more thoroughly *in vitro* rather than *in vivo*, and the physiological significance of many protein kinase phosphorylation reactions, particularly autophosphorylation, has not been fully established.

Myosin light chain kinases from skeletal, cardiac, and gizzard smooth muscles have been reported to undergo autophosphorylation. Skeletal- and cardiac-muscle myosin light chain kinases catalyze autophosphorylation in the presence of Ca^{2+} and calmodulin with the incorporation of about 1 mol phosphate/mol enzyme (*94*, *98*). In contrast, Ca^{2+} and calmodulin were reported to inhibit gizzard smooth-muscle myosin light chain kinase autophosphorylation, and significantly lower rates of autophosphorylation were observed with myosin light chain kinase from gizzard smooth muscle than from cardiac muscle (*163*). Myosin light chain kinase autophosphorylation has no apparent effect on catalytic activity or activation by calmodulin. The slow rate of autophosphorylation and lack of an effect on the biochemical properties of myosin light chain kinase indicates this reaction may not be biologically important.

Myosin light chain kinases from skeletal (*94*, *100*), cardiac (*98*, *100*), and smooth muscles (*86–89*, *108*, *109*, *117*, *164*) as well as from platelets (*87*, *165*) are phosphorylated *in vitro* by cAMP-dependent protein kinase. Cyclic AMP-dependent protein kinase catalyzes the maximal incorporation of about 1 mol phosphate per mol of skeletal or cardiac myosin light chain kinase (*94, 98*). However, a maximal incorporation of 2 mol phosphate per mol myosin light chain kinase is generally observed for the smooth-muscle enzymes in the absence of calmodulin. In the presence of calmodulin only 1 mol phosphate per mol of smooth-muscle myosin light chain kinase is incorporated. Cyclic GMP-dependent protein kinase was reported to phosphorylate tracheal and platelet myosin light chain kinases with the incorporation of approximately 1 and 1.5 mol phosphate per mol enzyme, respectively, in the absence of calmodulin (*87*). Peptide maps indicated that one common site was phosphorylated on platelet myosin light chain kinase by cAMP- and cGMP-dependent protein kinases, only in the presence of calmodulin (*87*).

The kinetic properties of phosphorylation of myosin light chain kinases by the cyclic nucleotide-dependent protein kinases have not been thoroughly investigated. The relative rates of phosphorylation for the two sites or sequence of phosphorylation, random or ordered, have not been established. However, some pertinent information is available. At 1 μM smooth-muscle myosin light chain kinase, the rate of phosphorylation is 0.04 to 0.2 μmol ^{32}P incorporated/min/mg of cAMP-dependent protein kinase catalytic subunit (*85, 87, 88, 166*). This concentration of myosin light chain kinase is similar to the concentration of the enzyme in smooth muscles. These rates of phosphorylation of smooth-muscle myosin light chain kinase *in vitro* are approximately 1% or lower than the rates of phosphorylation of other protein substrates known to be phosphorylated by cAMP-dependent protein kinase *in vivo* (*65*).

Phosphorylation of myosin light chain kinase isozymes by cAMP-dependent protein kinase has different effects on catalytic properties. Phosphorylation does not alter the catalytic activity of skeletal-muscle myosin light chain kinase (*94, 100*). Although Walsh and Guilleux (*100*) reported a 3.3-fold decrease in the apparent dissociation constant for calmodulin with phosphorylated cardiac-muscle myosin light chain kinase, Wolf and Hofmann (*98*) observed no significant change.

Calmodulin activation of smooth-muscle and platelet myosin light chain kinase is effected by phosphorylation with cAMP-dependent protein kinase *in vitro*. The concentration of calmodulin required for half-maximal activation is increased for diphosphorylated, but not monophosphorylated, myosin light chain kinases. In general, the increases are about 10-fold for myosin light chain kinases from platelets (*87, 165*) gizzard (*166*), trachea (*86*), stomach (*85*), myometrium (*89*), aorta (*88*), and carotid artery (*117*). Changes in calmodulin binding affinity with phosphorylation is not restricted to smooth-muscle and platelet myosin light chain kinases. Phosphorylation of myelin basic protein, troponin I, and histone H2A by cyclic AMP-dependent protein kinase decreases their ability to compete with myosin light chain kinase for calmodulin (*118*). This effect may be related to the increased acidic phosphate charge introduced into or near calmodulin-binding domains.

Adelstein and colleagues (*165–167*) proposed that, *in vivo,* cAMP-dependent phosphorylation of myosin light chain kinase decreases catalytic activity via a reduction in affinity for calmodulin which results in smooth-muscle relaxation. Although this is an intriguing hypothesis, there are numerous investigations that cast doubt on its physiological importance (*65*). Miller *et al.* (*86*) examined tracheal smooth-muscle myosin light chain kinase for a change in the calmodulin activation properties in muscles treated with the β-adrenergic agonist, iso-proterenol. Isoproterenol increased phosphorylase *a* formation approximately 3-fold within 5 min incubation, presumably due to stimulation of cAMP formation and activation of cAMP-dependent protein kinase. However, there was no change in the calmodulin activation properties of myosin light chain kinase (*86*). Thus, phosphorylation of the site associated with a decrease in calmodulin affinity was probably not effected. However, measurements of the extent of phosphorylation of tracheal myosin light chain kinase was not measured. De Lan-erolle *et al.* (*168*) examined cAMP accumulation, ^{32}P incorporation into myosin light chain kinase, and relaxation in tracheal muscle. Incorporation of ^{32}P was measured after immunoprecipitation of myosin light chain kinase in homogenates from radiolabeled smooth muscles. Under control conditions the calculated ^{32}P incorporation was 1.1 mol phosphate per mol myosin light chain kinase. This amount of incorporation is not expected to be due to phosphorylation by cAMP-dependent protein kinase because cAMP content is low, and other cAMP-dependent processes are at basal states (*86, 168, 169*). The addition of forskolin stimulates cAMP formation and increases the net phosphate incorporated by 0.6–0.8 mol phosphate per mol myosin light chain kinase. A similar amount of phosphate is incorporated whether the muscle is precontracted or relaxed. It is puzzling why the same amount of ^{32}P is incorporated with contracted and relaxed tissues since, in the former, calmodulin is presumably bound to myosin light chain kinase, whereas in the latter the enzyme is in the unbound state. The cAMP-dependent incorporation of only 1 mol phosphate per mol myosin light chain kinase would not be expected to change the calmodulin activation properties of smooth-muscle myosin light chain kinase if only one site is phosphorylated. However, in this study the phosphorylated sites were not identified nor were the calmodulin activation properties measured (*168*).

As pointed out previously (*65*), phosphorylation of smooth-muscle myosin P-light chain can be a transient event in relation to force development. It has been found that after prolonged incubation of tracheal smooth muscle with carbachol, the extent of phosphorylation of myosin P-light chain after reaching maximal values in a minute or less, decreases to control values after two hours while force is maintained. The addition of isoproterenol at this time results in relaxation (*86*).

Thus, β-adrenergic receptor stimulation may relax tracheal smooth muscle in the absence of significant phosphorylation of myosin P-light chain. A similar experimental approach has been reported with hog carotid artery (*170*). Under conditions where force is maintained and myosin P-light chain phosphorylation has returned to basal levels, forskolin-stimulated cAMP formation relaxes the muscles with no dephosphorylation of myosin P-light chain. Under similar conditions adenosine, 3-isobutyl-1-methylxanthine, sodium nitroprusside, and 8-bromo-cyclic GMP also relax the muscle without P-light chain dephosphorylation. Thus, dephosphorylation of myosin is not necessary for relaxation related to increases in cAMP formation. Jones *et al.* (*171*) measured changes in ion fluxes in relation to relaxation by forskolin for contractions produced by norepinephrine, angiotensin II, and KCl depolarization in rat aorta. These authors concluded that cAMP-dependent regulation of membrane ion fluxes represents a primary mechanism for relaxation and that the phosphorylation of myosin light chain kinase apparently functions, if at all, in a secondary capacity. These pharmacological studies in isolated smooth-muscle strips indicate that cAMP-dependent effects on sarcoplasmic Ca^{2+} concentrations may be the important process mediating smooth-muscle relaxation. Thus, although purified smooth-muscle myosin light chain kinases are phosphorylated by cAMP-dependent protein kinase, the physiological relevance is doubtful (*65*) when compared to the criteria of Krebs and Beavo (*146*) for establishing the physiological significance of a particular protein phosphorylation reaction.

F. BIOLOGICAL SIGNIFICANCE OF MYOSIN PHOSPHORYLATION

Calmodulin-dependent phosphorylation of myosin P-light chain by myosin light chain kinase plays a central role in the regulation of smooth-muscle contractility (*65*) (see Volume XVIII, Chapter 13). Studies with purified smooth-muscle actomyosin have demonstrated that phosphorylation of the myosin P-light chain increases actin activation of myosin Mg^{2+} ATPase activity which is viewed as an *in vitro* correlate of myosin cross-bridge cycling and smooth-muscle force generation *in vivo*.

The importance of myosin P-light chain phosphorylation in initiation of isometric force development has been demonstrated with a variety of smooth-muscle preparations. Stimulation of tissues with one of a number of contractile agents (i.e., electrical, KCl, pharmacological agonists), results in myosin P-light chain phosphorylation during isometric contraction (*65*). In skinned smooth-muscle fibers, a Ca^{2+}-independent contraction is produced by treatment of fibers with a calmodulin-independent fragment of chicken gizzard myosin light chain kinase (*172*). Moreover, irreversible thiophosphorylation of myosin in skinned smooth-muscle strips results in contractions that are maintained in the absence of Ca^{2+} (*173, 174*). Thus, myosin phosphorylation is sufficient to initiate and maintain force.

While Ca^{2+}-dependent myosin phosphorylation may be a primary event in smooth-muscle contraction, there is evidence that another Ca^{2+}-dependent regulatory system is involved in the regulation of force maintenance in smooth muscle. Myosin P-light chain phosphorylation may increase rapidly and then slowly decrease in smooth-muscle tissue while isometric force remains unchanged (65). However, force maintenance is Ca^{2+}-dependent (175). The transient phosphorylation of P-light chain observed in different smooth muscles are probably due to transient changes in intracellular Ca^{2+} concentration. After increasing to relatively high concentrations, the Ca^{2+} concentration declines to a range that remains above basal values. These low values are not sufficient to maintain activation of myosin light chain kinase by calmodulin but are sufficient to maintain force (65). The particular properties of the Ca^{2+} transient and resultant myosin phosphorylation probably depends upon the type of smooth-muscle or nonmuscle cell, as well as the specific agonist used. It should be emphasized that myosin P-light chain phosphorylation appears to be essential for smooth-muscle contraction, and the second site of Ca^{2+} regulation is activated following the transient phosphorylation. The other site of calcium regulation that is responsible for maintenance of tension in the absence of P-light chain phosphorylation remains unidentified.

In contrast to the obligatory role for myosin P-light chain phosphorylation in smooth-muscle contraction, skeletal-muscle myosin phosphorylation probably plays a modulatory role. It is generally accepted that in striated muscle the primary mode of Ca^{2+} regulation is via the thin filament-linked troponin–tropomyosin complex. Unlike smooth-muscle actomyosin, actin activation of skeletal-muscle myosin Mg^{2+}ATPase activity is not dependent upon myosin phosphorylation. However, an increase in the actin activation of skeletal-muscle myosin Mg^{2+}ATPase activity occurs as a result of phosphorylation due to a decrease in the apparent K_m for actin, with no significant change in the V_{max} value (154, 176).

Phosphorylation of skeletal-muscle myosin P-light chain in intact rat fast-twitch muscle correlates to potentiation of posttetanic isometric-twitch tension (177, 178). Isometric-twitch potentiation induced by low frequency repetitive stimulation in situ also correlates to phosphorylation of skeletal-muscle myosin P-light chain (179). Potentiation of isometric-twitch tension and P-light chain phosphorylation occurs in situ (under conditions of physiological stimulation frequency), primarily in white fast-twitch, but not red slow-twitch muscles (133). These results have been corroborated in human investigations (180).

Work on glycerinated rabbit psoas-muscle fibers demonstrate an effect of myosin P-light chain phosphorylation on tension generation at a submaximal Ca^{2+} concentration that does not fully activate the thin-filament regulatory proteins (181). At 0.6 μM Ca^{2+} isometric tension was only 10–20% of the maximal tension obtained at 10 μM Ca^{2+}. Under these conditions, addition of calmodulin and purified rabbit skeletal-muscle myosin light chain kinase results in an increase in isometric tension while the addition of a phosphoprotein phosphatase reverses the effect. There is no effect of phosphorylation on the normalized force–velocity curve, or the extrapolated maximal shortening velocity. At 10 μM Ca^{2+} there is no significant effect of P-light chain phosphorylation on the extent of isometric tension. Thus, P-light chain phosphorylation may increase isometric tension at submaximal Ca^{2+} concentrations by increasing the affinity of the myosin cross-bridge for the thin filament in a cooperative manner. These observations provide a mechanism for the potentiation of isometric-twitch tension by P-light chain phosphorylation.

III. Multifunctional Calmodulin-Dependent Protein Kinases

A. INTRODUCTION

Ca^{2+}-dependent protein kinase activity was first described for skeletal-muscle phosphorylase kinase (*12*) and myosin light chain kinase (*67*). Both protein kinases were recognized for their respective specificities for protein substrates, a property that contrasts markedly with cyclic nucleotide-dependent protein kinases. The observation was subsequently made by many investigators that Ca^{2+} stimulated protein kinase activity in homogenates prepared from a variety of tissues resulted in phosphorylation of proteins that did not correspond to myosin P-light chain or phosphorylase (*35, 182*). It has become apparent that this Ca^{2+}-dependent kinase activity is due to a class of calmodulin-dependent protein kinases that have similar biochemical properties. A calmodulin-stimulated glycogen synthase kinase has been purified from rabbit and rat liver (*183–185*) and rabbit skeletal muscle (*186*). Calmodulin-dependent protein kinase that phosphorylates synapsin I has been purified from rat brain tissue (*187–189*). Additional sources include *Torpedo* electric organ (*190*), turkey erythrocytes (*191*), bovine heart (*192*), rat pancreas (*193*), and neurons of *Aplysia* (*194*). In addition to binding calmodulin, these enzymes exhibit additional, similar biochemical properties including broad specificity for protein substrates, similar polypeptide composition (50 and 60 kDa) assembled in a oligomer of about 500 kDa, and autophosphorylatable subunits. The widespread distribution and broad substrate specificity of these calmodulin-dependent protein kinases suggest that, like calmodulin, they may have multiple roles in regulating cellular processes.

B. PHYSICOCHEMICAL PROPERTIES

Multifunctional calmodulin-dependent protein kinases have been identified and purified from mammalian brain in a number of laboratories. We first consider the properties of these kinases to allow for a discussion of the apparent discrepancies in subunit and native sizes, as well as substrate specificity, within a single tissue before addressing the problems of possible tissue- and species-specific differences.

1. Mammalian Brain Kinases

The structural properties of the multifunctional calmodulin-dependent protein kinases isolated from rat brain are summarized in Table VI. In general, the kinases are composed of a 50-kDa subunit comprising most of the protein (189, 195–201). In addition, most laboratories have also copurified a 60-kDa subunit (189, 196–198, 200, 201). Immunological crossreactivity suggests that the 50-kDa subunit is structurally related to the 60-kDa subunit (201). However, peptide maps of [125]I- or ^{32}P-labeled enzyme suggest that, although the 50- and 60-kDa subunits are structurally related, they represent distinct polypeptides (187, 196, 198).

In contrast, the doublet bands seen in both the 50- and 60-kDa subunits are presumably due to either limited proteolysis of the native subunit (187) or varying degrees of phosphorylation that cause mobility shifts upon electrophoresis (196). The problem of endogenous proteolysis and resultant kinase instability has been emphasized (196). McGuinness et al. (201) analyzed this problem with a monoclonal antibody to the brain kinase by immunoblots of freshly dissected tissue homogenized in boiling sodium dodecyl sulfate buffer. The 58-kDa (β') subunit was present in much lower amounts relative to the 60-kDa (β) subunit than the relative amounts observed in the purified kinase. Thus, the doublet bands are probably due to proteolysis during purification.

The sizes of the native brain kinases were determined by gel filtration and sucrose-density gradient centrifugation and, in general, are about 500 kDa (Table VI). Reported values for Stokes radius and $s_{20,w}$ range from 75–95 Å (189, 195, 197, 198, 201) and 7.3S–17.4S (189, 195, 197–201), respectively.

Although there are some differences in the reported size of the native enzymes isolated from brain (Table VI), the major discrepancy is the 165 kDa [$s_{20,w}$ = 7.3S] determined with enzyme immediately after purification (200). With longer times after purification, this purified kinase aggregated to 545 kDa [$s_{20,w}$ = 16.5S]. Differences in native mass become more obvious when the kinases from different animal species and tissues are compared. The possible reasons for these differences is discussed later. Obviously, the native mass of the brain kinases depends upon the relative ratio of the 50-kDa (α) subunit to the 60-kDa (β) subunit. For example, the difference in mass between the rat-cerebellar kinase ($1\alpha : 4\beta/\beta'$) and the forebrain kinase ($3\alpha : 1\beta/\beta'$) is 615 versus 550 kDa (Table VI).

A calmodulin-dependent protein kinase from rat brain that is composed of a single protein subunit of 50–55 kDa has been purified (195, 199). This 50-kDa

TABLE VI

Structural Properties of Multifunctional Calmodulin-Dependent Protein Kinases from Rat Brain

Tissue	Subcellular fraction	Subunit mass[a] (kDa)	Relative subunit composition[b]	Native mass (kDa)	Reference
Whole brain	Soluble	50 (α); 60/58 (β/β')	3α:1β/β'	650	(189)
Whole brain	Soluble	51; 60 doublet	51 > 60	460; 560	(198)
Whole brain	Soluble	52 doublet; 63 doublet	52 > 63	600	(196)
Whole brain	Soluble	51; 60	51 > 60	560	(197)
Whole brain	Particulate, "cytoskeletal" fraction	50 (α); 60 (β)	6α:1β	165; 545	(200)
Whole brain	Soluble	49	—	540; 640	(195)
Cerebral cortex	Soluble	55		540	(199)
Forebrain	Soluble and particulate	50 (α); 60/58 (β/β')	3α:1β/β'	550	(201)
Cerebellum	Soluble and particulate	50 (α); 60/58 (β/β')	1α:4β/β'	615	(201)

[a] The / indicates mass of subunits where the smaller subunit may be due to proteolysis or other modifications.
[b] The > indicates which subunit of the designated mass is more abundant.

subunit self-associates into an oligomeric kinase complex that has a mass of approximately 540 to 640 kDa. Bennett *et al.* (*189*) point out that an acid precipitation step (pH 6.1) used by Fukunaga *et al.* (*195*) causes precipitation of the 60-kDa (β) subunit of the brain calmodulin-dependent kinase. However, Yamauchi and Fujisawa (*199*) did not use an acid precipitation step and, therefore, the absence of a 60-kDa subunit cannot be readily explained.

The purification of an active kinase that lacks the 60-kDa subunit raises the question of the role of this subunit in relation to kinase activity. The antibody crossreactivity of the 50- and 60-kDa subunits with a monoclonal to the protein kinase provides evidence that the 60-kDa polypeptide is a subunit of the kinase (*201*). Bennett *et al.* (*189*) found that both the 50- and 60-kDa subunits coprecipitated with a monoclonal antibody to the partially purified kinase. Supporting evidence also comes from the partial peptide maps which show both subunits to be structurally related (*187, 196, 198*). The oligomeric structure has not been clarified in terms of whether the kinase exists in multiple heteropolymers of 50- and 60-kDa subunits or whether homopolymers exist in tissues where both subunits are found. There have been no studies involving cross-linking or nondenaturing polyacrylamide gels of the holoenzymes whereby the subunit compositions could be analyzed.

The multifunctional calmodulin-dependent protein kinases purified from brain exist as both soluble and particulate enzyme complexes. It appears that the 50- and 60-kDa subunits are similar, if not identical, regardless of subcellular localization (Table VI). Kennedy *et al.* (*187*) extracted the kinase from the particulate membrane fraction by reducing the ionic strength (keeping the solution isoosmotic) with recovery of 70% of the kinase activity. The kinases purified from the soluble and particulate fractions had identical specificities for substrates, effect of pH on kinase activity, and Ca^{2+} concentrations required for catalysis. McGuinness *et al.* (*201*) purified both particulate and soluble calmodulin-dependent protein kinases from rat forebrain and cerebellum. The forebrain enzyme was more highly concentrated in the soluble fraction (50% of total activity), and another 24–45% was extracted from the particulate fraction. In contrast, 20% of the total cerebellar kinase was present in the soluble fraction, and only an additional 10–15% of the activity could be extracted from the particulate fraction. However, the purified kinases were similar in biochemical properties except in the relative ratios of 50- to 60-kDa subunits (Table VI).

Several laboratories have identified the major postsynaptic density protein from rat brain as the 50-kDa subunit of the multifunctional calmodulin-dependent protein kinase (*188, 202, 203*). This membrane-bound protein has been determined to be similar to the 50-kDa subunit of the kinase on the basis of (*a*) [^{125}I]calmodulin binding; (*b*) peptide maps of ^{32}P-labeled and ^{125}I-labeled 50-kDa proteins; and (*c*) antibody crossreactivity. Goldenring *et al.* (*202*) suggested on the basis of two-dimensional tryptic peptide maps, that the membrane-bound

50- and 60-kDa subunits of the major postsynaptic density protein may have slightly different primary structure than the soluble 50- and 60-kDa kinase subunits. A calmodulin-dependent protein kinase isolated from rat brain neuronal nuclear matrix has 50- and 60-kDa autophosphorylated subunits which appear to be similar to the polypeptides of the postsynaptic densities (204). However, one-dimensional phosphopeptide maps and two-dimensional tryptic maps demonstrated that the membrane-bound 50-kDa major postsynaptic density protein was very similar, but not identical to the 50-kDa subunit of the calmodulin-dependent protein kinase from the nuclear matrix.

A calmodulin-dependent protein kinase has been purified from a postsynaptosomal cytoskeletal preparation from rat brain that is distinct from the postsynaptic density preparation (200). This cytoskeletal preparation is enriched 2- to 3-fold with kinase compared to the typical postsynaptic density preparation. In contrast to the particulate enzymes purified by Kennedy et al. (187), the kinase associated with the cytoskeleton required 8 M urea for extraction. The purified kinase has a 50-kDa subunit comprising most of the protein with a minor 60-kDa subunit. Both subunits bound [^{125}I]calmodulin and were autophosphorylated. The only apparent difference between this kinase and the more easily extractable kinases was the different relative ratio of subunits, $6\alpha : 1\beta$ (Table VI). It will be essential to determine the primary structure of these calmodulin-dependent kinases to establish whether there are significant differences in primary structure that are important for subcellular distribution.

There are specific forms of the multifunctional calmodulin-dependent protein kinase associated with particular brain regions (201). The α and β subunits appear to be identical in the different forms, but the relative ratio of $\alpha : \beta/\beta'$ subunits are distinct for the forebrain ($3\alpha : 1\beta/\beta'$) and cerebellum ($1\alpha : 4\beta'$), respectively (Table VI). These area-specific ratios exist shortly after tissue dissection and do not seem to arise as a result of the purification procedure. The purified kinases have different native masses, Stokes radii, and sedimentation coefficients, indicative of the relative ratios of $\alpha : \beta$ subunits. There were no differences in the ratio of α to β on either the ascending or descending sides of the protein peaks after gel filtration, thus arguing against homopolymers consisting of α or β subunits, respectively.

In summary, there may be subcellular and region-specific forms of the brain multifunctional calmodulin-dependent protein kinase that exist with different ratios of protein subunits. The biological significance of these differences may be related to different binding affinities for membranes, the cytoskeleton, or kinase substrates.

2. Other Tissues

The similarities in subunit composition and substrate specificity between the brain multifunctional calmodulin-dependent protein kinase and a number of other

TABLE VII

<small>STRUCTURAL PROPERTIES OF MULTIFUNCTIONAL CALMODULIN-DEPENDENT PROTEIN KINASES FROM DIFFERENT TISSUES</small>

Tissue and animal species	Subunit mass[a] (kDa)	Relative composition[b]	Native mass (kDa)	Reference
Brain, rat	50–55 60/58	Variable	165–650	(See Table VI)
Heart, steer[c]	55–57 73–75	55 > 73	100; 900	(192,207)
Skeletal muscle, rabbit	54 58/59	4(58);1(54)	696	(186)
Liver, rabbit	51/53 50/53	—	275; 500 300	(183) (184)
Rat[d]	56/57	—	—	(185)
Pancreas, rat[c]	51	—	300; 600	(209)
Erythrocyte, turkey[c]	50/54 58	58 > 50	—	(191)
Invertebrate neuronal, Aplysia[e]	50–51	—	—	(194)
Electric organ, Torpedo[c]	54 62	—	—	(190)

[a] The / indicates mass of subunit where the smaller subunit may be due to proteolysis or other modifications.

[b] The > indicates which subunit of the designated mass is more abundant.

[c] Not homogeneous.

[d] Comigrated with rabbit enzyme.

[e] Identified by monoclonal antibodies.

calmodulin-dependent protein kinases suggest that all of these enzymes may represent a class of homologous, multifunctional calmodulin-dependent protein kinases. In fact, several groups of investigators report antibody crossreactivity between the rat brain multifunctional calmodulin-dependent protein kinase and kinases isolated from other animal species and tissues. Antibodies to the rat brain enzyme have been shown to cross-react with the kinase from rabbit skeletal muscle (205, 206) and Aplysia californica (194). Very similar peptide maps were observed with the Aplysia 50-kDa polypeptide compared to the 50-kDa subunit from rat brain kinase (194). In addition, the kinases identified in other tissues bind [125I]calmodulin and are autophosphorylated. Table VII summarizes some of the structural properties of the multifunctional calmodulin-dependent protein kinase from a number of tissues and animal species.

In general, multifunctional calmodulin-dependent protein kinases from a wide variety of animal species (mammalian, avian, and invertebrate) are composed of either two subunits of approximately 50–55 and 60–75 kDa or the single lower-molecular-mass subunit, 50–55 kDa. Again it is necessary to consider whether differences in the purification procedures could account for the variable subunit compositions. In the case of kinases from rabbit and rat liver, rat pancreas, and *Aplysia* neuronal tissue, only the rabbit liver kinase purified by Ahmad *et al.* (*183*) was subjected to an acidic pH treatment during purification. As discussed previously, lowering the pH to 6.1 may cause precipitation of the larger subunit (*189*). The rabbit and rat liver kinases purified by Payne *et al.* (*184*) and Schworer *et al.* (*185*) were also reported to have a single subunit, and no acidic pH treatment was used. Therefore, the relative ratio of the two subunits in multifunctional calmodulin-dependent protein kinases appears to be dependent upon the type of tissue as well as animal species.

The multifunctional calmodulin-dependent protein kinases purified from nonneuronal tissue are present predominantly in the soluble cell fraction (*183, 184, 186, 192, 193, 207*). A significant portion of the kinase activity in *Torpedo* electric organ (*190*) and neuronal tissue of *Aplysia* (*194*) appears in the particulate fraction as well as in the soluble fraction, similar to the distribution of the enzyme in rat brain.

The variability in reported sizes of the smaller and larger subunit of nonneuronal multifunctional calmodulin-dependent protein kinases may be due simply to differences in relative electrophoretic mobility under different buffer systems. For example, multifunctional calmodulin-dependent protein kinase purified from rat liver was reported to have a molecular mass of 56–57 kDa (*185*) and, therefore, appears to be different from the rabbit liver kinase, 51–53 kDa. However, these authors reported that the two forms from rat and rabbit liver comigrated during polyacrylamide gel electrophoresis in the presence of sodium dodecyl sulfate using the discontinuous Laemmli (*208*) buffer system. Similarly, Woodgett *et al.* (*206*) reported that the multifunctional calmodulin-dependent protein kinase from rabbit skeletal muscle comigrates with the rat brain kinase, but that the apparent mass of the major component for both is either 50 or 55 kDa, depending upon the electrophoretic buffer system. Therefore, these minor differences in reported masses may reflect variable electrophoretic mobilities of the subunits under different conditions.

The differences in reported native masses may also be due to variable experimental conditions. For example, different ionic strengths used during the determination of the Stokes radius or sedimentation coefficient may have significant effects. Kloepper and Landt (*207*) reported two different native masses for the multifunctional kinase purified from bovine heart, either 100 or 900 kDa, as determined by sucrose-density gradient centrifugation. The relative proportion of total activity in either peak, as well as enzyme stability, was dependent upon the NaCl concentration. However, because this kinase preparation was only partially pure, it was not possible to rule out interactions with other proteins that may cause the great size differences. In contrast, Woodgett *et al.* (*186*) found that the Stokes radius determined by gel filtration of the rabbit skeletal-muscle kinase was unaffected by ionic strength up to 0.5 M NaCl. The native enzyme was shown by transmission electron microscopy to be a 10 nm diameter hexameric ring (*186*). Because the estimated mass was 700 kDa, the authors concluded that the structure was a dodecamer, composed of two hexameric rings stacked one upon the other. The suggestion was made that the 300-kDa form present in other tissues may be composed of a single hexameric ring. The enzyme from rabbit liver had an apparent native mass of either 500 or 250 kDa as determined by gel filtration or sucrose-density gradient centrifugation under conditions of identical ionic strength (*183*). Similar results were obtained with multifunctional calmodu- lin-dependent protein kinase from rat pancreas (*209*). Further characterization of the hydrodynamic properties of multifunctional calmodulin-dependent protein kinases needs to be performed to clarify the native structures.

C. CA^{2+} AND CALMODULIN ACTIVATION

The multifunctional protein kinases are dependent upon both Ca^{2+} and cal- modulin for activity with a variety of protein substrates (*35, 182*), but much less is known about the specific, biochemical properties of activation by calmodulin as compared to similar information for myosin light chain kinases. The enzyme activity is inhibited in a competitive fashion by various calmodulin antagonists (*55, 194*). Measurement of [^{125}I]calmodulin binding via the gel overlay method shows that binding polypeptides comigrate with both 50- and 60-kDa protein subunits in purified kinase preparations and binding is dependent upon the pres- ence of Ca^{2+} (*196, 200, 201*). These experimental approaches demonstrate qualitatively the Ca^{2+}·calmodulin dependence of the multifunctional calmodu- lin-dependent protein kinase activity and identify calmodulin binding subunits.

The concentration of calmodulin required for half-maximal activation is, in general, greater than the concentrations of calmodulin required for half-maximal activation of myosin light chain kinases (Table VIII). Reported values deter- mined under a variety of experimental conditions vary from 0.012 to 0.4 μM (Table VIII). Some investigators have also measured the concentration of Ca^{2+} required for half-maximal activation of the multifunctional calmodulin-depen- dent protein kinases. These experiments have been performed at a particular calmodulin concentration and $K_{0.5}$ Ca^{2+} values range from 0.8 to 4.0 μM (Table VIII). Thus, the concentrations of Ca^{2+} and calmodulin required for half- maximal activation of the multifunctional calmodulin-dependent protein kinases are within a range that is biologically relevant. However, additional studies need to be performed with the purified multifunctional calmodulin-dependent protein kinases to determine the stoichiometry of Ca_n^{2+}·calmodulin binding to the holo- enzyme required for activation and how many divalent cation binding sites in calmodulin need to be occupied by Ca^{2+} for enzyme activation. It will be interesting to compare the energy coupling values for Ca^{2+}, calmodulin, sub-

TABLE VIII

Ca^{2+} and Calmodulin Activation of Multifunctional
Calmodulin-Dependent Protein Kinases

$K_{0.5}Ca^{2+}$ at (μM)	[Calmodulin] (μM)	$K_{0.5}$Calmodulin at (μM)	$[Ca^{2+}]_{free}$ (μM)	Reference
4.0	0.6	0.40	300	(187)
0.8	6.0	0.16	5	(257)
1.9	0.3	0.012	200	(195)
		0.10	500	(54)
1.6	0.1	0.010	120	(199)
		0.080	400	(183)

strates, and multifunctional calmodulin-dependent protein kinase with the reported values for myosin light chain kinase from rabbit skeletal muscle.

D. Catalytic Properties

The multifunctional calmodulin-dependent protein kinase is present in a number of different tissues as discussed previously. Of the tissues examined, the greatest kinase activities are found in brain, and possibly constitute 0.1–0.4% of total brain protein (187, 189). Regional variations of calmodulin-dependent protein kinase activity are found in brain (210–212) and the enzyme comprises approximately 0.4 and 0.2% of total protein in rat forebrain and cerebellum regions, respectively (201). In brain and nonneuronal tissue, the relative catalytic activities are brain (100%), spleen (25%), heart (12%), adrenal (9%), skeletal muscle (4%), and liver and kidney (<0.1%) (55).

The multifunctional calmodulin-dependent protein kinases have a characteristic broad substrate specificity that distinguishes them from other calmodulin-regulated protein kinases (183, 184, 186, 189, 191, 205, 213). Apparent kinetic constants, K_m, and V_{max}, are available for a few substrates; however, most studies report relative rates of phosphorylation for different protein substrates. Good substrates are phosphorylated at catalytic rates ranging from 0.2–2.9 μmol/min/mg enzyme (Table IX). However, variation occurs in the reported rates of calmodulin-dependent phosphorylation with individual substrates as exemplified with synapsin I, 0.4 to 2.9 μmol/min/mg (187, 189, 191), and skeletal-muscle glycogen synthase, 0.2 to 2.2 μmol/min/mg (183, 184, 186, 198). Relative rates of phosphorylation of moderate and poor substrates listed in Table IX are 10–40% and less than 10%, respectively, of catalytic rates with good substrates. Obviously, these comparisons of relative phosphorylation rates can be made only within a given study because of the marked variation in kinase-specific activity for even a single substrate. The substrate properties of additional

TABLE IX

PROTEINS TESTED AS SUBSTRATES FOR MULTIFUNCTIONAL CALMODULIN-DEPENDENT PROTEIN KINASES

Substrate	Reference
Good substrates	
Synapsin I	(*187,189,206*)
Skeletal-muscle glycogen synthase	(*183,184,205,206*)
Gizzard myosin P-light chain	(*183,198*)
Microtubule-associated protein 2	(*196,232,233*)
Tau factor	(*232*)
Calmodulin-sensitive cyclic nucleotide phosphodiesterase	(*235*)
Tryptophan hydroxylase	(*199*)
Myelin basic protein	(*196*)
Tubulin	(*196,232*)
Moderate substrates	
Calmodulin-dependent protein kinase (autophosphorylation)	(*200,207*)
Histone H3	(*187,191,200*)
Histone H1	(*205,206*)
Tyrosine hydroxylase	(*236*)
Cardiac phospholamban	(*207*)
Microtubule-associated protein 2	(*205,207*)
Myelin basic protein	(*195*)
Poor substrates	
Casein	(*186,189,198,207*)
Phosvitin	(*186,198,205*)
Phosphorylase *b*	(*189,198,205,207,236*)
Phosphofructokinase	(*186*)
Acetyl-CoA carboxylase	(*186,205*)
ATP citrate lyase	(*186,205*)
Myelin basic protein	(*207*)
Protamine	(*195*)
Histone H2B	(*192,217*)
Gizzard myosin	(*184*)
Skeletal-muscle myosin P-light chain	(*186,191,205*)
Cardiac-muscle myosin P-light chain	(*183,207*)
Regulatory subunit of cAMP-dependent protein kinase	(*207*)
G-substrate	(*191*)
Phosphorylase *b* kinase	(*186*)
Histone H1	(*196,200*)

proteins cannot be classified because the kinetic properties or relative rates of phosphorylation are not available. These potential substrates include ribosomal protein S6 (*193, 209*), phenylalanine hydroxylase (*214*), vimentin (*198*), γ-aminobutyric acid modulin (*198, 215*), liver pyruvate kinase (*216*), high-mobility group 17 (*217*), liver glycogen synthase (*218*), and fodrin (*219*). The total list of proteins phosphorylated by multifunctional calmodulin-dependent protein kinases is overwhelming if considered as biologically relevant phosphorylation reactions. However, it is unlikely that all of these proteins are phosphorylated *in vivo*, and considerable efforts will be necessary to establish the functional significance of each reaction in addition to demonstrating that it occurs *in vivo*. As discussed by Krebs and Beavo (*146*) many proteins phosphorylated by cAMP-dependent protein kinase *in vitro* have not been shown to be phosphorylated in living cells.

There are significant differences in optimal conditions for phosphorylation of different protein substrates by multifunctional calmodulin-dependent protein kinases. For example, different pH optima were reported for phosphorylation of casein (pH 6.6) (*198*), skeletal-muscle glycogen synthase (pH 7.6–7.8) (*184, 198*), and synapsin I (pH 8–9) (*187*). The presence of 210 mM NaCl or 28 mM NaKHPO$_4$ inhibits the rate of casein phosphorylation by 50% (*198*). Synapsin I phosphorylation is inhibited 60% in the presence of 150 mM NaCl (*190*). The effects of salt inhibition are not cation-specific and are observed with Na$^+$, K$^+$, and Tris$^+$ (*190*). Thus, the comparison of phosphorylation rates with different proteins under identical reaction conditions may not necessarily reflect the maximal relative rates of phosphorylation *in vitro* due to different optimal conditions for different substrates.

Different phosphorylation rates with a particular protein substrate may also result from the method of protein substrate preparation or storage. Commercially obtained skeletal-muscle glycogen synthase is a poor substrate for multifunctional calmodulin-dependent protein kinase (*189, 201, 206*). In contrast, significant rates of calmodulin-dependent phosphorylation were obtained with skeletal-muscle glycogen synthase prepared by investigators (*183, 186, 220*). Kloepper and Landt (*207*) found that dephosphorylation of commercially obtained casein made it a better substrate for the multifunctional calmodulin-dependent protein kinase. Mixed histones at high concentrations of 2 mg/ml were phosphorylated at slow rates by partially activated kinase due to calmodulin binding to the basic histone proteins (*205*). However, the multifunctional calmodulin-dependent protein kinase phosphorylated lower concentrations of mixed histones at moderate rates (*205*).

The multifunctional calmodulin-dependent protein kinases have maximal catalytic rates with MgATP^{2-} as a metal ion–nucleotide substrate (*190, 195, 198*). Concentrations of free Mg^{2+} required for maximal protein kinase activity are influenced by the protein substrate (*195*). Synapsin I is phosphorylated at maximal rates in the presence of 1.9–4.9 mM free Mg^{2+}, but activity is diminished by higher Mg^{2+} concentrations (*190*). In contrast, approximately 9.9 mM free Mg^{2+} is required for maximal catalytic activity with histone H3 (*190*). Multifunctional calmodulin-dependent protein kinases can use Mn^{2+} as a metal cation cofactor (*190, 195, 198*). Half-maximal activation is obtained with lower concentrations of Mn^{2+} than Mg^{2+} (*195, 198*); however, substitution of Mn^{2+} for Mg^{2+} decreases the maximal rates of phosphorylation with either smooth-muscle myosin P-light chain (*195*) or synapsin I (*190*). Kuret and Schulman (*198*) reported Mg^{2+} and Mn^{2+} gave similar maximal rates of phosphorylation with casein. Thus, the effect of metal cations on multifunctional calmodulin-dependent protein kinase activity *in vitro* may be determined, in part, by the particular protein substrate.

Other nucleotides were examined for effects on catalytic activity. In the presence of $MgGTP^{2-}$, phosphorylation rates were 10% (*190*) and 41% of the rates obtained with $MgATP^{2-}$ with an apparent K_m value for $MgGTP^{2-}$ of 0.94 mM (*186*). $MgADP^-$ is a poor competitive inhibitor in respect to $MgATP^{2-}$, with an apparent inhibition constant of 0.47 mM in the presence of 0.1 mM $MgATP^{2-}$ (*198*).

Apparent K_m values for $MgATP^{2-}$ with brain multifunctional calmodulin-dependent protein kinases range from 3 to 114 μM with several different protein substrates (Table X). Kinases from nonneuronal tissues had apparent K_m values for $MgATP^{2-}$ in a more narrow range (5–45 μM) with different protein substrates. These variations may be related to different values obtained in different laboratories. Palfrey *et al.* (*190*) found similar values with kinases from turkey erythrocytes and *Torpedo* electric organ. In general, the apparent K_m values for ATP are lower with multifunctional calmodulin-dependent protein kinases in comparison with other calmodulin-dependent protein kinases such as phosphorylase kinase and myosin light chain kinase. Myosin light chain kinases have apparent K_m values for $MgATP^{2-}$ of 200–400 μM for skeletal-(*75, 115*) and

TABLE X

APPARENT K_m VALUES FOR $MgATP^{2-}$
WITH MULTIFUNCTIONAL CALMODULIN-DEPENDENT PROTEIN KINASES

Protein substrate	Kinase tissue source	$MgATP^{2-}$ K_m (μM)	Reference
Synapsin I	Brain	3–4	(*187*)
Tubulin	Brain	7	(*196*)
Casein	Brain	22	(*198*)
Skeletal-muscle glycogen synthase	Brain	114	(*220*)
Tryptophan hydroxylase	Brain	60	(*199,234*)
Gizzard myosin P-light chain	Brain	109	(*195*)
Synapsin I	Turkey erythrocytes	17	(*191*)
	Torpedo electric organ	15–20	(*190*)
Skeletal-muscle glycogen synthase	Skeletal muscle	45	(*186*)
	Liver	27	(*183*)
Ribosomal protein S6	Pancreas	5	(*209*)

cardiac-(97, 98) muscle kinases, and $50-120$ μM for smooth-muscle (80, 85, 109) and nonmuscle kinases (51, 221). Phosphorylase kinase has apparent K_m values, $200-260$ μM, similar to values reported for myosin light chain kinases from skeletal- and cardiac-muscle (222, 223). Multifunctional calmodulin-dependent protein kinases have K_m values for MgATP^{2-} that are similar to values obtained for cyclic AMP-dependent protein kinase, $3-15$ μM (224, 225), cyclic GMP-dependent protein kinase, $10-46$ μM (226, 227), and mammalian casein kinases, $4-22$ μM ($228-231$).

Kinetic data have been obtained for brain multifunctional calmodulin-dependent protein kinases with the protein substrates synapsin I, skeletal-muscle glycogen synthase, smooth-muscle myosin P-light chain, microtubule-associated protein 2 (MAP-2), tau factor, tryptophan hydroxylase, and calmodulin-stimulated cyclic nucleotide phosphodiesterase (Table XI). Most protein substrates are phosphorylated stoichiometrically with the incorporation of $1-2$ mol phosphate/mol protein. However, Yamamoto et al. (232) report the incorporation of approximately 19 and 15 mol phosphate/mol of tau factor and MAP-2, respectively. Serine is the major phosphoamino acid in most substrates.

Substrates from neuronal tissues such as synapsin I (187, 198, 201), MAP-2 ($232-234$), tau factor (232), and calmodulin-stimulated cyclic nucleotide phosphodiesterase (235) have low apparent K_m values of 0.2 to 5 μM. Tryptophan hydroxylase has an incredibly low K_m value of 0.3 nM (199). Apparent V_{max}

TABLE XI

CATALYTIC PROPERTIES OF MULTIFUNCTIONAL CALMODULIN-DEPENDENT
PROTEIN KINASES FROM BRAIN

Protein substrate	Maximal extent phosphorylation (mol phosphate/ mol substrate)	Apparent K_m (μM)	Apparent V_{max} (μmol/min/mg)	Reference
Synapsin I	2	0.4	1.0–4.4[a]	($189,198,201$)
Skeletal-muscle glycogen synthase	1.1–2.0	3–32	0.24	($198,205$)
Gizzard myosin P-light chain	1	30–60	0.19	($195,198$)
		30	0.32[a]	(232)
Microtubule-associated protein-2	3–14.8	0.2–1.6	0.22–0.74	($198,206,232–234$)
Tau factor	19.1	0.7	2.43	(232)
Tryptophan hydroxylase	—	0.0003	0.37	(199)
Calmodulin-sensitive cyclic nucleotide phosphodiesterase	0.7	5	0.21	(235)

[a] Specific activity, not V_{max} values.

TABLE XII

RELATIVE RATES OF PHOSPHORYLATION WITH MULTIFUNCTIONAL CALMODULIN-DEPENDENT
PROTEIN KINASES FROM NEURONAL AND NONNEURONAL TISSUES

	Brain (%)	Skeletal muscle (%)	Pheochromocytoma (%)
Glycogen synthase	100	100	100
Synapsin I (0.1 mg/ml)	620–760	500–710	487
Smooth-muscle P-light chain (0.8 mg/ml)	81–125	65–115	58
Phosphorylase b (1.5 mg/ml)	< 0.1	< 0.1	< 0.1
Phosvitin (2 mg/ml)	< 0.1	< 0.1	< 0.1

Data are from Refs. *205, 206,* and *236.*

values obtained with the protein substrates listed in Table XI were 0.19–2.4 μmol/min/mg with the highest value, 4.4 μmol/min/mg, obtained with synapsin I (*201*). The highest calculated turnover number for a 600-kDa multifunctional calmodulin-dependent protein kinase is 44 sec^{-1}. In comparison with other calmodulin-regulated protein kinases, calculated turnover numbers are 35 sec^{-1} for gizzard myosin light chain kinase (*147*), and 453 sec^{-1} for skeletal-muscle phosphorylase kinase (*223*).

Multifunctional calmodulin-dependent protein kinases from neuronal and nonneuronal tissues have similar catalytic properties with different protein substrates (Tables XI and XII). Catalytic activities seem to be the greatest with synapsin I for calmodulin-dependent protein kinase from brain (*187*), skeletal muscle (*205*), pheochromocytoma tumor (*236*), turkey erythrocytes (*191*), and *Torpedo* electric organ (*190*). Smooth-muscle myosin P-light chain and skeletal-muscle glycogen synthase were phosphorylated at slower but similar rates (*205*). A number of proteins are not phosphorylated at significant rates by either neuronal or nonneuronal multifunctional calmodulin-dependent protein kinases including phosphorylase b, skeletal-muscle myosin P-light chain, phosvitin, and gizzard smooth-muscle myosin P-light chain bound to myosin heavy chain (*205, 206, 236*).

The catalytic properties of brain multifunctional calmodulin-dependent protein kinases may be compared with those of other protein kinases, particularly cAMP-dependent protein kinase which phosphorylates several protein substrates in common with the calmodulin-dependent protein kinases. Synapsin I is a substrate for both kinases (*55, 237, 238*). Cyclic AMP-dependent protein kinase phosphorylates stoichiometrically synapsin I with the incorporation of 1 mol phosphate per mol synapsin I on site 1, identified in a 10-kDa peptide after *S. aureus* V8 protease digestion (*239*). Two mol phosphate per mol of synapsin I are incorporated into a distinct peptide by multifunctional calmodulin-dependent protein kinase (*239*). Thus, the two enzymes phosphorylate distinct sites in synapsin I.

Skeletal-muscle glycogen synthase is phosphorylated by calmodulin-dependent protein kinase activity at sites 1b and 2 (*186*). Several protein kinases in addition to calmodulin-dependent protein kinase phosphorylate skeletal-muscle glycogen synthase at sites 1b and/or 2 including cAMP-dependent protein kinase, phosphorylase kinase, cGMP-dependent protein kinase, and glycogen synthase kinase 4 (*186, 240*). Phosphorylation of glycogen synthase sites 1b and 2 by calmodulin-dependent protein kinase occurs at different rates, with a more rapid rate of phosphorylation at site 2 than at site 1b (*186*). Cyclic AMP-dependent protein kinase and cGMP-dependent protein kinase also have different phosphorylation rates with the fastest to slowest rates occurring as site 1a > 2 > 1b, respectively (*240*).

As discussed previously, specificity of protein kinases for protein substrates is significantly influenced by basic residues in the primary structure around the phosphorylated site (*156, 159, 160, 227*). Amino acid sequences of sites phosphorylated by the multifunctional calmodulin-dependent protein kinase are shown in Table XIII. All phosphorylation sites shown in Table XIII have one or more basic amino acids located two or three residues toward the amino terminus from the phosphorylatable serine; this property is similar to sites phosphorylated by cAMP-dependent protein kinase. Synthetic peptides of skeletal-muscle glycogen synthase or smooth-muscle myosin P-light chains were used as sub-

TABLE XIII

AMINO ACID SEQUENCES SURROUNDING SITES PHOSPHORYLATED BY MULTIFUNCTIONAL CALMODULIN-DEPENDENT PROTEIN KINASES[a]

Source	Sequence	Reference
Skeletal-muscle glycogen synthase site 2	P L S R T L <u>S</u> V S S L	(258,259)
Skeletal-muscle glycogen synthase site 1b	S G G S K R S N <u>S</u> V D T S	(260,261)
Smooth-muscle P-light chain	K K R P Q R A T <u>S</u> N V F S	(252)
Phenylalanine hydroxylase	S R K L <u>S</u> B F G Z Z	(262)
Liver pyruvate kinase	L R R A <u>S</u> L	(263)

[a] For amino acid abbreviations, see Table V.

strates for multifunctional calmodulin-dependent protein kinase (241). The apparent K_m value for a synthetic peptide representing glycogen synthase phosphorylation site 2 in Table XIII was 2-fold greater than the K_m value for native glycogen synthase (241). However, the V_{max} value for the synthetic peptide was 10-fold greater than the native protein substrate. A synthetic peptide identical with smooth-muscle myosin P-light chain was phosphorylated with a lower K_m value and a higher V_{max} value than values obtained with native gizzard-muscle myosin P-light chain. The arginine, three residues toward the amino terminus from the phosphorylatable serine in glycogen synthase and myosin P-light chain peptides, is an important determinant for phosphorylation by multifunctional calmodulin-dependent protein kinase.

E. PHOSPHORYLATION

The protein subunits of multifunctional calmodulin-dependent protein kinases are phosphorylated in the presence of Ca^{2+} and calmodulin (183, 186, 189, 195, 198, 204). Multiple sites in each protein subunit appear to be phosphorylated. Bennett et al. (189) reported the incorporation of 2 mol phosphate per mol of 50-kDa subunit and 3 mol phosphate per mol 60-kDa subunit with brain kinases. Approximately 4 mol phosphate are incorporated into the subunits of liver calmodulin-dependent protein kinase (183). Similarly, both protein subunits of skeletal-muscle calmodulin-dependent protein kinase are autophosphorylated with an estimated 4–5 mol phosphate per mol protein (186). Serine is the major amino acid autophosphorylated (196).

Autophosphorylation of multifunctional calmodulin-dependent protein kinases decreases subunit mobility in sodium dodecyl sulfate polyacrylamide gels (189, 198, 201, 204). After phosphorylation in vitro, the 50- and 60-kDa subunits change to 53–54 and 64–68 kDa, respectively (189, 198, 201, 204). Similar effects of phosphorylation on protein mobility in sodium dodecyl sulfate polyacrylamide gels are observed with type II regulatory subunit of cAMP-dependent protein kinase (242–244) and glycogen synthase kinase 3 (245). Thus, the reported differences in protein subunit mobility of multifunctional calmodulin-dependent protein kinases could be due, in part, to different amounts of phosphate incorporated into the polypeptides. However, there have been no chemical measurements of protein-bound phosphate in purified nonphosphorylated multifunctional calmodulin-dependent protein kinases.

Attempts have been made to identify the catalytic subunit of multifunctional calmodulin-dependent protein kinase. As a result of autophosphorylation studies and binding studies with 8-N_3-ATP (198, 246), both subunits were concluded to contain active sites. These studies cannot disregard, however, possible nonspecific interactions that may occur between the nitrene in 8-N_3-ATP and the protein (247). Furthermore, the data do not discount the possibility of a single type of catalytic subunit which autophosphorylates and phosphorylates another type of subunit.

The role of autophosphorylation on the biochemical properties of multifunctional calmodulin-dependent protein kinases is not clearly defined. Shields *et al.* (*246*) reported 50–70% increased binding of [^{125}I]calmodulin, measured by the gel overlay method, to both the 50- and 60-kDa polypeptides after autophosphorylation in synaptic junction fractions. LeVine *et al.* (*248*) examined calmodulin binding to a neuronal cytoskeleton preparation enriched with the kinase. Similar to the observations of Shields *et al.* (*246*), [^{125}I]calmodulin binding was greater to phosphorylated calmodulin-dependent protein kinase from enriched cytoskeleton by the gel overlay method (*248*). However, [^{125}I]calmodulin binding affinity to phosphorylated calmodulin-dependent protein kinase in enriched cytoskeleton decreased 2-fold when measurements were performed in a buffered solution (*248*). Results from these studies show opposite effects of calmodulin-dependent protein kinase autophosphorylation on calmodulin-binding properties. These differences may be a result of calmodulin binding measured with denatured or native calmodulin-dependent protein kinase by gel overlay or in solution, respectively. Additional studies are required for a thorough characterization of calmodulin-binding properties of autophosphorylated calmodulin-dependent protein kinase, as well as an evaluation of the biological significance.

F. BIOLOGICAL SIGNIFICANCE OF MULTIFUNCTIONAL
    CALMODULIN-DEPENDENT PROTEIN KINASES

There have been many reviews written about the roles Ca^{2+}- and cAMP-dependent protein phosphorylation play in the regulation of neuronal function (*182, 249–251*). For example, changes in the intracellular concentrations of Ca^{2+} and cyclic nucleotides have been implicated in the regulation of neurotransmitter biosynthesis and release, RNA transcription and protein translation, synaptic vesicle fusion, cytoskeletal protein organization, and ion channels, and it has been proposed that most, if not all, of these processes are regulated by cAMP or Ca^{2+}-dependent protein kinases. The reader should refer to the many chapters on control of specific enzymes and biological processes in this volume for more details.

Ouimet *et al.* (*212*) demonstrated by immunocytochemistry that the brain multifunctional calmodulin-dependent protein kinase is associated *in situ* with subcellular structures that contain known *in vitro* substrates, such as synapsin I and the microtubule-associated protein, MAP-2. Biochemical measurements suggest that calmodulin-dependent protein kinase activity is associated with isolated synaptosomes, neuronal nuclei and a cytoskeletal protein fraction. Clearly, the multifunctional calmodulin-dependent protein kinase is located within the brain at sites accessible to protein substrates identified *in vitro*.

Calcium regulation of biological processes is not limited to neuronal tissue. Neural stimulation of muscle and hormonal stimulation of liver and other tissues are examples of external signals that may be mediated intracellularly by the second messenger, Ca^{2+}. The multifunctional calmodulin-dependent protein kinase has been found in a large number of vertebrate tissues. The large number of protein substrates phosphorylated by this kinase suggests that it is a general protein kinase that mediates Ca^{2+}-regulated processes in many cell types. However, the specific role that this kinase plays in any particular cell depends on a number of factors including subcellular localization of the kinase and substrate availability. It will be necessary to satisfy the criteria of Krebs and Beavo (146) to establish the biological importance of phosphorylation for each protein phosphorylated by a multifunctional calmodulin-dependent protein kinase.

ACKNOWLEDGMENTS

The authors wish to express their sincere appreciation to many colleagues who sent reprints and preprints for this chapter. In addition, the authors acknowledge the dedicated, superb assistance of Nancy Bryant and Kathy Perdue in preparing this manuscript. Support has been generously provided from the National Institutes of Health (HL23990, HL26043, HL06296) and the Muscular Dystrophy Association for research described in this chapter.

REFERENCES

1. Ringer, S. (1883). *Practitioner* **31,** 81.
2. Ringer, S. (1883). *J. Physiol. (London)* **4,** 29.
3. Ringer, S. (1883). *J. Physiol. (London)* **4,** 222.
4. Locke, F. S. (1894). *Zentralbl. Physiol.* **8,** 166.
5. Campbell, A. K. (1983). "Intracellular Calcium—Its Universal Role as Regulator." Wiley, New York.
6. Kretsinger, R. H. (1983). *In* "Calcium and Cell Function" (W. Y. Cheung, ed.), Vol. 4, p. 212. Academic Press, New York.
7. Bailey, K. (1942). *BJ* **36,** 121.
8. Needham, D. M. (1942). *BJ* **36,** 113.
9. Heilbrunn, L. V., and Wiercinski, F. J. (1947). *J. Cell. Comp. Physiol.* **29,** 15.
10. Ebashi, S. (1963). *Nature (London)* **200,** 1010.
11. Ebashi, S., Kodama, A., and Ebashi, F. (1968). *J. Biochem. (Tokyo)* **64,** 465.
12. Meyer, W. L., Fischer, E. H., and Krebs, E. G. (1964). *Biochemistry* **3,** 1033.
13. Ozawa, E., Hosoi, K., and Ebashi, S. (1967). *J. Biochem. (Tokyo)* **61,** 531.
14. Sutherland, E. W., Robinson, G. S., and Butcher, R. W. (1968). *Circulation* **37,** 279.
15. Rasmussen, H. (1970). *Science* **170,** 404.
16. Rasmussen, H. (1983). *In* "Calcium and Cell Function" (W. Y. Cheung, ed.), Vol. 4, p. 1. Academic Press, New York.
17. Rasmussen, H., and Barrett, P. Q. (1984). *Physiol. Rev.* **64,** 938.
18. Maruta, H., and Korn, E. D. (1977). *JBC* **252,** 8329.
19. Kuczmarski, E. R., and Spudich, J. A. (1980). *PNAS* **77,** 7292.

20. Hammer, J. A., III, Albanesi, J. P., and Korn, E. D. (1983). *JBC* **258**, 10168.
21. Klee, C. B., Crouch, T. H., and Richman, P. G. (1980). *Annu. Rev. Biochem.* **49**, 489.
22. Cheung, W. Y. (1980). *Science* **207**, 19.
23. Van Eldik, L. J., Zendegui, J. G., Marshak, D. R., and Watterson, D. M. (1982). *Int. Rev. Cytology* **77**, 1.
24. Klee, C. B., and Vanaman, T. C. (1982). *Adv. Protein Chem.* **35**, 213.
25. Watterson, D. M., Sharief, F., and Vanaman, T. C. (1980). *JBC* **255**, 962.
26. Teo, T. S., and Wang, J. H. (1973). *JBC* **248**, 5950.
27. Lin, Y. M., Liu, Y. P., and Cheung, W. Y. (1974). *JBC* **249**, 588.
28. Watterson, D. M., Harrelson, W. G., Jr., Keller, P. M., Sharief, F., and Vanaman, T. C. (1976). *JBC* **251**, 4501.
29. Klee, C. B. (1977). *Biochemistry* **16**, 1017.
30. Wolff, D. J., Poirier, P. G., Broström, C. O., and Broström, M. A. (1977). *JBC* **252**, 4108.
31. Crouch, T. H., and Klee, C. B. (1980). *Biochemistry* **19**, 3692.
32. Dedman, J. R., Potter, J. D., Jackson, R. L., Johnson, J. D., and Means, A. R. (1977). *JBC* **252**, 8415.
33. Haiech, J., Klee, C. B., and Demaille, J. G. (1981). *Biochemistry* **20**, 3890.
34. Potter, J. D., Strang-Brown, P., Walker, P. L., and Iida, S. (1983). "Methods in Enzymology," Vol. 102, p. 135.
35. Manalan, A. S., and Klee, C. B. (1984). *Adv. Cyclic Nucleotide Protein Phosphorylation Res.* **18**, 227.
36. Cheung, W. Y. (1970). *BBRC* **38**, 533.
37. Kakiuchi, S., Yamazaki, R., and Nakajima, H. (1970). *Proc. Jpn. Acad.* **46**, 587.
38. Teshima, Y., and Kakiuchi, S. (1974). *BBRC* **56**, 489.
39. Kakiuchi, S., and Yamazaki, R. (1970). *BBRC* **41**, 1104.
40. Broström, C. O., Huang, Y.-C., Breckenridge, B. McL., and Wolff, D. J. (1975). *PNAS* **72**, 64.
41. Cohen, P., Burchell, A., Foulkes, J. G., Cohen, P. T. W., Vanaman, T. C., and Nairn, A. C. (1978). *FEBS Lett.* **92**, 287.
42. Epel, D., Patton, C., Wallace, R. W., and Cheung, W. Y. (1981). *Cell* **23**, 543.
43. Anderson, J. M., Charbonneau, H., Jones, H. P., McCann, R. O., and Cormier, M. J. (1980). *Biochemistry* **19**, 3113.
44. Nagao, S., Suzuki, Y., Watanabe, Y., and Nozawa, Y. (1979). *BBRC* **90**, 261.
45. Gopinath, R. M., and Vincenzi, F. F. (1977). *BBRC* **77**, 1203.
46. Jarrett, H. W., and Penniston, J. T. (1977). *BBRC* **77**, 1210.
47. Blum, J. J., Hayes, A., Jamieson, G. A., and Vanaman, T. C. (1980). *J. Cell Biol.* **87**, 386.
48. Tuana, B. S., and MacLennan, D. H. (1984). *JBC* **259**, 6979.
49. Wong, P. Y.-K., and Cheung, W. Y. (1979). *BBRC* **90**, 473.
50. Dabrowska, R., and Hartshorne, D. J. (1978). *BBRC* **85**, 1352.
51. Hathaway, D. R., and Adelstein, R. S. (1979). *PNAS* **76**, 1653.
52. Yazawa, M., Kuwayama, H., and Yagi, K. (1978). *J. Biochem (Tokyo)* **84**, 1253.
53. DeLorenzo, R. J., Freedman, S. D., Yohe, W. B., and Maurer, S. C. (1979). *PNAS* **76**, 1838.
54. Payne, M. E., and Soderling, T. R. (1980). *JBC* **255**, 8054.
55. Kennedy, M. B., and Greengard, P. (1981). *PNAS* **78**, 1293.
56. Amphlett, G. W., Vanaman, T. C., and Perry, S. V. (1976). *FEBS Lett.* **72**, 163.
57. Dedman, J. R., Potter, J. D., and Means, A. R. (1977). *JBC* **252**, 2437.
58. Korn, E. D. (1978). *PNAS* **75**, 588.
59. Stossel, T. P. (1978). *Annu. Rev. Med.* **29**, 427.
60. Taylor, E. W. (1979). *CRC Crit. Rev. Biochem.* **6**, 103.
61. Weingrad, S. (1979). In "*Handbook of Physiology*" (R. M. Berne, N. Sperelakis, and S. R.

Geiger, eds.), 2nd ed., Sect. 2, Vol. I, p. 393. Am. Physiol. Soc., Bethesda, Maryland.
62. Harrington, W. F., and Rodgers, M. E. (1984). *Annu. Rev. Biochem.* **53**, 35.
63. Leavis, P. C., and Gergely, J. (1984). *CRC Crit. Rev. Biochem.* **16**, 235.
64. Eisenberg, E., and Hill, T. L. (1985). *Science* **227**, 999.
65. Kamm, K. E., and Stull, J. T. (1985). *Annu. Rev. Pharmacol. Toxicol.* **25**, 593.
66. Means, A. R., and Dedman, J. R. (1980). *Nature (London)* **285**, 73.
67. Perrie, W. T., Smillie, L. B., and Perry, S. V. (1973). *BJ* **135**, 151.
68. England, P. J., Stull, J. T., and Krebs, E. G. (1972). *JBC* **247**, 5275.
69. Stull, J. T., Broström, C. O., and Krebs, E. G. (1972). *JBC* **247**, 5272.
70. Stull, J. T. (1980). *Adv. Cyclic Nucleotide Res.* **12**, 39.
71. Frearson, N., and Perry, S. V. (1975). *BJ* **151**, 99.
72. Frearson, N., Focant, B. W. W., and Perry, S. V. (1976). *FEBS Lett.* **63**, 27.
73. Adelstein, R. S., and Eisenberg, E. (1980). *Annu. Rev. Biochem.* **49**, 921.
74. Cooke, R., and Stull, J. T. (1981). *In* "Cell and Muscle Motility" (R. M. Dowben and J. W. Shay, eds.), Vol. 1, p. 99. Plenum, New York.
75. Pires, E. M. V., and Perry, S. V. (1977). *BJ* **167**, 137.
76. Yagi, K., Yazawa, M., Kakiuchi, S., Ohshima, M., and Uenishi, K. (1978). *JBC* **253**, 1338.
77. Daniel, J. L., and Adelstein, R. S. (1976). *Biochemistry* **15**, 2370.
78. Hathaway, D. R., Adelstein, R. S., and Klee, C. B. (1981). *JBC* **256**, 8183.
79. Dabrowska, R., Aromatorio, D., Sherry, J. M. F., and Hartshorne, D. J. (1977). *BBRC* **78**, 1263.
80. Adelstein, R. S., and Klee, C. B. (1981). *JBC* **256**, 7501.
81. Walsh, M. P., Hinkins, S., Dabrowska, R., and Hartshorne, D. J. (1983). "Methods in Enzymology," Vol. 99, p. 279.
82. Uchiwa, H., Kato, T., Onishi, H., Isobe, T., Okuyama, T., and Watanabe, S. (1982). *J. Biochem. (Tokyo)* **91**, 273.
83. Ngai, P. K., Carruthers, C. A., and Walsh, M. P. (1984). *BJ* **218**, 863.
84. Adachi, K., Carruthers, C. A., and Walsh, M. P. (1983). *BBRC* **115**, 855.
85. Walsh, M. P., Hinkins, S., Flink, I. L., and Hartshorne, D. J. (1982). *Biochemistry* **21**, 6890.
86. Miller, J. R., Silver, P. J., and Stull, J. T. (1983). *Mol. Pharmacol.* **24**, 235.
87. Nishikawa, M., de Lanerolle, P., Lincoln, T. M., and Adelstein, R. S. (1984). *JBC* **259**, 8429.
88. Vallet, B., Molla, A., and Demaille, J. G. (1981). *BBA* **674**, 256.
89. Higashi, K., Fukunaga, K., Matsui, K., Maeyama, M., and Miyamoto, E. (1983). *BBA* **747**, 232.
90. Nunnally, M. H., and Stull, J. T. (1984). *JBC* **259**, 1776.
91. Yazawa, M., and Yagi, K. (1978). *J. Biochem. (Tokyo)* **84**, 1259.
92. Blumenthal, D. K., and Stull, J. T. (1980). *Biochemistry* **19**, 5608.
93. Crouch, T. H., Holroyde, M. J., Collins, J. H., Solaro, R. J., and Potter, J. D. (1981). *Biochemistry* **20**, 6318.
94. Edelman, A. M., and Krebs, E. G. (1982). *FEBS Lett.* **138**, 293.
95. Mayr, G. W., and Heilmeyer, L. M. G., Jr. (1983). *Biochemistry* **22**, 4316.
96. Nagamoto, H., and Yagi, K. (1984). *J. Biochem. (Tokyo)* **95**, 1119.
97. Walsh, M. P., Vallet, B., Autric, F., and Demaille, J. G. (1979). *JBC* **254**, 12136.
98. Wolf, H., and Hofmann, F. (1980). *PNAS* **77**, 5852.
99. Sellers, J. R., and Harvey, E. V. (1984). *Biochemistry* **23**, 5821.
100. Walsh, M. P., and Guilleux, J. C. (1981). *Adv. Cyclic Nucleotide Res.* **14**, 375.
101. Guerriero, V., Jr., Rowley, D. R., and Means, A. R. (1981). *Cell* **27**, 449.
102. Nunnally, M. H., Rybicki, S. B., and Stull, J. T. (1985). *JBC* **260**, 1020.
103. Cleveland, D. W., Fischer, S. G., Kirschner, M. W., and Laemmli, U. K. (1977). *JBC* **252**, 1102.

104. Srivastava, S., and Hartshorne, D. J. (1983). *BBRC* **110**, 701.
105. Blumenthal, D. K., Takio, K., Edelman, A. M., Charbonneau, H., Titani, K., Walsh, K. A., and Krebs, E. G. (1985). *PNAS* **82**, 3187.
106. Cox, J. A., Comte, M., Fitton, J. E., and DeGrado, W. F. (1985). *JBC* **260**, 2527.
107. Malencik, D. A., and Anderson, S. R. (1984). *Biochemistry* **23**, 2420.
108. Adelstein, R. S., Conti, M. A., Hathaway, D. R., and Klee, C. B. (1978). *JBC* **253**, 8347.
109. Walsh, M. P., Dabrowska, R., Hinkins, S., and Hartshorne, D. J. (1982). *Biochemistry* **21**, 1919.
110. Foyt, H. L., Guerriero, V., Jr., and Means, A. R. (1985). *JBC* **260**, 7765.
111. Scordilis, S. P., and Adelstein, R. S. (1978). *JBC* **253**, 9041.
112. Bremel, R. D., and Shaw, M. E. (1978). *FEBS Lett.* **88**, 242.
113. Walsh, M. P., Cavadore, J. C., Vallet, B., and Demaille, J. G. (1980). *Can. J. Biochem.* **58**, 299.
114. Tanaka, T., Naka, M., and Hidaka, H. (1980). *BBRC* **92**, 313.
115. Nairn, A. C., and Perry, S. V. (1979). *BJ* **179**, 89.
116. Blumenthal, D. K., and Stull, J. T. (1982). *Biochemistry* **21**. 2386.
117. Bhalla, R. C., Sharma, R. V., and Gupta, R. C. (1982). *BJ* **203**, 583.
118. Malencik, D. A., Anderson, S. R., Bohnert, J. L., and Shalitin, Y. (1982). *Biochemistry* **21**, 4031.
119. Johnson, J. D., Holroyde, M. J., Crouch, T. H., Solaro, R. J., and Potter, J. D. (1981). *JBC* **256**, 12194.
120. Nishikori, K., Weisbrodt, N. W., Sherwood, O. D., and Sanborn, B. M. (1983). *JBC* **258**, 2468.
121. Huang, C. Y., Chau, V., Chock, P. B., Wang, J. H., and Sharma, R. K. (1981). *PNAS* **78**, 871.
122. Stull, J. T., Sanford, C. F., Manning, D. R., Blumenthal, D. K., and High, C. W. (1981). *In* "Protein Phosphorylation" (O. M. Rosen and E. G. Krebs, eds.), Vol. 8, p. 823. Cold Spring Harbor Press, Cold Spring Harbor, New York.
123. Cox, J. A., Malnoe, A., and Stein, E. A. (1981). *JBC* **256**, 3218.
124. Cox, J. A., Comte, M., Malnoe, A., Burger, D., and Stein, E. A. (1984). *In* "Metal Ions in Biological Systems" (H. Sigel, ed.), Vol. 17, p. 215. Dekker, New York.
125. Cox, J. A., Comte, M., and Stein, E. A. (1982). *PNAS* **79**, 4265.
126. Malnoe, A., Cox, J. A., and Stein, E. A. (1982). *BBA* **714**, 84.
127. Burger, D., Stein, E. A., and Cox, J. A. (1983). *JBC* **258**, 14733.
128. Keller, C. H., Olwin, B. B., LaPorte, D. C., and Storm, D. R. (1982). *Biochemistry* **21**, 156.
129. Olwin, B. B., Edelman, A. M., Krebs, E. G., and Storm, D. R. (1984). *JBC* **259**, 10949.
130. Cox, J. A. (1984). *FP* **43**, 3000.
131. Chau, V., Huang, C. Y., Chock, P. B., Wang, J. H., and Sharma, R. K. (1982). *In* "Calmodulin and Intracellular Ca^{++} Receptors" (S. Kakiuchi, H. Hidaka, and A. R. Means, eds.), p. 199. Plenum, New York.
132. Stull, J. T., Nunnally, M. H., Moore, R. L., and Blumenthal, D. K. (1985). *Adv. Enzyme Regul.* **23**, 123.
133. Moore, R. L., and Stull, J. T. (1984). *Am. J. Physiol.* **247**, C462.
134. LaPorte, D. C., Wierman, B. M., and Storm, D. R. (1980). *Biochemistry* **19**, 3814.
135. Tanaka, T., and Hidaka, H. (1980). *JBC* **255**, 11078.
136. Weiss, B., and Levin, R. M. (1978). *Adv. Cyclic Nucleotide Res.* **9**, 285.
137. Asano, M., and Hidaka, H. (1984). *In* "Calcium and Cell Function" (W. Y. Cheung, ed.), Vol. 5, p. 123. Academic Press, New York.
138. Zimmer, M., Gobel, C., and Hofmann, F. (1984). *FEBS Lett.* **139**, 295.
139. Maulet, Y., and Cox, J. A. (1983). *Biochemistry* **22**, 5680.

140. Barnette, M. S., Daly, R., and Weiss, B. (1983). *Biochem. Pharmacol.* **32**, 2929.
141. Newton, D. L., Olderwurtel, M. D., Krinks, M. H., Shiloach, J., and Klee, C. B. (1984). *JBC* **259**, 4419.
142. Ni, W.-C., and Klee, C. B. (1985). *JBC* **260**, 6974.
143. Newton, D. L., and Klee, C. B. (1984). *FEBS Lett.* **165**, 269.
144. Hartshorne, D. J., and Mrwa, U. (1980). *In* "Calcium Binding Proteins: Structure and Function" (F. L. Siegel, E. Carafoli, R. H. Kretsinger, D. H. MacLennan, and R. H. Wasserman, eds.), p. 255. Elsevier/North-Holland, Amsterdam.
145. Tawata, M., Kobayashi, R., and Field, J. B. (1983). *Endocrinology (Baltimore)* **112**, 701.
146. Krebs, E. G., and Beavo, J. A. (1979). *Annu. Rev. Biochem.* **48**, 923.
147. Persechini, A., and Hartshorne, D. J. (1983). *Biochemistry* **22**, 470.
148. Sellers, J. R., Chock, P. B., and Adelstein, R. S. (1983). *JBC* **258**, 14181.
149. Ikebe, M., Ogihara, S., and Tonomura, Y. (1982). *J. Biochem. (Tokyo)* **91**, 1809.
150. Sobieszek, A. (1985). *Biochemistry* **24**, 1266.
151. Scordilis, S. P., Anderson, J. L., Pollack, R., and Adelstein, R. S. (1977). *J. Cell Biol.* **74**, 940.
152. Bourguignon, L. Y. W., Nagpal, M. L., Balazovich, K., Guerriero, V., and Means, A. R. (1982). *J. Cell Biol.* **95**, 793.
153. Singh, T. J., Akatsuka, A., and Huang, K. P. (1983). *FEBS Lett.* **159**, 217.
154. Persechini, A., and Stull, J. T. (1984). *Biochemistry* **23**, 4144.
155. Kemp, B. E., Pearson, R. B., and House, C. (1982). *JBC* **257**, 13349.
156. Kemp, B. E., Pearson, R. B., and House, C. (1983). *PNAS* **80**, 7471.
157. Kemp, B. E., and Pearson, R. B. (1985). *JBC* **260**, 3355.
158. Michnoff, C. H., Stull, J. T., and Kemp, B. E. (1986). *JBC* (in press).
159. Chan, K. F. J., Hurst, M. O., and Graves, D. J. (1982). *JBC* **257**, 3655.
159a. Gauss, A., Mayr, G. W., and Heilmeyer, L. M. G. (1985). *EJB* **153**, 327.
160. Kemp, B. E., Bylund, D. B., Huang, T.-S., and Krebs, E. G. (1975). *PNAS* **72**, 3448.
161. Glass, D. B., and Krebs, E. G. (1982). *JBC* **257**, 1196.
162. Mrwa, U., and Hartshorne, D. J. (1980). *FP* **39**, 1564.
163. Foyt, H. L., and Means, A. R. (1985). *J. Cyclic Nucleotide Protein Phosphorylation Res.* **10**, 143.
164. Conti, M. A., and Adelstein, R. S. (1980). *FP* **39**, 1569.
165. Hathaway, D. R., Eaton, C. R., and Adelstein, R. S. (1981). *Nature (London)* **291**, 252.
166. Conti, M. A., and Adelstein, R. S. (1981). *JBC* **256**, 3178.
167. Adelstein, R. S., Sellers, J. R., Conti, M. A., Pato, M. D., and de Lanerolle, P. (1982). *FP* **41**, 2873.
168. de Lanerolle, P., Nishikawa, M., Yost, D. A., and Adelstein, R. S. (1984). *Science* **223**, 1415.
169. Silver, P. J., and Stull, J. T. (1982). *JBC* **257**, 6145.
170. Gerthoffer, W. T., Trevethick, M. A., and Murphy, R. A. (1984). *Circ. Res.* **54**, 83.
171. Jones, A. W., Bylund, D. B., and Forte, L. R. (1984). *Am. J. Physiol.* **246**, H306.
172. Walsh, M. P., Bridenbaugh, R., Hartshorne, D. J., and Kerrick, W. G. L. (1982). *JBC* **257**, 5987.
173. Hoar, P. E., Kerrick, W. G. L., and Cassidy, P. S. (1979). *Science* **204**, 503.
174. Peterson, J. W., III (1982). *Biophys. J.* **37**, 453.
175. Aksoy, M. O., Mras, S., Kamm, K. E., and Murphy, R. A. (1983). *Am. J. Physiol.* **245**, C255.
176. Pemrick, S. M. (1980). *JBC* **255**, 8836.
177. Manning, D. R., and Stull, J. T. (1979). *BBRC* **90**, 164.
178. Manning, D. R., and Stull, J. T. (1982). *Am. J. Physiol.* **242**, C234.

179. Klug, G. A., Botterman, B. R., and Stull, J. T. (1982). *JBC* **257**, 4688.
180. Houston, M. E., Green, H. J., and Stull, J. T. (1985). *Pfluegers Arch.* **403**, 348.
181. Persechini, A., Stull, J. T., and Cooke, R. (1985). *JBC* **260**, 7951.
182. Nairn, A. C., Hemmings, H. C., Jr., and Greengard, P. (1985). *Annu. Rev. Biochem.* **54**, 931.
183. Ahmad, Z., DePaoli-Roach, A. A., and Roach, P. J. (1982). *JBC* **257**, 8348.
184. Payne, M. E., Schworer, C. M., and Soderling, T. R. (1983). *JBC* **258**, 2376.
185. Schworer, C. M., Payne, M. E., Williams, A. T., and Soderling, T. R. (1983). *ABB* **224**, 77.
186. Woodgett, J. R., Davison, M. T., and Cohen, P. (1983). *EJB* **136**, 481.
187. Kennedy, M. B., McGuinness, T., and Greengard, P. (1983). *J. Neurosci.* **3**, 818.
188. Kennedy, M. B., Bennett, M. K., and Erondu, N. E. (1983). *PNAS* **80**, 7357.
189. Bennett, M. K., Erondu, N. E., and Kennedy, M. B. (1983). *JBC* **258**, 12735.
190. Palfrey, H. C., Rothlein, J. E., and Greengard, P. (1983). *JBC* **258**, 9496.
191. Palfrey, H. C., Lai, Y., and Greengard, P. (1984). *In* "The Red Cell: Sixth Ann Arbor Conference" (G. J. Brewer, ed.), p. 291. A. R. Liss, Inc., New York.
192. Palfrey, H. C. (1984). *FP* **43**, 1466 (abstr.).
193. Gorelick, F. S., Cohn, J. A., Freedman, S. D., Delahunt, N. G., Gershoni, J. M., and Jamieson, J. D. (1983). *J. Cell Biol.* **97**, 1294.
194. DeRiemer, S. A., Kaczmarek, L. K., Lai, Y., McGuinness, T. L., and Greengard, P. (1984). *J. Neurosci.* **4**, 1618.
195. Fukunaga, K., Yamamoto, H., Matsui, K., Higashi, K., and Miyamoto, E. (1982). *J. Neurochem.* **39**, 1607.
196. Goldenring, J. R., Gonzalez, B., McGuire, J. S., Jr., and DeLorenzo, R. J. (1983). *JBC* **258**, 12632.
197. Schworer, C. M., McClure, R. W., and Soderling, T. R. (1983). *FP* **43**, 1466 (abstr.).
198. Kuret, J., and Schulman, H. (1984). *Biochemistry* **23**, 5495.
199. Yamauchi, T., and Fujisawa, H. (1983). *EJB* **132**, 15.
200. Sahyoun, N., LeVine, H., III, Bronson, D., Siegel-Greenstein, F., and Cuatrecasas, P. (1985). *JBC* **260**, 1230.
201. McGuinness, T. L., Lai, Y., and Greengard, P. (1985). *JBC* **260**, 1696.
202. Goldenring, J. R., Casanova, J. E., and DeLorenzo, R. J. (1984). *J. Neurochem.* **43**, 1669.
203. Kelly, P. T., McGuinness, T. L., and Greengard, P. (1984). *PNAS* **81**, 945.
204. Sahyoun, N., LeVine, H., III, and Cuatrecasas, P. (1984). *PNAS* **81**, 4311.
205. McGuinness, T. L., Lai, Y., Greengard, P., Woodgett, J. R., and Cohen, P. (1983). *FEBS Lett.* **163**, 329.
206. Woodgett, J. R., Cohen, P., Yamauchi, T., and Fujisawa, H. (1984). *FEBS Lett.* **170**, 49.
207. Kloepper, R. F., and Landt, M. (1984). *Cell Calcium* **5**, 351.
208. Laemmli, U. K. (1970). *Nature (London)* **227**, 680.
209. Cohn, J. A., Gorelick, F. S., Delahunt, N. G., and Jamieson, J. D. (1984). *FP* **43**, 1466 (abstr.).
210. Walaas, S. I., Nairn, A. C., and Greengard, P. (1983). *J. Neurosci.* **3**, 291.
211. Walaas, S. I., Nairn, A. C., and Greengard, P. (1983). *J. Neurosci.* **3**, 302.
212. Ouimet, C. C., McGuinness, T. L., and Greengard, P. (1984). *PNAS* **81**, 5604.
213. Schworer, C. M., and Soderling, T. R. (1983). *BBRC* **116**, 412.
214. Doskeland, A. P., Schworer, C. M., Doskeland, S. O., Chrisman, T. D., Soderling, T. R., Corbin, J. D., and Flatmark, T. (1984). *EJB* **145**, 31.
215. Wise, B. C., Guidotti, A., and Costa, E. (1984). *Adv. Cyclic Nucleotide Protein Phosphorylation Res.* **17**, 511.
216. Schworer, C. M., El-Maghrabi, M. R., Pilkis, S. J., and Soderling, T. M. (1985). *FP* **44**, 705 (abstr.).
217. Sahyoun, N., LeVine, H., III, Bronson, D., and Cuatrecasas, P. (1984). *JBC* **259**, 9341.

218. Camici, M., Ahmad, Z., DePaoli-Roach, A. A., and Roach, P. J. (1984). *JBC* **259**, 2466.
219. Sobue, K., Kanda, K., and Kakiuchi, S. (1982). *FEBS Lett.* **150**, 185.
220. Iwasa, T., Fukunaga, K., Yamamoto, H., Tanaka, E., and Miyamoto, E. (1984). *ABB* **235**, 212.
221. Penn, E. J., Brocklehurst, K. W., Sopwith, A. M., Hales, C. N., and Hutton, J. C. (1982). *FEBS Lett.* **139**, 4.
222. Krebs, E. G., Love, D. S., Bratvold, G. E., Trayser, K. A., Meyer, W. L., and Fischer, E. H. (1964). *Biochemistry* **3**, 1022.
223. Tabatabai, L. B., and Graves, D. J. (1978). *JBC* **253**, 2196.
224. Yamamura, H., Nishiyama, K., Shimomura, R., and Nishizuka, Y. (1973). *Biochemistry* **12**, 856.
225. Sugden, P. H., Holladay, L. A., Reimann, E. M., and Corbin, J. D. (1976). *BJ* **159**, 409.
226. Inoue, M., Kishimoto, A., Takai, Y., and Nishizuka, Y. (1976). *JBC* **251**, 4476.
227. Glass, D. B., McFann, L. J., Miller, M. D., and Zeilig, C. E. (1981). *In* "Protein Phosphorylation" (O. M. Rosen and E. G. Krebs, eds.), Vol. 8, p. 267. Cold Spring Harbor Press, Cold Spring Harbor, New York.
228. Dahmus, M. E. (1976). *Biochemistry* **15**, 1821.
229. Dahmus, M. E. (1981). *JBC* **256**, 3319.
230. Dahmus, M. E., and Natzle, J. (1977). *Biochemistry* **16**, 1901.
231. Hathaway, G. M., and Traugh, J. A. (1979). *JBC* **254**, 762.
232. Yamamoto, H., Fukunaga, K., Goto, S., Tanaka, E., and Miyamoto, E. (1985). *J. Neurochem.* **44**, 759.
233. Schulman, H. (1984). *J. Cell Biol.* **99**, 11.
234. Fujisawa, H., Yamauchi, T., Nakata, H., and Okuno, S. (1984). *FP* **43**, 3011.
235. Fukunaga, K., Yamamoto, H., Tanaka, E., Iwasa, T., and Miyamoto, E. (1984). *Life Sci.* **35**, 493.
236. Vulliet, P. R., Woodgett, J. R., and Cohen, P. (1984). *JBC* **259**, 13680.
237. Ueda, T., Maeno, H., and Greengard, P. (1973). *JBC* **248**, 8295.
238. Huttner, W. B., and Greengard, P. (1979). *PNAS* **76**, 5402.
239. Huttner, W. B., DeGennaro, L. J., and Greengard, P. (1981). *JBC* **256**, 1482.
240. Embi, N., Parker, P. J., and Cohen, P. (1981). *EJB* **115**, 405.
241. Kemp, B. E., Pearson, R. B., Woodgett, J. R., and Cohen, P. (1985). *FP* **44**, 705 (abstr.).
242. Hofmann, F., Beavo, J. A., Bechtel, P. J., and Krebs, E. G. (1975). *JBC* **250**, 7795.
243. Rangel-Aldao, R., Kupiec, J. W., and Rosen, O. M. (1979). *JBC* **254**, 2499.
244. Scott, C. W., and Mumby, M. C. (1985). *JBC* **260**, 2274.
245. Hemmings, B. A., Yellowlees, D., Kernohan, J. C., and Cohen, P. C. (1981). *EJB* **119**, 443.
246. Shields, S. M., Vernon, P. J., and Kelly, P. T. (1984). *J. Neurochem.* **43**, 1599.
247. Zoller, M. J., and Taylor, S. S. (1979). *JBC* **254**, 8363.
248. LeVine, H., III, Sahyoun, N. E., and Cuatrecasas, P. (1985). *PNAS* **82**, 287.
249. Kennedy, M. B. (1983). *Annu. Rev. Neurosci.* **6**, 493.
250. Schulman, H. (1984). *Trends Pharmacol. Sci.* **5**, 188.
251. Nestler, E. J., Walaas, S. I., and Greengard, P. (1984). *Science* **225**, 1357.
252. Pearson, R. B., Jakes, R., John, M., Kendrick-Jones, J., and Kemp, B. E. (1984). *FEBS Lett.* **168**, 108.
253. Suzuyama, Y., Umegane, T., Maita, T., and Matsuda, G. (1980). *Hoppe-Seyler's Z. Physiol. Chem.* **361**, 119.
254. Collins, J. H. (1976). *Nature (London)* **259**, 699.
255. Matsuda, G., Maita, T., Suzuyama, Y., Setoguchi, M., and Umegane, T. (1977). *J. Biochem. (Toyko)* **81**, 809.
256. Matsuda, G., Maita, T., Kato, Y., Chen, J.-I., and Umegane, T. (1981). *FEBS Lett.* **135**, 232.

257. Burke, B. E., and DeLorenzo, R. J. (1981). *PNAS* **78,** 991.
258. Huang, T. S., and Krebs, E. G. (1979). *FEBS Lett.* **98,** 66.
259. Rylatt, D. B., and Cohen, P. (1979). *FEBS Lett.* **98,** 71.
260. Huang, T. S., and Krebs, E. G. (1977). *BBRC* **75,** 643.
261. Parker, P. J., Aitken, A., Bilham, T., Embi, N., and Cohen, P. (1981). *FEBS Lett.* **123,** 332.
262. Wretborn, M., Humble, E., Ragnarsson, U., and Engström, L. (1980). *BBRC* **93,** 403.
263. Hjelmquist, G., Andersson, J., Edlund, B., and Engström, L. (1974). *BBRC* **61,** 509.

5

Protein Kinase C

USHIO KIKKAWA • YASUTOMI NISHIZUKA

Department of Biochemistry
Kobe University School of Medicine
Kobe 650, Japan

I. Introduction

The biochemical basis of signal transduction across the cell membrane has long been a subject of interest, and protein kinase C has attracted great attention in the studies on the control of cellular functions and proliferation. Under physiological conditions the enzyme is activated by diacylglycerol in the presence of Ca^{2+} and membrane phospholipid. The diacylglycerol active in this role may arise in the plasma membrane only transiently from the receptor-mediated hydrolysis of inositol phospholipids, and this protein kinase appears to be indispensable for transmitting various extracellular informational signals from the cell surface into the cell interior. The signals relating to protein kinase C normally

THE ENZYMES, Vol. XVII

differ from those from a group of hormones and neurotransmitters that produce cyclic AMP as an intracellular messenger. Thus, protein kinase C and cyclic AMP-dependent protein kinase (protein kinase A) transduce distinctly different pieces of information into the cell through their own specific protein phosphorylation. This article describes some properties, mechanism of activation, and possible roles of protein kinase C in cell-surface signal transduction. Several aspects of this protein kinase system have been reviewed (1–6).

II. Properties

Protein kinase C was first found in 1977 as a proteolytically activated protein kinase that was capable of phosphorylating histone (7), and later found to be activated reversibly by association of membrane phospholipid in the presence of diacylglycerol and physiological concentration of Ca^{2+} (8, 9). This enzyme is ubiquitously distributed in tissues and organs, with the brain having highest activity (10, 11). In this tissue a large quantity of the enzyme is associated with synaptic membranes, whereas in most other tissues the enzyme is present mainly in the soluble fraction as an inactive form and recovered upon biochemical fractionation of subcellular components (12). The enzyme has been purified to near homogeneity from the soluble fraction of rat brain (12), bovine heart (13), pig spleen (14), bovine brain (15), rabbit renal cortex (16), and rabbit brain (17). The original method for purification from rat brain was later improved (18).

The brain enzyme shows a single band upon sodium dodecyl sulfate (SDS)-polyacrylamide gel electrophoresis with a molecular weight of about 82,000. The Stokes radius is 42Å, and the molecular weight is 87,000 as estimated by gel filtration analysis. The sedimentation coefficient is 5.1 S which corresponds to a molecular weight of 77,000. The frictional ratio of the enzyme is calculated to be 1.6, indicating an asymmetric nature of the molecule. Although the molecular weight mentioned above slightly varies with the methods employed for estimation, the enzyme is composed of a single polypeptide chain with no subunit structure. Neither calmodulin nor an antibody against calmodulin affects the enzymic activity. The isoelectric point of the enzyme is pH 5.6. The optimum pH range for activity is 7.5–8.0 with Tris acetate as a test buffer. Mg^{2+} is essential for the catalytic activity with the optimum range having about 5–10 mM. To some extent Mg^{2+} can be replaced by Mn^{2+} or Co^{2+}. The optimum concentrations for Mn^{2+} and Co^{2+} are 0.5–1 mM, with the maximum enzymic activity being approximately 50% of that with Mg^{2+}. The K_m value for ATP is about $6 \times 10^{-6} M$. Guanosine triphosphate does not serve as a phosphate donor. Protein kinase C utilizes ATP-γ-S as a phosphoryl donor, but the reaction velocity is very slow (19).

The enzymes obtained from various tissues appear to be similar and practically indistinguishable from one another at least in their kinetic and catalytic properties. However, there are minor differences; for example, the molecular weight of the heart enzyme is about 83,500 on SDS-polyacrylamide gel electrophoresis, 99,500 on sedimentation coefficient (5.6 S) and Stokes radius (42.9 Å), and 113,000 on gel filtration analysis (13). The optimum pH of heart enzyme is 6–7 with Pipes as a test buffer. It is not known at present whether these differences are due to intrinsic properties of the enzymes from different species and different tissues. Some heterogeneity of protein kinase C in brain and other tissues has been described, and the enzyme is very susceptible to proteolysis (7, 12).

III. Biochemical Activation

Protein kinase C per se is normally inactive. When assayed in a cell-free enzymic reaction, the protein kinase depends on Ca^{2+} as well as phospholipid for its activation. However, diacylglycerol, which is produced in membranes from inositol phospholipids in a signal-dependent fashion, dramatically increases the affinity of this enzyme to Ca^{2+}, and thereby renders it fully active without a net increase in the Ca^{2+} concentration (9, 20). Thus, the activation of this unique protein kinase is biochemically dependent on, but physiologically independent of, Ca^{2+}. Figure 1 shows some kinetic analyses indicating that the sensitivity of protein kinase C to this divalent cation is greatly increased by the addition of diacylglycerol.

Among various phospholipids tested, phosphatidylserine is absolutely required for the enzyme activation. At lower concentrations of Ca^{2+} other phospholipids such as phosphatidylethanolamine, phosphatidylinositol, phosphatidylcholine, and sphingomyelin are all inert. However, several of these phospholipids show positive or negative cooperativity for the activation of protein kinase C when supplemented to phosphatidylserine. For instance, phosphatidylethanolamine increases further the affinity of enzyme for Ca^{2+}, and makes it fully active at the range of 10^{-7} M of this cation. whereas both phosphatidylcholine and sphingomyelin show opposing effects (20). Thus, the asymmetric distribution of various phospholipids in the membrane phospholipid bilayer may take a part in the activation of enzyme, and the K_a value for Ca^{2+} depends on the phospholipid composition as well as on the presence of diacylglycerol. The enzyme activation is specific to Ca^{2+}, and none of other divalent cations tested is able to substitute for Ca^{2+}, except Sr^{2+} which is about 10% as active as Ca^{2+} at comparable concentrations.

In the enzymic reaction various diacylglycerols are capable of activating the enzyme. In physiological processes, 1-stearoyl-2-arachidonyl glycerol is likely the activator of this enzyme, since most of inositol phospholipids contain this diacylglycerol backbone (21). It was initially found that diacylglycerol active in

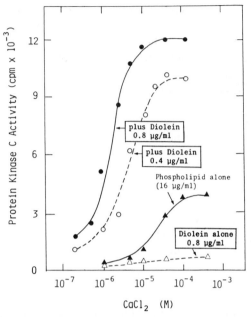

FIG. 1. Activation of protein kinase C by diacylglycerol in the presence of Ca^{2+} and phospholipid. Homogeneous protein kinase C obtained from rat brain was assayed in the presence of various concentrations of Ca^{2+} by measuring the incorporation of the radioactive phosphate of [γ-^{32}P]ATP into calf thymus H1 histone as a model phosphate acceptor. Where indicated, diolein and a mixture of phospholipids from human erythrocytes were added. Other detailed conditions are described elsewhere (9) [taken from Ref. (5)].

this role contains at least one unsaturated fatty acyl moiety at either 1 or 2 position (9). However, it became evident that when one fatty acyl moiety is replaced by a short chain the resulting diacylglycerols such as 1-palmitoyl-2-acetyl glycerol (22), dioctanoyl glycerol and dihexanoyl glycerol (23, 24) are also active to support the enzyme activation. More recently, the active diacylglycerol is found to be specific to the 1,2-sn-configuration, and other stereoisomers are totally inactive, suggesting that a highly specific lipid–protein interaction is needed for this enzyme activation (25). However, at higher concentrations of Ca^{2+}, the enzyme exhibits catalytic activity in the presence of phospholipid but without diacylglycerol as noted above. The detailed biochemical mechanism of this enzyme activation remains largely unknown.

Protein kinase C is alternatively activated by proteolysis with Ca^{2+}-dependent protease or trypsin (7, 26). When Ca^{2+}-dependent protease is employed, a limited proteolysis takes place, and a smaller component carrying enzymic activity is produced. The molecular weight is estimated to be about 51,000. The catalytically active component thus produced is totally independent of Ca^{2+}, phospholipid, and diacylglycerol. Protein kinase C, which is attached to the membrane, is more susceptible to limited proteolysis by Ca^{2+}-dependent protease (26). The protein kinases fully activated either proteolytically or non-proteolytically show similar kinetic and catalytic properties, and exhibit the same levels of enzymic activity. Although it has been proposed that such proteolytic activation of protein kinase C may occur in intact cell systems (27), the physiological significance of this proteolysis has not been defined.

IV. Physiological Activation

A wide variety of hormones, neurotransmitters, and many other biologically active substances activate cellular functions and proliferation through interaction with their specific cell surface receptors. It has been repeatedly shown that many of these signals provoke the breakdown of inositol phospholipids in the plasma membrane (28–31). Initially, phosphatidylinositol (PI) was regarded as a prime target (32), but further evidence seems to suggest that, after stimulation of the receptor, phosphatidylinositol-4,5-bisphosphate (PIP_2) rather than PI and phosphatidylinositol-4-phosphate (PIP) is degraded immediately to produce 1,2-diacylglycerol and inositol 1,4,5-trisphosphate (IP_3) (33–37). In general, this turnover of membrane phospholipids is associated with an increase in intracellular concentration of Ca^{2+}, which appears to mediate many of the subsequent physiological responses. Studies in this laboratory have provided evidence that the diacylglycerol produced in this way initiates the activation of protein kinase C, and thereby the information of extracellular signals is translated to protein phosphorylation as shown in Fig. 2.

It has been proposed that IP_3, the other product of PIP_2 breakdown, could serve as a mediator of Ca^{2+} release from intracellular stores (38). Evidence for this proposal has been obtained from the studies on the effect of IP_3 on various permeabilized cells, and this Ca^{2+}-releasing activity of IP_3 was first demonstrated in a preparation of rat pancreatic acinar cells (39). This possibility has been subsequently supported using several permeabilization and Ca^{2+}-measuring procedures in many other cell types and also using micro-injection of IP_3 into photoreceptors (38). If this proposal is correct, then the signal-dependent breakdown of a single molecule, PIP_2, may generate two intracellular mediators, diacylglycerol that induces protein kinase C activation and IP_3 that mobilizes Ca^{2+}. However, it is still not absolutely clear that only PIP_2 is hydrolyzed in response to external stimuli. It is possible that three inositol phospholipids are broken down by phospholipases C at different times at different rates or that these phospholipids are hydrolyzed at different sites within the cell (6). The phospholipases C in mammalian tissues, except for the enzyme of lysosomal origin, normally require high concentrations of Ca^{2+} when assayed in cell-free

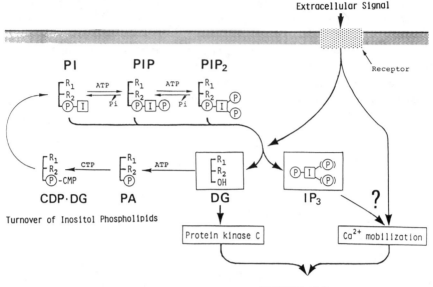

FIG. 2. Inositol phospholipid turnover and cell-surface signal transduction. PI, phos-phatidylinositol; PIP, phosphatidylinositol-4-phosphate; PIP₂, phosphatidylinositol-4,5-bisphos-phate; CDP·DG, CDP-diacylglycerol; PA, phosphatidic acid; DG, 1,2-diacylglycerol; IP₃, in-ositol-1,4,5-triphosphate; I, inositol moiety; and P, phosphoryl group [adapted from Ref. (3)].

systems. Although the regulatory mechanism of this enzyme is not clear, various results suggest that the breakdown of inositol phospholipids is Ca^{2+}-dependent but may not be regulated by Ca^{2+} (37). Evidence suggests that GTP may be involved in this signal-induced hydrolysis of inositol phospholipids (40–43).

Although it is generally accepted that the inositol phospholipid turnover is firmly linked to the activation of protein kinase C, the evidence for this signal transduction has primarily come from experiments with platelets as a model system, where many agonists and antagonists for the aggregation and release reaction are known (44, 45). When stimulated by thrombin, collagen, or platelet-activating factor (PAF), two endogenous platelet proteins with approximate molecular weights of 47,000 (47K protein) and 20,000 (20K protein) are heavily phosphorylated, and this phosphorylation reaction is normally associated with the release of their constituents such as serotonin (46, 47). The 20K protein is myosin light chain, and a calmodulin-dependent protein kinase is responsible for this phosphorylation reaction (48). Although the function of 47K protein remains unknown, the phosphorylation of this protein may be used as a marker of the activation of protein kinase C (45). When human platelets are stimulated, di-acylglycerol containing arachidonate is rapidly produced with the concomitant phosphorylation of 47K protein (44, 45, 49, 50).

Under appropriate conditions synthetic diacylglycerol such as 1-oleoyl-2-acetyl glycerol is permeable and directly activates protein kinase C without interaction with cell surface receptors (51, 52). This exogenously added diacylglycerol does not produce endogenous diacylglycerol nor does it induce inositol phospholipid breakdown. There is no indication of arachidonate release or damage of cell membranes. Instead, the exogenous diacylglycerol is rapidly converted *in situ* to the corresponding phosphatidate, that is, 1-oleoyl-2-acetyl-3-phosphoryl glycerol, probably through the action of diacylglycerol kinase. Furthermore, it has been shown that dioctanoyl glycerol and dihexanoyl glycerol are also permeable to cell membranes and activate protein kinase C directly (23, 24). Several lines of experimental evidence seem to indicate that protein kinase C is activated upon stimulation by various extracellular signals, and that the receptor-mediated hydrolysis of inositol phospholipids is a sign for the transmembrane control of protein phosphorylation. In this process protein kinase C may be reversibly attached to membranes, and presumably a quaternary complex consisting of the enzyme, phospholipid, diacylglycerol, and Ca^{2+} is produced. Diacylglycerol and Ca^{2+} may be synergistically effective to produce such a catalytically active complex, but the precise physiological picture of this unique lipid–protein interaction remains to be clarified.

V. Action of Tumor Promoters

Tumor-promoting phorbol esters, such as 12-*O*-tetradecanoyl-phorbol-13-acetate (TPA) first isolated from croton oil, elicit a variety of biological and biochemical actions in a manner very similar to hormones. In most cases Ca^{2+} is indispensable for causing such cellular responses, and a number of kinetic studies with various cell types suggest that their primary site of action is the cell surface membrane (53–57). Evidence is available that protein kinase C is a target for phorbol esters, since the tumor promoters directly activate this enzyme both *in vitro* and *in vivo*, and there is an approximate correlation between the ability of individual phorbol esters to promote tumor development and to activate the protein kinase (58, 59). Kinetic analysis indicates that TPA, which has a diacylglycerol-like structure, is able to substitute for diacylglycerol at extremely low concentrations. Like diacylglycerol, TPA dramatically increases the affinity of the enzyme for Ca^{2+} to the 10^{-7} M range, resulting in its full activation without detectable mobilization of Ca^{2+} when measured by an intracellular Ca^{2+}-indicator, quin 2.

Studies with a homogeneous preparation of protein kinase C indicate that [³H]phorbol-12,13-dibutyrate (PDBu), which is another potent tumor-promoting phorbol ester, may bind to the enzyme only when Ca^{2+} and phospholipid are present (60). This radioactive phorbol ester binds neither to protein kinase C nor to phospholipid *per se* irrespective of the presence or absence of Ca^{2+}, and all four components previously mentioned are needed simultaneously for the binding as well as for the enzyme activation. The apparent dissociation binding constant (K_d) of the tumor promoter is exactly identical with the activation constant (K_a) for the enzyme, and this value varies with the composition of phospholipids added to the incubation mixture. Again, phosphatidylserine is essential, and other phospholipids show positive or negative cooperativity for the binding. This may help explain the reported existence of apparent multiple binding sites of tumor promoter in broken cell preparations (61). In any case, the K_d values obtained with purified protein kinase C are remarkably similar to those previously described for the specific tumor-promoter-binding site on intact cell membranes (62–65). Diacylglycerols such as diolein compete with the radioactive phorbol ester for the binding, whereas neither monoolein, triolein, nor free oleic acid is active in this capacity under similar conditions.

Scatchard analysis indicates that roughly one molecule of PDBu binds to one molecule of protein kinase C in the presence of a physiological concentration of Ca^{2+} and an apparent excess of phospholipid (60). Presumably in intact cells, where phospholipid and Ca^{2+} are not limited, for each molecule of tumor-promoting phorbol ester intercalated into the membrane phospholipid bilayer, one molecule of protein kinase C moves to it and produces the quaternary complex by which the enzyme is activated. Studies in this and other laboratories (60–69) strongly suggest that protein kinase C is a receptor protein of tumor-promoting phorbol esters, and that many of the pleiotropic actions, if not all, of the tumor promoters may be mediated through the action of protein kinase C. It may be noted that mezerein (70), teleocidin, and debromoaplysiatoxin (71), which have no diacylglycerol-like structure but exhibit tumor-promoting activity, are all capable of activating protein kinase C presumably by causing membrane perturbation analogous to that phorbol esters do. However, it is still possible that these tumor promoters have additional actions on the membrane, particularly at higher concentrations. For instance, TPA may act as a weak Ca^{2+}-ionophore under certain conditions.

VI. Inhibitors

No inhibitor has been found that is specific for protein kinase C. Theoretically, at least three entities of inhibitors may exist: first, compounds that antagonize the action of diacylglycerol; second, phospholipid-interacting compounds that prevent the activation of the enzyme; and third, compounds that inhibit the catalytically active center of the enzyme. It is obviously important to develop specific inhibitors of this enzyme for several practical reasons. However, no inhibitor has been found that belongs to the first entity. Instead, R 59 022 (6-[2-[4-[(4-fluorophenyl)phenylmethylene]-1-piperidinyl]ethyl]-7-methyl-5H-thiazolo[3,2-a] pyrimidin-5-one) is shown to act as a selective inhibitor of diacylglycerol kinase, and thereby enhances the accumulation of diacylglycerol in membranes (72).

Most of the inhibitory compounds described appear to be in the second entity. For instance, protein kinase C is inhibited to various extents by psychotic drugs (trifluoperazine, chlorpromazine, fluphenazine, imipramine, etc.), local anesthetics (dibucaine, tetracaine, etc.), W-7 [N(6-aminohexyl)-5-chloro-1-naphthalenesulfonamide], verapamil, phentolamine, adriamycin, polyamines (spermine, spermidine, and putrescine), palmitoylcarnitine, melittin, heparin, polymixin B, and vitamin E (10, 73–77). The inhibition of protein kinase C by these drugs is not due to their interaction with the active center of the enzyme, since the catalytically active enzyme fragment, which is obtained by limited proteolysis (26), is not susceptible to any of these drugs. Kinetically, most of the drugs listed above interact with phospholipid, and inhibit the activation of the enzyme in a competitive manner. These phospholipid-interacting drugs usually also inhibit calmodulin-dependent protein kinases such as myosin light chain kinase by competing with calmodulin. In an experiment using intact platelets it was shown that some of these drugs such as chlorpromazine, dibucaine, and tetracaine do not inhibit thrombin-induced diacylglycerol formation, but in fact profoundly inhibit the activation of protein kinase C in a dose-dependent manner (44, 45). Cyclic nucleotide-dependent protein kinases are not susceptible to these drugs.

Some inhibitors that interact with the catalytically active center of protein kinases have been described (78). For instance, H-7 [1-(5-isoquinolinesulfonyl)-2-methylpiperazine] profoundly inhibits protein kinase C. However, protein kinase A is inhibited by this compound as well, but myosin light chain kinase is far less susceptible. It has been reported that some polypeptide cytotoxins effectively and specifically inhibit protein kinase C relative to protein kinase A and myosin light chain kinase, but the mode of their inhibitor actions is not known (79).

VII. Synergistic Roles with Calcium

Although bovine adrenal medullary cells (80) and some presynaptic muscarinic receptors (81) may be exceptions, it is generally the case that when stimulation of receptors leads to inositol phospholipid breakdown it simultaneously mobilizes Ca^{2+}. A series of studies indicate that the activation of protein kinase C appears to be a prerequisite but not a sufficient requirement for physiological responses of target cells, because the cellular responses to synthetic diacylglycerol or to tumor promoter per se are always incomplete. Under appropriate conditions it is possible to induce protein kinase C activation and Ca^{2+} mobilization selectively by the addition of permeable diacylglycerol or tumor promoter for the former and a Ca^{2+}-ionophore, such as A23187, for the latter. By using this procedure it is possible to demonstrate that protein kinase C activation and Ca^{2+} mobilization are both essential and act synergistically to elicit full physiological responses such as release reaction of serotonin (51, 52). In this experiment the concentration of A23187 (0.2–0.4 μM) is critical, because at higher concentrations of more than 0.5 μM this Ca^{2+}-ionophore itself will cause the phosphorylation of 47K protein as well as 20K protein, probably due to the nonspecific activation of phospholipase C that is accompanied by a large increase in Ca^{2+} concentration. Likewise, the synthetic diacylglycerol or tumor promoter alone at higher concentrations (more than 50 $\mu g/ml$ or 50 ng/ml, respectively) causes a significant release of platelet constituents. The precise reason for this enhanced release is unclear, but it is possible that these compounds can induce the release reaction by acting as membrane fusigens or weak Ca^{2+}-ionophores.

The involvement of the two pathways, protein kinase C activation and Ca^{2+} mobilization, in the signal transduction may explain, at least in part, the agonist-selectivity that is often observed in release reactions. For instance, again in platelets, serotonin and adenine nucleotides are released from dense bodies in response to a variety of signals such as thrombin, collagen, ADP, epinephrine, and PAF, while lysosomal enzymes are released only at higher concentrations of thrombin and collagen. By using permeabilized platelets it has been shown that such agonist-selectivity of release reactions is not related to Ca^{2+} concentrations, because there is no difference in their sensitivity to Ca^{2+} (82). Theoretically, it is possible that the two pathways mentioned may exert differential control over release reactions from different granules within a single activated platelet. In neuronal tissues, a single nerve ending frequently contains both peptides and classical transmitters presumably present in different stores (83). It is possible to imagine that the two synergistic routes may also be responsible for the frequency-selectivity of release reactions that is often observed during electrical stimulation. It is well known that depolarization of membranes by electrical stimulation or depolarizing agents induces inositol phospholipid turnover (84–87).

A potential role of protein kinase C for the activation of cellular functions has been suggested in many other systems. For instance, the synergistic roles of this enzyme with Ca^{2+} are proposed for the receptor-mediated release reactions and exocytosis of several endocrine as well as exocrine tissues as exemplified in Table I (88–109). It seems important to note that in both peripheral and central nervous tissues the two pathways appear to be essential for the neurotransmitter release from nerve endings. In fact, the combination of TPA and A23187 has been shown to induce full activation of acetylcholine release from the cholinergic nerve endings of guinea pig ileum (101). The physiological responses through these two pathways may include not only release reactions of various cell types but also many other cellular processes such as adrenal steroidogenesis (102), neutrophil superoxide generation (103–105), smooth muscle contraction (106, 107), and hepatic glycogenolysis (108, 109).

TABLE I

CELLULAR RESPONSES PROBABLY ELICITED BY SYNERGISTIC
ACTIONS OF PROTEIN KINASE C AND Ca^{2+}

Tissues	Responses	References
1. Release Reactions and Exocytosis		
Platelets	Serotonin	(51,52,88)
	Lysosomal enzyme	(88,89)
Mast cells	Histamine	(90)
Neutrophils	Lysosomal enzymes	(89,91,92)
Adrenal medulla	Catecholamine	(93)
Adrenal cortex	Aldosterone	(94)
Pancreatic islets	Insulin	(95–97)
Pancreatic acini	Amylase	(98)
Pituitary cells	Gonadotropin	(99)
	Thyrotropin	(100)
Ileal nerve endings	Acetylcholine	(101)
2. Metabolic Processes and Others		
Adrenal cortex	Steroidogenesis	(102)
Neutrophils	Superoxide generation	(103–105)
Smooth muscle	Contraction	(106,107)
Hepatocytes	Glycogenolysis	(108,109)

Alternatively, it has been proposed that the activation of protein kinase C *per se* is sufficient to induce some cellular responses such as serotonin release from platelets (*110*) and superoxide generation and exocytosis of neutrophils (*111*), and these reactions are shown to proceed without any detectable increase in Ca^{2+} concentrations only when protein kinase C is activated. The precise relationship between protein kinase C and Ca^{2+} actions during the activation of various cellular functions is a subject of great interest.

VIII. Growth Response and Down Regulation

The role of two pathways, protein kinase C activation and Ca^{2+} mobilization, is not only confined to the short-term responses described earlier in this chapter but is also extended to the long-term responses such as cell proliferation. It is possible to show with macrophage-depleted human peripheral lymphocytes that the two pathways are both essential and synergistically act for promoting DNA synthesis (*112, 113*). However, for long-term responses, a low concentration of some growth factor such as phytohemagglutinin is still needed, implying that another hitherto unknown signal pathway is involved in eliciting full activation of cell proliferation (*112*). In a similar set of experiments with Swiss 3T3 cell line, insulin is necessary for the growth response in addition to synthetic diacylglycerol or TPA

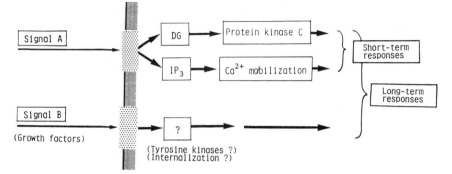

FIG. 3. Pathways of signal transduction for short-term and long-term cellular responses. DG, 1,2-diacylglycerol; and IP$_3$, inositol-1,4,5-trisphosphate.

(*114*). It has been known for some time that tumor promoter and growth factor such as epidermal growth factor (EGF) generally act in concert for cell proliferation (*115*). Platelet-derived growth factor (PDGF) or TPA is shown to induce messenger RNAs of some protooncogenes such as c-*fos* within 5–10 min (*116*), although the concentration of TPA employed for this experiment was extremely high. Figure 3 illustrates the hypothetical pathways of signal transduction for short-term and long-term cellular responses. Some of the receptors for growth factors are associated with a tyrosine-specific protein kinase activity. The role of protein kinase C and tyrosine-specific protein kinases in cell proliferation is another subject of current interest.

Very frequently protein kinase C appears to be related to "down regulation" or "negative feedback control." For instance, protein kinase C is shown to phosphorylate the receptor protein of some growth factors such as EGF with the concomitant decrease in both its tyrosine-specific protein kinase and growth factor-binding activities (*24, 117–122*). In EGF receptor, threonine-654 is shown to be phosphorylated by protein kinase C (*123*), which is located at the upstream close to its tyrosine-specific protein kinase domain (*124*). Although TPA markedly reduces EGF-binding to many mitogenically responsive cell types (*125–128*), the biological significance of this phosphorylation reaction remains to be explored. In an analogous fashion, it has been suggested that some receptors in human leukemic cell lines, such as the receptors of insulin (*129*), somatomedin C (*129*), transferrin (*130, 131*), and interleukin 2 (*132*), and also α_1-adrenergic receptor in rat hepatocytes (*133*), are phosphorylated by protein kinase C. However, a logical consequence of the phosphorylation reactions of these receptors has not been well substantiated.

The dual functions of protein kinase C described in the preceding section, apparently positive forward action and negative feedback action, are observed also in some other cellular processes. Myosin light chain is shown to be phosphorylated by protein kinase C, and the site of phosphorylation differs from that phosphory-lated by myosin light chain kinase (*134*). TPA and Ca^{2+}-ionophore are syn-ergistically involved in smooth muscle contraction (*106, 107*), but the phos-phorylation of myosin light chain by protein kinase C appears to suppress the actin-activated ATPase activity of myosin, resulting in the relaxation (*135*). It is possible that this reaction catalyzed by protein kinase C may constitute a feedback mechanism to prevent overresponse, since the reaction proceeds slowly compared with the contraction, which is completed within seconds.

A similar negative feedback control by protein kinase C may be possible for the signal-induced increase of Ca^{2+} (*110, 136–138*). Such a mechanism to decrease the concentration of Ca^{2+} has been described for Ca^{2+}-transport ATPase, which is activated by this divalent cation through a calmodulin-dependent mechanism (*139, 140*). Analogously, in cardiac systems, protein kinase C phosphorylates sarcoplasmic reticulum proteins, resulting in the enhanced Ca^{2+}-transport AT-Pase and, thus, leading to the decrease in the cytoplasmic Ca^{2+} concentration (*141–144*). The significance of such feedback control of the intracellular Ca^{2+} concentration is also not known. In physiological processes, both diacylglycerol and Ca^{2+} appear only transiently, and the informational signals pass through the membrane very quickly and elicit subsequent cellular responses. Perhaps, in biological systems, a positive signal may be immediately followed by a negative feedback mechanism. In some tissues with phorbol ester, it is possible that such a negative feedback phase may predominate under certain experimental conditions, where the tumor promoter stays in membranes and keeps protein kinase C active for an unusually long period of time.

Some other feedback mechanisms, such as the inhibition of protein kinase C by a calmodulin-dependent system (*145*) and the inhibition of inositol phospholipid breakdown by protein kinase C (*146*) have been suggested.

IX. Target Proteins and Catalytic Specificity

Although evidence is accumulating that protein kinase C is at times related to apparently negative feedback control as previously discussed, the enzyme plays positive roles in overall processes to activate many cellular functions and pro-liferation. In most tissues, however, crucial information for such physiological target proteins is unavailable. The enzyme has a broad substrate specificity *in vitro*, and phosphorylates seryl and threonyl residues but not tyrosyl residue of endogenous proteins. Protein kinase C is phosphorylated by itself in the simul-taneous presence of Ca^{2+}, phospholipid, and diacylglycerol (*12*). Approximately 2 mol of phosphate are incorporated into each mole of the enzyme, and both seryl and threonyl residues are phosphorylated. Table II (*147–167*) lists some proteins that may serve as phosphate acceptors *in vitro*. Although this list is expanding very

TABLE II

POSSIBLE PHOSPHATE ACCEPTOR PROTEINS OF PROTEIN
KINASE C

Phosphate acceptor proteins	References
Receptor proteins	
Epidermal growth factor receptor	(117–123)
Insulin receptor	(129)
Somatomedin C receptor	(129)
Transferrin receptor	(130,131)
Interleukin 2 receptor	(132)
Contractile proteins and cytoskeletous proteins	
Myosin light chain	(134,135,147)
Troponin T and I	(148)
Filamin	(149)
Vinculin	(149,150)
Microtubule-associated protein	(151)
Gap junction proteins	(152–154)
Membrane and nuclear proteins	
Histones and protamine	(7–9)
High mobility group proteins	(155)
Middle T antigen	(156)
Ca^{2+}-ATPase and phospholamban	(141–144)
Synaptic B-50 (F1) protein	(157,158)
Na^+–H^+ exchange protein	(137,159)
Enzymes and other proteins	
Glycogen phosphorylase kinase	(160)
Glycogen synthetase	(160,161)
Guanylate cyclase	(162)
Initiation factor 2 (β-subunit)	(163)
Fibrinogen	(164)
Myelin basic protein	(9,76,165)
Retinoid-binding protein	(166)
Ribosomal S6 protein	(167)

rapidly, most of these phosphorylation reactions remain to be explored for the physiological significance. For instance, the 47K protein in platelets described earlier in this chapter appears to be involved in the release reaction, but its definitive role is yet to be clarifed. In neuronal tissues, synapsin I, which is located specifically on the cytoplasmic surface of synaptic fesicles, appears to serve as a substrate for protein kinase A as well as for calmodulin-dependent protein kinase (168). Apparently, this protein is also phosphorylated by protein kinase C, suggesting that it is one of the possible targets of this protein kinase and plays some role in release reactions (4). B-50 protein (F1 protein), another brain protein which is associated specifically with presynaptic membranes (*169–172*), serves as a preferable substrate specific to protein kinase C (*157, 158*). This protein phosphorylation selectively increases after long-term potentiation in the hippocampal neural activity, and has been proposed to be related to the expression of synaptic plasticity (*173, 174*). It is noted that B-50 protein kinase previously found in brain tissues (*175*) has been later identified as protein kinase C (*176–178*). The role of protein kinase C may also be extended to the modulation of membrane conductance, channels and active transport, axoplasmic flow, neurotransmitter biosynthesis, and many other neuronal functions by phosphorylating the proteins involved (*4*). It has been shown that the activation of endogenous protein kinase C by TPA or the microinjection of this enzyme itself enhances the voltage-sensitive calcium current in bag cell neurons (*179*).

Although protein kinases C and A transduce distinctly different pieces of information into the cell as discussed earlier in this chapter, these two protein kinase pathways sometimes cause apparently similar cellular responses and often potentiate each other. Analysis indicates that protein kinases C and A often share the same phosphate acceptor proteins, even the same seryl and threonyl residues in a single protein molecule for phosphorylation. Extensive work by Krebs and his colleagues and others [reviewed in Ref. (*180*)] have shown that the primary structure of the vicinity of the aminoacyl residue to be phosphorylated is one of the determinant factors for the substrate recognition, and that protein kinase A reacts with the seryl and threonyl residues that are usually located at the downstream close to lysyl or arginyl residue. With some model substrate proteins such as myelin basic protein it is shown that, contrary to protein kinase A, protein kinase C appears to favor the hydroxylamino acids that are located at the upstream close to the basic aminoacyl residue. All seryl and threonyl residues that are phosphorylated commonly by these two enzymes have basic aminoacyl residues at both downstream and upstream locations (*181*). Although it is not known whether this principle can be extended to many other protein substrates, this approach appears to provide a clue to understand the reason why these protein kinases may show different but sometimes similar functions depending on the structure of substrate protein molecule.

X. Relation to Other Receptors

In receptor functions there may be dramatic heterogeneity and variations from tissue to tissue, but most tissues possess at least two major classes of receptors in transducing information across the membrane. One class is related to cyclic AMP formation, and the other to the inositol phospholipid turnover. In addition, the stimulation of the latter class of receptors normally releases arachidonic acid, and often increases cyclic GMP. Thus, the succeeding events of protein kinase C activation, Ca^{2+} mobilization, arachidonic acid release, and cyclic GMP formation appear to be integrated in a single receptor cascade.

The mode of physiological responses may be roughly divided into two groups. In bidirectional control systems in many tissues such as platelets, neutrophils, lymphocytes, and some neuronal cells, the signals that induce the inositol phospholipid turnover promote the activation of cellular functions, whereas the signals that produce cyclic AMP usually antagonize such activation. For instance, in platelets the signal-induced inositol phospholipid breakdown, diacylglycerol formation, 47K protein phosphorylation, and serotonin release are all blocked concurrently by prostaglandins that increase cyclic AMP as observed by dibutyryl cyclic AMP (2, 44, 45). This inhibitory action of cyclic AMP extends to the mobilization of Ca^{2+}, presumably through the decreased formation of IP_3 as previously mentioned and through the activation of protein kinase A. Protein kinase A has a potential to decrease cytosolic Ca^{2+} concentration by phosphorylating the regulatory components of Ca^{2+}-activated ATPase, thereby enhancing its catalytic activity (182). However, the molecular basis of the counteraction of inositol phospholipid breakdown by protein kinase A remains to be explored. It has been sometimes proposed that protein kinase C may inhibit agonist-stimulated adenylate cyclase presumably at the point of regulation by guanine nucleotides (183). However, such an interaction of the two receptor functions has not been unequivocally established.

In contrast, in monodirectional control systems in some tissues such as hepatocytes and many endocrine cells, the two classes of receptors appear not to interact with each other but to function independently or cooperatively. In hepatocytes, for instance, the inositol phospholipid turnover that is induced by α-adrenergic stimulators is not blocked by β-adrenergic stimulators nor by dibutyryl cyclic AMP (184). Both α- and β-stimulators are well known to cause glycogenolysis in the liver.

In some tissues such as pineal gland, the cellular responses to β-adrenergic stimulators are markedly potentiated by α-adrenergic stimulators that induce inositol phospholipid turnover and Ca^{2+} mobilization (185). Presumably, in such tissues protein kinase C potentiates the adenylate cyclase system or acts cooperatively with protein kinase A to induce full cellular responses. Considerable variations arise in receptor interaction, but further exploration of the biochemical basis of such interaction might be of great importance for understanding the basal mechanism of signal transduction.

Arachidonic acid is shown to be derived from inositol phospholipids through two consecutive reactions catalyzed by phospholipase C followed by diacylglycerol lipase (186). Inositol phospholipids in mammalian tissues contain mostly arachidonic acid at the position 2 (21). This fatty acid, however, may also be released from phosphatidylethanolamine as well as from phosphatidylcholine. Perhaps, when the receptor is stimulated, both phospholipases C and A_2 act in concert.

Although Ca^{2+} causes direct activation of guanylate cyclase in some tissues (*187*), it seems likely that arachidonic acid peroxide and prostaglandin endoperoxide serve as activators for this enzyme (*188, 189*). Cyclic GMP may have a function to act as a "negative" rather than a "positive" intracellular mediator, providing an immediate feedback control that prevents overresponse. It is known that sodium nitroprusside, which induces a marked elevation of cyclic GMP levels, is a powerful inhibitor of platelet activation (*190*). Indeed, analogous to cyclic AMP, 8-bromo cyclic GMP as well as sodium nitroprusside is shown to inhibit the signal-induced inositol phospholipid breakdown, and thereby counteracts the activation of protein kinase C (*191*). Although some functions of cyclic GMP and cyclic GMP-dependent protein kinase have been suggested for nervous tissues (*168*), crucial information on the role of this cyclic nucledotide is limited.

XI. Conclusion

This chapter summarizes our knowledge of protein kinase C, which is expanding very rapidly. The evidence available strongly suggests its crucial role in signal transduction for the activation of many cellular functions and proliferation, particularly at the early phase of responses. Perhaps, the signal-induced breakdown of inositol phospholipids initiates a cascade of events starting with Ca^{2+} mobilization and protein kinase C activation and ending with alterations of a variety of cellular processes including gene expression in long-term. However, it seems early to discuss the precise relationship between the roles of Ca^{2+} and protein kinase C. Several functions of this enzyme seem plausible, such as synergistic roles with Ca^{2+}, Ca^{2+}-sensitivity modulation, plasticity, down regulation, desensitization, or dual functions. Obviously, Ca^{2+} and protein kinase C each appear to play diverse roles in controlling cellular processes, and it is hoped that further exploration of the roles of this unique protein kinase may provide clues to the biochemical bases of signal transduction and cellular responses.

ACKNOWLEDGMENTS

The authors are grateful to Mrs. M. Furuta, Miss S. Fukase, and Mrs. S. Nishiyama for their skillful assistance for preparation of this manuscript.

REFERENCES

1. Nishizuka, Y. (1983). *Philos. Trans. R. Soc. London, Ser. B* **302**, 101–112.
2. Nishizuka, Y. (1983). *Trends Biochem. Sci.* **8**, 13–16.
3. Nishizuka, Y. (1984). *Nature (London)* **308**, 693–698.
4. Nishizuka, Y. (1984). *Science* **225**, 1365–1370.

5. Nishizuka, Y., Takai, Y., Kishimoto, A., Kikkawa, U., and Kaibuchi, K. (1984). *Recent Prog. Horm. Res.* **40**, 301–345.
6. Hirasawa, K., and Nishizuka, Y. (1985). *Annu. Rev. Pharmacol. Toxicol.* **25**, 147–170.
7. Inoue, M., Kishimoto, A., Takai, Y., and Nishizuka, Y. (1977). *JBC* **252**, 7610–7616.
8. Takai, Y., Kishimoto, A., Kikkawa, U., Mori, T., and Nishizuka, Y. (1979). *BBRC* **91**, 1218–1224.
9. Kishimoto, A., Takai, Y., Mori, T., Kikkawa, U., and Nishizuka, Y. (1980). *JBC* **255**, 2273–2276.
10. Kuo, J. F., Andersson, R. G. G., Wise, B. C., Mackerlova, L., Salomonsson, I., Brackett, N. L., Katoh, N., Shoji, M., and Wrenn, R. W. (1980). *PNAS* **77**, 7039–7043.
11. Minakuchi, R., Takai, Y., Yu, B., and Nishizuka, Y. (1981). *J. Biochem. (Tokyo)* **89**, 1651–1654.
12. Kikkawa, U., Takai, Y., Minakuchi, R., Inohara, S., and Nishizuka, Y. (1982). *JBC* **257**, 13341–13348.
13. Wise, B. C., Raynor, R. L., and Kuo, J. F. (1982). *JBC* **257**, 8481–8488.
14. Schatzman, R. C., Raynor, R. L., Fritz, R. B., and Kuo, J. F. (1983). *BJ* **209**, 435–443.
15. Parker, P. J., Stabel, S., and Waterfield, M. D. (1984). *EMBO J.* **3**, 953–959.
16. Uchida, T., and Filburn, C. R. (1984). *JBC* **259**, 12311–12314.
17. Inagaki, M., Watanabe, M., and Hidaka, H. (1985). *JBC* **260**, 2922–2925.
18. Kitano, T., Go, M., Kikkawa, U., and Nishizuka, Y. (1986). "Methods in Enzymology" **124**, 349–252
19. Wise, B. C., Glass, D. B., Chou, C. H. J., Raynor, R. L., Katoh, N., Schatzman, R. C., Turner, R. S., Kibler, R. F., and Kuo, J. F. (1982). *JBC* **257**, 8489–8495.
20. Kaibuchi, K., Takai, Y., and Nishizuka, Y. (1981). *JBC* **256**, 7146–7149.
21. Holub, B. J., Kuksis, A., and Thompson, W. (1970). *J. Lipid Res.* **11**, 558–564.
22. Kajikawa, N., Kikkawa, U., Itoh, K., and Nishizuka, Y. (1986). "Methods in Enzymology" (in press).
23. Lapetina, E. G., Reep, B., Ganong, B. R., and Bell, R. M. (1985). *JBC* **260**, 1358–1361.
24. Davis, R. J., Ganong, B. R., Bell, R. M., and Czech, M. P. (1985). *JBC* **260**, 1562–1566.
25. Rando, R. R., and Young, N. (1984). *BBRC* **122**, 818–823.
26. Kishimoto, A., Kajikawa, N., Shiota, M., and Nishizuka, Y. (1983). *JBC* **258**, 1156–1164.
27. Tapley, P. M., and Murray, A. W. (1984). *BBRC* **118**, 835–841.
28. Michell, R. H. (1975). *BBA* **415**, 81–147.
29. Hawthorne, J. N., and White, D. A. (1975). *Vitam. Horm. (N.Y.)* **33**, 529–573.
30. Berridge, M. J. *Mol. Cell. Endocrinol.* (1981). **24**, 115–140.
31. Michell, R. H., Kirk, C. J., Jones, L. M., Downes, D. P., and Creba, J. A. (1981). *Philos. Trans. R. Soc. London, Ser. B* **296**, 123–137.
32. Hokin, M. R., and Hokin, L. E. (1953). *JBC* **203**, 967–977.
33. Abdel-Latif, A. A., Akhtar, R. A., and Hawthorne, J. N. (1977). *BJ* **162**, 61–73.
34. Downes, P., and Michell, R. H. (1982). *Cell Calcium* **3**, 467–502.
35. Agranoff, B. W., Murthy, P., and Seguin, E. B. (1983). *JBC* **258**, 2076–2078.
36. Berridge, M. J., Dawson, R. M. C., Downes, C. P., Heslop, J. P., and Irvine, R. F. (1983). *BJ* **212**, 473–482.
37. Fisher, S. K., van Rooijen, L. A. A., and Agranoff, B. W. (1984). *Trends Biochem. Sci.* **9**, 53–56.
38. Berridge, M. J., and Irvine, R. F. (1984). *Nature (London)* **312**, 315–321.
39. Streb, H., Irvine, R. F., Berridge, M. J., and Schulz, I. (1983). *Nature (London)* **306**, 67–69.
40. Boyer, J. L., Garcia, A., Posadas, C., and Garcia-Sainz, J. A. (1984). *JBC* **259**, 8076–8079.
41. Haslam, R. J., and Davidson, M. M. L. (1984). *FEBS Lett.* **174**, 90–95.
42. Haslam, R. J., and Davidson, M. M. L. (1984). *J. Recept. Res.* **4**, 605–629.

43. Nakamura, T., and Ui, M. (1984). *FEBS Lett.* **173**, 414–418.
44. Kawahara, Y., Takai, Y., Minakuchi, R., Sano, K., and Nishizuka, Y. (1980). *BBRC* **97**, 309–317.
45. Sano, K., Takai, Y., Yamanishi, J., and Nishizuka, Y. (1983). *JBC* **258**, 2010–2013.
46. Lyons, R. M., Stanford, N., and Majerus, P. W. (1975). *J. Clin. Invest.* **56**, 924–936.
47. Haslam, R. J., and Lynham, J. A. (1977). *BBRC* **77**, 714–722.
48. Daniel, J. L., Holmsen, H., and Adelstein, R. S. (1977). *Thromb. Haemostasis* **38**, 984–989.
49. Rittenhouse-Simmons, S. (1979). *J. Clin. Invest.* **63**, 580–587.
50. Bell, R. L., and Majerus, P. W. (1980). *JBC* **255**, 1790–1792.
51. Kaibuchi, K., Sano, K., Hoshijima, M., Takai, Y., and Nishizuka, Y. (1982). *Cell Calcium* **3**, 323–335.
52. Kaibuchi, K., Takai, Y., Sawamura, M., Hoshijima, M., Fujikura, T., and Nishizuka, Y. (1983). *JBC* **258**, 6701–6704.
53. Van Duuren, B. L. (1969). *Prog. Exp. Tumor Res.* **11**, 31–68.
54. Hecker, E. (1971). *Methods Cancer Res.* **6**, 439–484.
55. Boutwell, R. K. (1974). *CRC Crit. Rev. Toxicol.* **2**, 419–443.
56. Weinstein, I. B., Lee, L. S., Mufson, A., and Yamasaki, H. (1979). *J. Suppramol. Struct.* **12**, 195–208.
57. Blumberg, P. M. (1980). *CRC Crit. Rev. Toxicol.* **8**, 153–197.
58. Castagna, M., Takai, Y., Kaibuchi, K., Sano, K., Kikkawa, U., and Nishizuka, J. (1982). *JBC* **257**, 7847–7851.
59. Yamanishi, J., Takai, Y., Kaibuchi, K., Sano, K., Castagna, M., and Nishizuka, Y. (1983). *BBRC* **112**, 778–786.
60. Kikkawa, U., Takai, Y., Tanaka, Y., Miyake, R., and Nishizuka, Y. (1983). *JBC* **258**, 11442–11445.
61. Blumberg, P. M., Jaken, S., Koning, B., Sharkey, N. A., Leach, K. L., Jeng, A. Y., and Yeh, E. (1984). *Biochem. Pharmacol.* **33**, 933–940.
62. Ashendel, C. L., Staller, J. M., and Boutwell, R. K. (1983). *BBRC* **111**, 340–345.
63. Sando, J. J., and Young, M. C. (1983). *PNAS* **80**, 2642–2646.
64. Horowitz, A. D., Greenebaum, E., and Weinstein, I. B. (1981). *PNAS* **78**, 2315–2319.
65. Solanki, V., and Slaga, T. J. (1981). *PNAS* **78**, 2549–2553.
66. Niedel, J. E., Kuhn, L. J., and Vandenbark, G. R. (1980). *PNAS* **80**, 36–40.
67. Leach, K. L., James, M. L., and Blumberg, P. M. (1983). *PNAS* **80**, 4208–4212.
68. Kraft, A. S., Anderson, W. B., Cooper, H. L., and Sando, J. J. (1982). *JBC* **257**, 13193–13196.
69. Shoyab, M., Todaro, G. J., and Tallman, J. F. (1982). *Cancer Lett.* **16**, 171–177.
70. Miyake, R., Tanaka, Y., Tsuda, T., Kaibuchi, K., Kikkawa, U., and Nishizuka, Y. (1984). *BBRC* **121**, 649–656.
71. Fujiki, H., Tanaka, Y., Miyake, R., Kikkawa, U., Nishizuka, Y., and Sugimura, T. (1984). *BBRC* **120**, 339–343.
72. de Chaffoy de Courcelles, D., Roevans, P., and Van Belle, H. (1985). *JBC* **260**, 15762–15770.
73. Mori, T., Takai, Y., Minakuchi, R., Yu, B., and Nishizuka, Y. (1980). *JBC* **255**, 8378–8380.
74. Hidaka, H., Sasaki, Y., Tanaka, T., Endo, T., Ohno, S., Fujii, Y., and Nagata, T. (1981). *PNAS* **78**, 4354–4357.
75. Schatzman, R. C., Wise, B. C., and Kuo, J. F. (1981). *BBRC* **98**, 669–676.
76. Wise, B. C., Glass, D. B., Chou, C. H. J., Raynor, R. L., Katoh, N., Schatzman, R. C., Turner, R. S., Kibler, R. F., and Kuo, J. F. (1982). *JBC* **257**, 8489–8495.
77. Katoh, N., Raynor, R. L., Wise, B. C., Schatzman, R. C., Turner, R. S., Helfman, D. M., Fain, J. N., and Kuo, J. F. (1982). *BJ* **202**, 217–224.

78. Hidaka, H., Inagaki, M., Kawamoto, S., and Sasaki, Y. (1984). *Biochemistry* **23**, 5036–5041.
79. Kuo, J. F., Raynor, R. L., Mazzei, G. J., Schatzman, R. C., Turner, R. S., and Kem, W. R. (1983). *FEBS Lett.* **153**, 183–186.
80. Swilem, A. M. F., Hawthorne, J. N., and Azila, N. (1983). *Biochem. Pharmacol.* **32**, 3873–3874.
81. Starke, K. (1980). *Trends Pharmacol. Sci.* **1**, 268–271.
82. Knight, D. E., Hallam, T. J., and Scrutton, M. C. (1982). *Nature (London)* **296**, 256–257.
83. Hökfelt, T., Johansson, O., Ljingdahl, Å., Lundberg, J. M., and Schultzberg, M. (1980). *Nature (London)* **284**, 515–521.
84. Yoshida, H., and Quastel, J. H. (1962). *BBA* **57**, 67–72.
85. Birnberger, A. C., Birnberger, K. L., Eliasson, S. G., and Simpson, P. C. (1971). *J. Neurochem.* **18**, 1291–1298.
86. Schacht, J., and Agranoff, B. W. (1972). *J. Neurochem.* **19**, 1417–1421.
87. Hawthorne, J. N., and Pichard, M. R. (1979). *J. Neurochem.* **32**, 5–14.
88. Knight, D. E., Niggli, V., and Scrutton, C. (1984). *EJB* **143**, 437–446.
89. Kajikawa, N., Kaibuchi, K., Matsubara, T., Kikkawa, U., Takai, Y., Nishizuka, Y., Itoh, K., and Tomioka, C. (1983). *BBRC* **116**, 743–750.
90. Katakami, Y., Kaibuchi, K., Sawamura, M., Takai, Y., and Nishizuka, Y. (1984). *BBRC* **121**, 573–578.
91. O'Flaherty, J. T., Schmitt, J. D., McCall, C. E., and Wykle, R. L. (1984). *BBRC* **123**, 64–70.
92. White, J. R., Huang, C. K., Hill, J. M., Jr., Naccache, P. H., Becker, E. L., and Sha'afi, R. I. (1984). *JBC* **259**, 8605–8611.
93. Knight, D. E., and Baker, P. E. (1983). *FEBS Lett.* **160**, 98–100.
94. Kojima, I., Lippes, H., Kojima, K., and Resmussen, H. (1983). *BBRC* **116**, 555–562.
95. Zawalich, W., Brown, C., and Rasmussen, H. (1983). *BBRC* **117**, 448–455.
96. Hubinont, C. J., Best, L., Sener, A., and Malaisse, W. J. (1984). *FEBS Lett.* **170**, 247–253.
97. Tamagawa, T., Niki, H., and Niki, A. (1985). *FEBS Lett.* **183**, 430–432.
98. de Pont J. J. H. H. M., and Flueren-Jacobs, A. M. M. (1984). *FEBS Lett.* **170**, 64–68.
99. Hirota, K., Hirota, T., Aguilera, G., and Catt, K. J. (1985). *JBC* **260**, 3243–3246.
100. Martin, T. F. J., and Kowalchyk, J. A. (1984). *Endocrinology (Baltimore)* **115**, 1527–1536.
101. Tanaka, C., Taniyama, K., and Kusunoki, M. (1984). *FEBS Lett.* **170**, 64–68.
102. Culty, M., Vilgrain, I., and Chambaz, E. M. (1984). *BBRC* **121**, 499–506.
103. Fujita, I., Irita, K., Takeshige, K., and Minakami, S. (1984). *BBRC* **120**, 318–324.
104. Robinson, J. M., Badway, J. A., Karnovsky, M. L., and Karnovsky, M. J. (1984). *BBRC* **122**, 734–739.
105. Dale, M. Maureen, and Penfield, A. (1984). *FEBS Lett.* **175**, 170–172.
106. Rasmussen, H., Forder, J., Kojima, I., and Scriabine, A. (1984). *BBRC* **122**, 776–784.
107. Baraban, J. M., Gould, R. J., Peroutka, G. J., and Snyder, S. H. (1985). *PNAS* **82**, 604–607.
108. Garrison, J. C., Johnsen, D. E., and Campanile, C. P. (1984). *JBC* **259**, 3283–3292.
109. Fain, J. N., Li, S. Y., Litosch, I., and Wallace, M. (1984). *BBRC* **119**, 88–94.
110. Rink, T. J., Sanchez, A., and Hallan, T. J. (1983). *Nature (London)* **305**, 317–319.
111. Di Virgilio, F., Lew, D. P., and Pozzan, T. (1984). *Nature (London)* **310**, 691–693.
112. Kaibuchi, K., Takai, Y., and Nishizuka, Y. (1985). *JBC* **260**, 1366–1369.
113. Truneh, A., Albert, F., Golstein, P., and Schmitt-Verhulst, A. M. (1985). *Nature (London)* **313**, 318–320.
114. Rozengurt, E., Rodriguez-Pena, A., Coombs, M., and Sinnett-Smith, J. (1984). *PNAS* **81**, 5748–5752.
115. Mastro, A. M., and Mueller, G. C. (1974). *Exp. Cell Res.* **88**, 40–46.
116. Kriuijer, W., Cooper, J. A., Hunter, T., and Verma, I. M. (1984). *Nature (London)* **312**, 711–716.

117. Cochet, C., Gill, G. N., Meisenhelder, J., Cooper, J. A., and Hunter, T. (1984). *JBC* **259**, 2553–2558.
118. McCaffrey, P. G., Friedman, B., and Rosner, M. R. (1984). *JBC* **259**, 12502–12507.
119. Iwashita, S., and Fox, C. F. (1984). *JBC* **259**, 2559–2567.
120. Brown, K. D., Blay, J., Irvine, R. F., Heslop, J. P.,and Berridge, M. J. (1984). *BBRC* **123**, 377–384.
121. Moon, S. O., Palfrey, H. C., and King, A. C. (1984). *PNAS* **81**, 2298–2302.
122. Friedman, B., Frackelton, A. R., Ross, A. H., Connors, J. M., Fujiki, H., Sugimura, T., and Rosner, M. R. (1984). *PNAS* **81**, 3034–3038.
123. Hunter, T., Ling, N., and Cooper, J. A. (1984). *Nature (London)* **311**, 480–483.
124. Ullrich, A., Coussens, L., Hayflick, J. S., Dull, T. J., Gray, A., Tam, A. W., Lee, J., Yarden, Y., Libermann, T. A., Schlessinger, J., Downword, J., Mayes, E. L. V., Whittle, N., Waterfield, M. D., and Seeburg, P. H. (1984). *Nature (London)* **309**, 418–430.
125. Lee, L. S., and Weinstein, I. B. (1978). *Science* **202**, 313–315.
126. Shoyab, M., DeLarco, J. E., and Todaro, G. J. (1979). *Nature (London)* **279**, 387–391.
127. Krupp, M. N., Connoly, D. T., Lane, M. D. (1982). *JBC* **257**, 11489–11496.
128. King, A. C., and Cuatrecasas, P. (1982). *JBC* **257**, 3053–3060.
129. Jacobs, S., Sahyoun, N. E., Saltiel, A. R., and Cuatrecasas, P. (1983). *PNAS* **80**, 6211–6213.
130. May, W. S., Jacobs, S., and Cuatrecasas, P. (1984). *PNAS* **81**, 2016–2020.
131. Klausner, R. D., Harford, J., and van Renswoude, J. (1984). *PNAS* **81**, 3005–3009.
132. Shackelford, D. A., and Trowbridge, I. S. (1984). *JBC* **259**, 11706–11712.
133. Corvera, S., and Garcia-Sainz, J. A. (1984). *BBRC* **119**, 1128–1133.
134. Nishikawa, M., Hidaka, H., and Adelstein, R. S. (1983). *JBC* **258**, 14069–14072.
135. Inagaki, M., Kawamoto, S., and Hidaka, H. (1984). *JBC* **259**, 14321–14323.
136. Tsien, R. Y., Pozzan, T., and Rink, T. J. (1982). *Nature (London)* **295**, 68–71.
137. Moolenaar, W. H., Tertoolen, L. G. J., and de Laat, S. W. (1984). *Nature (London)* **312**, 371–374.
138. Sagi-Eisenberg, R., Lieman, H., and Pecht, I. (1985). *Nature (London)* **313**, 59–60.
139. Carafoli, E., and Compton, M. (1978). *Curr. Top. Membr. Transp.* **10**, 151–216.
140. Vincenzi, F. F., and Hinds, T. R. (1980). *In* "Calcium and Cell Function" (W. Y. Cheung, ed.), Vol. 1, pp. 127–165. Academic Press, New York.
141. Limas, C. J. (1980). *BBRC* **96**, 1378–1383.
142. Lagast, H., Pozzan, T., Waldvogel, F. A., and Lew, P. D. (1984). *J. Clin. Invest.* **73**, 878–883.
143. Iwasa, Y., and Hosey, M. M. (1984). *JBC* **259**, 534–540.
144. Movsesian, M. A., Nishikawa, M., and Adelstein, R. S. (1984). *JBC* **259**, 8029–8032.
145. Albert, K. A., Wu, W. C. S., Nairn, A. C., and Greengard, P. (1984). *PNAS* **81**, 3622–3625.
146. Labarca, R., Janowsky, A., Patel, J., and Paul, S. M. (1984). *BBRC* **123**, 703–709.
147. Endo, T., Naka, M., and Hidaka, H. (1982). **105**, 942–948.
148. Katoh, N., Wise, B. C., and Kuo, J. F. (1983). *BJ* **209**, 189–195.
149. Kawamoto, S., and Hidaka, H. (1984). *BBRC* **118**, 736–742.
150. Werth, D. K., Niedel, J. E., and Pastan, I. (1983). *JBC* **258**, 11423–11426.
151. Takai, Y., Kikkawa, U., Kaibuchi, K., and Nishizuka, Y. (1984). *Adv. Cyclic Nucleotide Protein Phosphorylation Res.* **18**, 119–158.
152. Fitzgerald, D. J., Knowles, S. E., Ballard, F. J., and Murray, A. W. (1983). *Cancer Res.* **43**, 3614–3618.
153. Enomoto, T., Martel, N., Kanno, Y., and Yamasaki, H. (1984). *J. Cell. Physiol.* **121**, 323–333.
154. Gainer, H. St. C., and Murray, A. W. (1985). *BBRC* **126**, 1109–1113.

155. Ramachandran, C., Yan, P., Bradbury, E. M., Shyamala, G., Yasuda, H., and Walsh, D. A. (1984). *JBC* **259**, 13495–13503.
156. Hirata, F., Matsuda, K., Notsu, Y., Hattori, T., and del Carmine, R. *PNAS* **81**, 4717–4721.
157. Jolles, J., Zwiers, H., Dekker, A., Wirtz, K. W. A., and Gispen, W. H. (1981). *BJ* **194**, 283–291.
158. Aloyo, V. J., Zwiers, H., and Gispen, W. H. (1983). *J. Neurochem.* **41**, 649–653.
159. Lopez-Rivas, A., Stroobant, P., Waterfield, M. D., and Rozengurt, E. (1984). *EMBO J.* **3**, 939–944.
160. Kishimoto, A., Mori, T., Takai, Y., and Nishizuka, Y. (1978). *J. Biochem. (Tokyo)* **84**, 47–53.
161. Ahmad, Z., Lee, F. T., DePaoli-Roach, A., and Roach, P. J. (1984). *JBC* **259**, 8743–8747.
162. Zwiller, J., Revel, M. O., and Malviya, A. N. (1985). *JBC* **260**, 1350–1353.
163. Schatzman, R. C., Grifo, J. A., Merrick, W. C., and Kuo, J. F. (1983). *FEBS Lett.* **159**, 167–170.
164. Papanikolau, P., Humble, E., and Engström, L. (1982). *FEBS Lett.* **143**, 199–204.
165. Sulakhe, P. V., Petrali, E. H., Daris, E. R., and Thiessen, B. J. (1980). *Biochemistry* **19**, 5363–5371.
166. Cope, F. O., Staller, J. M., Mahsen, R. A., and Boutwell, R. K. (1984). *BBRC* **120**, 593–601.
167. Le Peuch, C. J., Ballester, R., and Rosen, O. M. (1983). *PNAS* **80**, 6858–6862.
168. Nestler, E. J., and Greengard, P. (1983). *Nature (London)* **305**, 583–588.
169. Ehrlich, Y. H., and Routtenberg, A. (1974). *FEBS Lett.* **45**, 237–243.
170. Routtenberg, A. (1979). *Prog. Neurobiol.* **12**, 85–113.
171. Sorenson, R. G., Kleine, L. P., and Mahler, H. R. (1981). *Brain Res. Bull.* **7**, 56–61.
172. Kristjansson, G. I., Zwiers, H., Oestreicher, A. B., and Gispen, W. H. (1982). *J. Neurochem.* **39**, 371–378.
173. Lovinger, D. M., Akers, R. F., Nelson, R. B., Barnes, C. A., McNaughton, B. L., and Routtenberg, A. (1985). *Brain Res.* **343**, 137–143.
174. Routtenberg, A., and Lovinger, D. (1985). *Behav. Neural Biol.* **43**, 3–11.
175. Zwiers, H., Schotman, P., and Gispen, W. H. (1980). *J. Neurochem.* **34**, 1689–1699.
176. Zwiers, H., Jolles, J., Aloyo, V. J., Oestreicher, A. B., and Gispen, W. H. (1982). *Prog. Brain Res.* **56**, 405–417.
177. Routtenberg, A. (1984). *In* "Neurobiology of Learning and Memory" (G. Lynch, J. McGaugh, and N. Weinberger, eds.), pp. 479–490. Guilford Press, New York.
178. Akers, R. F., and Routtenberg, A., (1985). *Brain Res.* **334**, 147–151.
179. DeRiemer, S. A., Strong, J. A., Albert, K. A., Greengard, P., and Kaczmarek, L. K. (1985). *Nature (London)* **313**, 313–316.
180. Krebs, E. G., and Beavo, J. A. (1979). *Annu. Rev. Biochem.* **48**, 923–959.
181. Kishimoto, A., Nishiyama, K., Nakanishi, H., Uratsuji, Y., Nomura, H., Takeyama, Y., and Nishizuka, Y. (1985). *JBC* **260**, 12492–12499.
182. Katz, A. M., Tada, M., and Kirschberger, M. A. (1975). *Adv. Cyclic Nucleotide Res.* **5**, 453–472.
183. Heyworth, C. M., Whetton, A. D., Kinsella, A. R., and Houslay, M. D. (1984). *FEBS Lett.* **170**, 38–42.
184. Kaibuchi, K., Takai, Y., Ogawa, Y., Kimura, S., Nishizuka, Y., Nakamura, T., Tonomura, A., and Ichihara, A. (1982). *BBRC* **104**, 105–112.
185. Sugden, D., Vanecek, J., Klein, D. C., Thomas, T. P., and Anderson, W. B. (1985). *Nature (London)* **314**, 359–361.
186. Bell, R. L., Kennerley, D. A., Stanford, N., and Majerus, P. W. (1979). *PNAS* **76**, 3238–3241.

187. Murad, F., Arnold, W. P., Mittal, C. K., and Braughler, J. M. (1979). *Adv. Cyclic Nucleotide Res.* **11**, 175–204.
188. Hidaka, H., and Asano, T. (1977). *PNAS* **74**, 3657–3661.
189. Graff, G., Stephenson, J. H., Glass, D. B., Haddox, M. K., and Goldberg, M. D. (1978). *JBC* **253**, 7662–7676.
190. Haslam, R. J., Salama, S. E., Fox, J. E. B., Lynham, L. A., and Davidson, M. M. L. (1980). *In* "Platelets; Cellular Response Mechanisms and their Biological Significance" (A. Rotman, F. A. Meyer, C. Gitler, and A. Silderberg, eds.), pp. 213–231. Wiley, New York.
191. Takai, Y., Kaibuchi, K., Matsubara, T., and Nishizuka, Y. (1981). *BBRC* **101**, 61–67.

6

Viral Oncogenes and Tyrosine Phosphorylation

TONY HUNTER • JONATHAN A. COOPER*

Molecular Biology and Virology Laboratory
The Salk Institute
San Diego, California 92138

*Present address: Fred Hutchinson Cancer Research Center, 1124 Columbia Street, Seattle, Washington 98104

191

THE ENZYMES, Vol. XVII

I. Introduction and Historical Perspective

For nearly thirty years phosphorylation has been recognized as a means of reversibly modulating the function of proteins. Over this period many protein kinases specific for serine and threonine residues in their substrates have been identified and characterized. However, not until the 1980s have enzymes with specificity for tyrosine, the third hydroxyamino acid, been reported. Tyrosine phosphorylating activity was originally detected in partially purified preparations of two viral transforming proteins (1, 2). Since that time it has become clear that there is a large family of protein-tyrosine kinases. Eight such cellular enzymes were first recognized in altered guises as parts of the transforming proteins of a series of acutely oncogenic retroviruses, while five other protein-tyrosine kinases are growth factor receptors. Over fifteen distinct protein-tyrosine kinase genes have been identified with the prospect of several more to follow.

The seminal discovery of a protein kinase activity associated with $pp60^{v\text{-}src}$ (3), the product of the Rous sarcoma virus (RSV) *src* gene, immediately suggested that transformation could be due to aberrant protein phosphorylation events that would modulate the functions of critical cellular proteins in an abnormal fashion. This idea was reinforced by the unexpected finding that $pp60^{v\text{-}src}$ phosphorylated tyrosine rather than serine or threonine (4). The subsequent demonstration that several other distinct retroviral oncogene products had similar tyrosine phosphorylating activities lent further credence to this notion. Out of twenty known viral oncogenes, there are eight whose products are protein-tyrosine kinases: namely, the v-*src*, v-*yes*, v-*fgr*, v-*fps* and v-*fes*, v-*abl*, v-*ros*, v-*erb*-B and v-*fms* oncogenes. Abnormal patterns of typrosine phosphorylation are manifest in cells transformed by most of the relevant viruses, and several substrates for the viral protein-tyrosine kinases have been identified. Nevertheless, despite extensive efforts, the proof that tyrosine phosphorylation is in fact critical for transformation by these viruses is not at hand. The problem has been to pinpoint substrates for the viral protein-tyrosine kinases whose functions are changed in ways that would explain the transformed phenotype.

The aim of this chapter is to summarize our knowledge of the viral protein-tyrosine kinases and their roles in viral transformation. For this purpose we have reviewed the properties of the individual viral protein-tyrosine kinases and contrasted them to the cognate cellular enzymes that are encoded by the cellular genes homologous to the viral oncogenes. We have included a discussion of a number of transformed cell types where altered tyrosine phosphorylation is evident, but which do not involve one of the aforementioned viral oncogenes. The general properties of protein-tyrosine kinases and their common structural features are described. We conclude with a progress report on the identification of substrates for the viral protein-tyrosine kinases. Throughout we have tried to address the question of how the viral protein-tyrosine kinases differ from their cellular counterparts, enzymes that clearly coexist peaceably with normal cells. In many places we have found it pertinent to compare and contrast the properties of the viral enzymes to those of the growth factor receptor protein-tyrosine kinases, particularly because of the abnormal growth state of transformed cells. For a more detailed review of the growth factor receptor protein-tyrosine kinases, however, the reader is referred to Chapter 7 by Morris White and C. Ronald Kahn.

II. Individual Viral Protein-Tyrosine Kinases and Their Cellular Homologues

A. pp60^{v-src} AND pp60^{c-src}

The complete sequence of the *src* gene of RSV predicts a primary translation product of 526 amino acids (*5–7*). In fact pp60^{v-src} is modified by the removal of the initiating methionine and the subsequent linkage of a myristyl group to the glycine which is exposed (*8, 9*), leaving a mature protein of 525 amino acids. The pp60^{v-src} was originally identified by immunoprecipitation from RSV-transformed cells with serum from an RSV tumor-bearing rabbit (TBR serum) (*10*). The protein kinase activity of pp60^{v-src} was also first detected in immunoprecipitates made with TBR serum, in this case leading to the phosphorylation of the immunoglobulin heavy chain (*3*). TBR sera contain variable amounts of antibodies to viral structural proteins in addition to those against pp60^{v-src}. Some TBR sera cross-react with pp60^{v-src} from many strains of RSV as well as pp60^{c-src}, while others are specific for pp60^{v-src}. Interestingly while immunoprecipitates of pp60^{v-src} made with TBR serum do not exhibit autophosphorylation or phosphorylation of exogenous substrates, similar immunoprecipitates of pp60^{c-src} will both autophosphorylate and phosphorylate added proteins in addition to phosphorylating heavy chain. A number of other reagents are available for immunoprecipitation of pp60^{v-src}, some of which recognize pp60^{c-src} as well. Antibodies against bacterially expressed pp60^{v-src} cross-react with pp60^{c-src} and recognize a population of pp60^{v-src} that is largely perinuclear, which is not detected with antitumor sera (*11, 12*). Several hybridomas derived from mice immunized with bacterially expressed pp60$^{v\text{-}src}$ produce monoclonal antibodies (*13, 14*) which have specificities similar to polyclonal antisera against this protein. The major antigenic determinants for both the polyclonal and monoclonal antibacterial pp60$^{v\text{-}src}$ are in the N-terminal half of pp60^{src}. Both types of antibody allow autophosphorylation of pp60^{src} and phosphorylation of exogenous substrates. Antipeptide antisera directed against amino acids 409–419 (*15, 16*), 498–512 (*17*) and against the C-terminal 6 residues (521–526) (*18*) immunoprecipitate pp60$^{v\text{-}src}$. The latter does not recognize pp60$^{c\text{-}src}$ due to its different C-terminus. Antibodies to the 409–419 (*16*) and 498–512 (*17*) *src* peptides inhibit pp60$^{v\text{-}src}$ protein kinase activity, while immunoprecipitates of pp60$^{v\text{-}src}$ made with antibodies to the 521–526 *src* allow phosphorylation of exogenous substrates (*18*). A number of antibodies specific for synthetic peptides corresponding to sequences from pp60$^{v\text{-}src}$ have been affinity purified with TBR sera (*19*).

Although pp60$^{v\text{-}src}$ is synthesized on soluble ribosomes, the bulk of the protein is found associated with cellular membranes, which include the plasma membrane and perinuclear membranes (*12, 13, 19–23*). However, pp60$^{v\text{-}src}$ is not an integral membrane protein but is bound in a peripheral fashion. Newly synthesized molecules of pp60$^{v\text{-}src}$ are associated in a complex with two cellular proteins, pp89 and pp50 (*24, 25*); pp89 has been identified as one of the major stress-induced proteins, but the function of pp50 is unknown. It has been proposed that this complex serves to transport pp60$^{v\text{-}src}$ to the membrane. A mutant pp60$^{v\text{-}src}$ in which the N-terminal glycine has been replaced by an alanine is not myristylated and is found to be largely soluble in the cell (*26*). Such pp60$^{v\text{-}src}$ mutants appear to be fully active as protein-tyrosine kinases but are nontransforming (*26, 27*). This implies that association of pp60$^{v\text{-}src}$ with the membrane, presumably via the myristyl group, is essential for transformation. The pp60$^{v\text{-}src}$ accounts for about 0.05% of total cell protein in cells transformed by RSV (*28*). The half-life of pp60$^{v\text{-}src}$, depending on the strain of RSV, ranges from 2 to 7 h (*28, 29*).

pp60$^{v\text{-}src}$ has two functional domains. The C-terminal 30,000 daltons (residues 270–526, numbering from Met 1) have been shown to contain the entire catalytic domain on the basis of limited proteolysis (*30, 31*). The sequence in this region of pp60$^{v\text{-}src}$ shows striking homology to that of the catalytic subunit of the cAMP-dependent protein kinase (*32*) (see Fig. 1). The homology includes a number of key sequence motifs, known to be critical for function, which will be discussed in Section IV. A similar catalytic domain is also recognizable in all the other protein-tyrosine kinases. As previously mentioned the very N-terminus of pp60$^{v\text{-}src}$ plays an essential role in the association of pp60$^{v\text{-}src}$ with cellular membranes. Two lines of evidence suggest that the rest of the N-terminal domain is involved in regulating catalytic activity. First, some of the point mutations that render pp60$^{v\text{-}src}$ temperature-sensitive for phosphotransferase activity map between residues 2 and 270

(*33*). In addition deletion mutations in this region can either render catalytic activity temperature-sensitive [e.g., deletion of residues 173–227 or 169–225 (*34, 35*)], affect the phenotype of the transformed cell [e.g., deletion of residues 16–82, 16–138 (*36*), 15–169 (*37*), or 135–236 (*38*)], or even abolish transforming activity altogether [e.g., deletion of residues 82–169 (*35*)]. Second, this domain of pp60^{v-src} also contains several phosphorylation sites, occupation of which can alter phosphotransferase activity.

A great deal of effort has gone in to determining both the sites and effects of

---

FIG. 1. (See pp. 196–197.) The amino acid sequences of the catalytic domains of 4 protein-serine kinases, myosin light chain kinase (MLCK) (*250*), phosphorylase γ-subunit (PhK-γ) (*249*), cGMP-dependent protein kinase (cGPK) (*248*), and cAMP-dependent protein kinase catalytic subunit (cAPK) (*247*) are aligned with the predicted sequences of 12 protein-tyrosine kinases: pp60^{v-src} (*src*) (Schmidt-Ruppin A strain of RSV) (*6*), chicken pp60^{c-src} (c-*src*) (*52*), P90$^{gag-yes}$ (*yes*) (*67*), P70$^{gag-fgr}$ (*fgr*) (*75*), P120$^{gag-abl}$ (*abl*) (*110*), P140$^{gag-fps}$ (*fps*) (*80*), P110$^{gag-fes}$ (*fes*) (Gardner-Arnstein strain) (*81*), P68$^{gag-ros}$ (*ros*) (*136*), the human insulin receptor (INS.R) (*245*), gp66/68$^{v-erb-B}$ (*erb*-B) (AEV-H strain) (*144*), human EGF receptor (EGF.R) (*159*), gP180$^{gag-fms}$ (*fms*) (*168*), and the human c-*fms* protein (c-*fms*) (only the C-terminal 97 residues are shown) (*183*). Finally the sequences of 4 other protein kinase-related proteins are given: P100$^{gag-mil}$ (*mil*) (*232*), P90$^{gag-raf}$ (*raf*) (the sequence shown starts at position 387; there are *gag* gene sequences upstream of this residue) (*233*), p37^{mos} (*mos*) (*235*), and its murine cellular counterpart (c-*mos*) (*235*). The sequences were aligned by eye for maximum homology. The single letter amino acid code has been used, with . representing a gap introduced to optimize homology and with - representing an identity between the v-*onc* sequence and the corresponding c-*onc* sequence (remember that while the chicken c-*src* and mouse c-*mos* sequences are strictly comparable to the chicken-derived v-*src* and mouse-derived v-*mos* genes respectively, the human EGF receptor is compared to the chicken-derived v-*erb*-B gene, and the human c-*fms* protein is compared to the cat-derived v-*fms* gene). Residue numbers are indicated: x/, starting number; /x, finishing number; (x) terminus. In the case of the MLCK and c-*fms* sequences, the precise position of the sequence shown in the protein is not known, but the C-terminal 368 residues of the MLCK sequence are presented (*250*). Residues that are common to all 4 protein-serine kinases are given underneath the cAPK lines in capital letters; residues that are found in 3 of the enzymes are shown in lower case letters, while residues that are highly conserved with regard to properties are indicated *. The same analysis for the protein-tyrosine kinase group is depicted above the *src* lines. No analysis was performed for the *mil, raf, mos,* and c-*mos* sequences, but residues that are conserved between all 20 sequences are indicated by vertical lines connecting the conserved residues between the protein-serine kinase and protein-tyrosine kinase groups. The lysine (K) which is modified by the ATP analog FSBA is in bold type, as are the threonine (T) in the cAMP-dependent protein kinase catalytic subunit which is autophosphorylated and the tyrosine (Y) which is autophosphorylated in the viral protein-tyrosine kinases. The C-termini of pp60^{v-src}, P90$^{gag-yes}$, P70$^{gag-fgr}$, P68$^{gag-ros}$, and gP180$^{gag-fms}$ are in bold type where they diverge from the corresponding cellular sequence (these cellular sequences are not known in the case of the c-*yes* and c-*fgr* proteins). Within the protein-tyrosine kinase family it should be noted that the v-*fps* and v-*fes* genes correspond to the same genetic locus in chickens and cats respectively (*82*). This is also true for the v-*mil* and v-*raf* genes which correspond to the same locus in chickens and mice respectively (*232–234*). For those interested in comparing additional sequences of protein kinase catalytic domains, the human and chicken c-*mos* gene sequences have been completed, while the *neu* gene (*209*) and p56^{lstra} (*228, 229*) sequences are also available. In addition there are at least two other *raf*-related genes, and a cDNA corresponding to one of these has been sequenced (*pks*).

This figure presents a multiple sequence alignment of protein kinase domains. The sequences are labeled (left to right groups): MLCK, PhK-γ, cGPK, cAPK, and the tyrosine/serine kinase family members src, c-src, yes, fgr, abl, fps, fes, ros, INS.R, erb-B, EGF.R, fms, mil, raf, mos, c-mos.

```
MLCK     ?/LCLPAREEDCFQ ILDDCPPPPAPPHRIVELRTGNVSSEFSMNSKEALGGGKFGAVCCTCEKSTGLK......LAAKVIKQTP.....KDKEMVMLEIEVMNQLN.HRNLIQLY
PhK-γ        (1)TRDAALPGSHSTHGFYENYEPKEILGRGVSSVVRRCIHKEPTCKE...YAVKIIDVTGGGSFSAEEVQ.ELREATLKEVDILRKVSGHPNIIQLK
cGPK    318/DSFPKHLIGGLDDVSNKAYEDAEAKAKYEAEAAFFANLKLSDFNIIDTLGVGGFGRVELVQLKSEESKT......FAMKILKRHIVDT.....RQQEHIRSEKQIMQGAH.SDFIVRLY
cAPK        (1)GNAAAAKKGSEQESVKEFLAKAKEDFLKKWENPAQNTAHLDQFERIKTLLGTSPGRVMLVKHMETGNH.....YAMKILDKQKVVKL.....KQIEHTLNEKRILQAVN.FPFFVKLE
                                                                                    *A Ki*           E E i    ** L
                                                                                    VA*K *                   L *
                                                                                                             L e*

src     226/LVAYYSKHADGLCHRLANVCPTSKPQTQGLAKDAWEIPRESLRLEAKLGQGCFGEVWMGTWNDTTR.....VAIKTLKPGTMSPE....AFLQEAQVMKKLRHEKLVQLYAVV.SE
c-src   226/-----------------T------------------------------------------------V-----------G------.......................................
yes     510/LVKHYREHADGLCHKLTTVCPTVKPQTQGLAKDAWEIPRESLRLEVKLGQGCFGEVWMGTWNGTTK.....VAIKTLKLGTMMPE....AFLQEAQIMKKLRHDKLVPLYAVV.SE
fgr     363/LVQHYREHVNDGLCHLLTAACTTMKPQTMGLAKDAWEISRSSITLLQRRLGTGCFGDVWLGMWNGSTK....VAVKTLKPGTMSPK....ASLEEAQIMKLLRHDKLVQLYAVV.PE
abl     323/HSTVADGLITTLHYPAPKRNKPTILYGVSPNYDKWEMERTDITMKHKLGGGQYGEVYEGVWKKYSLT....VAVKTLKEDTMEVE....EFLKEAAVMKEIKHPNLVQLLGVCTRE
fps     880/LPTIPLLIDHLLSQRPITRKSGIVLFRAVLRDKWVLNHEDVLLGERIGRGNFGEVFSGRLRADNTP.....VAVKSCRE.TLPPELKA....KFLQEARILKQCNHPNIVRLIGVCTQK
fes     656/FASIPLLVDHLLRSQQPLTKKSGIVLNRAVPRDKWVLNHEDLVLGEQIGRGNFGEVFSGRLRADNTL....VAVKSCRE.TLPPDIKA....KFLQEARILKQISHPNIVRLIGVCTQK
ros     217/KELAQLRGMAETVGLANACYAVSTLPSQAEIESLPAFPRDKLNLHKLLGSGAFGEVYEGTALDILADGSGESRVAVKTLKRGATDQEKE...EFLKEAHLMSKFDHPHILKLLGVCLLN
INS.R   943/RQPDGPLGPLYASSNREYLSASDVFPCSVVPDEWEVSREKITLLRELGQGSFGMVYEGNARDIIKGEAETR.VAVKTVNESASLRERI....EFLNEASVMKGFTCHHVVRLLGVVSKG
erb-B   98/HIVRKRTLRRLLQERELVEPLTPSGEAPNQAHLRILKETEFKKV.KVLGSGAFGTVYKGLWIPEGEKVTIP.VAIKELREATSPKANK....EILDEAYVMASVDNPHVCRLLGICLTS
EGF.R   649/-----------------------I--------------V------------------------------V---------------------..............................
fms    1108/KQKPKYGVRWKIIESYEGNSVTFIDPTQLPYNEWEFPRNNLQFGKTLGTGAFGKVVEATAFGIGKEDAVLK.VAVMLKSTAHADEKEALM.SELKIMSHLGQ..HENIVNLLGACTHG
                                                *G fG v g                              *G  K *
                                                 LG fG fg V                            VA*K *
                                                                                                         i

mil     515/PKTPVPAQRERAPGTNTQEKNKIRPRGQRDSSYYWEIEASEVLLSTRIGSGSFGTVYKGKWHGD.....AFRNEVAVLRKTRHVNI..LLFMGYMT
raf    gag-387/TQEKNKIRPRGQRDSSYYWKMEASEVMLSTRIGSGSFGTVYKGKWHGD.....AFRNEVAVLRKTRHVNI..LLFMGYMT
mos     53/RSCSIPLVAPRKAGKLFGTTPPRARGLPRRLAWFSIDWEQVCLMHRLGSGGFGSVYKATYHGVP....VAIKQVNKCTEDLRASQR.SFWAELNIAGL.RHDNIVRVVAASTAS
c-mos   22/--------------------------------------------------------------------------------R-
```

```
MLCK    AAIETHEIVLFMEYIEGGELFERIVDEDYH.............                         TEVDTMVFRQICDGILFM
PhK-γ   DFYETNTFFPVFDLMKKGELFDYLTEKVTLS..........                          EKETRKIMRALLEVICAL
cGPK    RTPKDSKYLYMLMEACLGGELWTILRDRGSPE.........                          DSTTRFYTACVVEAFAYL
cAPK    FSFKDNSNLYMVMEYVPGGEMFSHLRRIGRFS.........                          EPHARFYAAQIVLFEYL
                gGElf l                                                          l
        me                                                                    r
              G 11 *1*

src     EPIYIVIE......YMSKGSLLDFLKGEMGKYLR.....                            LPQIVDMAAQIASGMAYV
c-src   ------------------------------------------                        ---------------------
yes     EPIYIVTE......FMTKGSLLDFLKEGEGKFLK.....                            LPQIVDMAAQIADGMAYI
fgr     EPIYIVTE......FMCHGSLLEFLKDQEGQDLT.....                            LPQLVDMAAQVAEGMAYM
abl     PPFYIITE......FMTYGNLLDYLRECNRQEVS.....                            AVLLYMATQISSAMEYL
fps     QPIYIVME......LVQGGDFLSFLRSKGPRLK......                            MKKLIKMMENAAAGMEYL
fes     QPIYIVME......LVQGGDFLTFLRTEGARLR......                            MKTLLQMVGDAAAGMEYL
ros     EPQYLILE......LMEGGDLLSVLRGARKQFQSPLLT..                           LTDLLDICLDICKGCVYL
INS.R   QPTLVVME......LMAHGDLKSYLRSLRPEAENNGRPPPT.                         LQEMIQMAAEIADGMAYL
erb-B   TVQ.LITQ......LMPYGCLLDYIREHKDNIG.....                             SQYLLNWCVQIAKGMNYL
EGF.R   --------F----V---I---------------------
fms     THGGPVLVITE...YCCYGDLLNFLRRQAEAMPGPSLSVGQDPEAGAGYKNIHLEKKYVARDSGFSSQGVDTYVEMRPVSTSSSNDSFSEEDLGKEDGRPLELRDLLHFSSQVAQGMAFL
               *                                                                   gm y
                                                                                  LPQIVDMAAQIASGMAYV

mil     KDNLAIVTQ......WCEGSSLYKHLHVQETKFQ.....                            MFQLIDIARQTAQGMDYL
raf     KDNLAIVTQ......WCEGSSLYKHLHVQETKFQ.....                            MFQLIDIARQTAQGMDYL
mos     TRTPEDSNSLGTII.MEFGGNV.TLHQVIYDATRSPELSCRKQLS..                    LGKCLKYSLDVVNGLLFL
c-mos   22/---------------------------------E--
```

FIG. 1.

```
              [catalytic domain, continued]

MLCK   HKMRVLHLDLKPENILCVNTTGH....LVKIIDFGLARRYN.PNEKLKVN.....PGT...PE.FLSPEVNVD.....QISDKTDMWSLGVITYMLLS.GLSPFLGDDTETLNNVLS.GN
PhK-γ  HKLNIVHRDLKPENILLDDDM.....NIKITDFGFSCQLD.PGEKLREV..CGT...PS.YLAPEIIECSMNDNHPGYGKEVDMWSTGVIMYTLLA.GSPPFWHRKQMLMIRMIMS.GN
cGPK   HSKGIIYRDLKPENLILDHRG....KLKIIDFGFAKKIG.FGKRTWTF..CGT...PE.YLAPEIISLGHDISADYWSLGILMYELLT.GSPPFSCGPDPMKTYNIILR.GI
cAPK   HSLDLIYRDLKPENLLIDQQG.....YIQVTDFGFAKR..VKGRTWTL..CGT...PE.YLAPEIILSK.......GYNKAVDWWALGVLIYEMAA.GYPPFFADQPIQIYEKIVS.G.
       H       rDLKPEN** d                k    DFGfa          cGT        Pe  yLaPEii    g        D WslGv* Y ll  G  pPF       *  G

       e    *HRDl A N lv            K* DFG  r         Y  G       ApE        p kW ApE                SDvWsfG*ll E      g    Py
src    ERMNYIHRDLRAANILVGENL.....VCKIADFGLARL...IEDNEYTARQGAKF.PIKWTAPEAALYG....RFTIKSDVWSFGILLTELVTKGRVPYPGMVNREVLEQVER.GY
c-src  ERMNYIHRDLRAANILVGENL.....VCKIADFGLARL...IEDNEYTARQGAKF.PIKWTAPEAALYG....RFTIKSDVWSFGILLTELTTKGRVPYPGMVNREVLDQVER.GY
yes    ERMDYIHRDLRAANILVGERL.....VCKIADFGLARL...IEDNEYNPRQGAKF.PIKWTAPEAALPG....RFTIKSDVWSFGILLTELISKGRVPYPGMNNREVLEQVEH.GY
fgr    EKKNHIHRDLAARNCLVGENH.....LVKVADFGLSRL...MTGDTYTAHAGAKF.PIKWTAPESLAYN....KFSIKSDVWAFGVLLWEIATVGMSPYPGIDLSQVYELLEK.DY
abl    ESKHCIHRDLAARNCLVTEKN.....TLKISDFGMSRQ...EEDGVYASTGMKQIPVKWTAPEALNYG....WYSSESDVWSFGILLWEAFSLGAVPYANLSNQQTREAIEQ.GV
fps    ESKCCIHRDLAARNCLVTEKN.....VLKISDFGMSRE...EADGVYAASGLRLVPVKWTAPEALNYG....RYSSESDVWSFGILLWETFSLGASPWPNLSNQQTREFVEK.GG
fes    EKMRFIHRDLAARNCLVGENQ.....IYKNDYYRKREGEGLLPVRWMAPESLIDG....VFTNHSDVWAFGVLVWETLTLGQQPYEGLSNIEVLHHVRS.GG
ros    NAKKFVHRDLAARNCMVAHDF.....TVKIGDFGMTRD...IYETDYYRKGKGLLPVRWMAPESLKDG....VFTTSDMWSFGVVLWEITSLAEQPYQGLSNEQVLKFVMD.GG
INS.R  EERRLVHRDLAARNVLVKTPQ.....HVKITDFGLAKLLG.ADEKEYHAEGG.KV.PIKWMALESISSVLEK.GE  IYTHQSDVWSYGVTVWELMTFGSKPYDGIPASEISSVLEK.GE
erb-B  -D--..............................-E----------..-B--------..........-I---I---
EGF.R  ASKNCIHRDVAARNVLLTSGR.....VAKIGDFGLARDI..MNDSNYIVK.GNARLPVKWMAPESIFDC.....VTYVQSDVWSYGILLWEIFSLGLNPYPGILVNSKFYKLVKDGY

mil    HAKNIIHRDMKSNNIFLHEGL.....TVKIGDFGLATVKSRWSESQQVEQPTG....SILWMAPEVIRMQDSN...PFSFQSDVYSYGIVLYELMT.GELPYSHINNRDQIIFMVGRY
raf    HAKNIIHRDMKSNNIFLHEGL.....TVKIGDFGLATVKSRWSGSQQVEQPTG....SVLWMAPEVIRMQDDN...PFSPQSDVYSYGIVLYELMA.GELPYAHINNRDQIIFMVGRY
mos    HSQSIHLDLKPANILISEQD.....VCKISDFGCSQKLQDLRGRQASPPHIGG..TYTHQAPEILKGE......IATPKADIYSFGITLWQMTT.REVPYSGEPQYVQVAVAYNLR
c-mos  ...............................C---H---


              [carboxy-terminal region]

MLCK   WYFDEETFEAVSDE..AKDPVSNLIVKEQGARMSAAQCLAHPW.LNN.LAEKAKRCNRRLKSQILLKKYLMKRRWKKNFIAVSAANRFKKISSSGALM(?)
PhK-γ  YQPGSPEWDDYSDT..VKDLVSRFLVVQPQKRYTAEEALAHPWFFQQYVVEEVRHFSPRGKFKVICLTVLASVRIYYQYRRVKPVTREIVIRDPYALRPLRRLIDAYAFRIYGHWV/358
cGPK   DMIEFPPKIA.RN...AANLIKKLCRDNPSERLGNLKWGVKDIQKHKWFEG.FNWEGLRKGTLTPPIIPSVASPTDTSNFDSFPEDNDEPPPDDNSGWDI.DF(670)
cAPK   .KVRFPSHFS.SD...LKDLLRNLLQVDLTKRFGNLKDGVNDIKNHKWFATTDWIAIYQRKVEAPFIPKFKGPGDTSNFDDYEEEEIRVSINEKCGKEFSEF(349)
       *    p  sd            d1              R        e          r

       *    P  cp            *M  cW          RP F    *
src    RMPCPECPES....LHDLMCQCWRKDPEERPTFKYLQAQLLPACVLEVAB(526)
c-src  ...........R---------E----FLEDYFTSTEPQYQPGENL(533)
yes    RMPCPQGCPES....LHGLMKLCWKKDPDERPTFEYIQSFLEDYFTAABPSGY(812)
fgr    RMPCPPGCPAS....LYEAMEQTWRLDPEERPTFEYLQSPLEDYFNGPQQN(663)
abl    HMERPEGCPEK....VYELMRACWQWNPSDRPSFAEIHQAFETMFQESSISDEVKELGKRGTRGGAGSMLQAPELPTKTRCRRAAEQKDAPDTPELLHTKGLGESDALDSEP/687
fps    RLEPPEQCPED....VYRLMQRCWEYDPHRRPSSGAVHQDLIAIRKHR(1182)
fes    RLPCPELCPDA....VFRLMEQCWAYEPGQRPSFSAIYQELQSIRKHR(958)
ros    RLESPNNCPDD....IRDLMTRCWAQDPHNRPTFYYIQHKLQEIRHSPLCFSYFGLGDKESVAPLRIQTAPPQPL(552)
INS.R  YLDQPDNCPER....VTDLMRMCWQFNPNMRPTFLEIVNLLKDDLHPSFPEVSFFHSEENKAPESEELEMEPEDMENVPLDRSSHCQREEAGGRDGSSSLGFLRSYEEHIPYTH/1324
erb-B  RLPQPPICTID....VYMINVKCWMIDADSRPKFRELIAEFSKMARDPPRVIVIQGDERMHLPSPTDSKFYRTLMDEEDMEDIVDADEYLVPHQGFFNSPSTSRTPLLSSLSA/467
EGF.R  ...................N---A--D----D--V-----I-Q----S------/1016
fms    QMAQPAFAPFN....IYSIMQACWALEPTRRPTFQQICSLIQKQAQEDRRVPNYTNLPSSSSRLLRPWRGPPL(1511)
c-fms  ...................R--GGSGSSSELESSSEHLTCCEQGDIAQPLIQPNNYQF(?)

mil    ASPDLSKLYKNCP.KAMKRLVADCLKKVREERPLFPQILSSIALLQHSLPKINRSASEPSLHRASHTEDINSCTLTST.RLPVP(894)
raf    ASPDLSRLYKNCP.KAIKRLVADCVKKVKEERPLFPQILSSIELLQHSLPKINRSAPESLHRAAHTEDINACTLTTSPRLPVF(710)
mos    PSLAGAVFTASLTGKAIQNIIQSCWEARGLQRPSAELLQRDLKAFRGTLG(374)
c-mos  ...................A--(343)
```

FIG. 1. (Continued.)

197

phosphorylation on $pp60^{v-src}$. There is a major site of tyrosine phosphorylation at residue 416 (*39, 40*); other minor sites of tyrosine phosphorylation are located in the N-terminal half of the molecule but have not been mapped precisely (*41–44*). All these sites can apparently be phosphorylated by $pp60^{v-src}$ itself in a so-called "autophosphorylation" reaction. Whether this is a true intramolecular reaction or not is unresolved. $pp60^{v-src}$ is also phosphorylated on serine with the major site being Ser-17 (*36, 40*). This serine can be phosphorylated by the cAMP-dependent protein kinase *in vitro* (*45*), and there is evidence that elevation of cAMP levels in some RSV-transformed cell lines leads to increased N-terminal phosphorylation of $pp60^{v-src}$ (*46*), presumably at this site. Recently it has been shown that there are other sites of serine phosphorylation in the N-terminal half of the protein which can be phosphorylated by the Ca^{2+}-phospholipid-dependent diacylglycerol-activated protein kinase, protein kinase C (*47, 48*). Ser-12 and to a lesser extent Ser-48 are phosphorylated in $pp60^{v-src}$ isolated from cells treated with phorbol esters, compounds which activate protein kinase C (*48*). In RSV-transformed chick cells it has been estimated that about 60% of $pp60^{v-src}$ molecules are phosphorylated at Ser-17, while 30% are phosphorylated at Tyr-416 (*28*). Upon treatment with phorbol esters at least 50% of $pp60^{v-src}$ molecules are phosphorylated at Ser-12 (*48*).

The effects of these phosphorylations on the catalytic activity of $pp60^{v-src}$ have not been determined with certainty. Phosphorylation at Ser-17 may increase activity a few-fold (*46*), but deletion of this residue by site-directed mutation does not abolish either protein kinase activity or transforming ability (*36*). The effects of phosphorylation at Ser-12 have not yet been determined, but, by analogy with phosphorylation at the neighboring Ser-17, might be expected to increase catalytic activity. Phosphorylation at Tyr-416 has also been reported to increase the activiiy of $pp60^{v-src}$ (*49*), but once again a mutant $pp60^{v-src}$ in which Tyr-416 has been replaced with a phenylalanine is able to function as a protein kinase and can transform (*50, 51*). It appears that none of these phosphorylations is essential for activity but may provide a means of positive regulation.

In most respects $pp60^{c-src}$ is very similar to $pp60^{v-src}$, except at its C-terminus. From residues 1 to 514 chicken $pp60^{c-src}$ differs from $pp60^{v-src}$ by amino acid substitutions in a few scattered positions, the exact number depending on the strain of RSV. Starting at residue 515, however, the proteins diverge completely (*52*); $pp60^{c-src}$ has 19 additional amino acids encoded by contiguous sequences beyond those for residue 514, thus generating a protein with 533 amino acids. In contrast the 12 C-terminal amino acids of $pp60^{v-src}$ are derived from sequences about 1 kilobase downstream of the true C-terminus of the c-*src* gene as a consequence of a deletion that occurred during the genesis of RSV (*52*).

In primary fibroblasts and most cell lines $pp60^{c-src}$ accounts for about 0.005% of total cell protein, which is a considerably lower level than $pp60^{v-src}$ in RSV-transformed cells. There are cell types, however, where $pp60^{c-src}$ is more abundant. Certain neural cell types have about five times as much pp60$^{c\text{-}src}$ as fibroblasts (53, 54), while platelets have even higher levels (55). The majority of pp60$^{c\text{-}src}$ in rat neurons migrates more slowly than pp60$^{c\text{-}src}$ from rat fibroblasts on gel electrophoresis (53) and appears to be altered in its N-terminal half. The existence of a unique form of pp60$^{c\text{-}src}$ in neurons raises the possibility that multiple forms of pp60$^{c\text{-}src}$ could be produced from a single gene, for instance by alternate splicing, as occurs for the c-abl gene. Like pp60$^{v\text{-}src}$, pp60$^{c\text{-}src}$ is modified by myristylation of its N-terminal glycine (8), and most of pp60$^{c\text{-}src}$ appears to be membrane-associated in a fashion very similar to pp60$^{v\text{-}src}$ (21), although whether the microscopic distribution of the two proteins is the same is not certain.

pp60$^{c\text{-}src}$ is also phosphorylated on both serine and tyrosine. Tyr 416, however, is not a major phosphorylation site in pp60$^{c\text{-}src}$, even though the sequence around Tyr 416 is identical to that in pp60$^{v\text{-}src}$ (52). Another tyrosine in the C-terminal half is phosphorylated, which has been shown to be Tyr 527 in the unique C-terminal tail of pp60$^{c\text{-}src}$ (55a). Interestingly, pp60$^{c\text{-}src}$ autophosphorylates at Tyr 416 in vitro (39). Phosphorylation of Ser 17 in pp60$^{c\text{-}src}$ is observed in vivo (56), and this residue is phosphorylated in vitro by the cAMP-dependent protein kinase (57). The pp60$^{c\text{-}src}$ is also phosphorylated by protein kinase C at Ser 12 (48). The effects of these phosphorylations on the catalytic activity of pp60$^{c\text{-}src}$ are currently under investigation. There is mounting evidence that phosphorylation of Tyr 527 in pp60$^{c\text{-}src}$ acts as a negative regulator of its protein kinase activity (58). The c-yes and c-fgr proteins have a tyrosine in a homologous sequence near their C-termini, and these enzymes may also be negatively regulated by phosphorylation at this site.

Considerable effort has gone into trying to understand how pp60$^{v\text{-}src}$ causes transformation. Because of the similarity between pp60$^{c\text{-}src}$ and pp60$^{v\text{-}src}$ it was originally proposed that transformation might simply be due to the 10–50-fold greater levels of pp60$^{v\text{-}src}$ that are present in RSV-infected cells. This idea has now been rigorously tested by the expression of a cloned c-src gene in susceptible fibroblasts at levels equal to or greater than those usually achieved for pp60$^{v\text{-}src}$ (59–62). Such cells are at best partially transformed. Therefore it appears that there is a qualitative difference between pp60$^{v\text{-}src}$ and pp60$^{c\text{-}src}$. In vitro recombination experiments between cloned v-src and c-src genes show that the alteration at the C-terminus of pp60$^{v\text{-}src}$ is a major factor responsible for the transforming ability of this protein (59), although there are ancillary effects of some of the point mutations in the rest of the protein. For instance the substitution of Thr-338 by Ile in the pp60$^{c\text{-}src}$ catalytic domain is apparently sufficient to activate it for transformation (63).

The exact consequences of the altered C-terminus for the function of pp60$^{v\text{-}src}$, however, are not known. The presence of large amounts of pp60$^{c\text{-}src}$ in nondividing cells such as platelets and neurons suggests that pp60$^{c\text{-}src}$ may not normally be involved in proliferation. Instead $pp60^{c\text{-}src}$ could have a role in a membrane process such as secretion (64), which is a major activity in both platelets and nerve terminals. The apparently newly acquired ability of $pp60^{v\text{-}src}$ to stimulate cell growth could be accounted for if $pp60^{v\text{-}src}$ had different substrate specificities to $pp60^{c\text{-}src}$. In vitro, however, $pp60^{v\text{-}src}$ and $pp60^{c\text{-}src}$ display similar abilities to phosphorylate a wide variety of proteins and peptides, albeit few of which are physiological substrates (65, 66). The major difference is that the specific catalytic activity of $pp60^{c\text{-}src}$ is lower than that of $pp60^{v\text{-}src}$ in immunoprecipitates (65, 66). This activity difference is even more apparent in vivo. Cells expressing a level $pp60^{c\text{-}src}$ equal to that of $pp60^{v\text{-}src}$ in a v-src-transformed cell have a level of phosphotyrosine in protein the same as that in a normal cell (65), in contrast to the 10-fold elevation observed in a v-src-transformed cell. Thus the activity of $pp60^{c\text{-}src}$ is severely restricted in the cell. To what extent this is due to posttranslational modification(s), a regulatory protein(s), or limited access to suitable substrates is unclear, although as previously indicated the phosphorylation of Tyr-527 could play a key role. Certainly there is a good correlation between the occurrence of phosphate at Tyr-416 in the cell and the ability of a $pp60^{src}$ molecule to transform. This leaves open the question of whether $pp60^{v\text{-}src}$ has different substrate specificities in the cell compared to $pp60^{c\text{-}src}$, or whether transformation is a consequence of the cell being unable to regulate the activity of $pp60^{v\text{-}src}$.

B. $P90^{gag\text{-}yes}$ AND THE c-yes PROTEIN

The v-yes oncogene occurs in two distinct avian sarcoma viruses, Y73 virus and Esh sarcoma virus. In contrast to RSV, the v-yes sequences are expressed as part of a chimeric protein with viral structural gag gene sequences at its N-terminus. This gag-yes protein is predicted to contain 220 amino acids from the gag gene, 585 from v-yes sequences, and 7 C-terminal residues from an unused reading frame in the env gene (67). The expected 90-kDa protein is immunoprecipitated from Y73 virus-infected cells with anti-gag protein serum (68). Such immunoprecipitates have protein-tyrosine kinase activity detectable both by autophosphorylation of $P90^{gag\text{-}yes}$ and phosphorylation of added substrates. There are currently no yes-specific antibodies, although some of the monoclonal antibodies against $pp60^{v\text{-}src}$ (14) and the anti-src 498–512 peptide serum (69) immunoprecipitate $P90^{gag\text{-}yes}$.

Studies on biogenesis of $P90^{gag\text{-}yes}$ have shown that it is not modified by attachment of a lipid moiety (70). Instead the N-terminus is probably acetylated, like that of its progenitor $Pr76^{gag}$. $P90^{gag\text{-}yes}$ is distributed in the cell in very much the same manner as $pp60^{v\text{-}src}$, being largely membrane-associated (69). The nature of its attachment to membranes is not understood, but newly synthesized molecules of $P90^{gag\text{-}yes}$ are associated with pp89 and pp50 in a complex similar to that observed with $pp60^{v\text{-}src}$ (71, 72).

The location of the catalytic domain in P90$^{gag-yes}$ has not been defined by isolation of proteolytic fragments, but there is a striking sequence homology between residues 555 and 801 of P90$^{gag-yes}$ and residues 270–516 of pp60^{v-src}, where over 90% of the residues are identical (Fig. 1). pp60^{v-src} and P90$^{gag-yes}$ also have sequence homology upstream of the catalytic domain: greater than 80% of the residues between 365 and 554 of P90$^{gag-yes}$ are identical to residues 81–250 of pp60^{v-src}. There is no homology, however, between the N-terminal membrane-binding domain of pp60^{v-src} and P90$^{gag-yes}$. Despite this close relationship there is no doubt that the c-*src* and c-*yes* genes and their products are distinct entities.

P90$^{gag-yes}$ contains two sites of tyrosine phosphorylation and two sites of serine phosphorylation (73). One of the serine sites is in the *gag* region, and presumably corresponds to the site in p19^{gag}. Tyr-700, homologous to Tyr-416 in pp60^{v-src}, has been identified as one of the sites of tyrosine phosphorylation and is the major site of autophosphorylation *in vitro* (40). The effects of these phosphorylations on P90$^{gag-yes}$ protein kinase activity are unknown.

The product of the c-*yes* gene has not yet been identified, although transcripts of the gene are abundant in kidney and a variety of embryonic tissues (74). Because there are 585 v-*yes*-encoded amino acids in P90$^{gag-yes}$ and because the v-*yes* sequence is apparently truncated at both its N- and C-termini, it seems likely that the c-*yes* protein will be larger than 65 kDa. No mutations have yet been made in the v-*yes* gene to define which sequences are critical for transformation, and to determine how the c-*yes* gene has been activated. It is possible that, as is the case for pp60^{v-src}, the alteration at the C-terminus of P90$^{gag-yes}$ vis-à-vis the c-*yes* protein may be important.

C. P70$^{gag-fgr}$ AND THE c-*fgr* PROTEIN

The v-*fgr* gene is carried by the Gardner-Rasheed (GR) feline sarcoma virus. The transforming protein of GR-FeSV, P70$^{gag-fgr}$, is in fact a tripartite protein, predicted to have 663 amino acids of which, starting at the N-terminus, 118 are from the *gag* gene, 151 are from the 5′-end of a mRNA-encoding γ-actin (including the 128 N-terminal amino acids), 390 are from v-*fgr* sequences, and 5 are from the parental feline leukemia virus *env* gene in an unused reading frame (75). The 390 v-*fgr*-encoded amino acids are extremely homologous to the C-terminal region of the v-*yes* gene, but the v-*fgr* and v-*yes* genes are apparently derived from two distinct cellular genes (76). A protein of 70 kDa, which has a tyrosine-specific autophosphorylating activity, can be immunoprecipitated from GR-FeSV transformed cells using anti-feline p15^{gag} serum (77). There are no *fgr*-specific antibodies available.

Little work has been done on the biosynthesis of P70$^{gag-fgr}$, but it is modified by an N-terminal myristyl group, as is the feline *gag* gene precursor protein with which it shares N-terminal sequences (78). P70$^{gag\text{-}fgr}$ has been detected in the cytoplasm of transformed cells by immunofluorescence staining but there is no obvious coincidence of staining with membranes, despite the presence of the myristyl group (79). Subcellular fractionation studies suggest that P70$^{gag\text{-}fgr}$ is partially particulate and partially soluble.

By analogy with pp60$^{v\text{-}src}$ the catalytic domain of P70$^{gag\text{-}fgr}$ would be comprised of residues 407–516 (see Fig. 1), but this has not been tested experimentally. Like P90$^{gag\text{-}yes}$, P70$^{gag\text{-}fgr}$ shares sequences with pp60$^{v\text{-}src}$ upstream of the catalytic domain.

P70$^{gag\text{-}fgr}$ contains three sites of tyrosine phosphorylation and two sites of serine phosphorylation (78). Tyr-553, located in a position homologous to Tyr-416 in pp60$^{v\text{-}src}$, appears to be one of the sites of tyrosine phosphorylation and is the major site autophosphorylated *in vitro*. There are no indications whether any of these phosphorylations affect the activity of P70$^{gag\text{-}fgr}$.

The c-*fgr* protein has not been identified nor have any mutations of the v-*fgr* gene been created. The intriguing question of whether the actin sequences in P70$^{gag\text{-}fgr}$ play a role in its transforming activity remains unanswered. It is curious that the 3′ recombination sites for the v-*yes* and v-*fgr* genes are identical with respect to the acquired cellular sequences, although the recombination sites in the parental viruses are different. Thus both proteins are apparently truncated with respect to their cellular counterparts, but have different C-termini. This again implies that precise loss of C-terminal sequences may be important for activation.

D. P140$^{gag\text{-}fps}$ AND p98$^{c\text{-}fps}$; P85$^{gag\text{-}fes}$ AND p92$^{c\text{-}fes}$

The v-*fps* oncogene is found in a series of avian sarcoma viruses, including Fujinami sarcoma virus (FSV), PRCII, and PRCIV. These viruses contain overlapping v-*fps* sequences all of which have a common C-terminal region of 430 amino acids but which vary to some extent to the N-terminal side of this region due to deletions in the v-*fps* sequences. The different viruses induce distinguishable transformed phenotypes, probably due to these variations. All the viral transforming proteins contain N-terminal viral *gag* gene sequences joined to v-*fps* sequences. As an example, FSV P140$^{gag\text{-}fps}$ is a protein with 308 *gag*-derived amino acids and 874 v-*fps*-encoded amino acids (80).

The v-*fes* oncogene exists in two feline sarcoma viruses, Snyder-Theilin (ST) FeSV and Gardner-Arnstein (GA) FeSV. The two v-*fes* genes also have a common C-terminal region of 436 amino acids, and differ in the nature of the v-*fes* sequences nearer their N-termini. These sequences are expressed as *gag-fes* chimeras, and, for example, GA-FeSV P110$^{gag\text{-}fes}$ has 425 *gag*-derived amino acids and 609 *fes*-encoded amino acids (81). There proved to be striking sequence homologies between the v-*fps* and v-*fes* genes and their products suggesting that they were derived from genes that are equivalent in chicken and cats; this has been formally demonstrated (82). As a result one would expect the v-*fps* and v-*fes* proteins and their cellular counterparts to be similar in their properties. To a large extent this has turned out to be the case, and we therefore discuss the *fps* and *fes* gene products together.

Proteins of the expected sizes can be immunoprecipitated from v-*fps*-transformed cells either with anti-p19^{gag} serum or anti-*fps* tumor serum (83, 84). There are also monoclonal antibodies that recognize *fps* determinants mapping to a region N-terminal to the protein kinase domain (85). Immunoprecipitates of P140$^{gag\text{-}fps}$ have tyrosine-specific autophosphorylating activity and can phosphorylate exogenous substrates (86, 87). The *gag-fes* proteins can be immunoprecipitated from v-*fes*-transformed cells either with anti-feline p15^{gag} serum (88, 89) or anti-*fes* tumor serum (88). There is also a series of anti-*fes* peptide sera (90). In immunoprecipitates, *gag-fes* proteins will autophosphorylate and phosphorylate added substrates on tyrosine (88, 89).

Metabolic labeling studies have shown that newly synthesized P140$^{gag\text{-}fps}$ molecules are found in association with pp89 and pp50 (71, 72). Since the N-terminus of P140$^{gag\text{-}fps}$ is derived from the avian *gag* gene precursor, Pr76^{gag}, in all likelihood it is acetylated although this has not been formally shown. Half-lives of 3–5 h have been measured for several of the different *gag-fps* proteins (70). Unlike pp60$^{v\text{-}src}$ and P90$^{gag\text{-}yes}$, P140$^{gag\text{-}fps}$ is broadly distributed in the cytoplasm showing a diffuse staining pattern (91, 92). The absence of membrane association is corroborated by cell fractionation studies, where P140$^{gag\text{-}fps}$ behaves as a soluble protein at physiological ionic strength (91). P140$^{gag\text{-}fps}$, however, is found in the particulate membrane fraction at low ionic strength suggesting it does have some affinity for membranes (92).

In contrast to the *gag-fps* proteins, the N-termini of the *gag-fes* proteins are myristylated (93), like the feline *gag* gene precursor from which they stem. By immunofluorescence staining P85$^{gag\text{-}fes}$ has been found in the cytoplasm as well as in association with membrane structures (94). The myristyl group may predispose *gag-fes* proteins to this more pronounced membrane affiliation.

The catalytic domain of P140$^{gag\text{-}fps}$ has been defined by analysis of proteolytic fragments and lies at the C-terminal end of the protein (95). Comparison of the sequence of P140$^{gag\text{-}fps}$ with that of pp60$^{v\text{-}src}$ shows that residues 924–1174 of the v-*fps* sequence have striking homology with the catalytic domain of pp60$^{v\text{-}src}$ (residues 270–516) (Fig. 1). This region is essentially identical in all the strains of virus carrying the v-*fps* gene. Residues 700 to 950 form the equivalent catalytic domain in P110$^{gag\text{-}fes}$ (Fig. 1). There is also evidence for a second functional domain in P140$^{gag\text{-}fps}$ as discussed below.

P140$^{gag\text{-}fps}$ contains single minor sites of both serine and tyrosine phosphorylation in the *gag*-derived region, as well as one serine and two tyrosine sites in the v-*fps* region (95). The only identified site is Tyr 1073, which is the tyrosine homologous to Tyr-416 in pp60$^{v\text{-}src}$. Tyr-1073 is one of the major acceptors for the *in vitro* autophosphorylation reaction. Mutation of Tyr-1073, to Phe, Ser, Thr, and Gly has been accomplished (*96, 97*). In each case the mutant shows a diminished ability to phosphorylate added substrates and is noticeably weaker in its transforming activity. One presumes that these defects are due to the inability to phosphorylate this residue, although there is an outside chance that a tyrosine per se is required at this site for full activity. At the very least Tyr-1073 appears to be more important than Tyr-416 in pp60$^{v\text{-}src}$. There are preliminary indications that phosphorylation at this site increases the phosphotransferase activity of P140$^{gag\text{-}fps}$.

Both the c-*fps* and c-*fes* proteins have been identified using a variety of anti-v-*fps* and anti-v-*fes* antibodies. The c-*fps* protein is a 98-kDa polypeptide which is found predominantly in hematopoietic cells (*98*). The p98$^{c\text{-}fps}$ has protein-tyrosine kinase activity *in vitro*, both for autophosphorylation and towards exogenous substrates; p98$^{c\text{-}fps}$ apparently lacks phosphotyrosine when isolated from cells, but contains phosphoserine. In contrast to P140$^{gag\text{-}fps}$, p98$^{c\text{-}fps}$ is preodminantly soluble upon cell fractionation even at low ionic strength (*99*). The c-*fes* protein, somewhat smaller at 92 kDa, is also found in hematopoietic cells, particularly of the myeloid lineage (*100, 101*). Like p98$^{c\text{-}fps}$, p92$^{c\text{-}fes}$ autophosphorylates and phosphorylates added substrates. Interestingly a 94 kDa protein with tyrosine-specific autophosphorylating activity, is also found on immunoprecipitation of mammalian cells with rat anti-v-*fps* tumor serum (*100*) or anti-*fps* peptide serum (*101*). The distribution of this protein in different cell types is distinct from that of p92$^{c\text{-}fes}$, and their relationship is unclear. Although much of the sequence of the c-*fps* protein has been deduced from the nucleotide sequence of chicken c-*fps* genomic clones (*102*), the entire sequence of neither p98$^{c\text{-}fps}$ nor p92$^{c\text{-}fes}$ protein is known. From the fact that the v-*fps* sequences in FSV encode approximately 95 kDa, however, it is likely that they represent most of p98$^{c\text{-}fps}$.

The requirements for the transformation by *fps* sequences have been studied in detail. The *gag* sequences have been shown to be dispensible for transformation (*103*), although such mutant viruses are more weakly tumorigenic, as are viruses like PRCII which lack N-terminal v-*fps* sequences (*104, 105*). A virus in which cellular sequences homologous to the v-*fps* sequences in FSV have been substituted for the v-*fps* sequences is capable of transformation (*106*). The *gag*-derived sequences, however, are necessary for transformation by c-*fps* sequences (*106*). This suggests that there are subtle differences between v-*fps* and c-*fps* necessary for the transforming ability of v-*fps*, which remain to be determined. For instance there are at least 26 single amino acid differences between the c-*fps* protein and the v-*fps* protein in the shared region (*102*). The C-termini of the v-*fps* and c-*fps* proteins are the same, and so, in contrast to pp60$^{v\text{-}src}$, there is no indication that changes at the C-terminus are required. Linker insertion mutations have been created in many positions of the FSV *gag-fps* gene (*107*). Mutants with insertions into the catalytic domain are invariably nontransforming, as are those with insertions into the 100 residues upstream of this domain. Such mutations also define a region between residues 400 and 600 of P140$^{gag\text{-}fps}$ which is necessary for transformation. Mutants with insertions in between these two regions are fully active, suggesting that P140$^{gag\text{-}fps}$ may have a two-domain structure (*108*). A start has been made with identifying the *fes* sequences required for transformation; substitution of 80% of the v-*fes* sequences in GA-FeSV with the equivalent human c-*fes* sequences reduces but does not abolish transforming activity (*109*).

E.   P120$^{gag\text{-}abl}$ AND p150$^{c\text{-}abl}$

The v-*abl* gene was initially identified in Abelson murine leukemia virus (Ab-MuLV), but has subsequently been found in HZ-2 FeSV. Ab-MuLV induces B cell disease in mice, but can transform many other types of cell, including fibroblasts, under the right circumstances. P160$^{gag\text{-}abl}$ contains 236 *gag* gene-derived amino acids at its N-terminus, followed by 1008 v-*abl*-encoded residues (*110, 111*). P160$^{gag\text{-}abl}$ can be detected in Ab-MuLV-transformed cells by immunoprecipitation with either anti-murine P15^{gag} serum (*112, 113*) or an unusual type of antitumor sera from a C57-leaden mouse (*114*). Immunoprecipitates of P160$^{gag\text{-}abl}$ made with anti-*gag* sera display autophosphorylating activity and can also phosphorylate added proteins (*115, 116*). Recently several site-specific antibodies have been generated, both against bacterially expressed fragments of the v-*abl* sequence and against synthetic v-*abl* peptides (*117, 118*). The antibodies against sequences derived from the protein kinase domain inhibit the P160$^{gag\text{-}abl}$ protein-tyrosine kinase activity, while those against sequences outside the protein kinase domain have no effect (*117*). Variants of Ab-MuLV encoding smaller *gag-abl* proteins have arisen at a high frequency. A commonly used strain encodes P120$^{gag\text{-}abl}$ which has an internal deletion in the C-terminal half of the protein (*110, 111*).

A catalytic domain can be defined in P160$^{gag\text{-}abl}$ by comparison of its sequence with that of pp60$^{v\text{-}src}$. Residues 367–614 of P160$^{gag\text{-}abl}$ have strong homology with the catalytic domain of pp60$^{v\text{-}src}$ (Fig. 1). The positioning of this domain near the N-terminus of the v-*abl* sequences distinguishes P160$^{gag\text{-}abl}$ from the viral protein-tyrosine kinases discussed so far. An identical location for the catalytic domain has been determined experimentally by expression of fragments of the v-*abl* gene in *E. coli* and assaying the resultant *abl* proteins for protein-tyrosine kinase activity (*119*). As will be seen in the final paragraph of this section, much of the sequence C-terminal to the catalytic domain is dispensible for transformation.

P160$^{gag\text{-}abl}$ is myristylated like Pr65^{gag} (*93*), the *gag* gene precursor of

MuLV. Mutants in which the N-terminal region of P160$^{gag-abl}$ has been exchanged for one which cannot be myristylated are nontransforming suggesting that this modification is critical for transformation (*120*). Immunofluorescence staining localizes P160$^{gag-abl}$ in the cytoplasm of transformed fibroblasts, where it is associated with the plasma membrane and concentrated in adhesion plaques (*121*). An early study with antitumor serum apparently detected P160$^{gag-abl}$ on the surface of transformed cells (*114*), but later evidence suggests that the fluorescence may not have have been due to P160$^{gag-abl}$. Cell fractionation studies show that P160$^{gag-abl}$ is associated with particulate fractions and the cytoskeleton, but there is also some soluble protein (*122*).

P160$^{gag-abl}$ is multiply phosphorylated and contains four sites of serine phosphorylation, two of threonine phosphorylation, and two of tyrosine phosphorylation (*123*). Two of the serine sites are located in the *gag*-derived region. Tyr-514, one of the sites of tyrosine phosphorylation (*40, 124*), is homologous to Tyr-416 in pp60^{v-src}. The other site of tyrosine phosphorylation may correspond to Tyr-385 (*40, 117*). *In vitro* P160$^{gag-abl}$ undergoes autophosphorylation at multiple sites. One of these is Tyr-514, but the other major sites are not detected in P160$^{gag-abl}$ labeled metabolically (*123, 124*). The functional effects of these phosphorylations have not been ascertained.

The product of the murine c-*abl* gene has been identified as a 150-kDa protein (*125*). Recent evidence obtained by sequencing c-*abl* cDNA clones suggests that there are at least four distinct c-*abl* proteins which share their C-terminal 1010 residues but which differ at their N-termini, containing 20–40 unique amino acids as a result of the use of alternate 5' exons in their mRNAs (*126*). p150^{c-abl} is found predominantly in cells of hematopoietic origin, but occurs at low levels in many cell types. The product of the human c-*abl* gene (p145^{c-abl}), which is slightly smaller, also occurs in hematopoietic cells (*127*). Apparently neither protein contains phosphotyrosine, although both are phosphorylated on serine (*127, 128*). Originally p150^{c-abl} was reported to lack protein-tyrosine kinase activity (*128*), but by altering assay conditions it has been possible to show that p150^{c-abl} both autophosphorylates and phosphorylates added substrates (*129*). The major site of autophosphorylation corresponds to Tyr-514 in P160$^{gag-abl}$ (*120*).

Little is known about the function of p150^{c-abl} but the c-*abl* gene is found to have undergone a specific type of rearrangement in the leukemic cells of the majority of patients suffering from chronic myelogenous leukemia (*130, 131*), which results from a reciprocal translocation of part of chromosome 9 onto chromosome 22. In the translocation the c-*abl* gene on chromosome 9 is truncated at its 5' end and comes under the control of a transcription unit on chromosome 22 termed *bcr*. This results in a novel hybrid *bcr-abl* mRNA which generates a 210-kDa protein containing *bcr* sequences at its N-terminus followed by c-*abl* sequences (*118, 127*). The p210$^{bcr-abl}$ is active as a protein-tyrosine kinase *in vitro* and is phosphorylated on tyrosine in the cell (*127*). Presumably the enzymic activity of p210$^{bcr-abl}$ is important in the leukemic phenotype, but substrates specific for this protein-tyrosine kinase are still being sought.

The sequence requirements for transformation by v-*abl* have been rigorously analyzed. All except the N-terminal 14 amino acids of the *gag*-derived sequences have been deleted, leaving sufficient information to dictate myristylation (*120, 132*). This *gag*-deletion mutant has undiminished fibroblast transforming activity, but decreased ability to transform lymphoid cells. This decrease, however, is probably due to instability of the protein in lymphoid cells rather than an inherent deficiency (*133*). Provided there is an N-terminal sequence with a myristylation signal, the minimum region required for transformation of fibroblasts has been defined as residues 250–630 of P160$^{gag-abl}$ (*134*). This is somewhat longer than the minimal protein kinase domain. The additional residues are on the N-terminal side of this domain in a region of homology with pp60^{v-src} and P140$^{gag-fps}$. Although the residues from 630 on can be deleted without loss of fibroblast transforming activity, such mutants have attenuated ability to transform lymphoid cells (*132*).

The structures of the oncogenically activated *abl* proteins, P160$^{gag-abl}$ and P210$^{bcr-abl}$, are in many ways similar in that both have lost N-terminal sequences present in p150^{c-abl}. When isolated from the cell the specific activity of the p150^{c-abl} protein kinase appears to be lower than that of either P160$^{gag-abl}$ or p210$^{bcr-abl}$. This suggests that the N-terminal sequences of p150^{c-abl} may act as a negative regulatory domain. This idea is supported by the fact that there are different species of p150^{c-abl} with different N-terminal regions (*126, 135*), each of which could have a specific regulatory function. Elimination of this regulatory domain, like that of the C-terminal domain of pp60^{c-src}, may be critical for activation of the c-*abl* gene. This hypothesis remains to be tested directly.

F. P68$^{gag-ros}$ AND THE c-*ros* PROTEIN

The v-*ros* gene is carried by the UR2 avian sarcoma virus. The predicted structure of the v-*ros* gene product has 150 *gag* gene-derived amino acids at its N-terminus followed by 402 *ros*-encoded residues (*136*). A protein of 68 kDa can be immunoprecipitated by anti-p19^{gag} serum from UR2 virus-transformed chick cells (*137*). In such immunoprecipitates P68$^{gag-ros}$ autophosphorylates on tyrosine (*137*). One *ros*-specific antitumor serum has been generated (*138*).

P68$^{gag-ros}$ is not myristylated (*138*), but residues 158–175 are uncharged and could form a membrane attachment site. Immunofluorescence staining shows that P68$^{gag-ros}$ is present in the cytoplasm of transformed cells, but no pronounced concentration of the protein in adhesion plaques or cell junctions was noted (*139*). Cell fractionation studies, however, suggest that P68$^{gag-ros}$ is plasma membrane associated (*136*).

Comparison of the sequence of P68$^{gag\text{-}ros}$ with that of pp60$^{v\text{-}src}$ defines residues 251–518 as the catalytic domain of P68$^{gag\text{-}ros}$ (Fig. 1). The sequences on either side of this catalytic domain are relatively short and therefore probably not long enough to form separate functional domains, although the putative membrane attachment site could be critical for transformation.

P68$^{gag\text{-}ros}$ is phosphorylated in transformed cells predominantly on serine (137). Efforts to detect phosphotyrosine in P68$^{gag\text{-}ros}$ have been unsuccessful. Autophosphorylation of P68$^{gag\text{-}ros}$ occurs at a single major site. The identity of this tyrosine is unknown, but the properties of the phosphotyrosine-containing peptide are not consistent with its being Tyr-418, which is the tyrosine homologous to Tyr-416 in pp60$^{v\text{-}src}$ (140). In contrast to the viral protein-tyrosine kinases discussed so far P68$^{gag\text{-}ros}$ does not induce dramatic changes in the level of phosphotyrosine in proteins in transformed cells (141, 142). P68$^{gag\text{-}ros}$ may therefore have a more restricted substrate specificity than the other viral protein-tyrosine kinases.

The product of the c-ros gene has not been identified, although a 3.1 kb c-ros transcript has been detected exclusively in kidney (143). The sequences of the chicken c-ros gene and the v-ros gene diverge at residue 540 of P68$^{gag\text{-}ros}$ just downstream of the protein kinase domain. In this respect P68$^{gag\text{-}ros}$ is similar to pp60$^{v\text{-}src}$, P90$^{gag\text{-}yes}$, P70$^{gag\text{-}fgr}$, and gP180$^{gag\text{-}fms}$ (see Section II, H). Among the protein-tyrosine kinases the sequence of the catalytic domain of P68$^{gag\text{-}ros}$ is most closely related to that of the insulin receptor (see Fig. 1). This coupled with a second structural analogy to the EGF and insulin receptors, namely the existence of a putative transmembrane domain upstream of the protein kinase domain, strongly suggests that the c-ros protein will be a cell surface receptor with a ligand-regulated protein-tyrosine kinase activity. The nature of this receptor is of obvious interest.

G. gp68/72$^{v\text{-}erb\text{-}B}$ and the EGF Receptor

The v-erb-B oncogene is found in two distinct avian erythroblastosis viruses, AEV-ES4 and AEV-H. The former carries an additional and unrelated cell-derived oncogene, v-erb-A, which potentiates the effect of the v-erb-B gene. Although in inefected animals these viruses cause erythroblastosis, in culture both viruses can transform fibroblasts. The v-erb-B gene is of particular interest because it was derived from the chicken EGF receptor gene, of which it represents a part (see below). The complete sequence of the AEV-H v-erb-B gene has been determined (135), but the true N-terminus of the v-erb-B protein is unknown. The v-erb-B gene is expressed from a subgenomic RNA and therefore the N-terminus of the v-erb-B protein could either contain 6 amino acids of the gag gene product or be initiated within v-erb-B sequences. Assuming the former to be correct, then the v-*erb*-B protein has 6 *gag*-gene-encoded residues plus 609 from v-*erb*-B, predicting a protein of 61 kDa (*144*). In fact cells transformed by AEV contain multiple forms of v-*erb*-B protein (*145, 146*), which appear to represent differently glycosylated forms of a single polypeptide chain of about 61 kDa. The major v-*erb*-B gene product is gp66/68$^{v\text{-}erb\text{-}B}$. A small fraction of gp66/68$^{v\text{-}erb\text{-}B}$ is further modified to give proteins of 75–80 kDa (gp74$^{v\text{-}erb\text{-}rB}$) (*147–149*). The various v-*erb*-B proteins can be immunoprecipitated with anti-tumor serum raised in rats (*145*), or by antisera raised against a bacterially expressed protein corresponding to a fragment of the v-*erb*-B protein (*146*). A number of anti-*erb*-B peptide sera have also been reported (*150, 151*).

All the different forms of v-*erb*-B proteins are found in cellular membranes with which they become associated as a result of synthesis on membrane-bound ribosomes. It is unclear if a cleavable signal peptide is necessary for this interaction, since, as previously explained, the true N-terminus of the v-*erb* B proteins is unknown. Neither gp66/68$^{v\text{-}erb\text{-}B}$ nor the EGF receptor are modified by attachment of lipid. Assuming there is a *gag-erb*-B chimeric protein, and numbering from its N-terminus, the following structure can be deduced for gp74$^{v\text{-}erb\text{-}B}$. The first 74 amino acids (perhaps lacking those residues upstream of a signal peptide cleavage site) would lie outside the cell. This region contains three sites for N-linked glycosylation, and there is evidence for the presence of three oligosaccharide chains on gp66/68$^{v\text{-}erb\text{-}B}$. Residues 75–99 are uncharged and would constitute a transmembrane domain anchoring the protein in the membrane. Residues 100–615, which would then lie in the cytoplasm, include a protein kinase domain starting about 50 residues from the end of the transmembrane domain. Cell fractionation studies show that gp66/68$^{v\text{-}erb\text{-}B}$ is associated with intracytoplasmic membranes which appear to be perinuclear and may represent part of the Golgi apparatus (*147, 148*). In contrast gp74$^{v\text{-}erb\text{-}B}$ is largely exposed on the cell surface, and can be detected by immunofluorescence staining (*147*). From the slow rate of maturation of gp66/68$^{v\text{-}erb\text{-}rB}$ it seems as if the v-*erb*-B gene product is delayed in its transit through the Golgi apparatus. However, neither transport of the v-*erb* protein to the cell surface nor transformation is affected by the presence of the N-linked glycosylation inhibitor, tunicamycin.

Residues 147–399 of the predicted v-*erb*-B gene product show strong homology with the catalytic domain of pp60$^{v\text{-}src}$ (Fig. 1), yet initial attempts to demonstrate protein kinase activity associated with v-*erb*-B proteins were unsuccessful (*145, 146*). This was doubly surprising in light of the subsequent finding that the v-*erb*-B protein represented a part of the EGF receptor gene (*152*), whose product has bona fide protein-tyrosine kinase activity that is stimulated upon binding EGF (*153*). Recently, however, protein-tyrosine kinase activity has been detected in immunoprecipitates of v-*erb*-B proteins, by autophosphorylation and phosphorylation of exogenous substrates (*149, 150*), as well as in membranes from AEV-infected chick cells (*154*). The reason for this discrepancy is unclear;

some of the antibodies used in the earlier experiments may have inhibited the protein kinase activity. In addition the levels of *in vitro* phosphorylation are low by comparison with the other virally coded protein-tyrosine kinases and by comparison with the EGF receptor.

The v-*erb*-B proteins are only weakly phosphorylated in either AEV-transformed fibroblasts or erythroblasts (*145, 149*), containing predominantly phosphoserine and some phosphothreonine at sites which are unidentified. gp66/68$^{v-erb-B}$ and gp74$^{v-erb-B}$ may be substrates for protein kinase C, since all three forms of the v-*erb*-B protein show increased phosphorylation in response to 12-*O*-tetradecanoyl-phorbol-13-acetate (TPA) treatment of AEV-infected chick fibroblasts (*149*). One site of protein kinase C phosphorylation may be Thr-109, since the equivalent Thr in the EGF receptor (Thr-654) is phosphorylated by this enzyme (*155, 156*). Phosphorylation of Thr-654 diminishes the degree of stimulation of the EGF receptor protein-tyrosine kinase by EGF and decreases the affinity of the receptor for EGF (*157, 158*). The effect of the TPA-induced threonine phosphorylation on the v-*erb*-B protein-tyrosine kinase has not been determined, although TPA treatment of one AEV-transformed fibroblast clone causes growth arrest (*149*).

There is every reason to believe that the c-*erb*-B gene is the EGF receptor gene. The properties of the EGF receptor are reviewed in Chapter 7 by Morris White and C. Ronald Kahn. We discuss here only the features of the EGF receptor that are salient for an understanding of how the c-*erb*-B gene might be activated to transform. The complete sequence of the human EGF receptor has been deduced from the nucleotide sequence of overlapping cDNA clones (*159*). The EGF receptor has 1186 amino acids consisting of a 619-residue EGF-binding extracellular domain and a 542-residue intracellular domain separated by a single 26-amino acid transmembrane segment. The AEV-H v-*erb*-B protein corresponds to residues 551–1154 of the receptor, and thus is missing most of the extracellular domain (*144*). The v-*erb*-B protein is also truncated at the C-terminal end and lacks the last 32 residues of the EGF receptor. This region of the EGF receptor contains a major autophosphorylation site (Tyr-1173) which is thought to be important in the regulation of its protein-tyrosine kinase activity (*160*). The absence of this acceptor site may in part explain why the *in vitro* protein kinase activity of the v-*erb*-B protein was hard to detect.

A priori one might imagine that the loss of both the regulatory ligand-binding domain and the autophosphorylation site would lead to the constitutive activation of the v-*erb*-B protein as a protein kinase. In fact its specific activity seems to be lower than that of the unstimulated EGF receptor. This is borne out by the observation that AEV-transformed fibroblasts display only a very modest increase in the level of phosphotyrosine in protein (*154, 161*). Nevertheless the idea that the v-*erb*-B proteins mimic an occupied EGF receptor and deliver an unregulated growth signal is still very attractive. In keeping with this notion at least some of the phosphotyrosine-containing proteins unique to AEV-transformed fibroblasts are also observed following EGF treatment of fibroblasts (see Section V,B). An indication that the expression of gp74$^{v\text{-}erb\text{-}B}$ on the cell surface is critical for transformation comes from the properties of temperature-sensitive mutants of AEV (*147*). Mutant AEV-infected erythroblasts revert to a normal phenotype upon shifting to the restrictive temperature and this is accompanied by a loss of gp74$^{v\text{-}erb\text{-}B}$ from the surface.

A start has been made in defining the regions of the v-*erb*-B gene that are necessary for transformation of both fibroblasts and erythroblasts, as well as delineating which changes in the c-*erb*-B gene are required for activation. Deletions and insertions in the protein kinase domain of the v-*erb*-B gene abolish transforming activity (*162–164*). A two-amino acid insertion in the connecting region between the protein kinase domain and the transmembrane domain is without consequence (*164*). Likewise large deletions downstream of the protein kinase domain (*144, 163*) and a smaller deletion or insertions into the extracellular domain have no effect on transforming ability (*164*). Starting with the intact EGF receptor, truncations at both the N-terminus and C-terminus are required for the protein to have transforming activity for fibroblasts, although exactly how many amino acids have to be removed at either end has not been determined (*165*). In contrast the normal C-terminus may serve for erythroblast transformation, since the c-*erb*-B gene in transformed erythroblasts, which has been activated by the insertion of an avian leukosis virus genome, has an intact C-terminus (*166*). The temperature-sensitive mutants of AEV may also be instructive in this regard. For instance there are two mutations in the ts34 AEV-ES4 mutant both of which lie within the protein kinase domain (*167*). The change of a histidine (equivalent to His-270 in the AEV-H v-*erb* protein) to an aspartic acid is likely to be the crucial one. This mutation has a clear-cut effect on transport of the protein to the surface at the restrictive temperature, yet, despite its location in the protein kinase domain, the protein-tyrosine kinase activity of the ts34 v-*erb* protein does not appear to be temperature sensitive.

H. gP180$^{gag\text{-}fms}$ AND gp170$^{c\text{-}fms}$

The v-*fms* oncogene was originally characterized as the oncogene of the McDonough strain of feline sarcoma virus (SM-FeSV). It has subsequently been detected in a second feline sarcoma virus (HZ-5 FeSV). The nucleotide sequence of SM-FeSV predicts a chimeric product of 1511 amino acids, of which 536 are derived from the *gag* gene and 975 from the v-*fms* sequence (*168*). Several v-*fms* glycoproteins are generated from a single primary translation product (*169–171*). The first detectable product is a 180-kDa glycoprotein (gP180$^{gag\text{-}fms}$), which is processed to yield P65^{gag} and gp120$^{v\text{-}fms}$ (*172, 173*). The oligosaccharide side chains of the latter are then modified to yield a mature protein (gp140$^{v\text{-}fms}$). All the proteins containing v-*fms*-encoded sequences can be immunoprecipitated with rat antitumor serum (*171*, *173*). Rat monoclonal antibodies to v-*fms* have been reported and these like the antitumor serum recognize extracellular v-*fms* domains (*172*, *174*). Anti-*gag* protein sera will immunoprecipitate gP180$^{gag\text{-}fms}$ (*169–173*). Antisera against synthetic v-*fms* peptides and bacterially expressed proteins corresponding to v-*fms* have become available (*175*).

gp120$^{v\text{-}fms}$ is associated with intracellular membranes, while gp140$^{v\text{-}fms}$ is principally detectable on the cell surface both by immunofluorescence staining and by iodination (*173*, *176*). The majority of gp140$^{v\text{-}fms}$ appears to be located in coated pits (*176*). Since all the v-*fms* proteins are membrane-associated and indeed span the membrane, the N-terminus of the protein must provide a signal sequence. It is presumed that the synthesis of the *gag-fms* precursor is initiated at first in phase AUG codon which is normally used for the synthesis of the FeLV glycosylated *gag* gene product. The maturation of gp120$^{v\text{-}fms}$ to gp140$^{v\text{-}fms}$ is rather slow and incomplete (*173*), and, as is the case for the v-*erb*-B proteins, it appears that transit through the Golgi apparatus is impeded. Experiments with inhibitors of glucosidase I, which is involved in oligosaccharide maturation in the Golgi apparatus, show that maturation is required both for transport to the cell surface and for transformation (*177*). The *gag-fms* protein is modified by addition of myristate. This suggests that the signal peptide for insertion of the *gag-fms* protein into the membrane is internal and may be that normally used by p170$^{c\text{-}fms}$.

The structure of the v-*fms* proteins is reminiscent of that of the v-*erb*-B proteins. There are 1080 residues, containing 14 sites for N-linked glycosylation, upstream of a 26-residue transmembrane domain. Fifty residues beyond the transmembrane domain recognizable homology with the catalytic domain of pp60$^{v\text{-}src}$ is evident. Residues 1152–1478 are homologous to residues 270–516 of pp60$^{v\text{-}src}$ (Fig. 1). Note that this v-*fms* catalytic domain is 80 residues longer than that of pp60$^{v\text{-}src}$. This is largely due to a 68-amino acid insertion between residues 358 and 359 of pp60$^{v\text{-}src}$. An initial report that gp 120$^{v\text{-}fms}$ underwent autophosphorylation on tyrosine (*178*), was later confirmed (*179*). Like the v-*erb*-B proteins, the specific activity of the v-*fms* protein-tyrosine kinase appears to be low, and cells transformed by the v-*fms* oncogene show no increase in the level of phosphotyrosine in cellular protein (*178*).

When isolated from SM-FeSV transformed cells gp120$^{v\text{-}fms}$ was found to be phosphorylated on serine and threonine, but no phosphotyrosine was observed (*178*). The sites of phosphorylation have not been identified, but presumably lie in the cytoplasmic domain. The sites of autophosphorylation have not been mapped either. There is no indication whether phosphorylation of gp120$^{gag\text{-}fms}$ affects its activity.

Given the structure and properties of the v-*fms* proteins it seemed likely that the product of the c-*fms* gene would prove to be a growth factor receptor. This prediction has been substantiated by the finding that gp170$^{c\text{-}fms}$ is the receptor for a hematopoietic growth factor—colony stimulating factor 1 (CSF-1)—which stimulates the growth of cells in the myeloid–monocyte lineage (*175*). This identification is consistent with the nearly exclusive expression of c-*fms* mRNA in hematopoietic tissues, and the demonstration that a 170-kDa protein can be immunoprecipitated from cat spleen with anti-*fms* monoclonal antibodies, which becomes phosphorylated on tyrosine *in vitro* (*180*). Furthermore tyrosine phosphorylation is stimulated by the addition of CSF-1 to membranes prepared from cell lines expressing CSF-1 receptors (*181*).

The requirements for v-*fms* transformation have been investigated by site-directed mutagenesis. A deletion just upstream of the transmembrane domain abolishes transforming activity even though the mutant protein retains protein-tyrosine kinase activity (*179*). In the mutant-infected cells gp120$^{v\text{-}fms}$ accumulates in the Golgi apparatus, and no gp140$^{v\text{-}fms}$ is detectable. This implies that the expression of gp140$^{v\text{-}fms}$ on the cell surface is required for transformation, in keeping with the results obtained with glucosidase I inhibitors (*177*). A deletion which removes the transmembrane domain is also nontransforming (*174*). This mutant v-*fms* protein does not reach the surface either, and leaves no sequences exposed in the cytoplasm, apparently because the absence of the hydrophobic transmembrane domain allows transfer of the whole protein into the lumen of the endoplasmic reticulum. Deletion mutations in the protein kinase domain abrogate transforming activity (*182*). The requirements for activation of the c-*fms* gene have not yet been determined, but the v-*fms* and c-*fms* genes have different C-termini with the sequences diverging at residue 1498 of v-*fms* (*183*). From this point the c-*fms* protein has a further 39 amino acids while the v-*fms* protein has 13. By analogy with pp60$^{v\text{-}src}$ and pp60$^{c\text{-}src}$ this could prove to be an important difference. The v-*fms* protein apparently contains the normal c-*fms* N-terminal sequence (*183*), so it is unclear whether there are changes at the N-terminus of the gene, such as the addition of *gag* sequences, which are necessary for transformation. There are other scattered differences between the predicted v-*fms* and c-*fms* amino acid sequences, which might also be important.

With regard to the mechanism of transformation, while all the available evidence indicates that the protein-tyrosine kinase activity of the v-*fms* proteins is rather restricted, the results with site-directed mutations suggest that this activity is critical; gp140$^{v\text{-}fms}$ must be expressed on the surface, and its localization in coated pits (*176*) suggests that it is mimicking a growth factor receptor. It seems likely that gp140$^{v\text{-}fms}$ is constitutively activated as a protein-tyrosine kinase, since, although it can still bind CSF-1, CSF-1 does not stimulate autophosphorylation (*182*). It is interesting to note that the v-*fms* oncogene causes primarily sarcomas, whereas its cellular cognate regulates the growth of hematopoietic cells. Conversely, the v-*erb*-B oncogene transforms erythroid cells, while the EGF receptor acts in nonhematopoietic cells. The identification of the substrates for viral and cellular forms of these two protein-tyrosine kinases may yield a clue to this apparent paradox.

III. Other Transformation-Related Tyrosine Phosphorylation Systems

A. THE v-*sis* ONCOGENE, PDGF, AND THE PDGF RECEPTOR

The v-*sis* oncogene was initially found in simian sarcoma virus (SSV) and later in the Parodi-Irgens feline sarcoma virus (PI-FeSV). The v-*sis* gene of SSV has been intensively investigated, and indeed yielded the first direct connection between viral transformation and growth control, when it was discovered that the predicted sequence of the v-*sis* gene product was highly homologous to that of the B chain of human platelet-derived growth factor (PDGF) (*184–186*). The v-*sis* gene of SSV is expressed from a subgenomic mRNA in the form of an *env-sis* chimera containing 38 amino acids from the N-terminus of the viral *env* gene and 220 residues encoded by v-*sis* sequences. Homology with the B chain of PDGF starts at residue 99 and extends to residue 207. The retention of the N-terminus of the *env* gene supplies the *env-sis* protein with a signal sequence. The earliest detectable product of the v-*sis* gene is a 28-kDa protein which probably corresponds to the primary translation product minus the signal sequence with the addition of a single N-linked oligosaccharide chain at position 80 (*186–188*). The gp28$^{v\text{-}sis}$ is rapidly dimerized and then is trimmed by proteolytic processing at both its N- and C-termini. At the N-terminus the cleavage site is probably the Lys-Arg sequence (residues 97 and 98), which would generate an N-terminus identical to that of the B chain of PDGF. The C-terminal cleavage site could be at residue 207, the residue at which most of the human PDGF B chain molecules end (*189*).

There is evidence that the product of the v-*sis* gene is secreted and variable amounts of a 24-kDa dimer are found in the culture medium of SSV-transformed cells (*190, 191*). Both point mutations and small deletions within the signal sequence of the v-*sis* gene abolish transforming activity indicating that the v-*sis* products must enter the secretion pathway to be active (*192*). Since only cells bearing PDGF receptors can be transformed by SSV, there is every reason to believe that the v-*sis* proteins must function through interaction with the PDGF receptor, and thereby mimic PDGF. The PDGF receptor is a 180-kDa surface glycoprotein with an associated protein-tyrosine kinase activity which is stimulated severalfold upon binding PDGF (*193–195*). It seems likely that at least some aspects of the response of cells to PDGF are mediated through increased tyrosine phosphorylation of cellular proteins by activated PDGF receptor. This in turn suggests that transformation by the v-*sis* oncogene is also a result of tyrosine phosphorylation. No new phosphotyrosine-containing proteins, however, have been detected in v-*sis* transformed cells.

The functional similarity between the v-*sis* proteins and PDGF is underscored by the fact that SSV-transformed cells contain a mitogenic activity which is neutralized by anti-PDGF antibodies (*178, 190, 196*). Furthermore the growth of some types of SSV-transformed cell can be inhibited by these antibodies (*195*). It is not clear, however, whether the v-*sis* proteins actually have to be secreted from the cells to be active (*191*), since they could theoretically interact functionally with the PDGF receptor in some intracellular compartment. Cloned c-*sis* sequences are fully active in transformation when expressed in appropriate vectors (*197, 198*). Thus there are no critical structural alterations necessary for the transforming activity of the v-*sis* gene. However, since PDGF treatment of normal cells does not lead to transformation, this implies that there is a difference between autogenous generation of PDGF and its exogenous application to cells.

B. TGF-α AND THE EGF RECEPTOR

Certain transformed cells secrete growth factors that are able to cause the transient induction of the transformed phenotype in normal cells. For instance TGF-α and TGF-β in combination allow the growth of normal fibroblasts as colonies in agar. The sequence of TGF-α shows homology to EGF (*199*) and it has been shown that TGF-α interacts with the EGF receptor causing the same responses as EGF (*200–202*). Thus TGF-α stimulates the protein-tyrosine kinase activity of the EGF receptor as effectively as EGF, both *in vitro* and in intact cells. Since no other receptor has been identified for TGF-α, we can assume the biological effects of TGF-α are achieved by stimulating EGF receptor-mediated tyrosine phosphorylation.

In this context it is worth noting that a number of tumors have been isolated in which there is aberrant expression of the EGF receptor gene (*203*), commonly resulting in high levels of EGF receptor expression. Often the receptor gene is amplified and in a few cases there are alterations in gene structure, some of which may lead to a truncation similar to that giving rise to the v-*erb*-B protein. Whether these alterations in EGF receptor expression are in any way instrumental in the tumor phenotype is unclear.

C. THE *neu* ONCOGENE

The *neu* oncogene was first identified by transfection of DNA from a series of ethylnitrosourea-induced rat neuroglioblastomas onto NIH3T3 cells (*204*). The product of the *neu* oncogene was identified as a 185-kDa surface glycoprotein (p185^{neu}) through the use of antitumor serum (*205*). Subsequently monoclonal antibodies specific for p185^{neu} were isolated (206). p185^{neu} is phosphorylated on serine and threonine and displays tyrosine-specific autophosphorylating activity in immunoprecipitates (207). Hybridization studies showed the neu gene to be related to the v-erb-B gene (208) and this enabled the isolation of molecular clones of the neu gene (209). The nucleotide sequence of the neu gene shows that it has a protein kinase domain and in this region the neu gene is closely related to the c-erb-B gene (209). The neu gene, however, is clearly distinct from the c-erb-B gene. The neu gene is very likely to be the rat equivalent of the human c-erb-B-2 gene (210).

The size and properties of p185^{neu} strongly suggest that it is a growth factor receptor. A protein very similar to p185^{neu} can be detected on normal fibroblasts (208). So far, however, the true ligand for this normal cell protein has not been identified. Treatment of neu-transformed cells with monoclonal antibodies directed against p185^{neu} reverses the transformed phenotype (211). This reinforces the notion that p185^{neu} mimics a surface growth factor receptor. It seems likely that p185^{neu} is a constitutively activated form of a normal growth factor receptor which acts through tyrosine phosphorylation.

D. pp60$^{c\text{-}src}$ AND POLYOMA VIRUS MIDDLE T ANTIGEN

The initial report of a protein-tyrosine kinase activity was of that found in association with polyoma virus middle-sized tumor antigen (mT antigen) (1). Of the three polyoma virus T antigens, mT antigen is the one most intimately connected with tumorigenesis, being capable of transforming cell lines to a malignant state in the absence of the other two T antigens. The mT antigen is a membrane-associated protein but is synthesized on soluble ribosomes. It does not span the membrane but appears to interact with membranes via a hydrophobic sequence near its C-terminus. Studies with site-directed mutants have shown that mT antigen must interact with membranes to transform. Analysis of a variety of mT antigen mutants shows that there is a good but not perfect correlation between the presence of associated protein kinase activity and an ability to transform (1, 212, 213).

Despite strenous efforts, all attempts to demonstrate that mT antigen itself has enzymic activity have failed. Given the lack of primary sequence homology between mT antigen and other protein kinases perhaps this is not surprising. Measurement of the size of the protein kinase active fraction of mT antigen showed that it was considerably larger than the bulk of mT antigen, which behaved as a monomer (214). This suggested that mT antigen might be associated with a cellular protein kinase in the form of a complex. Subsequently it was shown that mT antigen is complexed with pp60$^{c\text{-}src}$, and that the majority of the associated protein-tyrosine kinase activity can be accounted for by pp60$^{c\text{-}src}$

(*215, 216*). Since pp60$^{c\text{-}src}$ is membrane-bound, and only mT antigens capable of binding to membranes are found in this complex (*216*), the interaction between the two proteins is likely to take place on the membrane. Only a small fraction of the pp60$^{c\text{-}src}$ and mT antigen populations are found in association. Under some circumstances there is evidence that the amount of mT antigen is limiting, but what regulates the formation of complex is unknown. It is worth noting that in polyoma virus-infected RSV-transformed cells pp60$^{v\text{-}src}$ is not found associated with mT antigen (*217*).

pp60$^{c\text{-}src}$ molecules which. are associated with mT antigen have severalfold greater activity towards exogenous substrates than free pp60$^{c\text{-}src}$ molecules (*218*). In some undefined way therefore the interaction with mT antigen increases the activity of pp60$^{c\text{-}src}$. As previously mentioned the activity of pp60$^{c\text{-}src}$ seems to be regulated in the cell (*65, 66*), possibly by virtue of a tyrosine phosphorylation in the unique C-terminal region of the molecule (*58*). Bound mT antigen appears to prevent phosphorylation of this site (*58, 218a*), and this may in part be responsible for the activation of the associated pp60$^{c\text{-}src}$ molecules. An affinity of mT antigen for the C-terminus of pp60$^{c\text{-}src}$ would explain its inability to associate with pp60$^{v\text{-}src}$. Despite the activation of pp60$^{c\text{-}src}$, polyoma virus-infected and transformed cells show no increase in the overall level of phosphotyrosine in protein and contain no new phosphotyrosine-containing proteins (*161*). In this regard they are similar to cells that express high levels of pp60$^{c\text{-}src}$ (*65*). The substrates for the activated pp60$^{c\text{-}src}$ await detection.

E.   p56^{lstra}

High levels of protein-tyrosine kinase activity were detected in membranes of the LSTRA mouse lyphoma cell line [originally isolated from a Moloney murine leukemia virus (Mo-MuLV)-induced thymoma] during a screen of suspension cells for this activity (*219*). Although membrane preparations from these cells can phosphorylate exogenous substrates, there is a major endogenous substrate of 56 kDa (*219–222*). This protein has a phosphorylation site contained in a tryptic peptide identical to that including Tyr-416 in pp60$^{v\text{-}src}$ (*223*). Partial purification of the protein-tyrosine kinase activity indicates that this protein, p56^{lstra}, is almost certainly the protein kinase itself labeled by autophosphorylation (*16*). LSTRA cells are not unique among mouse thymomas in having elevated levels of membrane-associated protein-tyrosine kinase activity. Several other Mo-MuLV-induced and Radiation-MuLV-induced thymomas contain a protein of similar size and structure (*222*). A protein closely related to p56^{lstra} is also found in normal thymocytes, albeit at lower levels, but is lacking from some cloned normal T cell lines (*221–226*). It is not detected in other types of hematopoetic cells, including B cells, nor in nonhematopoietic cells. B cells have a slightly smaller autophosphorylating membrane-associated protein-tyrosine kinase, which is structurally distinct (225, 226). Like pp60^{v-src}, p56^{lstra} is modified by the addition of a myristyl group to its N-terminus (222, 227).

cDNA clones corresponding to p56^{lstra} have been isolated and sequenced (228, 229) [the gene has been termed both *tck* (228) and *lsk*^{T} (229)]. The p56^{lstra} sequence is closely related to that of the v-*src* and v-*yes* genes but is distinct from both, as well as every other cloned protein-tyrosine kinase gene. Apparently p56^{lstra} is identical to the 56-kDa protein in normal thymocytes (228, 229), and the increased expression of p56^{lstra} is due to the insertion of an Mo-MuLV LTR (long terminal repeat) adjacent to the p56^{lstra} gene. The role, if any, of p56^{lstra} in the phenotype of LSTRA cells and the other thymomas remains unproven. LSTRA cells have higher levels of phosphotyrosine in protein than many T cell lines (221), but there is no true cognate control cell. The only phosphotyrosine-containing protein detectable in LSTRA cells is p56^{lstra} itself (222).

F.   THE v-*mil*, v-*raf*, AND v-*mos* ONCOGENES

The v-*mil* gene is part of the Mill Hill 2 avian acute leukemia virus genome and is expressed as a *gag-mil* chimera, P100$^{gag-mil}$ (230). The v-*raf* gene is the oncogene of the 3611-MuSV murine sarcoma virus and is also expressed as a *gag*-linked protein, P90$^{gag-raf}$ (231). Analysis of the *mil* and *raf* sequences shows that these two genes were almost certainly derived from a cellular gene which is equivalent in chicken and mouse (232–234). The v-*mos* gene is carried by a number of strains of Moloney murine sarcoma virus (Mo-MuSV), and encodes an *env-mos* protein, p37^{mos} (235–237). The sequences of the predicted products of these genes have regions with homology to that of the catalytic domain of the protein-tyrosine kinase family (see Fig. 1). P100$^{gag-mil}$, P90$^{gag-raf}$ (230), and p37^{mos} (236) have all been assayed in immunoprecipitates for protein kinase activity. None of them show any detectable protein-tyrosine kinase activity. All three proteins, however, have associated activities which will phosphorylate serine and to a lesser extent threonine either in exogenous substrates or in autophosphorylation reactions (238–240). Since these proteins themselves are all phosphorylated on serine, it is hard to rule out the possibility that the detected activity is due to an associated cellular protein-serine kinase. On balance, however, it seems likely that these proteins really do have intrinsic protein kinase activity. Bacterially synthesized v-*raf* (241) and v-*mos* (242) proteins have ATPase activities. Mutations in the ATP binding site of the v-*mos* gene abolish transforming activity (243). Cells transformed by these viruses do not display any perturbation of the normal tyrosine phosphorylation patterns (231, 244). It also seems probable that these enzymes are specific for serine rather than tyrosine. One indication of this specificity comes from an examination of the sequences in the region of these proteins which is homologous to the autophosphorylation site of the protein-tyrosine kinases. In no case is there a tyrosine residue, but instead there are one or more serine residues which could be autophosphorylated (Fig. 1).

IV. General Properties of Protein-Tyrosine Kinases

A. SEQUENCE HOMOLOGIES

The preceding account of the individual protein-tyrosine kinases has emphasized the existence of a comparable 30-kDa catalytic domain in each enzyme. In a few cases this domain has been defined on the basis of isolation of discrete proteolytic fragments. In the others, however, the existence of this domain has been deduced largely from a striking sequence homology with the catalytic domain of pp60^{v-src}. Figure 1 shows a comparison of the primary amino acid sequences in the relevant regions of all the protein-tyrosine kinases for which there are inferred sequences. This includes pp60^{v-src} (5–7), pp60^{c-src} (52), P90$^{gag-yes}$ (67), P70$^{gag-fgr}$ (75), P140$^{gag-fps}$ (80), P110$^{gag-fes}$ (81), P160$^{gag-abl}$ (110, 111), P68$^{gag-ros}$ (136), gp66/68$^{v-erb-B}$ (144), the EGF receptor (159), gP180$^{gag-fms}$ (168), and the insulin receptor (245, 246). These are compared to the sequences of four bona fide protein-serine kinases: cAMP-dependent protein kinase (247), cGMP-dependent protein kinase (248), the γ-subunit of phosphorylase kinase (249), and myosin light chain kinase (250). In addition the sequences of the putative viral protein-serine kinases P100$^{gag-mil}$ (232), P90$^{gag-raf}$ (233), and p37^{mos} (235) are presented. The sequences have been aligned to optimize homologies. (For another perspective of sequence homologies of protein-tyrosine kinases see Chapter 7 by White and Kahn.)

There are 16 residues within this domain that are absolutely conserved in all the protein kinases together with several other amino acids that are found in the great majority of protein kinases (Fig. 1). There are clusters of such highly conserved amino acids towards the N-terminus of the domain and in the C-terminal half. There is reason to believe that these may represent subdomains corresponding to an ATP binding site and a catalytic site respectively (see Section IV, B). Other regions of the domain are more variable and in some cases there are insertions of several amino acids. In addition there is the very large insertion found in the *fms* proteins. The homology among the protein-tyrosine kinases is greater than their homology with the group of protein-serine kinases. The converse is also true. Some comments on the specific features of these two types of protein kinase are made in Section IV, B.

There is another region of sequence homology between some of the protein-tyrosine kinases lying upstream of the catalytic domain. Residues 144 to 190 of pp60^{v-src} are clearly related to regions upstream of the catalytic domain in P90$^{gag-yes}$ (residues 428–474), P70$^{gag-fgr}$ (residues 281–327), P140$^{gag-fps}$ (residues 818–858), P85$^{gag\text{-}fes}$ (residues 594–634), and P160$^{gag\text{-}abl}$ (residues 244–288). This region is absent in P68$^{gag\text{-}ros}$, gp66/68$^{v\text{-}erb\text{-}B}$, gP180$^{gag\text{-}fms}$, and the insulin receptor among the protein-tyrosine kinases. It is also not found in any of the protein-serine kinases nor in p37^{mos}, P100$^{gag\text{-}mil}$, and P90$^{gag\text{-}raf}$. The function of this region is not yet defined although mutations in this region of pp60$^{v\text{-}src}$ affect its transforming activity (34–38), and it is necessary for P140$^{gag\text{-}fps}$ (107) and P160$^{gag\text{-}abl}$ (134) to transform.

The striking sequence similarities among the protein kinases imply that there was a single ancestral catalytic domain which diverged to give rise to the two classes of protein kinase specific for serine and tyrosine. It is interesting to note, however, that even among the most closely related protein-tyrosine kinases, the intron–exon structure has not been conserved.

B. CATALYTIC DOMAIN

The precise limits of a competent protein-tyrosine kinase catalytic domain have not been defined, and could vary slightly from enzyme to enzyme. On the C-terminal side the leucine corresponding to Leu-516 in pp60$^{v\text{-}src}$ appears to be critical. All the protein-tyrosine kinases have a hydrophobic residue at this site (either Leu or Phe) (Fig. 1). There is no corresponding hydrophobic residue in the protein-serine kinases. A two-amino acid deletion of residues 502–504, as well as all substitutions of the C-terminus of pp60$^{v\text{-}src}$ which include Leu-516 abolish both protein kinase and transforming activities (34, 251). Substitution of the nine residues from Pro-518 with nine amino acids from SV40 had no effect on transforming activity (252). In addition, although the sequences of pp60$^{v\text{-}src}$ and pp60$^{c\text{-}src}$ diverge at residue 514, both proteins have a leucine at position 516. The removal of sequences beyond the residue equivalent to Leu-516 in gP180$^{gag\text{-}fms}$ (182) and p37^{mos} (253) also abolishes transforming activity. The effects of antibodies directed against specific peptide sequences provide another way of deducing regions important for catalytic activity. As predicted, antibodies against the v-src peptide 521–526 (18) or the c-src peptide 527–533 (216) are permissive for protein kinase activity, while antibodies to the src peptide 498–512 inhibit (17). Antibodies against the v-mos peptide 362–374 which ends adjacent to the leucine equivalent to Leu-516 inhibit the p37^{mos}-associated protein-serine kinase activity (254). If Leu-516 is the terminal residue of the catalytic domain this puts the end of the domain very close to the C-terminus of many of the protein-tyrosine kinases (e.g., 10 residues away for pp60$^{v\text{-}src}$; 17 for pp60$^{c\text{-}src}$; 12 for P90$^{gag\text{-}yes}$; 10 for P70$^{gag\text{-}fgr}$; 8 for P140$^{gag\text{-}fps}$ and P110$^{gag\text{-}fes}$; 34 for gP180$^{gag\text{-}fms}$). There may be a requirement for a small C-terminal extension, but from the positioning of the catalytic domain in the EGF receptor and P160$^{gag\text{-}abl}$ it is clear there is no necessity for a C-terminal location within the protein.

The boundary on the N-terminal side is less clear-cut. Homology with some of the other protein-tyrosine kinases is evident from position 260 onwards in pp60^{v-src} (Fig. 1). The fact that the N-terminus of the γ-subunit of phosphorylase kinase lies at a position equivalent to residue 249 in pp60^{v-src} suggests that the catalytic domain in the protein-tyrosine kinases probably does not extend much beyond this point. A deletion extending to position 264 greatly reduces the protein-tyrosine kinase activity of pp60^{v-src} (37), but on the other hand an N-terminal deletion mutant of p37^{mos} in which the N-terminus of the truncated *mos* protein lies at a position equivalent to 270 in pp60^{v-src} can transform (253).

One critical region of the catalytic domain has been identified by labeling with the ATP affinity analog, *p*-fluorosulfonylbenzoyladenosine (FSBA). In the cAMP-dependent protein kinase this reagent reacts with Lys-72 (255), while in pp60^{v-src} it reacts with the equivalent Lys-295 (256), and in the EGF receptor with Lys-721 (257). In all cases the derivatization inhibits protein kinase activity. A lysine in this position is absolutely conserved in all the protein kinases, being found in the sequence X-Ala-X-Lys, where both X residues are nonpolar. Based on its reaction with FSBA it seems likely that this lysine is in the vicinity of either the β or γ phosphate of the substrate ATP. Site-directed mutagenesis of this position has been achieved for pp60^{v-src} (258, 259), p37^{mos} (243), and P140$^{gag-fps}$ (260). Any substitution, including arginine, completely abolishes both protein kinase and transforming activity in all three cases. Some specific attribute of the ϵ-amino group of lysine in this position must be critical. For instance if the ϵ-amino group were simply needed to neutralize a phosphate negative charge, one would have anticipated that the guanido group of arginine would be able to substitute. Since this is not the case it seems possible that this ϵ-amino group is directly involved in phosphorylation, perhaps mediating proton transfer. Antibodies to synthetic peptides corresponding to this region in the EGF receptor do inhibit protein kinase activity.

The other residues that create the ATP binding site have not been so clearly defined. About 20 residues to the N-terminal side of this lysine, there is a cluster of glycines in the sequence Gly-X-Gly-X-X-Gly which is also highly conserved. The only protein kinase lacking even one of these glycines is the γ-subunit of phosphorylase kinase. Such a glycine-rich region is a common feature of nucleotide binding sites in many types of protein including ATPases, GTPases, and dehydrogenases. In the proteins of this type for which three-dimensional structures are available the Gly-X-Gly-X-X-Gly sequence forms an elbow around the nucleotide, with the first glycine making contact with the ribose ring and the second lying close to the terminal pyrophosphate (261–263). Mutation of these glycines has so far not been reported. A similar glycine-rich stretch is found, in several GTP binding proteins with GTPase activity, such as the transducin α-subunit, EFTu, and p21^{ras} (residues 10–15), but there is no following lysine (264, 265). There are three additional regions of primary sequence homology between all these proteins which have been implicated in GTP binding based on the tertiary structure of EFTu (for p21^{ras} these are residues 58–73, 111–120, and 148–155). There are no obvious homologues of these sequences in the protein kinases. Nevertheless we can guess that some of the highly conserved residues in the C-terminal half of the catalytic domain of the protein kinases will interact with ATP, and participate in phosphate transfer.

In this regard the conserved acidic residues (4 out of the 11 absolutely conserved residues in this region are acidic) seem likely to play an important role (263). The sequence Arg-Asp-Leu (residues 385–387 in pp60$^{v\text{-}src}$) defines the N-terminal end of the region with the greatest homology among the protein kinases (Fig. 1). All of the protein kinases contain aspartic acid at the position equivalent to Asp-386 in pp60$^{v\text{-}src}$, and have an Asp-Phe-Gly sequence corresponding to residues 404–406 in pp60$^{v\text{-}src}$. To our knowledge neither of these sequences has been mutagenized. The sequence Ala-Pro-Glu (residues 430–432 in pp60$^{v\text{-}src}$) is found in all protein kinases except gp66/68$^{v\text{-}erb\text{-}B}$ and the EGF receptor, which have Ala-Leu-Glu, and the myosin light chain kinase, which has Ser-Pro-Glu. This sequence has been mutagenized in pp60$^{v\text{-}src}$ by substitution of individual residues at the three sites (264, 265). All the replacements tested so far abolish both protein kinase and transforming activities, but there has not been a systematic survey. It seems probable that the glutamic acid will be essential, but that there will be some flexibility at the other two sites.

To the C-terminal side of this region (i.e., from residue 485 in pp60$^{v\text{-}src}$) there are sequences that are highly conserved among the protein-tyrosine kinases, which for the most part are not represented in the protein-serine kinases (Fig. 1). It is possible that this region is involved in part in determining substrate and amino acid specificity, but there are clearly other regions of the catalytic domain involved in substrate recognition. For instance an affinity peptide substrate analog, which has been devised for the cAMP-dependent protein kinase, reacts with Cys-198 (266), a cysteine which is conserved in three of the protein-serine kinases. No equivalent cysteine is found in the protein-tyrosine kinases, but Cys-198 lies next to a glycine (residue 421 in pp60$^{v\text{-}src}$) which is conserved in all the protein kinases. This region is of interest because it is adjacent to the autophosphorylation sites in the protein-tyrosine kinases and in the cAMP-dependent protein kinase. The auto-phosphorylation site in pp60$^{v\text{-}src}$ is Tyr-416. The equivalent tyrosine is also autophosphorylated in P90$^{gag\text{-}yes}$, P70$^{gag\text{-}fgr}$, P140$^{gag fps}$, P110$^{gag\text{-}fes}$, and P160$^{gag\text{-}abl}$, but is not detectably phosphorylated in gp66/68$^{v\text{-}erb\text{-}B}$, the EGF receptor, or P68$^{gag\text{-}ros}$. Thr-196 in the cAMP-dependent protein kinase is auto-phosphorylated (247). On the basis of the affinity peptide labeling studies we can deduce that this region lies very near the catalytic center. Possibly in the absence of an exogenous substrate this region forms a loop which can fold over into the active center and receive phosphate in a true intramolecular reaction. It is interesting to note that the sequences upstream of the autophosphorylation sites bear some resemblance to those favored by the different enzymes in their exogenous substrates (see Section IV, C).

It is evident that further progress in defining critical residues in the catalytic domains of protein kinases and their functions in phosphate transfer will require a knowledge of their three-dimensional structure. To date there are no X-ray crystallographic data for a protein kinase, but progress is being made with the catalytic subunit of the cAMP-dependent protein kinase. The ability to produce large quantities of several of the protein-tyrosine kinases in bacteria should also be an advantage in this context (267–271), although so far such proteins have been largely insoluble and therefore mostly inactive. The single exception is pt*abl*50, which corresponds to a fragment of the v-*abl* gene containing the catalytic domain (270). This protein has been purified to homogeneity and retains a high level of protein-tyrosine kinase activity. In contrast another bacterially expressed *abl* protein has protein-tyrosine kinase activity, but at only one-hundredth the level of pt*abl*50 (271).

The mechanism of phosphate transfer by the protein-tyrosine kinases has not been investigated in detail. Rather few of them have been purified to homogeneity, which is a prerequisite for proper kinetic analysis. Highly purified preparations of pp60^{v-src} (272–277), pt*abl*50 (270), and the EGF receptor (278–280) have been reported. In general there does not appear to be a phosphoenzyme intermediate, and the phosphate linked to the autophosphorylation site in the catalytic domain is not turned over during phosphate transfer (91). A detailed kinetic analysis for the EGF receptor protein-tyrosine kinase using a synthetic peptide substrate suggests that the enzyme works via a sequentially ordered bi bi reaction where the peptide is the first substrate to bind and ADP is the last product to be released (281). Such a mechanism does not require a phosphoenzyme intermediate. The only residue we know for certain that is in the active center is the lysine lying close to the β or γ phosphate of the ATP.

All the protein-tyrosine kinases require Mg^{2+} or Mn^{2+} for activity, and presumably this is bound to the incoming ATP. The preference of many of the protein-tyrosine kinases for Mn^{2+} when assayed in the partially purified state may not exist with the purified enzymes (270). It has been suggested that this is a result of the ability of Mn^{2+} ions to inhibit phosphotyrosine-specific phosphatases in crude systems. In the case of the cAMP-dependent protein kinase there is evidence for a separate metal ion binding site. The fact that the optimum metal ion concentration for the protein-tyrosine kinases is considerably higher than their K_m for the metal ion–ATP complex suggests that the protein-tyrosine kinases may also have a separate metal ion binding site (270, 275). Nothing is known about the chirality of the ATP required for the phosphate transfer by the protein-tyrosine kinases. Although many of these enzymes have strict requirement for ATP (e.g., P110$^{gag-fes}$; the insulin receptor), others will use GTP instead of ATP (e.g., pp60$^{v\text{-}src}$; pt*abl*50), but the K_m for GTP (circa 100 μM) is considerably higher than for ATP (10–30 μM) in the cases where it has been measured (49, 270).

Initial experiments suggested that the turnover numbers of the protein-tyrosine kinases might be rather low compared with those of the protein-serine kinases casting some doubt on the authenticity of this phosphotransferase activity. It appears, however, that the low turnover numbers were due to a combination of inappropriate assay conditions and partial inactivation of the enzymes, which are hard to purify because of their scarcity and lipophilic nature. With the availability of better substrates in the form of synthetic tyrosine-containing peptides and larger quantities of purified enzymes, the activities prove to be comparable to those of the protein-serine kinases. For instance pt50^{abl} has a turnover number of 170 μmol/min/μmol using angiotensin as a substrate (270), while that for the EGF receptor is 7–55 μmol/min/μmol using synthetic peptide substrates (280–282). These values compare with a V_{max} of 150–200 μmol/min/μmol for the cAMP-dependent protein kinase using histone as a substrate (283).

The protein-tyrosine kinases that have been tested will also catalyze transfer of phosphate from phosphotyrosine-containing proteins to ADP forming ATP (270, 284). From the kinetics of formation of ATP, the free energy of the phosphate ester linkage to tyrosine in proteins can be calculated to be about 10 kcal, which is essentially the same as for the β-γ phosphodiester bond of ATP (284). In the case of pt*abl*50 the K_m for ADP for this reaction appears to be surprisingly low (270). In the absence of peptide substrate the cAMP-dependent protein kinase has ATPase activity (285). It is not clear to what extent this is true for the protein-tyrosine kinases; the pt*abl*50 protein has very low ATPase activity (<0.1 nmol/min/mg) (270).

There have been reports that individual protein-tyrosine kinases may possess other catalytic activities. The ATP-dependent DNA nicking activity associated with purified preparations of the EGF receptor (286), however, has proved to be due to a contaminating protein rather than an inherent property (287). Phosphatidylinositol (PI) kinase activities have been detected in purified preparations of pp60$^{v\text{-}src}$ (288) and in immunoprecipitates of both P68$^{gag\text{-}ros}$ (289) and middle T antigen-associated pp60$^{c\text{-}src}$ (290). The most recent evidence indicates that this activity is probably also due to associated enzymes rather than being intrinsic. For example pt*abl*50 lacks this PI kinase activity (270), while the activity copurifying with the EGF receptor is separable (291). Removal of pp60$^{v\text{-}src}$ and p150$^{c\text{-}abl}$ from transformed cell extracts did not reduce PI kinase activity (270, 292). Moreover membranes from cells infected with RSV temperature-sensitive mutants show elevated PI kinase activity which is as thermostable as that of wild-type RSV infected cells, even though the pp60$^{v\text{-}src}$ protein-tyrosine kinase activity of the mutant cell membranes is more thermolabile than that of wild type (293). On balance it seems unlikely that the protein kinase domain of the protein-tyrosine kinases will have other enzymatic activities, but given the large sizes of many of these proteins, however, one should not rule out the possibility that individual proteins can carry out additional reactions which would be catalyzed by another region of the protein.

C. REQUIREMENTS FOR SUBSTRATE SELECTION

Although a description of the known intracellular substrates for protein-tyrosine kinases is given in Section V, it is pertinent at this stage to discuss on what basis these enzymes select their substrates. The recognition of serine or threonine residues by protein-serine kinases in their substrates depends at least in part on the primary sequence surrounding the target hydroxyamino acid. In every case the canonical recognition sequence contains charged residues in the vicinity of the serine or threonine. For example the cAMP-dependent protein kinase has a specificity for the sequence Arg-Arg-X-Ser-Y where X can be any amino acid and Y is a hydrophobic amino acid (294). This specificity was originally deduced from the examination of sequences of natural sites of phosphorylation and then refined through the use of synthetic peptide substrates.

In the case of the protein-tyrosine kinases few sequences for sites of phosphorylation in physiological substrates are known. The sequences of several sites of autophosphorylation are available, however, and these are characterized by the presence of neighboring negatively charged residues on the N-terminal side of the tyrosine, with glutamic acid predominating over aspartic acid (see Fig. 1). Several of the autophosphorylation sites lie in the catalytic domain and one should therefore consider the possibility that they have been conserved because they serve some purpose in addition to being phosphorylation sites. Nevertheless the autophosphorylation sites that lie outside the catalytic domain, such as those in the EGF receptor (160), also contain glutamic acid residues. The two substrate phosphorylation site sequences which have been determined are those for enolase and lactate dehydrogenase (LDH) (295). These sequences also lack basic residues and have glutamic acids in the vicinity of the tyrosine. The enolase site, however, only has glutamic acids on the C-terminal side.

In contrast to many of the protein-serine kinases the issue of whether the acidic residues are critical for recognition is not so readily settled, since the K_m for synthetic peptides based on the known phosphorylation sites are at least two orders of magnitude higher than for the equivalent residues in the intact proteins (295). The best peptide substrates yet found for protein-tyrosine kinases contain multiple glutamic acid residues; for example, gastrin with five glutamic acid residues upstream of the tyrosine (296) and a peptide based on the N-terminus of the red cell anion transport protein (Band 3) that also has multiple acidic residues upstream of the target tyrosine (297), both of which have a K_m of about 100 μM. A peptide corresponding to the sequence in enolase, which has only glutamic acid residues on the C-terminal side of the tyrosine, however, also has a similar K_m (295). The deleterious effects of replacement of individual glutamic acids in peptides based on the autophosphorylation site in pp60^{v-src} suggests that the presence of these acidic residues is beneficial (298, 299). It should also be noted that synthetic random polyaminoacids containing glutamic acid, alanine, and tyrosine are excellent substrates for most protein-tyrosine kinases (300). Nevertheless acidic residues are not obligatory for phosphorylation since there are reasonably good peptide substrates such as Val^5-angiotensin (K_m circa 1 mM) that lack acidic residues altogether (301, 302). In conclusion, while acidic residues in the vicinity of target tyrosines probably play some role, it seems likely that there are other factors in addition to primary sequence, such as tertiary structure, which are important for recognition of substrates by protein-tyrosine kinases.

By and large the different protein-tyrosine kinases have rather similar specificities *in vitro* toward peptide and protein substrates, and in general are rather promiscuous, phosphorylating many proteins which are clearly not physiological substrates. It would appear that some type of control over substrate selection has to be exerted in the cell if the different protein-tyrosine kinases are to have meaningful and distinct specificities. One contributory factor could be the relative locations of the protein-tyrosine kinases and potential substrates in the cell. All of the protein-tyrosine kinases that have been identified are located in the cytoplasm and many of them are associated with cytoskeletal structures and/or membranes. This positioning may restrict the mobility of the enzymes and potentially limit their access to soluble cytoplasmic proteins.

D. REGULATION OF PROTEIN-TYROSINE KINASES

Apart from the growth factor receptor protein-tyrosine kinases whose activities are regulated by their cognate ligands, there are no known physiological regulatory molecules for these enzymes. None of the well-characterized enzymes displays dependence on cyclic nucleotides, nor on Ca^{2+} or calmodulin, although there is one report of a calmodulin-regulated enzyme (303) and another of a polyamine-stimulated tyrosine phosphorylating activity (304), which need to be substantiated. There are strong indications, however, that the activities of the protein-tyrosine kinases can be regulated by covalent modification.

All the enzymes examined have at least one site of serine and/or threonine phosphorylation and often several. Most of the protein-tyrosine kinases are also phosphorylated on tyrosine in the cell, and all of them autophosphorylate *in vitro*. The autophosphorylation sites may lie within the catalytic domain or in other cytoplasmic regions of the protein. It is not clear whether these autophosphorylation reactions are strictly intramolecular. The accepted way to prove that a reaction is intramolecular is to demonstrate that it is dilution-independent. One of the difficulties in doing such experiments is the lipophilic nature of most of these enzymes, which often leads to aggregation that cannot be reversed by dilution. In the case of the EGF receptor the autophosphorylation reaction occurs in a dilution-independent fashion and therefore appears to be intramolecular ($279, 305$). This does not mean, however, that the autophosphorylation sites might not be accessible to exogenous protein-tyrosine kinases. In the case of P140$^{gag\text{-}fps}$ there is some evidence that this is indeed possible at least *in vitro* (260).

In general, autophosphorylation results in augmented enzymic activity, usually as a consequence of an increase in V_{max}. Among the viral protein-tyrosine kinases a fewfold increase in activity has been observed for pp60$^{v\text{-}src}$ (49) and P140$^{gag\text{-}fps}$ (306). Autophosphorylation of the insulin receptor also enhances its activity (307), although this may in part be accounted for by a decrease in the K_m for ATP. Autophosphorylation of the EGF receptor may elevate its activity towards exogenous substrates (308), but there is some disagreement over this matter (309). It is clear from the mutations in which the tyrosine in the autophosphorylation sites has been replaced by a phenylalanine that the presence of a phosphate moiety at this site is not an obligatory requirement for enzymatic activity ($50, 96$). One speculation is that the autophosphorylation reactions are truly intramolecular, and that the tyrosine in question, whether it be in the catalytic domain or elsewhere, competes with exogenous proteins for the substrate binding site. When phosphorylated the part of the protein containing the tyrosine would be displaced from the substrate binding site, thus allowing access to other proteins.

There are also sites of tyrosine phosphorylation which may not result from autophosphorylation. For instance Tyr-527 in pp60$^{c\text{-}src}$ is not phosphorylated when preparations of pp60$^{c\text{-}src}$ are incubated with ATP *in vitro* ($39, 56$). This site therefore may be phosphorylated by another protein-tyrosine kinase. The nature of this protein-tyrosine kinase is unknown, but it may play an important regulatory role since the phosphorylation of pp60$^{c\text{-}src}$ at this site appears to diminish its catalytic activity (58). There is also a tyrosine phosphorylation site in P140$^{gag\text{-}fps}$ which must be phosphorylated by another protein-tyrosine kinase (260).

Serine and threonine phosphorylation of the protein-tyrosine kinases may exert either positive or negative regulatory effects. For instance phosphorylation of pp60$^{v\text{-}src}$ at Ser-17 by the cAMP-dependent protein kinase appears to increase its activity (46). As is the case for the tyrosine autophosphorylation site, however, phosphorylation of Ser-17 is not required for enzymic activity (36). Threonine phosphorylation of the EGF receptor at Thr-654 by protein kinase C decreases the EGF-dependent stimulation of the receptor protein-tyrosine kinase ($157, 158$). The identities of most of the protein kinases phosphorylating the protein-tyrosine kinases have not been ascertained. As a consequence the effects of the majority of the nontyrosine phosphorylations have not been adequately assessed. From the precedents with the protein-serine kinases, we can anticipate, however, that there will be sophisticated regulatory interactions both among the protein-tyrosine kinases themselves and between the protein-tyrosine kinases and the protein-serine kinases.

An interesting general principle regarding the regulation of the normal cellular protein-tyrosine kinases is beginning to emerge. This is exemplified by pp60$^{c\text{-}src}$ whose activity is normally suppressed, apparently by tyrosine phosphorylation at a site distinct from the autophosphorylation site in the catalytic domain. Upon isolation, however, pp60$^{c\text{-}src}$ has detectable protein-tyrosine kinase activity, as manifested by autophosphorylation at Tyr-416, although its specific activity towards exogenous substrates is considerably lower than that of pp60$^{v\text{-}src}$ (65, 66). The c-fps and c-abl proteins appear to be similar in this regard. Neither p98$^{c\text{-}fps}$ nor p150$^{c\text{-}abl}$ are phosphorylated on tyrosine in the intact cell, yet both have measurable protein-tyrosine kinase activity in vitro, which leads to phosphorylation at the major tyrosine acceptor site in their catalytic domains (98, 129). Furthermore their specific activities are lower than those of their viral counterparts. What regulates the activities of p98$^{c\text{-}fps}$ and p150$^{c\text{-}abl}$ is unclear, but it could also involve reversible covalent modification. We can extend this negative regulatory principle to the growth factor receptor protein-tyrosine kinases. The EGF receptor is not detectably phosphorylated on tyrosine in fibroblasts until it binds EGF when it undergoes phosphorylation at sites which are autophosphorylated in vitro even in the absence of ligand (310). Thus all the cellular protein-tyrosine kinases that have been characterized are subject to negative regulation. Their activation is correlated with the appearance of phosphate at the major in vitro autophosphorylation site. The alterations in structure that generate the viral protein-tyrosine kinases appear to abrogate the normal negative regulatory control mechanisms.

V. Cellular Substrates for Protein-Tyrosine Kinases

Considerable effort has been expended in a search for substrates of both the viral and the growth factor receptor protein-tyrosine kinases, in the hope that a knowledge of these proteins and their functions would enable a molecular explanation of the transformed phenotype and mitogenesis. A number of proteins have been identified which are likely to be intracellular substrates for one or more of these enzymes.

For protein phosphorylation to act as a mechanism for control of protein function, it must be reversible. Protein-serine phosphorylation is demonstrably reversible and there are several well-characterized protein phosphatases specific for phosphoserine-containing proteins. In the cell the average half-life for phosphate linked to tyrosine is short, being on the order of 30 min. This implies the existence of phosphotyrosine phosphatases, and indeed there are several reports of such activities (311). For the purposes of our discussion of substrates it should be borne in mind that the activities and regulation of the phosphotyrosine phosphatases will be critical in determining the rate and extent of phosphorylation of a given protein-tyrosine kinase substrate in the cell.

Because of the infidelity of the protein-tyrosine kinases *in vitro,* substrates have largely been sought by trying to identify proteins from virally transformed or growth factor-treated cells that contain elevated levels of phosphotyrosine compared to the appropriate control cell. This has proved to be a difficult task because of the paucity of phosphotyrosine relative to phosphoserine and phosphothreonine, even in cells containing activated protein-tyrosine kinases. In normal cells phosphotyrosine accounts for about 0.05% of the phosphate linked to protein in acid-stable fashion, with the remainder being phosphoserine (90%) and phosphothreonine (10%) *(161)*. Upon transformation with viruses containing the v-*src,* v-*yes,* v-*fgr,* v-*fps/fes,* and v-*abl* oncogenes or treatment of cells bearing high levels of EGF or PDGF receptors with the cognate growth factor, the level of phosphotyrosine increases up to tenfold but rarely rises above 0.5% *(161, 312, 313)*. The most commonly used techniques for identifying phosphotyrosine-containing proteins have been either immunoprecipitation of specific proteins or one- or two-dimensional gel analysis of total proteins from ^{32}P-labeled cells. Treatment of gels with strong alkali has been a helpful adjunct to both approaches because the phosphate ester linkage of phosphotyrosine is stable to alkaline hydrolysis while that of the much more abundant phosphoserine is rather labile *(314, 315)*. Ultimately, however, the proof that a protein contains phosphotyrosine has to be obtained by analysis of either acid or protease hydrolysates. More recently the use of antiphosphotyrosine antibodies has begun to offer an alternative method of identifying phosphotyrosine-containing proteins *(316, 317)*.

In general the substrate specificities of the viral protein-tyrosine kinases apparent in virally transformed cells are rather similar to one another *(244)*. Likewise the specificities of the EGF and PDGF receptor protein-tyrosine kinases are similar to each other *(313)*, with rather little overlap between the two classes of enzyme. For this reason we discuss their substrates separately. In assessing the relevance of a given substrate there are a number of factors to consider: (1) Is the same tyrosine(s) phosphorylated by the protein-tyrosine kinase in question *in vitro?* (2) What is the stoichiometry of phosphorylation in the cell? (3) What is the function of the protein? (4) Is its activity changed by phosphorylation in a meaningful sense?

A. Substrates for the Viral Protein-Tyrosine Kinases

The substrates for the viral protein-tyrosine kinases can be divided into two main categories: glycolytic enzymes and cytoskeletal proteins. Enolase, phosphoglycerate mutase (PGM) and lactate dehydrogenase (LDH) were found to contain phosphotyrosine in RSV-transformed chick cells *(318)*. None of these enzymes contains any phosphotyrosine when isolated from normal cells. They do contain phosphotyrosine in cells transformed by viruses containing the v-*yes,* v-*fgr,* v-*fps/fes,* and v-*abl* oncogenes, but not the v-*ros,* v-*erb*-B, and v-*fms* oncogenes (*244, 318*). Enolase and LDH are phosphorylated at a single identified tyrosine residue (*295*). In both cases the same tyrosine is phosphorylated *in vitro* by a variety of purified protein-tyrosine kinases (*295*). The fact that glycolytic flux is increased 3- to 4-fold in transformed cells suggested that the phosphorylation of these glycolytic enzymes might have some significance. The stoichiometry of phosphorylation of these enzymes, however, is low, ranging from 1–10%, and there are no detectable changes in the activities of enolase, PGM, or LDH in RSV-tansformed cells (*319*). Furthermore none of these enzymes is normally considered to be rate-limiting for glycolysis. Phosphofructokinase (PFK), which is the key control enzyme in glycolysis, does not appear to contain phosphotyrosine in RSV-transformed cells (*320*). Therefore it seems likely that there is another mechanism for increasing glycolytic flux which does not involve direct tyrosine phosphorylation of glycolytic enzymes. The phosphorylation of enolase, PGM, and LDH is probably a gratuitous event consequent upon their relatively high concentrations and the elevated levels of active protein-tyrosine kinase in the cell. Presumably the cell can tolerate this degree of abnormal tyrosine phosphorylation. Phosphotyrosine phosphatases must recognize these unusual substrates, thus limiting the total level of phosphorylated protein that can accumulate.

In contrast to the glycolytic enzymes which are cytosolic, the second type of substrate is characterized by its association with particulate structures. Vinculin, which is localized in adhesion plaques, was the first cytoskeletal protein to be identified as a substrate for pp60$^{v\text{-}src}$ (*321*). Subsequently two other phosphotyrosine-containing proteins, first detected by two-dimensional gel analysis, p36 (*322–329*) and p81 (*330*), have been found to be part of a submembraneous cortical skeleton. It is obviously attractive to speculate that phosphorylation of one or more cytoskeletal proteins might be instrumental in the morphological changes evident in transformed cells. For instance the proposed function of vinculin is to act as a linker between actin filament bundles and a receptor on the inner face of the plasma membrane present in adhesion plaques (*331*). If phosphorylation of vinculin were to abrogate its function as a linker, then this could explain the observed disarray of actin filaments in RSV-transformed cells. Since pp60$^{v\text{-}src}$ is concentrated in adhesion plaques (*22*) it would be in an ideal position to phosphorylate vinculin. There is, however, also a soluble pool of vinculin, and currently it is not clear which population is phosphorylated by pp60$^{v\text{-}src}$. The phosphorylation of vinculin on tyrosine is stimulated about tenfold in cells transformed by the v-*src,* v-*yes,* and v-*abl* oncogenes but not in cells transformed by the v-*fps* oncogene (*321*). Vinculin contains several sites of serine and threonine phosphorylation in both normal and transformed cells. There is one major site of tyrosine phosphorylation that shows increased occupancy upon transformation (*321*). Vinculin can be phosphorylated *in vitro* with purified pp60$^{v\text{-}src}$ in a reaction stimulated by certain phospholipids (*332*).

Is the phosphorylation of vinculin likely to be significant in the morphological phenotype of RSV-transformed cells? Against this idea is the fact that only 1% of vinculin molecules are phosphorylated on tyrosine at steady state (321). Moreover there are mutants of RSV that cause phosphorylation of vinculin without a gross reorganization of the actin filaments being apparent (333). In favor of a physiological role for vinculin phosphorylation is the fact that phosphotyrosine can be detected in vinculin in normal cells transformed with the v-fps gene (321) and the morphology of these cells is much less markedly changed than that of v-src-transformed cells. This question may not be resolved until it can be shown whether tyrosine phosphorylation of vinculin alters its function in vitro in a meaningful fashion.

p36 and p81 are not conventional cytoskeletal proteins but form part of a submembraneous cortical skeleton which is associated peripherally with the inner face of the plasma membrane. Although p36 is diffusely distributed throughout this whole structure (327–329), p81 is localized to microvillar regions forming part of the microvillar core (330); p36 is highly conserved and is expressed in many cell types (334, 335). It is abundant in intestinal epithelial cells where it forms part of the terminal web structure which underlies the microvillar array (334). It is the major phosphotyrosine-containing protein in RSV-transformed cells (336), and is also phosphorylated on tyrosine in cells transformed by the v-yes, v-fgr, v-fps/fes, and v-abl oncogenes (244). It is weakly phosphorylated in v-erb-B-transformed fibroblasts (154, 322), but not detectably phosphorylated in v-fms- or v-ros-transformed cells. At steady state up to 25% of p36 molecules contain phosphotyrosine at a single site (324–326). In vitro p36 is phosphorylated at this tyrosine, which maps near the N-terminus of the molecule (323, 337). In transformed cells p36 is also found to contain phosphoserine at about the same level as phosphotyrosine (324–326). Since p36 from normal cells contains very little phosphate, we can conclude that this serine phosphorylation is induced by transformation. The provenance of the protein-serine kinase(s) that phosphorylate p36 is unknown, but it is noteworthy that different pp36 molecules contain phosphoserine and phosphotyrosine. In fibroblasts and intestine more than half of p36 exists as a heterotetramer in the form $(p36)_2(p10)_2$ (338, 339). p10 has been sequenced and has homology to the brain Ca^{2+}-binding protein S-100 (337). The tetrameric form of p36, purified from gut, associates with actin and fodrin in a Ca^{2+}-dependent fashion, although the concentration of Ca^{2+} required for association are much higher than physiological (339). In vitro the phosphorylation of p36 is dramatically reduced by the presence of p10, but like vinculin its phosphorylation is stimulated by certain phospholipids (340). Perhaps the converse is true and tyrosine phosphorylation prevents the association of p10 thus altering the function of p36. The subcellular distribution of p36, however, is apparently not changed by phosphorylation (341). No adequate tests of the functions of tyrosine-phosphorylated pp36 compared to p36 have been conducted. It remains possible that the phosphorylation of p36 is gratuitous and is a consequence of its membrane location and relatively abundant nature (0.1–0.3% of total cell protein).

pp81 was first recognized as a phosphotyrosine-containing protein in A431 cells following treatment with EGF (*312*). It later became apparent that its unphosphorylated form, p81, was closely related to and possibly identical with a previously characterized minor constituent of the microvillar cores of intestinal epithelial cells (*330*). Like p36, p81 is a highly conserved protein, and it is expressed in most of the same cell types as p36; p81 is phosphorylated on tyrosine in cells transformed by the v-*src* and v-*fes* oncogenes (*330, 342*). Other transformed cell types have not been tested. At steady state about 10% of the p81 population contains phosphotyrosine, which is predominantly located at a single site (*330*). The same site is phosphorylated *in vitro* by pp60^{v-src}. Protein pp81 also contains phosphoserine and phosphothreonine at multiple sites. Immunofluorescence staining of A431 cells shows that p81 is localized to microvilli (*330*). Upon cell lysis a fraction of p81 is found to be particulate, but no direct association with another protein has been detected. The soluble population of p81 behaves as a monomer; pp81 distributes during cell fractionation indistinguishably from p81. Thus like p36 there is no indication whether the tyrosine phosphorylation of p81 is gratuitous or functionally significant.

A number of other substrates for one or more of the viral protein-tyrosine kinases have been reported. pp50, which together with pp89 is associated with the newly synthesized products of the v-*src*, v-*yes*, and v-*fps* genes, contains phosphotyrosine (*2, 25, 71, 72*) and is thereby presumed to be a substrate for the respective protein-tyrosine kinases. Protein p50 is not phosphorylated on tyrosine in normal cells but is found associated with pp89 (*343*). Its function is unknown, but since the protein-tyrosine kinase activity of the complexed viral enzymes is lower than that of the free forms, pp50 could act as a negative regulator of phosphotransferase activity. pp42B is a phosphotyrosine-containing protein found in v-*src*-, v-*yes*-, and v-*fps*-transformed chick cells (*244, 314*). It is not found in the equivalent transformed mammalian cells. Since pp42B is identical to one of the two major phosphotyrosine-containing proteins observed in growth factor-treated cells, it is discussed further in Section V, B. Yet other putative substrates have been detected using antiphosphotyrosine antibodies (*317*). A 130-kDa protein of unknown nature is the most prominent protein among those detectable with these antibodies in RSV-transformed cells (*344*).

One can estimate that the phosphotyrosine content of the substrates described in the preceding paragraphs do not account for the total increment in phosphotyrosine in the transformed cell. It seems likely that most of the abundant viral protein-tyrosine kinase substrates have been identified. Presumably other substrates are scarce proteins, possibly with regulatory functions. What might they be? Mutants of pp60^{v-src} which cannot be myristylated have full protein-tyrosine kinase activity but fail to transform (26). Phosphotyrosine levels in protein are elevated to nearly the same extent in mutant-infected cells as in wild-type RSV-transformed cells. Among the proteins that are detectably phosphorylated at tyrosine in the mutant-infected cells are enolase, LDH, p36, and p81, although the latter two are more weakly phosphorylated than in wild-type transformed cells (26). For reasons that are not clear vinculin shows a dramatic increase in phosphotyrosine content in the mutant-infected cells. Clearly the phosphorylation of these substrates is not sufficient to induce transformation. The phosphorylation of other as yet unidentified substrates must be required. The concomitant failures of the nonmyristylated pp60^{v-src} to associate with membranes and to transform implies that some of the critical substrates might be membrane-associated. In keeping with this notion new phosphotyrosine-containing proteins have been detected among the glycoproteins of RSV-transformed cells (345). A number of membrane proteins would make interesting candidate substrates. For example, the phosphatidylinositol kinases, the phospholipases, and the guanine nucleotide-binding regulatory proteins of adenylyl cyclase would all be targets whose phosphorylation could have physiological consequences.

There is a striking difference in the patterns of substrates identified in cells transformed by viral oncogenes whose products are related to growth factor receptors (v-erb-B, v-fms, and possibly v-ros) and those which are not (v-src, v-yes, v-fgr, v-fps/fes, and v-abl). The former type of transformed cell display either a very small or no increase in the level of phosphotyrosine in cellular protein and few if any individual substrates are detected. In contrast the latter type of transformed cell exhibits large increases in phosphotyrosine levels and several individual substrates are detectable. This could reflect differences in substrate specificities for the two types of enzyme. Alternatively it could suggest that the latter type of protein-tyrosine kinase is simply more promiscuous, and that the critical substrates for transformation are the same in both cases.

B. SUBSTRATES FOR THE GROWTH FACTOR RECEPTOR PROTEIN-TYROSINE KINASES

The subject of substrates for the receptor protein-tyrosine kinases is dealt with in more depth in the Chapter 7 by Morris White and C. Ronald Kahn. We therefore simply compare the identities and properties of such substrates with those of the viral protein-tyrosine kinases. Attempts to identify substrates for the growth factor receptor protein-tyrosine kinases in growth factor-treated cells have not been rewarded with a large number of candidates, and in most cases the growth factor receptor itself proves to be the major tyrosine-phosphorylated protein in growth factor-treated cells (317, 346). In addition to the relevant receptors, EGF- and PDGF-treated quiescent fibroblasts contain several new phosphotyrosine-containing proteins in the 40–45 kDa size class. These have been detected either on one-dimensional gels (*347*) or two-dimensional gels where these proteins form a cluster in the p*I* 6.5–7 range (*313, 348*). Two-dimensional gel analysis shows that there are two pair of proteins (pp45A and pp45B; pp42A and pp42B) in which the members of each pair are closely related and are probably charge isomers (*349*). A phosphotyrosine-containing protein of 41 kDa (p41) is also observed with some cell types. A similar protein is phosphorylated on tyrosine in human chronic myelogenous leukemia cell lines which contain p210$^{bcr-abl}$ (*350*). A low level of p36 phosphorylation is induced by PDGF in some but not all mouse fibroblast lines (*313*). Enolase, PGM, and LDH are not found to contain phosphotyrosine in growth factor-treated cells despite the observable increase in glycolytic flux. Phosphorylation of vinculin on tyrosine is not detectable either, even though there are rapid shape changes in growth factor-treated cells.

In the A431 epidermoid human tumor cell line, which has unusually high numbers of EGF receptors, EGF treatment stimulates tyrosine phosphorylation of the EGF receptor, p81, p42, and p36 (*314*), as well as a protein of 35 kDa (p35) (*351*). Protein p35 is apparently distinct from p36. but it shares properties with p36 being a membrane-bound protein whose association is dependent on divalent cations (*351*). Protein p35 is of interest because its phosphorylation occurs with somewhat slower kinetics than the other proteins in the EGF-treated A431 cells, and it has been suggested that internalized receptor molecules may catalyze this reaction (*352*). Tyrosine phosphorylation of p35 has also been detected in permeabilized human fibroblasts treated with EGF (*353*). In considering the results obtained with A431 cells it should be borne in mind that some of the observed phosphorylations may be abnormal, because instead of responding mitogenically to normal concentrations of EGF A431 cells stop growing.

All the growth factor-stimulated tyrosine phosphorylations, except that of p35, occur extremely rapidly being maximal within 5–10 minutes or earlier (*313, 314*). In general they are also transient, with the levels of phosphorylation returning to near baseline within 2 h. The decline in phosphorylation probably reflects the internalization of the receptor–growth factor complex and the consequent inactivation of the receptor protein-tyrosine kinase. Continuous presence of growth factor is required to maintain pp42 in its fully phosphorylated state during the first hour of treatment.

Of the various putative growth factor receptor protein-tyrosine kinase substrates pp42 has received the greatest attention largely because it is the most commonly observed, and because tyrosine phosphorylation of p42 can also be stimulated by mitogens, such as phorbol esters, whose receptors are not protein-tyrosine kinases (*34, 349, 354, 355*). In fibroblasts the unphosphorylated form of pp42 is a minor cellular protein accounting for about 0.002% of total cell protein (*356*). Upon treatment with growth factors at least 50% of p42 becomes phosphorylated. An increase in serine phosphorylation of p42 is observed concomitant with the increase in tyrosine phosphorylation. Since p42 has not yet been purified, direct tests of the ability of the purified growth factor receptor protein-tyrosine kinases to phosphorylate p42 have not been conducted. The majority of both p42 and pp42 appear to be soluble (*356*); p42 is a highly conserved protein, and its tyrosine phosphorylation has been observed in species ranging from avian to human (*313, 330*). A number of incorrect guesses as to its function have been made based on its size and p*I*, and as a result we can say that p42 is not the catalytic subunit of either the cAMP-dependent protein kinase or the casein kinases, nor is it one of the components of the GTP-binding factors that regulate adenylyl cyclase.

There is no direct evidence that the phosphorylation of any of the identified substrates is important in the mitogenic response. The transience of the observed phosphorylations are in contrast to the requirement for the continued presence of the growth factor for several hours for mitogenesis. In the case of the EGF receptor there are conditions under which the increase in protein-tyrosine kinase activity can apparently be dissociated from subsequent mitogenesis (*357*). It seems possible that p42 phosphorylation is required for some of the early responses to growth factors but is not sufficient for mitogenesis. It is also of interest that pp42 is detected in some but not all of the transformed cells which contain active viral protein-tyrosine kinases. One might have imagined that the unregulated growth of such cells could be maintained if the viral protein-tyrosine kinase phosphorylated one or more of the critical targets in a mitogenic pathway that used tyrosine phosphorylation. This may still be the case, but p42 is apparently not such a protein.

VI. Conclusions

A large family of cellular genes encoding protein-tyrosine kinases has been identified since 1980. The true functions of most of these cellular enzymes remain obscure. A subset of the protein-tyrosine kinases are growth factor receptors which are activated by ligand binding. Given this commonality of receptor function, it is appealing to speculate that tyrosine phosphorylation of cellular proteins is involved in the response of cells to growth factors. The demonstrable functional effects of protein-serine and threonine phosphorylation suggest that it would be possible to generate suitable intracellular signals by protein-tyrosine phosphorylation. To play devil's advocate, however, one could argue that the only tyrosine phosphorylation event that is critical is the autophosphorylation of the receptor. If this were the case, then the nature of the signal generated by growth factor binding would be unresolved. To settle this question it is obviously vital to identify further substrates for the growth factor receptor protein-tyrosine kinases and to determine what effect phosphorylation has on their functions. One would like to focus on proteins involved in the mitogenic response, but the lack of genetics has handicapped the identification of the proteins in this pathway.

The functions of the nonreceptor protein-tyrosine kinases are an enigma. Some like pp60$^{c\text{-}src}$ are most abundant in nondividing cells, and therefore may play no part in proliferation. We must presume that they have roles in other cellular processes. Given the membrane localization of many of these enzymes it seems likely that they will be involved in membrane function. For instance they could form part of response systems to external stimuli by acting as coupling factors for receptors which themselves lack signalling mechanisms. Another possibility is that they are part of membrane fusion systems and could regulate secretion and vesicle transport.

There is an excellent correlation between the protein-tyrosine kinase activity of the viral transforming proteins and their ability to transform, but there is no direct proof that tyrosine phosphorylation of cellular proteins is responsible for the transformed phenotype. As is the case for the growth factor receptor protein-tyrosine kinases and mitogenesis, no protein-tyrosine kinase substrates critical for the process of transformation are known. The identification of further substrates for the viral enzymes is obviously crucial if we are to prove this matter. We can anticipate that such substrates will be minor cellular proteins, which are membrane-associated and which may have regulatory functions in cellular physiology. In this regard cellular mutants with altered abilities to be transformed by the viral protein-tyrosine kinases would be enormously useful.

Assuming that tyrosine phosphorylation of cellular proteins is required for transformation, we are left with the question of how acquisition by a retrovirus can cause such a dramatic change in the function of a cellular protein-tyrosine kinase. The activation of cellular protein-tyrosine kinases into transforming proteins could simply involve changes which cause these enzymes to be recalcitrant to normal control. For example the receptor protein-tyrosine kinases could then transmit an unregulated and continuous signal to proliferate. Since transformed cells differ from normal cells in many respects other than uncontrolled growth, however, activation could also have subtle effects on the substrate specificity of the enzymes. For instance the activation of the nonreceptor protein-tyrosine kinases may involve changes that allow them to mimic other protein-tyrosine kinases, such as the receptor protein-tyrosine kinases. Nevertheless it seems reasonable to conclude that the critical difference between the normal cellular enzymes and the viral protein-tyrosine kinases is the inability of the latter to be regulated in a normal fashion by the cell.

ACKNOWLEDGMENTS

We wish to thank numerous colleagues providing results prior to their publication.

REFERENCES

1. Eckhart, W., Hutchinson, M. A., and Hunter, T. (1979). *Cell* **18,** 925.
2. Witte, O. N., Dasgupta, A., and Baltimore, D. (1980). *Nature (London)* **283,** 826.
3. Collett, M. S., and Erikson, R. L. (1978). *PNAS* **75,** 2021.
4. Hunter, T., and Sefton, B. M. (1980). *PNAS* **77,** 1311.
5. Schwartz, D., Tizard, R., and Gilbert, W. (1983). *Cell* **32,** 853.
6. Takeya, T., Feldman, R. A., and Hanafusa, H. (1982). *J. Virol.* **44,** 1.
7. Czernilofsky, A. P., Levinson, A. D., Varmus, H. E., Bishop, J. M., Tischler, E., and Goodman, H. M. (1983). *Nature (London)* **301,** 736.
8. Buss, J. E., and Sefton, B. M. (1985). *J. Virol.* **53,** 7.
9. Schultz, A. M., Henderson, L. E., Oroszlan, S., Garber, E. A., and Hanafusa, H. (1985). *Science* **227,** 427.
10. Brugge, J. S., and Erikson, R. L. (1977). *Nature (London)* **269,** 346.
11. Gilmer, T. M., and Erikson, R. L. (1983). *J. Virol.* **45,** 462.
12. Resh, M. D., and Erikson, R. L. (1985). *J. Cell Biol.* **100,** 409.
13. Parsons, S. J., McCarley, D. J., Ely, C. M., Benjamin, D. C., and Parsons, J. T. (1984). *J. Virol.* **51,** 272.
14. Lipsich, L. A., Lewis, A. J., and Brugge, J. S. (1983). *J. Virol.* **48,** 352.
15. Sefton, B. M., and Hunter, T., unpublished results.
16. Casnellie, J. E., Gentry, L. E., Rohrschneider, L. R., and Krebs, E. G. (1984). *PNAS* **81,** 6676.
17. Gentry, L. E., Rohrschneider, L. R., Casnellie, J. E., and Krebs, E. G. (1983). *JBC* **258,** 11219.
18. Sefton, B. M., and Walter, G. (1982). *J. Virol.* **44,** 467.
19. Tamura, T., Bauer, H., Birr, C., and Pipkorn, R. (1983). *Cell* **34,** 587.
20. Willingham, M. C., Jay, G., and Pastan, I. (1979). *Cell* **18,** 125.
21. Courtneidge, S. A., Levinson, A. D., and Bishop, J. M. (1980). *PNAS* **77,** 3783.
22. Rohrschneider, L. R. (1980). *PNAS* **77,** 3514.
23. Kreuger, J. G., Wang, E., and Goldberg, A. R. (1980). *Virology* **101,** 25.
24. Oppermann, H., Levinson, A. D., Levintow, L., Varmus, H. E., Bishop, J. M., and Kawai, S. (1981). *Virology* **113,** 736.
25. Brugge, J., Erikson, E., and Erikson, R. L. (1981). *Cell* **25,** 363.
26. Kamps, M. P., Buss, J. E., and Sefton, B. M. (1985). *PNAS* **82,** 4625.
27. Cross, F. R., Garber, E. A., Pellman, D., and Hanafusa, H. (1984). *Mol. Cell. Biol.* **4,** 1834.
28. Sefton, B. M., Patschinsky, T., Berdot, C., Hunter, T., and Elliott, T. (1982). *J. Virol.* **41,** 813.
29. Ziemiecki, A., Friis, R. R., and Bauer, H. (1982). *Mol. Cell. Biol.* **2,** 355.
30. Levinson, A. D., Courtneidge, S. A., and Bishop, J. M. (1981). *PNAS* **78,** 1624.
31. Brugge, J. S., and Darrow, D. (1984). *JBC* **259,** 4550.
32. Barker, W. C., and Dayhoff, M. O. (1982). *PNAS* **79,** 2836.
33. Stoker, A. W., Enrietto, P. J., and Wyke, J. A. (1984). *Mol. Cell. Biol.* **4,** 1508.
34. Bryant, D. L., and Parsons, J. T. (1982). *J. Virol.* **44,** 683.

35. Parsons, J. T., Bryant, D. L., Wilkerson, V., Gilmartin, G., and Parsons, S. J. (1984). "Cancer Cells," Vol. 2, p. 37. Cold Spring Harbor Lab., Cold Spring Harbor, New York.
36. Cross, F. R., and Hanafusa, H. (1983). *Cell* **34**, 597.
37. Cross, F. R., Garber, E. A., and Hanafusa, H. (1985). *Mol. Cell. Biol.* **5**, 2789.
38. Kitamura, N., and Yoshida, M. (1983). *J. Virol.* **46**, 985.
39. Smart, J. E., Oppermann, H., Czernilofsky, A. P., Purchio, A. F., Erikson, R. L., and Bishop, J. M. (1981). *PNAS* **78**, 6013.
40. Patschinsky, T., Hunter, T., Esch, F. S., Cooper, J. A., and Sefton, B. M. (1982). *PNAS* **79**, 973.
41. Collett, M. S., Wells, S. K., and Purchio, A. F. (1983). *Virology* **128**, 285.
42. Purchio, A. F., Wells, S. K., and Collett, M. S. (1983). *Mol. Cell. Biol.* **3**, 1589.
43. Brown, D. J., and Gordon, J. A. (1984). *JBC* **259**, 9580.
44. Collett, M. S., Belzer, S. K., and Purchio, A. F. (1984). *Mol. Cell. Biol.* **4**, 1213.
45. Collett, M. S., Erikson, E., and Erikson, R. L. (1979). *J. Virol.* **29**, 770.
46. Roth, C. W., Richert, N. D., Pastan, I., and Gottesman, M. M. (1983). *JBC* **258**, 10768.
47. Purchio, A. F., Shoyab, M., and Gentry, L. E. (1985). *Science* **229**, 1393.
48. Gould, K. L., Woodgett, J. R., Buss, J. E., Cooper, J. A., Shalloway, D., and Hunter, T. (1985). *Cell* **42**. 849.
49. Graziani, Y., Erikson, E., and Erikson, R. L. (1983). *JBC* **258**, 6344.
50. Snyder, M. A., Bishop, J. M., Colby, W. W., and Levinson, A. D. (1983). *Cell* **32**, 891.
51. Snyder, M. A., and Bishop, J. M. (1984). *Virology* **136**, 375.
52. Takeya, T., and Hanafusa, H. (1983). *Cell* **32**, 881.
53. Brugge, J. S., Cotton, P. C., Queral, A. E., Barrett, J. N., Nonner, D., and Keane, R. W. (1985). *Nature (London)* **316**, 554.
54. Sorge, L. K., Levey, B. T., and Maness, P. F. (1984). *Cell* **36**, 249.
55. Golden, A., Nemeth, S. P., and Brugge, J. S. (1986). *PNAS* **83**, 852.
55a. Cooper, J. A., Gould, K. L., Cartwright, C. A., and Hunter, T. (1986). *Science* **231**, 1431.
56. Karess, R. E., and Hanafusa, H. (1981). *Cell* **24**, 155.
57. Purchio, A. F., Erikson, E., Collett, M. S., and Erikson, R. L. (1981). *Cold Spring Harbor Conf. Cell Proliferation* **8**, 1203.
58. Courtneidge, S. A. (1985). *EMBO J.* **4**, 1471.
59. Iba, H., Takeya, T., Cross, F. R., Hanafusa, T., and Hanafusa, H. (1984). *PNAS* **81**, 4424.
60. Shalloway, D., Coussens, P. M., and Yaciuk, P. (1984). *PNAS* **81**, 7071.
61. Parker, R. C., Varmus, H. E., and Bishop, J. M. (1984). *Cell* **37**, 131.
62. Johnson, P. J., Coussens, P. M., Danko, A. V., and Shalloway, D. (1985). *Mol. Cell. Biol.* **5**, 1073.
63. Hanafusa, H., personal communication.
64. Parsons, S. J., and Creutz, C. E. (1986). *BBRC* **134**, 736.
65. Coussens, P. M., Cooper, J. A., Hunter, T., and Shalloway, D. (1985). *Mol. Cell. Biol.* **5**, 3304.
66. Iba, H., Cross, F. R., Garber, E. A., and Hanafusa, H. (1985). *Mol. Cell. Biol.* **5**, 1058.
67. Kitamura, N., Kitamura, A., Toyoshima, K., Hirayama, Y., and Yoshida, M. (1982). *Nature (London)* **297**, 205.
68. Kawai, S., Yoshida, M., Segawa, K., Sugiyama, H., Ishizaki, R., and Toyoshima, K. (1980). *PNAS* **77**, 6199.
69. Gentry, L. E., and Rohrschneider, L. R. (1984). *J. Virol.* **51**, 539.
70. Beemon, K., and Mattingly, B., personal communication.
71. Lipsich, L. A., Cutt, J. R., and Brugge, J. S. (1982). *Mol. Cell. Biol.* **2**, 875.
72. Adkins, B., Hunter, T., and Sefton, B. M. (1982). *J. Virol.* **43**, 448.
73. Patschinsky, T., and Sefton, B. (1981). *J. Virol.* **39**, 104.

74. Gessler, M., and Barnekow, A. (1984). *Biosci. Rep.* **4**, 757.
75. Naharro, G., Robbins, K. C., and Reddy, E. P. (1984). *Science* **223**, 63.
76. Nishizawa, M., Semba, K., Yoshida, M. C., Yamamoto, T., Sasaki, M., and Toyoshima, K. (1986). *Mol. Cell. Biol.* **6**, 511.
77. Naharro, G., Dunn, C. Y., and Robbins, K. C. (1983). *Virology* **125**, 502.
78. Sefton, B. M., personal communication.
79. Rohrschneider, L. R., personal communication.
80. Shibuya, M., and Hanafusa, H. (1982). *Cell* **30**, 787.
81. Hampe, A., Laprevotte, I., Galibert, F., Fedele, L. A., and Sherr, C. J. (1982). *Cell* **30**, 775.
82. Groffen, J., Heisterkamp, N., Shibuya, M., Hanafusa, H., and Stephenson, J. R. (1983). *Virology* **125**, 480.
83. Lee, W.-H., Bister, K., Pawson, A., Robins, T., Moscovici, C., and Duesberg, P. H. (1980). *PNAS* **77**, 2018.
84. Hanafusa, T., Wang, L.-H., Anderson, S. M., Karess, R. E., Hayward, W. S., and Hanafusa, H. (1980). *PNAS* **77**, 3009.
85. Ingman-Baker, J., Hinze, E., Levy, J. G., and Pawson, T. (1984). *J. Virol.* **50**, 572.
86. Pawson, T., Guyden, J., Kung, T.-H., Radke, K., Gilmore, T., and Martin, G. S. (1980). *Cell* **22**, 767.
87. Feldman, R. A., Hanafusa, T., and Hanafusa, H. (1980). *Cell* **22**, 757.
88. Barbacid, M., Beemon, K., and Devare, S. G. (1980), *PNAS* **77**, 5158.
89. Reynolds, F. H., Van de Ven, W. J. M., Blomberg, J., and Stephenson, J. R. (1981). *J. Virol.* **37**, 643.
90. Sen, S., Houghten, R. A., Sherr, C. J., and Sen, A. (1980). *PNAS* **80**, 1246.
91. Feldman, R. A., Wang, E., and Hanafusa, H. (1983). *J. Virol.* **45**, 782.
92. Moss, P., Radke, K., Carter, C., Young, J., Gilmore, T., and Martin, G. S. (1984). *J. Virol.* **52**, 557.
93. Schultz, A., and Oroszlan, S. (1984). *Virology* **133**, 431.
94. Rohrschneider, L. R., and Gentry, L. E. (1984). *Adv. Viral Oncol.* **4**, 269.
95. Weinmaster, G., Hinze, E., and Pawson, T. (1983). *J. Virol.* **46**, 29.
96. Weinmaster, G., Zoller, M. J., Smith, M., Hinze, E., and Pawson, T. (1984). *Cell* **37**, 559.
97. Weinmaster, G., and Pawson, T. (1985). *JBC* **261**, 328.
98. Mathey-Prevot, B., Hanafusa, H., and Kawai, S. (1982). *Cell* **28**, 897.
99. Young, J. C., and Martin, G. S. (1984). *J. Virol.* **52**, 913.
100. MacDonald, I., Levy, J., and Pawson, T. (1985). *Mol. Cell. Biol.* **5**, 2543.
101. Feldman, R. A., Garbilove, J. L., Tam, J. P., Moore, M. A. S., and Hanafusa, H. (1985). *PNAS* **82**, 2379.
102. Huang, C.-C., Hammond, C., and Bishop, J. M. (1985). *JMB* **181**, 175.
103. Foster, D. A., and Hanafusa, H. (1983). *J. Virol.* **48**, 744.
104. Guyden, J. C., and Martin, G. S. (1982). *Virology* **122**, 71.
105. Duesberg, P. H., Phares, W., and Lee, W.-H. (1983). *Virology* **131**, 144.
106. Foster, D. A., Shibuya, M., and Hanafusa, H. (1985). *Cell* **42**, 105.
107. Stone, J. C., Atkinson, T., Smith, M., and Pawson, T. (1984). *Cell* **37**, 549.
108. Stone, J. C., and Pawson, T. (1985). *J. Virol.* **55**, 721.
109. Sodrowski, J., Goh, W. C., and Haseltine, W. A. (1984). *PNAS* **81**, 3039.
110. Reddy, E. P., Smith, M. J., and Srinivasan, A. (1983). *PNAS* **80**, 3623.
111. Lee, R., Paskind, M., Wang, J. Y. J., and Baltimore, D. (1985). *In* "RNA Tumor Viruses" (R. Weiss, N. Teich, H. Varmus, and J. Coffin, eds.), p. 861. Cold Spring Harbor Lab., New York.
112. Witte, O. N., Rosenberg, N., Paskind, M., Shields, A., and Baltimore, D. (1978). *PNAS* **75**, 2488.

113. Reynolds, F. H., Sacks, T. L., Deobaghar, D. N., and Stephenson, J. R. (1978). *PNAS* **75**, 3974.
114. Witte, O. N., Rosenberg, N., and Baltimore, D. (1979). *J. Virol.* **31**, 776.
115. Witte, O. N., Ponticelli, A., Gifford, A., Baltimore, D., Rosenberg, N., and Elder, J. (1981). *J. Virol.* **39**, 870.
116. Van de Ven, W. J. M., Reynolds, F. H., and Stephenson, J. R. (1980). *Virology* **101**, 185.
117. Konopka, J. B., Davis, R. L., Watanabe, S. M., Ponticelli, A., Schiff-Maker, L., Rosenberg, N., and Witte, O. N. (1984). *J. Virol.* **51**, 223.
118. Kloetzer, W., Kurzrock, R., Smith, L., Talpaz, M., Spiller, M., Gutterman, J., and Arlinghaus, R. (1985). *Virology* **140**, 230.
119. Wang, J. Y. J., and Baltimore, D. (1985). *JBC* **260**, 64.
120. Mathey-Prevot, B., and Baltimore, D. (1985). *EMBO J.* **4**, 1769.
121. Rohrschneider, L. R., and Najita, L. M. (1984). *J. Virol.* **51**, 547.
122. Rotter, V., Boss, M. A., and Baltimore, D. (1981). *J. Virol.* **38**, 336.
123. Sefton, B. M., Hunter, T., and Raschke, W. C. (1981). *PNAS* **78**, 1552.
124. Reynolds, F. H., Oroszlan, S., and Stephenson, J. R. (1982). *J. Virol.* **44**, 1097.
125. Witte, O. N., Rosenberg, N., and Baltimore, D. (1979). *Nature (London)* **281**, 396.
126. Ben-Neriah, Y., Bernards, A., Paskind, M., Daley, G. Q., and Baltimore, D. (1986). *Cell* **44**, 577.
127. Konopka, J. B., Watanabe, S. M., and Witte, O. N. (1984). *Cell* **37**, 1035.
128. Ponticelli, A. S., Whitlock, C. A., Rosenberg, N., and Witte, O. N. (1982). *Cell* **29**, 953.
129. Konopka, J. B., and Witte, O. N. (1985). *Mol. Cell. Biol.* **5**, 3116.
130. Collins, S. J., and Groudine, M. T. (1983). *PNAS* **80**, 4813.
131. Heisterkamp, N., Stephenson, J. R., Groffen, J., Hansen, P. F., de Klein, A., Bartram, C. R., and Grosfeld, G. (1983). *Nature (London)* **306**, 239.
132. Prywes, R., Foulkes, J. G., Rosenberg, N., and Baltimore, D. (1983). *Cell* **34**, 569.
133. Prywes, R., Hoag, J., Rosenberg, N., and Baltimore, D. (1985). *J. Virol.* **54**, 123.
134. Prywes, R., Foulkes, J. G., and Baltimore, D. (1985). *J. Virol.* **54**, 114.
135. Wang, J. Y. J., personal communication.
136. Neckameyer, W. S., and Wang, L.-H. (1985). *J. Virol.* **53**, 879.
137. Feldman, R. A., Wang, L.-H., Hanafusa, H., and Balduzzi, P. C. (1982). *J. Virol.* **42**, 228.
138. Sefton, B. M., and Patschinsky, T., unpublished results.
139. Notter, M. F. D., and Balduzzi, P. C. (1984). *Virology* **136**, 56.
140. Hunter, T., and Cooper, J. A. (1983). *UCLA Symp. Mol. Cell. Biol.* [N.S.] **6**, 369.
141. Antler, A. M., Greenberg, M. E., Edelman, G. M., and Hanafusa, H. (1985). *Mol. Cell. Biol.* **5**, 263.
142. Sefton, B. M., Cooper, J. A., and Patschinsky, T., unpublished results.
143. Shibuya, M., Hanafusa, H., and Balduzzi, P. C. (1982). *J. Virol.* **42**, 143.
144. Yamamoto, T., Hihara, H., Nishida, T., Kawai, S., and Toyoshima, K. (1983). *Cell* **34**, 225.
145. Hayman, M. J., Ramsay, G. M., Savin, K., Kitchener, G., Graf, T., and Beug, H. (1983). *Cell* **32**, 579.
146. Privalsky, M. L., Sealy, L., Bishop, J. M., McGrath, J. P., and Levinson, A. D. (1983). *Cell* **32**, 1257.
147. Hayman, M. J., and Beug, H. (1984). *Nature (London)* **309**, 460.
148. Privalsky, M. L., and Bishop, J. M. (1984). *Virology* **135**, 356.
149. Decker, S. J. (1985). *JBC* **260**, 2003.
150. Kris, R. M., Lax, I., Gullick, W., Waterfield, M. D., Ullrich, A., Fridkin, M., and Schlessinger, J. (1985). *Cell* **40**, 619.
151. Akiyama, T., Yamada, Y., Ogawara, H., Richert, N., Pastan, I., Yamamoto, T., and Kasuga, M. (1984). *BBRC* **123**, 797.

152. Downward, J., Yarden, Y., Mayes, E., Scrace, G., Totty, N., Stockwell, P., Ullrich, A., Schlessinger, J., and Waterfield, M. D. (1984). *Nature (London)* **307,** 521.
153. Buhrow, S. A., Cohen, S., and Staros, J. V. (1982). *JBC* **257,** 4019.
154. Gilmore, T., DeClue, J. E., and Martin, G. S. (1985). *Cell* **40,** 609.
155. Hunter, T., Ling. N., and Cooper, J. A. (1984). *Nature (London)* **311,** 480.
156. Davis, R. J., and Czech, M. P. (1985). *PNAS* **82,** 1974.
157. Cochet, C., Gill, G. N., Meisenhelder, J., Cooper, J. A., and Hunter, T. (1984). *JBC* **259,** 2553.
158. Friedman, B., Frackelton, A. R., Ross, A. H., Connors, J. M., Fujiki, H., Sugimura, T., and Rosner, M. R. (1984). *PNAS* **81,** 3034.
159. Ullrich, A., Coussens, L., Hayflick, J. S., Dull, T. J., Gray, A., Tam, A. W., Lee, J., Yarden, Y., Liberman, T. A., Schlessinger, J., Downward, J., Mayes, E. L. V., Waterfield, M. D., Whittle, N., and Seeburg, P. H. (1984). *Nature (London)* **309,** 418.
160. Downward, J., Parker, P., and Waterfield, M. D. (1984). *Nature (London)* **311,** 483.
161. Sefton, B. M., Hunter, T., Beemon, K., and Eckhart, W. (1980). *Cell* **20,** 807.
162. Frykberg, L., Palmieri, S., Beug, H., Graf, T., Hayman, M. J., and Vennstrom, B. (1983). *Cell* **32,** 227.
163. Sealy, L., Privalsky, M. L., Moscovici, G., Moscovici, C., and Bishop, J. M. (1983). *Virology* **130,** 155.
164. Privalsky, M. L., Bassiri, M., and Ng, M., personal communication.
165. Riedel, H., and Ullrich, A., personal communication.
166. Nilsen, T., Maroney, P. A., Goodwin, R. G., Rottman, F. M., Crittenden, L. B., Raines, M. A., and Kung, H.-J. (1985). *Cell* **41,** 719.
167. Scotting, P. J., and Hayman, M. J., personal communication.
168. Hampe, A., Gobet, M., Sherr, C. J., and Galibert, F. (1984). *PNAS* **81,** 85.
169. Barbacid, M. L., Lauver, A. V., and Devare, S. G. (1980). *J. Virol.* **33,** 196.
170. Van De Ven, W. J. M., Reynolds, F. H., Nalewaik, R. P., and Stephenson, J. R. (1980). *J. Virol.* **35,** 165.
171. Ruscetti, S. K., Turek, L. P., and Sherr, C. J. (1980). *J. Virol.* **35,** 259.
172. Anderson, S. J., Furth, M., Wolff, L., Ruscetti, S. K., and Sherr, C. J. (1982). *J. Virol.* **44,** 696.
173. Anderson, S. J., Gonda, M. A., Rettenmeier, C. W., and Sherr, C. J. (1984). *J. Virol.* **51,** 730.
174. Rettenmeier, C. W., Roussel, M. F., Quinn, C. O., Kitchingman, G. R., Look, A. T., and Sherr, C. J. (1985). *Cell* **40,** 971.
175. Sherr, C. J., Rettenmeier, C. W., Sacca, R., Roussel, M. F., Look, A. T., and Stanley, E. R. (1985). *Cell* **41,** 665.
176. Manger, R., Najita, L., Nichols, E. J., Hakomori, S.-I., and Rohrschneider, L. (1984). *Cell* **39,** 327.
177. Nichols, E. J., Manger, R., Hakomori, S., Herscovics, A., and Rohrschneider, L. R. (1985). *Mol. Cell. Biol.* **5,** 3467.
178. Barbacid, M., and Lauver, A. V. (1981). *J. Virol.* **40,** 812.
179. Roussel, M. R., Rettenmeier, C. W., Look, A. T., and Sherr, C. J. (1984). *Mol. Cell. Biol.* **4,** 1999.
180. Rettenmeier, C. W., Chen, J. H., Roussel, M. F., and Sherr, C. J. (1985). *Science* **228,** 320.
181. Yeung, Y. G., Jubinsky, P. T., and Stanley, E. R., personal communication.
182. Sherr, C. J., personal communication.
183. Coussens, L., Van Beveren, C., Smith, D., Chen, E., Mitchell, R. L., Isacke, C. M., Verma, I. M., and Ullrich, A. (1986). *Nature (London)* **320,** 277.
184. Waterfield, M. D., Scrace, G. T., Whittle, N., Stroobant, P., Johnsson, A., Wasteson, A., Westermark, B., Heldin, C.-H., Huang, J. S., and Deuel, T. F. (1983). *Nature (London)* **304,** 35.

185. Doolittle, R. F., Hunkapiller, M. W., Hood, L. E., Devare, S. G., Robbins, K. C., Aaronson, S. A., and Antoniades, H. N. (1983). *Science* **221**, 275.

186. Devare, S. G., Reddy, E. P., Robbins, K. C., Andersen, P. R., Tronick, S. R., and Aaronson, S. A. (1983). *PNAS* **80**, 731.

187. Robbins, K. C., Antoniades, H. N., Devare, S. G., Hunkapiller, M. W., and Aaronson, S. A. (1983). *Nature (London)* **305**, 605.

188. Deuel, T. F., Huang, J. S., Huang, S. S., Stroobant, P., and Waterfield, M. D. (1983). *Science* **221**, 1348.

189. Johnsson, A., Heldin, C.-H., Wasteson, A., Westermark, B., Deuel, T. F., Huang, J. S., Seeburg, P. H., Gray, A., Ullrich, A., Scrace, G., Stroobant, P., and Waterfield, M. D. (1984). *EMBO J.* **3**, 2963.

190. Owen, A. J., Pantazis, P., and Antoniades, H. N. (1984). *Science* **225**, 54.

191. Robbins, K. C., Leal, F., Pierce, J. H., and Aaronson, S. A. (1985). *EMBO J.* **4**, 1783.

192. Hannink, M., and Donoghue, D. J. (1984). *Science* **226**, 1197.

193. Ek, B., Westermark, B., Wasteson, A., and Heldin, C.-H. (1982). *Nature (London)* **295**, 419.

194. Nishimura, J., Huang, J. S., and Deuel, T. F. (1982). *PNAS* **79**, 4303.

195. Frackelton, A. R., Tremble, P. M., and Williams, L. T. (1984). *JBC* **259**, 7909.

196. Huang, J. S., Huang, S. S., and Deuel, T. F. (1984). *Cell* **39**, 79.

197. Josephs, S. F., Ratner, L., Clarke, M. F., Westin, E. H., Reitz, M. S., and Wong-Staal, F. (1984). *Science* **225**, 636.

198. Gazit, A., Igarishi, H., Chiu, I.-M., Srinivasan, A., Yaniv, A., Tronick, S. R., Robbins, K. C., and Aaronson, S. A. (1984). *Cell* **39**, 89–97.

199. Marquadt, H., Hunkapiller, M. W., Hood, L. E., Twardzik, D. R., De Larco, J. E., Stephenson, J. R., and Todaro, G. J. (1983). *PNAS* **80**, 4684.

200. Reynolds, F. H., Todaro, G. J., Fryling, C., and Stephenson, J. R. (1981). *Nature (London)* **292**, 259.

201. Carpenter, G., Stoscheck, C. M., Preston, Y. A., and De Larco, J. E. (1983). *PNAS* **80**, 5627.

202. Pike, L. J., Marquadt, H., Todaro, G. J., Gallis, B., Casnellie, J. E., Bornstein, P., and Krebs, E. G. (1982). *JBC* **257**, 14628.

203. Libermann, T. A., Nusbaum, H. R., Razon, N., Kris, R., Lax, I., Soreq, H., Whittle, N., Waterfield, M. D., Ullrich, A., and Schlessinger, J. (1985). *Nature (London)* **313**, 144.

204. Shih, C., Padhy, L. C., Murray, M., and Weinberg, R. A. (1981). *Nature (London)* **290**, 261.

205. Padhy, L. C., Shih, C., Cowing, D., Finkelstein, R., and Weinberg, R. A. (1982). *Cell* **28**, 865.

206. Drebin, J. A., Stern, D. F., Link, V. C., Weinberg, R. A., and Greene, M. I. (1984). *Nature (London)* **312**, 545.

207. Stern, D. F., Heffernan, P. A., and Weinberg, R. A. (1986). *Mol. Cell. Biol.* **6**, 1729.

208. Schechter, A. L., Stern, D. F., Vaidyanathan, L., Decker, S. J., Drebin, J. A., Greene, M. I., and Weinberg, R. A. (1984). *Nature (London)* **312**, 513.

209. Bargmann, C. I., Hung, M.-C., and Weinburg, R. A. (1986). *Nature (London)* **312**, 513.

210. Yamamoto, T., Ikawa, I., Akiyama, T., Semba, K., Nomura, N., Miyajima, N., Saito, T., and Toyoshima, K. (1986). *Nature (London)* **319**, 230.

211. Drebin, J. A., Link, V. C., Stern, D. F., Weinberg, R. A., and Greene, M. I. (1985). *Cell* **41**, 695.

212. Smith, A. E., Smith, R., Griffin, B., and Fried, M. (1979). *Cell* **18**, 915.

213. Schaffhausen, B. S., and Benjamin, T. L. (1979). *Cell* **18**, 935.

214. Walter, G., Hutchinson, M. A., Hunter, T., and Eckhart, W. (1982). *PNAS* **79**, 4025.

215. Courtneidge, S. A., and Smith, A. E. (1983). *Nature (London)* **303**, 435.

216. Courtneidge, S. A., and Smith, A. E. (1984). *EMBO J.* **3**, 585.

217. Kaplan, P. L., personal communication.

218. Bolen, J. B., Thiele, C. J., Israel, M. A., Yonemoto, W., Lipsich, L. A., and Brugge, J. S. (1984). *Cell* **38**, 767.
218a. Cartwright, C. A., Kaplan, P. L., Cooper, J. A., Hunter, T., and Eckhart, W. (1986). *Mol. Cell. Biol.* **6**, 1562.
219. Casnellie, J. E., Harrison, M. L., Pike, L. J., Hellstrom, K. E., and Krebs, E. G. (1982). *PNAS* **79**, 282.
220. Gacon, G., Gisselbrecht, S., Piau, J. P., Boissel, J. P., Tolle, J., and Fisher, S. (1982). *EMBO J.* **1**, 1579.
221. Casnellie, J. E., Harrison, M. L., Hellstrom, K. E., and Krebs, E. G. (1983). *JBC* **258**, 10738.
222. Voronova, A. F., Buss, J. E., Patschinsky, T., Hunter, T., and Sefton, B. M. (1985). *Mol. Cell. Biol.* **4**, 2705.
223. Casnellie, J. E., Harrison, M. L., Hellstrom, K. E., and Krebs, E. G. (1982). *JBC* **257**, 13877.
224. Swarup, G., Dasgupta, J. D., and Garbers, D. L. (1983). *JBC* **258**, 10341.
225. Earp, H. S., Austin, K. S., Buessow, S. C., Dy, R., and Gillespie, G. Y. (1984). *PNAS* **81**, 2347.
226. Earp, H. S., Austin, K. S., Gillespie, G. Y., Buessow, S. C., Davies, A. A., and Parker, P. J. (1985). *JBC* **260**, 4351.
227. Marchildon, G. A., Casnellie, J. E., Walsh, K. A., and Krebs, E. G. (1984). *PNAS* **81**, 7679.
228. Voronova, A. F., and Sefton, B. M. (1986). *Nature (London)* **319**, 682.
229. Marth, J. D., Peet, R., Krebs, E. G., and Perlmutter, R. M. (1985). *Cell* **43**, 393.
230. Hu, S. S. F., Moscovici, C., and Vogt, P. K. (1978). *Virology* **89**, 162.
231. Rapp, U. R., Reynolds, F. H., and Stephenson, J. R. (1983). *J. Virol.* **45**, 914.
232. Kan, N. C., Flordellis, C. S., Mark, G. E., Duesberg, P. H., and Papas, T. S. (1983). *PNAS* **80**, 3000.
233. Mark, G. E., and Rapp, U. R. (1984). *Science* **224**, 285.
234. Kan, N. C., Flordellis, C. S., Mark, G. E., Duesberg, P. H., and Papas, T. S. (1984). *Science* **223**, 813.
235. Van Beveren, C., Galleshaw, J. A., Jonas, V., Berns, A. J. M., Doolittle, R. F., Donoghue, D. J., and Verma, I. M. (1981). *Nature (London)* **289**, 258.
236. Papkoff, J., Verma, I., and Hunter, T. (1982). *Cell* **29**, 417.
237. Papkoff, J., Nigg, E. A., and Hunter, T. (1983). *Cell* **33**, 161.
238. Moelling, K., Heimann, B., Beimling, P., Rapp, U. R., and Sander, T. (1984). *Nature (London)* **312**, 558.
239. Kloetzer, W. S., Maxwell, S. A., and Arlinghaus, R. B. (1983). *PNAS* **80**, 412.
240. Maxwell, S. A., and Arlinghaus, R. B. (1985). *Virology* **143**, 321.
241. Rapp, U., personal communication.
242. Seth, A., and Vande Woude, G. (1985). *J. Virol.* **56**, 144.
243. Hannink, M., and Donoghue, D. J. (1985). *PNAS* **82**, 7894.
244. Cooper, J. A., and Hunter, T. (1981). *Mol. Cell. Biol.* **1**, 394.
245. Ullrich, A., Bell, J. R., Chen, E. Y., Herrara, R., Petruzzelli, L. M., Dull, T. J., Gray, A., Coussens, L., Liao, Y.-C., Tsubokawa, M., Mason, A., Seeburg, P. H., Grunfeld, C., Rosen, O. M., and Ramachandran, J. (1985). *Nature (London)* **313**, 756.
246. Ebina, Y., Ellis, L., Jarnagin, K., Edery, M., Graf, L., Clauser, E., Ou, J.-H., Masiarz, F., Kan, Y. W., Goldfine, I. D., Roth, R. A., and Rutter, W. J. (1985). *Cell* **40**, 747.
247. Shoji, S., Parmelee, D. C., Wade, R. D., Kumar, S., Ericsson, L. H., Walsh, K. A., Neurath, H., Long, G. L., Demaille, J. G., Fischer, E. H., and Titani, K. (1981). *PNAS* **78**, 848.
248. Takio, K., Wade, R. D., Smith, S. B., Krebs, E. G., Walsh, K. A., and Titani, K. (1984). *Biochemistry* **23**, 4207.
249. Reimann, E. M., Titani, K., Ericsson, L. H., Wade, R. D., Fischer, E. H., and Walsh, K. A. (1984). *Biochemistry* **23**, 4185.

250. Takio, K., Blumenthal, D. K., Edelman, A. M., Walsh, K. A., Krebs, E. G., and Titani, K. (1985). *Biochemistry* **24,** 6028.
251. Wilkerson, V. W., Bryant, D. L., and Parsons, J. T. (1985). *J. Virol.* **55,** 314.
252. Shalloway, D., Coussens, P. M., and Yaciuk, P. (1984). "Cancer Cells," Vol. 2, p. 9. Cold Spring Harbor Lab., Cold Spring Harbor, New York.
253. Donoghue, D. J., personal communication.
254. Arlinghaus, R. B., personal communication.
255. Zoller, M. J., Nelson, N. C., and Taylor, S. S. (1981). *JBC* **256,** 10837.
256. Kamps, M. P., Taylor, S. S., and Sefton, B. M. (1984). *Nature (London)* **310,** 589.
257. Russo, M. W., Lukas, T. J., Cohen, S., and Staros, J. V. (1985). *JBC* **260,** 5205.
258. Kamps, M. P., and Sefton, B. M. (1986). *Mol. Cell. Biol.* **6,** 751.
259. Snyder, M. A., Bishop, J. M., NcGrath, J. P., and Levinson, A. D. (1985). *Mol. Cell. Biol.* **5,** 1772.
260. Weinmaster, G., Zoller, M. J., and Pawson, A. J. (1986). *EMBO. J.* **5,** 69.
261. Wierenga, R., and Hol, W. (1983). *Nature (London)* **302,** 258.
262. Rossman, M., Garavito, R., and Eventoff, W. (1977). *FEBS Symp.* **49,** 3.
263. Sternberg, M. J., and Taylor, W. R. (1984). *FEBS Lett.* **175,** 387.
264. Bryant, D. L., and Parsons, J. T. (1983). *J. Virol.* **45,** 1211.
265. Bryant, D. L., and Parsons, J. T. (1984). *Mol. Cell. Biol.* **4,** 862.
266. Bramson, H. N., Thomas, N., Matsueda, R., Nelson, N. C., Taylor, S. S., and Kaiser, E. T. (1982). *JBC* **257,** 10575.
267. Gilmer, T. M., and Erikson, R. L. (1981). *Nature (London)* **294,** 771.
268. McGrath, J. P., and Levinson, A. D. (1982). *Nature* **295,** 23.
269. Wang, J. Y. J., Queen, C., and Baltimore, D. (1982). *JBC* **257,** 13181.
270. Foulkes, J. G., Chow, C., Gorka, C., Frackelton, A. R., and Baltimore, D. (1985). *JBC* **260,** 8070.
271. Ferguson, B., Pritchard, M. L., Feild, J., Rieman, D., Greig, R. G., Poste, G., and Rosenberg, M. (1985). *JBC* **260,** 3652.
272. Erikson, R. L., Collett, M. S., Erikson, E., and Purchio, A. F. (1979). *PNAS* **76,** 6260.
273. Levinson, A. D., Oppermann, H., Varmus, H. E., and Bishop, J. M. (1980). *JBC* **255,** 11973.
274. Blithe, D. L., Richert, N. D., and Pastan, I. H. (1982). *JBC* **257,** 7135.
275. Richert, N. D., Blithe, D. L., and Pastan, I. H. (1982). *JBC* **257,** 7143.
276. Fukami, Y., and Lipmann, F. (1985). *PNAS* **82,** 321.
277. Sugimoto, Y., Erikson, E., Graziani, Y., and Erikson, R. L. (1985). *JBC* **260,** 18383.
278. Cohen, S., Ushiro, H., Stoschek, C., and Chinkers, M. (1982). *JBC* **257,** 1523.
279. Weber, W., Bertics, P. J., and Gill, G. N. (1984). *JBC* **259,** 14631.
280. Parker, P. J., Young, S., Gullick, W. J., Mayes, E. L. V., Bennett, P., and Waterfield, M. D. (1984). *JBC* **259,** 9906.
281. Erneux, C., Cohen, S., and Garbers, D. L. (1983). *JBC* **258,** 4137.
282. Pike, L. J., Kuenzel, E. A., Casnellie, J. E., and Krebs, E. G. (1984). *JBC* **259,** 9913.
283. Bechtel, P. J., Beavo, J. A., and Krebs, E. G. (1977). *JBC* **252,** 2691.
284. Fukami, Y., and Lipmann, F. (1983). *PNAS* **80,** 1872.
285. Armstrong, R. N., Kondo, H., and Kaiser, E. T. (1979). *PNAS* **76,** 722.
286. Mrockowski, B., Mosig, G., and Cohen, S. (1984). *Nature (London)* **309,** 270.
287. Cohen, S., Schlessinger, J., and Das, M., personal communications.
288. Sugimoto, Y., Whitman, M., Cantley, L. C., and Erikson, R. L. (1984). *PNAS* **81,** 2117.
289. Macara, I. G., Marinetti, G. V., and Balduzzi, P. C. (1984). *PNAS* **81,** 2728.
290. Whitman, M., Kaplan, D. R., Schaffhausen, B., Cantley, L., and Roberts, T. M. (1985). *Nature (London)* **315,** 239.
291. Thompson, D. M., Cochet, C., Chambaz, E. M., and Gill, G. N. (1985). *JBC* **260,** 8824.

292. Sugano, S., and Hanafusa, H. (1985). *Mol. Cell Biol.* **5**, 2399.
293. Sugimoto, Y., and Erikson, R. L. (1985). *Mol. Cell. Biol.* **5**, 3914.
294. Kemp, B. E., Graves, D. J., Benjamini, E., and Krebs, E. G. (1977). *JBC* **252**, 4888.
295. Cooper, J. A., Esch, F., Taylor, S. S., and Hunter, T. (1984). *JBC* **259**, 7835.
296. Baldwin, G. S., Knesel, J., and Monckton, J. M. (1983). *Nature (London)* **301**, 435.
297. Dekowski, S. A., Rybicki, A., and Drickamer, K. (1983). *JBC* **258**, 2750.
298. Hunter, T. (1982). *JBC* **257**, 4843.
299. Pike, L. J., Gallis, B., Casnellie, J. E., Bornstein, P., and Krebs, E. G. (1982). *PNAS* **79**, 1443.
300. Braun, S., Raymond, W. E., and Racker, E. (1984). *JBC* **259**, 2051.
301. Wong, T. W., and Goldberg, A. R. (1983). *JBC* **258**, 1022.
302. Wong, T. W., and Goldberg, A. R. (1984). *JBC* **259**, 3127.
303. Migliaccio, A., Rotondi, A., and Auricchio, F. (1984). *PNAS* **81**, 5921.
304. Kuehn, G. (1984). *Mol. Aspects Cell. Regul.* **3**, 185.
305. Yarden, Y., and Schlessinger, J. (1986). *Biochemistry* (in press).
306. Pawson, T., personal communication.
307. Rosen, O. M., Herrera, R., Olowe, Y., Petruzzelli, L. M., and Cobb, M. H. (1983). *PNAS* **80**, 3237.
308. Bertics, P. J., and Gill, G. N. (1985). *JBC* **260**, 14642.
309. Downward, J., Waterfield, M. D., and Parker, P. J. (1985). *JBC* **260**, 14538.
310. Decker, S. J. (1984). *Mol. Cell. Biol.* **4**, 1718.
311. Foulkes. J. G. (1983). *Curr. Top. Microbiol. Immunol.* **107**, 163.
312. Hunter, T., and Cooper, J. A. (1981). *Cell* **24**, 741.
313. Cooper, J. A., Bowen-Pope, D., Raines, E., Ross, R., and Hunter, T. (1982). *Cell* **31**, 263.
314. Cooper, J. A., and Hunter, T. (1981). *Mol. Cell. Biol.* **1**, 165.
315. Cheng, Y.-S. E., and Chen, L. B. (1981). *PNAS* **78**, 2388.
316. Ross, A. H., Baltimore, D., and Eisen, H. (1981). *Nature (London)* **294**, 654.
317. Frackelton, A. R., Ross, A. H., and Eisen, H. N. (1983). *Mol. Cell. Biol.* **3**, 1343.
318. Cooper, J. A., Reiss, N. A., Schwartz, R. J., and Hunter, T. (1983). *Nature (London)* **302**, 218.
319. Bissell, M. J., White, R. C., Hatie, C., and Bassham, J. A. (1973). *PNAS* **70**, 2951.
320. Alexander, C., Cooper, J. A., and Hunter, T., unpublished results.
321. Sefton, B. M., Hunter, T., Ball, E. H., and Singer, S. J. (1981). *Cell* **24**, 165.
322. Radke, K., and Martin, G. S. (1979). *PNAS* **76**, 5212.
323. Erikson, E., and Erikson, R. L. (1980). *Cell* **21**, 829.
324. Radke, K., Gilmore, T., and Martin, G. S. (1980). *Cell* **21**, 821.
325. Cooper, J. A., and Hunter, T. (1983). *JBC* **258**, 1108.
326. Courtneidge, S., Ralston, R., Alitalo, K., and Bishop, J. M. (1983). *Mol. Cell. Biol.* **3**, 340.
327. Greenberg, M. E., and Edelman, G. M. (1983). *Cell* **33**, 767.
328. Nigg, E. A., Cooper, J. A., and Hunter, T. (1983). *J. Cell Biol.* **97**, 1601.
329. Radke, K., Carter, V. C., Moss, P., Dehazya, P., Schliwa, M., and Martin, G. S. (1983). *J. Cell Biol.* **97**, 1601.
330. Gould, K. L., Cooper, J. A., Bretscher, A., and Hunter, T. (1986). *J. Cell Biol.* **102**, 660.
331. Geiger, B. (1979). *Cell* **18**, 193.
332. Ito, S., Richert, N., and Pastan, I. (1982). *PNAS* **79**, 4628.
333. Rohrschneider, L. R., and Rosok, M. J. (1983). *Mol. Cell. Biol.* **3**, 731.
334. Gould, K. L., Cooper, J. A., and Hunter, T. (1984). *J. Cell Biol.* **98**, 487.
335. Greenberg. M. E., Brackenbury, R., and Edelman, G. M. (1984). *J. Cell Biol.* **98**, 473.
336. Martinez, R., Nakamura, K. D., and Weber, M. J. (1982). *Mol. Cell. Biol.* **2**, 653.
337. Glenney, J. R., and Tack, B. F. (1985). *PNAS* **82**, 7884.

338. Erikson, E., Tomasiewicz, H. G., and Erikson, R. L. (1984). *Mol. Cell. Biol.* **4,** 77.
339. Gerke, V., and Weber, K. (1984). *EMBO J.* **3,** 227.
340. Glenney, J. R. (1985). *FEBS Lett.* **192,** 79.
341. Cooper, J. A., and Hunter, T. (1982). *J. Cell Biol.* **94,** 287.
342. Cooper, J. A., Scolnick, E. M., Ozanne, B., and Hunter, T. (1983). *J. Virol.* **48,** 752.
343. Brugge, J., and Darrow, D. (1982). *Nature (London)* **295,** 250.
344. Comoglio, P. M., Di Renzo, M. F., Tarone, G., Giancotti, F. G., Naldini, L., and Marchisio, P. C. (1984). *EMBO J.* **3,** 483.
345. Monteagudo, C. A., Williams, D. L., Crabb, G. A., Tondravi, M., and Weber, M. J. (1984). "Cancer Cells," Vol. 2; p. 69. Cold Spring Harbor Lab., Cold Spring Harbor, New York.
346. Ek, B., and Heldin, C.-H. (1984). *JBC* **259,** 11145.
347. Nakamura, K. D., Martinez, R., and Weber, M. J. (1983). *Mol. Cell. Biol.* **3,** 380.
348. Kohno, M. (1985). *JBC* **260,** 1771.
349. Cooper, J. A., Sefton, B. M., and Hunter, T. (1984). *Mol. Cell. Biol.* **4,** 30.
350. Freed, E., and Hunter, T., unpublished results.
351. Fava, R., and Cohen, S. (1984). *JBC* **259,** 2636.
352. Sawyer, S. T., and Cohen, S. (1985). *JBC* **260,** 8233.
353. Giugni, T. D., James, L. C., and Haigler, H. T. (1985). *JBC* **260,** 15081.
354. Bishop, R., Martinez, R., Nakamura, K. D., and Weber, M. J. (1983). *BBRC* **115,** 536.
355. Gilmore, T., and Martin, G. S. (1983). *Nature (London)* **306,** 487.
356. Cooper, J. A., and Hunter, T. (1985). *Mol. Cell. Biol.* **5,** 3304.
357. Schreiber, A. B., Yarden, Y., and Schlessinger, J. (1981). *BBRC* **101,** 517.

7

The Insulin Receptor and Tyrosine Phosphorylation

MORRIS F. WHITE • C. RONALD KAHN

Research Division
Joslin Diabetes Center and
Department of Medicine
Brigham and Women's Hospital
Harvard Medical School
Boston, Massachusetts 02215

THE ENZYMES, Vol. XVII
Copyright © 1986 by Academic Press, Inc.

I. Introduction and Scope

Cellular growth and metabolism is tightly regulated by a variety of hormones and growth factors that bind specifically to their target cells via surface membrane receptors (1, 2). In many cases, these interactions cause the production of a distinctive second messenger at the inner face of the plasma membrane that transmits the regulatory signal to intracellular sites. For example, biogenic amines, many polypeptide hormones, and prostaglandins stimulate the adenylate cyclase system which elevates the intracellular concentration of cAMP and activates cAMP-dependent protein kinases (3, 4). Cellular substrates for these cAMP-dependent kinases undergo phosphorylation on serine or threonine residues which alters their activity (4, 5). Similarly, α-adrenergic agents elevate the cytoplasmic concentration of inositol trisphosphate and Ca^{2+}, and the level of diacylglycerol in the plasma membrane (6). These second messengers initiate a series of events including the activation of calmodulin-dependent kinases (6, 7) or the protein kinase C (7, 8). Although the mobilization of cellular Ca^{2+} (9, 10) or the generation of a second messenger, either cAMP (11, 12) or some other mediator (13, 14), has been implicated at times in the response of cells to insulin and other growth factors, none of these classic second-messenger systems seem to adequately account for the cellular action of these polypeptides (15).

Phosphorylation of serine and threonine residues in cellular proteins is quantitatively significant and an important regulatory mechanism for cellular metabolism. In contrast, phosphorylation of tyrosine residues is extremely rare, and thus phosohotyrosine is nearly 3000-fold less abundant than phosphothreonine and phosphoserine combined (16). The search for a molecular explanation for the action of hormones and growth factors has lead to the observation that the plasma membrane receptors for epidermal growth factor (EGF) (17, 18), insulin (19, 20), platelet-derived growth factor (PDGF) (21–23) and insulin-like growth factor (IGF-I) (24–26) are tyrosine-specific protein kinases. These receptors were identified as tyrosine kinases because, in each case, the appropriate hormone stimulated tyrosine phosphorylation of a membrane protein that was subsequently identified as its receptor.

The physiology of the receptor-related tyrosine kinases is not completely understood, but they are likely to play a central role in the signal transmission of the receptors and the regulation of cellular growth and metabolism (1). A strong connection between tyrosine phosphorylation and cellular growth was originally suggested when cells infected with the Rous sarcoma virus (RSV) were found to contain elevated levels of phosphotyrosine (16, 27). This increase is due to autophosphorylation of a protein of $M_r = 60,000$ (pp60^{src}) which is the gene product responsible for cellular transformation by RSV (28), and tyrosine phosphorylation of other cellular proteins (29). Normal cells also contain tyrosine kinases which cross-react with antibodies against oncogene products suggesting that these cellular homologs may be important in normal growth regulation as well (28).

This review focuses on one of the hormone-regulated tyrosine kinases found in nearly all cells, the insulin receptor kinase. We consider its structure and enzymic properties, and the characteristics of tyrosine phosphorylation that occur in intact cells during their initial response to insulin. Finally, we compare the insulin receptor to other tyrosine kinases and summarize evidence supporting the notion that tyrosine phosphorylation is a key event regulating cellular differentiation, growth, and metabolism. Figure 1 shows schematic drawings of the general structural features of a few tyrosine kinases that are discussed in this review.

II. Structure of the Insulin Receptor and Its Relation to Other Tyrosine-Specific Protein Kinases

A. INTRODUCTION

A variety of experimental approaches have been used to determine the structure of plasma membrane receptors in mammalian cells and tissues. The insulin receptor has been studied using affinity chromatography, SDS-polyacrylamide

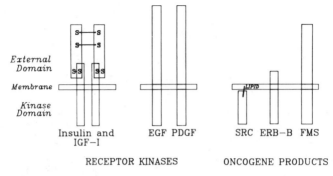

FIG. 1. A schematic comparison of the structure of the receptors for insulin, IGF-I, EGF, and PDGF, and the oncogene products pp60^{src}, gp65^{erb-B}, and gp140^{fms}.

gel electrophoresis (SDS-PAGE) of receptors labeled covalently with [^{125}I]insulin by cross-linking or photoreactive [^{125}I]insulin analogs, immunoprecipitation of biosynthetically labeled receptors with antiinsulin receptor antibody and analysis by SDS-PAGE, and nucleotide sequencing of cDNA that contains the coding region for the insulin receptor precursor. In this section, we describe the insulin receptor structure and compare it in general terms to the structure of the types I and II IGF receptors, and the receptors for PDGF and EGF. We also identify some specific amino acid sequence domains that serve as structural landmarks in tyrosine-specific receptor kinases and oncogene products, and serine–threonine kinases.

B.   INSULIN RECEPTOR

1. *Purification*

Affinity purification of the insulin receptor solubilized from rat liver using insulin-Sepharose was first achieved by Cuatrecasas (*30–32*), and improved by Jacobs *et al.* (*33–35*), Rosen *et al.* (*35, 36*) and Fujita-Yamaguchi *et al.* (*37, 38*), to give sufficient quantities of receptor for structural analysis, amino acid composition, and limited sequence analysis. These studies show that the receptor consists of two distinct proteins termed the α- and β-subunits (*39*). By gel filtration, ultracentrifugation, and SDS-PAGE on nonreducing gels, the receptor migrates as an oligomer with an estimated molecular mass of about 350,000. In some cases, larger and smaller oligomers are also detected (Fig. 2, lanes a and b). Upon reduction with dithiothreitol, only the α and β-subunits are obtained from the 350-kDa protein suggesting a stoichiometry of (αβ)$_2$ for the nonreduced form (*38, 40*). A minor protein of lower molecular mass (52 kDa) has also been observed in many of these studies (*32–34*). By tryptic peptide mapping this appears to be a degradation product of the β-subunit (*38*); however, Fehlmann *et*

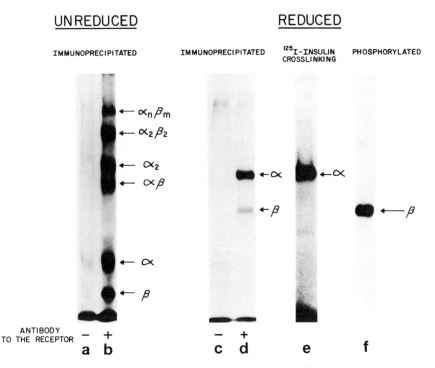

FIG. 2. An autoradiogram showing an SDS-PAGE separation of the insulin receptor under non-reducing (a and b) and reducing conditions (c–f). The receptor was surface-labeled with ^{125}I and immunoprecipitated (a–d) with control ($-$) or antiinsulin receptor serum ($+$), labeled with [^{125}I]insulin by crosslinking (e), or phosphorylated *in vitro* with [γ-^{32}P]ATP (f).

al. have suggested that it may be a major histocompatibility component that is associated with the receptor during immunoprecipitation (*41*). The purified receptor retains insulin-binding properties similar to those of the membrane-bound protein. The stoichiometry of insulin binding has been controversial (*42*), but in some preparations of receptor, 1.5–1.6 insulin molecules bind to the $M_r = 350,000$ oligomer suggesting that this structure contains two insulin binding sites (*37*).

2. *Biosynthesis and Glycosylation*

In parallel with attempts at receptor purification, studies of receptor structure and biosynthesis were undertaken by Kahn and co-workers using human sera containing antibodies against the insulin receptor (*43, 44*). Immunoprecipitation of the receptor following cell surface or biosynthetic labeling (Fig. 2, lanes c and d) also indicate that the insulin receptor in the plasma membrane is composed of two glycoprotein subunits linked covalently by disulfide bonds (*40, 45–47*).

Pulse-chase labeling studies with cultured human lymphocytes indicate that the α- and β-subunits of the insulin receptor are obtained from a single protein precursor (48). This was confirmed in 3T3-L1 cells (49). The first polypeptide detected with antireceptor antibody during biosynthetic labeling has a molecular mass by SDS-PAGE of 180,000 (49). N-linked high mannose-type core oligosaccharides are added cotranslationally to give a 190-kDa form. At this stage, the precursor binds to immobilized lentil lectin owing to its high mannose content, and removal of mannose from the precursor with endoglycosidase H regenerates the 180-kDa precursor (49). The 190-kDa proreceptor contains both an insulin-binding site (50) and the insulin-stimulated tyrosine kinase (50, 51), that is, it possess properties of both the α- and β-subunits (see the following sections).

In some, but not all cells, the 190-kDa proreceptor is converted to an $M_r = 210,000$ species before cleavage to form the α- and β-subunits (48). This larger protein is partially resistant to endoglycosidase H (49) and binds to wheat germ agglutinin (WGA)-agarose suggesting that the apparent increase of molecular mass is due to the addition of complex-type oligosaccharide that contain N-acetylglucosamine (48). However, the 210-kDa precursor is insensitive to neuraminidase suggesting that it has not been sialated. Cleavage of this protein does not yield mature α- and β-subunits, but rather a disulfide-link oligomer with subunit molecular masses of 120,000 and 80,000, respectively (49). Prior to insertion into the plasma membrane, the receptor subunits appear to be further glycosylated, acquire sialic acid and migrate on SDS-PAGE with molecular masses of 135,000 (α) and 95,000 (β). This presumably occurs in the Golgi region of the cell. The role of glycosylation in transport of the mature receptor from the Golgi to the plasma membrane is unknown.

Complete deglycosylation of the mature insulin receptor by treatment with trifluoromethane sulfonic acid yields aglyco α- and β-subunits with molecular masses of 98,500 and 80,000, respectively (52). This is consistent with the molecular mass of the aglycoprecursor which contains a single copy of each of the subunits (49). Based on digestion with endoglycosidase F, endoglycosidase H, and chemical deglycosylation, both subunits appear to contain complex N-linked oligosaccharides, as well as some polymannose units. The β-subunit probably also contains O-linked sugars since its molecular weight after endoglycosidase F treatment is still greater than that after chemical deglycosylation (52).

The cDNA for the insulin receptor has been cloned by two groups using partial amino acid sequences to synthesize oligonucleotide probes (53, 54). The amino acid sequence deduced from the nucleotide sequence of this DNA is consistent with the biosynthetic studies and indicates that the insulin receptor arises from a single chain precursor that contains both α- and β-subunits. During processing, this precursor is cleaved at a cationic domain, Arg-Lys-Arg-Arg, to yield both subunits (53, 54). The α- and β-subunits have calculated molecular masses of

82,000 and 69,000, respectively, and both contain several potential sites for glycosylation. Presumably, the discrepancy of the molecular masses measured after deglycosylation and that calculated from the amino acid sequence is due to anomalous migration of the subunits during SDS-PAGE.

3. Oligomeric Structure

Before reduction of the disulfide linkages between the α- and β-subunits, the insulin receptor extracted from human lymphocytes and rat hepatoma cells is composed of several oligomeric forms (40, 55, 56). These species have molecular masses by SDS-PAGE of 210,000, 270,000, 350,000, and 520,000 (Fig. 2). Two-dimensional SDS-PAGE reveals that these proteins correspond to $\alpha\beta$ heterodimers, α_2 homodimers, and two forms of $\alpha\beta$ heterodimers of high molecular weight, respectively (40, 55). Fujita-Yamaguchi suggests that the $\alpha\beta$ heterodimer of the purified receptor exhibits the highest kinase activity (57), although our studies suggest that the higher-molecular-weight forms are equally active in the intact cell. When receptors are extracted from adipocytes and placenta one disulfide-linked complex with a M_r = 350,000 is predominant (39). A model hypothesized frequently to account for this predominant species was proposed by Massague et al. and consists of two $\alpha\beta$ heterodimers linked by disulfide bridges (58, 59). A diagram of this oligomeric form shown in Fig. 3, is based on information derived from several sources including specific domains predicted from the presumed amino acid sequence.

4. Membrane Orientation of the Functional Domains

The transmembrane orientation of the α- and β-subunits was determined by iodination and specific immunoprecipitation of the receptor from plasma membrane vesicles (60). The β-subunit is iodinated in both right-side-out and inside-out plasma membrane vesicles from rat adipocytes, whereas the α-subunit is iodinated on right-side-out vesicles only. These experiments suggest that the α-subunit, or at least all of its tyrosine residues, are located at the extracellular face of the plasma membrane. Conversely, the β-subunit consists of external, internal, and transmembrane domains. This structural hypothesis is supported by the deduced amino acid sequence information which suggests that a single hydrophobic transmembrane region probably exists only in the β-subunit (Fig. 3) (53, 54). During translation in the rough endoplasmic reticulum, the insulin receptor precursor is apparently inserted once into the lipid bilayer such that the entire α-subunit and an N-terminal portion of the β-subunit are ultimately situated at the outer face of the plasma membrane.

The differences in membrane orientation of the two subunits is consistent with differences in their specialized function. Affinity-labeling of the insulin receptor with photoreactive [125I]insulin analogs (61, 62) or covalent cross-linking of [125I]insulin to its receptor with disuccinimidyl suberate (63, 64) identifies the α-

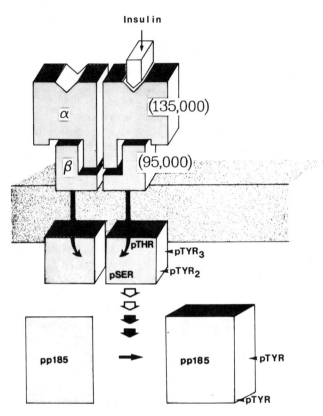

FIG. 3. A schematic drawing of the insulin receptor and phosphorylation of a soluble cellular protein of molecular mass 185 kDa (see Sections II, B and IV, F, 3 for details).

subunit as the primary insulin-binding site (Fig. 2). In contrast, the β-subunit is poorly labeled by these techniques, but is uniquely labeled when insulin receptors are immunoprecipitated from [^{32}P]orthophosphate-labeled cells (19, 20) (Fig. 2). In addition, the β-subunit is specifically labeled by ATP affinity reagents suggesting that the catalytic domain of the receptor kinase is located in the 95-kDa component (65–67). Thus, the regulatory subunit is located at the extracellular face and the catalytic subunit is located at the inner face of the plasma membrane.

5. Turnover and Transport

The rate of insulin receptor biosynthesis and turnover has been estimated by measuring changes in insulin binding after inhibition of protein synthesis (68) or glycosylation (69), tryptic digestion of surface receptors (70), labeling newly synthesized receptors with heavy isotopes (71, 72), and by immunoprecipitation of radioactively labeled receptors (47). These experiments suggest that the half-life of the insulin receptor in the absence of insulin is 9 to 12 h in chick liver cells (71, 72) and IM-9 lymphocytes (47). This steady state is due to a balance between receptor synthesis and degradation; the degradation rate appears to be regulated and involves receptor endocytosis and fusion with lysosomes.

In the presence of insulin, the receptor enters the cell at an accelerated rate and its half-life is decreased. For example, after incubation of rat adipocytes with insulin, the intracellular subset of receptors rapidly increases within 30 s (73). These internalized receptors initially escape degradation and are recycled back to the plasma membrane (73, 74). However, during prolonged exposure to insulin, the degradation rate of the receptor increases 2- to 3-fold (47, 75). Since there is no change in the rate of biosynthesis (47), this results in down-regulation of the receptors and a concomitant desensitization of the cell to insulin (55, 75, 76). Whether internalization is necessary to transport the activated tyrosine kinase to a specific site in the cell where it is more or less active is unknown.

C. OTHER MEMBERS OF THE LIGAND-REGULATED TYROSINE KINASE FAMILY

1. Types I and II Insulin-Like Growth Factor Receptors

In addition to the insulin receptor, several other growth factor receptors have an associated ligand-stimulated tyrosine kinase domain (1). One of these that is closely related to the insulin receptor is the type I insulin-like growth factor receptor (Fig. 1). IGFs are chemically and immunologically distinct polypeptide hormones that mimic the biological effects of insulin, but exhibit different potencies (77, 78). Two IGFs have been identified, called IGF-I and IGF-II (79, 80) which bind to two distinct membrane receptors (81). The type I IGF receptor reacts preferentially with IGF-I as compared to IGF-II, and reacts only weakly with insulin. Likewise, the type II IGF receptor binds IGF-II with a higher affinity than IGF-I; it does not recognize insulin at all (82). Insulin receptors have a weak affinity for both types of insulin-like growth factors (77).

IGF receptors have been demonstrated on a wide variety of cells (83–86). Typically, both subtypes of IGF receptors coexist on the same cell; however, some cells possess only the type I or type II IGF receptor. For example, cultured human lymphocytes contain only IGF-I and insulin receptors (84), liver cells and some cultured hepatomas cell lines (FAO and H35) possess only IGF-II insulin receptors (87), whereas fibroblasts and muscle cells possess all three (88).

In addition to their independent expression on cells, receptors for insulin and IGF-I can be distinguished by their distinct immunoreactivity. Kull et al. have prepared monoclonal antibodies that specifically recognize the type I IGF receptor (89). Kasuga et al. observed that most sera from patients with autoantibodies to the insulin receptor also react with the type I receptor; however, one serum was selective for the insulin receptor and did not recognize the type I IGF receptor (90). Thus, although the IGF-I and insulin receptors are related, they can be distinguished by their preferential binding to the homologous hormone, their independent occurrence on cells, and their distinct immunoreactivity.

The structure of the type I IGF receptor is generally homologous to that of the insulin receptor. It is composed of two subunits with molecular masses by SDS-PAGE of 132,000 (α-subunit) and 92,000–98,000 (β-subunit) (89, 91, 92). The molecular mass of the nonreduced receptor has been estimated by SDS-PAGE to be greater than 300,000. Both subunits are detected after specific immunoprecipitation of the receptor from surface iodinated or [^{35}S]methionine-labeled cells (91, 92). Like the insulin receptor, only the α-subunit is detected when the receptor is labeled with [^{125}I]IGF-I by cross-linking (92–95), and only the β-subunit is observed in cells labeled with [^{32}P]orthophosphate. IGF-I stimulates this phosphorylation in proportion to its affinity for the receptor (24–26, 93, 96, 97). In the presence of monensine, an inhibitor of protein processing, the α- and β-subunits are not detected, but a new protein of M_r = 180,000 is immunoprecipitated from [^{35}S]methionine-labeled cells which presumably represents the precursor of the type I IGF receptor (91). By analogy with the more extensively studied insulin receptor, the type I IGF receptor consists of a complex of the type, $\beta\alpha$-$\alpha\beta$, linked covalently by disulfide bonds in a similar way to that shown for the insulin receptor in Figs. 1 and 3 (92). Thus, the α-subunit contains the extracellular binding domain, and the β-subunit possess an intracellular tyrosine kinase and is linked to the α-subunit through disulfide bonds.

The type II IGF receptor is not structurally homologous to the type I IGF or the insulin receptors (81). It is composed of a single polypeptide chain with a molecular mass by SDS-PAGE before reduction with dithiothreitol of about 220,000 (92). It appears to have intramolecular, but not intermolecular disulfide bonds since its mobility on SDS-PAGE decreases after reduction with dithiothreitol. In H35 cells, the type II receptor is synthesized as a 230-kDa precursor that is glycosylated with high-mannose side chains to form a 245-kDa glycoprotein (98). This is processed to a mature form (250 kDa) by removal of the mannose oligosaccharides and attachment of neuraminic acid. The type II receptor is phosphorylated in the intact cell (99), but in contrast to the insulin and IGF-I receptors, this is due primarily to phosphoserine; it has not been shown to undergo ligand-stimulated tyrosine phosphorylation (97). Thus, the type II receptor does not appear to possess an intrinsic tyrosine kinase activity and the nature of the signal created by this receptor during ligand binding is unknown.

2. Epidermal Growth Factor Receptor and Platelet-Derived Growth Factor Receptor

The membrane receptors for EGF and PDGF also possess a ligand-stimulated tyrosine kinase (1, 100). The EGF receptor is a single glycoprotein of 170 kDa that undergoes tyrosine autophosphorylation after EGF binding (*17, 18, 101, 102*). The EGF-binding domain, phosphorylation sites, and phosphotransferase activity remain associated with the $M_r = 170,000$ monomer after EGF affinity chromatography, electrophoresis in nondenaturing gels, and immunoprecipitation suggesting that each domain is intrinsic to the EGF receptor (*18, 101*). Limited tryptic digestion of the EGF receptor purified from the human epidermal carcinoma cell, A431, after affinity labeling with [^{125}I]EGF or biosynthetically labeled with [^{35}S]methionine, [^{3}H]glucosamine, or [^{32}P]orthophosphate suggests that the EGF-binding site and oligosaccharide-attachment sites are located in an $M_r = 125,000$ N-terminal portion of the monomer (*103*); this fragment has a similar molecular size to the α-subunit of the insulin receptor. In contrast, the ATP-binding site and phosphorylation sites are located in an $M_r = 55,000$ C-terminal fragment which is slightly larger than the intracellular portion of the β-subunit of the insulin receptor. The EGF-binding site and the catalytic site are separated by an hydrophobic transmembrane domain, but unlike the insulin receptor, there is no cationic domain that is susceptible to proteolytic processing (*104*). Thus, the structure of the mature EGF receptor resembles the uncleaved precursor forms of the insulin and type I IGF receptors.

Although less well studied, the PDGF receptor structure appears to resemble that of the EGF receptor (Fig. 1). [^{125}I]PDGF specifically cross-links to a plasma membrane glycoprotein of $M_r = 180,000$ (*105*) which contains a PDGF-stimulated tyrosine kinase (*21–23, 106, 107*). Purification of the PDGF receptor by affinity chromatography with immobilized PDGF has not been reported; however, the receptor has been purified to near homogeneity by affinity chromatography on Sepharose-immobilized monoclonal antiphosphotyrosine antibody (*108*). This antibody provides an affinity reagent that can be used to purify phosphotyrosine-containing proteins from crude cell homogenates. In this approach, intact BALB/3T3 cells were stimulated with PDGF at 4°C for 3 h and then the cells were lysed and solubilized in the presence of phosphatase inhibitors. The phosphotyrosine-containing proteins bound to the immobilized antiphosphotyrosine antibody and upon elution with phenylphosphate, a 6000-fold purification of the PDGF receptor was obtained. The specific binding activity of this preparation, 0.3 nmol/mg protein, is substantially less than the theoretical specific binding of PDGF receptor (5.6 nmol/mg protein). This difference may be due to inactivation that occurs during the experiment, or a regulatory phenomenon mediated by receptor aggregation like that reported during purification of insulin (*57*) and EGF (*109*) receptors. SDS-PAGE separation of the eluted protein and identification of the receptor by silver stain or autoradiography of [^{35}S]methionine-labeled protein reveals a single protein of $M_r = 180,000$. This approach to the purification of membrane receptors which takes advantage of tyrosine phosphorylation should prove to be very useful for this system as well as other phosphotyrosine-containing proteins that have not yet been described.

D. AMINO ACID SEQUENCE HOMOLOGIES BETWEEN MEMBRANE
    RECEPTORS AND TYROSINE KINASE ONCOGENE PRODUCTS

1. *Introduction*

The structural relationship between cell surface receptors, and viral and pro-tooncogene products was first directly demonstrated by the discovery that the amino acid sequence of the internal domain of the EGF receptor is nearly identical to the transforming protein of the avian erythroblastosis virus, gp65$^{v\text{-}erb}$ *(104, 110)*. Presumably, the avian erythroblastosis virus acquired the gene that codes most of the intracellular domain of the EGF receptor but not the regulatory domain that contains the EGF-binding site and is ordinarily found at the outer face of the plasma membrane *(110)*. The phenotype of cells transformed by the avian erythroblastosis virus appears to result from the unregulated tyrosine kinase of v-*erb*-B which contains a constitutively activated EGF receptor catalytic site *(110, 112)*. This relationship has been strengthened further by the high degree of homology between the catalytic domains of the insulin receptor and several other oncogene products *(53, 54)*. These discoveries suggest that two types of tyrosine kinases exist in cells: One is regulated by peptide hormones and growth factors and the other is composed of viral oncogene products which may be inaccurate and unregulated copies of normal cellular genes.

The amino acid sequences of the receptors for insulin and EGF reveal certain conserved domains which serve provisionally to categorize cell surface membrane receptors. For example, they contain cysteine-rich sequences in the extracellular domain, a membrane-spanning sequence, and a carboxyterminal signal transducing function containing a tyrosine kinase. In addition, some viral oncogene products, in particular, v-*erb*-B *(110)*, and the v-*fms* glycoprotein (gp140^{fms}) *(113, 114)* share some of these features. This section describes these characteristics.

2. *Extracellular Cysteine-Rich Domain
    of the Membrane Receptor Kinases*

The extracellular portion of the insulin and EGF receptors contains a high percentage of cysteine residues in certain extracellular domains (Fig. 4). The α-subunit of the insulin receptor contains one N-terminal cysteine-rich region *(53, 54)*, whereas the EGF receptor of human *(104)* and drosophila *(115)* origin contains two of these domains of about equal size, one at the N-terminus and one close to the membrane-spanning domain. This extracellular domain is not unique to tyrosine kinases since the low-density lipoprotein receptor also possess such a region near its N-terminus (Fig. 4) *(116)*. The function of these Cys residues is unknown, but they could stabilize the ligand binding site through intramolecular disulfide bonds, or mediate receptor aggregation through formation of inter-receptor disulfide bonds.

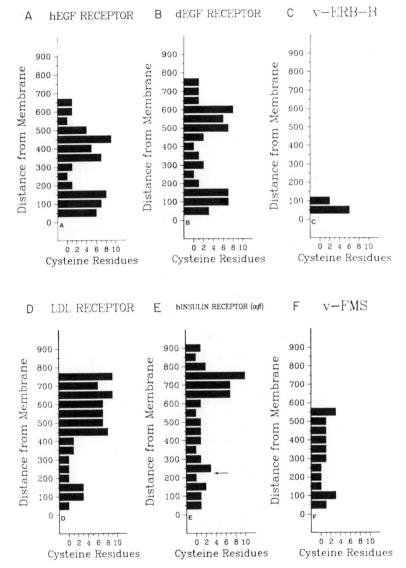

FIG. 4. The distribution of cysteine residues in the extracellular domain of the insulin receptor (E) and some other receptors and tyrosine-specific protein kinases. The arrow in panel E indicates the position of cleavage between the α- and β-subunits.

LDLR

Ala·Leu·Ser·Ile·Val·Leu·Pro·Ile·Val·Leu·Leu·Val·Phe·Leu·Cys·Leu·Gly·Val·Phe·Leu·Leu·Trp————————————Lys·Asn·Trp·Arg

HIR

Ile·Ile·Ile·Gly·Pro·Leu·Ile·Phe·Val·Phe·Leu·Phe·Ser·Val·Val·Ile·Gly·Ser·Ile·Tyr·Leu·Phe·Leu————————————Arg·Lys·Arg

HEGFR

Ile·Ala·Thr·Gly·Met·Val·Gly·Ala·Leu·Leu·Leu·Leu·Val·Val·Ala·Leu·Gly·Ile·Gly·Leu·Phe·Met————————————Arg·Arg·Arg

v-ERB-B

Ile·Ala·Ala·Gly·Val·Val·Gly·Gly·Leu·Leu·Cys·Leu·Val·Val·Val·Gly·Leu·Gly·Ile·Gly·Leu·Tyr·Leu————————————Arg·Arg·Arg

v-FMS

Leu·Leu·Phe·Thr·Pro·Val·Leu·Leu·Thr·Cys·Met·Ser·Ile·Met·Ala·Leu·Leu·Leu·Leu·Leu·Leu·Leu·Leu·Leu·Tyr————————————Lys·Tyr·Lys

v-ROS

Ile·Thr·Ala·Ile·Val·Ala·Val·Ile·Gly·Ala·Val·Val·Leu·Gly·Leu·Thr·Ser·Leu·Thr·Ile·Ile·Ile·Leu·Phe·Gly·Phe·Val·Trp—His·Gln·Arg·Trp

FIG. 5. The amino acid sequence of the membrane spanning regions of the insulin receptor and some other membrane receptors and tyrosine kinases. The shaded regions emphasizes the cationic domain which marks the end of the hydrophobic region and the beginning of the intracellular domain.

The $gp75^{erb-B}$ possesses a short extracellular domain of about 75 amino acids beyond the transmembrane region (110, 117). This fragment contains 8 cysteine residues at positions similar to those of the EGF receptor (Fig. 4), but due to truncation completely lacks the analogous region near the N-terminus (117). The v-fms coded protein retains an extracellular amino-terminal segment that contains nearly 450 amino acids, but no cysteine-rich region is predicted (Fig. 4). Possibly, $gp140^{fms}$ is still dependent on an interaction with its ligand (113). However, it is likely that the v-fms gene product (140 kDa) is a truncated version of c-fms (170 kDa), the latter being the membrane receptor for murine colony stimulating factor, CSF-I (114). Both proteins undergo tyrosine autophosphorylation in immunoprecipitates (118), and CSF-I stimulates phosphorylation of c-fms (114). Thus, by comparison with the insulin and EGF receptors, a large cysteine-rich domain necessary for ligand binding may be missing in v-fms leading to a ligand-independent activation of its associated tyrosine kinase in a way similar to that observed for the v-erb-B (111, 112).

3. Membrane-Spanning Region

The insulin and EGF receptors contain presumed membrane-spanning regions based on the existence of a sequence domain with a high proportion of hydrophobic amino acid residues. Figure 5 shows these sequences for a collection of receptors and some tyrosine kinase oncogene products. None of these sequences are exactly identical, not even those for the human and *Drosophila* EGF receptors. However, each domain has a similar length and hydrophobic feature, and is flanked at the C-terminal end by a short, strongly hydrophilic or cationic sequence that presumably identifies the beginning of the cytoplasmic domain. For the insulin receptor, the membrane-spanning region is in the β-subunit, and thus is separated from the cysteine-rich region of the α-subunit (53, 54). In contrast, the cysteine region of the EGF receptor and its transmembrane domain are part of the same molecule (104). These hydrophobic domains presumably represent the only transmembrane region in these proteins, suggesting that the conformational change induced by the ligand binding must be transmitted through this domain to regulate the function of the tyrosine kinase.

4. Intracellular Catalytic Domain

The cytoplasmic domains of the EGF and insulin receptors have a high degree of sequence homology to the src-related oncogene products that contain a tyrosine kinase activity and lesser but important homology to serine–threonine kinases. Ullrich et al. have compared a 275 amino acid interval between residues 990 and 1265 of the precursor of the insulin receptor to the corresponding domain in the EGF receptor and the oncogene products v-abl, v-src, v-fes, and v-fms (53). Nearly 40% of the amino acid residues in this region are in similar positions in four of these tyrosine kinases. In particular, each tyrosine kinase, including the insulin receptor, contains an amino acid sequence that has been previously identified as a site of ATP-binding in cAMP-dependent protein kinases (Sequence A). In the case of the receptors for insulin and EGF, v-erb-B and v-fms, this domain begins exactly 50 residues from the end of the trans-membrane-spanning region. It can be identified in cytoplasmic serine–threonine kinases, oncogene products that possess a tyrosine or serine–threonine kinase, and membrane receptor tyrosine kinases (Fig. 6). It has the general form shown in Sequence A

$$Leu\cdot Gly\cdot X\cdot Gly\cdot X\cdot X\cdot Gly\cdot X\cdot Val$$

SEQUENCE A

and occurs between 13 and 20 residues before a lysine. This lysine residue is covalently labeled with the ATP affinity reagent, p-FSO$_2$BzADO, in the cAMP-dependent protein kinase (133), pp60^{src} (134), and the EGF receptor (135) suggesting that it is involved in ATP-binding for each enzyme. The corresponding residue in the human insulin receptor precursor is Lys-1018 and it is probably also labeled by ATP affinity reagents (65–67). The ε-amino group of the lysine residue is thought to be conserved in protein kinases because it facilitates the nucleophilic reaction between the γ-phosphate group of ATP and the hydroxyl group of the substrate (134).

Since the discovery that pp60^{src} is a tyrosine kinase, its major phosphoaccep-tor site at Tyr-416 and the adjacent amino acids have served as a domain for comparison with other tyrosine kinases (53, 100, 104, 127). This tyrosine resi-

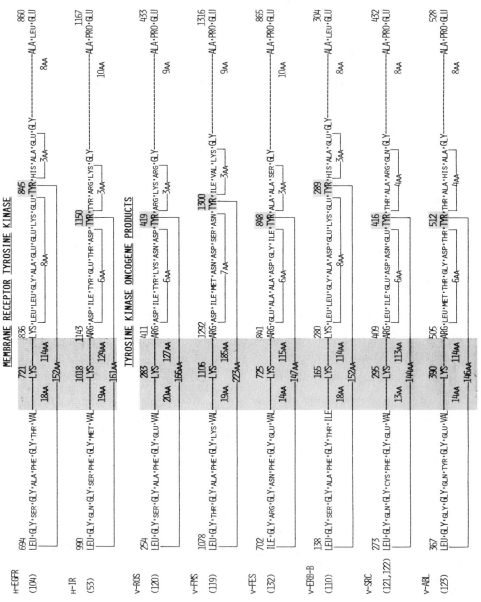

MEMBRANE RECEPTOR TYROSINE KINASE

TYROSINE KINASE ONCOGENE PRODUCTS

FIG. 6 (see legend on p. 264.)

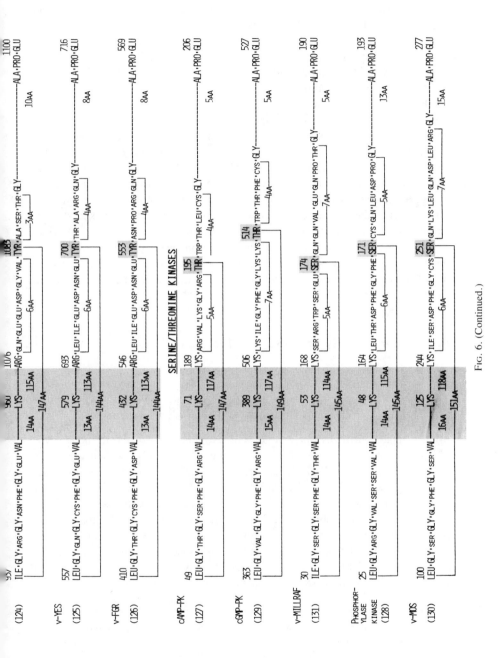

FIG. 6. (Continued.)

263

due occurs 121 amino acids away from the lysine residue in the ATP-binding site of pp60^{src}. It is followed by 5 residues terminating in a Gly, and preceded by an arginine residue 6 positions to its N-terminal side and two negatively charged amino acids, Asp-413 and Glu-415 (Fig. 6). Tyr-416 is followed by a sequence of five amino acids that ends with a Gly residue; eight amino acids beyond this point, an Ala·Pro·Glu sequence occurs. These characteristic domains surround a similar tyrosine residue in both the EGF and insulin receptors which occurs 124 and 132 amino acid residues from the catalytic lysine, respectively (53, 54, 104). A similar tyrosine is also found in the other oncogene products, but there is considerable variability in the surrounding sequence suggesting that there must be some flexibility in this domain with respect to the putative phosphoacceptor site. Tyrosine is conspicuously absent from the serine–threonine kinases shown in Fig. 6. However, a corresponding sequence can be identified which contains a serine or threonine residue in a relatively similar position; it is flanked at its N-terminus by an arg residue and at its C-terminus by Ala·Pro·Glu (Fig. 6). Thus, receptors for insulin and EGF, tyrosine kinase oncogene products, and serine–threonine kinases possess similar characteristic landmarks. (For another perspective of sequence homologies of protein-tyrosine kinases see Chapter 6 by Hunter and Cooper.)

III. Properties of Tyrosine-Specific Protein Kinases

A. Introduction

Following the original observation of Kasuga *et al.* that the insulin receptor was a tyrosine kinase (20), several groups began to characterize this activity using receptor purified from various sources including isolated rat hepatocytes (136–139), skeletal muscle (140), adipocytes, (141, 142) and hepatoma cells (139, 143); 3T3-L1 cells (144), and cultured mouse melanoma cells (145); and human erythrocyte (146, 147), fibroblasts (147), lymphocytes (19), and placental membranes (36–38, 148–150). In both crude detergent extracts (20, 143) and highly purified preparations (36, 149, 150), insulin stimulates tyrosine phosphorylation of the β-subunit of the insulin receptor. Using these experimental systems, several enzymic characteristics of the insulin receptor have been described. Some of these characteristics are briefly summarized in Table I and certain aspects are discussed in detail in this section.

FIG. 6. The relative positions of the amino acid sequences near the proposed ATP-binding site (including and to the left of the shaded Lys residue) and the *src*-related autophosphorylation site (to the right of the shaded Lys) in the insulin and EGF receptor, several oncogene products, and some serine–threonine kinases. Similar amino acids are shown in large letters.

TABLE I

Some Characteristics of Tyrosine-Specific Protein Kinases[a]

	pp60src	EGF	Insulin	PDGF	Type-I IGF
			KINETIC PROPERTIES		
Intramolecular autophosphorylation	Yes (100)	Yes (101,109,178)	Yes (36,143,177)	—	Yes (26)
Autophosphorylation activates phosphotransferase	Yes (100)	—	Yes (155,165,195)	—	Yes (26)
Ligand stimulation (K_d, nM)	—	10 (17,18)	5 (246)	20 (182)	20 (26) / 3 (199)
Nucleotide preference	ATP–GTP (278)	ATP–GTP (17)	ATP (20)	ATP–GTP (182)	ATP (26)
K_m, μM (ATP)	23 (278)	0.2 (178)	20 (143)	2 (182)	20 (26)
K_m, μM (GTP)	30 (278)	—	—	2 (182)	—
Cation activators					
Autophosphorylation	Mn^{2+}	Mn^{2+}	Mn^{2+}	Mn^{2+}	Mn^{2+}
Phosphorylation	Mg^{2+}	Mn^{2+}	Mn^{2+}–Mg^{2+}	—	Mg^{2+}
Cation inhibitors	—	Zn^{2+} (185)	Ni^{2+},Zn^{2+},Ca^{2+} (181)	—	—
		SUBSTRATES FOR TYROSINE KINASES			
Metabolic enzymes					
1,6-Diphosphofructokinase	—	—	[Ref. (222) and Fig. 12]	—	—
Lactate dehydrogenase	(206,223)	—	[Ref. (222) and Fig. 12]	—	—
Enolase	(206,223)	—	[Ref. (222) and Fig. 12]	—	—
Phosphoglycerate matose	(206,223)	—	[Ref. (222) and Fig. 12]	—	—
Cytosolic malic dehydrogenase	(279)	—	—	—	—
Structural proteins					
Vinculin	(280)	—	—	—	—
Microtubule proteins	—	—	(228)	—	—
Tubulin	—	—	(228)	—	—

(continued)

TABLE I (*Continued*)

	pp60*src*	EGF	Insulin	PDGF	Type-I IGF
Regulatory proteins					
Calmodulin	—	—	(216)	—	—
Gastrin	—	(203)	—	—	—
Angestensin II	(206)	—	(201)	—	—
Growth hormone	—	(281)	—	—	—
Progesterone receptor	—	(282)	—	—	—
Myosin regulatory light chain	—	(231)	(201)	—	—
Other polypeptides					
(Glu-4–Tyr)$_n$	—	(283)	(283,284)	—	(26)
src-like peptide	(100)	(197)	(149)	(199)	—
Histone 2B	(163)	(199)	(149)	(182)	—
Anti-*src* Ab	(100)	(214)	(149)	—	—
Casein	—	(17)	(97,285)	—	—

^{a} Numbers in parentheses are references.

B. PURIFICATION OF MEMBRANE RECEPTORS FOR TYROSINE
   KINASE ASSAYS

On most cells, membrane receptors for insulin, EGF, and PDGF are very minor proteins. Thus, nearly all of our current information about membrane receptor kinases has been obtained from a few cell types that contain a relatively high concentration of the particular receptor. The best example is the human epidermal carcinoma cell line, A431, which has been used for most studies of the EGF receptor and possesses 2 to 3 million receptors per cell (100, 151). Similarly, the rat hepatoma cell FAO and the mouse adipocyte-like 3T3-L1 cells have been valuable tools for insulin receptor studies since they are insulin responsive and contain about 50,000 to 100,000 receptors per cell (152, 153). The plasma membrane from human placenta is also a rich source of both EGF and insulin receptors, and this tissue has been used as a source for both receptor purification and mRNA for cloning the corresponding receptor cDNAs (53, 54, 104).

Initial attempts to demonstrate receptor kinase activity have relied on isolation of plasma membranes from homogenized cells. This approach was successfully applied to studies with the EGF and PDGF receptors; however, the kinase activity of the insulin receptor is weak in the intact plasma membrane and detergent solubilization is necessary to increase in vitro autophosphorylation to a detectable level (154). Thus, most in vitro studies of the insulin receptor kinase have been done with detergent-solubilized receptor purified by various methods.

Affinity chromatography using immobilized wheat germ agglutinin has been employed often to partially purify the solubilized insulin receptor. This method also enriches for the receptors of EGF, PDGF, and IGF-I. Further purification of the kinase activity of the EGF and insulin receptors has been achieved by the use of immunoprecipitation or ligand-affinity chromatography. In the latter case, determining proper conditions for elution is critical to retain an active receptor kinase. For example, elution of the insulin receptor from insulin-Sepharose with 6 M urea, as originally proposed for purification of binding activity (35), did not yield an active tyrosine kinase (144). However, milder elution conditions (1 M NaCl, pH 5.5) recovered purified insulin receptors with tyrosine kinase activity and maximum insulin binding (150, 155). The inclusion of dithiothreitol during purification was also found to improve the recovery of the kinase activity (36). These highly purified preparations have been used by several investigators to study the kinetic characteristic of the insulin receptor kinase.

Insulin-affinity purified receptor is not always optimal for studies of insulin receptor function. For many studies, we have successfully utilized the insulin receptor purified partially by affinity chromatography on WGA-agarose (156). The receptor can be eluted from this affinity column at neutral pH by N-acetylglucosamine with essentially 100% recovery. Using FAO cells, the eluate has a binding capacity of 100 pmol insulin/mg protein, corresponding to about 1%

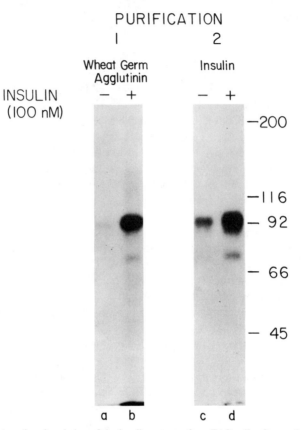

FIG. 7. *In vitro* phosphorylation of the insulin receptor from FAO cells after partial purification on immobilized WGA (lanes a and b) and after further purification by elution from insulin-Sepharose (lanes c and d).

pure insulin receptor (*143*). This mixture of glycoproteins, however, contains only a single phosphoprotein after incubation with [γ-^{32}P]ATP, Mn^{2+}, and insulin which corresponds to the β-subunit of the insulin receptor. After phosphorylation, the autoradiogram of the WGA-purified receptor is indistinguishable from that obtained with the same receptor that undergoes further purification on insulin-Sepharose (Fig. 7). Thus, purification of the receptor by WGA chromatography provides two advantages in comparison with insulin-Sepharose: It uses mild elution conditions that are more likely to retain the physiological state of the receptor, and it avoids insulin so the receptor remains in an inactive state until the initiation of the experiment.

C. AUTOPHOSPHORYLATION OF THE INSULIN RECEPTOR

1. *Introduction*

Nearly all protein kinases undergo a self-catalyzed autophosphorylation reaction (*157–159*). This is especially fortunate for the study of tyrosine kinases because it provides a unique molecular marker for enzymes with this rare specificity. For example, the kinase activity of the insulin receptor was first detected when Kasuga *et al.* discovered that insulin stimulated phosphorylation of the β-subunit of the insulin receptor (*19*).

In the intact cell, the insulin receptor, like other tyrosine kinases, contains phosphoserine and phosphothreonine. Only after insulin stimulation is phosphotyrosine detected (*20*). Following purification and *in vitro* phosphorylation, the membrane receptors for insulin (*150, 160*), EGF (*161*), PDGF (*108*), and certain oncogene products (*162, 163*) contain exclusively phosphotyrosine (*100*); however, there are some exceptions. For example, phosphorylation of the partially purified receptor from rat liver plasma membranes occurs on both tyrosine and serine residues *in vitro* (*164*). Similarly, insulin receptors from placenta immobilized on insulin-Sepharose exhibit tyrosine, as well as a small amount of serine phosphorylation (*165*). Serine phosphorylation, however, has never been detected in the β-subunit of the highly purified insulin receptor eluted from an insulin-affinity column (*36, 150*). These results suggest that the insulin receptor and other growth factor receptors and oncogene products are tyrosine kinases, but are also substrates for serine and threonine kinases in the intact cell which may remain associated with these proteins during partial purification.

2. *Ligand Specificity and Stimulation of Tyrosine Kinases in Vitro*

Stimulation of insulin receptor autophosphorylation by insulin analogs corresponds exactly to their relative binding affinities (*20, 146, 166*). Half-maximal stimulation of the insulin receptor kinase occurs at about 5 nM insulin which corresponds to the concentration of insulin for half-maximal receptor occupancy. IGF-I binds to the insulin receptor with 100-fold lower affinity than insulin and displays equally decreased stimulation of insulin receptor phosphorylation. On the other hand, IGF-I stimulates phosphorylation of its own receptor with a dose similar to its binding (*26*). Similar specificity is also observed with both the EGF (*17, 18*) and PDGF (*21–23*) receptors. Thus, activation of the tyrosine kinase of growth factor receptors is strictly regulated by an interaction with a highly specific ligand.

Several agents mimic insulin action in the intact cell and some of these compounds have been tested for their ability to stimulate the kinase activity of the insulin receptor. Mild tryptic digestion has been shown to mimic insulin action on glucose transport and glycogen synthase in adipocytes (*167*). Trypsin treatment of the purified insulin receptor also stimulates tyrosine phosphorylation of the β-subunit and a proteolytic product of $M_r = 72,000$ (*168*). Similarly, concanavalin A (*169*), vanadate (*142*), and some antiinsulin receptor antibodies (*170*) stimulate tyrosine phosphorylation of the receptor and mimic the insulin response (*171–173*). However, all antibodies that mimic insulin action do not stimulate phosphorylation (*170, 174*). The reason that all antiinsulin receptor sera that mimic insulin action do not stimulate phosphorylation is unclear but may relate to differences in relative potency, antibody heterogeneity, or differences in antibody and insulin action. Clearly, further study of these agents may provide an explanation for the role of tyrosine kinases in insulin action.

One of the most striking differences between the receptor kinases and the tyrosine kinase of the oncogene products is that the latter lacks a ligand-binding regulatory domain. As a result, the oncogene kinases appear to be constitutively activated. For example, the v-*erb*-B gene product is homologous to the catalytic domain of the EGF receptor, but is an active kinase in the absence of EGF (*111, 112*). Presumably this is because the extracellular regulatory site is missing but may also be due to the C-terminal deletion (*110*). Removal of the binding domain of the EGF receptor by tryptic digestion yields a 42-kDa C-terminal fragment that also contains an activated tyrosine kinase (*175*). Thus, the loss of the regulatory domain of the EGF receptor whether by proteolytic processing or genetic mutation appears to activate the kinase. Similarly, mild trypsin digestion of the insulin receptor activates the tyrosine kinase (*169*), and Goren *et al.* have shown that 85-kDa and 70kDa fragments of the β-subunit also retain tyrosine kinase activity.

3. *Tyrosine Autophosphorylation in Membrane Receptors Is an
   Intramolecular Reaction*

The mechanism of autophosphorylation has been studied with solubilized insulin and EGF receptors. In detergent solution, phosphorylation of the β-subunit during insulin stimulation is a rapid reaction reaching steady state in about 2 min (Fig. 8A). The exact rate depends on the experimental conditions including temperature, and ATP, insulin, and divalent metal ion concentration (*138, 143, 176*); however, the rate of this reaction is not dependent on the concentration of the receptor protein in solution suggesting that autophosphorylation occurs by an intramolecular reaction (Fig. 8B) (*36, 143, 177*). Based on similar experiments, autophosphorylation of the EGF receptor, as well as other tyrosine kinases also occurs through an intramolecular reaction (*100, 109, 178*). The intramolecular reaction could be due to aggregation of hydrophobic proteins in a detergent micelle; however phosphorylation of other proteins like histone H2B by the insulin receptor follows the predictions of Michaelis–Menton kinetics and is sensitive to the enzyme concentration suggesting that phosphorylation

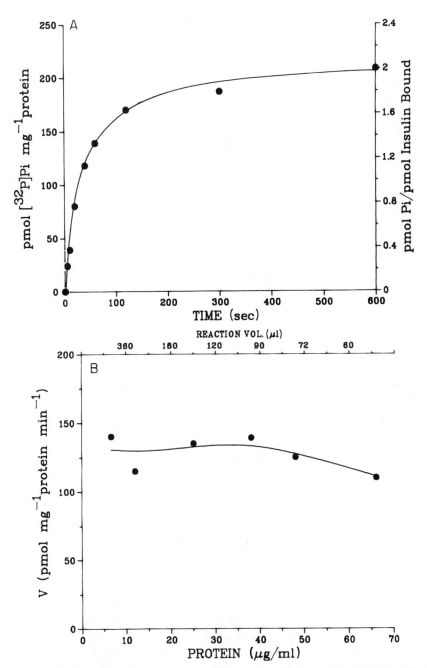

FIG. 8. (A) The time course of *in vitro* phosphorylation of the β-subunit of the WGF-purified insulin receptor from FAO cell at 22°C, 50 μM ATP and 100 nM insulin. (B) Using similar conditions, the effect of diluting only the protein concentration on the initial velocity (V) of autophosphorylation measured during a 60-sec time interval.

of exogenous substrates occurs through a bimolecular reaction in detergent solution (36). In addition, autophosphorylation of soluble serine–threonine kinases
also occurs by an intramolecular reaction suggesting that this mechanism may be
of general occurrence (157).

The stoichiometry of insulin receptor autophosphorylation has been estimated
in vitro with solubilized receptors. Using insulin receptors partially purified from
FAO cells on wheat germ agglutinin agarose, White et al. estimated that at
steady state, about 2 mol of phosphate are associated with the β-subunit per 1
mol of insulin-binding sites (143). Assuming a 1 : 1 ratio of α- to β-subunits and
one competent insulin-binding site in each α-subunit, this corresponds to about 2
molecules of phosphate per β-subunit (Fig. 8A). Petruzzelli et al. arrived at the
same conclusion using insulin Sepharose-purified receptor from human placenta
(36). These measurements are consistent with the identification of two major
tryptic phosphopeptides from the β-subunit by HPLC (see Section II, C, 5). In
contrast, 0.8–1 mol of phosphate is incorporated into 1 mole of the monomeric
EGF receptor during the initial period of phosphorylation (151, 178). This
stoichiometry further increases, however, during extended periods of phosphorylation.

4. Kinetics of Autophosphorylation of the Insulin Receptor

Kinetic characteristics of insulin receptor autophosphorylation have been studied in a few experimental systems (143, 165, 176). Using the partially purified
insulin receptor from FAO cells, at 22°C, incorporation of $[\gamma\text{-}^{32}P]ATP$ (50 μM)
is half-maximal at 30 sec and nearly 90% complete after 2 min (Fig. 8A). Initial
velocity curves of insulin-stimulated autophosphorylation are sigmoidal with
respect to the ATP concentration, and as expected, the Lineweaver–Burk plot of
these data is parabolic (143). Insulin stimulates the V_{max}, with no effect on the
K_m, but the exact values are highly dependent on the reaction conditions. Using
100 nM insulin, the K_m for ATP and the V_{max} are 36 μM and 0.86 pmol/binding
sites/min, respectively.

The sigmoidal kinetic curves may arise from the fact that multiple tyrosine
residues are phosphorylated during this reaction. In addition, there are indications that the phosphorylation at the first sites (see Section III, C, 5, pY2 and
pY3) may activate the receptor for phosphorylation of others which could give
rise to apparent positive cooperativity. Possibly, when phosphorylation is limited
to a single tyrosine residue, the kinetic characteristic of this system will be
simpler. For example, only one mole of phosphate is incorporated into each mole
of the EGF receptor during the initial incubation at low temperature and low ATP
levels (0.1 to 0.5 μM), and in contrast to the insulin receptor, its Lineweaver–
Burk plot is linear suggesting no cooperativity (178). However, these kinetic
data must be viewed cautiously because the usual steady-state assumption that
the concentration of the enzyme–substrate complex is constant during the mea-

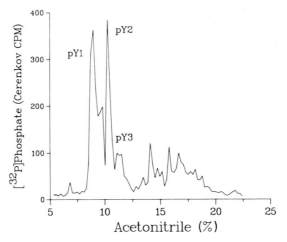

FIG. 9. Reverse-phase HPLC elution profile of tryptic phosphopeptides obtained from the β-subunit of the insulin receptor that was purified from FAO cells on WGA-agarose and phosphorylated *in vitro* for 10 min with 50 μ*M* ATP and 100 n*M* insulin.

surement does not apply since the autophosphorylation reaction is intramolecular and the concentration of the catalytic site and the phosphate acceptor sites are rapidly changing (*179*).

5. Identification and Kinetics of Tyrosine Autophosphorylation Sites in the EGF and Insulin Receptor

When the phosphorylated β-subunit is digested with trypsin, several phosphopeptides can be separated by reverse phase HPLC and the kinetic properties of phosphorylation at each site can be quantified. Although different sources of receptor have been used at different stages of purification, a similar set of tryptic phosphopeptides has been described by several investigators (*36, 143, 160, 165*). Figure 9 shows an HPLC profile that is representative of the *in vitro* phosphorylation of the β-subunit from FAO cells. Two major phosphotyrosine-containing peptides are obtained that are labeled pY1 and pY2, and a third peptide that is slightly smaller *in vitro* is labeled pY3.* Other peptides are observed to elute at higher concentrations of acetonitrile especially after long phosphorylation intervals (*143*). Tryptic peptides pY2 and pY3 migrate at similar

---

*These phosphopeptides have been labeled differently in other published reports, and in fact their corespondence is only speculative since the HPLC column and solvent systems were different in each case. However, based on their relative elution positions we suggest the following relationship: pY1—Peptide 2 in Ref. (*143*), 1 in Ref. (*165*), and I in Ref. (*36*); pY2—Peptide 4 in Ref. (*143*), 2 in Ref. (*165*), and II in Ref. (*36*); pY3—Peptide 5 in Ref. (*143*) and 3 in Ref. (*165*).

positions as the phosphotyrosine-containing peptides that occur during labeling of intact cells, but pY1 is not detected *in vivo* [see following discussion and Ref. (*160*)]. In our hands, all of the peptides contain phosphotyrosine only (*36, 160*) which is consistent with the fact that no phosphoserine or phosphothreonine has been found in the highly purified insulin receptor or the partially purified receptor from FAO cells. Some reports from other laboratories under somewhat different conditions have found that these phosphopeptides also contain a small amount of phosphoserine as well (*165*).

The exact positions of pY1, pY2, and pY3 in the β-subunit of the insulin receptor have not yet been determined by amino acid sequence analysis because of the difficulty in obtaining sufficient amounts of the purified tryptic phosphopeptides. This problem has been overcome for the EGF receptor using the A431 cells (*161*). Thus, the positions of the phosphotyrosine residues in the EGF receptor have been determined and their sequences are shown in Fig. 10. Each of

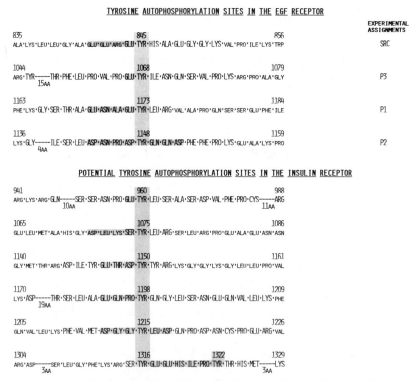

FIG. 10. The shaded tyrosine residues indicate experimentally determined autophosphorylation sites in the EGF receptor and possible sites of autophosphorylation in the insulin receptor. The capitalized residues indicate the tryptic peptide expected from these domains.

the three *in vitro* sites (P1, P2, and P3) are in the C-terminal region of this receptor. Only one of the tyrosine residues (P1) is observed during *in vivo* labeling. Interestingly, the *src*-related site (Tyr-845) is apparently not phosphorylated in the EGF receptor. The phosphopeptides migrate on reverse-phase HPLC relative to their total hydrophathy calculated according to Kyte and Doolittle (*180*). Peptide P3 yields two peaks on the HPLC owing to variable digestion of its C-terminal Lys-Arg pair (*161*).

The insulin receptor contains a number of tyrosine residues in sequences on the intracellular portion of the β-subunit that might favor phosphorylation. Goren *et al.* have made some tentative assignments of the sites of phosphorylation by subjecting the receptor to limited digestion with a series of proteases followed by tryptic peptide mapping on HPLC (unpublished results, this laboratory) (Fig. 10). Two major *in vitro* sites of phosphorylation, pY2 and pY3 appear to be located within a 10-kDa terminal fragment of the β-subunit and can be removed by mild digestion with a variety of enzymes. These are most likely Tyr-1316 or -1322, both of which appear in the same tryptic peptide. Peptide pY1 is more centrally located in the β-subunit since it is found in the 70-kDa tryptic fragment, and could be Tyr-1150 which is in the *src*-related peptide, or Tyr-1075 or 960. Sequencing of the peptides will be required for definitive assignment.

The time course, and the effects of insulin and ATP concentration on the phosphorylation of the various sites of the β-subunit has been studied with the receptor from FAO cells (*143*). Increasing the concentration of insulin increases the phosphorylation of pY1, pY2, and pY3, except at 1 μM insulin where the intensity of pY1 decreases (Fig. 11A). Autophosphorylation of the β-subunit is usually inhibited slightly by concentrations of insulin greater than 100 nM (*141, 143*) and this appears to be entirely due to the 50% decrease in pY1. Half-maximal autophosphorylation occurs at about 5 nM insulin and no differences are detected in the peptide maps obtained within this interval. The unstimulated receptor yields an identical tryptic peptide map suggesting that insulin stimulates phosphorylation at all of the sites (*36*).

The rate of autophosphorylation at the various sites in the β-subunit is not equal. During insulin stimulation, only pY2 and pY3 are detected after 5 sec of incubation with [γ-^{32}P]ATP, whereas pY1 shows a definite lag (Fig. 11B). This time course is consistent with the finding that pY2 and pY3 but not pY1 occur *in vivo* suggesting that pY2 and pY3 may be the preferred phosphorylation sites (*160*). During 30 sec of incubation, the rate of autophosphorylation increases sigmoidally with the concentration of ATP and this relation is reflected in each of the phosphorylation sites (Fig. 11C).

6. Regulation of Autophosphorylation by Divalent Metal Ions

Divalent cations are essential for the full activity of protein kinases. Mn^{2+} is the most potent activator *in vitro* for autophosphorylation of the receptors for

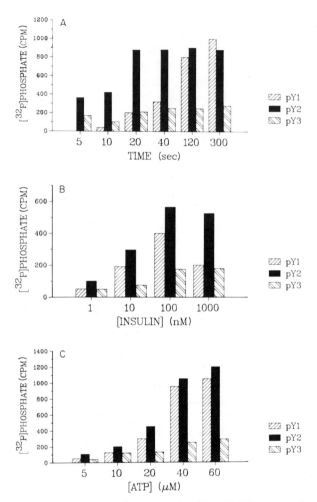

FIG. 11. The effect of time (A), insulin concentration (B), and ATP concentration (C) on the *in vitro* phosphorylation of pY1, pY2, and pY3 in the β-subunit of the insulin receptor [See Ref. (*143*) for details].

EGF (*181*), PDGF (*182*), and insulin (*143, 176*), and for pp60^{src} (*183*). This is particularly true at micromolar ATP concentration used in most of the *in vitro* kinase assays. However, when the ATP concentration is increased toward physiologic levels (1 or 2 m*M*), Mg^{2+} or Mn^{2+} almost equally support autophosphorylation of the insulin receptor (*176*). In contrast, Cu^{2+}, Zn^{2+}, Cd^{2+}, and Ni^{2+} are potent inhibitors of the insulin receptor kinase (*184*), and Zn^{2+} partially inhibits the kinase of EGF receptors in plasma membranes (*185*). Inhibition by Zn^{2+} has not been reported consistently (*186*), probably because Zn^{2+} is chelated by components in the *in vitro* assays (*184*). In contrast, Zn^{2+} appears to satisfy the ion requirement of pp60^{src} (*187*) suggesting that the catalytic site or regulation of pp60^{src} may be different than that of the insulin receptor.

Mn^{2+} stimulates insulin receptor autophosphorylation *in vitro* by decreasing the K_m for ATP (*143*). The effect is observed at concentrations of Mn^{2+} (0.1 mM to 0.5 mM) which are in excess of that necessary to form the presumed substrate, $MnATP^{2-}$. Thus, it is unlikely that Mn^{2+} stimulation occurs solely through chelation of ATP. Since Mn^{2+} and Mg^{2+} form similar complexes with ATP (*188, 189*), the lack of effect of Mg^{2+} in the presence or absence of Mn^{2+} suggests that activation of the receptor kinase requires binding of the free Mn^{2+} to a specific regulatory site on the receptor (*143*). These results are analogous to the selective stimulation by divalent cations of the hepatic adenylate cyclase system (*190*) and phosphoenolpyruvate carboxykinase (*191*). The physiologic role of this regulatory mechanism for the insulin receptor is unknown. However, free Mn^{2+} levels in hepatocytes change between 0.25 μM and 0.71 μM during starvation and feeding, respectively (*192*), suggesting that Mn^{2+} could increase the sensitivity of hepatocytes to insulin after meals.

Our kinetic studies suggest that the K_m for Mn^{2+} approaches zero as the ATP concentration increases (*143*). In contrast, the V_{max} is independent of Mn^{2+} suggesting that Mn^{2+}-binding may precede $MnATP^{2-}$-binding by an equilibrium-ordered mechanism (*143*). Thus, the insulin receptor will not bind ATP in the absence of this divalent cation suggesting that Mn^{2+} is an essential activator of the insulin receptor. Since the ATP concentration is nearly 2 mM in the intact cell, the autophosphorylation reaction could be supported by the micromolar levels of cellular Mn^{2+} (*192*) or by a parallel decrease of the K_m for Mg^{2+} (*176*). Thus, in the intact cell, Mn^{2+}, and less effectively Mg^{2+}, probably bind to the receptor at an allosteric site and activate the insulin receptor kinase.

7. Regulation of the Insulin Receptor Phosphotransferase by Autophosphorylation

One of the important issues about autophosphorylation is whether it regulates the protein kinase activity. Several studies have suggested that autophosphorylation may activate the phosphotransferase of protein kinases. For example, autophosphorylation at serine residues in the regulatory subunit of the cAMP-dependent protein kinase decreases the rate at which the active catalytic subunit reassociates with the regulatory subunit, thus preserving the more active form of the enzyme (*159, 193, 194*). The activity of other serine kinases is also regulated by autophosphorylation (*157*).

In a parallel fashion, Rosen *et al.* have suggested that the phosphorylated insulin receptor still catalyzes phosphorylation of histone at the insulin-stimulated rate even after the bound insulin is removed by acidic elution (*155*). This result suggests that insulin-binding may be necessary to stimulate autophosphorylation, but then the phosphorylated receptor possesses an active phosphotransferase which is no longer dependent on bound insulin. Yu and Czech have suggested that autophosphorylation of tryptic peptide 2 (pY2 by our numbering scheme) may be responsible for this activation (165). Pang et al., using an antiphosphotyrosine antibody to purify phosphotyrosine-containing receptor, have also confirmed that the autophosphorylated receptor appears to remain activated after the removal of insulin (195).

D. Substrate Specificity and Kinetic Mechanism of Tyrosine-Specific Receptor Kinases

1. Use of Synthetic Peptides to Study Substrate Specificity

In addition to the autophosphorylation reactions, tyrosine-specific receptor kinases and oncogene products catalyze a phosphotransferase reaction between ATP and tyrosine residues in other polypeptides. Both natural and synthetic tyrosine-containing peptides have been used to study substrate specificity and some of these proteins are listed in Table I. The amino acid sequences of the synthetic peptides are often based on the major site of tyrosine phosphorylation between amino acid residues 414 and 424 in pp60^{src} as shown in Sequence B. (163).

Ile·Glu·Asp·Asn·Glu·**TYR**·Thr·Ala·Arg·Glu·Gly

SEQUENCE B

Casnellie et al. have synthesized a peptide analog, shown in Sequence C, with only a single phosphorylation site at tyrosine and a cationic N-terminal extension to enhance its adsorption to phosphocellulose paper (196).

Arg·Arg·Leu·Ile·Glu·Asp·Ala·Glu·**TYR**·Ala·Ala·Arg·Gly

SEQUENCE C

Receptors for insulin, EGF and PDGF phosphorylate this peptide and the K_m values are near 1 mM (149, 197–199). In each case, the appropriate hormone stimulates the phosphorylation 3- to 10-fold by increasing the V_{max} of the kinase.

Important features of substrate peptides have been studied by several investigators. Alone, tyramine and acetyltyramine are very low affinity substrates for tyrosine kinases (200), and the phosphorylation of Tyr-Arg by the insulin receptor is not detected (201) suggesting that adjacent amino acids are important for recognition at the catalytic site. Hunter reported that a glutamyl residue preceding the tyrosine by one to five amino acids is necessary for phosphorylation (202), and it is generally true that a tyrosine residue located to the C-terminal side of glutamic acid residues is phosphorylated by tyrosine kinases. Gastrin

(203) and a random copolymer of glutamate and tyrosine (97) are good examples. The systematic modification shown in Sequence D

$$\text{Leu·Ile·X}_1\text{·X}_2\text{·Ala·X}_3\text{·TYR·Thr·Ala}$$

SEQUENCE D

at the variable positions suggests that Glu residues in place of X_1, X_2, and X_3 yields the best substrates, with position X_3 being most critical (204). Arg residues added to either end of this peptide cause an increase in the K_m suggesting that anionic residues enhance and cationic residues inhibit interactions of peptides with the EGF receptor. This contrasts with serine kinases (205) which typically recognize a serine residue preceded by cationic amino acids, as shown in Sequence E.

$$\text{Leu·Arg·Arg·Ala·SER·Leu·Gly}$$

SEQUENCE E

Substitution of the serine residue in this peptide with tyrosine yields a poor substrate for the insulin receptor (149).

Factors other than the primary amino acid sequence are probably important for substrate recognition by tyrosine kinases. For example, enolase has a K_m of 2 μM for phosphorylation by pp60^{src}, but a synthetic peptide based on the sequence of the phosphorylation site has a K_m = 150 μM (206). Furthermore, the presence of anionic amino acids at the N-terminal side of a tyrosine residue is not obligatory for phosphorylation: Glutamate residues are located at the C-terminal side of the phosphorylated tyrosine in enolase (206); [val⁵]angiotensin II, a substrate for both the EGF and insulin receptor, does not contain any anionic residues (201, 207). Thus, it is likely that the tertiary structure of a substrate is also important for its recognition at the catalytic site of tyrosine kinases.

Steady-state kinetic analysis of the EGF receptor kinase using a synthetic peptide, shown in Sequence F suggests that catalysis occurs by an ordered bi bi reaction (208).

$$\text{Leu·Glu·Asp·Ala·Glu·TYR·Ala·Ala·Arg·Arg·Arg·Arg·Gly}$$

SEQUENCE F

In this kinetic model (209), the peptide binds to the catalytic site first followed by ATP; the phosphopeptide is the first and ADP is the last product to be released (208). Thus, both peptide and ATP must bind to the receptor before the products are formed. The sequential mechanism suggests that phosphorylation of exogenous substrates does not occur through a phosphoryl intermediate such as an autophosphorylation site (208). This conclusion is also supported by experiments using [γ-³⁵S]ATP [adenosine-5′-O-(3-[³⁵S]thriotriphosphate)] as alternative phosphate donors (210). EGF binding has no effect on K_m for ATP or the substrate (Sequence F), but stimulates phosphorylation by increasing the V_{max}. Thus, ligand-binding probably stimulates the catalytic step of the EGF receptor kinase with no effect on the binding of ATP or peptide. The kinetic mechanisms of hexokinase and phosphorylase kinase have also been shown to be sequential (211, 212) and this may be true for the insulin receptor since similar mechanisms of enzymatic phosphorylation seem to occur in various kinases whether the acceptor site is a seryl or tyrosyl residue (208). However, the possibility that one of the autophosphorylation sites of the insulin receptor serves as a catalytic intermediate has not been ruled out.

2. *In Vitro Identification of Protein Substrates*
   for the Insulin Receptor

One approach to understanding the role of a tyrosine kinase in hormone action is to identify potential substrates using an *in vitro* assay system. Several purified cellular proteins have been found to undergo tyrosine phosphorylation by pp60^{src} and the receptors for insulin, EGF and PDGF (163). Some of these are listed in Table I. However, this approach suffers from the limitation that the physiologic significance of the phosphorylation remains entirely unknown until *in vivo* correlations can be established. Nevertheless, even though the immediate physiological significance is lacking, several of these substrates are of particular biochemical interest.

Phosphorylation of the IgG in the immunocomplex of pp60^{src} has become a standard assay for pp60^{src} and other tyrosine kinases. When the purified insulin receptor is incubated with antisera to pp60^{src}, insulin stimulates tyrosine phosphorylation of the IgG heavy chains (149), and a similar effect has been reported with receptors for EGF and PDGF (213, 214). The phosphorylation of this population of IgG by several tyrosine kinase suggests a similar recognition between the kinases and the anti-pp60^{src}-IgG. In contrast, phosphorylation was not detected with four different antibodies containing antisera to the insulin receptor or with three different nonimmune sera (149). Addition of immobilized protein A to the reaction mixture results in precipitation of the phosphorylated IgG but not the insulin receptor suggesting that the antibody to pp60^{src} interacts enzymically with the insulin receptor, but that this is not sufficient for immunoprecipitation (149). So far, only one antibody to pp60^{src} has been shown to immunoprecipiate the insulin receptor (215).

Haring *et al.* recently surveyed several serine kinases as possible substrates for the insulin receptor (216). Of the five rat liver proteins evaluated [phosphorylase kinase, glycogen synthase kinase, casein kinase I, casein kinase II, and calmodulin dependent (CDR) kinase], only the CDR kinase was phosphorylated by the insulin receptor under the assay conditions used. In the presence of the CDR kinase, insulin also stimulated phosphorylation of calmodulin on tyrosine residues (216). Although the exact site of tyrosine phosphorylation in calmodulin was not determined, calmodulin contains only two tyrosine residues (Tyr-99 and Tyr-138) (*217*). Both occur in Ca^{2+}-binding domains but the latter is surrounded by anionic residues shown in Sequence G which are similar to those observed in other proteins that undergo phosphorylation; this is probably the site of phosphorylation.

Glu·Val·Asn·**TYR**·Glu·Glu

SEQUENCE G

Calmodulin is not regarded as a protein that is regulated by phosphorylation (*217*). However, serine phosphorylated forms have been isolated recently from chicken brain and muscle (*218*). McDonald *et al.* showed that the insulin receptor binds to calmodulin, possibly through its β-subunit (*219*). Calcium and calcium ionophores have also been shown to modify insulin receptor phosphorylation (*220*) and evidence has been presented for a Ca^{2+}-binding site on the receptor itself (*221*). Although further studies are necessary to clarify the relation between these proteins, these observations may form the basis of a mechanism that links the insulin receptor to certain postreceptor events.

Recently, we surveyed the *in vitro* phosphorylation of several glycolytic enzymes by the WGA-purified insulin receptor from FAO cells (*222*). Phosphofructokinase (PFK), the enzyme that catalyzes the rate-limiting step in glycolysis, and phosphoglycerate mutase (PGM) are most strongly phosphorylated by the insulin receptor (Fig. 12). The K_m for PFK phosphorylation is about 6 μM which is one of the lowest for the insulin receptor and near the hepatic concentration of the enzyme (*222a*). However, under our *in vitro* assay conditions, the steady-state stoichiometry is less than 0.01 mol phosphate/mol enzyme. Cooper *et al.* have shown that PGM, lactate dehydrogenase (LDH), and enolase, but not PFK, are phosphorylated on tyrosine residues in cells infected with the Rous sarcoma virus (RSV) (*223*) and *in vitro* during incubation with pp60[src] and [γ-^{32}P]ATP (*206*). In contrast, LDH and enolase are phosphorylated relatively weakly by the insulin receptor *in vitro* suggesting a difference between the substrate specificity of these kinases (Fig. 12). None of the glycolytic enzymes have been detected to undergo tyrosine phosphorylation in the intact cell during insulin stimulation. Whether phosphorylation of the glycolytic enzymes is important for insulin action or cellular transformation through the tyrosine kinase signal mechanism is unknown (*206*).

Certain cytoskeletal proteins are cellular targets for tyrosine kinases (*224*). For example, vinculin is phosphorylated by the *src* kinase (*224, 225*). Vinculin is concentrated at the inner face of the plasma membrane at sites of actin-membrane interaction and cell substratum contact, and it may be involved in the attachment of microfilaments to the cytoplasmic side of the plasma membrane (*226*). Since the organization of microfilaments is disrupted in RSV-transformed cells (*227*), it is possible that pp60[src] phosphorylation of vinculin could mediate this effect.

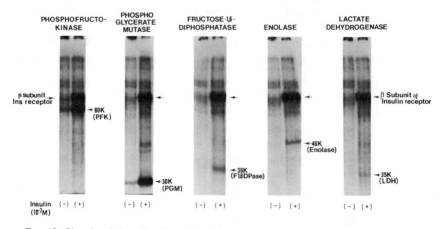

FIG. 12. Phosphorylation of various glycolytic enzymes *in vitro* by the WGA-purified insulin receptor from FAO cells in the absence (−) and presence (+) of 100 n*M* insulin. Each glycolytic enzyme was tested at 0.1 mg/ml, 22°C, and 50 μ*M* ATP.

Tubulin and microtubule-associated proteins (MAPs) have also been shown to undergo tyrosine phosphorylation by the purified insulin receptor kinase (*228*). The K_m for these substrates is less than 10 μ*M*. Tubulin and MAPs are phosphorylated on serine and threonine residues by cyclic AMP-dependent and calmodulin-dependent kinases, and these phosphorylations appear to regulate microtubule assembly and inhibit MAP-actin associations (*229, 230*). Whether the tyrosine phosphorylation occurs *in vivo* and also alters their function must be addressed in future work. In addition, the myosin regulatory light chain is phosphorylated by the EGF receptor kinase (*231*). Thus, several cytoskeletal proteins have been shown to undergo tyrosine phosphorylation, but the exact physiologic function of these phosphorylation events are not understood.

Tyrosine kinases have also been shown to phosphorylate phosphoinositides such as phosphotidyl inositol (PI) *in vitro*. This was originally demonstrated for v-*src* (*232*) and v-*ros* (*233*), and we have observed this for the highly purified insulin receptor (*233a*). In the case of insulin, this effect can be competitively inhibited by polypeptide substrates for the insulin receptor. Immunoprecipiation of the reaction solution with antiinsulin receptor antibodies depletes by greater than 90% the phosphotidyinositol kinase activity; however, the phosphorylation of PI to PIP is at best weakly insulin-stimulated. It has been shown that the EGF receptor can be separated from almost all PI kinase activity. Thus, the association of PI kinase with tyrosine kinases will require further study.

E. PHOSPHORYLATION OF THE INSULIN AND EGF RECEPTOR
   BY OTHER KINASES

1. *Casein Kinase I*

In the intact cell, tyrosine kinases also contain phosphoserine and phosphothreonine (*20, 100, 160*). Since the purified insulin receptor possess only a tyrosine kinase activity (*143, 160, 234*), these observations suggest that *in vivo* the receptor is a substrate for serine–threonine kinases that are independent of the receptor. Tuazon *et al.* have shown that casein kinase I phosphorylates partially purified insulin receptor from human placenta (*235*). Similar experiments were attempted without success by Haring *et al.* (*216*); however, the inclusion of 10 m*M* Mn^{2+} in the standard assays probably inhibited the activity of casein kinase I (*234*). Casein kinase I catalyzes serine phosphorylation primarily on residues in the β-subunit of the insulin receptor but some phosphorylation is also detected in the α-subunit in the absence of added insulin. Whether serine phosphorylation of the insulin receptor is catalyzed *in vivo* by this kinase has not been established and the effect of this serine phosphorylation on the activity of the receptor tyrosine kinase has not been reported, but these findings provide a clue for future *in vivo* studies.

2. pp60^{src}

In collaboration with Wirth and Pastan, we have shown that pp60^{src} catalyzes phosphorylation of the WGA-purified insulin receptor from FAO cells (*236*). This phosphorylation occurs on tyrosine residues of the β-subunit and is dependent on the concentration of pp60^{src} added to the reaction (Fig. 13). At a relatively low concentration of pp60^{src} there is an additive effect between insulin-stimulated and pp60^{src}-catalyzed phosphorylation of the β-subunit, whereas, at high concentrations of pp60^{src}, no further effect of insulin could be detected (*236*). Separation of the tryptic phosphopeptides by reverse-phase HPLC suggests that pp60^{src} phosphorylates the same or very similar sites in the β-subunit of the receptor (*237*). Yu *et al.* using the same preparation of pp60^{src} have provided evidence suggesting that phosphorylation of the β-subunit by pp60^{src} activates the receptor phosphotransferase in a way similar to insulin-stimulated autophosphorylation (*237*). Thus, tyrosine phosphorylation of the insulin receptor by the *src*-related oncogenes could provide a mechanism of cellular transformation.

In A431 cells infected with RSV, Cooper and Hunter showed that both the EGF receptor and pp60^{src} could phosphorylate an M_r-34,000-protein, but pp60^{src} did not phosphorylate the EGF receptor (*238*). The pp60^{src} binds to the plasma membrane which puts it in close proximity to the insulin receptor. However, the phosphorylation of the insulin receptor by pp60^{src} in the intact cell has

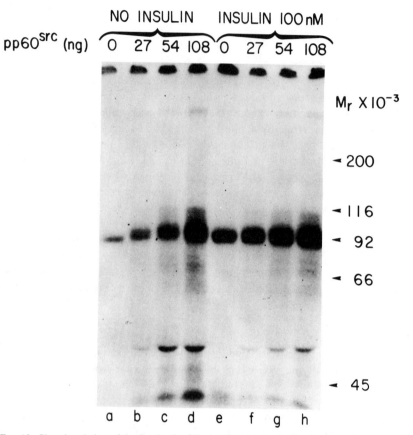

FIG. 13. Phosphorylation of the β-subunit of the insulin receptor in the absence and presence of insulin by the indicated amount of pp60^{src}. The total reaction volume was 50 μl and each sample was immunoprecipitated by antiinsulin receptor antibody before SDS-PAGE.

not been determined. Thus, the physiologic relevance of our observations demonstrating an interaction between pp60^{src} and the β-subunit remain untested and must await further studies using intact cells infected RSV.

3. Protein Kinase C

Based on our experience with phorbol esters, the calcium and phospholipid-dependent protein kinase C may phosphorylate the insulin receptor in an intact cell [See Ref. (239) and Sec. V, B.]; however, the purified C-kinase has not been shown to phosphorylate the insulin receptor *in vitro* (unpublished results, this laboratory). In contrast, purified C-kinase from rat brain catalyzes phosphorylation of the EGF receptor in isolated plasma membranes of A431 cells (240).

Tryptic peptide mapping of the EGF receptor reveals a distinct increase in the phosphorylation of three peptides, all of which contain phosphothreonine (*241*). A small increase in phosphoserine also occurs but this is not associated with distinct tryptic peptides and is probably distributed in many nonspecific sites. The major site of threonine phosphorylation is Thr-654. This was deduced from the migration of synthetic peptides with the authentic tryptic peptides from the EGF receptor (*241, 242*). Thr-654 is surrounded by cationic amino acids, shown in Sequence H, and is nine amino acid residues from the C-terminal side of the presumed transmembrane domain (*104*); a similar Thr residue does not occur in the insulin receptor.

Arg·Arg·Arg·His·Ile·Val·Arg·Lys·Arg·**THR**·Leu·Arg·Arg

SEQUENCE H

Since the tyrosine kinase domain presumably begins about 50 amino acid residues from the membrane, Thr-654 is located between the EGF-binding domain and the catalytic domain. This phosphothreonine could have an ionic effect on the conformation of the EGF receptor and alter the transmission of the EGF signal. In fact, threonine phosphorylation of the EGF receptor decreases the EGF-stimulated tyrosine autophosphorylation and thus may represent an important regulatory mechanism for EGF receptor function.

IV. Tyrosine Phosphorylation of Insulin Receptors and Other Proteins in the Intact Cell

A. INTRODUCTION

Enzymatic characterization of the purified insulin receptor was necessary to understand the biochemistry of phosphorylation and receptor kinase activity. However, to appreciate the physiologic role of receptor kinases we must focus on their function in the intact cell. For the insulin receptor, this has been studied most extensively in the well-differentiated, insulin-sensitive rat hepatoma, FAO. These cells possess a variety of insulin responses such as activation of glycogen synthase, induction of tyrosine amino transferase and amino acid transport, and stimulation of DNA synthesis (*152*). The FAO cell also has a high concentration of cell-surface receptors, most of which undergo tyrosine phosphorylation during insulin stimulation (unpublished results in this laboratory).

Successful study of *in vivo* phosphorylation experiments depends on the use of methods that completely block dephosphorylation during purification and immunoprecipitation of the phosphotyrosine-containing proteins. We employ two steps, (*a*) rapid freezing of the cells with liquid nitrogen at the end of an experiment to terminate the labeling procedure, and (*b*) thawing of the cells into a solution containing sodium vanadate (2 mM) and other phosphatase and protease inhibitors. Vanadate is a potent inhibitor of phosphotyrosine-specific phosphatases (*185, 243*) and dramatically improves the recovery of phosphotyrosine-containing proteins from the FAO cell (*160*). Applying these steps to our experiments and using high-affinity antiinsulin receptor and antiphosphotyrosine antibodies, we can study insulin receptor phosphorylation in the intact cell and begin to identify phosphotyrosine-containing proteins that occur immediately after insulin stimulation.

B. PHOSPHORYLATION OF THE INSULIN RECEPTOR IS AN EARLY
   CELLULAR RESPONSE TO INSULIN BINDING

Several insulin responses, such as stimulation of glucose transport, occur within seconds after insulin binding. Thus, the primary molecular events in the mechanism of insulin action that precede these changes must occur very rapidly after insulin binding. To determine if phosphorylation meets this criteria, FAO cells were labeled with [^{32}P]orthophosphate and then incubated with or without insulin. The labeled insulin receptors were solubilized, purified by affinity chromatography on immobilized wheat germ agglutinin agarose, immunoprecipitated with an antiinsulin receptor antibody and separated by SDS-PAGE. Using this procedure, we have found that tyrosine phosphorylation of the insulin receptor is stimulated within 5 sec after insulin binding, reaches steady state within 20 sec, and remains constant for at least 60 min (Fig. 14) (*160*). At no time is any phosphorylation of the α-subunit detected. This finding is consistent with the notion that tyrosine residues of the α-subunit are located entirely at the external face of the plasma membrane and inaccessible to the catalytic site of the β-subunit which is inside of the cell.

In contrast to the *in vitro* situation, phosphoamino acid analysis of the β-subunit extracted from the M_r = 95,000 region of the SDS-polyacrylamide gels reveals that in the basal state, receptor phosphorylation is due entirely to phosphoserine and phosphothreonine (*20, 160, 244*). Incubation of FAO cells with insulin for as little as 20 sec stimulates tyrosine phosphorylation of the β-subunit to a level approximately equal to that of phosphoserine (*160, 244*). Thus, insulin-stimulated phosphorylation of the β-subunit quickly reaches a steady state in FAO cells and the earliest new sites of phosphorylation are tyrosine residues.

The extent of change in serine and threonine phosphorylation is unclear. In FAO cells, some increase in phosphoserine occurs after longer incubations with insulin (*20, 160, 244*), whereas in freshly isolated hepatocytes, this appears to be the predominant phosphorylation stimulated in intact cells (*136, 137*). An exact quantitative analysis of these results is difficult since limited acid hydrolysis produces only partial cleavage of the peptide linkages and the phosphate bonds of serine, threonine, and tyrosine are hydrolyzed partially and unequally during this procedure (*245*).

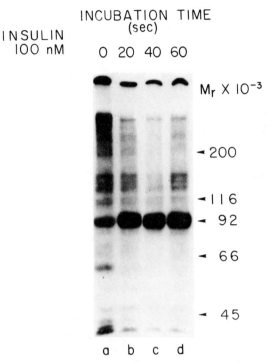

INSULIN
100 nM

INCUBATION TIME
(sec)

0 20 40 60

$M_r \times 10^{-3}$

◄ 200

◄ 116

◄ 92

◄ 66

◄ 45

a b c d

FIG. 14. A time course of phosphorylation of the insulin receptor β-subunit in [32P]ortho-phosphate-labeled FAO cells treated without insulin (lane a) or with 100 nM insulin for 20 sec, 40 sec, or 60 sec (lanes b–d). Before SDS-PAGE, the receptor was purified by affinity chromatography on WGA-agarose and immunoprecipitated with antiinsulin receptor antibodies.

C. QUANTITATIVE ANALYSIS OF BASAL AND INSULIN-STIMULATED PHOSPHORYLATION SITES

Trypsin digestion of the insulin receptor provides a method to characterize in greater quantitative detail the pattern of phosphorylation of the β-subunit in the intact cell. The phosphorylated β-subunit purified from $^{32}P_i$-labeled FAO cells yields several tryptic phosphopeptides which can be separated by reverse-phase HPLC and analyzed for phosphoamino acid content (*160*). In the absence of insulin, a single major [32P]phosphopeptide is detected that contains only phosphoserine (pS) (*160*) (Fig. 15A). Other minor tryptic peptides are also detected which contain phosphoserine, as well as one which contains phosphothreonine (pT in Fig. 15A). Although, the minor phosphoserine sites are somewhat variable in appearance, pS and pT are always detected in the basal receptor (*160*). The predominance of pS in the insulin-free receptor is consistent with the total phosphoamino acid content of the β-subunit, and as expected, no phosphotyrosine-containing peptides are detected in the β-subunit before insulin stimulation.

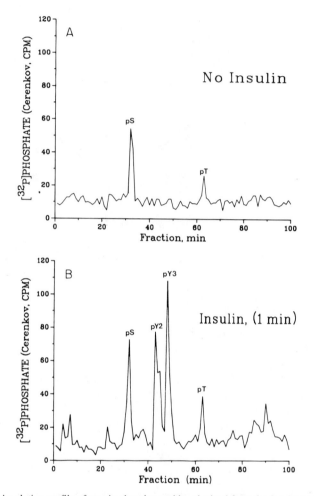

FIG. 15. An elution profile of tryptic phosphopeptides obtained from the β-subunit of the insulin receptor labeled in the intact FAO cell with [³²P]orthophosphate for 2 h. The β-subunit was purified on WGA-agarose and by specific immunoprecipitation from basal cells or cells stimulated with 100 n*M* insulin for 1 min. After SDS-PAGE, the gel fragments containing the β-subunit were digested with trypsin and the peptides were separated by reverse-phase HPLC.

After incubation of cells with insulin for only 20 sec, two additional phosphopeptides (pY2 and pY3) are detected in the tryptic digests of the β-subunit (Fig. 15B). Within the limits of detection, both peptides contain only phosphotyrosine (*160*). The phosphoamino acid composition of the other phosphopeptides is identical to that obtained in the absence of insulin. Thus, in less than 1 min, insulin stimulates phosphorylation at two tyrosine residue in the β-subunit of the insulin receptor in intact cells.

TABLE II

COMPARISON OF INSULIN-STIMULATED PHOSPHORYLATION IN THE INTACT FAO
CELL BY AUTORADIOGRAPHY OF THE β-SUBUNIT AND TRYPTIC PEPTIDE MAPPING

	Tryptic phosphopeptides					Relative intensity of β-subunit by autoradiography (arbitrary units)
	pS	pY2	pY3 (CPM)	pT	Total	
No insulin	81	—	—	23	101	350
Insulin (100 nM)	102	154	158	43	457	1617
Stimulation	1.2	—	—	1.7	4.5[a]	4.6

[a] 88% of this stimulation is due to phosphorylation of pY2 and pY3.

Table II summarizes the radioactivity associated with the four major peptides (pS, pY2, pY3, and pT). The relative increase in phosphorylation calculated from these phosphopeptides is 4.5-fold which equals exactly the increase predicted by densitometic scanning of the β-subunit on the corresponding autoradiogram. The phosphotyrosine-containing peptides (pY2 and pY3) are responsible for nearly 90% of the insulin-stimulated phosphorylation detected during the 1 min incubation, and the remaining 10% is distributed between phosphoserine- and phosphothreonine-containing peptides, pS and pT, respectively.

It is worth noting that in our earlier reports, insulin-stimulated phosphorylation in FAO cells was slower and the phosphoserine content of the β-subunit always predominated significantly over the amount of phosphotyrosine (234), whereas in current experiments the two are nearly equal phosphotyrosine predominates (160). Similarly, experiments in other laboratories with freshly isolated hepatocytes suggest that phosphoserine is the predominant phosphoamino acid in the β-subunit after insulin stimulation (137). The early results in the FAO cell, and the lack of phosphotyrosine in the hepatocytes may be due to underestimation of phosphotyrosine due to dephosphorylation that occurs during receptor extraction and purification (160). In fact, further studies with isolated rat hepatocytes in our laboratory using appropriate phosphatase inhibitors show that the initial effect of insulin on the β-subunit is the stimulation of tyrosine phosphorylation. However, regulatory mechanisms which are not understood are also likely to affect the ratio between phosphoserine and phosphotyrosine in the insulin-stimulated β-subunit.

D. COMPARISON OF THE SITES OF PHOSPHORYLATION IN VIVO AND IN VITRO

Comparison by reverse-phase HPLC of the tryptic phosphopeptides obtained from the insulin-stimulated β-subunit after phosphorylation in vitro and in vivo

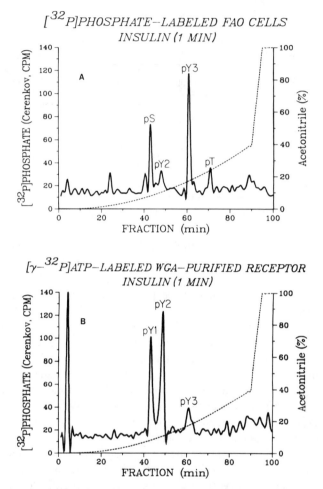

FIG. 16. A comparison of the tryptic phosphopeptides obtained from the insulin-stimulated β-subunit of the insulin receptor labeled *in vivo* (A) or *in vitro* (B).

reveals distinctly different sets of phosphopeptides (*160*). As previously noted after *in vitro* phosphorylation, three tryptic phosphopeptides are detected in the β-subunit, all of which contain only phosphotyrosine (pY1, pY2, pY3 in Fig. 9). When this is compared to the profile obtained from the β-subunit phosphorylated *in vivo*, only two of the phosphopeptides (pY2, pY3) were found to elute at identical positions (Fig. 16) (*160*). The major phosphotyrosine-containing peptide (pY3) from the β-subunit labeled *in vivo* is a minor peptide in the receptor phosphorylated *in vitro*. pY2 is a major site *in vitro* but variable *in vivo* and is sometimes less than pY3. Peptide pY1 is obtained only during *in vitro* phosphorylation, but migrates on a µBondapac C18 column in the same position as pS obtained from the $^{32}P_i$-labeled cells.

The reasons for the difference between the phosphorylation sites that occur *in vivo* and *in vitro* are not known, but such differences are often observed. For example, the EGF receptor contains three distinct tyrosine phosphorylation sites *in vitro*, but only one is strongly phosphorylated *in vivo* (*161*). For the insulin receptor, the appearance of pY1 may be due to an alteration in the conformation or accessibility of the catalytic or phosphate acceptor sites secondary to solubilization of the receptor. Peptides pY2 and pY3 are phosphorylated most rapidly *in vitro*, and may be the physiologically occurring sites. Caution should be used during the interpretation of results obtained with the purified receptor since regulatory effects resulting from tyrosine autophosphorylation at nonphysiological sites may not reflect exactly events that occur in the intact cell. However, it appears that pY2 which occurs both *in vitro* and *in vivo* may be partly responsible for regulation of the insulin receptor phosphotransferase (*165*).

E. DEPHOSPHORYLATION OF THE INSULIN RECEPTOR

When insulin is removed from its receptor the cellular metabolic responses are rapidly lost. To determine the relation between the loss of the insulin response and receptor phosphorylation, the reversibility of insulin-stimulated phosphorylation of the insulin receptor was studied (*246*). Incubation of insulin-stimulated FAO cells with insulin antibody to facilitate hormone dissociation reduced receptor phosphorylation almost to the level found in the absence of insulin (Fig. 17A). The half-time of the dephosphorylation of receptor was about 8 min and similar to the dissociation rate of insulin from the receptor. These data are consistent with the notion that phosphorylation and dephosphorylation of the β-subunit parallel the onset and termination of insulin action.

During the continued presence of insulin, the phosphorylation of the β-subunit is constant for at least 60 min. To determine if this represents a dynamic steady state, an isotope exchange experiment was carried out (*246*). The insulin receptor in $^{32}P_i$-labeled FAO cell was stimulated with 10^{-8} M insulin. When the radioactivity was chased with unlabeled phosphate, $^{32}P_i$ was lost from the β-subunit with a half-life of 20 min (Fig. 17B). The slower rate of dephosphorylation in this chase experiment compared to antibody-induced dephosphorylation was due presumably to the time required for dilution of the specific activity of the cellular [γ-^{32}P]ATP. Thus, the level of receptor phosphorylation represents a steady state between autophosphorylation and dephosphorylation catalyzed by an exogenous phosphatase (*246*).

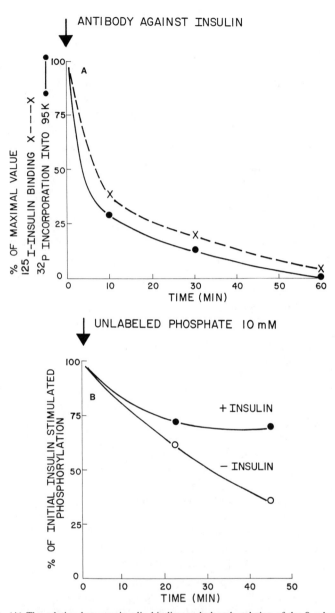

FIG. 17. (A) The relation between insulin-binding and phosphorylation of the β-subunit. Phosphorylation and binding were allowed to reach steady state in the intact FAO cell and then insulin dissociation was facilitated by addition of antiinsulin antibodies. The amount of [^{125}I]insulin and [^{32}P]phosphate remaining associated with the insulin receptor at the indicated times is plotted as a percentage of the maximum measured before the addition of antibody. (B) A pulse–chase experiment of phosphorylation of the β-subunit. FAO cells were labeled for 2 h with [^{32}P]P_i, incubated without (−) or with (+) insulin for 15 additional min, and then unlabeled phosphate was added to the cells for the indicated time intervals. The percentage of the label remaining in the β-subunit after the chase is plotted in the figure.

F. USE OF ANTIPHOSPHOTYROSINE ANTIBODIES TO ISOLATE
    RECEPTOR KINASES AND THEIR SUBSTRATES

1. *Introduction*

Sequential protein phosphorylation is an attractive mechanism for the transmission of an extracellular signal from a membrane-bound receptor kinase to
intracellular sites. Since the purified insulin receptor catalyzes tyrosine phosphorylation of various substrates, it is reasonable to hypothesize that the second
event during the cellular insulin response is a tyrosine phosphorylation of another
cellular protein. The identification of these substrates in the intact cell is difficult
because of the extremely low level of phosphotyrosine-containing proteins and
the overwhelming amount of phosphoserine and phosphothreonine. Two-dimensional polyacrylamide gel electrophoresis has been applied successfully to this
problem for pp60^{src} and the EGF receptor (*100, 163, 238*), but it may miss some
phosphoproteins with p*I* values below 5.5 or phosphoproteins of high molecular
weight due to the background phosphorylation that occurs in these regions (*163*).
[For a detailed discussion of the substrates identified by this technique see Ref.
(*163*)].

Antiphosphotyrosine antibodies coupled with one-dimensional SDS-PAGE
provide an alternative approach to identify tyrosine phosphorylation events. This
technique was first described by Ross *et al.* who produced a monoclonal antibody to phosphotyrosine by immunization with aminobenzyl phosphonate coupled by diazotization to Keyhole Limpet hemocyanin (KLH) (*247*). Polyclonal
antiphosphotyrosine antibodies have been prepared successfully in rabbits using
phenylphosphate derivatives (*181, 195*). Antiphosphotyrosine antibodies have
been used to isolate phosphoproteins from retrovirus-transformed cells and cells
stimulated with EGF, PDGF, or insulin. In this section, we describe our experience with antiphosphotyrosine antibodies.

2. *Preparation and Purification*
   of Antiphosphotyrosine Antibodies

We have prepared antiphosphotyrosine antibodies in New Zealand and Dutch
Belted rabbits using the method of Pang *et al.* (*195*). The antigen is prepared by
reaction of phosphotyramine with bromoacetyl bromide to yield bromoacetyl
phosphotyramine which couples to KLH by reacting with the ε-amino groups of
its lysine residues. This method of coupling the hapten appears to be superior
because it is carried out under mild conditions and introduces about 20 to 30
molecules of phenylphosphate into each molecule of KLH though weakly antigenic amide bonds. The antibodies obtained after three months of cutaneous
inoculation are easily purified by affinity chromatography on immobilized phosphotyramine. About 4 mg of IgG are obtained from 20 ml of serum. The average
titer is about 1 : 50,000 by a radioimmunoassay using immobilized BSA-phos-
photyramine and [^{125}I]protein A (unpublished results, this laboratory). Quan-

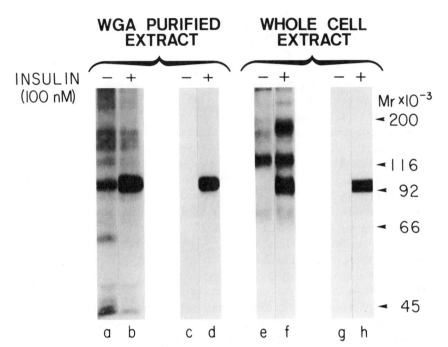

FIG. 18. Purification of phosphotyrosine-containing proteins from [32P]orthophosphate-labeled FAO cells by immunoprecipitation with antiinsulin receptor and antiphosphotyrosine antibodies. FAO cells were labeled for 2 h. Insulin (100 nM) was omitted (−) or added (+) for 1 min and then a supernatant of the whole cell detergent extract was prepared; half of it was purified on immobilized WGA-agarose. The WGA-purified extract was immunoprecipitated with antiinsulin receptor antibody (lanes a and b) or antiphosphotyrosine antibody (lanes c and d). The whole cell extract was immunoprecipitated with antiphosphotyrosine antibody and the precipitated phosphoproteins were either separated by PAGE (lanes e and f) or immunoprecipitated a second time with antireceptor antibodies before SDS-PAGE (lanes g and h). The proteins were reduced with dithiothreitol.

titative immunoprecipitation of labeled insulin receptors can be achieved with <3 μg/ml of this IgG per 10 pmol of receptor. It is specific for phosphotyrosine and bound proteins can only be eluted with phenyl phosphate derivatives and not by phosphate (200 mM) or alkyl phosphate-containing compounds (244); however, it cross-reacts to a small degree with sulfotyrosine (244a).

3. Immunoprecipitation with Antiphosphotyrosine Antibodies of the Insulin Receptor and a Possible in Vivo Substrate

When the glycoprotein fraction purified from 32P-labeled FAO cells in the absence of insulin was immunoprecipitated with an antiphosphotyrosine antibody, no phosphotyrosine-containing proteins were identified (Fig. 18, lane c). Incubation of FAO cells with 100 nM insulin for 1 min results in the stimulation of a single [^{32}P]phosphoprotein in the WGA agarose-purified cell extract that specifically precipitates with the antiphosphotyrosine antibody (Fig. 18, lane d). This protein corresponds to the β-subunit of the insulin receptor on SDS-PAGE, and is completely immunoprecipitated with antiinsulin receptor antibodies. Phosphotyrosine, but not phosphoserine or phosphothreonine completely inhibits immunoprecipitation of the β-subunit.

In contrast to the single protein in the WGA eluate, this phosphotyrosine antibody specifically immunoprecipitates several labeled proteins from whole-cell extracts of FAO cells (248). Before incubation of cells with insulin, two phosphoproteins (M_r = 120,000 and 75,000) were eluted from the antiphosphotyrosine antibody immunocomplex (Fig. 18, lane e). After insulin stimulation, two additional proteins of M_r = 95,000 (pp95) and M_r = 185,000 (pp185) were immunoprecipitated suggesting that they undergo *de novo* tyrosine phosphorylation in response to insulin (Fig. 18, lane f). The pp95 is the β-subunit of the insulin receptor since it is immunoprecipitated completely by antiinsulin receptor antibodies (Fig. 18, lanes g and h), whereas pp185 is not recognized by this antibody.

The insulin dose response curve for phosphorylation of pp185 is identical to the curve for the β-subunit, and both proteins reach maximum phosphorylation during 30 sec incubation with insulin (248). The pp185 is not retained by wheat germ agglutinin agarose and can be extracted from cells without the use of detergents (unpublished results in this laboratory). This suggests that pp185 is a soluble protein that is located in the cytoplasm or is weakly associated with membranes. Both the β-subunit and pp185 contain phosphoserine and phosphothreonine in addition to phosphotyrosine. The tryptic peptide map of pp185 does not resemble that of the β-subunit suggesting that pp185 is an insulin-stimulated phosphotyrosine-containing protein that is not structurally related to the insulin receptor. Thus, pp185 fulfills basic criteria expected for a substrate of the insulin receptor that transmits the insulin signal from the plasma membrane to other intracellular sites.

4. Immunoprecipitation of [^{32}P]Phosphoproteins from EGF- and PDGF-Stimulated Cells

Antiphosphotyrosine antibodies have been used to study phosphotyrosine-containing proteins in Rous sarcoma virus and Abelson murine leukemia virus-transformed cells (247, 249, 250), in A431 cells, and in fibroblasts after stimulation of the cells with PDGF or EGF (21, 249). In each case, the major proteins precipitated from whole-cell extracts correspond to the transforming gene product of the retrovirus or the phosphorylated receptor. For example, the major PDGF-stimulated phosphoprotein immunoprecipitated from BALB/3T3 and human fibroblasts have an M_r = 185,000. Several physical characteristics suggest that this protein is the PDGF receptor (and not the 185-kDa protein described in the previous section) including its molecular weight, isoelectric point ($pI = 4.2$), binding to DEAE cellulose, and wheat germ agglutinin (21). Similarly, these antibodies precipitate the EGF receptor from stimulated cells (249).

Other phosphoproteins have also been detected in fibroblasts following hormone stimulation. Using a monoclonal antiphosphotyrosine antibody, Frackelton et al. observed a cytosolic 74-kDa protein following PDGF stimulation of the BALB/3T3 cell that may represent a substrate of the activated PDGF receptor (22). Ek and Heldin have used a polyclonal antiphosphotyrosine antibody to immunoprecipitate several phosphoproteins from PDGF-stimulated human fibroblasts (23). Further studies are necessary to determine the nature of these proteins and whether they are involved in transmission of the PDGF signal.

G. INSULIN RECEPTOR SUBSETS SEPARATED BY ANTIRECEPTOR
     AND ANTIPHOSPHOTYROSINE ANTIBODIES

In collaboration with J. Shafer at the University of Michigan, we have used antiinsulin receptor and antiphosphotyrosine antibodies as a pair of reagents to separate the phosphotyrosine-containing subset of receptors from the total receptor pool (244). Before insulin stimulation, the β-subunit of the receptor does not contain phosphotyrosine so the receptor is precipitated only with antireceptor antibodies. In contrast, following insulin stimulation, an increasing portion of receptors precipitate with the antiphosphotyrosine antibody. Using a second precipitation with antiinsulin receptor antibody after clearing the solution with the antiphosphotyrosine antibody, however, an apparently constant amount of phosphorylated receptors remain in the cell extract (244). By phosphoamino acid analysis, this receptor subset contains little or no phosphotyrosine suggesting that it is not accessible to insulin, is not a functional tyrosine kinase, or undergoes dephosphorylation during the incubation. The role of this subset of receptors in insulin action is unknown.

V. Regulation of Hormone Receptors by Multisite Phosphorylation

A. INTRODUCTION

One of the cellular responses to a specific hormone is to alter the interaction of other different receptors with their ligand. For example, insulin binding on adipocytes or hepatoma cells increases nearly 10-fold the binding of IGF II to its receptor (251–253). This effect is rapid, temperature dependent, not present in isolated membranes, and mimicked by antiinsulin receptor antibodies (251). This appears to be due in part to insulin-stimulated translocation of IGF-II receptors to the plasma membrane (*253a*). Relaxin stimulates insulin binding to cell surface receptors in certain cases (*254*) and isoproterenol decreases insulin binding in rat adipocytes (*255*). The phorbol ester PMA (phorbol 12-mytristate 13-acetate) rapidly decreases insulin binding in some (*256*) but not all cells (*239*). Similarly, EGF binding is decreased by vasopressin (*257*), PDGF, fibroblast derived growth factor, isoproterenol (*255*), and PMA (*258*). These effects do not occur through competition with insulin or EGF at the binding site.

To determine the mechanism involved in these regulatory events, several investigators have studied protein phosphorylation and the role of the Ca^{2+}-dependent and phospholipid-activated protein kinase C (*259*). Protein kinase C is one of the two known, operationally distinct branches of the Ca^{2+} messenger system that exists in mammalian cells (the other branch is calmodulin-dependent) (*6, 7*). Activation of the Ca^{2+} messenger system by hormones like vasopressin or isoproterenol causes a transient rise in cytosolic Ca^{2+} and a prolonged increase in the diacylglycerol content of the plasma membrane. Both of these factors lead to activation of protein kinase C (*7, 8, 259, 260*). Hormonal stimulation may be mimicked by phospholipase C which increases the cellular content of diacylglycerol (*262*), A23187 which elevates intracellular Ca^{2+} (*263*), or incubation with exogenous diacylglycerols or PMA (*8, 261, 264*). Neidel *et al.* have presented evidence that the protein kinase C is the PMA receptor (*260, 261*). Thus, the regulation of the EGF and insulin receptors by PMA or hormones that activate the Ca^{2+} messenger system could occur through serine–threonine phosphorylation catalyzed by the protein kinase C.

B. EFFECT OF PMA-STIMULATED PHOSPHORYLATION ON THE ACTIVITY OF EGF AND INSULIN RECEPTOR

The dose response for the inhibition of EGF binding by PMA is very close to that of its ability to stimulate the Ca^{2+} and phospholipid-dependent protein kinase C or to act as a tumor promoter (*259*). Recent reports have shown that PMA stimulates serine–threonine phosphorylation of the EGF receptor in A431 cells, fibroblasts, and KB cells possibly by activating the protein kinase C (*240, 241, 242, 263, 265*). The major threonine phosphorylation site is located in a domain that is very close to the membrane-spanning region of the EGF receptor which could affect the binding of EGF at the outer face of the membrane (*266*). PMA also inhibits EGF-stimulated tyrosine autophosphorylation (*240, 266*), but how this occurs is not known. These results suggest that protein kinase C may regulate both binding and catalytic activity of EGF receptor by producing a threonine phosphorylation in the EGF receptor.

Jacobs *et al.* showed that PMA stimulates the phosphorylation of the β-subunit of the insulin receptor in IM-9 lymphocytes (*24*) and we observed a similar but quantitatively smaller effect on FAO cells (*239*). In FAO cells, this phosphoryla-

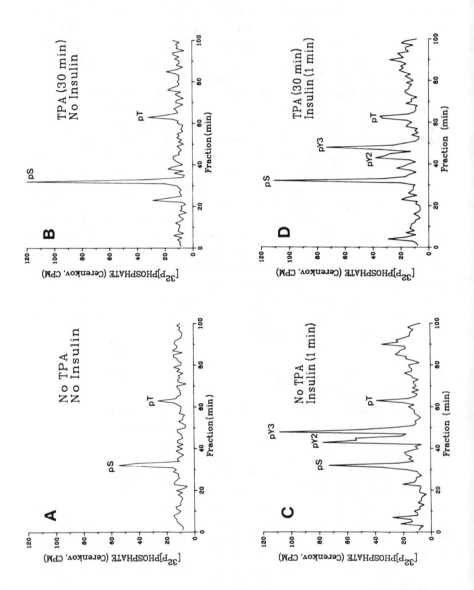

tion occurs on serine and threonine residues that migrate in identical positions on reverse-phase HPLC as the sites phosphorylated before the addition of the phorbol ester (i.e., in the basal state) (Fig. 19, A and B). PMA has no detectable effect on insulin binding in FAO cells; however, insulin-stimulated phosphorylation of the β-subunit is reduced when compared to control cells (239). Insulin stimulation of both glycogen synthase and tyrosine amino transferase is also largely inhibited by PMA treatment and this correlates with the decreased response of the cells to insulin-stimulated autophosphorylation (239). Separation of the tryptic phosphopeptides from the β-subunit of insulin-stimulated cells indicates that PMA inhibits tyrosine autophosphorylation of pY2 and pY3 (Fig. 19, C and D). These data show that phorbol esters stimulate insulin receptor phosphorylation on serine residues and reduce insulin-stimulated tyrosine phosphorylation and insulin response suggesting a regulatory role for serine phosphorylation and protein kinase C in insulin action.

VI. Evidence That Tyrosine Phosphorylation Is Physiologically Important

A. THE RELATIONSHIP BETWEEN TYROSINE PHOSPHORYLATION AND CELLULAR TRANSFORMATION

A growing list of observations suggest that tyrosine kinases regulate early events in cellular metabolism, growth, and differentiation. About half of the known oncoviruses contain an oncogene that codes for a protein tyrosine kinase (267). Transformation of cells with these viruses leads to an increased level of phosphotyrosine in proteins (16, 27). However, the physiological relevance of tyrosine phosphorylation, although strongly implied, is still not proven.

The significance of tyrosine kinases has been best studied with the RSV. To establish the relationships between tyrosine phosphorylation and cellular transformation, mutant viruses have been prepared which contain altered or impaired tyrosine kinases. For example, a mutant strain of the RSV that is temperature-sensitive for induction of transformation also yields a thermally labile tyrosine kinase (268–271). These results suggest that the kinase activity of pp60^{src} is necessary for cellular transformation.

FIG. 19. An elution profile of tryptic phosphopeptides obtained from the GGB-subunit of the insulin receptor labeled in the intact FAO cell with [^{32}P]orthophosphate for 2 h. The β-subunit was purified on WGA-agarose and by immunoprecipitation with antiinsulin receptor antibodies from basal cells (A); cells treated for 30 min with 1 μM PMA (B); cells stimulated with 100 nM insulin for 1 min (C); or cells treated with 1 μM PMA for 30 min before insulin stimulation (D). After SDS-PAGE, the gel fragments containing the β-subunit were digested with trypsin and the peptides were separated by reverse-phase HPLC.

To determine whether autophosphorylation of the tyrosine kinase is essential, mutant RSV and feline sarcoma virus have been prepared by site-directed mutagenesis which contain a Phe residue in place of Tyr-416 of pp60^{src} (272) or Tyr-1073 of p140$^{gag\text{-}fps}$ (273), respectively. This change does not affect the phosphotransferase activity of pp60^{src} nor the ability of RSV mutant to transform mouse fibroblasts (262). However, the mutant virus shows an impaired ability to induce tumors (274). A similar mutation in p140$^{gag\text{-}fps}$ impairs both fibroblast transformation and the phosphotransferase activity (273). Thus, under some circumstances there appears to be a physiologic relationship between autophosphorylation, the tyrosine-specific phosphotransferase, and cellular transformation.

For the insulin receptor, insulin-resistant patients and mutant cells, some of which appear to express receptors with an altered tyrosine kinase, may offer an opportunity to determine the physiologic significance of receptor kinase activity. These are discussed in the following section.

B. EVIDENCE THAT TYROSINE PHOSPHORYLATION IS IMPORTANT
    FOR INSULIN ACTION

1. *Insulin Receptor Phosphorylation in Mutant Melanoma Cells*

Insulin inhibits the growth of Cloudman melanoma cells in culture. Taking advantage of this characteristic, Pawlek *et al.* have selected several variants from this cell line which show an altered insulin growth response (275, 276). We have characterized insulin binding and receptor phosphorylation in three of these cell lines: the wild-type Cloudman melanoma, 1A, a variant called 111 which is insulin-resistant, that is, insulin neither stimulates nor inhibits growth, and another variant, 46, that requires insulin for growth (145). [^{125}I]Insulin binding to intact cells is similar for the wild type 1A and insulin-stimulated variant; however, the insulin-resistant variant 111 shows approximately a 30% decrease of receptor affinity with no apparent difference in receptor number. A similar pattern of insulin binding is observed when the insulin receptor from each melanoma cell line is partially purified on wheat germ agglutinin-agarose, suggesting that the slightly altered binding in type (111) is intrinsic to the receptor (145). Since binding is altered only at low hormone concentrations, however, it cannot account for the total insulin resistance observed.

Phosphorylation was studied during incubation of the WGA-purified receptor from each cell line with insulin and [γ-^{32}P]ATP in an *in vitro* assay. Insulin stimulated the autophosphorylation of the β-subunit of the receptor from all three cell types with similar kinetics. However, the amount of phosphate incorporated into the β-subunit in the insulin-resistant cell line 111 is decreased approximately 50% from that observed with the two other cell lines. This difference is reflected in the entire insulin dose response curve. Phosphopeptide analysis by trypsin digestion of the β-subunit and separation of the phosphopeptides by reverse-phase HPLC indicates that at least three sites of autophosphorylation exist in the receptor of these cells (*145*). These may correspond to pY1, pY2, and pY3 as described previously. Only pY1 and pY3 were detected in the insulin-resistant variant suggesting that a specific alteration in the properties of autophosphorylation of the insulin receptor exist in this insulin resistant cell line and this may affect the transmission of growth promoting signals in the cell. From the *in vitro* studies of Yu and Czech (*165*), pY2 may be important for activating the kinase, thus, its absence from the β-subunit in 111 cells may be relevant.

2. *Insulin Receptor Phosphorylation in Cells from Insulin-Resistant Patients*

Human monocytes, erythrocytes, and fibroblasts contain specific insulin receptors which have binding characteristics similar to those found in other classical insulin target cells (*146, 277*). Although less is known about the receptor in these cells, they are useful for human studies because they are readily accessible from the periphereal circulation, and the number of cells ordinarily collected is sufficient to supply receptors for phosphorylation studies (*146*). Insulin stimulates 4- to 5-fold the tyrosine phosphorylation of the β-subunit of the insulin receptor contained in a Triton X-100 solubilized membrane fraction from human erythrocytes. We have used these systems to characterize the binding and kinase properties of the insulin receptor in patients with Type A insulin resistance and acanthosis nigricans.

The Type A syndrome is characterized by severe insulin resistance due to a defect in insulin action. In some patients, there is a striking decrease in insulin binding, whereas in others, insulin binding is normal. To better understand the pathophysiology of this syndrome, both the binding and kinase activity of the insulin receptor have been characterized in cells from patients with this syndrome (*147, 278*).

Most of the patients show a marked decrease in binding in both intact cells and in the soluble receptor. In most of these cases, this alteration is due to a decrease in receptor number, although an occasional patient exhibits a decrease in receptor affinity for insulin. In all of these patients, maximal phosphorylation was reduced by 60% to 90%. In patients with decreased number of receptors, this is due to decrease in β-subunits. In the patients with altered affinity, there was also a rightward shift of the insulin dose-response curve. The most interesting patients, however, are those in whom insulin binding is normal or nearly normal, but phosphorylation is decreased. Two such patients have been studied in detail. In these, receptor autophosphorylation in monocytes, erythrocytes, and fibroblasts is decreased by about 60% (*147, 278*). This appears to be due to a decrease in tyrosine kinase activity. Additional molecular studies will be required to established the cause of these changes; however, these data provide an important clue as to the role of receptor kinase activity in insulin action.

VII. Conclusions

Phosphotyrosine-containing proteins and tyrosine kinases are minor components of normal cells that are closely related to the regulation of cellular growth and metabolism. The insulin receptor is a member of this family of tyrosine kinases, which also includes the receptors for epidermal growth factor, insulin-like growth factor-I, platelet-derived growth factor and murine colony stimulating factor, and the gene products of certain cellular oncogenes and transforming retroviruses. In this review, we have focused on the structure and function of the insulin receptor to establish the hypothesis that tyrosine kinases play a key role in cellular regulation and tranmission of the insulin signal. The insulin receptor is composed of two functional domains: an extracellular regulatory subunit that binds insulin (the α-subunit) and a transmembrane subunit that contains tyrosine kinase activity (the β-subunit). The tyrosine kinase is intrinsic to the insulin receptor and is present in all cells studied so far that contain the receptors. One of the earliest detectable responses to insulin binding in an intact cell is activation of this kinase and tyrosine phosphorylation of intracellular domains on the β-subunit. From studies with the purified insulin receptor, autophosphorylation activates the phosphotransferase in the β-subunit which catalyzes tyrosine phosphorylation of other proteins on tyrosine residues. The search for substrates in the intact cell has been difficult due to their very low concentration; however, phosphotyrosine antibodies have provided a very sensitive tool to begin identifying the cellular substrates for the insulin receptor and other tyrosine kinases. For example, a protein of 185 kDa (pp185) is immunoprecipited from FAO cells immediately after insulin stimulation, suggesting that it is an endogenous substrate for the insulin receptor kinase. One of the most important problems to address in the future is the identification of other cellular substrates and determination of their role in the transmission of the insulin signal.

ACKNOWLEDGMENTS

This work has been supported in part by grants to C.R.K. (AM31036 and AM29770) and to M.F.W. (AM07163 and AM35988) from the Institute of Health and Human Development, National Institute of Health, United States Public Health Service, and a career development award to MFW from the American Diabetes Association.

REFERENCES

1. Heldin, C. H., and Wastermark, B. (1984). *Cell* **37**, 9–20.
2. Kahn, C. R., Baird, K. L., Flier, J. S., Grunfeld, C., Harmon, J. T., Harrison, L. C., Karlsson, F. A., Kasuga, M., King, G. L., Lang, U., Podskalny, J. M., and Van Obberghen, E. (1981). *Recent Prog. Horm. Res.* **37**, 447–538.
3. Ross, E. M., and Gilman, A. G. (1980). *Annu. Rev. Biochem.* **49**, 533–564.

4. Rubin, C. S., and Rosen, O. M. (1975). *Annu. Rev. Biochem.* **44**, 831–887.
5. Krebs, E. G., and Beavo, J. A. (1979). *Annu. Rev. Biochem.* **48**, 923–959.
6. Exton, J. H. (1981). *Mol. Cell. Endocrinol.* **23**, 233–265.
7. Rasmussen, H., and Barrett, P. Q. (1984). *Physiol. Rev.* **64**, 938–984.
8. Nishizuka, Y. (1984). *Nature (London)* **308**, 693–698.
9. Czech, M. P. (1977). *Annu. Rev. Biochem.* **46**, 359–384.
10. Denton, R. M., Brownsey, R. W., and Belsham, G. J. (1981). *Diabetologia* **1**, 347–362.
11. Chalapowski, F. J., Kelly, L. A., and Butcher, R. W. (1975). *Adv. Cyclic Nucleotide Res.* **6**, 245–338.
12. Pastan, I., Johnson, G. S., and Anderson, W. B. (1975). *Annu. Rev. Biochem.* **44**, 491–522.
13. Czech, M. P. (1981). *Am. J. Med.* **70**, 142–150.
14. Cheng, K., and Larner, J. (1985). *Annu. Rev. Physiol.* **47**, 405–424.
15. Kahn, C. R. (1979). *Trends Biochem. Sci.* **4**, 263–266.
16. Hunter, T., and Sefton, B. M. (1980). *PNAS* **77**, 1311–1315.
17. Carpenter, G., King, L., Jr., and Cohen, S. (1979). *JBC* **254**, 4884–4891.
18. Cohen, S., Carpenter, G., and King, L., Jr. (1980). *JBC* **255**, 4834–4842.
19. Kasuga, M., Karlsson, F. A., and Kahn, C. R. (1982). *Science* **215**, 185–187.
20. Kasuga, M., Zick, Y., Blithe, D. L., Crettaz, M., and Kahn, C. R. (1982). *Nature* **298**, 667–669.
21. Nishimura, J., Huang, J. S., and Deuel, T. F. (1982). *PNAS* **79**, 4303–4307.
22. Frackelton, A. R., Tremble, P. M., and Williams, L. T. (1984). *JBC* **259**, 7909–7915.
23. Ek, B., and Heldin, C. H. (1984). *JBC* **259**, 11145–11152.
24. Jacobs, S., Sahyoun, N. E., Saltiel, A. R., and Cuatrecasas, P. (1983). *PNAS* **80**, 6211–6213.
25. Rubin, J., Shia, M. A., and Pilch, P. (1983). *Nature (London)* **305**, 440.
26. Sasaki, N., Ress-Jones, R. W., Zick, Y., Nissley, S. P., and Rechler, M. M. (1985). *JBC* **260**, 9793–9804.
27. Cooper, J. A., and Hunter, T. (1981). *J. Cell Biol.* **91**, 878–893.
28. Bishop, M. (1983). *Annu. Rev. Biochem.* **52**, 301–354.
29. Sefton, B. M., Hunter, T., Beeman, K., and Eckhart, W. (1980). *Cell* **20**, 807–816.
30. Cuatrecasas, P. (1972). *PNAS* **69**, 318–322.
31. Cuatrecasas, P. (1972). *JBC* **247**, 1980–1991.
32. Cuatrecasas, P. (1972). *PNAS* **69**, 1277–1281.
33. Jacobs, S., Shechter, Y., Bissell, K., and Cuatrecasas, P. (1977). *BBRC* **77**, 981–988.
34. Jacobs, S., Hazum, E., Shechter, Y., and Cuatrecasas, P. (1979). *PNAS* **76**, 4918–4921.
35. Siegel, T. W., Ganguly, S., Jacobs, S., Rosen, O. M., and Rubin, C. (1983). *JBC* **256**, 9266–9273.
36. Petruzzelli, L., Herrera, R., and Rosen, O. M. (1984). *PNAS* **81**, 3327–3331.
37. Fujita-Yamaguchi, Y., Choi, S., Sakamoto, Y., and Itakura, K. (1983). *JBC* **258**, 5045–5049.
38. Fujita-Yamaguchi, Y. (1984). *JBC* **259**, 1206–1211.
39. Massague, J., Pilch, P. F., and Czech, M. P. (1981). *PNAS* **77**, 7137–7141.
40. Kasuga, M., Hedo, J. A., Yamada, K. M., and Kahn, C. R. (1982). *JBC* **257**, 10392–10399.
41. Fehlmann, M., Peyron, J. F., Samson, M., Van Obberghen, E., Brandenburg, D., and Brossette, N. (1985). *PNAS* **82**, 8634–8637.
42. Pang, D. T., and Shafer, J. A. (1984). *JBC* **259**, 8589–8596.
43. Kahn, C. R., Flier, J. S., Bar, R. S., Archer, J. A., Gorden, P., Martin, M., and Roth, J. (1976). *N. Engl. J. Med.* **294**, 739–745.
44. Flier, J. S. (1982). *Clin. Immunol. Rev.* **1**, 215–256.
45. Van Obberghen, E., Kasuga, M., LeCam, A., Hedo, J. A., Itin, A., and Harrison, L. C. (1981). *PNAS* **78**, 1052–1056.
46. Hedo, J. A., Kassuga, M., Van Obberghen, E., Roth, J., and Kahn, C. R. (1981). *PNAS* **78**, 4791–4795.

47. Kasuga, M., Kahn, C. R., Hedo, J. A., Van Obberghen, E., and Yamada, K. M. (1981). *PNAS* **78**, 6917–6921.
48. Hedo, J. A., Kahn, C. R., Hayoshi, M., Yamada, K. M., and Kasuga, M. (1983). *JBC* **258**, 10020–10026.
49. Ronnett, G. V., Knutson, V. P., Kohanski, R. A., Simpson, T. L., and Lane, M. D. (1984). *JBC* **259**, 4566–4575.
50. Blackshear, P. J., Nemenoff, R. A., and Avruch, J. (1983). *FEBS Lett.* **158**, 243–246.
51. Rees-Jones, R., Hedo, H. A., Zick, Y., and Roth, J. (1983). *BBRC* **116**, 417–422.
52. Herzberg, V. L., Grigorescu, F., Edge, A. S. B., Spiro, R. G., and Kahn, C. R. (1985). *BBRC* **129**, 789–796.
53. Ullrich, A., Bell, J. R., Chen, E. Y., Herrera, R., Petruzzelli, L. M., Dull, T. J., Gray, A., Coussens, L., Liao, Y.-C., Tsubokawa, M., Mason, A., Seeburg, P. H., Grunfeld, C., Rosen, O. M., and Ramachandran, J. (1985). *Nature (London)* **313**, 756–761.
54. Ebina, Y., Ellis, L., Jarnagin, K., Edery, M., Graf, L., Clauser, E., Ou, J.-H., Masiar, F., Kan, Y. W., Goldfine, I. D., Roth, R. A., and Rutter, W. J. (1985). *Cell* **40**, 747–758.
55. Crettaz, M., Jialal, I., Kassuga, M., and Kahn, C. R. (1984). *JBC* **259**, 11543–11549.
56. Chratchko, Y., Gazzano, H., Van Obberghen, E., and Fehlmann, M. (1984). *Mol. Cell. Endocrinol.* **36**, 59–65.
57. Fujita-Yamaguchi, Y., and Katuria, S. (1985). *PNAS* **82**, 6095–6099.
58. Massague, J., Pilch, P. F., and Czech, M. P. (1981). *JBC* **256**, 3182–3190.
59. Massague, J., Czech, M. P. (1982). *JBC* **257**, 6729–6738.
60. Hedo, J. A., and Simpson, I. A. (1984). *JBC* **259**, 11083–11089.
61. Yip, C. C., Yeung, C. W. T., and Moule, M. L. (1978). *JBC* **253**, 1743–1745.
62. Yip, C. C., Yeung, C. W., and Moule, M. L. (1980). *Biochemistry* **19**, 70–76.
63. Pilch, P. F., and Czech, M. (1979). *JBC* **254**, 3375–3381.
64. Pilch, P. F., and Czech, M. (1980). *JBC* **255**, 1722–1731.
65. Roth, R. A., and Cassell, M. P. (1983). *Science* **219**, 299–301.
66. Van Obberghen, E., Rossi, B., Kowalski, A., Gazzano, H., and Ponzio, G. (1983). *PNAS* **80**, 945–949.
67. Shia, M. A., and Pilch, P. F. (1983). *Biochemistry* **22**, 717–721.
68. Kosmakos, F. C., and Roth, J. (1980). *JBC* **255**, 9860–9869.
69. Rosen, O. M., Chia, G. H., Fung, C., and Rubin, C. S. (1979). *J. Cell. Physiol.* **99**, 37–42.
70. Marshall, S. (1983). *Diabetes* **32**, 319–325.
71. Reed, B. C., and Lane, M. D. (1980). *PNAS* **77**, 285–289.
72. Krupp, M., and Lane, M. D. (1981). *JBC* **256**, 1689–1694.
73. Marshall, S. (1985). *JBC* **260**, 4136–4144.
74. Fehlmann, M., Carpentier, J.-L., Van Obberghen, E., Freychet, P., Thamm, P., Saunders, D., Brandenburg, D., and Orci, L. (1982). *PNAS* **79**, 5921–5925.
75. Ronnett, G. V., Knutson, V. P., and Lane, M. D. (1982). *JBC* **257**, 12257–12262.
76. Crettaz, M., and Kahn, C. R. (1984). *Diabetes* **33**, 477–485.
77. King, G. L., and Kahn, C. R. (1984). *In* "Growth and Maturation Factors" (G. Guroff, ed.), Vol. 2, pp. 223–265. Wiley & Sons Inc. New York.
78. Froesch, E. R., Schmid, C., Schwander, J., and Zapf, J. (1985). *Annu. Rev. Physiol.* **47**, 443–467.
79. Rinderknecht, E., and Humbel, R. E. (1978). *JBC* **253**, 2769–2776.
80. Rinderknecht, E., and Humbel, R. E. (1978). *FEBS Lett.* **89**, 283–286.
81. Rechler, M. M., and Nissley, S. P. (1985). *Annu. Rev. Biochem.* **47**, 425–442.
82. Rechler, M. M., Zpaf, J., Nissley, S. P., Froesch, E. R., and Moses, A. C. (1980). *Endocrinology (Baltimore)* **107**, 1451–1459.
83. Rechler, M. M., Nissley, S. P., Podskalny, J. M., Moses, A. C., and Fryklund, L. (1977). *J. Clin. Endocrinol. Metab.* **44**, 820–831.

84. Marshall, R. N., Underwood, L. E., Voina, S. J., Foushee, D. B., and Van Wyk, J. J. (1974). *J. Clin. Endocrinol. Metab.* **39**, 283–292.
85. Rosenfeld, R. G., and Hintz, R. L. (1980). *Endocrinology (Baltimore)* **107**, 1841–1848.
86. Polychronakos, C., Guyda, H. J., and Posner, B. I. (1983). *J. Clin. Endocrinol. Metab.* **57**, 436–438.
87. Megyesi, K., Kahn, C. R., Roth, J., Neville, D. M., Jr., Nissley, S. P., Humbel, R. E., and Froesch, E. R. (1975). *JBC* **250**, 8990–8996.
88. Nissley, S. P., and Rechler, M. M. (1978). *Natl. Cancer Inst. Monogr.* **48**, 167.
89. Kull, F. C., Jacobs, S., Su, Y.-F., Suoboda, M. E., Van Wyk, J. J., and Cuatrecasas, P. (1983). *JBC* **258**, 6561–6566.
90. Kasuga, M., Sasaki, N., Kahn, C. R., Nissley, S. P., and Rechler, M. M. (1983). *J. Clin. Invest.* **72**, 1459–1469.
91. Jacobs, S., Kull, F. C., and Cuatrecasas, P. (1983). *PNAS* **80**, 1128–1131.
92. Massague, J., and Czech, M. P. (1982). *JBC* **257**, 5038–5041.
93. Bhaumick, B., Bala, R. M., and Hollenberg, M. D. (1981). *PNAS* **78**, 4279–4283.
94. Chernausek, S. D., Jacobs, S., and Van Wyk, J. J. (1981). *Biochemistry* **20**, 7345–7350.
95. Kasuga, M., Van Obberghen, E., Nissley, S. P., and Rechler, M. M. (1981). *JBC* **256**, 5305–5308.
96. Jacobs, S., Kull, F. C., Earp, H. S., Svoboda, M., Van Wyk, J. J., and Cuatrecasas, P. (1983). *JBC* **258**, 9581–9584.
97. Zick, Y., Sasaki, N., Rees-Jones, R. W., Grunberger, G., Nissley, S. P., and Rechler, M. M. (1984). *BBRC* **119**, 6–13.
98. MacDonald, R. G., and Czech, M. P. (1985). *JBC* **260**, 11357–11365.
99. Haskell, J. F., Nissley, S. P., Rechler, M. M., Sasaki, N., Greenstein, L. A., Lee, L. (1984). *Int. Congr. Ser.—Exerpta Med.* **652**, 726 (Abstr. 931).
100. Hunter, T., and Cooper, J. A. (1985). *Annu. Rev. Biochem.* **54**, 897–930.
101. Cohen, S., Ushiro, H., Stosckeck, C., and Chinkers, M. (1982). *JBC* **257**, 1523–1531.
102. Buhrow, S. A., Cohen, S., and Staros, J. V. (1982). *JBC* **257**, 4019–4002.
103. Chinkers, M., and Brugge, J. S. (1984). *JBC* **259**, 11534–11542.
104. Ullrich, A., Coussens, L., Hayflick, J. S., Dull, T. J., Gray, A., Tam, A. W., Lee, J., Yarden, Y., Liberman, T. A., Schlessinger, J., Downward, J., *et al.* (1984). *Nature (London)* **309**, 418–425.
105. Glenn, K., Bowen-Pope, D. F., and Ross, R. (1982). *JBC* **257**, 5172–5176.
106. Heldin, C.-H., Ek, B., and Ronnstrand, L. (1983). *JBC* **258**, 10054–10061.
107. Ek, B., Westermark, B., Wasteson, A., and Heldin, C.-H. (1982). *Nature (London)* **295**, 419–420.
108. Daniel, T. O., Tremble, P. M., Frackelton, A. R., and Williams, L. T. (1985). *PNAS* **82**, 2684–2687.
109. Biswas, R., Basu, M., Sen-Majumdar, A., and Das, M. (1985). *Biochemistry* **24**, 3795–3802.
110. Downward, J., Yarden, Y., Mayes, E., Scrace, G., Totty, N., Stockwell, P., Ullrich, A., Schlessinger, J., and Waterfield, M. D. (1984). *Nature (London)* **307**, 521–527.
111. Decker, S. J. (1985). *JBC* **260**, 2003–2006.
112. Gilmore, T., DeClue, J. E., and Martin, G. S. (1985). *Cell* **40**, 609–618.
113. Rettenmier, C. W., Roussel, M. F., Quinn, C. O., Kitchingman, G. R., Look, A. T., and Sherr, C. J. (1985). *Cell* **40**, 971–981.
114. Sherr, C. J., Rettenmier, C. W., Sacca, R., Roussel, M. F., Look, A. T., and Stanley, E. R. (1985). *Cell* **41**, 665–676.
115. Liuneh, E., Glazer, L., Segal, D., Schlessinger, J., and Shilo, Ben-Zion. (1985). *Cell* **40**, 599–607.

116. Yamamoto, T., Davis, C. G., Brown, M. S., Schneider, W. J., Casey, M. L., Goldstein, J. L., and Russel, D. W. (1984). *Cell* **39**, 27–38.
117. Yamamoto, T., Nishida, T., Miyajima, N., Kawai, S., Ooi, T., Toyoshima, K. (1983). *Cell* **35**, 71–78.
118. Retternmier, C. W., Chen, J. H., Roussel, M. F., and Shorr, C. J. (1985). *Science* **228**, 320–322.
119. Hampe, A., Gobet, M., Sherr, C. J., and Galibert, F. (1984). *PNAS* **81**, 85–89.
120. Neckameyer, W. S., and Wang, L.-H. (1985). *J. Virol.* **53**, 879–884.
121. Czernilofsky, A. P., Levinson, A., Varmus, H., Bishop, J., Tischer, E., and Goodman, H. (1980). *Nature (London)* **287**, 198–203.
122. Czernilofsky, A. P., Levinson, A., Varmus, H., Bishop, J., Tischer, E., and Goodman, H. (1983). *Nature (London)* **301**, 736.
123. Reddy, E. P., Smith, M. J., and Srinivasan, A. (1983). *PNAS* **80**, 3623–3627.
124. Shibuya, M., and Hanafusa, H. (1982). *Cell* **30**, 787–795.
125. Kitamura, N., Kitamura, A., Toyoshima, K., Hirayama, Y., and Yoshida, M. (1982). *Nature (London)* **297**, 205–208.
126. Naharro, G., Robbins, K. C., and Reddy, E. P. (1984). *Science* **223**, 63–66.
127. Shoji, S., Parmellee, D., Wade, R. D., Kumar, S., Ericsson, L., Walsh, K. A., Neurath, H., Long, G., Demaille, J., Fischer, E. H., and Titami, K. (1981). *PNAS* **78**, 848–851.
128. Reimann, E. M., Titani, K., Ericsson, L. H., Wade, R. D., Fischer, E. H., and Walsh, K. A. (1984). *Biochemistry* **23**, 4185–4192.
129. Takio, K., Wade, R. D., Smith, S. B., Krebs, E. G., Walsh, K. A., and Titani, K. (1984). *Biochemistry* **23**, 4207–4218.
130. Van Beueren, C., Galleshaw, J. A., Jonas, V., Berns, A. J. M., Doolittle, R. F., Donaghue, D. J., and Verma, I. M. (1981). *Nature (London)* **289**, 258–262.
131. Kan, N. C., Flordellis, C. S., Mark, G. E., Duesberg, P. H., and Papas, T. S. (1984). *Science* **223**, 813–816.
132. Hampe, A., Laprevotte, I., Galibert, F., Fedele, L. A., and Sherr, C. J. (1982). *Cell* **30**, 775–785.
133. Zoller, M. J., Nelson, N. C., and Taylor, S. S. (1981). *JBC* **256**, 10837–10842.
134. Kamps, M. P., Taylor, S. S., and Sefton, B. M. (1984). *Nature (London)* **310**, 589–592.
135. Russo, M. W., Lukas, T. J., Cohen, S., and Staros, J. V. (1985). *JBC* **260**, 5205–5208.
136. Van Obberghen, E., and Kowalski, A. (1982). *FEBS Lett.* **143**, 179–182.
137. Gazzano, H., Kowalski, A., Fehlmann, M., and Van Obberghen, E. (1983). *Biochem. J.* **216**, 575–582.
138. Zick, Y., Whittaker, J., and Roth, J. (1983). *JBC* **258**, 3431–3434.
139. Blackshear, P. J., Nemenoff, R. A., and Avruch, J. (1984). *Endocrinology (Baltimore)* **114**, 141–152.
140. Burant, C. F., Treutelaar, M. K., Landreth, G. E., and Buse, M. G. (1984). *Diabetes* **33**, 704–708.
141. Haring, H. U., Kasuga, M., and Kahn, C. R. (1982). *BBRC* **108**, 1538–1545.
142. Tamura, S., Brown, T. A., Whipple, J. H., Fujita-Yamaguchi, Y., Dubler, R. E., Cheng, K., and Larner, J. (1984). *JBC* **259**, 6650–6658.
143. White, M. F., Haring, H. U., Kasuga, M., and Kahn, C. R. (1984). *JBC* **259**, 255–264.
144. Petruzzelli, L. M., Gangaly, S., Smith, C. R., Cobb, M. H., Rubin, C. S., and Rosen, O. M. (1982). *PNAS* **79**, 6792–6796.
145. Haring, H. U., White, M. F., Kahn, C. R., Kasuga, M., Lauris, V., Fleischmann, P., Murray, M., and Pawelek, J. (1984). *J. Cell Biol.* **99**, 900–908.
146. Grigorescu, F., White, M. F., and Kahn, C. R. (1983). *JBC* **258**, 13708–13716.
147. Grigorescu, F., Flier, J. S., and Kahn, C. R. (1984). *JBC* **259**, 15003–15006.

148. Avruch, J., Nemenoff, R. A., Blackshear, P. J., Pierce, M. W., and Osathanondh, R. (1982). *JBC* **257**, 15162–15166.
149. Kasuga, M., Fujita-Yamaguchi, Y., Blithe, D. L., White, M. F., and Kahn, C. R. (1983). *JBC* **258**, 10973–10980.
150. Kasuga, M., Fujita-Yamaguchi, Y., Blithe, D. L., and Kahn, C. R. (1983). *PNAS* **80**, 2137–2141.
151. Cohen, S. (1983). "Methods in Enzymology," Vol. 99, pp. 379–387.
152. Crettaz, M., and Kahn, C. R. (1983). *Endocrinology (Baltimore)* **113**, 1201–1209.
153. Karlsson, F. A., Grunfeld, C., Kahn, C. R., and Roth, J. (1979). *Endocrinology (Baltimore)* **104**, 1383–1392.
154. Hammerman, M. R., and Gavin, J. R., II (1984). *Am. J. Physiol.* **247**, F408–F417.
155. Rosen, O. M., Herrera, R., Olowe, Y., Petruzzelli, L. M., and Cobb, M. H. (1983). *PNAS* **80**, 3237–3240.
156. Kasuga, M., White, M. F., and Kahn, C. R. (1985). "Methods in Enzymology," Vol. 109, pp. 609–621.
157. Flockhart, D. A., and Corbin, J. D. (1982). *CRC Crit. Rev. Biochem.* **12**, 133–186.
158. Rangel-Aldao, R., and Rosen, O. M. (1976). *JBC* **251**, 7526–7529.
159. Erlichman, J., Rangel-Aldao, R., and Rosen, O. M. (1983). "Methods in Enzymology," Vol. 99, pp. 176–186.
160. White, M. F., Takayama, S., and Kahn, C. R. (1985). *JBC* **260**, 9470–9478.
161. Downward, J., Parker, P., and Waterfield, M. D. (1984). *Nature (London)* **311**, 483–485.
162. Smart, J. E., Opperman, H., Czernilofsky, A. P., Purchio, A. S., Erikson, R. L., and Bishop, J. M. (1981). *PNAS* **78**, 6013–6017.
163. Cooper, J. A., and Hunter, T. (1983). *Curr. Top. Microbiol. Immunol.* **107**, 125–159.
164. Zick, Y., Grunberger, G., Podskalny, J. M., Moncada, V., Taylor, S., Gorden, P., and Roth, J. (1983). *BBRC* **116**, 1129–1135.
165. Yu, K.-T., and Czech, M. (1984). *JBC* **259**, 5277–5286.
166. Zick, Y., Kasuga, M., Kahn, C. R., and Roth, J. (1983). *JBC* **258**, 75–80.
167. Kono, T., and Barham, F. W. (1971). *JBC* **246**, 6204–6209.
168. Tamura, S., Fujita-Yamaguchi, Y., and Larner, J. (1983). *JBC* **258**, 14749–14752.
169. Roth, R. A., Cassell, D. J., Maddux, B. A., and Goldfine, I. D. (1983). *BBRC* **115**, 245–252.
170. Zick, Y., Rees-Jones, R. W., Taylor, S. I., Gorden, P., and Roth, J. (1984). *JBC* **259**, 4396–4400.
171. Cuatrecasas, P., and Tell, G. P. E. (1973). *PNAS* **70**, 485–489.
172. Kahn, C. R., Baird, K. L., and Van Obberghen, E. (1981). *FEBS Lett.* **129**, 131–134.
173. Heyliger, C. E., Tahiliani, A. G., and McNeill, J. H. (1985). *Science* **227**, 1474–1477.
174. Simpson, I. A., and Hedo, J. A. (1984). *Science* **223**, 1301–1304.
175. Basu, M., Biswas, R., and Das, M. (1984). *Nature (London)* **311**, 477–480.
176. Nemenoff, R. A., Kwok, Y. C., Shulman, G. I., Blackshear, P. J., Osathanondh, R., and Avruch, J. (1984). *JBC* **259**, 5058–5065.
177. Shia, M. A., Rubin, J. B., and Pilch, P. F. (1983). *JBC* **258**, 14450–14455.
178. Weber, W., Bertics, P. J., and Gill, G. N. (1984). *JBC* **259**, 14631–14636.
179. Segel, I. H. (1975). "Enzyme Kinetics," pp. 227–272. Wiley, New York.
180. Kyte, J., and Doolittle, R. F. (1982). *JMB* **157**, 105–132.
181. King, Jr. L. E., Carpenter, G., and Cohen, S. (1980). *Biochemistry* **19**, 1524–1528.
182. Ek, B., and Heldin, C.-H. (1982). *JBC* **257**, 10486–10492.
183. Richert, N. D., Blithe, D. L., and Pastan, I. (1982). *JBC* **257**, 7143–7150.
184. Pang, D. T., and Shafer, J. A. (1985). *JBC* **260**, 5126–5130.
185. Swarup, G., Cohen, S., and Garbers, D. L. (1982). *BBRC* **107**, 1104–1109.
186. Brautigan, D. T., Bornstein, P., and Gallis, B. (1981). *JBC* **256**, 6519–6522.

187. Ziemiecki, A., Hennig, D., Gardner, L., Ferdinand, F.-J., Friis, R. R., Bauer, H., Pedersen, N. C., Johnson, L., and Theilen, G. H. (1984). *Virology* **138**, 324–331.
188. Cleland, W. W. (1967). *Annu. Rev. Biochem.* **36**, 77–112.
189. O'Sullivan, W. J., and Smithers, G. W. (1979). "Methods in Enzymology," Vol. 63, pp. 294–336.
190. Londos, C., and Preston, M. S. (1977). *JBC* **252**, 5951–5961.
191. Schramm, V. L., Fullin, F. A., and Zimmerman, D. M. D. (1981). *JBC* **256**, 10803–10808.
192. Ash, D. E., and Schramm, V. L. (1982). *JBC* **257**, 9261–9264.
193. Rangel-Aldao, R., and Rosen, O. M. (1976). *JBC* **251**, 3375–3380.
194. Rangel-Aldao, R., and Rosen, O. M. (1977). *JBC* **252**, 7140–7145.
195. Pang, D., Sharma, B., and Shafer, J. A. (1985). *ABB* **242**, 176–186.
196. Casnellie, J. E., Harrison, M. L., Pike, L. J., Hellström, K. E., and Krebs, E. G. (1982). *PNAS* **79**, 282–286.
197. Pike, L. J., Gallis, B., Casnellie, J. E., Bornstein, P., and Krebs, E. G. (1982). *PNAS* **79**, 1443–1447.
198. Pike, L. J., Bower-Pope, D. F., Ross, R., and Krebs, E. G. (1983). *JBC* **258**, 9383–9390.
199. Pike, L. J., Kuenzel, E. A., Casnellie, J. E., and Krebs, E. G. (1984). *JBC* **259**, 9913–9921.
200. Braun, S., Ghang, M. A., and Racker, E. (1983). *Anal. Biochem.* **135**, 369–378.
201. Stadtmauer, L. A., and Rosen, O. M. (1983). *JBC* **258**, 6682–6685.
202. Hunter, T. (1982). *JBC* **257**, 4843–4848.
203. Baldwin, G. S., Kresel, J., and Menckton, J. M. (1983). *Nature (London)* **301**, 435–437.
204. House, C., Baldwin, G. S., and Kemp, B. E. (1984). *EJB* **140**, 363–367.
205. Kemp, B., Graves, D. J., Benjamin, E., and Krebs, E. G. (1977). *JBC* **252**, 4888–4894.
206. Cooper, J. A., Esch, F. S., Taylor, S. S., and Hunter, T. (1984). *JBC* **259**, 7835–7841.
207. Wong, T. W., and Goldberg, A. R. (1983). *JBC* **258**, 1022–1025.
208. Erneux, C., Cohen, S., and Garbers, D. L. (1983). *JBC* **258**, 4137–4142.
209. Segel, I. H. (1975). "Enzyme Kinetics: Behavior and Analysis of Rapid Equilibrium and Steady-state Enzyme Systems," pp. 560–590. Wiley, New York.
210. Cassel, D., Pike, L. J., Grant, G. A., Krebs, E. G., and Glasser, L. (1983). **258**, 2945–2950.
211. Fromm, H. J., and Zewe, V. (1962). *JBC* **237**, 3027–3032.
212. Tabakabai, L., and Graves, D. J. (1978). *JBC* **253**, 2196–2202.
213. Chinkers, N., and Cohen, S. (1981). *Nature (London)* **290**, 516–519.
214. Kudlow, J. E., Buss, J. E., and Gill, G. N. (1981). *Nature (London)* **290**, 519–521.
215. Perrotti, N., Taylor, S. I., Richert, N. D., Rapp, U. R., Pastan, I., and Roth, J. (1985). *Science* **227**, 761–763.
216. Haring, H. U., White, M. F., Kahn, C. R., Ahmad, Z., DePaoli-Roach, A. A., and Roach, P. J. (1985). *J. Cell. Biochem.* **28**, 171–182.
217. Means, A. R. (1979). *Recent Prog. Horm. Res.* **37**, 333–367.
218. Plancke, Y. D., and Lazarides, E. (1983). *Mol. Cell. Biol.* **3**, 1412.
219. McDonald, J. M., Goewent, R. R., and Graves, C. B. (1984). *Diabetes* **33**, 10A.
220. Plehure, W. E., Williams, P. F., Caterson, I. D., Harrison, L. C., and Turtle, J. R. (1983). *BJ* **214**, 361–366.
221. Williams, P. F. (1984). *Diabetes* **33**, 10A.
222. White, M. F., and Sale, E. M. (1984). *Diabetes* **33**, 121
222a. Sale, E. M., White, M. F., and Kahn, C. R. *J. Cell. Biol.* (submitted for publication).
223. Cooper, J. A., Reiss, N. A., Schwartz, R. J., and Hunter, T. (1983). *Nature (London)* **302**, 219–223.
224. Sefton, B. M., and Hunter, T. (1981). *Cell* **24**, 165–174.
225. Ito, S., Richert, N. D., and Pastan, I. (1983). *JBC* **258**, 14626–14631.
226. Hynes, R. (1982). *Cell* **28**, 437–438.

227. Wang, E., and Goldberg, A. R. (1976). *PNAS* **73**, 4065–4069.
228. Kadawaki, T., Fujita-Yamaguchi, Y., Nishida, E., Takaku, F., Akiyama, T., Kathuria, S., Akanuma, Y., and Kasuga, M. (1985). *JBC* **260**, 4016–4020.
229. Jameson, L., Frey, T., Zeeberg, B., Dalldorf, F., and Caplow, M. (1980). *Biochemistry* **19**, 2472–2479.
230. Goldenring, J. R., Gonzalez, B., McGuire, J. S., Jr., and DeLorenzo, R. J. (1983). *JBC* **258**, 12632–12640.
231. Gallis, B., Edelman, A. M., Casnellie, J. E., and Krebs, E. G. (1983). *JBC* **258**, 13089–13093.
232. Sugimoto, Y., Whitman, M., Cantley, L. C., and Erikson, R. L. (1984). *PNAS* **81**, 2117–2121.
233. Macara, I. G., Marinetti, G. V., and Balduzzi, P. C. (1984). *PNAS* **81**, 2728–2732.
233a. Sale, G. J., Fujita-Yamaguchi, Y., and Kahn, C. R. (1986). *BJ* **155**, 345–351.
234. Kasuga, M., Zick, Y., Blithe, D. L., Karlsson, F. A., Haring, H. U., and Kahn, C. R. (1982). *JBC* **257**, 9891–9894.
235. Tuazon, P. T., Pang, D. T., Shafer, J. A., and Traugh, J. A. (1985). *J. Cell. Biochem.* **28**, 159–170.
236. White, M. F., Werth, D. K., Pastan, I., and Kahn, C. R. (1984). *J. Cell. Biochem.* **28**, 171–182.
237. Yu, K.-T., Werth, D. K., Pastan, I.. and Czech, M. P. (1985). *JBC* **260**, 5838–5846.
238. Cooper, J. A., and Hunter, T. (1981). *J. Cell Biol.* **91**, 878.
239. Takayama, S., White, M. F., Lauris, V., and Kahn, C. R. (1984). *PNAS* **81**, 7797–7801.
240. Cochet, C., Gill, G. N., Meisenhelder, J., Cooper, J. A., and Hunter, T. (1984). *JBC* **259**, 2553–2558.
241. Hunter, T., Ling, N., and Cooper, J. A. (1984). *Nature (London)* **311**, 480–483.
242. Davis, R., and Czech, M. P. (1984). *JBC* **259**, 8545–8549.
243. Swarup, G., Cohen, S., and Bargers, D. (1981). *JBC* **256**, 8197–8201.
244. Pang, D. T., Sharma, B., Shafer, J. A., White, M. F., and Kahn, C. R. (1985). *JBC* **260**, 7131–7136.
244a. Shafer, J. A., personal communication.
245. Cooper, J. A., Sefton, B. M., and Hunter, T. (1983). "Methods in Enzymology," Vol. 99, pp. 387–405.
246. Haring, H.-U., Kasuga, M., White, M. F., Crettaz, M., and Kahn, C. R. (1984). *Biochemistry* **23**, 3298–3306.
247. Ross, A. A., Baltimore, D., and Eisen, H. N. (1981). *Nature (London)* **294**, 654–656.
248. White, M. F., Maron, R., and Kahn, C. R. (1985). *Nature (London)* **318**, 183–186.
249. Frackelton, A. R., Ross, A. H., and Eisen, H. N. (1983). *Mol. Cell. Biol.* **3**, 1343–1352.
250. Comoglio, P. M., DiRenzo, M. F., Tarone, G., Giancotti, F. G., Naldini, L., and Marchisio, P. C. (1984). *EMBO J.* **3**, 483–489.
251. King, G. L., Rechler, M. M., and Kahn, C. R. (1982). *JBC* **257**, 10001–10006.
252. Oppenheimer, C. L., Pessin, J. E., Massague, J., Gitomer, W., and Czech, M. P. (1983). *JBC* **258**, 4824–4830.
253. Oka, Y., Mottola, C., Oppenheimer, C. L., and Czech, M. P. (1984). *PNAS* **81**, 4028–4032.
253a. Corvera, S., and Czech, M. P. (1985). *PNAS* **82**, 7314–7318.
254. Oleffsky, J. M., Saekow, M., and Kroc, R. L. (1982). *Ann. N.Y. Acad. Sci.* **380**, 220.
255. Pessin, J. E., Gitomer, W., Oka, Y., Oppenheimer, C. L., and Czech, M. P. (1983). *JBC* **258**, 7386–7394.
256. Homopoulus, P., Testa, U., Gourdin, M.-F., Hervy, C., Titeux, M., and Vainchenker, W. (1982). *EJB* **129**, 389–393.
257. Rozengurt, E., Brown, K. D., and Pettican, P. (1981). *JBC* **256**, 716–722.

258. Magnun, B. E., Matrisian, L. M., and Bowden, G. T. (1980). *JBC* **255**, 6373–6381.
259. Castagna, M., Takai, Y., Kaibuchi, K., Sano, K., Kikkawa, U., and Nishizuka, Y. (1982). *JBC* **257**, 7847–7851.
260. Niedel, J. E., Kuhn, L. J., and Niedel, J. E. (1983). *PNAS* **80**, 36–40.
261. Vandenbark, G. R., Kuhn, L. J., and Niedel, J. E. (1984). *J. Clin. Invest.* **73**, 448–457.
262. Jetten, A. M., Ganong, B. R., Vandenbark, G. R., Shirley, J. E., and Bell, R. M. (1985). *PNAS* **82**, 1941–1945.
263. Fearn, J. C., and King, A. C. (1985). *Cell* **40**, 991–1000.
264. Davis, R. J., Ganong, B. R., Bell, R. M., and Czech, M. P. (1985). *JBC* **260**, 5315–5322.
265. Iwashita, S., and Fox, C. F. (1984). *JBC* **259**, 2559–2567.
266. Hunter, T. (1984). *Nature (London)* **311**, 414–416.
267. Bishop, J. M. (1985). *Cell* **42**, 23–38.
268. Erikson, R. L., Collett, M. S., Erickson, E., and Purchio, A. F. (1979). *PNAS* **76**, 6260–6264.
269. Levinson, A. D., Oppermann, H., Varmus, H. E., and Bishop, J. M. (1980). *JBC* **255**, 11973–11980.
270. Collet, M. S., Erikson, R. L. (1978). *PNAS* **75**, 2021–2024.
271. Sefton, B. M., Hunter, T., and Beeman, K. (1980). *J. Virol.* **33**, 220–229.
272. Snyder, M. A., Bishop, J. M., Colby, W. W., and Levinson, A. D. (1983). *Cell* **32**, 891–901.
273. Weinmaster, G., Zoller, M. J., Smith, M., Hinze, E., and Pawson, T. (1984). *Cell* **37**, 559–568.
274. Snyder, M. A., and Bishop, J. M. (1984). *Virology* **136**, 375–386.
275. Pawelek, J., Murray, M., and Fleischmann. (1982). *Cold Spring Harbor Conf. Cell Proliferation* **9**, 911–920.
276. Kahn, C. R., Murray, M., and Pawlek, J. (1980). *J. Cell. Physiol.* **103**, 109–119.
277. Gammeltoft, S. (1984). *Physiol. Rev.* **64**, 1321–1378.
278. Graziani, Y., Erikson, E., and Erikson, R. L. (1983). *JBC* **258**, 2126–2129.
279. Rubsamen, H., Saltenberger, K., Frus, R. R., and Eigenbrodt, E. (1982). *PNAS* **79**, 228–232.
280. Sefton, B. M., Hunter, T., Ball, E. H., and Singer, S. J. (1978). *Cell* **24**, 165–174.
281. Baldwin, G. S., Greco, B., Hearn, M. T. W., Knesel, J. A., Morgan, F. J., and Simpson, R. J. (1983). *PNAS* **80**, 5276–5280.
282. Ghosh-Dastidar, P., Coty, W. A., Griest, R. E., Woo, D. D. L., and Fox, C. F. (1981). *PNAS* **81**, 1654–1658.
283. Braun, S., Raymond, W. E., and Racker, E. (1984). *JBC* **259**, 2051–2054.
284. Rees-Jones, R. W., Hendricks, S. A., Quarum, M., and Roth, J. (1984). *JBC* **259**, 3470–3474.
285. Rees-Jones, R. W., and Taylor, S. I. (1985). *JBC* **260**, 4461–4467.

8

Phosphoprotein Phosphatases

LISA M. BALLOU • EDMOND H. FISCHER

Department of Biochemistry
University of Washington
Seattle, Washington 98195

311

THE ENZYMES, Vol. XVII

I. Introduction and Historical Overview: The "PR Enzyme"

Aside from a few reports in the late 1940s on enzyme activities catalyzing the dephosphorylation of phosphoproteins such as casein and phosvitin (*1, 2*), the real importance of protein phosphatases in the control of metabolic processes became evident only when their participation in the regulation of the interconvertible enzyme phosphorylase was demonstrated. Indeed, the history of protein phosphatases is intimately linked to that of phosphorylase, the first enzyme shown by Parnas and Ostern in Poland (*3*) and Carl and Gertie Cori in the United States (*4, 5*) to be involved in the metabolism of glycogen.

When first discovered, phosphorylase was found to have an absolute requirement for adenylic acid (*6*). In 1939, there was a brief report that a form of the enzyme from yeast did not require AMP for activity (*7*). This was discounted until, in 1943, phosphorylase was crystallized in the Cori's laboratory in a form that did not require AMP (*8–10*). The active enzyme was called phosphorylase *a;* it was assumed that this represented the native molecule, since in crude muscle extracts it was rapidly converted to the form that required AMP, which was termed phosphorylase *b.* The logical hypothesis was advanced that in native phosphorylase, AMP was covalently bound to the protein as a prosthetic group. Conversion of phosphorylase *a* to *b* would then be catalyzed by a "prosthetic group-removing enzyme" or "PR enzyme" that would release the cofactor (*9*). But this hypothesis had to be abandoned when no AMP was found in the supernatant or dialysate of the reaction; furthermore, no AMP, adenosine, or ribose could be detected in hydrolysates of phosphorylase *a* using the most sensitive microbiological procedures available at that time (*11*). This unexpected result indicated that "PR enzyme" could not be a prosthetic group-removing enzyme responsible for the cleavage of adenylic acid. In 1943, Cori and Green discovered that spleen extracts with PR enzyme activity also displayed proteolytic activity (*9*). This suggested that PR enzyme might be a protease, an assumption that seemed to be confirmed when it was found that trypsin could also convert phosphorylase a into an AMP-requiring species. The possibility that PR enzyme might be a protease seemed further supported when it was shown that conversion of phosphorylase a to b was accompanied by a near halving of the molecule, from M_r ~500,000 to 250,000 (12). Although it was understood that the term "prosthetic group-removing enzyme" was a misnomer, it was proposed to retain the abbreviation for this enzyme (now well entrenched in the biochemical literature) under the new designation of "phosphorylase-rupturing enzyme."

In the mid-1950s, it was shown both in muscle (13, 14) and in liver (15, 16) that the conversion of phosphorylase b to a consisted of a phosphorylation of the molecule catalyzed by a specific phosphorylase kinase. A particular serine residue in each subunit was phosphorylated; therefore, the reverse reaction, namely, the conversion of phosphorylase a to b, had to involve an enzymic dephosphorylation of the protein. For the first time, then, it was clear that the PR enzyme was in reality a phosphorylase phosphatase. It is interesting to note that as early as 1943, Cori and Green (9) showed the presence of phosphate in phosphorylase a; in 1956, Wosilait and Sutherland reported that phosphate was liberated by PR enzyme in an amount that increased with increasing conversion of phosphorylase a to b (17).

When protein phosphorylation–dephosphorylation became recognized as a widespread mechanism of intracellular regulation, the question arose as to whether each protein was phosphorylated by a specific kinase and dephosphorylated by a corresponding phosphatase, or whether the interconverting enzymes reacted with many different protein substrates. Earlier studies led to rather confusing conclusions because none of the phosphatase preparations used was homogeneous. Since then, several of the enzymes have been obtained in essentially pure form; while some of these are rather similar and show overlapping substrate specificities, others are clearly distinct and dephosphorylate a narrow range of substrates. This chapter deals with the cytoplasmic phosphoseryl-, phosphothreonyl-, as well as the phosphotyrosyl-protein phosphatases. The mitochondrial pyruvate dehydrogenase phosphatase is discussed in Volume XVIII, Chapter 3. There have been several comprehensive reviews of the protein phosphatases (18–21).

II. Classification of Protein Phosphatases

Whereas a considerable amount of information has been gained on the structure, mechanism of action, subunit composition, functional sites, and regulation of a variety of protein kinases, until the last decade very little was known about the protein phosphatases. Therefore, it was difficult to devise a system of classification for these enzymes. In general, enzymes are distinguished from one another and categorized according to their (*a*) substrate specificity; (*b*) dependence on activator molecules or metal ions, etc.; (*c*) inhibitors; and (*d*) molecular (e.g., size) and immunological characteristics, cellular distribution, etc. These criteria could not be readily applied to the protein phosphatases. First, to measure their activity, one cannot simply add [^{32}P]ATP and look at counts incorporated as is done with the kinases. One must first isolate the protein substrates, phosphorylate them with the appropriate kinases, and then follow the release of counts. Some of the substrates are difficult to purify; others have multiple sites of phosphorylation that are not equally susceptible to a given enzyme. Furthermore, even when a good substrate is available, it might be dephosphorylated by a number of different phosphatases. Enzyme preparations capable of dephosphorylating several of the enzymes involved in glycogen synthesis and breakdown have led to the concept of a multifunctional protein phosphatase regulating glycogen metabolism (*19*). The use of phosphopeptide substrates is fraught with the same disadvantages as observed with protein kinases: Both types of enzymes appear to be far more promiscuous when acting on small peptides rather than on intact proteins.

Second, most kinases have been classified on the basis of their dependency on modulator compounds such as cyclic nucleotides, Ca^{2+} and calmodulin (CaM), diacylglycerol, hemin, or double-stranded RNA. Except for a CaM-dependent protein phosphatase, most of these enzymes appear to act independently of such activator molecules, though the activity of some is modulated by heat-stable protein inhibitors. Others are dependent on Mn^{2+}, Mg^{2+}, or Ni^{2+} under certain circumstances, but this requirement varies with the substrate used, the pH of the reaction, or the form of the enzyme present. For instance, while type 1 phosphorylase phosphatase shows no Mg^{2+} dependence when acting on protein substrates, the metal ion is absolutely required for the dephosphorylation of *p*-nitrophenylphosphate (*p*NPP).

Third, while the interaction of protein phosphatases with heat-stable inhibitors has been exploited for classification purposes (*22*), this approach has its own pitfalls. Both inhibitor-1 and -2 are modulated by phosphorylation–dephosphorylation (see Section III); therefore, erroneous conclusions may be drawn if one works with a system containing contaminating kinases or phosphatases. To further complicate the matter, association of the enzyme with other components (e.g., the deinhibitor protein) can render it insensitive to the inhibitors (*23–25*). Finally, the state of aggregation and enzymic properties of phosphatases can vary during their isolation. Such behavior is particularly noticeable in the case of the major phosphorylase phosphatases in skeletal muscle and liver, which appear to exist in tissue extracts in the form of high-molecular-weight complexes which can be disrupted in the course of purification. Data on the immunological properties of phosphatases are now starting to appear and may soon be relied upon for the purpose of enzyme classification. No information on the genetic structure of the proteins is yet available.

A breakthrough in the area of phosphatase categorization occurred when Lee *et al.* found that treatment of the high-molecular-weight enzyme with 80% ethanol converted it to a low-molecular-weight form of M_r ~35,000 (*26*). This, along with metal ion sensitivity, formed the basis of some early attempts to subdivide the enzymes (*27–29*). However, it soon became apparent that the M_r ~35,000 phosphatase preparations contained two distinct enzymes that could be distinguished by their susceptibilities to the heat-stable inhibitors as well as their relative activities toward the α- and β-subunits of phosphorylase kinase (*30*). The parent high-molecular-weight enzymes could also be differentiated in this way (*31, 32*).

These properties have provided the grounds for a new classification system for the phosphoseryl- and phosphothreonyl-protein phosphatases as proposed by Cohen and Ingebritsen (*18, 22*), in which the type 1 enzymes are those that are susceptible to inhibitor-1 and -2 and preferentially dephosphorylate the β-subunit of phosphorylase kinase, whereas the type 2 enzymes are not inhibited and display higher activity toward the α-subunit (Table I). Types 1 and 2A seem to be the most closely related; they have broad substrate specificities and attack essentially the same phosphate groups but at different rates. Phosphatase-1 has a specific activity toward phosphorylase *a* approximately 10-fold higher than that of type 2A. Phosphatase-2B (calcineurin) is clearly different because of its stimulation by Ca^{2+}–CaM. The type 2C enyzme shows an absolute dependence on divalent metal ions such as Mg^{2+}; it acts poorly on phosphorylase *a* but is very active toward hydroxymethylglutaryl (HMG)-CoA reductase and HMG-CoA reductase kinase. Finally, not considered in Cohen's scheme are the phosphotyrosyl-protein phosphatases, for which a limited amount of information is available.

As a test of this classification system, Cohen *et al.* have fractionated tissue extracts by ion exchange and gel filtration chromatographies and analyzed the fractions using close to twenty phosphoserine- and phosphothreonine-containing protein substrates (*33–38*). Making allowances for the potential dangers inherent in this kind of study, they concluded that all of the phosphatase activities present could be ascribed to the type 1 and 2 enzymes previously identified. We have followed the classification scheme given in Table I in discussing the individual phosphatases.

III. Phosphatase Type 1 (Phosphorylase Phosphatase)

A. Introduction

Because of the great variety of forms in which type 1 phosphorylase phosphatase can exist and the multiplicity of factors that can regulate its activity, we have defined in Table II the main species involved together with some of their characteristics.

TABLE I

CLASSIFICATION OF PROTEIN PHOSPHATASES

Type	Protein phosphatases — Some common names	Inhibition by inhibitor-1 and -2	Activators	Phosphorylase kinase α:β activity ratio	Specificity	Catalytic subunit (M_r)	Some preferred natural protein substrates
1	Phosphorylase phosphatase; MgATP-dependent phosphatase; F_C	Yes	F_A-MgATP Mn^{2+}	0.01–0.05	Broad	38,000	Phosphorylase a Phosphorylase kinase β Glycogen synthase Histones H1 and H2B HMG-CoA reductase kinase
2A	—	No	Basic proteins and polyamines Mn^{2+}	~ 5	Broad	36,000	Myosin light chain Protamine Phosphorylase a Histones H1 and H2B Inhibitor-1 and -2
2B	Calcineurin	No	Ca^{2+}–CaM Mn^{2+} Ni^{2+}	~ 100	Narrow	61,000	Inhibitor-1 Phosphorylase kinase α Myosin light chain Protein K.-F. DARPP-32
2C	Mg^{2+}-dependent phosphatase	No	Mg^{2+} Mn^{2+}	~ 10–20	Broad	46,000	Myosin light chain Histone H2B HMG-CoA reductase HMG-CoA reductase kinase
3	Phosphotyrosyl-protein phosphatase	No	?	—	?	?	Glycogen synthase EGF receptor ?

TABLE II

GLOSSARY FOR THE TYPE 1 PHOSPHORYLASE PHOSPHATASE SYSTEM

Names of enzymes, activators, or inhibitors	Size (M_r)	Properties
High-molecular-weight complex	$\sim 250,000^a$	High-molecular-weight form of undefined composition present in tissue extracts
I_G; Glycogen-bound phosphatase	$260,000^a$ $137,000^b$	Spontaneously active enzyme obtained from glycogen particles. A 1:1 complex between the catalytic subunit and the G component
MgATP-dependent phosphatase; $F_C \cdot M$; inactive $M_r = 70,000$ complex	$70,000^{a,b}$ (or $140,000^a$)	A 1:1 complex between the catalytic subunit and a regulatory subunit (inhibitor-2). Isolated in an inactive form. (May also contain an additional $M_r = 62,000$ subunit.)
Catalytic subunit; F_C	$38,000^c$	Can exist in an inactive and various active conformations
E_i	$38,000^c$	Inactive form of the catalytic subunit
E_a^{Mn}	$38,000^c$	Catalytic subunit activated by Mn^{2+}
$E_a^{F_A}$	$38,000^c$	Active catalytic subunit obtained following F_A-MgATP activation of the $M_r = 70,000$ complex
E_i^{Tr} and E_a^{Tr-Mn}	$33,000^c$	Nicked inactive or Mn^{2+}-activated catalytic subunits obtained after partial trypsinolysis
F_A; Glycogen synthase kinase-3	$47,000^b$ $51,000^c$	Activating factor for the $M_r = 70,000$ complex. Phosphorylates inhibitor-2
Inhibitor-1	$60,000^a$ $26,000^c$ $18,640^d$	Heat-stable protein that is inhibitory after phosphorylation by cAMP-dependent protein kinase. Dephosphorylated when bound to the catalytic subunit and by phosphatases 2A and 2B
Inhibitor-2; regulatory subunit; modulator protein (M)	$42,000^a$ $25,500^{b,e}$ $31,000^c$	Heat-stable protein inhibitory in its dephospho form. Phosphorylated by F_A and autodephosphorylated when bound to the catalytic subunit. Also dephosphorylated by phosphatases 2A and 2B
G component	$103,000^c$	A subunit associated with the glycogen-bound phosphatase that interferes with inhibition by inhibitor-1 and -2
Deinhibitor protein	$17,500^a$ $8,900^b$ $8,300^c$	A polypeptide isolated from liver glycogen particles that interferes with inhibition by inhibitor-1 and -2. Could be a proteolytic fragment of the G component

[a] By gel filtration.
[b] By ultracentrifugation analysis.
[c] By sodium dodecyl sulfate (SDS)-gel electrophoresis.
[d] By sequence analysis.
[e] By amino acid analysis.

B. PURIFICATION AND SUBUNIT STRUCTURE

Type 1 phosphatase can probably be considered as the classical phosphorylase phosphatase of skeletal muscle. In this tissue it is the major enzyme affecting the activity of not only phosphorylase, but also glycogen synthase and phosphorylase kinase (*36, 39*). The concentration in skeletal muscle is ~0.5 μ*M* (skeletal muscle > liver > brain > heart > adipose tissue), with activity levels of 25 nmol P_i/min/g of tissue using phosphorylase *a* as a substrate (*36, 40*). This value does not represent the total amount of enzyme present, but only that which is already spontaneously active: A significant portion exists as an inactive complex that can be detected only after incubation with MgATP and the kinase F_A or with trypsin in the presence of Mn^{2+}. The complex is difficult to detect in crude extracts because of interference from metabolites such as AMP, the heat-stable inhibitors, and other phosphatases that can be activated by divalent cations or inhibited by ATP (*41, 42*). In addition to these cytosolic forms, approximately 50–60% of the muscle enzyme and 20–30% of that from the liver is associated with the protein–glycogen complex, where type 1 is virtually the only phosphatase species present (*34, 39, 43*).

Three forms of the enzyme have been purified to a state of homogeneity: the free catalytic subunit, the inactive MgATP-dependent complex, and the glycogen-bound species. Most of these preparations have been obtained from rabbit skeletal muscle following separation by ion exchange chromatography, gel filtration, and affinity chromatography on Sepharose linked to ligands such as polylysine (*42–45*), aminohexane (*25, 46*), histone (*47*), and glucosamine 6-phosphate (*24*). Blue Sepharose has also been useful, especially for the separation of the inactive complex from its free subunits (*42, 44, 46, 48*). It is important to include a cocktail of protease inhibitors at least during the early stages of purification to prevent digestion by endogenous proteases. Denaturing treatments such as precipitation with 80% ethanol allow for the recovery of a fully active type 1 catalytic subunit; this species has been obtained in amounts of 30–50 μg per 100 g of skeletal muscle (*40, 44, 49*) and has also been purified from the liver (*30, 47, 50*). The protein varies in size from M_r = 32,000–38,000, with the lower-molecular-weight species apparently resulting from proteolysis. It has been proposed on the basis of Western blotting that the native enzyme could exist as an even larger polypeptide of M_r = 70,000 (*51*). There is no evidence for the presence of free catalytic subunit in tissue extracts since all of the phosphorylase phosphatase activity elutes during gel filtration at a volume corresponding to M_r ≈ 250,000 (*26, 51, 52*). The exact nature of the enzyme(s) in this fraction is not known; it could comprise a mixture of phosphatases $2A_0$ and $2A_1$, activated MgATP-dependent enzyme, and some of the spontaneously active phosphatase that has been released from glycogen particles into the cytosol.

The presence of a MgATP-stimulated phosphorylase phosphatase was first demonstrated by Merlevede and Riley in bovine adrenal cortex (53) and then further described in a variety of avian and mammalian tissues (41, 42, 54–57) as well as in *Neurospora crassa* (58). The system could be resolved by DEAE chromatography into two components: F_C, a totally inactive phosphatase; and F_A, a factor that activated the enzyme in the presence of MgATP (59, 60). A heat-stable inhibitor later recognized as inhibitor-2 copurified with F_C and was also shown to be essential for activation (42, 45, 46, 61, 62). F_A has been purified to homogeneity; all evidence suggests that it is identical to glycogen synthase kinase-3 (56, 63, 64). It activates the MgATP-dependent enzyme through phosphorylation of the regulatory subunit (44, 45, 62, 65, 66).

A form of the inactive phosphatase has been isolated in a state of homogeneity from rabbit skeletal muscle (45, 46) with yields of 60–145 µg per 100 g of tissue. It has also been partially purified from bovine cardiac muscle (67). Procedures avoiding organic solvents (46) or including precipitation with 50% acetone (45) have yielded enzyme preparations displaying an M_r of ~70,000 by gel filtration and consisting of a 1 : 1 complex between the $M_r = 38,000$ catalytic subunit and $M_r = 31,000$ inhibitor-2. The latter protein is very susceptible to proteolysis and in some cases may be completely lost (45, 61, 68). Peptide maps of the type 1 catalytic subunit as isolated after ethanol treatment and the $M_r = 38,000$ component from the complex are identical (46); indeed, reconstitution of the free subunit with inhibitor-2 leads to the formation of a MgATP-dependent complex with properties virtually identical to those of the native enzyme (46, 65). However, there is one important difference between the two proteins: While the enzyme isolated after ethanol precipitation is fully active, the catalytic subunit present in the purified $M_r = 70,000$ complex is in an inactive conformational state, as discussed in Section III,E.

Early preparations of F_C contained material displaying an M_r of 140,000 by gel filtration and 70,000 by sucrose density centrifugation; SDS-gels showed a major protein of $M_r = 70,000$ thought to represent the phosphatase and some diffuse bands in the $M_r = 30,000$–40,000 range (42, 69). It now appears that the $M_r = 70,000$ protein may have been a contaminant, while the faint $M_r = 30,000$–40,000 bands were partially degraded forms of the catalytic and inhibitor moieties. Subsequently, a highly purified preparation of F_C was reported to contain, in addition to the catalytic and regulatory subunits, a third component of $M_r = 62,000$ believed to be a part of the complex but whose function has not been investigated (66). This MgATP-dependent enzyme migrated in sucrose gradients with an apparent M_r of 70,000, suggesting that the $M_r = 62,000$ component dissociates under these conditions.

The spontaneously active glycogen-bound phosphatase coprecipitates with glycogen at pH 6.1 and is then released into the supernatant by α-amylase digestion (24, 25, 43, 70). The enzyme purified from rabbit skeletal muscle (60 µg per 100 g of tissue) has been designated as 1_G and is said to be a 1 : 1 complex between the type 1 catalytic subunit and an $M_r = 103,000$ G component (see Table II) (25); inhibitor-2 is usually not detected in the glycogen fraction. The free catalytic moiety has also been prepared from liver (48) and skeletal muscle (43) glycogen particles; again, peptide maps of this protein are virtually identical to those of the soluble enzymes (25). The G component, which mediates binding to glycogen, is extremely sensitive to proteolysis and stains poorly with Coomassie Blue and silver. Inhibitor-2 and 1_G combine to form an inactive F_A–MgATP-dependent complex. However, in contrast to the usual instantaneous inhibition seen with the other type 1 enzymes, inhibition of 1_G by both inhibitor-1 and -2 is time- and temperature-dependent, with a $t_{1/2}$ of several minutes. Sensitivity to inhibitor-1 is restored by phosphorylation of the G subunit by cAMP-dependent protein kinase to 0.9 mol P_i/mol of protein; in this sense, the G component acts like the deinhibitor protein and in fact could be its precursor (25). Limited trypsinolysis produces the same effect, presumably by degrading the large subunit (24, 70).

C. Substrate Specificity

The type 1 phosphatases show a broad substrate specificity but are particularly active toward phosphorylase a, as shown in Table III (71–74). Specific activities

TABLE III

Substrate Specificity of Various Phosphatase Type 1 Species

Substrate	Enzyme species	Specific activity (nmol P_i/min/mg)	K_m (μM)	Reference
Phosphorylase a	Catalytic subunit	15,000–25,000	5–10	(38, 40, 43, 48, 49, 71)
	F_C activated by F_A-MgATP	5,700–20,000	10	(42, 46, 69, 72)
	F_C activated by trypsin-Mn^{2+}	7800	—	(45)
	1_G	5700	5	(24, 25)
Phosphorylase kinase β-subunit	Catalytic subunit	~3400	—	(49, 73)
	F_C activated by F_A-MgATP	200	—	(42)
Glycogen synthase	Catalytic subunit	1100	—	(49)
	1_G	—	6	(24)
Inhibitor-1	Catalytic subunit	—	0.19	(74)
Inhibitor-2	Catalytic subunit	—	0.04	(38)
Lysine-rich histone	Catalytic subunit	490	—	(49)
pNPP	Catalytic subunit	3740	9000	(49)
	F_C activated by trypsin-Mn^{2+}	2600	3000	(45)

have been reported to be as high as 50,000–60,000 nmol P_i/min/mg after partial proteolysis of the free catalytic subunit (40, 44, 65). Values for the MgATP-dependent and glycogen-bound complexes are somewhat lower due to their greater molecular masses but correspond to about 15,000 nmol P_i/min/mg for the isolated catalytic component. The K_m for phosphorylase obtained with the liver MgATP-dependent phosphatase is 18 μM, approximately twice that of the muscle enzyme (72). All of these phosphatases dephosphorylate the β-subunit of phosphorylase kinase 20–100 times faster than the α (24, 25, 46, 49). Some investigators have measured greater relative activities toward the kinase than toward phosphorylase a (33, 38), while others have shown the opposite (24, 32, 72). The enzyme also dephosphorylates a number of other proteins not listed in Table III, including histones H1 and H2B, myosin light chain, HMG-CoA reductase kinase, and troponin I. Type 1 phosphatases require Mn^{2+} in order to attack inhibitor-1 (32, 74) or the nonprotein substrate p-nitrophenylphosphate (pNPP) (24, 45, 49); in the latter case, Mg^{2+} is also effective ($Mg^{2+} > Mn^{2+} > Ca^{2+}$) and pH optima of both 7.5 (as is observed with protein substrates) and 8.5 have been reported. There is good evidence that the pNPP phosphatase (pNPPase) activity is actually due to the type 1 phosphatase rather than a contaminating enzyme; it is generated simultaneously with the phosphorylase phosphatase activity during trypsin-Mn^{2+} activation of the inactive $M_r = 70,000$ complex (45).

D. INHIBITOR-1, INHIBITOR-2, AND THE DEINHIBITOR PROTEIN

Among the known protein phosphatases, the type 1 enzymes are unique in their sensitivity to two small inhibitory polypeptides termed inhibitor-1 and inhibitor-2. A heat-stable, trypsin-labile inhibitor of phosphorylase phosphatase was first detected by Lee et al. in rabbit liver (75). Huang and Glinsmann later separated two inhibitors from rabbit skeletal muscle; the first, inhibitor-1, was active only after phosphorylation by cAMP-dependent protein kinase (76, 77) whereas the second, inhibitor-2, was active in its dephospho form and copurified with an inactive phosphatase (77, 78). Inhibition by both proteins is usually immediate, although a slow reaction can be observed in concentrated tissue extracts and with the active complex purified from glycogen particles (25, 36). This is apparently due to the presence of the deinhibitor protein or the G component. Some type 1 phosphatase preparations have been reported to lose their sensitivity to the inhibitors after prolonged storage of the enzymes, but the reason for this phenomenon is unknown (74).

1. Inhibitor-1

Inhibitor-1 has been obtained in a state of homogeneity from rabbit skeletal muscle (~150 μg per 100 g of tissue), where its concentration is estimated to be 1.5–1.8 μM, or 3–4 times higher than that of the type 1 phosphatase (36, 79).

Rabbit reticulocytes, by contrast, contain the enzyme but no detectable inhibitor-1 (*80*). The complete amino acid sequence shows that the $M_r = 18,640$ polypeptide chain consists of 165 residues with an acetylated N-terminal methionine (*81*). The behavior of the inhibitor during gel filtration ($M_r = 60,000$) and SDS-gel electrophoresis ($M_r = 26,000$) indicates that it possesses an asymmetric structure (*77, 79*). Inhibitory activity appears after phosphorylation of Thr-35 by cAMP-dependent protein kinase (*81*); the cGMP-dependent enzyme probably phosphorylates the same residue (*82*). The isolated inhibitor contains 0.5–0.7 mol P_i/mol of protein on Ser-67 even in preparations where Thr-35 is completely dephosphorylated, but this modification does not appear to influence activity (*81*). The phosphorylation state of inhibitor-1 increases in the presence of isoproterenol or adrenalin and decreases in response to insulin, suggesting an important role in the hormonal control of cellular processes in which phosphatase type 1 is implicated (*83–88*).

Inhibition of the free catalytic subunit is noncompetitive with respect to phosphorylase *a* when examined at high phosphatase concentrations but mixed ($K_i = 1.5–7.5$ nM) at low concentrations; this difference is due to the fact that the K_i is similar to the concentration of phosphatase used in the assays so analysis by an unmodified Michaelis-Menten equation might be invalid (*32, 74, 77*). Proteolytic fragments of the protein are also effective inhibitors; the smallest one found having full activity consists of residues 9–54 (*89*). The F_A-activated complex, although it already contains inhibitor-2 as its regulatory subunit, is also sensitive to inhibitor-1 with half-maximal effects at 1 nM (*42, 46, 72, 90*). Inhibition can be abolished by limited proteolysis with trypsin, by dilution (*32*), or by removal of the phosphate from Thr-35. Dephosphorylation by protein phosphatase-1 requires Mn^{2+} and is inhibited competitively by inhibitor-2 ($K_i = 8$ nM); however, the K_m of inhibitor-1 as a substrate (190 nM) is much higher than its K_i as an inhibitor (*74*). Phosphatase-2A and -2B are also very effective against the protein (*36*).

2. *Inhibitor-2*

Unlike inhibitor-1, inhibitor-2 is active in its dephospho form. It has been purified to homogeneity from rabbit skeletal muscle with yields of 50–100 μg per 100 g of tissue (*38, 65, 91, 92*) and its concentration is estimated to be about the same as that of the type 1 phosphatase (*38*). It is not known whether any free inhibitor exists *in vivo* or whether it is entirely bound within the $M_r = 70,000$ complex (*93*). Like inhibitor-1, it has little ordered structure and displays high molecular sizes by gel filtration and SDS-gel electrophoresis (see Table II). The amino acid composition, which is different from that of inhibitor-1, gives a size of $M_r = 25,500$ (*46, 77, 78, 92*). Immunoblotting of crude tissue extracts using a polyclonal antibody against this protein has revealed an $M_r = 60,000$ polypeptide that may be related to inhibitor-2, perhaps as an unprocessed precursor (*94*).

Inhibitor-2 displays mixed inhibition at high phosphatase concentrations and competitive inhibition with respect to phosphorylase a at low concentrations, with a K_i of 3.1 nM (74). The catalytic subunit may possess two sites for binding inhibitor-2: The first (K_d ~0.1 nM) positions the protein in the correct orientation to produce the species that can be activated by F_A. At higher concentrations (K_d ~5 nM) the protein binds to the second site and produces an inhibited form that cannot be activated by the kinase (38, 44, 62, 66). In fact, the F_A-activated MgATP-dependent complex can be inhibited by additional inhibitor-2, probably because of binding to the latter site (42, 90). Inhibitory activity is lost upon phosphorylation of a specific threonyl residue by F_A. The sequence surrounding this site (Pro-Ser-Thr(P)-Pro-Tyr) is similar to the one found in glycogen synthase site 3abc, also phosphorylated by F_A, except that the synthase is phosphorylated on seryl residues (95). Even after prolonged incubation of the isolated inhibitor with F_A, only $\leq$ 0.5 mol P_i/mol of protein is incorporated; the reason for this low level of phosphorylation is not understood (38, 50, 95). The protein is dephosphorylated by phosphatase-2A and -2B as well as by the type 1 enzyme (K_m = 40 nM); the latter reaction is not blocked by inhibitor-1 or -2 (38).

Casein kinase II introduces at least 3 mol P_i/mol of inhibitor-2 on seryl residues without activating the phosphatase complex, and apparently this allows an enhanced phosphorylation and activation by F_A (50). cAMP-dependent protein kinase also phosphorylates the protein but without affecting its activity (50, 62).

3. Deinhibitor Protein

The deinhibitor protein is a trypsin-labile, heat-stable polypeptide first detected in liver glycogen particles as an activity that protected the type 1 phosphatases from inhibition by inhibitor-1 and -2 (23, 96). It has been obtained in homogeneous form from canine liver (14 μg per 100 g of tissue) and displays an M_r of 8300 on SDS-gels (48). The deinhibitor not only protects the spontaneously active enzyme from inhibitor-1, but is also able to reverse the inhibition in a time-dependent manner under conditions where dephosphorylation of the inhibitor is minimal (97). This is in contrast to the effect seen with inhibitor-2, where inhibition cannot be reversed (48). The deinhibitor also enhances the extent of activation brought about by F_A (98) and increases the activity of the phosphatase toward phosphoinhibitor-1, just as Mn^{2+} does (23). The protein is said to be phosphorylated and inactivated by cAMP-dependent protein kinase and reactivated by a histone H1-stimulated (type 2A?) phosphatase, but the level of phosphorylation is very low (0.02 mol P_i/mol of protein) (99). The presence of the deinhibitor or other proteins of similar function explains why some type 1 phosphatase preparations from glycogen particles are relatively insensitive to inhibitor-1 and -2 (23–25, 70).

E. REGULATION OF THE M_r = 70,000 MgATP-DEPENDENT
    COMPLEX

1. *Activation by Trypsin and Mn²⁺*

Divalent metal ions, particularly Mn^{2+}, have been known to stimulate phosphorylase phosphatase from the time of the first reported preparation (*100*). However, attempts to isolate the Mn^{2+}-sensitive enzyme were initially unsuccessful—the level of activation was highly variable and was especially low in those preparations that had been isolated the most rapidly. It was finally recognized that sensitivity to Mn^{2+} was a property that the enzyme acquired over time, apparently as a result of the action of endogenous proteases. This assumption was confirmed when it was demonstrated that the phosphatase was strongly activated by the combined action of trypsin and Mn^{2+} (*52, 68*).

Some of the characteristics of the activation process could be elucidated after the enzyme had been obtained in pure form (*45*). When isolated in the presence of protease inhibitors, the M_r = 70,000 complex is totally inactive and can be fully activated by incubation with trypsin in the presence of 1 mM Mn^{2+}; by contrast, virtually no activity appears following treatment with either Mn^{2+} or trypsin alone. During trypsinolysis the regulatory subunit is rapidly destroyed, while the catalytic moiety is partially degraded to a species of M_r = 33,000 that is still inactive until exposed to the divalent cation; this trypsin-Mn^{2+}-activated enzyme is termed "E_a^{Tr-Mn}." The catalytic subunit possesses several trypsin-sensitive sites that give rise to a series of large fragments, one of which (M_r = 33,000) is remarkably stable to further proteolysis, especially when Mn^{2+} is present. Cleavage of the protein must occur toward the carboxyl end of the molecule, since the remaining core has a blocked N-terminus. Results similar to these have been obtained using chymotrypsin (*46*).

It was not at first understood why trypsin alone does not activate the phosphatase, even though it degrades the inhibitory subunit. There were three possible explanations for this observation.

1. Trypsin converts the catalytic subunit to an inactive form that must be reactivated by Mn^{2+}.
2. A small fragment of the inhibitor remains bound to the active site and is released upon addition of the metal ion.
3. The catalytic subunit in the M_r = 70,000 complex already exists in an inactive state.

This matter was resolved by separating the two subunits by ion-exchange fast protein liquid chromatography (FPLC) under nondenaturing conditions (*62*). The free, intact catalytic subunit was recovered in an inactive state (E_i) that could be fully activated by Mn^{2+}. Therefore, the phosphatase complex is different from the cAMP-dependent protein kinase, where dissociation by cAMP releases active catalytic subunits; in this case, dissociation yields only an inactive enzyme.

There is no evidence that the regulatory and catalytic components of the $M_r = 70,000$ complex dissociate spontaneously, since no activity can be measured even after a 24 h incubation in the presence of 1 mM Mn^{2+}. In fact, any activity seen in the presence of the divalent metal ion may be taken as an indication that the complex has sustained some proteolysis.

Activation of the free catalytic subunit by divalent cations is highly specific for Mn^{2+} and Co^{2+}, with half-maximal response at ~40 μM Mn^{2+} (*61, 62*); Ca^{2+} and Mg^{2+} at up to 10 mM are without effect. Mn^{2+} activation appears to proceed in two steps, according to Scheme I (*62*):

$$E_i \underset{\text{EDTA}}{\overset{\text{Mn}^{2+}}{\rightleftharpoons}} [E \cdot Mn^{2+}]_a \overset{\text{Mn}^{2+}}{\xrightarrow{\hspace{1em}}} E_a Mn$$

SCHEME I

First, there is an immediate activation that can be reversed by EDTA, presumably due to the formation of an enzyme–metal complex. This is followed by a slow conversion to an EDTA-resistant form ($E_a{}^{Mn}$) in which no metal ion seems to be present, as demonstrated by the lack of radioactivity when ^{54}Mn^{2+} is used. It is not known whether Mn^{2+} induces a stable change in conformation or whether it brings about a covalent modification of the protein, such as an SH–SS interconversion.

2. Activation by F_A and MgATP

The physiological regulation of the $M_r = 70,000$ phosphatase undoubtedly involves the kinase F_A and MgATP, as first described by Merlevede *et al.* (*41, 56, 59*). The mechanism of activation by F_A has been clarified only in part. In early studies, no incorporation of ^{32}P from [γ-^{32}P]ATP into the phosphatase could be detected, nor was there any evidence for an adenylylation reaction using ^{14}C-labeled ATP (*42, 101*). Because of the metal ion dependence of certain forms of the enzyme, it was proposed that activation might result from an F_A-mediated transfer of Mg ions from the MgATP complex [(*69, 101*); see also Ref. (*67*)]. It became known that activation of the phosphatase involves a phosphorylation of the regulatory subunit (*44, 45, 62, 65, 66*). The phosphate incorporated is released following addition of EDTA to block the kinase reaction, indicating that the phosphatase, once activated, catalyzes its own dephosphorylation (*45, 62*). Because of these two competing reactions, activation is strongly dependent on F_A concentration; at low F_A-to-phosphatase ratios the activity reaches a low plateau whose level increases with increasing amounts of the kinase (*41, 45, 56*).

Activation cannot be explained simply on the basis of a dissociation of the phosphorylated complex, since this would release the catalytic subunit in its inactive state. Obviously, a conversion of the enzyme to its active form must occur sometime during the phosphorylation–dephosphorylation cycle. This was directly demonstrated by subjecting the F_A-activated phosphatase to FPLC: the catalytic subunit emerged in a completely active state (designated $E_a{}^{F_A}$), as seen by the fact that addition of Mn^{2+} caused no further activation (62). Even when the phosphatase was activated only 10 to 15% by F_A, all of the catalytic subunit was recovered in the active form. The most likely explanation for the presence of such high levels of masked activity is that the catalytic subunit becomes inhibited following autodephosphorylation but retains its active conformation.

Several models describing the activation cycle of the MgATP-dependent complex have been proposed (62, 66, 67, 102); however, it is not yet certain at which point the enzyme becomes catalytically active. One complication is that phosphatase activity is not proportional to the amount of phosphate present in the regulatory subunit: The total amount of ^{32}P incorporated is substoichiometric, usually between 0.1 and 0.3 mol P_i/mol of inhibitor-2 (62, 65, 66). One way to account for this discrepancy would be to assume that activation actually occurs during the dephosphorylation step. If this were so, the enzyme would have to be able to catalyze its own intramolecular dephosphorylation while remaining inactive toward external substrates. In favor of this hypothesis is the observation that the regulatory subunit is phosphorylated to a much higher extent in the presence of [^{35}S]adenosine-5'-(γ-thio)triphosphate (ATPγS) than with ATP, but the phosphatase is activated to a lower level. Furthermore, ^{35}S is released in the presence of Mg^{2+}, with concomitant appearance of phosphorylase phosphatase activity. Thus, it has been suggested that the active enzyme is directly produced from the phosphorylated intermediate by a Mg^{2+}-dependent intramolecular dephosphorylation reaction (67). It should be noted that others have found no activation using ATPγS, even when the thiophosphorylation stoichiometry was as high as 0.8 mol/mol of inhibitor-2 (65, 101). In this case, thiophosphorylation prevented subsequent activation by F_A-MgATP.

An alternative explanation for the lack of correlation between phosphatase activity and ^{32}P incorporation could be that the entire system displays hysteretic characteristics; that is, both the catalytic and regulatory subunits might undergo slow changes in conformation as depicted in Fig. 1 (102). According to this scheme, phosphorylation by F_A would be accompanied by synchronous changes in both the regulatory and catalytic components, the former losing its inhibitory activity and the latter being placed in its active conformation (I → II). When the regulatory subunit undergoes autodephosphorylation (II → II*) it remains in the noninhibitory state for some time before switching back to its inhibitory form (II* → III). Thus, complex III is inactive, even though the catalytic subunit is still in the active conformation. This species is similar to the cAMP-dependent protein kinase in that any dissociation of the enzyme would liberate active catalytic subunits. Form III can either be reactivated by phosphorylation or it can return to the original inactive state by a slow conversion of E_a to E_i (III → I). If the rate of autodephosphorylation (II → II*) were greater than the rate at which

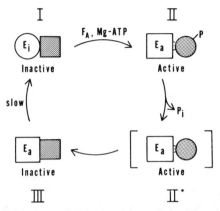

FIG. 1. Model depicting the hysteretic behavior of the catalytic and regulatory subunits during F_A treatment. The open symbols represent the catalytic subunit in the inactive (circle) and active (square) conformations. The shaded symbols designate inhibitor-2 in its noninhibitory (circle) and inhibitory (square) states. Form I is the native inactive enzyme. From Ballou *et al.* (*102*).

inhibitor-2 regains its inhibitory conformation (II* → III), most of the active phosphatase would exist as the dephosphorylated complex II*. Therefore, activation would be rapid and initially determined by the rate of phosphorylation, but the final level of activity would be independent of the amount of phosphate incorporated. There is no evidence that the active $M_r = 70,000$ phosphatase dissociates following phosphorylation (*46, 62, 66*); however, the MgATP-dependent complex reconstituted from the nicked $M_r = 33,000$ form of the catalytic subunit does dissociate upon activation (*65*).

The forms depicted in Fig. 1 can be identified and measured by the use of trypsin with or without Mn^{2+}, as previously described. In each case, trypsin-Mn^{2+} treatment gives full activity, whereas trypsin treatment alone allows expression of all catalytic subunits in the active conformation, whether present in the active or inactive complexes. Assay without any treatment measures only the active enzyme (forms II and II* according to Fig. 1). When this type of analysis was carried out at a low concentration of F_A, it was found that within 30 min the catalytic subunit was virtually entirely converted to its active conformation, although only 10–15% of the enzyme was activated (*62*). Therefore, under these conditions the phosphatase must exist predominantly as the inactive form III.

3. Conversion of E_a to E_i

The active catalytic subunit E_a does not spontaneously return to the inactive state E_i; this conversion seems to have an absolute requirement for the regulatory subunit (*62*). Furthermore, there are distinct differences in the rate at which inactivation occurs, depending on the nature of the enzyme used. The reaction proceeds much faster with $E_a^{F_A}$ ($t_{1/2}$ about 12 min) than with E_a^{Mn}, indicating some difference in behavior between these two species. E_a^{Tr-Mn} cannot be reconverted to E_i by incubation with inhibitor-2, despite the fact that it is strongly inhibited by the regulatory subunit. Therefore, the C-terminal portion of the molecule that is cleaved during tryptic attack might be involved in the inactivation process, either by masking the catalytic site or by participating in a rearrangement of the protein structure. Because inhibitor-2 is required for both the activation and inactivation processes, it is clear that its function goes beyond that of a simple inhibitor; for this reason it has been termed a modulator of enzyme activity (61).

That the various active forms of the phosphatase are not identical is further demonstrated by their substrate specificities (103). While the regulatory subunit of type II cAMP-dependent protein kinase (R_{II}) is readily dephosphorylated by all activated forms of the enzyme, synthetic peptides patterned according to the phosphorylation site of R_{II} are attacked only by E_a^{Mn} and E_a^{Tr-Mn}, but hardly at all by the F_A-activated complex or the active catalytic subunit derived from it (E_a^{FA}). Similar data were obtained with peptides covering the phosphorylation sites of the α- and β-subunits of phosphorylase kinase and phosphorylase a. These results cannot be ascribed to differences in affinity in view of the lack of response even at the highest substrate concentration used (i.e., 200 μM peptide). Rather, they suggest that enzyme specificity is affected by factors other than simply the amino acid sequence surrounding the phosphorylated site.

4. Involvement of Sulfhydryl Groups in Enzyme Activity

When the two components of the $M_r = 70,000$ complex are separated by FPLC in the absence of reducing agent, the catalytic subunit is recovered in an inactive state that no longer responds to Mn^{2+} treatment (103). This form of the enzyme must first be exposed to high concentrations of reducing agent in order to regain its sensitivity to the metal ion. The reaction is time-, temperature-, and concentration-dependent with respect to thiols [$t_{1/2}$ ~1 min with 50 mM dithiothreitol (DTT) at 30°C] and has a pH optimum of about 9. Maximal activity is obtained when the enzyme is incubated with Mn^{2+} together with the reducing agent. Several SH compounds are effective (DTT > 2-mercaptoethylamine > 2-mercaptoethanol > reduced glutathione > cysteine) while ascorbic acid, NADH, or NADPH are without effect. These data seem to indicate that the enzyme can undergo an SH–SS interchange and that Mn^{2+} activation requires a reduced form of the catalytic subunit. This assumption is supported by the observation that disulfides such as oxidized glutathione (GSSG) can inhibit all active forms of the enzyme, probably through formation of a mixed disulfide (104, 105). The inactive species can be reactivated by excess DTT even in the absence of Mn^{2+}, indicating that GSSG inactivation is not accompanied by a conversion of E_a to E_i. N-Ethylmaleimide (NEM) is also a strong inhibitor of the phosphatase; prior incubation with GSSG does not protect the enzyme from NEM inactivation.

These data might indicate that there are two sets of SH groups present in the phosphatase; while they seem to differ in chemical reactivity, they both appear to contribute to expression of enzymatic activity.

F. OTHER INHIBITORS AND ACTIVATORS

Each of the various inactive and active forms of the type 1 phosphatase responds differently to divalent cations. As previously mentioned, the free inactive catalytic subunit is fully activated by 1 mM Mn^{2+} or Co^{2+}, but the same enzyme bound to inhibitor-2 in the MgATP-dependent complex is not (45, 61, 62). By contrast, active species of the phosphatase are inhibited $\leq$ 50% by Mn^{2+}, particularly when phosphorylase a is used as a substrate (36, 44, 49, 62, 72, 106), whereas activity against several other phosphoproteins (e.g., phosphorylase kinase and glycogen synthase) is slightly stimulated (36, 72). Mg^{2+} and Ca^{2+} at low millimolar levels usually have no effect (36, 62, 106), although partial activation by Mg^{2+} or Ni^{2+} has been observed under some conditions (67, 68, 101). The enzymes can be inactivated by incubation with 50 mM NaF, 1 mM PP_i, or ATP and then partially reactivated by Mn^{2+} (36, 44, 69, 70); this phenomenon is addressed in detail in Section IV,D. The nicked catalytic subunit obtained after trypsin-Mn^{2+} activation is resistant to ATP even after prolonged incubation (68); likewise, these reagents do not appear to affect the M_r = 70,000 complex (44).

Basic compounds such as polylysine or protamine are strongly inhibitory when phosphorylase a is used as a substrate (43, 107). The free catalytic subunit is inhibited 50% by 100 μg/ml histone H1 or 3 μg/ml M_r = 17,000 polylysine, and about 90% by 30 μg/ml protamine. Polyamines are also inhibitory, with half-maximal effects at 0.04 mM spermine, 0.9 mM spermidine, and 9.0 mM putrescine (106). These substances act as activators with most other protein substrates: For instance, dephosphorylation of glycogen synthase site 3abc is stimulated more than 2-fold by 2 mM spermine and 13-fold by 100 μg/ml histone H1.

Heparin is a potent noncompetitive inhibitor of the type 1 catalytic subunit (K_i = 8 μg/ml) but has no effect on the type 2A enzyme at up to 50 μg/ml (71, 108). Inhibition can be reversed by polybrene (a cationic polymer), protamine, and histones H1 and H3; polybrene was also said to reverse the effect of inhibitor-1 and -2. Another specific inhibitor of the type 1 enzyme is DARPP-32, a dopamine- and cAMP-regulated neuronal phosphoprotein of M_r = 32,000 (82, 109, 110). Many of its properties are similar to those of inhibitor-1, although the two proteins are distinct gene products. DARPP-32 is effective only in its phospho form and is phosphorylated on a threonyl residue by cAMP-dependent protein kinase at a site homologous to that of inhibitor-1. It displays mixed inhibition with respect to phosphorylase a, with a K_i of 0.5–2.2 nM. The regulatory subunit of type II cAMP-dependent protein kinase (111, 112) as well as the heat-stable inhibitor of the kinase (PKI) (*103*) have also been reported to block activity of the phosphatase. PKI acts as a competitive inhibitor with respect to phosphorylase *a* (K_i = 30 n*M*); synthetic peptides corresponding to the inhibitory site also inhibit the phosphatase but with K_i values two orders of magnitude larger. Finally, the type 1 enzymes are activated by SH compounds and inactivated by cystamine, 5,5'-dithiobis(2-nitrobenzoic acid) (DTNB) and GSSG, but are relatively resistant to iodoacetic acid and iodoacetamide (*43*).

G. HORMONAL CONTROL

Protein phosphorylation–dephosphorylation is probably the most prominent mechanism by which cellular events can be regulated. Many protein kinases, such as those dependent on cAMP, cGMP, Ca^{2+}–CaM, or diacylglycerol and the tyrosine protein kinases, become activated in response to extracellular signals, principally hormones and growth factors. It would therefore be very surprising if the protein phosphatases that catalyze the reverse reactions were not also under hormonal control. For instance, it is well known that glycogen degradation is initiated when a number of enzymes become phosphorylated following hormonal stimulation of adenylate cyclase. To bring about an arrest of glycogenolysis and a stimulation of glycogen synthesis, as observed under the influence of insulin, either the protein kinases must be inhibited or the protein phosphatases activated. The type 1 phosphatase is a good candidate as a target for insulin activation since it acts on phosphorylase kinase, phosphorylase, and glycogen synthase.

There is evidence that hormonal control of the enzyme might be exerted at least in part at the level of inhibitor-1. Any dephosphorylation of the inhibitor mediated by the action of insulin would result in an increase in phosphatase activity; this has indeed been observed (*85, 86*). The process is all the more complex in that dephosphorylation of the protein can be catalyzed by the "inhibited" enzyme itself, as well as by some of the other protein phosphatases. Another possible target might be the kinase F_A. The complication here is that activation of this enzyme by insulin would favor both the inactivation and activation of glycogen synthase; as glycogen synthase kinase-3, F_A would promote the phosphorylation of site 3abc, while at the same time it would bring about the dephosphorylation of this site by activating the type 1 phosphatase. No activation of F_A by the insulin receptor–kinase or the EGF receptor–kinase has yet been demonstrated.

Alternatively, insulin (or perhaps a second messenger of the hormone) could block the autodephosphorylation of complex II (Fig. 1), thereby freezing the phosphatase in its active form, or it could allow expression of the potential activity hidden in the inactive form III by causing the complex to relax or dissociate, just as cAMP causes the cAMP-dependent protein kinase to dissociate. However, no such direct effect of insulin on the purified system has been demonstrated. A growth factor-dependent phosphorylation of the regulatory subunit does occur upon incubation of the phosphatase with insulin or EGF receptors, but this does not affect enzyme activity. Further understanding of the structural and regulatory properties of phosphorylase phosphatase should help clarify the role of hormones in this system.

IV. Phosphatase Type 2A

A. LOCALIZATION AND PURIFICATION

Protein phosphatase 2A can be recovered from tissue extracts in three subtypes, termed $2A_0$, $2A_1$, and $2A_2$, in order of their elution from DEAE-cellulose (*34*). Phosphatase $2A_0$ may be largely inactive under most assay conditions; its basal level of activity toward phosphorylase is low and somewhat variable (*113*). Phosphatase $2A_2$ appears to be derived from $2A_0$ and/or $2A_1$ through the loss of a loosely bound subunit (*113–115*). One important characteristic that distinguishes these enzymes from the other protein phosphatases is their ability to be activated by basic proteins.

A survey of rabbit tissues using phosphorylase *a* as a substrate showed that liver is an abundant source of phosphatase 2A, at 26 nmol P_i/min/g of tissue (liver > brain > heart > skeletal muscle > adipose tissue) (*36*). Virtually all of the enzyme is cytosolic. The different subtypes have been purified to homogeneity from many sources using numerous purification procedures. Investigators have been particularly imaginative in their choice of affinity chromatography media. These have included hydrophobic supports such as butyl-, phenyl-, and aminohexyl-Sepharose (*116–119*); unphosphorylated or phosphorylated proteins such as myosin-, histone-, and myosin light chain-Sepharose (*114, 120, 121*); thiophosphorylated substrates such as phosphorylase- or myosin light chain-Sepharose (*113, 119*); and other supports such as reactive red-120 agarose (*122*) and polylysine-Sepharose (*123, 124*). Large yields have been obtained from smooth muscle [200 μg $2A_1$/100 g of turkey gizzard (*125*) and 90 μg $2A_2$/100 g of bovine aorta (*122*)] as well as from pig heart (84 μg $2A_1$/100 g of tissue) (*115*). Rabbit muscle and rat liver usually give about 10–25 μg of each 2A species per 100 g of tissue (*113, 114, 118*).

B. SIZE AND SUBUNIT COMPOSITION

Type 2A phosphatases are multisubunit enzymes consisting of a catalytic subunit of M_r ~36,000 bound to various other components. The catalytic moiety is common to each 2A subtype (*113*); it is a distinct gene product from the type 1 catalytic subunit as shown by peptide mapping (*40, 49*). It is most likely, though, that the two proteins are structurally homologous; their amino acid compositions

are similar and several cross-reacting monoclonal antibodies have been reported
(49, 126). By contrast, polyclonal antibodies against the skeletal muscle 2A
subunit did not cross-react with the purified type 1 phosphatase (40), and they
inhibited the activities of the type 1, 2B, and 2C enzymes by only 10–15% in
liver extracts (39).

Phosphatase $2A_0$ purified from rabbit skeletal muscle consists of three subunits termed A (M_r = 60,000), B' (M_r = 54,000), and C (M_r = 36,000) and is
recovered as an AB'C_2 complex (Table IV) (127–131); for the sake of clarity,
the subunit nomenclature introduced by Cohen et al. (113) is used. The same A
subunit is also present in subtypes $2A_1$ and $2A_2$; the apparent molecular size of
this component varies considerably (M_r ~60,000–70,000), depending upon the
SDS-polyacrylamide gel electrophoretic system used (131). The type 2A phosphatases are asymmetric proteins with frictional ratios as high as 1.7 (115); they
therefore display anomolously high molecular sizes as determined by gel filtration chromatography.

Phosphatases $2A_1$ and $2A_0$ are quite similar in structure (Table IV); they share
the same A and C subunits but, at least in rabbit skeletal muscle, the B and B'
moieties (M_r ~55,000) are different (113). The subunit stoichiometry has usually been reported to be ABC or ABC_2; it is not certain whether this difference is
a reflection of tissue or species diversity. There is some question as to whether
the isolated $2A_1$ phosphatases actually represent the forms present in tissue
extracts, since a decrease in size has been observed during purification (115).
This could be due to the loss of a subunit, to a conformational change, or to
partial proteolysis of the protein during the preparative procedure. The $2A_1$
species from turkey gizzard can associate to some degree to produce dimers and
tetramers (125).

Loss of the B' or B subunit from phosphatase $2A_0$ and/or $2A_1$ leads to the
formation of subtype $2A_2$. It is possible to fractionate muscle extracts under
conditions that prevent the formation of this species; in this case, 25–30% of the
total 2A activity appears as $2A_0$ and the remainder as $2A_1$ (113). Most preparations of $2A_2$ phosphatase are said to occur in a 1 : 1 complex between the A and C
subunits (Table IV). Unlike the loosely held B subunit, the A and C components
usually remain tightly bound to one another, although partial dissociation has
been observed during gel filtration in the presence of 2% glycerol (114, 118) or
0.2 M salt (132). Free C can be generated by harsh procedures such as addition of
40–80% ethanol (26, 133), freeze–thawing in the presence of 0.2 M 2-mercap-
toethanol (114, 134), or treatment with 6 M urea or proteases (135, 136).

C. Substrate Specificity

Type 2A phosphatases have broad substrate specificities as seen in Tables I and
V. The reported specific activities toward a given substrate can be highly vari-

TABLE IV

SIZE AND SUBUNIT COMPOSITION OF TYPE 2A PHOSPHATASES

Name	Source	Holoenzyme $(M_r)^a$	Designation and size of subunits[b]		Subunit stoichiometry	Reference
Enzymes classified as $2A_0$						
$2A_0$	Rabbit skeletal muscle	181,000	A	60,000	1:1:2	(113)
			B′	54,000		
			C	36,000		
Enzymes classified as $2A_1{}^c$						
SMP-I	Turkey gizzard	165,000		60,000	1:1:1	(119, 125, 127)
				55,000		
				38,000		
—	Pig heart	171,000	β′	69,000	1:1:1	(115)
			γ′	56,000		
			α′	34,000		
IB	Rat liver	185,000	β	69,000	1:1:2	(114)
			γ	58,000		
			α	35,000		
Myosin light chain phosphatase	Chicken gizzard	—		67,000	1:1.8:0.6	(121)
				54,000		
				34,000		
$2A_1$	Rabbit skeletal muscle	202,000	A	60,000	1:1:2	(113)
			B	55,000		
			C	36,000		
Enzymes classified as $2A_2$						
SMP-IV	Turkey gizzard	—		58,000	—	(120)
				40,000		
eIF-2 phosphatase	Rabbit reticulocytes	100,000		60,000	1:1	(127, 128)
				38,000		
II	Rat liver	154,000	β	69,000	2:1	(118)
			α	35,000		
I (3C)	Bovine heart	95,000	β	63,000	1:1	(29, 129)
			α	35,000		
Myosin phosphatase	Bovine aorta	—		67,000	1:1	(122)
				38,000		
IV	Human erythrocytes	104,000	β	69,000	1:1	(130)
			α	32,000		
H-II	Rabbit skeletal muscle	140,000		70,000	1:1	(131)
				35,000		
$2A_2$	Rabbit skeletal muscle	107,000	A	60,000	1:1	(113)
			C	36,000		

[a] M_r calculated by ultracentrifugation analysis.
[b] Subunit M_r determined by SDS-gel electrophoresis.
[c] These enzymes have been classified as $2A_1$ as opposed to $2A_0$ because during purification by ion exchange chromatography they were retained on the column after a salt wash that would elute the A_0 fraction.

able, sometimes differing by one order of magnitude or more. In most cases these differences are due to the use of substrates at less-than-saturating concentrations, the presence or absence of activating compounds such as metal ions, caffeine or basic proteins, variations in the assay conditions and the source, and, most importantly, the degree of purity of the preparation. An example of this can be seen when comparing the activities toward phosphorylase kinase and phosphorylase a; a partially purified type $2A_1$ enzyme assayed in the presence of 1 mM Mn^{2+} displayed an activity ratio of 3.4 (34), whereas a value of 0.1 was obtained with the homogeneous enzyme measured in the presence of caffeine instead of the metal ion (106).

All type 2A phosphatases examined show about a 5-fold preference for the dephosphorylation of the α- rather than the β-subunit of phosphorylase kinase (33, 49, 113, 127). Activity toward this substrate is lower (49, 106, 131) than toward myosin light chain (Table V) (137–140), protamine, or histone H2B (specific activities of 5300 and 1800 nmol P_i/min/mg, respectively, using the free catalytic subunit) (115, 141). Similarly, activities of 960, 2670 and 2190 nmol P_i/min/mg for the $2A_0$, $2A_1$, and $2A_2$ enzymes, respectively, have been reported toward phosphorylase a when measured in the presence of caffeine and protamine (113); these values are lower than those obtained with phosphatase type 1 (49, 106). Aside from the autodephosphorylation reaction described in Section III,E for the type 1 enzyme, the type 2A phosphatases appear to be the major enzymes responsible for the dephosphorylation of inhibitor-2 in skeletal muscle (38); they are also said to play a major role in controlling the enzymes involved in glycolysis, glycogen metabolism, gluconeogenesis, and aromatic amino acid breakdown in the liver (36, 37).

The activity of the free catalytic subunit changes substantially when it recombines with the A and B subunits (115, 135, 142). The A and C components form an AC complex (i.e., phosphatase $2A_2$), with a concomitant 7-fold increase in activity against histone H1. Addition of subunit B to this complex causes about a 50% inhibition of the dephosphorylation of both phosphorylase a and glycogen synthase. The free M_r ~36,000 catalytic moiety is also more active than the ABC complex (i.e., the $2A_1$ species) toward myosin, myosin light chain, myosin light chain kinase, and phosphorylase kinase (125, 143). Removal of the inhibitory B subunit explains why type 2A preparations of high molecular weight display increases in activity toward phosphorylase upon dilution (36, 116) or dissociation by the harsh treatments discussed above; activation by basic proteins does not involve such a dissociation (107).

The $2A_2$ phosphatase from bovine heart was reported to possess a divalent cation-dependent activity toward (P)Tyr-IgG and (P)Tyr-casein (144). Maximal activity was only about 5% of that measured with phosphorylase a as a substrate. It is questionable whether this activity is intrinsic to the enzyme because it has a different thermal stability and responds differently to inhibitors such as pNPP

TABLE V

KINETIC CONSTANTS OF VARIOUS PHOSPHATASE 2A PREPARATIONS

Phosphatase	Source	Substrate	Kinetic constants K_m (μM)	Kinetic constants V_{max} (nmol P_i/min/mg)	Reference
Enzymes classified as 2A$_1$					
SMP-I	Turkey gizzard	Myosin light chain	10	7,700	(*125*)
—	Pig heart	Histone H2B	24	—	(*137*)
		Histone H2B heptapeptide	187	—	(*137*)
B	Canine heart	Mixed histones	4–10	—	(*138*)
Enzymes classified as 2A$_2$					
SMP-IV	Turkey gizzard	Heavy mero-myosin	6	—	(*139*)
eIF-2 phosphatase	Rabbit reticulocytes	eIF-2	30	1,100	(*128*)
IV	Human erythrocytes	Spectrin	2	153	(*130*)
		Histone H2B	48	1,860	(*130*)
H-II	Rabbit skeletal muscle	Lysine-rich histones	58	3,800	(*131*)
		Phosphorylase *a*	4–21	1,160–1,290	(*131*)
Free catalytic subunit					
—	Turkey gizzard	Heavy mero-myosin	19	—	(*139*)
		Myosin light chain	50	—	(*125*)
—	Pig heart	Histone H2B	81	—	(*137*)
		Histone H2B heptapeptide	90	—	(*137*)
C-II	Rabbit skeletal muscle	Phosphorylase *a*	6	—	(*49*)
		*p*NPP	7000	—	(*49*)
2A$_c$	Rabbit skeletal muscle	Phosphorylase *a*	5	—	(*38*)
		Inhibitor-2	1	—	(*38*)
—	Bovine heart	Kemptide	113	—	(*140*)
S	Canine heart	Mixed histones	10	—	(*138*)

and P_i. The type 2A phosphatases from chicken brain are also said to have a small but readily detectable activity against phosphotyrosyl proteins (*145*).

High levels of an alkaline phosphatase-like activity (20,000–25,000 nmol P_i/min/mg as measured by the hydrolysis of *p*NPP) are associated with some preparations of the enzyme (*129, 131*). Though 95% of it can be separated by hydrophobic chromatography (*146*), some low level of activity against nonprotein substrates might be intrinsic to the protein phosphatase. The activity is stimulated by histone H1 (*147*) and unaffected by the classical alkaline phosphatase inhibitors L-homoarginine and L-phenylalanine (*146*). β-naphthylphosphate can also serve as a substrate, but not phosphoserine, phosphothreonine, α-naphthylphosphate, or ATP. The *p*NPPase activity differs from that against phosphoproteins in its high optimum pH (8.5 versus 7.5) (*148*), lower thermal stability (*49, 131, 144*), lack of inactivation by ATP or PP_i (*117*), and requirement for both divalent cations ($Mg^{2+} > Mn^{2+} > Co^{2+}$) and SH compounds (*117*). The K_m for both Mg^{2+} and *p*NPP is in the low millimolar range, while that for Mn^{2+} is about 5 μM (*117, 144, 148*).

D. ACTIVATORS AND INHIBITORS

1. *Effect of Divalent Cations*

All forms of the enzyme can be activated by Mn^{2+} at concentrations of 0.1–5 m*M*. The effect is usually small (≤3-fold) and has been observed with almost all substrates tested. The concentration of Mn^{2+} required for half-maximal activation differs considerably from substrate to substrate, ranging from 0.03 m*M* with phenylalanine hydroxylase to ~1 m*M* with glycogen synthase sites 2 and 3 (*106, 131*). Dephosphorylation of histone H2B by the $2A_1$ enzyme is stimulated 4- to 5-fold, with optimal activity obtained at 20–30 m*M* Mn^{2+} (*137*). The largest increases in activity (5- to 17-fold) have been seen with phenylalanine hydroxylase, inhibitor-1, and glycogen synthase (site 3abc) as substrates (*106*). Activity against thiophosphorylated phosphorylase is also stimulated by Mn^{2+} (*149*). By contrast, the effects of 1–5 m*M* Mg^{2+} can be highly variable: there might be no effect (*128, 150*), slight (≤2-fold) stimulation (*118, 130*), or slight inhibition (*118, 120*), depending upon which enzyme and substrate are used. Ca^{2+} at up to 5 m*M* is either without effect (*128, 150*) or slightly inhibitory (*125*), while Zn^{2+} and Fe^{2+} are always strong inhibitors (*128, 137*).

2. *Are Phosphatase-1 and Phosphatase-2A Metalloenzymes?*

It has long been known that certain phosphatase preparations can be inactivated by ATP, PP_i, or KF and then reactivated by Mn^{2+} or Co^{2+} (*53, 151–153*). Many of the earlier studies examining these effects were carried out on partially purified systems containing a mixture of different phosphatases so the data were difficult to interpret. However, we know that most of these preparations contained predominantly phosphatase-1 and -2A.

The most effective inactivating agents are pyrophosphoryl compounds such as PP_i, ATP, CTP, GTP, and ADP (*53, 125, 151, 154*). Other reagents include AMP-P(NH)P and 50 m*M* KF (*27, 152*). These do not cause a simple inhibition of the enzymic reaction since the inactivation is not reversed by dilution or gel filtration and is time-, temperature-, and concentration-dependent. For instance, 20 μM PP_i inactivated a phosphatase $2A_2$ preparation 57% after 10 min at 30°C (118). Under similar conditions, millimolar concentrations of P_i, AMP, EDTA, and other chelators have no effect (53, 117, 151–154), though in one instance (152) inactivation was observed after a 16 h incubation with 5 mM P_i and 1 mM EDTA, and in another case a preparation of phosphatase $2A_2$ purified in the presence of EDTA and EGTA was recovered in a state that required Mn^{2+} or Co^{2+} for activity (122). Only these two cations are effective, and not Mg^{2+}, Ca^{2+}, or Mg^{2+}ATP (27, 117, 122, 153, 154). The reaction also requires the presence of a sulfhydryl reagent (153, 155). Half-maximum activation with Co^{2+} occurs in the 20–40 μM range and that for Mn^{2+} is about 10-fold higher (153, 154). As usual, these values depend upon the type of enzyme and substrate used. The free catalytic subunit is generally more susceptible to inactivation than are the 2A complexes (34, 118, 125, 138) or the type 1 catalytic subunit (30).

These observations have led to the hypothesis that these phosphatases might be metalloproteins (27, 151, 152, 154, 156). ATP, PP_i, and F^-, by virtue of some special structural feature, would be able to bind to or enter the catalytic site and remove an essential metal ion that would generally be inaccessible to EDTA or EGTA. On the other hand, there are some analytical and kinetic data that do not support this model. First, metal analysis by plasma atomic emission spectroscopy of active M_r ~35,000 catalytic subunit did not reveal significant amounts of Ca, Cd, Co, Cu, Fe, Mg, Mn, Ni, Sn, or Zn (155). Second, activation of the pure type 1 catalytic subunit (62) or of a preparation containing a high-molecular-weight form of the enzyme (52) by $^{54}Mn^{2+}$ resulted in no significant incorporation of radioactivity into the protein. Finally, the effects of Mn^{2+}, ATP, and F^- were examined from a kinetic viewpoint using an enzyme preparation consisting of approximately 88% type 2A and 12% type 1 catalytic subunits (140). Activity on certain peptide substrates was stimulated up to 100-fold by Mn^{2+}, in contrast to only a 2-fold stimulation when acting on phosphorylase a or phosphohistones. The data indicated that the phosphatase is a metal-sensitive enzyme that displays full activity upon binding of a single Mn^{2+} ion. ATP and F^- were inhibitory but apparently not through chelation of an essential metal ion. Rather, a model was presented in which Mn^{2+}, F^-, and ATP affected activity by binding to individual sites on the enzyme.

Obviously, the relationship between divalent cations and phosphatase activity is highly complex and remains to be fully understood. It has been suggested that Mn^{2+} and Co^{2+} may induce a conformational change, perhaps involving thiol groups on the enzyme (62, 140, 155). In any case, more structural studies are clearly needed.

3. Stimulation by Basic Proteins and Polyamines

In contrast to the type 1 enzyme whose activity toward phosphorylase a is inhibited by micromolar concentrations of basic proteins or polyamines, the 2A phosphatases are stimulated 5- to 10-fold (107, 124, 150, 157). Half-maximal stimulation of rabbit muscle $2A_0$, $2A_1$, and $2A_2$ is observed at 0.04 μM poly-lysine (M_r = 17,000), 0.3 μM histone H1 and protamine (113), and 0.2 mM spermine (106). When polylysines of varying chain length were used, it was found that both the extent of activation and the optimal concentration decreased with increasing degree of polymerization, so that maximal activity was reached at 1–3 μM lysyl residues (158).

These basic compounds act by binding to the phosphatase molecule, thereby lowering the K_m for substrate (158–160); therefore, it is not surprising that the stimulatory effects vary from enzyme to enzyme. The 2A phosphatase complexes from a given tissue respond similarly, while the free catalytic subunit is activated to a lesser extent and requires 5 to 10 times more of the basic protein to display half-maximal activity. In comparison to the muscle enzymes, those from the liver require 3- to 7-fold higher concentrations of protamine, polylysine, or histone H1 (107, 113). Although activation is mainly phosphatase-directed, effects on the substrate have also been observed. Activation is especially large (10- to 185-fold) with phosphorylase kinase, glycogen synthase (site 3abc), and phenylalanine hydroxylase (107, 160). By contrast, dephosphorylation of the acidic myosin light chain is inhibited because polylysine binds to this substrate (158). As previously discussed, dephosphorylation of glycogen synthase and phenylalanine hydroxylase is also greatly accelerated by Mn^{2+}. In fact, there appears to be some correlation between the effects of basic proteins and Mn^{2+}: Enzyme preparations that have been inactivated by "ageing" in the presence of EGTA can be reactivated by histone H1 as well as by the divalent cation (150). It is possible that such basic compounds could serve as the physiological counterparts to Mn^{2+} or Co^{2+} and that they could regulate phosphatase activity in vivo. It has been suggested that polyamines could act as second messengers of insulin action (106, 161).

4. Other Inhibitors and Activators

P_i is a competitive inhibitor (half-maximal effect at 1.4 mM) of both $2A_1$ and the isolated catalytic subunit (125). Reagents such as NEM (0.5 mM), p-chloromercuribenzoate (0.1 mM), and GSSG (2 mM) are also inhibitory, although 10 mM iodoacetate is without effect (130). As mentioned previously, inhibitor-1 and -2 do not inhibit the 2A species. These enzymes are relatively resistant to tryptic digestion (118, 131); limited proteolysis causes a slight activation and conversion to a polylysine-insensitive form (158). Insulin at low micromolar concentrations binds to and activates 4-fold both phosphatase $2A_2$ and the isolated catalytic subunit; the hormone also forms a complex with the type 1 catalytic subunit, but in this instance there is no change in activity. There is no evidence that this interaction is of physiological significance (162).

V. Phosphatase Type 2B (Calcineurin)

Calcineurin was independently discovered by the groups of Wang (*163, 164*), Klee (*165, 166*), and Cheung (*167*) as a high-molecular-weight, heat-labile calmodulin-binding protein from brain acting as a powerful inhibitor of the Ca^{2+}-dependent cyclic nucleotide phosphodiesterase. It was shown to consist of two polypeptide chains—an $M_r = 61,000$ calmodulin-binding A or α-subunit (*168, 169*) and an $M_r = 15,000$–$16,000$ B or β-subunit possessing calcium-binding properties (*166, 170*). In 1982, calcineurin was found to be identical to a highly purified type 2B phosphatase displaying restricted specificity toward the α-subunit of phosphorylase kinase and inhibitor-1 (*171–173*). The enzyme requires Ca^{2+} or Mn^{2+} for activity and is further stimulated up to 10-fold by calmodulin.

A. LOCALIZATION

Calcineurin (also referred to as the $M_r = 80,000$ calmodulin-binding protein or $CaM\text{-}BP_{80}$) (*167, 174*) is primarily found in brain and other nervous tissues, although lower levels have been detected in other organs by radioimmunoassay or an immunoblotting technique using polyclonal antibodies to the brain protein (*36, 174–176*). Concentrations varying from ~20 mg/kg of tissue in chick and fish brain to 140 mg/kg in rat cerebrum have been measured, with most of the material present in the caudate nucleus and the putamen (*174, 176*). Immunocytochemical localization in mouse basal ganglia showed that both calcineurin and calmodulin exist mainly in the postsynaptic densities and dendritic microtubules (*177*), but a direct relationship between the two proteins is not evident: The level of calmodulin in the testis is approximately twice that in the brain, yet the calmodulin-binding protein is barely detectable by radioimmunoassay (*174*). Such results should be viewed with caution, however, since calcineurin might be immunologically distinct in the two organs. This situation exists in skeletal muscle, where its concentration as determined by radioimmunoassay is at least ten times less than in the brain (*174*), but at approximately the same level (12.5–25 mg/kg) by activity and direct isolation (*36, 172*). Lower amounts appear to be present in bovine heart (*174, 175, 178*) but none has been detected in crude extracts of bovine gastric smooth muscle (*175*). Because of the differences in immunological cross-reactivity, it is likely that a family of structurally and functionally related Ca^{2+}-dependent phosphatases will be found in various tissues.

Levels of calcineurin increase significantly during embryonic development of rat cerebrum and cerebellum and chick brain and retina, while those of calmodulin remain essentially constant. These studies indicate a close relationship between the synthesis of calcineurin and synaptogenesis (*176*).

B. PURIFICATION

Ca^{2+}-dependent, calmodulin-regulated phosphatases have been purified to homogeneity from brain (*165, 169, 179*), skeletal muscle (*172*), and heart (*178, 180*). A difficulty in following the enzyme during purification has been to distinguish it from the many other protein phosphatases present. Two assays have been used, one enzymatic and the other based on direct visualization of the two polypeptide chains. In the first, the release of ^{32}P from appropriate substrates such as myosin light chain, phosphorylase kinase, or inhibitor-1 is measured, with the activity taken as the difference between the value obtained in the presence of calcium and that obtained in the presence of EGTA. Inhibition by trifluoperazine (150 μM) can also be utilized, since the other phosphatases are unaffected by this compound; however, highly purified preparations of the type 2B enzyme may be relatively insensitive to the inhibitor (*172*). In the second approach, the small subunit, which exhibits a characteristic increase in mobility in the presence of Ca^{2+}, can be quantitated by densitometric analysis of SDS-slab gels stained with Coomassie Blue (*181*). The large subunit can be distinguished from the many other proteins of similar size and quantitated by utilizing either immunoblotting (*175*) or a gel overlay technique in which the protein is electrophoretically transferred from SDS-polyacrylamide gels onto nitrocellulose, then detected with [^{125}I]calmodulin which still interacts with the protein (*175, 181*).

All purification procedures include an affinity step in addition to the usual separation by chromatofocusing or by ion exchange, gel filtration, or Affi-Gel Blue chromatography (*171, 172, 179–182*). Klee et al. (*181*) and Tallant et al. (*182*) utilized affinity chromatography on calmodulin-Sepharose in the presence of Ca^{2+} followed by elution in the presence of EGTA. A substrate affinity column of thiophosphorylated myosin light chain has also been used successfully; again, absorption is carried out in the presence of Ca^{2+} and elution with EGTA (*173*). The enzyme is obtained in amounts of 0.3–1.0 mg/100 g of tissue from bovine brain (*181, 182*) but only 7 μg/100 g from rabbit skeletal muscle (*172*). The phosphatase in muscle extracts is extremely labile, especially in the presence of Ca^{2+}.

Specific activities of $\geqslant 400$ nmol P_i/min/mg of protein have been reported for the brain phosphatase (*181*) and about 800–2000 for the skeletal muscle enzyme (*172*) using myosin light chain and inhibitor-1 as substrates, respectively. Tenfold differences in specific activity can easily be observed depending on the substrate and the assay conditions used.

C. SUBUNIT STRUCTURE

All type 2B phosphatases are heterodimers consisting of a $1:1$ complex between the large A polypeptide chain ($M_r = 61,000$) and the small hydrophobic B subunit ($M_r = 19,200$ by sequence analysis or $15,000–16,000$ by SDS-gel electrophoresis). The A chain of the heart and skeletal muscle enzymes appears on gels as a doublet (A + A') of $M_r \sim 61,000$ and $58,000$, respectively, probably as a result of limited proteolysis (171, 175). The dimeric structure has been confirmed by cross-linking with dimethylsuberimidate (165, 166); sedimentation analysis gives a value of $s_{20,w} = 4.7–4.96$, consistent with a molecular weight of $85,000–90,000$ for the $1:1$ complex (168, 169). The holoenzyme obtained from brain has an extinction coefficient of $\epsilon_{278}^{1\%} = 9.6$ (165).

It appears that the A subunit constitutes the catalytic and regulatory core of the enzyme: it contains the catalytic site and sites of binding for the B subunit, calmodulin, and transition metal ions. The strongest evidence that the catalytic site resides in the A subunit was provided by separation of the two components by immunoabsorbent chromatography: The free $M_r = 61,000$ protein displayed phosphatase activity that was stimulated by Ca^{2+}–CaM (183). The isolated subunit has also been prepared by gel filtration in the presence of 6 M urea (183, 184). This material is inactive with Ca^{2+}–CaM alone and shows an absolute requirement for Mn^{2+} (Mn^{2+}–CaM > Mn^{2+}). In addition, it has been reported that prolonged exposure of the holoenzyme to trypsin leads to the production of $M_r = 43,000$ and $41,000$ fragments of the A subunit and destruction of the B component. Such a preparation still retains full phosphatase activity, implying that the A subunit possesses the catalytic site (185). A danger of experiments such as these is that low levels of contamination by intact holoenzyme could vitiate the data.

The sequence of the B subunit has recently been determined (170). The polypeptide chain contains 168 residues; it is rich in tyrosine ($\epsilon_{277}^{1\%} = 3.1$) but lacks cysteine and tryptophan and, like the catalytic subunit of cAMP-dependent protein kinase, is blocked by a myristyl group at the N-terminus (170, 186). The molecule shows a high degree of homology with calmodulin and troponin C, particularly in the four regions thought to bind Ca^{2+} on the basis of the X-ray crystallographic structure of carp parvalbumin (187). Addition of Ca^{2+} to the B subunit induces a blue shift of the absorption spectrum similar to that observed with calmodulin (170).

The purified enzyme apparently contains about 0.3 equivalents of covalently bound phosphate as well as near stoichiometric amounts (0.6–0.9 mol/mol) of very tightly bound Zn and Fe, as revealed by atomic absorption spectroscopy; little other metal ion has been detected (*188*). Phosphate content does not vary in the course of activation–deactivation of the enzyme, presenting no evidence for autodephosphorylation. Likewise, a correlation between Zn and Fe content and activity has not been established. Activation by Mn^{2+} and other metal ions has suggested that the enzyme could be a Mn^{2+}, Co^{2+}, or Ni^{2+} metalloenzyme (*184*).

D. SUBSTRATE SPECIFICITY

Studies on the substrate specificity of phosphatase 2B have been greatly complicated by the complex nature of divalent metal ion activation of the enzyme. First, calcineurin requires Ca^{2+} to render the B subunit functional and to allow the binding of calmodulin that further stimulates enzyme activity. In addition, the A subunit is itself activated by direct interaction with divalent metal ions such as Mn^{2+} or Ni^{2+}. The extent of metal ion activation varies with the purity and state of activation of the enzyme preparation, the nature of the substrate (particularly when low-molecular-weight compounds are used) and the pH at which the assays are carried out, so that data from different laboratories cannot be readily compared (see, for instance, values obtained for casein in Table VI) (*189–194*). Finally, contamination by other phosphatases may sometimes be a problem.

Phosphatase 2B does not exhibit the broad specificity that most other protein phosphatases seem to have. The purified enzyme dephosphorylates quite specifically the α-subunit of phosphorylase kinase (*33, 171*), inhibitor-1 and -2 (*38*), smooth muscle myosin light chain (*33, 172, 181*), the regulatory subunit R_{II} of cAMP-dependent protein kinase (*181*), and a number of brain proteins such as protein K.-F., DARPP-32, and G-substrate (*190*) (Table VI). It is interesting to note that G-substrate, like DARPP-32, shares a number of physicochemical features with inhibitor-1; all three proteins can act as phosphatase inhibitors, they are phosphorylated on threonyl residues, and the sequences around the phosphorylation sites are similar. The activity of the enzyme is 1–3 orders of magnitude lower on synapsin I, histones, phosphorylase *a*, and the β-subunit of phosphorylase kinase ($\alpha/\beta \approx 100$).

Calcineurin is also reported to dephosphorylate some phosphotyrosyl residues in proteins (*195*). Using (P)Tyr-casein as a substrate, the K_m and V_{max} with Mg^{2+} alone were 2.2 μM and 0.4 nmol P_i/min/mg, respectively. Addition of Ca^{2+}–CaM decreased the K_m to 0.6 μM and increased V_{max} to 4.6 nmol P_i/min/mg. These values are far below the V_{max} of 121 nmol P_i/min/mg obtained with (P)Ser-casein. By comparison, a specific phosphotyrosyl-protein phosphatase from rabbit kidney dephosphorylates (P)Tyr-casein with a V_{max} of 2000 nmol P_i/min/mg (See Section VII,D). The type 2B phosphatase also attacks low-molecular-weight phosphoesters such as β-naphthylphosphate >

TABLE VI

SPECIFIC ACTIVITY OF CALCINEURIN TOWARD VARIOUS PROTEIN SUBSTRATES

Source of enzyme	Substrate	K_m (μM)		Specific activity (nmol P_i/min/mg)		Reference
		−CaM	+CaM	−CaM	+CaM	
Rabbit skeletal muscle	Inhibitor-1	2.5	2.5	170[a]	2080[a]	(172)
	Phosphorylase kinase α-subunit	5.9	—	1040[a]	—	(172)
	Myosin light chain	—	3.7	—	790[a]	(172)
Bovine brain	Inhibitor-1	2.5	5	1[a]	286[a]	(189)
	Protein K.-F.[c]	—	~3[b]	18	320	(190)
	DARPP-32[d]	—	1.6	3.8	260	(190)
	G-Substrate[e]	—	3.8	1.3	80	(190)
	Casein	10.8	4.2	4.9[a]	72[a]	(185)
		1	1.3	5[a]	120[a]	(189)
		42	32	0.01[a]	0.13[a]	(191)
	Myelin basic protein	19	14	5[a]	56[a]	(191)

[a] V_{max} value.

[b] $S_{0.5}$ rather than K_m.

[c] An M_r = 18,000 protein resembling myelin basic protein that is phosphorylated by a cyclic nucleotide–Ca^{2+}-independent kinase (192).

[d] An M_r = 32,000 protein present in the cytosol of dopaminoceptive neurons in the basal ganglia and phosphorylated by cAMP-dependent protein kinase (193).

[e] An M_r = 23,000 protein present in the cytosol of cerebellar Purkinje cells serving as a specific substrate for cGMP-dependent protein kinase (194).

pNPP > α-naphthylphosphate >> phospho-DL-tyrosine (196, 197). The specific activity on these substrates is of the same order of magnitude as observed with phosphoproteins. Phosphothreonine is dephosphorylated at a very low rate, while phosphoserine, NADP, glucose 6-phosphate, AMP, IMP, GMP, and ATP are untouched.

E. ROLE OF REGULATORY SUBUNITS AND METAL IONS

Calcineurin, through its B subunit, binds 4 mol Ca^{2+}/mol of protein with an affinity in the low micromolar range (166). This stimulates the phosphatase to less than 5% of its maximum activity. Calmodulin forms a 1 : 1 complex with the enzyme as measured by a continuous variation Job plot (188), with a very low dissociation constant of 4×10^{-9} M (165). Binding has little effect on K_m but increases V_{max}, as seen in Table VI. As for other calmodulin-dependent enzymes, limited proteolysis activates calcineurin and converts it to a form that is insensitive to Ca^{2+}–CaM (183, 185, 198). This treatment is accompanied by a decrease in K_m from 10.8 to 1.2 μM and an increase in V_{max} from 4.9 to 30.9 nmol P_i/min/mg using casein as a substrate (*185*). Although the modified enzyme no longer binds calmodulin and displays full activity in the absence of the regulatory protein, it can still be activated by Ca^{2+} (half-maximal effect at 0.4 μM), since the B subunit is still present (*185, 198*). Millimolar concentrations of Ca^{2+}, however, are inhibitory (*198, 199*), whereas addition of 1 mM Mn^{2+} causes further activation (*185, 199*). Unlike the trypsinized holoenzyme, which shows some activity in the absence of any metal ions, partial trypsinolysis of the free A subunit prepared in the presence of urea yields a species that still has an absolute requirement for Mn^{2+} (*184*).

All investigators agree that the enzyme exhibits multiple levels of activity corresponding to multiple conformational states. The native AB dimer is inactive in the absence of Ca^{2+}. Saturation of the B subunit with Ca^{2+} yields a partially active enzyme that can be further stimulated by Ca^{2+}–CaM. However, the activity of each of these three states can be further increased—sometimes considerably—by binding of transition metal ions, principally Mn^{2+} (*171, 196, 200*), Co^{2+} (*196, 200*), and Ni^{2+} (*200–202*), to the A subunit. These six states of activity are depicted schematically in Fig. 2 (*197*).

Pallen and Wang (*202*) found that divalent cations could be subdivided into four groups according to their effects on enzyme activity. The first category includes Ni^{2+} and Mn^{2+}, which can activate the enzyme in the absence of calmodulin, though calmodulin further stimulates activity. The second is composed of Co^{2+}, Ca^{2+}, Sr^{2+}, and Ba^{2+}, cations that activate in the presence of

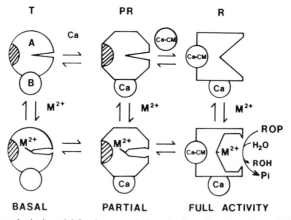

FIG. 2. A hypothetical model for the synergistic activation of calcineurin by Ca^{2+}, calmodulin, and divalent metal ions. The symbols used are A and B, the A and B subunits of calcineurin; T, PR, and R, tight, partially relaxed, and relaxed conformations; *shaded* and *open* areas on the A subunit, regulatory and catalytic domains, respectively; CM, calmodulin; M^{2+}, divalent cation; ROP, phosphoester. Reprinted with permission from Li (*197*).

calmodulin but not in its absence. The third, Be^{2+}, Mg^{2+}, Cd^{2+}, and Fe^{2+}, which are ineffective with or without calmodulin. And the fourth, Zn^{2+}, which inhibits the activation by ions of classes one and two (*195, 202*). There is no absolute rule as to the effectiveness of one metal ion over the other. For instance, while Ni^{2+}–CaM was reported to have five times the stimulatory power of Mn^{2+}–CaM with histone H1 as substrate (and 8-fold that of Ca^{2+}–CaM) (*200*), it was only one-fifth as effective as Mn^{2+}–CaM using myelin basic protein (*191*).

While in most studies Mg^{2+} is essentially without effect, Li *et al.* (*195, 197*) have reported strong stimulation of enzyme activity, particularly when low-molecular-weight substrates are used. At pH 7.4 the order of effectiveness of metal ions was $Ni^{2+} > Mn^{2+} > Mg^{2+} >> Co^{2+}$, whereas at pH 8.6 Mg^{2+} was just as effective as Ni^{2+}, if not more so. It should be noted that at this high pH contamination by alkaline phosphatase could be a problem. The metal ion present appears to be the main factor in determining the optimum pH of the reaction (*189*).

F. MECHANISM OF ACTIVATION AND DEACTIVATION

Activation of calcineurin by metal ions such as Mn^{2+} and Ni^{2+} is highly complex: It is slow and appears to proceed through several steps corresponding to slow changes in conformation (*188, 202*). At least two forms of the Ni^{2+}-activated enzyme have been postulated (*202*), the first being susceptible to chelating agents while the latter, obtained after a period of incubation, is resistant. This behavior was ascribed to a very tight binding of the metal ion, as accepted by most investigators (*188, 199, 202, 203*). Similar observations were made with Mn^{2+}, though in this case the fully activated enzyme was inhibited by EDTA, suggesting a less tight binding. It is interesting to note that this behavior is reminiscent of that seen with the inactive type 1 catalytic subunit, which is activated by Mn^{2+} in two steps, the first being reversed by EDTA while the second is not (*62*). However, the lack of incorporation of $^{54}Mn^{2+}$ into this protein indicates that the interconversions may be solely due to changes in conformation or a covalent modification of the molecule (see Section III,E).

Mn^{2+} and Ni^{2+} apparently compete for binding to calcineurin at the same site. However, once one of the cations is bound the other cannot displace it. In order to account for these data, it has been postulated that binding of one cation converts the enzyme into a form with a high affinity for that cation so, as a result, it cannot bind the other (*202*). One could also assume that each cation specifically induces a distinct conformational state. Convincing evidence in support of this hypothesis has been obtained by the use of a monoclonal antibody specific for the B subunit (*203*). The antibody inhibited the Ni^{2+}-stimulated but not the Mn^{2+}-stimulated form of the enzyme. This study also suggested that the B subunit plays a critical role in the expression of the Ni^{2+}-induced activity of the phosphatase.

That the control of calcineurin by various effector molecules is not a simple process is further demonstrated by the fact that Ca^{2+}–CaM, an activator, also participates in a deactivation reaction (188). A first indication of this was obtained by Stewart et al. (171), who observed that the enzyme emerged from a column of immobilized calmodulin in a form that had a very low specific activity in the presence of Ca^{2+}–CaM but could be reactivated by incubation with Mn^{2+}. The first-order deactivation rate constant is increased by substrates or competitive inhibitors such as PP_i, ADP, and GDP; a mathematical treatment of this process in the presence or absence of different ligands has been presented (188). The behavior of phosphatase 2B with respect to activation–deactivation is similar to that of the type 1 enzyme, where the modulator protein is required for both reactions.

VI. Phosphatase Type 2C

A. LOCALIZATION AND PURIFICATION

The type 2C enzyme was originally described as a glycogen synthase phosphatase notable for its absolute dependence on Mg^{2+} or Mn^{2+} and its low activity toward phosphorylase a (27, 204, 205). It has been detected in all mammalian tissues that have been examined. Liver appears to be the most abundant source, followed by heart and brain, skeletal muscle, and adipose tissue; activity levels in rat liver are approximately 1.6 nmol P_i/min/g of tissue using phosphorylase kinase as a substrate (36). The liver and skeletal muscle enzymes are cytosolic, with none found associated with glycogen particles or microsomes (36).

Many different preparations of what appears to be the same Mg^{2+}-dependent phosphatase have been described and given a variety of designations. Most may be classified as type 2C, though some might have been contaminated with Mn^{2+}-stimulated type 1 and/or 2A enzymes. These include phosphatase IA (205) and fraction IIB (27) from rat liver; fraction II of human erythrocytes (130); smooth muscle phosphatase II or SMP-II from turkey gizzard (119); bovine heart Peak-C (129); and phosphatase U-2 or 4 from canine heart (29, 129).

Purification procedures generally involve ammonium sulfate fractionation followed by gel filtration, anion exchange, and affinity chromatography. Phosphatases 2C and $2A_1$ coelute from DEAE-cellulose but can be easily separated by gel filtration (34). When ion-exchange chromatography is carried out in the presence of Mg^{2+}, the 2C enzyme elutes at a lower salt concentration (205).

Affinity columns of histone (205) or thiophosphorylated myosin light chain (119) coupled to Sepharose have been used successfully; in the latter case, Mg^{2+} is required for binding of the enzyme and elution is carried out with EDTA. Approximately 50–100 μg of purified phosphatase 2C can be obtained from 100 g of rat liver (205, 206) and ~40 μg from turkey gizzard (207).

B. SIZE AND SUBUNIT COMPOSITION

Type 2C phosphatases display sizes of M_r = 46,000 ± 6,000 as determined by SDS-gel electrophoresis, gel filtration chromatography, or sedimentation equilibrium centrifugation. The enzyme therefore appears to be monomeric, although one report suggests that the rat liver phosphatase might exist as a dimer of M_r = 90,000 on the basis of sucrose gradient centrifugation and chromatography on Sephacryl S-200 (206).

C. SUBSTRATE SPECIFICITY

These enzymes attack a wide variety of protein substrates (Table VII). In general, they have low activity toward the phosphorylase kinase β-subunit and phosphorylase a and exhibit a high histone H2B : H1 activity ratio. Good substrates include the α-subunit of phosphorylase kinase, pyruvate kinase, glycogen synthase (apparent K_m = 0.6 μM) (204), and myosin light chain (apparent K_m = 7.9 μM, V_{max} = 53,000 nmol/min/mg); by contrast, intact myosin is not dephosphorylated (207). The enzyme from bovine heart has been reported to dephosphorylate pNPP (129) and phosphotyrosyl residues in IgG (208). These reactions were cation-dependent, inhibited by Zn^{2+}, NaF, PP$_i$, and ATP, and displayed a pH optimum of 8.5–9.0, whereas that for phosphoseryl-protein substrates is closer to neutrality (204, 206, 207). The specific activity of the partially purified enzyme using pNPP as a substrate was 90 nmol P$_i$/min/mg as compared to 1,330 for myosin light chain (129).

The question as to the physiological importance of phosphatase 2C in regulating cellular processes has been addressed by measuring its activity in extracts against a large number of protein substrates (36, 37, 209). In liver, the enzyme represents a small but still significant fraction of the activity toward glycogen synthase, 6-phosphofructo-2-kinase/fructose-2,6-bisphosphatase, and pyruvate kinase, and is responsible for a sizeable percentage of the activity toward 6-phosphofructo-1-kinase. It is also the major enzyme acting on HMG-CoA reductase and the reductase kinase; however, since the phosphatase is not found in the microsomal fraction with which the reductase is tightly associated, its involvement in the regulation of cholesterol synthesis is uncertain (36). Likewise, although the smooth-muscle enzyme exhibits considerable activity against isolated myosin light chain, there is some question as to whether it is important as a light

TABLE VII

Relative Specificity of Protein Phosphatase 2C from Various Sources

	Relative activity (%)[a]			
	Rabbit liver (33)	Rat liver (37)	Rat liver (206)	Turkey gizzard(119,127)
Phosphorylase kinase (α)	100	100	100	100
Phosphorylase kinase (β)	5	5–10	14	5–10
Phosphorylase a	2	2	14	2
Glycogen synthase (site 1a)	21	—	—	24
Glycogen synthase (site 2)	86	—	—	84
Glycogen synthase (site 3abc)	8	—	—	8
Inhibitor-1	19	—	—	22
Pyruvate kinase	58	400	286	66
ATP-citrate lyase	15	—	—	18
Acetyl-CoA carboxylase	28	—	—	—
Myosin light chain	191	—	—	200
Histone H1	9	—	—	16
Histone H2B	224	—	71	280
Casein	—	—	14	13
6-Phosphofructo-1-kinase	—	260	286	—
6-Phosphofructo-2-kinase/ fructose-2,6- bisphosphatase	—	230	171–229	—
Fructose-1,6-bisphosphatase	—	77	314	—
Phenylalanine hydroxylase	—	54	—	—

[a] Activity is expressed as % P_i released relative to the dephosphorylation of the α-subunit of phosphorylase kinase, which was set at 100%. Different assay conditions were used by the different investigators. The values in most cases do not represent activities obtained under saturating conditions. Less than 5% activity was seen with glycogen synthase (site 5), myosin, myosin light chain kinase, and eIF-2(α).

chain phosphatase *in vivo* since it does not bind to myosin, thiophosphorylated myosin, or actin (*139*).

D. Activators and Inhibitors

Phosphatase 2C is completely dependent on divalent metal ions for activity with all substrates that have been tested. Maximum activity is usually obtained with 5–10 mM Mg^{2+} (*206, 207*), with half-maximal effects at ~1 mM (*204, 205*). Mn^{2+} and Co^{2+} can substitute for Mg^{2+} ($Mg^{2+} \geqslant Mn^{2+} > Co^{2+}$) but Ca^{2+} has no effect.

The enzyme is potently inhibited by NaF, P_i, and PP_i, with K_i values of 1.0,

1.0, and 4.2 mM, respectively, as well as by ATP, ADP, and AMP (K_i = 5, 3.2, and 4.0 mM, respectively) (206, 207). Unlike phosphatase types 1 and 2A, incubation of type 2C with 50 mM NaF/1 mM EDTA does not convert it to a form that now shows a specific requirement for Mn^{2+}; Mg^{2+} is still just as effective in activating the enzyme (27). Likewise, precipitation with 80% ethanol does not convert the protein to a lower-molecular-weight species, although there is a partial loss of activity (130, 205). Inhibitor-1 and -2 and trifluoperazine are without effect (33, 130, 172, 206).

The phosphorylation–dephosphorylation of liver pyruvate kinase (PK-L) is affected by ligands that bind to the enzyme. PK-L is phosphorylated and inactivated by cAMP-dependent protein kinase and dephosphorylated by protein phosphatase-1, -2A, and -2C (33). Positive allosteric effectors of PK-L such as phosphoenolpyruvate (PEP) and fructose 1,6-bisphosphate (F-1,6-P_2) inhibit the phosphorylation reaction, whereas negative effectors such as alanine and phenylalanine counteract the effects of PEP and F-1,6-P_2 and allow an increase in phosphorylation (210–212). There has been one report that these ligands also affect the dephosphorylation of PK-L by a Mg^{2+}-stimulated phosphatase present in hepatocyte lysates; PEP and F-1,6-P_2 partially inhibited this reaction, an effect that was antagonized by alanine (213). If this were so, then one could assume that the conformational changes induced in PK-L by the positive allosteric effectors would make the serine residues less accessible to both protein kinases and phosphatases. It should be noted that others have not observed a substrate-specific effect of PEP or F-1,6-P_2 on the dephosphorylation of PK-L; furthermore, no unequivocal evidence for a role of metabolites or effectors in regulating the dephosphorylation of several other glycolytic or gluconeogenic enzymes has been obtained (37, 209).

VII. Phosphotyrosyl-Protein Phosphatases

A. BACKGROUND

The phosphorylation of proteins on tyrosyl residues was first discovered in connection with the transforming proteins of certain tumor viruses (214). It was found that these proteins, the most well-known of which is pp60[v-src], are tyrosyl-protein kinases (215, 216) [for review, see Ref. (217)]. Normal cells contain low levels of an enzyme analogous to the viral src gene product designated as pp60[c-src]. Likewise, receptors for certain growth factors (e.g., epidermal growth factor (EGF), platelet-derived growth factor, insulin, and insulin-like growth factor I) possess tyrosyl-protein kinase activity [reviewed in Ref. (218)]. Enzyme activity of the receptors is activated by ligand binding; they undergo autophosphorylation and act on both endogenous membrane proteins and exogenous substrates.

Dephosphorylation of tyrosyl residues in proteins was first detected in membranes of A-431 human epidermoid carcinoma cells, a line having a high density of EGF receptors (*219, 220*). When A-431 membranes were incubated with EGF and ^{32}P-ATP, there was a rapid incorporation of counts into membrane protein followed by a slow loss of radioactivity due to the action of phosphatases. Similar results were obtained by Sefton *et al.* (*221*), who examined tyrosine phosphorylation in chick cells infected with a Rous sarcoma virus mutant temperature-sensitive for both transformation and pp60^{v-src} activity. When the transformed cells were shifted from the permissive to the nonpermissive temperature, the level of phosphotyrosine decreased substantially, again demonstrating the existence of phosphotyrosyl-protein phosphatases. These two early studies suggested that the level of phosphotyrosyl residues in cells would reflect the relative activities of the competing kinases and phosphatases.

B.   PURIFICATION

Phosphotyrosyl-protein phosphatase activity has been detected in a variety of tissues and cell lines (*222*). In the rabbit, it occurs in the order kidney > liver > brain > lung = heart > skeletal muscle. The level of 39 nmol P_i/min/g of kidney tissue using (P)Tyr-albumin as a substrate is in the same order of magnitude as that of the type 1 phosphorylase phosphatase in skeletal muscle (*223*). The enzyme seems to be unequally distributed between the particulate and soluble fractions, with the majority being found in the latter (*224, 225*). Many of the problems inherent in purifying the other protein phosphatases have been encountered with the tyrosine-specific enzymes: they also exist in a variety of forms, they are usually recovered in very low yield, many are unstable even when stored at $-20°C$, and most importantly, their substrates and the kinases needed to phosphorylate them are far more difficult to prepare in large quantity. Indeed, aside from the receptors already mentioned and some undefined membrane proteins, no natural substrates for these enzymes are known. As a result, proteins such as casein, tubulin, reduced and alkylated albumin, and carboxymethylated and succinylated phosphorylase have been used to measure enzyme activity (Table VIII) (*223–233*). These substrates are less than ideal because though they often possess multiple tyrosyl groups that can be phosphorylated, the stoichiometry of ^{32}P incorporation is usually low (<0.1 mol P_i/mol of protein). Furthermore, some of the modified proteins are highly insoluble and can be used only at low concentrations unless they are dissolved in 10 N NaOH (*223*), where they undergo partial hydrolysis. Most ideal would be to use a synthetic peptide corresponding to one of the phosphorylation sites of a known natural substrate. A single attempt to do so utilizing a phosphotyrosyl peptide derived from pp60^{src}

TABLE VIII

SUBSTRATES OF PHOSPHOTYROSYL-PROTEIN PHOSPHATASES

Substrate	Kinase	Reference
EGF receptor–kinase	EGF receptor–kinase	(226–229)
IgG	pp60$^{v\text{-}src}$	(29, 144, 225, 230)
Histones	EGF receptor–kinase	(195, 227, 228, 231)
Carboxymethylated and succinylated phosphorylase	EGF receptor–kinase	(226)
Casein	EGF receptor–kinase	(29, 195, 226, 232)
	pp60$^{v\text{-}src}$	(144, 145)
Reduced, alkylated bovine serum albumin	EGF receptor–kinase	(223, 232)
Tubulin	pp60$^{v\text{-}src}$	(145)
Glutamine synthetase	—	(224)
67K, 50K, 37K (membrane proteins)	?	(233)
Myosin light chain	EGF receptor–kinase	(232)
pNPP	—	(144, 145, 223, 227, 231)

failed in that it was not dephosphorylated (232). Because of these obstacles, none of these enzymes has been obtained in a homogeneous state.

In all purifications described, the enzyme does not seem to behave as a single entity, but can be separated into at least three fractions. In the first purification from detergent lysates of Ehrlich ascites tumor cells, 60% of the activity flowed through a column of DEAE-Sephadex and was not further considered. The material that was retained was eluted with NaCl and then subjected to affinity chromatography on Zn^{2+} chelated to iminodiacetic acid-agarose. Seventy-five percent of the activity emerged with 20 mM histidine (pool 1) and the remainder with 60 mM histidine (pool 2). After further purification by gel filtration chromatography, the pool 1 enzyme displayed a specific activity of 1.2 nmol P_i/min/mg versus 29 for the pool 2 material using chemically modified (P)Tyr-phosphorylase as a substrate (226). A similar scheme was used to purify the rabbit kidney enzyme (223). In this instance, 70% of the phosphatase did not bind to the DEAE column and elution from Zn^{2+}-iminodiacetic acid-agarose was carried out with EDTA instead of histidine. After further ion exchange and gel filtration steps, two enzyme species were obtained with specific activities of 213 (Peak I) and 281 nmol P_i/min/mg (Peak II) using chemically modified (P)Tyr-albumin as a substrate. Three forms of the phosphatase have also been resolved by DEAE chromatography from bovine heart (termed Y-1, Y-2, and Y-3) (29), chicken embryo fibroblasts (pTPI, pTPII, and pTPIII) (225) and chicken brain (T_1, T_2, and T_3) (145).

C. Physical Properties

Little is known about the structure of phosphotyrosyl-protein phosphatases. Several of the enzymes display sizes in the M_r = 35,000–40,000 range (223, 226), similar to that of the catalytic subunits of the type 1 and 2A protein phosphatases. A comparison of α-chymotryptic fragments from the tyrosyl phosphatase Peak I and the type 1 catalytic subunit was made using reverse-phase HPLC; preliminary evidence suggested that the two enzymes are not identical but will most likely prove to be homologous (234). The peptide maps of Peak I (M_r = 34,000) and II (M_r = 37,000) phosphotyrosyl-protein phosphatases have been compared by one-dimensional polyacrylamide gel electrophoresis and it appears that these two proteins are also distinct enzymes (223).

Phosphatases of larger size have also been reported, including Y-2 (M_r = 65,000) (29); T_1 (M_r = 30,000–100,000), T_2 (M_r = 43,000), and T_3 (M_r = 95,000) from chicken brain (145); and pTPI (M_r = 55,000), pTPII (M_r = 50,000), and pTPIII (M_r = 95,000) from fibroblasts (225). It is not yet known whether these species represent multisubunit complexes.

D. Enzymic Properties

A number of (P)Tyr-protein substrates that have been used to study phosphotyrosyl-protein phosphatases are listed in Table VIII. These have usually been phosphorylated by $pp60^{v-src}$ or the EGF receptor–kinase. The nonprotein substrate pNPP has also been employed because of its structural similarity to phosphotyrosine, although, as previously discussed, it is also attacked by the phosphoseryl- and phosphothreonyl-protein phosphatases. As expected, the main distinguishing characteristic of these enzymes is their near exclusive specificity toward tyrosyl versus seryl phosphate residues. The pool 1 phosphatase of Ehrlich ascites tumor cells, for instance, dephosphorylated chemically modified (P)Tyr-phosphorylase with a V_{max} of 0.17 nmol P_i/min/mg (K_m = 0.8 μM) but had essentially no activity toward phosphorylase a (226).

The purified cytosolic Peak I and II phosphatases were reported to readily dephosphorylate acidic proteins such as serum albumin, casein, and myosin light chain but did not act on basic substrates such as histone or various peptides that were rapidly dephosphorylated by calf intestinal alkaline phosphatase. It was suggested that if one were to use (P)Tyr-histone as a substrate, one would measure only the activity of alkaline phosphatase, not that of the phosphotyrosyl-protein phosphatase (232).

Kinetic constants for the two phosphatases from rabbit kidney are listed in Table IX. The K_m and V_{max} values are in the same order of magnitude as those reported for phosphoseryl-protein phosphatases with their preferred substrates. The Peak I enzyme had a pH optimum of 7–7.5 for both pNPP and protein

TABLE IX

KINETIC CONSTANTS OF PEAKS I AND II PHOSPHOTYROSYL-PROTEIN PHOSPHATASES TOWARD
VARIOUS SUBSTRATES[a]

	Peak I		Peak II	
Substrate	K_m (μM)	V_{max} (nmol P_i/min/mg)	K_m (μM)	V_{max} (nmol P_i/min/mg)
(P)Tyr-bovine serum albumin (reduced & alkylated)	12.5	18,200	1.2	240
(P)Tyr-casein	3.6	2,000	6.6	540
(P)Tyr-myosin light chain	1.4	5,600	—	—
pNPP	200	—	2,400	—

[a] Data are from Refs. (223, 232).

substrates, as opposed to pH 5–5.5 for the Peak II material (223). Considering that acid phosphatase attacks phosphotyrosyl residues in proteins, contamination by this enzyme cannot be discounted. Several other tyrosyl-protein phosphatases have been reported to operate best around neutrality (224–226).

E.   ACTIVATORS AND INHIBITORS

Because there are several different forms of phosphotyrosyl-protein phosphatase and they have been assayed under a variety of conditions, it is difficult to make definite statements about their effectors. However, some general rules seem to be appropriate:

1. Orthovanadate at micromolar levels is a strong inhibitor of the phosphotyrosyl-phosphatases as opposed to the phosphoseryl enzymes (29, 223, 225, 227, 228). This effect was first observed with the membrane-associated activity of A-431 cells, where 50% inhibition was obtained at 1 μM VO_4^{-3} (228). It has been suggested that this inhibitory action might be linked to the insulin-like effect of vanadate.
2. Brautigan et al. (229) found that 10 μM Zn^{2+} completely inhibited the dephosphorylation of A-431 cell membrane proteins by the endogenous phosphatase. In other systems substantial inhibition is seen with 5–100 μM Zn^{2+} (223–227, 230, 233). This property has led to the use of Zn^{2+} affinity columns to purify the enzyme (223, 226). Mn^{2+}, Mg^{2+}, Co^{2+}, and Ca^{2+} at up to 1 mM are somewhat less inhibitory (145, 225, 229).
3. pNPP is inhibitory in the low millimolar range (29, 226, 227).

4. Phosphotyrosine is a competitive inhibitor that gives 50% inhibition at 100–400 µM *(224, 226)*; phosphoserine and phosphothreonine, however, are not inhibitory.

5. A-431 cell membrane proteins thiophosphorylated with ATPγS were attacked at one-twentieth to one-fortieth the rate observed with normally phosphorylated proteins *(235)*.

6. F$^-$, PP$_i$, P$_i$, and ATP are less effective inhibitors of the phosphotyrosyl- than the phosphoseryl-protein phosphatases. Fluoride at 50 mM either has little effect *(145, 230)* or inhibits up to 45% *(29, 227)*; high concentrations of PP$_i$ (5–30 mM) or P$_i$ (10 mM) inhibit 70% or more *(145, 226)*, whereas 1 mM ATP causes almost complete inhibition of the enzyme *(225)*.

7. Inhibitor-1 and -2 are without effect *(29, 230)*.

8. Tetramisole, a powerful inhibitor of alkaline phosphatases, does not affect the activity of the tyrosine protein phosphatase of Ehrlich ascites tumor cells *(226)*.

9. EDTA at concentrations up to 5 mM either activates or is without effect *(145, 226, 229)*. The Y-2 enzyme of bovine heart was activated 5- to 10-fold by EDTA, with half-maximum activation at 15 µM. EGTA was a poor activator, while other chelators such as desferrioxamine or hydroxylquinoline were as effective as EDTA *(29)*.

10. Reducing agents such as DTT are potent stimulators of some of the enzymes *(223, 224)* but others, such as the Peak II material from rabbit kidney, remain unaffected.

F. OTHER ENZYMES WITH PHOSPHOTYROSYL-PROTEIN PHOSPHATASE ACTIVITY

Several other enzymes that dephosphorylate phosphotyrosyl groups in proteins have been reported. In view of their low levels of activity as compared with the tyrosyl-specific enzymes, their physiological importance is questionable. Nonetheless, they are mentioned here for the sake of completeness.

Alkaline phosphatases from calf intestine, bovine liver, and E. coli will dephosphorylate (P)Tyr-histone and the EGF receptor–kinase in A-431 cell membranes with activities approximately 6 orders of magnitude lower than those obtained with *p*NPP *(231)*. The reactions were strongly inhibited by 2 mM *p*NPP and 20 mM EDTA but not by 50 mM F$^-$. Similar properties have been reported for bovine kidney alkaline phosphatase using (P)Tyr-IgG as substrate *(144)*. An acid phosphatase in membranes of the human tumor astrocytoma dephosphorylated (P)Tyr-histones and phosphotyrosine. This enzyme activity was said to be different from classical acid phosphatases in that it did not attack the usual substrates such as β-glycerophosphate and was not inhibited by 10 mM L(+)-

tartrate. Histone dephosphorylation was stimulated by 5 mM EDTA and poorly inhibited by 100 μM Zn^{2+} or 10 mM pNPP, phosphotyrosine, or P$_i$. VO$_4^{-3}$, on the other hand, brought about a 50% inhibition at 0.5 μM (236). Li et al. (237) have suggested that the predominant phosphotyrosyl-protein phosphatase activity of human prostate is due to the prostatic acid phosphatase. The enzyme, a dimer of identical M_r = 50,000 subunits, dephosphorylated (P)Tyr-IgG and (P)Tyr-casein, though at very low rates (0.5 nmol P$_i$/min/mg) because the substrate concentration (1.56 nM) was so low. Typical acid phosphatase inhibitors such as molybdate and vanadate were strongly inhibitory at micromolar concentrations as were 5 mM L(+)-tartrate, F$^-$, and P$_i$. As discussed in the previous sections, all type 2 phosphatases have been said to dephosphorylate phosphotyrosyl-proteins at very low rates. In many cases the two activities have different properties and may very well be due to contaminating enzymes.

ACKNOWLEDGMENTS

We thank Carmen Westwater and Pamela Holbeck for their assistance in typing the manuscript. Our work on the phosphatases was supported by grants from the NIH (AM 07902) and the Muscular Dystrophy Association.

REFERENCES

1. Harris, D. L. (1946). *JBC* **165**, 541–550.
2. Feinstein, R. N., and Volk, M. E. (1949). *JBC* **177**, 339–346.
3. Parnas, J. K., and Ostern, P. (1935). *Biochem. Z.* **279**, 94.
4. Cori, C. F., and Cori, G. T. (1937). *Proc. Soc. Exp. Biol. Med.* **36**, 119–122.
5. Cori, G. T., Colowick, S. P., and Cori, C. F. (1938). *JBC* **123**, 375–380.
6. Cori, G. T., Colowick, S. P., and Cori, C. F. (1938). *JBC* **123**, 381–389.
7. Kiessling, W. (1939). *Biochem. Z.* **302**, 50.
8. Green, A. A., and Cori, G. T. (1943). *JBC* **151**, 21–29.
9. Cori, G. T., and Green, A. A. (1943). *JBC* **151**, 31–38.
10. Cori, C. F., Cori, G. T., and Green, A. A. (1943). *JBC* **151**, 39–55.
11. Velick, S. F., and Wicks, L. F. (1951). *JBC* **190**, 741–751.
12. Keller, P. J., and Cori, G. T. (1953). *BBA* **12**, 235–238.
13. Fischer, E. H., and Krebs, E. G. (1955). *JBC* **216**, 121–132.
14. Krebs, E. G., Kent, A. B., and Fischer, E. H. (1958). *JBC* **231**, 73–83.
15. Sutherland, E. W., and Wosilait, W. D. (1955). *Nature (London)* **175**, 169–171.
16. Rall, T. W., Sutherland, E. W., and Wosilait, W. D. (1956). *JBC* **218**, 483–495.
17. Wosilait, W. D., and Sutherland, E. W. (1956). *JBC* **218**, 469–481.
18. Cohen, P. (1978). *Curr. Top. Cell. Regul.* **14**, 117–196.
19. Lee, E. Y. C., Silberman, S. R., Ganapathi, M. K., Petrovic, S., and Paris, H. (1980). *Adv. Cyclic Nucleotide Res.* **13**, 95–131.
20. Li, H.-C. (1982). *Curr. Top. Cell. Regul.* **21**, 129–174.
21. Merlevede, W., Vandenheede, J. R., Goris, J., and Yang, S.-D. (1984). *Curr. Top. Cell. Regul.* **23**, 177–215.

22. Ingebritsen, T. S., and Cohen, P. (1983). *Science* **221**, 331–338.
23. Goris, J., Camps, T., Defreyn, G., and Merlevede, W. (1981). *FEBS Lett.* **134**, 189–193.
24. Khatra, B. S., and Soderling, T. R. (1983). *ABB* **227**, 39–51.
25. Stralfors, P., Hiraga, A., and Cohen, P. (1985). *EJB* **149**, 295–303.
26. Brandt, H., Killilea, S. D., and Lee, E. Y. C. (1974). *BBRC* **61**, 548–554.
27. Mackenzie, C. W., Bulbulian, G. J., and Bishop, J. S. (1980). *BBA* **614**, 413–424.
28. Li, H.-C., Hsiao, K.-J., and Chan, W. W. S. (1978). *EJB* **84**, 215–225.
29. Chernoff, J., and Li, H.-C. (1983). *ABB* **226**, 517–530.
30. Ingebritsen, T. S., Foulkes, J. G., and Cohen, P. (1980). *FEBS Lett.* **119**, 9–15.
31. Antoniw, J. F., Nimmo, H. G., Yeaman, S. J., and Cohen, P. (1977). *BJ* **162**, 423–433.
32. Nimmo, G. A., and Cohen, P. (1978). *EJB* **87**, 353–365.
33. Ingebritsen, T. S., and Cohen, P. (1983). *EJB* **132**, 255–261.
34. Ingebritsen, T. S., Foulkes, J. G., and Cohen, P. (1983). *EJB* **132**, 263–274.
35. Ingebritsen, T. S., Blair, J., Guy, P., Witters, L., and Hardie, D. G. (1983). *EJB* **132**, 275–281.
36. Ingebritsen, T. S., Stewart, A. A., and Cohen, P. (1983). *EJB* **132**, 297–307.
37. Pelech, S., Cohen, P., Fisher, M. J., Pogson, C. I., El-Maghrabi, M. R., and Pilkis, S. J. (1984). *EJB* **145**, 39–49.
38. Tonks, N. K., and Cohen, P. (1984). *EJB* **145**, 65–70.
39. Alemany, S., Tung, H. Y. L., Shenolikar, S., Pilkis, S. J., and Cohen, P. (1984). *EJB* **145**, 51–56.
40. Tung, H. Y. L., Resink, T. J., Hemmings, B. A., Shenolikar, S., and Cohen, P. (1984). *EJB* **138**, 635–641.
41. Goris, J., Dopere, F., Vandenheede, J. R., and Merlevede, W. (1980). *FEBS Lett.* **117**, 117–121.
42. Yang, S.-D., Vandenheede, J. R., Goris, J., and Merlevede, W. (1980). *JBC* **255**, 11759–11767.
43. Gratecos, D., Detwiler, T. C., Hurd, S., and Fischer, E. H. (1977). *Biochemistry* **16**, 4812–4817.
44. Resink, T. J., Hemmings, B. A., Tung, H. Y. L., and Cohen, P. (1983). *EJB* **133**, 455–461.
45. Ballou, L. M., Brautigan, D. L., and Fischer, E. H. (1983). *Biochemistry* **22**, 3393–3399.
46. Tung, H. Y. L., and Cohen, P. (1984). *EJB* **145**, 57–64.
47. Khandelwal, R. L., Vandenheede, J. R., and Krebs, E. G. (1976). *JBC* **251**, 4850–4858.
48. Goris, J., Waelkens, E., Camps, T., and Merlevede, W. (1984). *Adv. Enzyme Regul.* **22**, 467–484.
49. Silberman, S. R., Speth, M., Nemani, R., Ganapathi, M. K., Dombradi, V., Paris, H., and Lee, E. Y. C. (1984). *JBC* **259**, 2913–2922.
50. DePaoli-Roach, A. A. (1984). *JBC* **259**, 12144–12152.
51. Brautigan, D. L., Shriner, C. L., and Gruppuso, P. A. (1985). *JBC* **260**, 4295–4302.
52. Brautigan, D. L., Picton, C., and Fischer, E. H. (1980). *Biochemistry* **19**, 5787–5794.
53. Merlevede, W., and Riley, G. A. (1966). *JBC* **241**, 3517–3524.
54. Merlevede, W., Goris, J., and DeBrandt, C. (1969). *EJB* **11**, 499–502.
55. Chelala, C. A., and Torres, H. N. (1970). *BBA* **198**, 504–513.
56. Vandenheede, J. R., Yang, S.-D., Goris, J., and Merlevede, W. (1980). *JBC* **255**, 11768–11774.
57. Cohen, P., Yellowlees, D., Aitken, A., Donella-Deana, A., Hemmings, B. A., and Parker, P. J. (1982). *EJB* **124**, 21–35.
58. Tellez de Inon, M. T., and Torres, H. N. (1973). *BBA* **297**, 399–412.
59. Goris, J., Defreyn, G., and Merlevede, W. (1979). *FEBS Lett.* **99**, 279–282.
60. Yang, S.-D., Vandenheede, J. R., Goris, J., and Merlevede, W. (1980). *FEBS Lett.* **111**, 201–204.

61. Yang, S.-D., Vandenheede, J. R., and Merlevede, W. (1981). *JBC* **256**, 10231–10234.
62. Villa-Moruzzi, E., Ballou, L. M., and Fischer, E. H. (1984). *JBC* **259**, 5857–5863.
63. Hemmings, B. A., Yellowlees, D., Kernohan, J. C., and Cohen, P. (1981). *EJB* **119**, 443–451.
64. Woodgett, J. R., and Cohen, P. (1984). *BBA* **788**, 339–347.
65. Hemmings, B. A., Resink, T. J., and Cohen, P. (1982). *FEBS Lett.* **150**, 319–324.
66. Jurgensen, S., Shacter, E., Huang, C. Y., Chock, P. B., Yang, S.-D., Vandenheede, J. R., and Merlevede, W. (1984). *JBC* **259**, 5864–5870.
67. Li, H.-C., Price, D. J., and Tabarini, D. (1985). *JBC* **260**, 6416–6426.
68. Brautigan, D. L., Ballou, L. M., and Fischer, E. H. (1982). *Biochemistry* **21**, 1977–1982.
69. Vandenheede, J. R., Yang, S.-D., and Merlevede, W. (1981). *JBC* **256**, 5894–5900.
70. Khatra, B. S. (1984). *Proc. Soc. Exp. Biol. Med.* **177**, 33–41.
71. Erdodi, F., Csortos, C., Bot, G., and Gergely, P. (1985). *BBA* **827**, 23–29.
72. Stewart, A. A., Hemmings, B. A., Cohen, P., Goris, J., and Merlevede, W. (1981). *EJB* **115**, 197–205.
73. Ganapathi, M. K., Silberman, S. R., Paris, H., and Lee, E. Y. C. (1981). *JBC* **256**, 3213–3217.
74. Foulkes, J. G., Strada, S. J., Henderson, P. J. F., and Cohen, P. (1983). *EJB* **132**, 309–313.
75. Brandt, H., Lee, E. Y. C., and Killilea, S. D. (1975). *BBRC* **63**, 950–956.
76. Huang, F. L., and Glinsmann, W. H. (1975). *PNAS* **72**, 3004–3008.
77. Huang, F. L., and Glinsmann, W. H. (1976). *EJB* **70**, 419–426.
78. Huang, F. L., and Glinsmann, W. H. (1976). *FEBS Lett.* **62**, 326–329.
79. Nimmo, G. A., and Cohen, P. (1978). *EJB* **87**, 341–351.
80. Foulkes, J. G., Ernst, V., and Levin, D. H. (1983). *JBC* **258**, 1439–1443.
81. Aitken, A., Bilham, T., and Cohen, P. (1982). *EJB* **126**, 235–246.
82. Hemmings, H. C., Nairn, A. C., and Greengard, P. (1984). *JBC* **259**, 14491–14497.
83. Nemenoff, R. A., Blackshear, P. J., and Avruch, J. (1983). *JBC* **258**, 9437–9443.
84. Foulkes, J. G., and Cohen, P. (1979). *EJB* **97**, 251–256.
85. Foulkes, J. G., Jefferson, L. S., and Cohen, P. (1980). *FEBS Lett.* **112**, 21–24.
86. Foulkes, J. G., Cohen, P., Strada, S. J., Everson, W. V., and Jefferson, L. S. (1982). *JBC* **257**, 12493–12496.
87. Khatra, B. S., Chiasson, J.-L., Shikama, H., Exton, J. H., and Soderling, T. R. (1980). *FEBS Lett.* **114**, 253–256.
88. Chang, L. Y., and Huang, L. C. (1980). *Acta Endocrinol. (Copenhagen)* **95**, 427–432.
89. Aitken, A., and Cohen, P. (1982). *FEBS Lett.* **147**, 54–58.
90. Goris, J., Defreyn, G., Vandenheede, J. R., and Merlevede, W. (1978). *EJB* **91**, 457–464.
91. Yang, S.-D., Vandenheede, J. R., and Merlevede, W. (1981). *FEBS Lett.* **132**, 293–295.
92. Foulkes, J. G., and Cohen, P. (1980). *EJB* **105**, 195–203.
93. Yang, S.-D., Vandenheede, J. R., and Merlevede, W. (1983). *BBRC* **113**, 439–445.
94. Gruppuso, P. A., Johnson, G. L., Constantinides, M., and Brautigan, D. L. (1985). *JBC* **260**, 4288–4294.
95. Aitken, A., Holmes, C. F. B., Campbell, D. G., Resink, T. J., Cohen, P., Leung, C. T. W., and Williams, D. H. (1984). *BBA* **790**, 288–291.
96. Defreyn, G., Goris, J., and Merlevede, W. (1977). *FEBS Lett.* **79**, 125–128.
97. Waelkens, E., Goris, J., and Merlevede, W. (1984). *Biochem. Soc. Trans.* **12**, 827.
98. Goris, J., Waelkens, E., and Merlevede, W. (1983). *BBRC* **116**, 349–354.
99. Goris, J., Parker, P. J., Waelkens, E., and Merlevede, W. (1984). *BBRC* **120**, 405–410.
100. Cori, G. T., and Cori, C. F. (1945). *JBC* **158**, 321–332.
101. Yang, S.-D., Vandenheede, J. R., and Merlevede, W. (1981). *FEBS Lett.* **126**, 57–60.
102. Ballou, L. M., Villa-Moruzzi, E., and Fischer, E. H. (1985). *Curr. Top. Cell. Regul.* **27**, 183–192.

103. Ballou, L. M., Villa-Moruzzi, E., McNall, S. J., Scott, J. D., Blumenthal, D. K., Krebs, E. G., and Fischer, E. H. (1985). *Adv. Protein Phosphatases* **1,** 21–37.
104. Shimazu, T., Tokutake, S., and Usami, M. (1978). *JBC* **253,** 7376–7382.
105. Usami, M., Matsushita, H., and Shimazu, T. (1980). *JBC* **255,** 1928–1933.
106. Tung, H. Y. L., Pelech, S., Fisher, M. J., Pogson, C. I., and Cohen, P. (1985). *EJB* **149,** 305–313.
107. Pelech, S., and Cohen, P. (1985). *EJB* **148,** 245–251.
108. Gergely, P., Erdodi, F., and Bot, G. (1984). *FEBS Lett.* **169,** 45–48.
109. Hemmings, H. C., Williams, K. R., Konigsberg, W. H., and Greengard, P. (1984). *JBC* **259,** 14486–14490.
110. Hemmings, H. C., Greengard, P., Tung, H. Y. L., and Cohen, P. (1984). *Nature (London)* **310,** 503–508.
111. Gergely, P., and Bot, G. (1977). *FEBS Lett.* **82,** 269–272.
112. Jurgensen, S. R., Chock, P. B., Taylor, S., Vandenheede, J. R., and Merlevede, W. (1985). *FP* **44,** 1052 (Abstr. 3750).
113. Tung, H. Y. L., Alemany, S., and Cohen, P. (1985). *EJB* **148,** 253–263.
114. Tamura, S., and Tsuiki, S. (1980). *EJB* **111,** 217–224.
115. Imaoka, T., Imazu, M., Usui, H., Kinohara, N., and Takeda, M. (1983). *JBC* **258,** 1526–1535.
116. Lee, E. Y. C., Mellgren, R. L., Killilea, S. D., and Aylward, J. H. (1978). *FEBS Symp.* **42,** 327–346.
117. Li, H.-C. (1979). *EJB* **102,** 363–374.
118. Tamura, S., Kikuchi, H., Kikuchi, K., Hiraga, A., and Tsuiki, S. (1980). *EJB* **104,** 347–355.
119. Pato, M. D., and Adelstein, R. S. (1980). *JBC* **255,** 6535–6538.
120. Pato, M. D., and Kerc, E. (1984). *Biophys. J.* **45,** 354a.
121. Onishi, H., Umeda, J., Uchiwa, H., and Watanabe, S. (1982). *J. Biochem. (Tokyo)* **91,** 265–271.
122. Werth, D. K., Haeberle, J. R., and Hathaway, D. R. (1982). *JBC* **257,** 7306–7309.
123. DiSalvo, J., Waelkens, E., Gifford, D., Goris, J., and Merlevede, W. (1983). *BBRC* **117,** 493–500.
124. Yang, S.-D., Vandenheede, J. R., and Merlevede, W. (1984). *BBRC* **118,** 923–928.
125. Pato, M. D., and Adelstein, R. S. (1983). *JBC* **258,** 7047–7054.
126. Speth, M., Alejandro, R., and Lee, E. Y. C. (1984). *JBC* **259,** 3475–3481.
127. Pato, M. D., Adelstein, R. S., Crouch, D., Safer, B., Ingebritsen, T. S., and Cohen, P. (1983). *EJB* **132,** 283–287.
128. Crouch, D., and Safer, B. (1980). *JBC* **255,** 7918–7924.
129. Li, H.-C. (1981). *Cold Spring Harbor Conf. Cell Proliferation* **8,** 441–457.
130. Usui, H., Kinohara, N., Yoshikawa, K., Imazu, M., Imaoka, T., and Takeda, M. (1983). *JBC* **258,** 10455–10463.
131. Paris, H., Ganapathi, M. K., Silberman, S. R., Aylward, J. H., and Lee, E. Y. C. (1984). *JBC* **259,** 7510–7518.
132. Khandelwal, R. L., Zinman, S. M., and Ng, T. T. S. (1980). *BBA* **626,** 486–493.
133. Imazu, M., Imaoka, T., Usui, H., and Takeda, M. (1978). *BBRC* **84,** 777–785.
134. Kobayashi, M., Kato, K., and Sato, S. (1975). *BBA* **377,** 343–355.
135. Imazu, M., Imaoka, T., Usui, H., Kinohara, N., and Takeda, M. (1981). *J. Biochem. (Tokyo)* **90,** 851–862.
136. Killilea, S. D., Mellgren, R. L., Aylward, J. H., Metieh, M. E., and Lee, E. Y. C. (1979). *ABB* **193,** 130–139.
137. Imaoka, T., Imazu, M., Ishida, N., and Takeda, M. (1978). *BBA* **523,** 109–120.
138. Li, H.-C., and Hsiao, K.-J. (1977). *EJB* **77,** 383–391.

139. Sellers, J. R., and Pato, M. D. (1984). *JBC* **259**, 7740–7746.
140. Shacter-Noiman, E., and Chock, P. B. (1983). *JBC* **258**, 4214–4219.
141. Titanji, V. P. K., Zetterqvist, O., and Engström, L. (1980). *FEBS Lett.* **111**, 209–213.
142. Imaoka, T., Imazu, M., Usui, H., Kinohara, N., and Takeda, M. (1980). *BBA* **612**, 73–84.
143. Ganapathi, M. K., and Lee, E. Y. C. (1984). *ABB* **233**, 19–31.
144. Chernoff, J., Li, H.-C., Cheng, Y.-S. E., and Chen, L. B. (1983). *JBC* **258**, 7852–7857.
145. Foulkes, J. G., Erikson, E., and Erikson, R. L. (1983). *JBC* **258**, 431–438.
146. Li, H.-C., Hsiao, K.-J., and Sampathkumar, S. (1979). *JBC* **254**, 3368–3374.
147. Schlender, K. K., and Mellgren, R. L. (1984). *Proc. Soc. Exp. Biol. Med.* **177**, 17–23.
148. Li, H.-C., and Chan, W. W. S. (1981). *ABB* **207**, 270–281.
149. Tabarini, D., and Li, H.-C. (1980). *BBRC* **95**, 1192–1199.
150. Mellgren, R. L., and Schlender, K. K. (1983). *BBRC* **117**, 501–508.
151. Kato, K., Kobayashi, M., and Sato, S. (1975). *J. Biochem. (Tokyo)* **77**, 811–815.
152. Khatra, B. S., and Soderling, T. R. (1978). *BBRC* **85**, 647–654.
153. Khandelwal, R. L., and Kamani, S. A. S. (1980). *BBA* **613**, 95–105.
154. Hsiao, K.-J., Sandberg, A. R., and Li, H.-C. (1978). *JBC* **253**, 6901–6907.
155. Yan, S. C. B., and Graves, D. J. (1982). *Mol. Cell. Biochem.* **42**, 21–29.
156. Burchell, A., and Cohen, P. (1978). *Biochem. Soc. Trans.* **6**, 220–222.
157. Wilson, S. E., Mellgren, R. L., and Schlender, K. K. (1983). *BBRC* **116**, 581–586.
158. DiSalvo, J., Gifford, D., and Kokkinakis, A. (1984). *Proc. Soc. Exp. Biol. Med.* **177**, 24–32.
159. Wilson, S. E., Mellgren, R. L., and Schlender, K. K. (1982). *FEBS Lett.* **146**, 331–334.
160. Mellgren, R. L., Wilson, S. E., and Schlender, K. K. (1984). *FEBS Lett.* **167**, 291–294.
161. Huang, L. C., and Chang, L. Y. (1980). *BBA* **613**, 106–115.
162. Speth, M., and Lee, E. Y. C. (1984). *JBC* **259**, 4027–4030.
163. Wang, J. H., and Desai, R. (1976). *BBRC* **72**, 926–932.
164. Wang, J. H., and Desai, R. (1977). *JBC* **252**, 4175–4184.
165. Klee, C. B., and Krinks, M. H. (1978). *Biochemistry* **17**, 120–126.
166. Klee, C. B., Crouch, T. H., and Krinks, M. H. (1979). *PNAS* **76**, 6270–6273.
167. Cheung, W. Y., Lynch, T. J., and Wallace, R. W. (1978). *Adv. Cyclic Nucleotide Res.* **9**, 233–251.
168. Richman, P. G., and Klee, C. B. (1978). *JBC* **253**, 6323–6326.
169. Sharma, R. K., Desai, R., Waisman, D. M., and Wang, J. H. (1979). *JBC* **254**, 4276–4282.
170. Aitken, A., Klee, C. B., and Cohen, P. (1984). *EJB* **139**, 663–671.
171. Stewart, A. A., Ingebritsen, T. S., Manalan, A., Klee, C. B., and Cohen, P. (1982). *FEBS Lett.* **137**, 80–84.
172. Stewart, A. A., Ingebritsen, T. S., and Cohen, P. (1983). *EJB* **132**, 289–295.
173. Tonks, N. K., and Cohen, P. (1983). *BBA* **747**, 191–193.
174. Wallace, R. W., Tallant, E. A., and Cheung, W. Y. (1980). *Biochemistry* **19**, 1831–1837.
175. Manalan, A. S., Krinks, M. H., and Klee, C. B. (1984). *Proc. Soc. Exp. Biol. Med.* **177**, 12–16.
176. Tallant, E. A., and Cheung, W. Y. (1983). *Biochemistry* **22**, 3630–3635.
177. Wood, J. G., Wallace, R. W., Whitaker, J. N., and Cheung, W. Y. (1980). *J. Cell Biol.* **84**, 66–76.
178. Krinks, M. H., Haiech, J., Rhoads, A., and Klee, C. B. (1984). *Adv. Cyclic Nucleotide Protein Phosphorylation Res.* **16**, 31–47.
179. Wallace, R. W., Lynch, T. J., Tallant, E. A., and Cheung, W. Y. (1979). *JBC* **254**, 377–382.
180. Wolf, H., and Hofmann, F. (1980). *PNAS* **77**, 5852–5855.
181. Klee, C. B., Krinks, M. H., Manalan, A. S., Cohen, P., and Stewart, A. A. (1983). "Methods in Enzymology," Vol. 102, pp. 227–244.

182. Tallant, E. A., Wallace, R. W., and Cheung, W. Y. (1983). "Methods in Enzymology," Vol. 102, pp. 244–256.
183. Winkler, M. A., Merat, D. L., Tallant, E. A., Hawkins, S., and Cheung, W. Y. (1984). PNAS 81, 3054–3058.
184. Merat, D. L., Hu, Z. Y., Carter, T. E., and Cheung, W. Y. (1984). BBRC 122, 1389–1396.
185. Tallant, E. A., and Cheung, W. Y. (1984). Biochemistry 23, 973–979.
186. Aitken, A., Cohen, P., Santikarn, S., Williams, D. H., Calder, A. G., Smith, A., and Klee, C. B. (1982). FEBS Lett. 150, 314–318.
187. Kretsinger, R. H., and Nockolds, C. E. (1973). JBC 248, 3313–3326.
188. King, M. M., and Huang, C. Y. (1984). JBC 259, 8847–8856.
189. Li, H.-C., and Chan, W. W. S. (1984). EJB 144, 447–452.
190. King, M. M., Huang, C. Y., Chock, P. B., Nairn, A. C., Hemmings, H. C., Chan, K.-F. J., and Greengard, P. (1984). JBC 259, 8080–8083.
191. Wolff, D. J., and Sved, D. W. (1985). JBC 260, 4195–4202.
192. Greengard, P., and Chan, K.-F. J. (1983). FP 42, 2048 (Abstr. 1701).
193. Walaas, S. I., Aswad, D. W., and Greengard, P. (1983). Nature (London) 301, 69–71.
194. Aswad, D. W., and Greengard, P. (1981). JBC 256, 3487–3493.
195. Chernoff, J., Sells, M. A., and Li, H. C. (1984). BBRC 121, 141–148.
196. Pallen, C. J., and Wang, J. H. (1983). JBC 258, 8550–8553.
197. Li, H.-C. (1984). JBC 259, 8801–8807.
198. Manalan, A. S., and Klee, C. B. (1983). PNAS 80, 4291–4295.
199. Gupta, R. C., Khandelwal, R. L., and Sulakhe, P. V. (1984). FEBS Lett. 169, 251–255.
200. King, M. M., and Huang, C. Y. (1983). BBRC 114, 955–961.
201. Pallen, C. J., and Wang, J. H. (1983). Can. Fed. Biol. Soc. Proc. 26, 52 (abstr.).
202. Pallen, C. J., and Wang, J. H. (1984). JBC 259, 6134–6141.
203. Matsui, H., Pallen, C. J., Adachi, A.-M., Wang, J. H., and Lam, P. H.-Y. (1985). JBC 260, 4174–4179.
204. Binstock, J. F., and Li, H.-C. (1979). BBRC 87, 1226–1234.
205. Hiraga, A., Kikuchi, K., Tamura, S., and Tsuiki, S. (1981). EJB 119, 503–510.
206. Mieskes, G., Brand, I. A., and Soling, H.-D. (1984). EJB 140, 375–383.
207. Pato, M. D., and Adelstein, R. S. (1983). JBC 258, 7055–7058.
208. Li, H.-C., Tabarini, D., Cheng, Y.-S., and Chen, L. B. (1981). FP 40, 1539 (Abstr. 5).
209. Mieskes, G., and Soling, H.-D. (1985). BJ 225, 665–670.
210. Feliu, J. E., Hue, L., and Hers, H.-G. (1977). EJB 81, 609–617.
211. Berglund, L., Ljungström, O., and Engström, L. (1977). JBC 252, 613–619.
212. El-Maghrabi, M. R., Haston, W. S., Flockhart, D. A., Claus, T. H., and Pilkis, S. J. (1980). JBC 255, 668–675.
213. Mojena, M., and Feliu, J. E. (1983). Mol. Cell. Biochem. 51, 103–110.
214. Erickson, R. L., Collett, M. S., Erickson, E. L., and Purchio, A. F. (1979). PNAS 76, 6260–6264.
215. Hunter, T., and Sefton, B. M. (1980). PNAS 77, 1311–1315.
216. Collett, M. S., Purchio, A. F., and Erickson, R. L. (1980). Nature (London) 285, 167–169.
217. Sefton, B. M., and Hunter, T. (1984). Adv. Cyclic Nucleotide Protein Phosphorylation Res. 18, 195–226.
218. Pike, L. J., and Krebs, E. G. (1985). In "The Receptors" (P. Cohen, ed.). Academic Press, New York (in press).
219. Carpenter, G., King, L., and Cohen, S. (1979). JBC 254, 4884–4891.
220. Ushiro, H., and Cohen, S. (1980). JBC 255, 8363–8365.
221. Sefton, B. M., Hunter, T., Beemon, K., and Eckhart, W. (1980). Cell 20, 807–816.
222. Foulkes, J. G. (1983). Curr. Top. Microbiol. Immunol. 107, 163–180.

223. Shriner, C. L., and Brautigan, D. L. (1984). *JBC* **259,** 11383–11390.
224. Martensen, T. M. (1982). *FP* **41,** 443 (Abstr. 1015).
225. Nelson, R. L., and Branton, P. E. (1984). *Mol. Cell. Biol.* **4,** 1003–1012.
226. Horlein, D., Gallis, B., Brautigan, D. L., and Bornstein, P. (1982). *Biochemistry* **21,** 5577–5584.
227. Swarup, G., Speeg, K. V., Cohen, S., and Garbers, D. L. (1982). *JBC* **257,** 7298–7301.
228. Swarup, G., Cohen, S., and Garbers, D. L. (1982). *BBRC* **107,** 1104–1109.
229. Brautigan, D. L., Bornstein, P., and Gallis, B. (1981). *JBC* **256,** 6519–6522.
230. Foulkes, J. G., Howard, R. F., and Ziemiecki, A. (1981). *FEBS Lett.* **130,** 197–200.
231. Swarup, G., Cohen, S., and Garbers, D. L. (1981). *JBC* **256,** 8197–8201.
232. Sparks, J. W., and Brautigan, D. L. (1985). *JBC* **260,** 2042–2045.
233. Gallis, B., Bornstein, P., and Brautigan, D. L. (1981). *PNAS* **78,** 6689–6693.
234. Moberg, E. A., and Brautigan, D. L., personal communication.
235. Cassel, D., and Glaser, L. (1982). *PNAS* **79,** 2231–2235.
236. Leis, J. F., and Kaplan, N. O. (1982). *PNAS* **79,** 6507–6511.
237. Li, H.-C., Chernoff, J., Chen, L. B., and Kirschonbaum, A. (1984). *EJB* **138,** 45–51.

Section II

Control of
Specific Enzymes

9

Glycogen Phosphorylase

NEIL B. MADSEN

Department of Biochemistry
University of Alberta
Edmonton, Alberta
Canada T6G 2H7

THE ENZYMES, Vol. XVII

I. Introduction

Glycogen phosphorylase from skeletal muscle was the first enzyme found to be metabolically interconvertible between two discrete forms that exhibit different modes of allosteric control (*1*). The interconversion was later shown to involve the phosphorylation of a single serine residue by a specific protein kinase (*2–4*) and its dephosphorylation by protein phosphatase (*5*). The wealth of accumulated detail available for phosphorylase and for its interconverting enzymes exceeds that for the other interconvertible enzymes and amply justifies a separate chapter. The interconverting enzymes themselves are dealt with in other chapters and the historical aspects are also expanded elsewhere in this volume. Although glycogen phosphorylase has not been reviewed in this treatise since 1972 (*6*), a number of other reviews are available (*7–12*) which deal with various aspects of the structure–function relationships of the enzyme and with its regulation. In this review, emphasis is placed on the glycogen phosphorylases from skeletal muscle because these are the only ones for which adequate structural information is available and all comments are directed to this system unless explicitly stated otherwise.

This chapter begins by briefly reviewing the state of our knowledge of the enzymic characteristics of phosphorylase. This should remind us that regulation of this enzyme by phosphorylation is superimposed on the evolutionary earlier control by feedback inhibition (glucose and glucose 6-phosphate) and by the cell's energy charge (inhibition by ATP; activation by AMP and P_i). The molecular structure of the enzyme is reviewed because it is the best delineated of any protein dealt with in this volume and knowledge of this structure has provided us with at least partial explanations of the effects of phosphorylation and of its control.

Regulation of the phosphorylation state of glycogen phosphorylase has been touched on in other chapters but here it is appropriate to deal with substrate-directed controls. Especially interesting in this regard is the glucose-mediated activation of phosphorylase phosphatase in the liver because here we can see that an ancient feedback control, operating via allosteric conformational changes, has been adapted to serve as one of the modern physiological mechanisms for glucose homeostasis. The structural consequences of phosphorylating serine-14 appear to arise chiefly from the binding of the N-terminus across the subunit interface with consequent strengthening of the intersubunit interactions within the dimer. One of the functional results of this phenomenon is to change the allosteric behavior from the sequential to the concerted model, thus resolving an old controversy w1thin a single enzyme. Phosphorylase *a,* by having a covalently bound activating ligand in the form of the phosphoserine, has had its allosteric equilibrium constant (ratio between inactive and active conformers) decreased by three orders of magnitude. Therefore it has "escaped" allosteric control in that it is no longer dependent on AMP for activity nor is the inhibition by ATP or glucose-6-P of significance. In practical physiological terms, phosphorylase activity in the muscle can be turned on and off by its phosphorylation–dephosphorylation. Nevertheless many of the basic characteristics of the enzyme are unaltered by its phosphorylation state, including the catalytic mechanism and substrate specificity, the binding to the glycogen particle, and the inhibition by glucose binding to the active site and compounds binding to the negative effector site.

II. Enzymic Characteristics of Phosphorylase

A. BIOLOGICAL ROLE

Glycogen phosphorylase (1,4-α-D-glucan:orthophosphate α-D-glycosyltransferase, E.C. 2.4.1.1) catalyzes the first step in the intracellular degradation of glycogen as shown in Scheme I.

$$(\alpha\text{-1,4-Glucoside})_n + P_i \rightleftharpoons (\alpha\text{-1,4-glucoside})_{n-1} + \alpha\text{-D-glucose-1-P}$$

SCHEME I

when n represents the number of glucosyl residues. Although the measured equilibrium constant is somewhat less than one at neutral pH (13), the reaction *in vivo* proceeds as written because the intracellular concentration of P_i greatly exceeds that of glucose-1-P (14). Glycogen is a highly branched molecule and phosphorylase can degrade only to within four glucosyl units of an α-1,6-linked side chain. A second enzyme, colloquially known as the debranching enzyme, is needed to deal with the branch points. It is not known to be under any control, but is has been suggested to become rate limiting in muscle when phosphorylase is fully activated (15).

The role of phosphorylase, then, is to supply the glycolytic pathway with a regulated amount of phosphorylated glucose units derived from tissue glycogen stores. In muscle this function is regulated according to the energy needs of contraction, whereas in liver the enzyme's activity is modulated to provide enough glucose to maintain the fasting blood-sugar level. In other tissues the enzyme may respond to emergency situations such as anoxia. Glycogen is distributed widely in nature and its utilization via phosphorylase appears to be related to energy needs in growth, development, or starvation, as well as for the purposes previously discussed.

Electron microscopy reveals the presence of large-molecular-weight aggregates of glycogen in both muscle (16) and liver (17). These "glycogen particles," when isolated under mild conditions, contain the majority of the phosphorylase in both tissues, as well as the other enzymes associated with glycogen metabolism (*18, 19*). Phosphorylase (and glycogen synthase) are rather unique because, while quite soluble and easily extracted in aqueous media, they are rendered particulate by binding to their substrate via separate "glycogen storage" sites. Certainly for phosphorylase, and possibly for the synthase, this mode of binding orients their "control faces" toward the cytosol and permits metabolic interconversion while still bound to the glycogen.

B.   KINETICS AND MECHANISM

The kinetic mechanism of glycogen phosphorylates from liver, muscle, and some other tissues has been established as rapid-equilibrium random bi bi through the use of conventional kinetic studies (*20–22*) and, in some cases, isotope exchange at equilibrium (*23, 24*). The latter technique provides convincing evidence that the kinetic mechanism is not altered by the interconversion between the *a* and *b* forms, nor by changing the normal Michaelian kinetics to the allosteric type in the presence of ATP or glucose (*25*). This type of kinetic mechanism is useful for studying an allosteric enzyme because the apparent Michaelis constants are, in principle, equivalent to dissociation constants. It does not mean, however, that a covalently bound reaction intermediate might not occur during the catalytic process.

The catalytic mechanism has proven to be very elusive but the key to it must be the presence in the active site of pyridoxal phosphate. Since the possible roles of this coenzyme in the mechanism have been reviewed (*12*) as well as extensively discussed in the literature (*26, 27*), only a cursory summary is given here. One may remember, too, that a complete understanding of catalytic mechanisms is not necessary for an elucidation of their biological regulation. The role of pyridoxal phosphate in catalysis by phosphorylase is recognized as being unconventional because years of painstaking research have eliminated consideration of all parts of the coenzyme except the phosphate moiety, and that is essential (*12*). Both crystallographic (*28, 29*) and chemical (*30, 31*) studies suggest that the phosphate of the coenzyme may interact with the phosphate of the substrate but the precise involvement remains unclear. Discussion revolves around the possibility that the phosphate acts as a proton donor (acid catalyst) (*27, 32*) or as an electrophile (*26, 33, 34*).

Since the active site is buried deeply within the monomer, it is not surprising that conventional chemical modification studies have not been very rewarding. Although modification of a single carboxyl group causes inactivation (*35*), there is no evidence that this group is within the active site, although a carboxyl group is an attractive candidate for involvement in an enzyme-bound glucosyl intermediate. Lys-573 and Arg-568 are found within the active site by crystallographic analysis and are also located there by chemical means (*36, 37*). The former may bind the phosphate of the coenzyme while the latter may interact with the phosphate substrate. Other groups within the active site that may be involved with catalysis include Asp-283, Tyr-572, and His-570. The latter may account for the alkaline pK of 7.1 in the pH-activity curve (*38*). Finally, several lines of research provide indirect evidence for a glucosyl–enzyme intermediate in the catalytic reaction (*39*).

C. ALLOSTERIC CONTROL

Manifestations of allosteric control were discovered very early, consisting of the requirement for AMP for the activity of phosphorylase *b* in 1936 (*40*), the inhibition by glucose in 1940 (*13*), and the inhibition of phosphorylase *a* by glucose in 1943 (*41*). In the latter case, Cori, Cori, and Green showed clearly that the kinetics for the substrate, glucose 1-P, change from first to second order. Phosphorylase *a*, of course, was shown to have no requirement for AMP although its activity was stimulated by 50% and its dissociation constant for this effect was 1.5 μM instead of 30 to 50 μM for the *b* form (in the presence of saturating substrates). Parmeggiani and Morgan discovered the inhibition of phosphorylase *b* by ATP in 1962 (*42*) and from then on a number of laboratories have published many papers dealing with the allosteric phenomena in this enzyme. Several reviews may be consulted for details (*7, 8, 11*) and further discussion of allostery are deferred until the effects of serine-14 phosphorylation are discussed in Section VI.

III. Molecular Structure

A. SUBUNIT RELATIONSHIPS

Glycogen phosphorylase was the first enzyme shown to exist in oligomeric form. The first step in this exposition was the discovery that the interconverting enzyme, which we now know is a protein phosphatase, not only caused a change in the enzymic characteristics when converting the *a* form to the *b* form, but also caused a concomitant halving of the molecular weight (*43*). Not long afterward it was discovered that titration of sulfhydryl groups with mercurials resulted in a correlated loss of activity and an additional halving of the molecular weight, so that phosphorylase *a* was shown to be a tetramer of identical subunits whereas phosphorylase *b* was a dimer (*44*). The initial assumption of identical subunits, a reasonable interpretation of the then available data, was questioned more than once and cannot be considered to have been proved until much later when sufficient amino acid sequence data became available. The reader should be reminded that the difference in oligomeric structure between the *a* and *b* forms is a rather unique feature of some skeletal muscle systems, since the phosphorylase *a* enzymes of liver (*45*), lobster muscle (*46*), and rabbit heart (isozyme I) (*47*) are dimers. While AMP, especially in the presence of magnesium, may promote a tetrameric form of phosphorylase *b* (*48*), the functional oligomer for both forms of the muscle enzyme appears to be the dimer, since the latter has a higher affinity for glycogen and a higher specific activity (*49*).

B. PRIMARY STRUCTURE AND COMPARATIVE STUDIES

The amino acid sequence of phosphorylase was announced in 1977 (*50*), just in time to play a key role in the crystallographic studies. Each monomer contains 841 amino acids and together with the coenzyme and the phosphate on serine-14, yields a molecular weight of 97,412 for phosphorylase *a*. Since then, almost complete sequences have been obtained for the maltodextrin phosphorylase of *E. coli* (*51*) and for the phosphorylase found in potato (*52*). In addition, sequences surrounding the phosphorylatable site and the coenzyme site have been determined for phosphorylases from yeast (*53*) and dogfish muscle (*54*). Although these authors have commented on the significance of their findings, a thorough review and assessment is still awaited. I have reviewed these findings briefly (*12*) and reiterate the main points of interest in this chapter. There is almost 100% homology for the residues surrounding the coenzyme for all these phosphorylases. In addition, where the residues comprising the active site are known (rabbit muscle, *E. coli,* and potato enzymes) the homology also approaches 100%. It is only slightly less for residues comprising the twisted beta-sheet cores of the N-terminal and C-terminal domains but considerably less for the outer helices and loops. This suggests, as noted by Palm *et al.* (*51*), that the basic structure of all phosphorylases is remarkably similar. The greatest dissimilarities occur in the residues comprising the glycogen storage site of the rabbit muscle enzyme (see below), since the *E. coli* and potato enzymes have no such sites, and the N-terminal residues associated with regulation, both phosphorylation and AMP binding. Thus the *E. coli* enzyme lacks the first 17 residues of the muscle phosphorylase, and hence the phosphorylatable serine. There is virtually no homology until residue 80, so that the nucleoside subsite for AMP binding cannot be located while the arginine residues comprising the subsite for the phosphate moiety are present, suggesting a common primitive phosphoryl binding site which was extended in some eukaryotes. The potato phosphorylase also shows little homology until residue 80 of the muscle enzyme sequence is reached. In the dogfish enzyme, 12 of 14 residues surrounding the phosphorylatable serine are identical with those of the muscle enzyme, and the interconvertible enzymes are freely interchangeable. However, the phosphorylatable residue in yeast phosphorylase is a threonine in a sequence that is not similar to those that contain the phosphorylatable serine in other phosphorylases. The sequence studies available present a consistent picture of an ancient enzyme that has shown very strong conservation of the sequences concerned with the catalytic mechanism and basic structure of the protein. The major evolutionary divergence has been concerned with the N-terminal residues where most of the control elements are located, including interconversion by phosphorylation, AMP binding and activation, and subunit interactions.

C. THREE-DIMENSIONAL STRUCTURE

Crystallographic studies in Oxford and Edmonton (later San Francisco) have developed structures of phosphorylase b at 2.0 Å resolution (55–58) and phosphorylase a at 2.1 Å resolution (10, 59, 60) although full descriptions of the molecules at high resolution have not been published. The crystals of phosphorylase a were grown in the presence of glucose, which is found in the glucose-binding subsite of the active site, and the structure is therefore that of the inactive T conformation. This presents severe problems in trying to assess the catalytic mechanism but the general structure, features of the phosphorylatable serine site, and the ligand-binding sites can all be determined. The phosphorylase b crystals, grown in the presence of IMP, are isomorphous with those of the a form and also contain the enzyme in its T-state conformation, although one or two features of an expected active R-state are apparent. In general the structures of the a and b forms are quite similar except at the control sites on the N-terminus, as discussed in Section V.

The structure of phosphorylase is typical of the α and β class of proteins. Two major domains each consist of a core of twisted β-sheet surrounded by α-helices. The domains are associated closely together, with their interface containing the pyridoxal phosphate and the large cavity of the active site, the latter being some 15 Å from the surface of the protein. The two subunits of the dimer are closely associated through contacts between the N-terminal domains and it is via this interface region that both homotropic allosteric interactions between subunits and heterotropic interactions within a subunit are transmitted between ligand-binding sites (58, 61, 62). Furthermore, the effects of serine-14 phosphorylation are manifested through changes in these intersubunit contacts, discussed in Section V.

Although the two subunits of the phosphorylase dimer are related by a twofold axis of symmetry, the dimer is remarkably asymmetric, as we illustrated by computer-drawn space-filling models (62). One can define a concave "catalytic face" which is bound to the glycogen molecule by the glycogen storage sites on the periphery of this face. Closer to the diad axis is found the active site cavity. Ligand-binding and/or kinetic studies have shown that glucose or stereochemically similar compounds bind in the glucose subsite of the active site and promote the T conformation in both *a* and *b* forms of phosphorylase (*60, 63–66*). Glucose-1-P and analogues such as UDPglucose and glucose 1,2-cyclic phosphate bind in much the same location but promote the R conformation (*55, 67–69*). On the solvent side of a protein loop that forms part of the active site is found an inhibitor site that binds purine derivatives such as inosine, adenosine, and caffeine. The effects of these compounds are synergistic with those of glucose because they and glucose stabilize each other's binding sites. Thus Asn-284 forms a hydrogen bond with the 2-hydroxyl of glucose whereas, on the outside of the loop, the phenyl ring of Phe-285 sandwiches the base between itself and Tyr-612 (*60, 70*). Because the loop containing Asp-283, Asn-284, and Phe-285 must move to accommodate glucose-1-P or phosphorylated glucose inhibitors (*69*), glucose and caffeine exhibit competitive inhibition with respect to glucose-1-P and are synergistic in stabilizing the T-state conformation for both *a* and *b* forms of the enzyme.

On the side of the dimer opposite the catalytic face one can define a convex "control face." Here are found the two binding sites for AMP, each consisting of a phosphoryl site of two or three arginines, while the nucleoside portion interacts with several residues including some for the symmetry-related subunit (*71*). Important among these residues is Tyr-75 because chemical studies indicate it is involved in AMP binding (*72, 73*). It does not interact with AMP in the T-state phosphorylase *a* (*63*) but it does interact with the purine base of AMP in the phosphorylase *b* crystals (*71*). This suggests a vaguely defined mechanism for the movement of helix 49 to 75 which is seen upon the addition of substrate to the phosphorylase *a* crystals (*61*), as well as a means for communicating homotropic cooperativity between AMP-binding sites since the above mentioned helix connects these. Also binding to the AMP site are the inhibitors ATP and glucose-6-P, showing slightly different binding modes and causing different conformations of adjacent protein groups, as has been well defined for the physiologically significant case of phosphorylase *b* (*58, 74*). Although the binding sites for AMP and glucose-6-P do not overlap, their respective phosphates both bind to Arg-309, which occupies different positions in the two cases and explains the exclusivity of binding observed previously.

Close by the AMP-binding site is the site for phosphorylation, Ser-14. The phosphate of the latter, as seen in crystalline phosphorylase *a*, forms a salt bridge to Arg-69 of helix 49–75, just to the solvent side of AMP, thus forming an obvious link between the two control sites. Since residues 1–18 cannot be visualized in crystalline phosphorylase *b* and are assumed to be flexible (*75*) whereas all but the first 4 are seen in phosphorylase *a*, one of the obvious effects of the phosphorylation of phosphorylase *b* is to create a covalently attached ligand which now lies across the intersubunit interface and strengthens it by the formation of new salt bridges and other interactions (*76*). Therefore among the ligand-binding sites on phosphorylase *a* may be placed the two for the serine-14 phosphate (Arg-69 from the same subunit and Arg-43 from the symmetry-related subunit), and one for each of Arg-10 and Lys-11, which form hydrogen bonds to Gly-116 and Tyr-113, respectively, from the symmetry-related subunit.* The effects of this intersubunit binding of the N-terminal residues of phosphorylase a are discussed in Section V.

IV. Substrate-Directed Control of Phosphorylation and Dephosphorylation

A. PHOSPHORYLASE b KINASE

Krebs *et al.* (*77*) showed that glycogen decreases the K_m of phosphorylase b for nonactivated kinase by a factor of 10, and that this was not likely to be due to its stimulation of kinase activation by direct binding. Tabatabai and Graves (*78*) found that glycogen had no significant effect on the kinase action on the synthetic tetradecapeptide containing residues 5 to 18. We may therefore assume that when phosphorylase b is bound to glycogen, leaving the control face exposed, there are sufficient conformational changes to improve the binding of the kinase. With a rapid-equilibrium random bi bi mechanism (*78*), the K_m should be a measure of binding. Presumably, with a mobile N-terminus, the effect of glycogen is on residues surrounding Ser-14 other than the first 20.

The promoter of the allosteric R-state, AMP, does not have a significant effect on kinase activity (*77*), nor does another substrate, glucose-1-P. The allosteric inhibitors, glucose-6-P and ATP, do inhibit kinase activity on phosphorylase b (*77*). The ATP effect is difficult to analyse since there is inhibition of the kinase when acting on the synthetic peptide (*78*) but the inhibition by glucose-6-P is unequivocally substrate-directed because kinase action on the synthetic peptide is not affected (*79*). The latter study also showed that inhibition by glucose-6-P is competitive with respect to phosphorylase b, noncompetitive to MgATP, suggesting that the sugar phosphate, when bound to the activator site, promotes a conformation of phosphorylase that binds poorly to the kinase. Covalently bound AMP prevents the inhibition. A study employing cross-linking reagents suggests that AMP does not affect the motility of the N-terminus (*80*). Although this conclusion has not been confirmed by definitive protein chemistry, it would be consistent with the general concept that substrate-directed effects on phosphorylase kinase activity are not mediated via the N-terminus but through changes in the adjacent protein structure, which constitutes the kinase-binding

*Personal communications from Dr. R. J. Fletterick, reported also in his address before "The Robert A. Welch Foundation Conferences on Chemical Research," XVII, Houston, Texas (1983).

site and which lowers the K_m by 70-fold in comparison to the synthetic peptide (78).

B. Phosphorylase a Phosphatase

The major portion (~75%) of the phosphorylase a phosphatase activity in skeletal muscle is accounted for by protein phosphatase-1, as designated by Cohen and his colleagues (81). Furthermore, protein phosphatase-1 is virtually the only activity found in the glycogen particle (81, 82). Therefore the preparation of phosphorylase a phosphatase utilized in the studies of Martensen et al. (83, 84) was most likely composed chiefly of protein phosphatase-1 and, as discussed in (82), that studied by Detwiler et al. (85) almost certainly was. Martensen et al. (83, 84) utilized both phosphorylase a and the phosphorylated tetradecapeptide (residues 5–18 of phosphorylase a) to determine whether ligands affected the phosphatase directly or via interaction with the substrate. Surprisingly, inorganic phosphate, a substrate for phosphorylase, was a competitive inhibitor for either phosphatase substrate, a result compatible with the idea that this phosphatase may have the ordered uni bi kinetic mechanism commonly found for hydrolytic enzymes. AMP, as shown earlier in (86), caused inhibition that was substrate directed, as was true also for glucose-1-P because there was no inhibition with the phosphorylated peptide as substrate. Both AMP (83, 85) and glucose-1-P (83) caused inhibition which was superficially competitive with phosphorylase a, which was true also for UDPglucose (85). Since all these compounds promote the R-state of phosphorylase a, one may assume that the latter either binds poorly or not at all to the phosphatase or is a poor substrate.

Martensen et al. (84) also studied the activation of the phosphatase by metabolites and demonstrated that glucose, glucose-6-P, and glycogen all acted by binding to the substrate phosphorylase a. The first two caused large increases in maximal velocity with no significant change in the K_m. Glycogen did not affect the V_{max} but lowered the K_m significantly.

Detwiler et al. (85) examined the effects of metabolites on the kinetics of muscle phosphorylase a phosphatase acting on the natural substrate. They concluded that the AMP–phosphorylase a complex was a poor substrate with a V_{max} only 5% of the unliganded substrate but with a K_m increased by only 50%. Even for this kinetic mechanism, the results suggested that the affinity between phosphatase and substrate has not been altered significantly and that the inhibition affects primarily the velocity of the catalytic reaction. Similarly, the activation by glucose or glucose-6-P resulted in a three- or twofold increase in V_{max} with little change in K_m. Caffeine, long known to be an activator (87), was also shown to act through the substrate (86). With the exception of ADP and ATP (85), which cause inhibition, all the metabolites that stabilize the inactive T conformation are activators, whereas all the compounds that promote the active R configuration of phosphorylase *a* are inhibitors of the phosphatase. The allosteric control is therefore reinforced or superceded with a compatible control of covalent interconversion; ATP and ADP may not be an exception because they inhibit phosphorylase *a* only slightly, as does glucose-6-P. These compounds, while binding at the activator site, may cause local changes around the serine-14 phosphate without affecting the active site.

The substrate-directed activations and inhibitions of phosphorylase phosphatase have been interpreted at the molecular level from considerations of the X-ray-derived crystal structure of phosphorylase and the effects on that structure of allosteric effectors (*61, 62*). We suggested that AMP, and/or the active-site ligands glucose-1-P and UDPglucose, stabilize the active R conformation of phosphorylase *a* to which the phosphatase may bind but in which the serine phosphate is firmly bound and unavailable to the active site of phosphatase. This presumed firm binding of the serine phosphate and adjacent peptide chain is consistent with the crystallographic studies (*61*), NMR data (*88*), cross-linking studies (*89*), and protection of the N-terminal chain against proteolytic attack (*90*). Glucose and caffeine, either singly or acting synergistically together, stabilize the inactive T conformation in which the serine phosphate is exposed to catalytic action by the phosphatase but in which the surrounding binding site for the modifying enzyme is not altered significantly.

Parallel with these developments in the muscle system, Hers and Stalmans and their collaborators developed the hypothesis that phosphorylase *a* is the glucose receptor in the liver cell (*91–93*). They demonstrated *in vivo*, in isolated hepatocytes, and *in vitro*, that glucose activates phosphorylase *a* phosphatase and that phosphorylase *a* inhibits the phosphatase for glycogen synthase. *In vivo*, a glucose load results in the conversion of phosphorylase *a* to *b* and, after a lag period of approximately the two minutes needed to reduce the proportion of phosphorylase *a* to less than 10 to 20% of the total, the activation of glycogen synthase and deposition of glycogen. The molecular basis for the *in vitro* behavior of the muscle system may also be applied to the liver system. Even if there are separate phosphatases for phosphorylase and synthase, inhibition of the one for synthase by phosphorylase *a* until the concentration of phosphorylase *a* is reduced below a minimal value is consistent with the mechanism of glucose activation. The very poor binding of phosphorylase *b* to the phosphorylase phosphatase (K_m = 140 μ*M*) is consistent if this phosphatase is the one that also acts on the synthase in response to the physiological stimulus to glucose.

Work by Cohen and collaborators has shown that the collective activities of protein phosphatases-2A in rabbit liver account for 60% of the activity with phosphorylase *a* as substrate and 40 to 70% with the various phosphorylated sites of glycogen synthase (*81*). Therefore control of a single phosphatase, inhibited by phosphorylase *a* in the presence of AMP and substrates, activated by glucose, and then released from the phosphorylase system to act on glycogen synthase and other enzymes when there is no longer sufficient phosphorylase a left to bind it, provides the simplest explanation for the observed effects of glucose. This does not preclude additional controls for the same or other protein phosphatases. Kasvinsky et al. (94) showed that caffeine increases the glucose-induced phosphatase action on phosphorylase a in intact heptocytes, but it is not known if there is a natural metabolite in liver that utilizes the inhibitor site on phosphorylase a. Monanu and Madsen (95) have studied the substrate-directed regulation of liver protein phosphatases-$2A_1$ and -$2A_2$ acting on liver phosphorylase a and have confirmed that the essential elements elucidated for the muscle system are present. Both AMP and ATP are particularly potent inhibitors that require glucose and caffeine acting together for reversal. Ligands have much greater effects on maximal velocity than on K_m.

In assessing the physiological significance of these substrate-directed metabolite regulations of phosphorylase a phosphatase activity, one may note, as discussed by Martensen et al. (84), that the concentration of glucose-6-P in frog sartorius muscle subjected to tetanic stimulation may rise to 3.6 mM. The concentration level and the timing suggest that this metabolite may play a role in the decline of phosphorylase a observed after prolonged stimulation. The activation of the phosphatase is complemented by the inhibition of phosphorylase kinase (79). Since little if any free glucose is observed in muscle, it can play no role in the regulation of protein phosphatase activity in that tissue. In liver, however, with an in vivo glucose concentration equivalent to that of the serum, and with the wealth of physiological evidence provided by Hers and his colleagues, we can assume that there is a major physiological role for glucose as a regulatory metabolite.

The coenzyme, pyridoxal phosphate, probably plays no physiological role in the regulation of phosphorylase a phosphatase in muscle, but a comprehensive study demonstrated that its absence or alteration affected phosphatase activity and regulation significantly (96). Apophosphorylase a was dephosphorylated more rapidly than the native enzyme but glucose or caffeine did not increase the rate, whereas AMP was still able to inhibit. Examination of dephosphorylation rates of phosphorylases reconstituted with various pyridoxal phosphate analogs showed that, in general, only analogs that restore catalytic activity to the phosphorylase would restore the effects of glucose and caffeine and, in particular, this meant having two ionizable hydrogens at the 5'-position. Therefore the coenzyme is needed for the transmission of conformational changes in at least the case of caffeine, which can dissociate pyridoxal-reconstituted phosphorylase a from tetramer to dimer without increasing the phosphatase activity. This example, as well as others given, show that the activating effects of glucose and caffeine cannot be explained completely by suggesting that it is due to the dissociation of tetramers to dimers (97). Furthermore, the idea that the phosphatase dephosphorylates only the dimeric form (97) was contradicted by the finding that glucose increases by 6-fold the phosphatase action on phosphorylase *a* reconstituted with the phosphoethenyl analog of pyridoxal phosphate without causing it to dissociate (96). We should remember, too, that the substrate-directed control of liver phosphorylase *a* phosphatase cannot operate via alterations in the oligomeric structure.

V. Structural Consequences of Serine-14 Phosphorylation

A. TIGHTER SUBUNIT INTERACTIONS

As mentioned in Section III,C, crystallographic studies on phosphorylase *b* failed to locate the first 18 residues of the N-terminus (75), whereas all but the first 4 are well defined in the structure of crystalline phosphorylase *a* (59) where it lies across the subunit interface. These differences between the *a* and *b* forms in the crystalline state are consistent with solution studies published before and after the structures became available. For example, it was shown that subtilisin carries out a limited proteolysis of phosphorylase *b*, cleaving on the carboxy side of residues 16 and 264, whereas with phosphorylase *a* the activity is confined largely to residue 264 (50, 98). The slight activity of subtilisin on the N-terminus of the *a* form is decreased even further by AMP and increased only slightly by glucose or glycogen (99). The latter finding appears to rule out the tetrameric state of phosphorylase *a* as a reason for differences in the proteolytic action and focuses attention on the conformation of the N-terminus. The marked acceleration of subtilisin action at residue 16 by glucose-6-P (99) agrees with the effect of this ligand on phosphatase activity, as previously discussed. It is interesting to note that all ligands tested, whether promoters of the inactive T-state or of the active R-state, decreased the inactivation rate by subtilisin (99), suggesting a tighter association of the "tower loop" (containing the Gln-264–Ala-265 bond) with the symmetry-related subunit.

The structure of the N-terminus and its association with the main body of the protein was described at a resolution of 2.5 Å (59, 62), and the extra stabilizing effect on the stimuli interactions of the dimer has been discussed extensively (76). Figure 1 depicts interactions of the N-terminus with residues from both subunits, drawn from more highly refined coordinates at a resolution of 2.1 Å. The phosphate makes on intrasubunit salt bridge to Arg-69 of the helix containing the Tyr-75 which may well bind to the adenine moiety of AMP in the active R conformation. The role of this salt bridge in communicating between the serine phosphate and the activating AMP is discussed in Section III,C (61). A new intersubunit salt bridge is formed, between the serine phosphate and Arg-43. In addition, two new hydrogen bonds are formed between Arg-10 and Lys-11 to

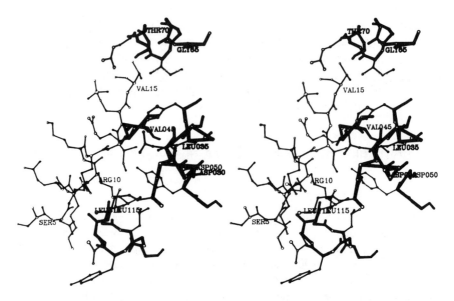

FIG. 1. Stereo diagram to illustrate the interactions of the phosphorylated N-terminus with residues from both subunits. Residues 65–70 and 5–15 (the N-terminus) are from one subunit; all others are from the symmetry-related subunit. The tetrahedral phosphate on Ser-14 (to the left of Val-15) interacts with Arg-69 from above and Arg-43' from the other subunit (both distances = 2.6 Å). An —NH₂ of Arg-10 interacts with the carbonyl oxygen of Gly-116' (2.9 Å) (just below the caption for Leu-115'). The ε-amino group of Lys-9 forms a similar hydrogen bond with the carbonyl oxygen of Tyr-113' (3.0 Å) near the bottom of the diagram. (The side chain of Tyr 113' is the lowest feature shown.) Figure 1 was drawn by Dr. R. J. Fletterick from coordinates of phosphorylase a at 2.1 Å resolution refined to an R-factor of 0.19 (S. Sprang and R. J. Fletterick, unpublished data).

Gly-116 and Tyr-113, respectively, of the symmetry-related subunit, and steric complementarity appears favorable for van der Waals contacts.* The two new salt bridges are in addition to the two found in both a and b forms, the latter being inaccessible to solvent and hence stronger. The two new salt bridges, while relatively accessible to solvent, are surrounded by polar and nonpolar contacts which may increase their stability. An NMR study indicated that the phosphate is titratable, hence accessible to solvent, and mobile (at least in the glucose-liganded protein) but behaves as though interacting with positive charges (*100*). One may assume that these salt bridges and their supporting environment, as well as the hydrogen bonds and sterically favorable van der Waals contacts, are the chief causes of the considerably increased stability and subunit interactions displayed by the phosphorylase a dimer as compared to dimer b.

Figure 2 illustrates a striking feature of the structure in the area of the N-

*Personal communications from Dr. R. J. Fletterick, reported also in his address before "The Robert A. Welch Foundation Conferences on Chemical Research," XVII, Houston, Texas (1983).

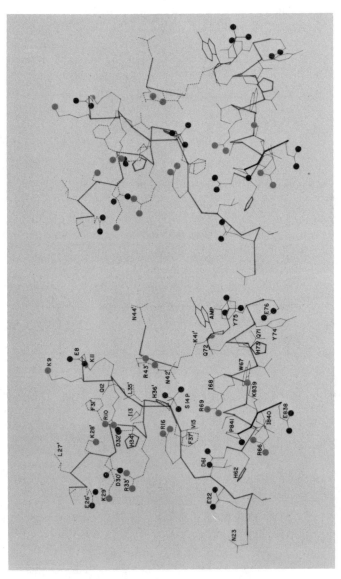

FIG. 2. Stereo diagram of the surface layer of residues surrounding Ser-14-P. Residues from the symmetry-related subunit are drawn in broken lines and the residue numbers are primed. This diagram was drawn from coordinates at a resolution of 2.5 Å and was published in different form in Ref. (62).

terminus. The negative charges of the serine phosphate are surrounded by positive charges derived from both subunits, while all but one of the nearby negative charges (Asp-32) are either oriented away from the surface or are outside the ring of positive charges. The eleven positive charges are His-73, Lys-839, Arg-69, Arg-66, Arg-16, Arg-10, Arg-33′, Lys-29′, Lys-28′, Arg-43′, and His-36′, where primed numbers indicate the symmetry-related subunit. The possible significance of these positive charges in providing a binding site for the phosphatase has been discussed thoroughly (62) and the analogy to similar asymmetric positive charge distributions in cytochrome c was pointed out, the latter providing binding sites for the enzymes that interconvert cytochrome c between its oxidized and reduced forms. Long before we knew the three-dimensional structure adjacent to the serine phosphate, the presence of four positive charges in the contingent amino acid sequence suggested to Fischer *et al.* (4) that the neutralization of these charges by the covalently attached phosphate permitted interactions at this locus which were previously electrostatically repulsed, thus leading to the activation of the enzyme and its association into tetramers. Sealock and Graves (101) studied the effects of various salts on phosphorylase activity and concluded that the interaction of the covalently bound phosphate with groups at a specific site is very sensitive to the ionic environment. The predictions of both these groups are consistent with the available structure.

Pertinent to our discussion of the location of the N-terminus is an immunological study carried out by Janski and Graves (102). They showed that antibodies specific for the N-terminal region of phosphorylase a are highly specific for only the first four amino-terminal residues of the enzyme. This agrees with the failure to locate these residues in the crystal structure. Furthermore, only one molecule of antibody binds per mole of dimer, causing a strong inhibition of both kinase and phosphatase activity but with less effect on the catalytic phosphorylase activity or K_m for glucose-1-P and AMP, and a decreased K_m for glycogen. Interestingly, the antibody stabilized phosphorylase b against resolution of the pyridoxal phosphate under the usual conditions, thus making it more like phosphorylase a and suggesting that the antibody is cross-linking the two monomers and preventing the dissociation necessary for resolution.

As discussed with regard to subtilisin, the N-terminal region is susceptible to proteases and it has long been known that trypsin will cause the formation of a pseudophosphorylase b species termed phosphorylase b' (103, 104). This species is active in the presence of AMP but it does not exhibit homotropic cooperativity for AMP or glucose-1-P and heterotropic interactions are greatly reduced (90). It is known that phosphorylase b' lacks the first 16 residues (4) and will bind the phosphorylated tetradecapeptide (residues 5–18) (105), with the resultant induction of enzymic properties similar to those of phosphorylase a. The same result was obtained with phosphorylase b, indicating that the dephosphorylated N-terminus could not prevent the phosphorylated peptide from binding, again emphasizing the difference between the *a* and *b* forms with respect to the conformation of the N-terminus. Finally, Janski and Graves (*106*) showed that nucleotide improved markedly the binding of a synthetic phosphorylated peptide (residues 1–18) to phosphorylase *b*. Surprisingly, in view of the crystal structure, studies with various lengths of synthetic peptide suggested that residues 1–4 improve the binding, whereas residues 5–6 (Ser-Asp) are very important. This is consistent with unpublished data from crystallographic studies which suggest that the N-terminus binds more extensively as well as more strongly in the R-state than in the T-state.*

Although the actual amount of extra energy of stabilization of the phosphorylase *a* dimer (compared to the *b* dimer) that has been conferred by the two new intersubunit salt bridges is the subject of some debate (*76*), the literature provides ample qualitative evidence for the existence of this stabilization. Thus phosphorylase *a* is more stable to thermal denaturation (*107*). Evidence has been provided that phosphorylase *b* dimers dissociate to a small extent and, at low concentrations, are completely dissociated in 2 *M* NaCl, whereas phosphorylase *a* is merely dissociated from tetrameric to dimeric form in that concentration of salt (*108*). The *b* forms of the heart-muscle phosphorylase isozymes hybridize readily whereas the *a* forms do not (*47*), and this result is pertinent to our present discussion because heart isozyme III appears to be identical to the single isozyme of skeletal muscle. Interestingly, enough, AMP or glucose-6-P were, like phosphorylation of serine-14, able to block hybridization. In another relevant study, Shaltiel *et al.* showed that 0.4 *M* imidazole citrate causes the dissociation of phosphorylase *b* into monomers (*109*), from which the coenzyme is readily resolved with cysteine, whereas this deforming salt does not dissociate phosphorylase *a* (*110*). Again, either AMP or phosphorylation of serine-14 prevents resolution of the pyridoxal phosphate in the imidazole citrate buffer (*104*). The stabilization of subunit interactions by AMP or glucose-6-P may be explained by their interaction with residues in both subunits (*58, 71*).

One more example of the subunit association in phosphorylase may be given to indicate that there remain unresolved ambiguities. Hedrick *et al.* (*111*) showed that apophosphorylase *b* exists as a tetramer at 0°, a dimer at 23°, and a monomer at 35°. A later study showed much the same behavior for apophosphorylase *a* (*112*), so that we are unable to distinguish between the phosphorylated and unphosphorylated forms of apophosphorylase with respect to intersubunit stability. This appears to be the only exception to evidence suggesting the greater stability of the phosphorylase *a* dimer. While removal of the coenzyme destabilizes the intersubunit associations, it has little effect on some other parameters since AMP can still bind (*111*), glycogen binds as well as to the holoenzyme (*113*), and phos-

*Personal communications from Dr. R. J. Fletterick, reported also in his address before "The Robert A. Welch Foundation Conferences on Chemical Research," XVII, Houston, Texas (1983).

phorylase kinase exhibits the same rate of reaction on the apo form of phosphorylase *b* as on the holo form (*111*). Thus the loss of the coenzyme does not appear to alter significantly the structure of the N-terminus or of the glycogen storage site. However, the accessibility of the serine-14 phosphate may be greater in apophosphorylase *a* as judged by the increased activity of the protein phosphatase with this substrate (*96*). As previously mentioned, AMP is able to inhibit phosphatase action on the apophosphorylase *a* whereas there is no effect of glucose or caffeine, again emphasizing the importance of intersubunit contacts for the transmital of conformational changes between the active site region and the regulatory region containing the sites for serine phosphate and AMP.

The two serine-14 phosphates of the phosphorylase *a* dimer are separated by approximately 40 Å around the curved surface of the protein, and it was suggested that the huge multimeric protein kinase might be able to phosphorylate both serine-14s simultaneously (*62*). Similarly, the "holophosphatases," presumably containing more than one catalytic subunit, may also be able to carry out a simultaneous dephosphorylation. Nevertheless, convincing evidence was obtained that partially phosphorylated intermediates occur during the interconversion of phosphorylase *a* and *b* (*114, 115*). Thus, during the phosphatase reaction, 50% of the phosphate could be released from serine-14 with no loss of activity as measured at high concentrations of glucose-1-P whereas, when measured at the usual low concentrations of substrate, or in the presence of glucose-6-P, the loss of activity coincided with dephosphorylation. The reverse phenomenon was observed during the conversion of phosphorylase *b* to *a* by the kinase. The results were interpreted as indicative of the formation of phosphorylase *a*/*b* hybrids having properties intermediate with respect to the parent forms, so that the affinity for glucose-1-P was less than that for *a* but greater than for *b*. Similarly, the inhibition by glucose-6-P is reminiscent of the *b* form, not the *a*. Two serine phosphates per tetramer stabilized the tetrameric state in the presence of glucose-1-P (*115*) while the hybrid dimer showed a reversal of caffeine inhibition by AMP which the phosphorylase *b* dimer does not (*116*).

These results are reasonably interpreted, in light of the structures of phosphorylase *a* and *b*, as suggesting that the phosphorylase *a*/*b* dimer has one phosphorylated N-terminus bound across the subunit interface, providing part of the increased stabilization and site–site interaction observed upon full conversion to the phosphorylated form. Randomization (reshuffling) of the two types of subunits in the oligomers, as well as formation of hybrids from the homogeneous forms, has also been observed and analyzed (*115*).

The suggestion that hybrid forms may occur *in vivo* during the interconversions of the *a* and *b* forms and thus lead to increased sensitivity to glucose-6-P-mediated regulation of glycogenolysis has received some support by the finding of these hybrids during the "flash activation" of phosphorylase bound to glycogen particles (*117*). In addition, the conversion of phosphorylase *b* to *a* in rabbit and frog muscle in response to hormonal and electrical stimulation was investigated by using activity measurements sensitive to the hybrid species (*118*). The assay for the hybrid species was based on these workers' previous finding that AMP increases its activity in the presence or absence of caffeine whereas phosphorylase *b* activity in the presence of AMP is completely inhibited by caffeine (*116*). They also showed that AMP could reverse the ATP inhibition of the hybrid under conditions where this is not true for phosphorylase *b*. Radda and his colleagues at Oxford employed electron spin resonance techniques with spin-labelled phosphorylase *b* to demonstrate the transient formation of the *a/b* hybrid as an intermediate in the *in vitro* conversion (*8*). However. they were unable to detect the intermediate during conversion of phosphorylase *b* to *a* in the isolated glycogen particle.

It is apparent that the formation of phosphorylase *a/b* hybrids under normal physiological conditions is a distinct possibility, and this species may exhibit different control characteristics from either of the more stable forms, as suggested by the two groups most active in this area (*114, 118*). One feels that further investigation of these phenomena is warranted, both to confirm the formation of the hybrids under physiological conditions by using more direct chemical methods, and to explore in more detail their significance with respect to regulation. This is an intriguing subject with considerable interest both for the regulation of glycogen phosphorylase and as a model for other more complicated protein phosphorylation systems.

B. OTHER MEASURABLE STRUCTURAL EFFECTS

One of the most notable results of the phosphorylation of serine-14 is the association of dimers to form tetramers in the case of phosphorylases from most skeletal muscles, as discussed in Section III,A. This association must be related to the N-terminal region but no information is available as to the actual residues involved, and there is sufficient evidence for the long-range transmission of conformational changes in this molecule to caution us against assuming that the N-terminus forms the interface. Phosphorylase *b* will form tetramers in the presence of AMP, especially with Mg, while phosphorylase *b'*, lacking the first 16 residues, will not (*105*). Huang and Graves (*119*) determined the dissociation constant for the tetramer to dimer dissociation and showed that it increased markedly with temperature, yielding standard enthalpy changes of 60 kcal/mol and an entropy change of 170 units.

Many of the differences observed in the reactivities of functional groups of the two forms of phosphorylase may well be due to the difference in oligomeric state. The rates for both the inhibition and the subsequent dissociation of phosphorylase *b* as a consequence of mercurials reacting with sulfhydryl groups is slower than those observed with phosphorylase *a* (*44, 119, 120*). The same phenomenon was observed when iodoacetamide was employed (*110, 121*). In this case specific rate constants for Cys-108 and Cys-142 were measured and those in phosphorylase *a* reacted faster than those in phosphorylase *b*. Access to Cys-142 is via the internal cavity between the two monomers, and requires some flexing or "breathing." Because the cavity is closed to the external solvent in the static T-state (*76*), the cysteine in the less tightly associated *b* dimer would be expected to have a greater reactivity than that in the more rigid *a* tetramer. Clearly, we do not yet understand all the factors involved in these structure–function relationships.

VI. Functional Results of the Phosphorylation of Serine-14

A. Changes in the Allosteric Constant, *L*, and in AMP Binding

Although it is generally considered to be an over simplification to apply the concerted model for allosteric transitions to phosphorylase *b*, nevertheless it is useful to use the concept of a "final" inactive T-state conformation and an active R-state conformation. The allosteric constant, *L*, which is the equilibrium constant for the molar ratio of T to R, has been estimated to be at least 3300 (*122, 123*). The same constant *L* for phosphorylase *a*, where the two-state model is more applicable, has been estimated at between 3 and 13 (*8, 124*). For discussion purposes we may assign nominal values of 3000 and 10, indicating that the energy required for the $T \rightarrow R$ transition has been reduced by at least 3.5 kcal. This reduction in energy must come from the interaction of the serine-14 phosphate with the main body of the dimer, as discussed in Section V,A. The strength of the newly formed salt bridges, plus other possible interactions, remains a matter for lively discussion (*76*), with values for salt bridges estimated at anywhere from 1 kcal/mol in hemoglobin (*125*) to 3 for an internal salt bridge in chymotrypsin (*126*). The maximal value for a salt bridge formed from a protein-bound serine phosphate that is accessible to solvent has been estimated at 5 kcal (*100*). For purposes of discussion, a nominal value of 4 kcal has been assigned to represent the net energy of stabilization afforded by the interaction of the phosphorylated N-terminus across the subunit interface of the phosphorylase dimer.

In Fig. 3 an attempt has been made to analyze the thermodynamic consequences of the phosphorylation of serine-14. As previously pointed out the effect on the allosteric constant *L*, reduces the energy of this transition from +4.9 to +1.4 kcal. This reduction coincides almost exactly with the increased energy of binding of AMP. Thus the latter binds 200 times more tightly to phosphorylase *a* than to *b* ($K_m = 2 \mu M$ versus $400 \mu M$), increasing the binding energy by 3.2 kcal

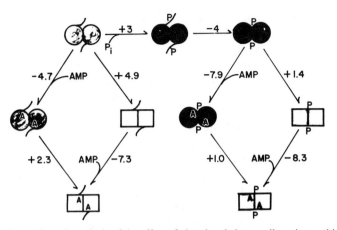

FIG. 3. Thermodynamic analysis of the effect of phosphorylation on allosteric transitions in the phosphorylase system. Phosphorylase *b* dimers are shown on the left, phosphorylation of serine-14 and the subsequent binding of the N-terminus across the subunit interface is depicted on top horizontal line whereas phosphorylase *a* dimers are shown on the right. Circles represent the T conformation and squares the R conformation, but dimers liganded with AMP will have a different conformation than unliganded dimers. Numbers are estimates for the free energy changes (in kcal) for each equilibrium.

(-4.7 to -7.9 kcal), a value close to but somewhat smaller than the change in the energy of the allosteric transition. It is obvious that, in phosphorylase *b*, a large fraction of the binding energy of the activating nucleotide is used to cause a conformational change leading from the T- to the R-state. In the case of phosphorylase *a*, although not symbolized on the diagram, the conformation of the AMP-liganded enzyme is very close to the R-state, and, conversely, substrate alone is able to stabilize a conformation similar to the R-state, as shown by identical maximal velocities in the presence and absence of AMP. On the contrary, it was first shown by Radda and colleagues, reviewed in (8), that glucose-1-P induced a major conformational change in AMP-liganded phosphorylase *b*. Part of the energy for the T → R transition in phosphorylase *b* must therefore be provided by the binding of the substrate and the crude estimate for the second part of this transition, shown in the figure as $+2.3$ kcal, is based, perhaps naively, on the limiting K_m for glucose-1-P for phosphorylase *b* being 10 times that for phosphorylase *a* (22). Therefore, 1.3 kcal was added to the 1.0 estimated as the maximal energy involved in the similar transition of AMP-liganded phosphorylase *a* to the fully activated R-state saturated with substrates.

The diagram shown in Fig. 3 is an oversimplification and the values given for the various free energies should not be taken as anything but estimates. The purpose is to illustrate how, as Sprang and Fletterick suggest, "we can imagine the N-terminus to behave as an intramolecular allosteric effector of the R-state organizing the subunit surface. This 'effector' must be phosphorylated to bind at the dimer surface'' (76). The resultant symmetrical alterations in compensatory free energy changes are more than just coincidential.

The considerable increase in subunit interactions within the dimer that is brought about by the binding of the phosphorylated N-terminus across the subunit interface was discussed previously. Not surprisingly, this has made the existence of intermediate conformations involving different conformations of the two subunits much more difficult in phosphorylase a, so that symmetry must be conserved, as required by the concerted transition model of Monod et al. (127). Thus Griffiths et al. (128) have shown that the binding of AMP to phosphorylase a follows the symmetry model rather than the simple sequential model. On the other hand, members of the same group, as reviewed by Busby and Radda (8), have shown that the binding of most ligands to phosphorylase b follows the sequential model of Koshland et al. (129). As reviewed by Madsen et al. (7) other laboratories have produced data that tend to agree, for the most part, with these conclusions, although there are some rather untidy discrepancies. As Koshland pointed out, the concerted transition or symmetry model is a limiting case of the general sequential model and may be brought about when the intersubunit interactions are too strong to permit stable intermediate forms (130). The phosphorylase system represents, at least to a first approximation, an example of how a simple chemical modification provides an enhanced subunit interaction sufficient to transform sequential allosteric transitions to the concerted mode.

B.   ESCAPE FROM ALLOSTERIC CONTROL

This heading is meant to dramatize the fundamental biological significance of the conversion of phosphorylase b to a under physiological conditions, and not to imply that the a form does not exhibit allostery. Figure 4 demonstrates that phosphorylase a has lost its requirement for AMP under conditions where phosphorylase b exhibits a virtually obligate requirement for activity (Shechosky and Madsen, unpublished research). It may be seen, too, that ATP causes severe inhibition of phosphorylase b, while accentuating the homotropic interaction of the AMP-binding sites. On the other hand, ATP inhibits only that extra activity exhibited by phosphorylase a in the presence of AMP. The data illustrate also the much greater binding affinity of phosphorylase a for AMP. Thus the phosphorylation of serine-14 eliminates both the need for an activating nucleotide and the inhibition by metabolites that bind at the activator site, namely ATP, ADP, and glucose-6-P.

Lowry et al. pointed out that since the main effect of AMP on phosphorylase a was to decrease the K_m for the substrates, glucose-1-P, and inorganic phosphate, the activation would be important at the low P_i concentrations found in muscle (131). However, the apparent K_m for P_i in the absence of AMP is 3 mM so that

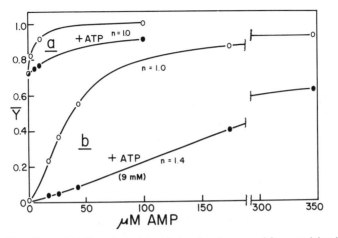

FIG. 4. The effects of AMP on the activity of phosphorylases a and b, as modulated by ATP. Activity was measured at 30° with 16 mM glucose-1-P and 1% glycogen and is represented by $\bar{Y}$, the ratio to maximal activity. When present, ATP was 9 mM; a indicates results with phosphorylase a, b represents phosphorylase b, and n is the Hill coefficient (from S. Shechosky and N. B. Madsen, unpublished experiments).

the enzyme would have partial activity with this substrate in the millimolar range (22). Busby and Radda addressed the problem of control of phosphorylase activity in activated muscle and conclude that if it were all converted to the a form, half would be liganded with AMP, the remainder with ATP and ADP (8). While maximal activity would not be expected under these conditions, the enormous concentration of phosphorylase in muscle would be more than adequate to account for the observed activities, the latter being directly related to phosphate concentration. On the other hand, phosphorylase b in the resting muscle would be expected to be almost totally liganded with glucose-6-P, ATP, and ADP, with almost no bound AMP (8), thus accounting for the very low rate of glycogenolysis under these conditions (132).

In 1962, Danforth et al. (133) developed a method to "freeze" instantly the interconverting system for phosphorylase in muscle, and to extract the enzymes and measure their activities under conditions which precluded any further changes. They were thus able to establish that the phosphorylase a activity of resting frog sartorius muscle is less than 5% of the total potential activity, thereby resolving the previous unsatisfactory relationship between the rate of glycogenolysis and the phosphorylase a content (132). They were able to establish the kinetics of the activation of phosphorylase a in response to either electrical stimulation or adrenalin, the effect of the latter being much slower, as well as the kinetics for the decay of the phosphorylase a activity after stimulation. These studies have been summarized by Helmreich and Cori (134).

A major contribution to the studies of the interconversion of phosphorylase b and a in muscle was made possible by the discovery of a strain of mice that appear to lack phosphorylase b kinase (*135*). Danforth and Lyon (*136*) demonstrated that while the proportion of phosphorylase in the a form in resting muscles of these I-strain mice was the same as that in the muscles of normal C57 mice, approximately 5%, electrical stimulation of the latter resulted in a rapid transformation of phosphorylase to 70% a (with a half-time of one second), whereas no effect was seen in the muscles from the I-strain mice. Furthermore, there was a pronounced lag in the onset of glycogenolysis in the muscles from the I-strain mice compared to the normal situation, as measured by the production of glucose-6-P or lactate. The rate of formation of these two compounds was half that by the normal muscles, and the total produced was also half. Therefore the importance of the conversion of phosphorylase b to a was well illustrated but the question of a mechanism that would allow the stimulation of phosphorylase b during muscle contraction was also raised by these studies.

The most obvious explanation, an increase in the concentration of AMP sufficient to activate phosphorylase b, was discounted by the studies of Griffiths and his colleagues, who showed significant changes in the AMP levels of muscles from normal or phosphorylase kinase-deficient mice upon 15–30 min of forced exercise by swimming (*137*). Furthermore, while the total AMP concentration was of the order of 0.1 mM, they suggested on the basis of the adenylate kinase equilibrium that the free AMP concentration might not exceed 0.006 mM, too low to activate phosphorylase b. On the other hand, the IMP concentrations rose from 0.73 mM in the muscles of the resting kinase-deficient mice to 1.65 mM after exercise. Since the K_m of IMP for phosphorylase b is less than 1 mM (*138*), this could provide sufficient activation to account for the increased glycogenolysis. One problem with scenario revolves around the earlier finding of Black and Wang (*138*) that IMP does not improve the K_m for the substrate, glucose-1-P, which remains at at least 32 mM at all IMP concentrations tested. We have confirmed (unpublished experiments) that in the direction of phosphorolysis at 1 mM IMP, the concentration of P_i required to reach half-maximal velocity is approximately 40 mM, the Hill coefficient is 1.7 and the maximal velocity (at 10 μmol/min/mg) is only one-third that observed with AMP. Furthermore, at 5 mM P_i, the specific activity was only 0.27 μmol/min/mg.

C.   CHARACTERISTICS REMAINING UNCHANGED

It may be unnecessary to belabor the obvious fact that most of the basic enzymic characteristics of phosphorylase are not altered by phosphorylation of serine-14, but since this is the prototype interconvertible enzyme system, principles established with it may act as guides for evaluating other systems. There is no reason to believe that the nature of the catalytic mechanism has been altered in any fundamental aspect. While the maximal activity of phosphorylase *a* is slightly more than that of the *b* form, this probably does not reflect any change in the catalytic mechanism. The energy of activation of the two forms, measured in the presence of AMP, is identical within experimental error (7). The two forms bind to glycogen with much the same dissociation constant, indicating a limited effect of the N-terminus on the glycogen storage site and thereby ensuring that phosphorylase remains bound to the glycogen particle regardless of its state of activity.

Another characteristic which varies only in detail between the two forms is the synergistic inhibition by glucose and caffeine. The kinetically derived dissociation constants for glucose with the *a* and *b* forms are approximately 3 and 1 mM, respectively, while those for caffeine are 0.2 and 0.08 mM, respectively (64, 66). The interaction constants are quite similar, being calculated at 0.3 and 0.2 for the *a* and *b* forms. Because of the uncertainties in these calculations, there is no significant difference in the extent to which one of these ligands improves the binding of the other. Although the ligands bind somewhat more tightly to phosphorylase *b* than to *a*, the conversion has not eliminated the inhibitions or modified the allosteric interactions. Any metabolic controls exerted at or near the catalytic site may possibly be affected only slightly by covalent interconversion. It will be interesting to see if this is a general principle applicable to other metabolically interconvertible enzymes.

VII. Concluding Remarks

The control of skeletal muscle glycogen phosphorylase by the reversible phosphorylation of its serine-14 provides us with a model system in which we should be able to discern principles of structure–function relationships applicable to other metabolically interconvertible enzymes. We have seen that after phosphorylation the N-terminal 18 residues bind across the subunit interface and strengthen the subunit interactions. It is suggested that the extra energy afforded by this interaction reduces the energy required for allosteric transitions and thereby allows phosphorylase *a* to escape from the allosteric controls to which phosphorylase *b* is subject. Many other interconvertible enzyme systems exhibit a similar release from allosteric restraints in their active forms, even though activation may involve dephosphorylation rather than phosphorylation of critical residues, and we may look forward, in due course, to the delineation of structural changes similar in principle to those discovered for the phosphorylase system. Increasing complexity will be observed in most cases, however, because, while phosphorylase is complicated enough, it is a "clean" and simple enzyme compared to some other systems. Just to take examples from the area of glycogen metabolism, glycogen synthase exhibits the phosphorylation of seven serines arranged in three groups with various phosphorylation patterns exhibited as the result of action by three separate kinases. Phosphorylase b kinase is composed of four different types of subunits, two of which undergo phosphorylation.

While progress in the structure–function relationships in the phosphorylase system may seem impressive, we are at a rather superficial level in our understanding and considerable work is required to clarify our working hypothesis. Comparison of the refined high-resolution structures of the a and b forms of phosphorylase is planned and should be very useful. Since both structures are of the inactive T conformation, we need the structure of at least one R-form so we can understand how serine-14 phosphorylation facilitates the T $\rightarrow$ R transition. We have only the slightest knowledge of how the enzymes catalyzing the interconversion interact with their phosphorylase substrates. As pointed out in Section VI,B, research on mice that lack phosphorylase b kinase has illuminated the physiological benefits of the conversion from the b to a form, but the mechanism by which these mice can still carry out glycogenolysis remains a matter for further investigation. The reader will note that I have emphasized the skeletal muscle phosphorylase system while saying little about the equally important system in liver, let alone those in other tissues. Aside from feeling more comfortable with the muscle system because of its wealth of structure–function information, I consider that the state of research on the liver system is in great flux, making a definitive treatment difficult. One should encourage a concerted effort to determine structures for the phosphorylases of liver because this would provide a firm base for defining the physiological role and mechanism for their control by phosphorylation.

ACKNOWLEDGMENTS

I am grateful to Dr. R. J. Fletterick for helpful discussions and for providing Fig. 1. I wish to thank Mrs. P. McDonald for her skillful typesetting through several revisions.

Research reported from this laboratory was supported by Grant MT-1414 from the Medical Research Council of Canada.

REFERENCES

1. Green, A. A., Cori, G. T., and Cori, C. F. (1942). *JBC* **142**, 447–448.
2. Fischer, E. H., and Krebs, E. J. (1955). *JBC* **216**, 121–132.
3. Krebs, E. G., and Fischer, E. H. (1956). *BBA* **20**, 150–157.
4. Fischer, E. H., Graves, D. J., Crittenden, E. R. S., and Krebs. E. H. (1959). *JBC* **234**, 1698–1704.
5. Fischer, E. H., Graves, D. H., and Krebs, E. G. (1957). *FP* **16**, 180.
6. Graves, D. J., and Wang, J. H. (1972). "The Enzymes," 3rd ed., Vol. 7, pp. 435–482.

7. Madsen, N. B., Avramovic-Zikic, O., Lue, P. F., and Honkel, K. O. (1976). *Mol. Cell Biochem.* **11**, 35–50.
8. Busby, S. J. W., and Radda, G. K. (1976). *Curr. Top. Cell. Regul.* **10**, 89–160.
9. Helmreich, E. J., and Klein, H. W. (1980). *Angew. Chem.* **19**, 441–455.
10. Fletterick, R. J., and Madsen, N. B. (1980). *Annu. Rev. Biochem.* **49**, 31–61.
11. Dombradi, V. (1981). *Int. J. Biochem.* **13**, 125–139.
12. Madsen, N. B., and Withers, S. G. (1986). *In* "Coenzymes and Cofactors. Pyridoxal Phosphate and Derivatives" (D. Dolphin, R. Poulson, and O. Avramovic, eds.), Vol. 1B, pp. 355–385. Wiley, New York.
13. Cori, G. T., and Cori, C. F. (1940). *JBC* **135**, 733–756.
14. Larner, J., Villar-Palasi, C., and Rechman, D. J. (1960). *ABB* **86**, 56–60.
15. Taylor, C., Cox, A. J., Kevrohan, J. C., and Cohen, P. (1975). *EJB* **51**, 105–115.
16. Wanson, J.-C., and Drochmans, P. (1968). *JBC* **38**, 130–150.
17. Porter, K. R., and Bruni, C. (1959). *Cancer Res.* **19**, 997–1009.
18. Meyer, F., Heilmeyer, L. M. G., Haschke, R. H., and Fischer, E. H. (1970). *JBC* **245**, 6642–6647.
19. Caudwell, B., and Cohen, P. (1978). *EJB* **86**, 511–518.
20. Maddaiah, V. T., and Madsen, N. B. (1966). *JBC* **241**, 3873–3881.
21. Engers, H. D., Bridger, W. A., and Madsen, N. B. (1969). *JBC* **244**, 5936–5942.
22. Engers, H. D., Shechosky, S., and Madsen, N. B. (1970). *Can. J. Biochem.* **48**, 746–754.
23. Engers, H. D., Bridger, W. A., and Madsen, N. B. (1970). *Can. J. Biochem.* **48**, 755–758.
24. Gold, A. M., Johnson, R. M., and Tseng, J. K. (1970). *JBC* **245**, 2564–2572.
25. Engers, H. D., Bridger, W. A., and Madsen, N. B. (1970). *Biochemistry* **9**, 3281–3284.
26. Madsen, N. B., and Withers, S. G. (1984). *In* "Chemical and Biological Aspects of Vitamin B6 Catalysis" (A. E. Evangelopolous, ed.), Part A, pp. 117–126. Alan R. Liss, Inc., New York.
27. Klein, H. W., Im, M. J., and Helmreich, E. J. M. (1984). *In* "Chemical and Biological Aspects of Vitamin B6 Catalysis (A. E. Evangelopolous, ed.), pp. 147–160.
28. Sygusch, J., Madsen, N. B., Kasvinsky, P. J., and Fletterick, R. J. (1977). *PNAS* **74**, 4757–4761.
29. Weber, I. T., Johnson, L. N., Wilson, K. S., Yeates, D. G. R., and Wild, D. J. (1978). *Nature (London)* **274**, 433–436.
30. Parrish, T., Uhing, R. J., and Graves, D. J. (1977). *Biochemistry* **16**, 4824–4831.
31. Withers, S. G., Madsen, N. B., Sykes, B. D., Takagi, M., Shimomura, S., and Fukui, T. (1981). *JBC* **256**, 10759–10762.
32. Klein, H. W., Palm, D., and Helmreich, E. J. M. (1982). *Biochemistry* **21**, 6675–6684.
33. Takagi, M., Fukui, T., and Shimomura, S. (1982). *PNAS* **79**, 3716–3719.
34. Withers, S. G., Madsen, N. B., Sprang, S. R., and Fletterick, R. J. (1982). *Biochemistry* **21**, 5372–5382.
35. Avramovic-Zikic, O., Breidenbach, W. C., and Madsen, N. B. (1974). *Can. J. Biochem.* **52**, 146–148.
36. Shimomura, S., Nakano, K., and Fukui, T. (1978). *BBRC* **82**, 462–468.
37. Dreyfus, M., Vandenbunder, B., and Buc, H. (1980). *Biochemistry* **19**, 3634–3642.
38. Kasvinsky, P. J., and Meyer, W. L. (1977). *ABB* **181**, 616–631.
39. Klein, H. W., Schiltz, E., and Helmreich, E. J. M. (1981). *In* "Protein Phosphorylation" (O. M. Rosen and E. G. Krebs, ed.), pp. 305–320. Cold Spring Harbor Lab., Cold Spring Harbor, New York.
40. Cori, C. F., and Cori, G. T. (1936). *Proc. Soc. Exp. Biol. Med.* **34**, 702–705.
41. Cori, C. F., Cori, G. T., and Green, A. A. (1943). *JBC* **151**, 39–55.

42. Parmeggiani, A., and Morgan, H. E. (1962). *BBRC* **9,** 252–256.
43. Keller, P. J., and Cori, G. T. (1953). *BBA* **12,** 235–238.
44. Madsen, N. B., and Cori, C. F. (1956). *JBC* **223,** 1055–1065.
45. Appleman, M. M., Krebs, E. G., and Fischer, E. H. (1966). *Biochemistry* **5,** 2101–2107.
46. Assaf, S. A., and Graves, D. J. (1969). *JBC* **244,** 5544–5555.
47. Davis, C. H., Schlisefield, L. H., Wolf, D. P., Leavitt, C. A., and Krebs, E. G. (1967). *JBC* **242,** 4824–4833.
48. Kent, A. B., Krebs, E. G., and Fischer, E. H. (1958). *JBC* **232,** 549–558.
49. Huang, C. Y., and Graves, D. J. (1970). *Biochemistry* **9,** 660–671.
50. Titani, K., Koide, A., Ericsson, L. H., Kumar, S., Wade, R., Walsh, K. A., Neurath, H., and Fisher, E. (1977). *PNAS* **74,** 4762–4766.
51. Palm, D., Goerl, R., Burger, K. J., Buhner, M., and Schwartz, M. (1984). *In* "Chemical and Biological Aspects of Vitamin B6 Catalysis" (A. E. Evangelopoulos, ed.), pp. 209–221. Alan R. Liss, Inc., New York.
52. Nakano, K., Kikumoto, Y., and Fukui, T. (1984). *In* "Chemical and Biological Aspects of Vitamin B6 Catalysis" (A. E. Evangelopoulos, ed.), pp. 171–180. Alan R. Liss, Inc., New York.
53. Lerch, K., and Fischer, E. H. (1975). *Biochemistry* **14,** 2009–2014.
54. Cohen, P., Saari, J. C., and Fischer, E. H. (1971). *Biochemistry* **10,** 5233–5241.
55. Jenkins, L. N., Stuart, D. I., Stura, E. A., Wilson, K. S., and Zanotti, G. (1981). *Philos. Trans. R. Soc. London, Ser. B* **293,** 23–41.
56. Lorek, A., Wilson, K. S., Stura, E. A., Jenkins, J. A., Zanotti, G., and Johnson, L. N. (1980). *JMB* **140,** 565–580.
57. Sansom, M. S. P., Stura, A., Babu, Y. S., McLaughlin, P., and Johnson, L. N. (1984). *In* "Chemical and Biological Aspects of Vitamin B6 Catalysis" (A. E. Evangelopoulos, ed.), pp. 127–146. Alan R. Liss, Inc., New York.
58. Lorek, A., Wilson, K. S., Sansom, M. S. P., Stuart, D. I., Stura, E. A., Jenkins, J. A., Hajdu, J., and Johnson, L. N. (1984). *BJ* **218,** 45–60.
59. Sprang, S. R., and Fletterick, R. J. (1979). *JMB* **131,** 523–551.
60. Sprang, S. R., Goldsmith, E. J., Fletterick, R. J., Withers, S. G., and Madsen, N. B. (1982). *Biochemistry* **21,** 5364–5371.
61. Madsen, N. B., Kasvinsky, R. J., and Fletterick, P. J. (1978). *JBC* **253,** 9097–9101.
62. Fletterick, R. J., Sprang, S., and Madsen, N. B. (1979). *Can. J. Biochem.* **57,** 789–797.
63. Kasvinsky, P. J., Madsen, N. B., Sygusch, J., and Fletterick, R. J. (1978). *JBC* **253,** 3343–3351.
64. Kasvinsky, P. J., Shechosky, S., and Fletterick, R. J. (1978). *JBC* **253,** 9102–9106.
65. Withers, S. G., Sykes, B. D., Madsen, N. B., and Kasvinsky, P. J. (1979). *Biochemistry* **24,** 5342.
66. Madsen, N. B., Shechosky, S., and Fletterick, R. J. (1983). *Biochemistry* **22,** 4460–4465.
67. Sygusch, J., Madsen, N. B., and Fletterick, R. J. (1977). *PNAS* **74,** 4757–4761.
68. Johnson, L. N., Jenkins, J. A., Wilson, K. S., Stura, E. A., and Zanotti, G. (1980). *JMB* **140,** 565–580.
69. Withers, S. G., Madsen, N. B., Sprang, S. R., and Fletterick, R. J. (1982). *Biochemistry* **21,** 5372–5382.
70. Sprang, S. R., Fletterick, R. J., Stern, M., Yang, G., Madsen, N. B., and Sturtevant, J. M. (1982). *Biochemistry* **21,** 2036–2048.
71. Stura, E. A., Zanotti, G., Babu, Y. S., Sansom, M. S. P., Stuart, D. I., Wilson, K. S., and Johnson, L. N. (1983). *JMB* **170,** 529–565.
72. Lee, Y. M., and Benisek, W. F. (1976). *JBC* **251,** 1553–1560.
73. Lee, Y. M., and Benisek, W. F. (1978). *JBC* **253,** 5460–5463.

74. Johnson, L. N., Stura, E. A., Wilson, K. S., Sansom, M. S. P., and Weber, I. T. (1979). *JMB* **134,** 639–653.
75. Weber, I. T., Johnson, L. N., Wilson, K. S., Yeates, D. G. R., Wild, D. L., and Jenkins, J. A. (1978). *Nature (London)* **274,** 433–437.
76. Sprang, S., and Fletterick, R. J. (1980). *Biophys. J.* **32,** 175–192.
77. Krebs, E. G., Love, D. S., Bratvold, G. E., Traysey, K. A., Meyer, W. L., and Fischer, E. H. (1964). *Biochemistry* **3,** 1022–1033.
78. Tabatabai, L. B., and Graves, D. J. (1978). *JBC* **253,** 2196–2202.
79. Tu, J. I., and Graves, D. J. (1973). *BBRC* **53,** 59–65.
80. Gusev, N. B., and Hajdu, J. (1979). *BBRC* **90,** 70–77.
81. Ingebritsen, T. S., Stewart, A. A., and Cohen, P. (1983). *EJB* **132,** 297–307.
82. Ingebritsen, T. S., Foulkes, J. G., and Cohen, P. (1983). *EJB* **132,** 263–274.
83. Martensen, T. M., Brotherton, J. E., and Graves, D. J. (1973). *JBC* **248,** 8323–8328.
84. Martensen, T. M., Brotherton, J. E., and Graves, D. J. (1973). *JBC* **248,** 8329–8336.
85. Detwiler, T. C., Gratecos, D., and Fischer, E. H. (1977). *Biochemistry* **16,** 4818–4823.
86. Nolan, C., Nova, W. B., Krebs, E. G., and Fischer, E. H. (1964). *Biochemistry* **3,** 542–551.
87. Sutherland, E. W. (1951). *In* "Phosphorus Metabolism" (W. S. McElroy and B. Glass, eds.), pp. 53–61. Johns Hopkins Press, Baltimore, Maryland.
88. Withers, S. G., Madsen, N. B., and Sykes, B. D. (1981). *Biochemistry* **20,** 1748–1756.
89. Dombradi, V., Toth, B., Bot, G., Hajdu, J., and Friedrich, P. (1982). *Int. J. Biochem.* **14,** 277–284.
90. Graves, D. J., Mann, S. A. S., Philip, G., and Oliveira, R. J. (1968). *JBC* **243,** 6090–6098.
91. Stalmans, W., Laloux, M., and Hers, H. G. (1974). *EJB* **49,** 415–427.
92. Hers, H. G. (1976). *Annu. Rev. Biochem.* **45,** 167–189.
93. Stalmans, W. (1976). *Curr. Top. Cell. Regul.* **11,** 51–97.
94. Kasvinsky, P. J., Fletterick, R. J., and Madsen, N. B. (1981). *Can. J. Biochem.* **59,** 387–395.
95. Monanu, M. O., and Madsen, N. B. (1985). *Can. J. Biochem. Cell Biol.* **63,** 115–121.
96. Yan, S. C. B., Uhing, R. J., Parrish, R. F., Metzler, D. E., and Graves, D. I. (1979). *JBC* **254,** 8263–8269.
97. Bot, G., Kovacs, E., and Gergely, P. (1977). *Acta Biochim. Biophys. Acad. Sci. Hung.* **12,** 335–341.
98. Raibaud, O., and Goldberg, M. E. (1973). *Biochemistry* **12,** 5154–5161.
99. Dombrádi, B., Tóth, B., Gergely, P., and Bot, G. (1983). *Int. J. Biochem.* **15,** 1329–1336.
100. Vogel, H. J., and Bridger, W. A. (1983). *Can. J. Biochem. Cell Biol.* **61,** 363–369.
101. Sealock, R. W., and Graves, D. J. (1967). *Biochemistry* **6,** 201–207.
102. Janski, A. M., and Graves, D. J. (1979). *JBC* **254,** 1644–1652.
103. Cori, G. T., and Cori, C. F. (1945). *JBC* **158,** 321–332.
104. Keller, P. J. (1955). *JBC* **214,** 135–141.
105. Carty, T. J., Tu, J.-I., and Graves, D. J. (1975). *JBC* **250,** 4980–4985.
106. Janski, A. M., and Graves, D. J. (1979). *JBC* **254,** 4033–4039.
107. Graves, D. J., Sealock, R. W., and Wang, J. H. (1965). *Biochemistry* **4,** 290–296.
108. Cohen, P., Duewer, T., and Fischer, E. H. (1971). *Biochemistry* **10,** 2683–2694.
109. Shaltiel, S., Hedrick, J. L., and Fischer, E. H. (1966). *Biochemistry* **5,** 2108–2116.
110. Avramovic, O., Smillie, L. B., and Madsen, N. B. (1970). *JBC* **245,** 1558–1565.
111. Hedrick, J. L., Shaltiel, L., and Fischer, E. H. (1966). *Biochemistry* **5,** 2117.
112. Shaltiel, S., Hedrick, J. L., Pocker, A., and Fischer, E. H. (1969). *Biochemistry* **8,** 5189–5196.
113. Kastenschmidt, L. L., Kastenschmidt, J., and Helmreich, E. (1968). *Biochemistry* **7,** 3590–3608.
114. Hurd, S. S., Teller, D., and Fischer, E. H. (1966). *BBRC* **24,** 79–84.

115. Fischer, E. H., Hurd, S. S., Koh, P., Seery, V. L., and Teller, D. C. (1968). *In* "Control of Glycogen Metabolism" (W. J. Whelan, ed.), pp. 19–33. Academic Press, New York.
116. Bot, Y., Kovács, E. F., and Gergely, P. (1974). *BBA* **370**, 78–84.
117. Heilmeyer, L. M. G., Meyer, F., Haschke, R. H., and Fischer, E. H. (1970). *JBC* **245**, 6649–6656.
118. Gergely, P., Bot, G., and Kovács, E. F. (1974). *BBA* **370**, 78–84.
119. Huang, C. Y., and Graves, D. J. (1970). *Biochemistry* **9**, 660–671.
120. Madsen, N. B. (1956). *JBC* **223**, 1067–1074.
121. Battell, M. L., Zarkadas, C. G., Smillie, L. B., and Madsen, N. B. (1968). *JBC* **243**, 6202–6209.
122. Madsen, N. B., and Shechosky, S. (1967). *JBC* **242**, 3301–3307.
123. Kastenschmidt, L. L., Kastenschmidt, J., and Helmreich, E. (1968). *Biochemistry* **7**, 4543–4550.
124. Helmreich, E., Michaelides, M. C., and Cori, C. F. (1967). *Biochemistry* **6**, 3695–3710.
125. Perutz, M. F. (1978). *Science* **201**, 1187–1191.
126. Fersht, A. R. (1971). *Cold Spring Harbor Symp. Quant. Biol.* **36**, 71–73.
127. Monod, J., Wyman, J., and Changeux, J.-P. (1965). *JMB* **12**, 88–118.
128. Griffiths, J. R., Price, N. C., and Radda, G. K. (1974). *BBA* **358**, 275–280.
129. Koshland, D. E., Nemethy, G., and Filmer, D. (1966). *Biochemistry* **5**, 365–385.
130. Koshland, D. E. (1969). *Curr. Top. Cell. Regul.* **1**, 1–27.
131. Lowry, O. H., Schulz, D. W., and Passonneau, J. V. (1964). *JBC* **239**, 1947–1953.
132. Cori, C. F. (1956). *In* "Enzymes: Units of Biological Structure and Function" (O. H. Gaebler, ed.), pp. 573–583. Academic Press, New York.
133. Danforth, W. H., Helmreich, E., and Cori, C. F. (1962). *PNAS* **48**, 1191–1199.
134. Helmreich, E., and Cori, C. F. (1965). *Adv. Enzyme Regul.* **3**, 91–107.
135. Lyon, J. B., and Porter, J. (1963). *JBC* **238**, 1–11.
136. Danforth, W. H., and Lyon, J. B. (1964). *JBC* **239**, 4047–4050.
137. Rahim, Z. H. A., Perrett, D., and Griffiths, J. R. (1976). *FEBS Lett.* **69**, 203–205.
138. Black, W. J., and Wang, J. H. (1968). *JBC* **243**, 5892–5898.

10

Phosphorylase Kinase

CHERYL A. PICKETT-GIES • DONAL A. WALSH

Department of Biological Chemistry
School of Medicine
University of California, Davis
Davis, California 95616

THE ENZYMES, Vol. XVII

I. Introduction

Protein phosphorylation and dephosphorylation are well recognized as a major metabolic control mechanism in eucaryotic cells. In mammalian systems, phosphorylation plays a role in the regulation of protein function in such diverse processes as carbohydrate and lipid metabolism, gene regulation and macromolecular synthesis, membrane transport, ionic homeostasis, muscle contraction, cytoskeletal organization, neural transmission, oncogenesis, and many others. Many avenues that have led to our understanding of regulation by protein phosphorylation have been obtained from studies of liver and muscle glycogen metabolism, and this area of investigation has often served as the initiation point for the elucidation of the primary principles that govern such regulatory phenomena. It was during studies of muscle glycogenolysis that the phosphorylation of glycogen phosphorylase, the first example of enzyme regulation by phosphorylation, was discovered. By the 1940s, through work in the laboratories of Cori and others, the basic biochemistry of glycogen breakdown had been established with the characterization of the phosphorylase-catalyzed conversion of glycogen to glucose 1-phosphate. At the same time, it was well known that exposure of liver cells to epinephrine caused a rapid breakdown of glycogen and an increase in free glucose, suggesting that epinephrine was somehow acting to increase the activity of phosphorylase (*1*). During the 1940s, workers in the Cori laboratory (*2*) demonstrated that phosphorylase existed in two molecular forms, but the nature of the difference and the mechanism by which these forms were interconverted remained unclear for more than a decade. Soon after the first protein kinase was described in 1954 (*3*), it was established that the two forms of phosphorylase were interconverted by phosphorylation–dephosphorylation (*4*). In 1950, Sutherland (*1*) demonstrated that epinephrine elicited the production of a heat-stable factor that was capable of activating phosphorylase in cell homogenates. This factor was soon identified as adenosine 3′,5′-monophosphate or cAMP (*5*), and was shown to be produced from ATP by the action of a membrane-bound adenylate cyclase (6). Meanwhile, characterization of the kinase responsible for the phosphorylation of phosphorylase was initiated in the laboratory of E. H. Fischer and E. G. Krebs. It became clear that this enzyme was also subject to phosphorylation and this modification increased its catalytic activity toward phosphorylase (7, 8). Addition of cAMP to purified phosphorylase kinase in the presence of Mg^{2+} and ATP stimulated the activation of the enzyme, thus the ground work was laid for the hypothesis of a "cascade" system in the hormonal regulation of glycogen metabolism. For a time it was thought that this sequence of events was complete; however, further investigation demonstrated that cAMP did not act directly on phosphorylase kinase but rather upon another enzyme contaminating the preparations. Identification of this enzyme, the cAMP-dependent protein kinase, by Walsh et al. (9) in 1968, completed the basic sequence of steps in the β-adrenergic regulation of phosphorylase.

Continuing studies on the enzymes of glycogenolysis have upheld the basic scheme of events as initially presented for the regulation of glycogen breakdown, but equally important, have shown that the system is much more complex than originally envisioned. From the studies of these enzymes, and of other enzymes that have since been shown to be regulated by protein phosphorylation, it has become abundantly clear that regulation by protein phosphorylation can often involve highly intricate interrelationships of multiple protein phosphorylation control mechanisms. Some of these complexities include the following:

1. Protein phosphorylation may either activate or inactivate the biological function of a protein.
2. Changes in activity may be reflected by alterations in the affinity for substrates, the affinity for allosteric ligands, the maximum velocity of the reaction, or combinations of some or all of these.
3. Protein phosphorylation can occur in multiple sites on a single polypeptide chain, catalyzed either by the same or by different enzymes.
4. The status of phosphorylation of a site can regulate not only the activity of the protein, but also the regulation by phosphorylation of other sites.
5. The system for protein dephosphorylation is as deeply complex as the system for phosphorylation. Protein phosphatases are, like the protein kinases, subject to a variety of regulations including protein phosphorylation and dephosphorylation.
6. Specific regulation by the phosphatases and kinases does not appear to follow a simple pattern. Different sites phosphorylated by the same kinase might require different phosphatases for their dephosphorylation and, vice versa, the same phosphatase can catalyze dephosphorylation of sites phosphorylated by different kinases.

The above are but a few of the complexities that have become apparent in systems regulated by protein phosphorylation; in fact, from the simple schemes of the glycogenolytic cascade presented in the 1960s and early 1970s, which appeared to adequately describe the physiological manifestation of such regulation, we have reached a stage in which it is very difficult to put into a physiological context the complexities of the processes that are now apparent. It is within this framework that this chapter on phosphorylase kinase is presented. Here, we attempt to at least indicate the extensive network of regulatory phenomena that impinge upon the control of phosphorylase kinase activity. For many, however, a full physiological rationale cannot be completely presented. While we have attempted to be complete, not all of the information that is available about phosphorylase kinase can be given adequate treatment; such can only be obtained from the cited original work. Three notable reviews on phosphorylase kinase have been presented by Chan and Graves (10), Carlson et al. (11), and Malencik and Fischer (12). The understanding of the control of glycogenolysis has served well for many years as a guideline for how other metabolic control systems may function. Similarly, an understanding of phosphorylase kinase can serve as a model to understand the properties of other regulatory enzymes. An attempt is made in this chapter to provide this understanding.

II. Physicochemical Properties

A. SOURCES AND ISOLATION

Phosphorylase kinase was first isolated from rabbit skeletal muscle (7) where it constitutes approximately 0.5% of the soluble protein. The commonly employed purification procedures involve the sequential use of (1) acid precipitation of the glycogen pellet and associated enzymes (including phosphorylase kinase) from crude extracts and/or, (2) two-step differential ultracentrifugation, (3) ammonium sulfate precipitation, and (4) gel filtration (13, 14). In addition to these procedures or as alternatives to various steps, several investigators have utilized purification techniques such as DEAE chromatography (14), affinity chromatography on immobilized calmodulin (15) or phosphorylase (16), hydrophobic chromatography (17), and sucrose density-gradient ultracentrifugation (18). Using a variety of these methods, the enzyme can be obtained in an essentially homogeneous form, although often the only criterion of homogeneity applied has been SDS-polyacrylamide gel electrophoresis (SDS-PAGE). Minor contaminants ($\sim$1%) are frequently present; whether such contaminants might have affected the outcome of results has often been ignored, perhaps inappropriately. In addition to the rabbit skeletal muscle enzyme, phosphorylase kinase has been purified from mouse (19), dogfish (20), and red bovine (21) skeletal muscle; from bovine cardiac muscle (22); from chicken gizzard (22a); and from rat liver (23). All of these enzymes appear to share many very similar characteristics with respect to physical properties, enzymic activities, and modes of regulation. Since by far the predominant number of *in vitro* studies have utilized the rabbit skeletal muscle enzyme, these results constitute most of what is reported in this review. Where there are established differences for the enzyme from other sources, they are noted. One phosphorylase kinase that appears to be quite distinct in molecular size, subunit composition, and enzymic characteristics, is that obtained from yeast (*24*); the studies so far on this enzyme, however, have been quite limited (*24, 25*).

B. SUBUNIT STRUCTURE

The isolated, homogeneous rabbit skeletal-muscle phosphorylase kinase has a molecular weight of approximately 1.3×10^6 (*13, 14*) and is composed of four types of subunits in the stoichiometry α $(\alpha')_4\beta_4\gamma_4\delta_4$. Two isozymes have been identified that differ in the size of the largest subunit, designated α or α'. Molecular weights of these subunits, based on their electrophoretic mobility in SDS-polyacrylamide gels, have been reported as α = 145,000 (*14*), 136,000 (*26*), 127,000 (*27*), 118,000 (*13*); α' = 133,000 (*26*), 134,000 (*22*), 140,000 (*14*); β = 128,000 (*14*), 120,000 (*26*), 113,000 (*27*), 108,000 (*13*); with the first listed ones being the most generally accepted. Absolute values for γ = 44,673 (*28*) and δ = 16,680 (*29*) have been determined from sequence analysis and are in primary agreement with prior values determined by SDS-PAGE.

C. ISOZYMES

When first isolated from mixed muscle types the minor α' band seen in SDS-gels was suspected to be a proteolytic product from the more abundant α band. However, further purification or treatment of extracts to inhibit proteolysis did not alter the relative amount of α' observed. Jennissen and Heilmeyer (*26*) and later Burchell *et al.* (*30*) prepared phosphorylase kinase from red (slow-twitch) and white (fast-twitch) skeletal muscle. Characterization of the enzyme suggested that the α' enzyme was predominant in slow-twitch muscle whereas the α enzyme was predominant in fast-twitch muscle. This evidence, coupled with the lack of α subunit in purified cardiac phosphorylase kinase (*22*), led to the generally accepted hypothesis that two isozyme forms of phosphorylase kinase exist, one form containing four α-subunits and the other four α'-subunits, and that each isozyme is specific for a particular type of muscle fiber. Separation of these isozymes appears to be best accomplished by calmodulin-affinity chromatography (*15*).

There are several questions about phosphorylase kinase isozymes that need to be addressed. Phosphorylase kinase exists in most nonmuscle tissues but for only one of these, liver, has the isozyme form been evaluated. The isozyme that has been purified from liver appears most likely to be the α type (*23*), but this protein was purified starting from the proteins associated with the glycogen complex, which for liver phosphorylase kinase represents only 5% of the total enzyme present in the cell. Whether this isozyme also constitutes the bulk of phosphorylase kinase present in liver is not known. (Studies of this type have been severely hampered by problems of proteolysis during purification, to which the phosphorylase kinase appears particularly susceptible. Often even extensive use of protease inhibitors does not negate the problem.) It is not known for any tissues whether within a single cell type one or both isozymes are expressed and/or whether heterologous enzyme containing both α- and α'-subunits in the same molecule can occur.

Whether the two isozymes of phosphorylase kinase have unique physiological functions is not known. While this would appear most reasonable and there are clear differences in *in vitro* properties, it is not apparent what different physiological purposes the two isozymes might serve. The most plausible, albeit speculative, proposal is that each is differentially regulated in response to Ca^{2+}, this is discussed in more detail in Sections V,A and VIII,A. The original correlation (26, 30) between phosphorylase kinase isozyme-type and muscle fiber contractile property (i.e., fast- or slow-twitch) should be expanded. Rat skeletal *flexor digitorum brevis* muscle,which is 90% FOG fibers (fast-twitch oxidative glycolytic), contains almost entirely the α' isozyme (30a), and high levels of α' isozyme have also been reported in other fast-twitch muscles containing high proportions of FOG fibers (30b). Thus phosphorylase kinase isozyme distribution in muscle is best related to metabolic rather than contractile properties with the α isozyme being present in cells that rely primarily on glycolytic activity [FG (fast-twitch glycolytic) fibers] and the α' isozyme in tissues with higher oxidative capacity [FOG and SO (slow-twitch oxidative) fibers and cardiac muscle]. Recently, Lawrence et al. (30b) have reported that repetitive long-term stimulation ($\sim$10 weeks) of muscle prompted a change in isozyme distribution with an elevation of the α' isozyme. This could well correspond to an increased proportional mass of oxidative fiber types.

It has been suggested that differences in phosphorylase kinase forms (isozymes) may also be related to their subcellular distribution. Studies by Schwartz et al. (31), Horl and Heilmeyer (32), and Le Peuch et al. (33) have indicated that an endogenous "phosphorylase kinase" was associated with the sarcoplasmic reticulum, that added phosphorylase kinase modulated sarcoplasmic reticulum Ca^{2+} transport, and that this appeared to be due, at least in part, to the phosphorylation of the protein "phospholamban." The substrate specificity of this sarcoplasmic reticulum phosphorylase kinase and the standard enzyme isolated from cytosol (and glycogen particle) are apparently identical but the two enzymes appear to be distinct based upon immunogenicity, the requirement for calmodulin for activity, inhibition by fluphenazine, and the retention of the sarcoplasmic reticulum enzyme in I-strain mice in which the cytosol-glycogen particle enzyme is genetically absent. The relationship between these two enzyme forms, in terms of structure, subunit composition, etc., has not been evaluated. Most likely, the membrane associated enzyme is not truly a "phosphorylase kinase."

Multiple forms of phosphorylase kinase might also exist based upon size and possibly subunit stoichiometry. In one of the original studies of the purification of phosphorylase kinase (13), it was recognized that the holoenzyme ($M_r = 1.3 \times 10^6$) tended to aggregate with the formation of complexes that were large enough to result in turbidity. Upon further characterization (12), this turbid fraction appeared to be rather contaminated by another polypeptide, suggested possibly to be α-actinin. This turbid phosphorylase kinase fraction has generally been regarded to be artifactual, arising as a consequence of pressure-dependent aggregation. Subsequently, Hessova et al. (34) observed that the higher-molecular-weight fraction also occurs as a consequence of heparin treatment. Of interest, heparin treatment promotes the formation of the phosphorylase kinase activity that these authors designate A_o (see Section V,B), a form that is calcium-independent and that they suggest plays an important role in the cell in permitting Ca^{2+}-independent phosphorylase a formation. Concomitant with a decreased sensitivity to Ca^{2+}, this turbid fraction appears to contain a decreased level of the δ-subunit (34). Determination of the stoichiometry of phosphorylase kinase does have some inherent difficulties but the possibility that the enzyme might exist in other than the α (or α')$_4$ $\beta_4\gamma_4\delta_4$ form definitely merits further consideration (12, 22). The actual data demonstrating a precise subunit stoichiometry is neither extensive nor definitive, especially for the documentation of the ratio of γ to either α or β.

D. SPATIAL ARRANGEMENT OF SUBUNITS

Clues to the potential spatial arrangement of phosphorylase kinase have come from electron microscopy (35, 35a), cross-linking studies (36–39), and isolation of partial structures (40). By either conventional negative staining or scanning transmission EM (35a), phosphorylase kinase appears as a bilobal structure resembling two opposing parentheses held together by two short cross bridges. This butterfly-like nonglobular structure has the molecular mass of the hexadecamer holoenzyme. Selective chymotryptic cleavage of the α subunit did not destroy the overall structure except that the individual lobules were misshapen. Restricted tryptic cleavage of both α and β subunits tended to convert the molecule to two equal halves, suggesting that the native molecule is comprised of bridged octamers each of structure $\alpha_2\beta_2\gamma_2\delta_2$. Interestingly, even with tryptic cleavage of a high percentage of α and β subunits, an unexpected proportion of molecules remained in the butterfly structure of the original mass of the holoenzyme. These patterns of effects suggest that both the β subunit and strong noncovalent forces contribute to the general conformation of the cross bridges in the native molecule. Earlier studies have also provided important insights into subunit–subunit interactions. Chan and Graves (40), using lithium bromide–promoted dissociation, obtained two partial complexes, $\alpha\gamma\delta$ and $\gamma\delta$, both of which retained catalytic activity; the characteristics of expression of activity would suggest that these two complexes retained some (or most) of the same interactions between the subunits as exist in the native molecule. From cross-linking studies, using dimethyl suberimidate, dimeric pairs of $\alpha\alpha$, $\beta\beta$, $\alpha\gamma$, and $\beta\gamma$, and trimers of $\alpha\beta\gamma$ and $\beta\gamma_2$ have been obtained (36–37), indicating close interactions of these various combinations of subunits in the native molecule. Also of interest, similar cross-linking approaches by Fitzgerald and Carlson (38) using difluorodinitrobenzene have shown that on enzyme activation by a variety of means an increased β–β interaction is observed. The current 3-dimensional structure of phosphorylase kinase is unknown; whatever eventual structure the holoenzyme proves to have, it will need to account for the various observations listed above. One early schematic presented by Picton et al. (39) to account for most of the cross-linking observations and the general "butterfly" EM observations is depicted in Fig. 1, but this does not readily account for the most recent EM observations (35a). Also shown in the schematic model is the interaction of four additional calmodulin molecules (δ') which by cross-linking studies (37) have been shown to be associated with α- and β-subunits. The role of these additional calmodulin interactions is discussed in Section V,C).

E.   SUBUNIT ISOLATION

The individual subunits α, α', and β have yet to be isolated in a form that maintains their full structural integrity. The α-, β-, and γ-subunits can be iso-

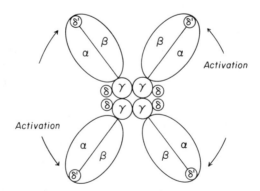

FIG. 1. Pictoral arrangement of the subunits of phosphorylase kinase. The arrows depict one type of conformational change suggested to occur upon activation. Adapted from Picton et al. (39).

lated in a denatured form by high-performance liquid chromatography (HPLC) (27) and the γ-subunit also by Sephadex G-200 gel filtration in the presence of SDS (13). Initially, the γ-subunit was reported to have been isolated with retained catalytic activity (41), however, the form isolated was subsequently identified as the γδ dimer. Pretreatment of the γδ dimer with EGTA, followed by sucrose density-gradient ultracentrifugation resulted in the formation of a Ca^{2+}-independent kinase activity which, although not specifically tested, may well be an isolated γ-subunit with full structural integrity (40). The γ-subunit, isolated by HPLC, has been obtained following renaturation, in an active form (42). The δ-subunit can be readily isolated with retained function by heat denaturation of the holoenzyme (43).

F. *In Vivo* Subunit Synthesis

The synthesis of subunits α and β from γ *in vivo* has been reported (43a). In that study the turnover of the α and β subunits in skeletal muscle appeared to be ~1.6 fold greater than that of the γ subunit. Clearly more data of this type is needed.

III. Subunit Function and Interaction between Subunits

A. The Catalytic Subunit(s)

The function of the individual subunits of phosphorylase kinase is only partially understood. Several reports indicate that the γ-subunit has catalytic activity. This was first proposed by Hayakawa *et al.* (44) when it was observed that activation of the enzyme by trypsin treatment occurred with proteolysis of the α- and β-subunits, whereas the γ-subunit remained intact. More direct evidence has come from the work of Skuster *et al.* (41), Chan and Graves (40, 45, 46), and Kee and Graves (42) who have isolated a γδ complex that retained full catalytic activity for the phosphorylation of both phosphorylase and phosphorylase kinase. Furthermore, treatment of the γδ complex with EGTA followed by ultracentrifugation led to a Ca^{2+}-independent phosphorylase kinase activity, but Ca^{2+}-dependency could be restored with addition of exogenous δ-subunit (46). Equally strong evidence that the γ-subunit contains a catalytic site comes from the recognition that there is substantial homology between the sequence of the γ-subunit and that of the catalytic subunit of the cAMP-dependent protein kinase (28).

There is also evidence, albeit less direct, that the β-subunit might contain another catalytic site (distinct from that on the γ-subunit). This is supported by several lines of experimentation. One approach has been the use of ATP analogs to covalently label ATP binding sites on phosphorylase kinase. Gulyaeva *et al.* (*47*), utilizing alkylating ATP analogs modified in the triphosphate moiety, found that both β- and γ-subunits were labeled. Modification of the β-subunit correlated well with inactivation of the enzyme while γ-subunit modification appeared to have little effect on activity. Subsequently, King *et al.* (*48, 48a*), demonstrated preferential labeling of the β-subunits of phosphorylase kinase with two photoaffinity analogs of ATP, 8-azido ATP and its 2',3'-dialdehyde derivative, both of which serve as the phosphoryl-donor substrate and thus must interact at the catalytic site. With either of these, labeling of the β subunit was accompanied with loss of activity, but correlations were inexact and greater (faster) derivatization occurred than inactivation. Both labeling of the β subunit and inactivation (on a percentage basis) were equally protected by ADP; there were, however, disparate effects with addition of divalent cations in that labeling was depressed in the presence of Mg^{2+} or Ca^{2+}, whereas inactivation was unaffected by Mg^{2+} and enhanced by Ca^{2+}. Affinity labeling has also been studied by this group using 5'-*p*-fluorosulfonyl benzoyl adenosine (*48b*); again, the β subunit was preferentially labeled and the enzyme inactivated, but because other subunits (α and γ) were also derivatized, no conclusions could be reached about exact correlations. One difficulty in interpreting such data is that phosphorylase kinase has been shown to have eight binding sites for ADP (see Section IV,A) and this leaves open two possibilities. The first is that there are four catalytic sites (one on each γ) and four allosteric sites (on β?); the second is that the latter are also catalytic sites. ADP, however, is clearly an allosteric regulator. The presence of a catalytic site on the β-subunit has also been suggested by Fischer *et al.* (*49*), who reported isolation of a catalytically active phosphoprotein after proteolysis of phosphorylase kinase phosphorylated predominantly in the β-subunit; a full report of these findings has not been presented. Compatible with this observation, however, Killilea and Ky (*50*) observed that following extended trypsin treatment of cardiac phosphorylase kinase, only one polypeptide remained, which corresponded to the β-subunit, but catalytic activity had been retained. Similar indirect evidence also exists suggesting that the α subunit might contain a catalytic site. Crabb and Heilmeyer (*27*) have shown that there is some sequence homology between the N-terminal region of the α-subunit and that of the transforming protein from Rous sarcoma virus, which is a tyrosine protein kinase. ATP-dependent derivatization of the α subunit has also been reported using fluorescein isothiocyanate (*50a*), which with several proteins binds at or near an ATP catalytic site. Very selective derivatization of the α (and α') subunit was seen that could be blocked by ATP addition (*50a, 50b*), but as with the data described above for the β subunit this could possibly represent binding to a regulatory rather than a catalytic site. The sequence of the FITC derivatization site has been reported (*50b*).

Evidence for potential multiple catalytic sites also stems from studies of phosphorylase kinase-catalyzed phosphorylation of substrates other than phosphorylase b. Carlson and Graves (51) suggested the possible existence of two separate catalytic sites to explain the enhanced autophosphorylation they observed when autocatalytic reactions were carried out in the presence of phosphorylase or peptide analogs of its phosphorylated region. They also observed that troponin did not inhibit the phosphorylase b to a conversion at pH 8.2, and in fact, accelerated the reaction at pH 6.8. One would normally predict that addition of one substrate (phosphorylase in the former case, troponin in the latter) would competitively inhibit the phosphorylation of another substrate (phosphorylase kinase and phosphorylase b, respectively). Dickneite et al. (52) reported that antibodies to phosphorylase kinase inhibited phosphorylase b phosphorylation in an uncompetitive manner but inhibited troponin phosphorylation in a competitive manner. This type of differential inhibition of activity toward various substrates was later described by King and Carlson (53) using an ATP analog to affinity-label phosphorylase kinase. Affinity-labeling in the presence or absence of Ca^{2+} and Mg^{2+} allowed them to distinguish three different classes of substrates based on their reactivity with the partially inactivated enzyme. In explaining their results, the authors proposed a model in which glycogen synthase and phosphorylase b are preferentially phosphorylated at one type of catalytic site, whereas troponin I and troponin T are phosphorylated at another. Further evidence supporting the idea of multiple catalytic sites has been reported by Kilimann and Heilmeyer $(54, 55)$ who have distinguished three separate activities of phosphorylase kinase towards phosphorylase b by their dependence on Ca^{2+}, Mg^{2+}, NH_4Cl, and pH. These three types of enzymic activity have different apparent specificities toward the protein substrates phosphorylase b, troponin I and T, and phosphorylase kinase; albeit that the proposed different substrate specificities from the two reports $(53–55)$ do not match.

Although the several lines of evidence presented here suggest that the γ-subunit and another subunit, potentially β, both contain catalytic sites, the data are still ambiguous. There can be little doubt, especially with the work of Chan and Graves $(40, 45, 46)$ and Reimann et al. (28), that the γ-subunit has a catalytic site. However, many of the observations suggesting that there is a second site can be rationalized if one assumes complex interactions within the phosphorylase kinase molecule. That phosphorylase kinase might exhibit complex interactions would hardly be surprising in the light of what is already known concerning its regulation. Chan and Graves (45) have potentially provided the most important clue that it is quite likely only the γ-subunit that contains a catalytic site (at least for phosphorylase b). They have shown that the molar activities of the holoenzyme, the $\alpha\gamma\delta$ complex and the $\gamma\delta$ complex (plus additional calmodulin) are, respectively, 99.3, 91.4, and 104 molecules/sec with phosphorylase as substrate. Thus, the $\gamma\delta$ complex (devoid of α and β) and the $\alpha\gamma\delta$ complex (devoid of β) exhibit the full catalytic competence of the holoenzyme. This would obviate the need to evoke a second catalytic site, in particular for the phosphorylase b to a conversion. However, other substrates might be phosphorylated by other catalytic sites, and it has been reported that phosphorylase kinase exhibits a low level of phosphatidyl inositol kinase activity which constitutes quite a different type of substrate (55a). Also of interest in a study of monoclonal antibodies directed against phosphorylase kinase, one clonal antibody was found that interacted equally with the α, β, and γ subunits (55b). If each contained a similar catalytic site to which the antibody was directed, this may be as would be expected.

B.   THE δ-SUBUNIT AND NATURE OF ITS INTERACTIONS
IN THE HOLOENZYME

The existence of the δ-subunit of phosphorylase kinase was not demonstrated until 1978 (43) due in large part to its small size and poor staining with typical protein stains. The identity of the δ-subunit and calmodulin, a calcium-binding protein first identified in the brain, was suggested by several physicochemical properties and was confirmed by its amino acid composition and ability to reactivate calmodulin-dependent enzymes (43). The amino acid sequence of the δ-subunit has been found to be identical to that of bovine uterus calmodulin and to differ only in amide assignments at two residues from that of bovine brain calmodulin (56). Calmodulin acts as a Ca^{2+}-dependent modulator of a wide variety of enzymes and the mediator of the control of these enzymes in response to physiological fluxes of Ca^{2+}.

If Ca^{2+} binding to the holoenzyme occurs exclusively through the δ-subunit, one might expect that phosphorylase kinase would show similar binding properties to those observed with calmodulin; any observed differences may give an indication of restrictions placed upon the δ-subunit as a consequence of it being an integral component of the holoenzyme. This analysis has been made by Heilmeyer et al. (57–59) who presented a comparison of the Ca^{2+}-binding properties of the holoenzyme and its isolated δ-subunit. Their data are summarized briefly in a simplified form in Table I. Analyses of binding were performed under three conditions: low ionic strength, high ionic strength, and high ionic strength plus Mg^{2+}. At low ionic strength, the holoenzyme binds 3–4 mol of Ca^{2+} per $\alpha\beta\gamma\delta$ with high affinity ($K_d = 20$–1000 nM). At high ionic strength, in the absence of Mg^{2+}, two of these high-affinity sites are retained but two are lost. In the presence of Mg^{2+}, two effects occur—the affinity of the two retained sites is diminished, but two other high-affinity sites are now detectable. Kohse and Heilmeyer (59) classified the sites as Ca^{2+}–Mg^{2+} and Ca^{2+}-specific. The Ca^{2+}/Mg^{2+} sites bind either ion so that in the presence of Mg^{2+} the apparent affinity for Ca^{2+} is depressed. The Ca^{2+}-specific sites bind only Ca^{2+}

TABLE I

COMPARISON OF CA^{2+} BINDING TO CALMODULIN OR THE δ-SUBUNIT OF
PHOSPHORYLASE KINASE[a]

Conditions	n	Isolated calmodulin K_d (M)	n^b	Phosphorylase kinase K_d (M)	Site
Low ionic strength	4	5.8×10^{-8}	3	2.0×10^{-8}	Ca^{2+}/Mg^{2+} plus
			1	6.0×10^{-6}	Ca^{2+}-specific
High ionic strength	2	4.0×10^{-6}	2	2.0×10^{-8}	Ca^{2+}/Mg^{2+}
	2	4.0×10^{-5}	—	—	Ca^{2+}-specific
High ionic strength	2	6.6×10^{-5}	2	2.5×10^{-7}	Ca^{2+}/Mg^{2+}
plus Mg^{2+}	2	2.8×10^{-5}	2	3.0×10^{-6}	Ca^{2+}-specific

[a] Adapted from Kohse and Heilmeyer (59).
[b] Per (αβγδ).

but, in the case of the holoenzyme at high ionic strength, the Ca^{2+}-specific sites require the presence of Mg^{2+} for Ca^{2+} binding. The designations for these sites are indicated in Table I. With the isolated δ-subunit, at low ionic strength, 4 mol of Ca^{2+} are bound per mol with high affinity, with both stoichiometry and affinity similar to what is observed with holoenzyme. At high ionic strength, the isolated δ-subunit retains four Ca^{2+}-binding sites, but their affinity is reduced. Addition of Mg^{2+} reduces the affinity of two of these sites (the Ca^{2+}–Mg^{2+} sites) further but does not affect binding to the Ca^{2+}-specific sites. Cooperativity with Hill values of ~2 was obtained for Ca^{2+} binding to both types of sites in both the isolated δ-subunit and holoenzyme. From these results, Kohse and Heilmeyer (59) drew the following conclusions:

1. The similarities of binding of Ca^{2+} to the isolated δ-subunits and holoenzyme, especially with respect to number of sites, designation of types of sites, and the cooperativity of Ca^{2+} binding, indicates that Ca^{2+}-binding by phosphorylase kinase can be fully accounted for as occurring through the δ-subunits.
2. The fact that only in isolated δ-subunits is the affinity of Ca^{2+} for the Ca^{2+}–Mg^{2+} sites depressed by an increase in ionic strength suggests that the integration of calmodulin into the holophosphorylase kinase stabilizes it in a conformation that is similar to that of the isolated subunit at low ionic strength.
3. Subunit–subunit interaction in the holoenzyme, most likely involving heterologous subunits, modifies the conformation of the environment of the Ca^{2+}-specific sites so that binding occurs only in the presence of Mg^{2+}.

The specific Mg^{2+}-binding sites may be on the δ-subunit or elsewhere on the phosphorylase kinase molecule. The interaction of the δ-subunit with the other subunits of phosphorylase kinase differs from that of calmodulin with other calmodulin-regulated enzymes. As exemplified by the regulation of myosin light chain kinase by calmodulin and presented in Chapter 4 in this volume (60), the typical mode of calmodulin interaction can be described by the two-step reaction as shown in Scheme I.

$$\text{Calmodulin} + Ca^{2+} \rightarrow Ca^{2+}\text{-calmodulin} \xrightarrow{\text{enzyme}} Ca^{2+}\text{-calmodulin-enzyme}$$

SCHEME I

That is, calmodulin binds only in the presence of Ca^{2+}, and when Ca^{2+} is removed, calmodulin dissociates from the enzyme. In contrast to this, the δ-subunits are tightly bound integral components of phosphorylase kinase holoenzyme, and are not readily dissociated by such agents as the Ca^{2+}-chelators, EDTA and EGTA, or by high concentrations of urea (12, 37, 43). [A slow rate of exchange (15% per week) of ^{14}C-labeled exogenous calmodulin with the δ-subunit can occur (37)]. Cross-linking studies (36, 37) have indicated that the δ-subunit is primarily bound to γ-subunit. This interaction is maintained during the lithium bromide-promoted dissociation to form the partial complexes $\alpha\gamma\delta$ and $\gamma\delta$ (40); however, it appears that at the level of the $\gamma\delta$ dimer the interaction is more closely analogous to other calmodulin-regulated enzymes since the dimer can apparently be dissociated by treatment with EGTA (46).

C. THE α- AND β-SUBUNITS AS REGULATORS

In addition to the possible functions previously discussed, several lines of evidence suggest that the α- and β-subunits serve a regulatory function. Phosphorylation of the α- and β-subunits by the cAMP-dependent protein kinase (13, 14) or by autophosphorylation (61, 62) results in activation, as does limited proteolytic degradation of these subunits (14, 63). Activation of the enzyme also results from dissociation of the holoenzyme by LiBr (45), which has led to the suggestion that the activity of the enzyme is inhibited by the regulatory subunit(s) (α and β), and that this inhibition can be relieved by phosphorylation, limited proteolysis, or dissociation.

In consideration of potential roles of phosphorylase kinase subunits, a possible indicator of unique function is the observation that whereas, for those enzymes tested, the β-, γ- (and δ-) subunits appear identical (by SDS-PAGE), this is not true for the α-subunit. Not only is it clear that within species there are two forms of α-subunit (α and α') but between species, the primary subunit that appears different is the α (or α') (64). Two recent reports have described the preparation of subunit specific antibodies which promise to be of some assistance in further defining the specific roles that each subunit serves (*55b, 64a*).

D. DOES THE γ-SUBUNIT HAVE ADDITIONAL ROLES?

In a single report by Fischer *et al.* (*49*), it was indicated that there was considerable homology between the γ-subunit of dogfish phosphorylase kinase and dogfish actin, even to the extent of interaction with myosin. None of this, however, was commented on in the subsequent follow-up full-length paper (*20*) and there is clearly no homology between equivalent rabbit muscle proteins (*28*). The reason for the high propensity of phosphorylase kinase to aggregate (*13*) (it does even with pressure which normally promotes disaggregation) is not known, and a similarity of the γ-subunit with actin would be an attractive explanation. The polymerized form has been reported to contain another component of the contractile apparatus, α-actinin (*12*).

IV. Catalytic Properties

A. CHARACTERISTICS OF THE PHOSPHORYLASE *b* TO *a* REACTION

The major reaction thought to be catalyzed by phosphorylase kinase *in vivo* is the phosphorylation of phosphorylase *b*. In this reaction, phosphorylase *b*, a dimer, is phosphorylated at each of two identical serine residues in the presence of Mg^{2+} and ATP. Mg^{2+} added in excess of that required to form the substrate, $MgATP^{2-}$, results in stimulation of phosphorylase kinase (*7, 65, 66*). This may be via additional binding sites for free Mg^{2+} (*67*), although free ATP^{4-} has been suggested to be inhibitory and thus some uncertainty exists as to whether free Mg^{2+} is stimulatory or free ATP^{4-} inhibitory (*7, 68, 69*). In addition to Mg^{2+}, Ca^{2+} is required for the activity of both the activated and nonactivated form of the enzyme (*7, 70*). The allosteric effects of Ca^{2+} and Mg^{2+} are discussed in more detail in Section V,B. Cheng *et al.* (*71*) have studied the interaction of ADP with phosphorylase kinase. In addition to being a product, ADP is an allosteric activator; 8 mol are bound per $(αβγδ)_4$ with K_d values in the range of 0.26 to 17 μ*M*. ADP stimulates both phosphorylase conversion and autophosphorylation and inhibits β-subunit dephosphorylation. Binding at this allosteric site is highly specific for the ADP moiety and many ADP analogs could not substitute for it (*71*).

Nonactivated phosphorylase kinase isolated from resting muscle (in the presence of divalent cations and in the absence of phosphatase inhibitors) has little activity at physiological pH (pH 6.8–7.0), but has considerable activity at pH values greater than 7.6 (7). Following phosphorylation, catalyzed by one of several protein kinases, or following limited proteolysis, the activity at pH 6.8–7.0 increases markedly and much more so than that measured at higher pH values (e.g., pH 6.8 activity increases as much as 50-fold with phosphorylation by the cAMP-dependent protein kinase). These properties are maintained during purification of the protein to homogeneity. Since dissociation of the enzyme into partial complexes ($\alpha\gamma\delta$ and $\gamma\delta$) also results in a marked activation at pH 6.8 with little change in activity at pH 8.2, it is most likely that the regulatory subunits (α and β) in nonactivated enzyme inhibit the catalytic site from exhibiting maximum catalytic potential; this inhibition is relieved either by a conformational change induced by pH or covalent modification, or by removal of the inhibitory subunits by dissociation and/or proteolysis. The changes in activity that occur either with pH or covalent modification can be attributed almost entirely to changes in affinity for phosphorylase.

This change in pH dependency has been exploited as a means to express the activation status of phosphorylase kinase, especially for an evaluation of enzyme activation occurring in intact tissues. Thus, nonactivated phosphorylase kinase has a ratio of activity at pH 6.8 to that at 8.2 of ~0.04–0.08, and activation by pH change, covalent modification, or proteolysis increases the activity ratio to ~0.2–0.9. Although this measurement has gained wide acceptance, there are some inherent problems with its use. For example, in studies of phosphorylase kinase activation in guinea pig hearts, Hayes and Mayer (72) could detect no changes in the ratio of activity at pH 6.8 to that at 8.2, despite a readily observable change in the pH 6.8 specific activity that was clearly a consequence of cAMP-dependent activation. The differences were shown to be attributable to differences between the kinetic parameters of guinea pig cardiac phosphorylase kinase and those of either the rat cardiac or rabbit skeletal muscle enzymes (72). It is important to note that the activity at pH 8.2 is not static but also changes with phosphorylation and proteolysis albeit, in most cases, less dramatically than the activity at pH 6.8 (44). Several endogenous factors also affect the pH 8.2 activity measurement and a greater variation in its quantitative value is often experienced (73). Some caution is therefore necessary in the interpretation of pH 6.8–8.2 activity ratios, and, we have repeatedly found that the measurement of pH 6.8 specific activity gives a more reliable index of phosphorylase kinase activation state.

B. KINETICS

Kinetic studies of phosphorylase kinase have been difficult because of an unusual lag in its catalytic reaction. This lag is pH-dependent, being more marked at pH values near neutrality but not so apparent at pH 8.2 (11, 66). It also appears to be dependent upon buffers, preincubation with substrates, and enzyme concentration. Several initial observations suggested that the lag may have been due to autophosphorylation of phosphorylase kinase (discussed in Section VI, D) with consequent activation (62). Subsequently, King and Carlson (74, 75) reported that the lag seen in phosphorylase conversion (and in autophosphorylation) at pH 6.8 can be diminished by preincubating the kinase with Mg^{2+} and Ca^{2+}. This process, which was termed "synergistic activation," requires the presence of both ions at half-maximal concentrations of 5 μM Ca^{2+} and 4 mM Mg^{2+}. Activation, which is maximal in 2 min, is reversed by chelators and decreased by both ATP and phosphorylase b. King and Carlson (75) concluded that this synergistic activation by Ca^{2+} and Mg^{2+} is the primary cause of the lag in the phosphorylase kinase reaction and that autophosphorylation, if it occurs, is secondary. Presumably, Ca^{2+} and Mg^{2+} promote a slow conformational change in the phosphorylase kinase structure, a situation that has been termed "hysteresis" (66, 75). This Ca^{2+} plus Mg^{2+}-dependent synergistic activation has also been shown to occur within the relatively physiological milieu of the glycogen particle (75a).

Table II presents a summary of reported kinetic constants for phosphorylase kinase (7, 45, 76); for simplicity, the data are presented in two parts, A and B, reflecting assays done by different laboratories. The trends in each are similar, and whether the apparent differences reflect minor differences in assay conditions or enzyme preparation is not known. The data in part A represent best what has been explored of the nature of activation by pH or phosphorylation; the data in part B are directed at what occurs upon dissociation of phosphorylase kinase into partial complexes. As previously noted, nonactivated enzyme has a pH 6.8–8.2 activity ratio of ~0.05 and activation by phosphorylation results in a marked change in the pH 6.8 activity but minimal changes at pH 8.2. These activations, either by pH or covalent modification, are reflected in the kinetic constants (Table II, part A). Between pH 7.0 and 8.5 there is a 10-fold decrease in the K_m for phosphorylase with essentially no change in either the K_m for ATP or the V_{max} of the reaction. Similarly, phosphorylation of phosphorylase kinase decreases the K_m for phosphorylase, at lower pH, without modification of the other kinetic parameters. Thus, activation of phosphorylase kinase, either by an increase in pH or by covalent modification, is attributable to a change in affinity for phosphorylase.

The dissociation of phosphorylase kinase into the partial complexes $\alpha\gamma\delta$ and $\gamma\delta$ occurs concomitantly with an increase in catalytic activity which is minimal at pH 8.2 but marked at pH 6.8 [i.e., the ratio of activity at pH 6.8 to that at 8.2 of holoenzyme, $\alpha\gamma\delta$, and $\gamma\delta$ are, respectively, 0.04–0.07, 0.50–0.60, and 0.9–1.00 (45)]. These changes in activity upon dissociation were likewise reflected by changes in kinetic constants (Table II, part B). As with increasing pH or covalent modification, activation by dissociation is accompanied by an increase in affinity for phosphorylase, reflected primarily at pH 6.8 rather than at pH 8.2.

TABLE II

KINETIC CONSTANTS FOR PHOSPHORYLASE KINASE

| | | Phosphorylase | | | ATP | | | |
| | | | Conditions[a] | | | Conditions | | |
Enzyme	pH	K_m (μM)	ATP (mM)	Mg^{2+} (mM)	K_m (mM)	Phos (μM)	Mg^{2+} (mM)	Ref.
A. Effects of activation								
Nonactivated	8.5	33	3	10				(8)
	8.2	40	3	10	0.31	33	10	(8)
	7.6	125–250	3	10				(8)
	7.4				0.24	33	10	(8)
	7.0	370	NS					(76)
Activated	8.2	17	3	10				(8)
	7.5	37	3	10	0.38	33	10	(8)
	7.0	20	NS					(76)
B. Effects of dissociation								
Nonactivated	8.2	250	2.8	10	0.22	100	10	(45)
	8.2	230	—	10	0.26	—	10	(45)
Activated	8.2	80	2.8	10				
$\alpha\gamma\delta$	8.2	110	2.8	10	0.50	100	10	(45)
	8.2	91	—	10	0.58	—	10	(45)
$\gamma\delta$	8.2	94	2.8	10	0.95	100	10	(45)
	6.8	83	2.8	10	0.86	100	10	(45)

[a] NS, not specifically stated.

In the case of dissociation to the $\gamma\delta$ dimer, there appears also to be a loss in affinity for ATP, presumably reflecting some role for the α and β subunits in the interactions of ATP with the holoenzyme.

Kinetic studies with phosphorylase b as substrate are potentially subject to interpretative errors since effectors may be enzyme and/or substrate directed. For this reason, alternate substrates (which do not bind such factors as metal ions, nucleotides, buffers, or glycogen) have been sought. Tessmar and Graves (77) have studied a tetradecapeptide composed of the same amino acid sequence that surrounds the phosphorylated serine in phosphorylase. This peptide is phosphorylated at the same site as the native substrate. Also, the reaction with the peptide is similar to that with phosphorylase in several important aspects; the reaction shows the same type of lag in product formation, a similar pH dependence, and essentially the same Ca^{2+} and $MgATP^{2-}$ requirement. Although the K_m for peptide is considerably greater than that for phosphorylase b (suggesting the involvement of a greater region of phosphorylase in binding to phosphorylase kinase than simply the 14 amino acids at the phosphorylation site or the requirement for a precise peptide chain conformation), once bound, the peptide is readily phosphorylated.

Using this substance, Tabatabai and Graves (69) studied the kinetic mechanism of the phosphorylase kinase reaction. With activated phosphorylase kinase (phosphorylated by the cAMP-dependent protein kinase) and either the tetradecapeptide or phosphorylase b, initial rate data suggest a sequential-type mechanism. Competitive inhibition patterns with analogs of the tetradecapeptide or of ATP are consistent with a random bi bi mechanism.

The reversibility of the phosphorylase b to a reaction (catalyzed by phosphorylase kinase) has also been studied. Early reports suggested that the reaction was irreversible (78) but later studies indicate that reversal can take place in the presence of glucose which tends to dissociate tetrameric phosphorylase a to a dimer (79). Interestingly, the latter report indicates that the pH dependence of the reverse reaction differs from that of the forward reaction and that phosphorylation of phosphorylase kinase by the cAMP-dependent protein kinase does not affect the rate of the reverse reaction.

C. PEPTIDE SUBSTRATE SPECIFICITY

In early studies of phosphorylase kinase (80) it was shown that a tetradecapeptide, isolated by chymotryptic digestion of phosphorylase and containing the seryl residue that was phosphorylated in the native molecule, could be readily phosphorylated by phosphorylase kinase, albeit with a fivefold lower V_{max} and a fivefold higher K_m. This peptide has served as the initiation point for studies of phosphorylase kinase substrate specificity (81–83). Phosphorylation of the peptide shares many of the characteristics of that of the native substrate. The reaction requites Ca^{2+}, and exhibits a low pH 6.8–8.2 activity ratio with nonactivated phosphorylase kinase; the peptide is phosphorylated at a faster rate by enzyme activated either by phosphorylation or proteolysis, and the reaction shows the characteristic initial lag in reaction rate. Subsequently, it has been shown that glycogen synthase is also phosphorylated by phosphorylase kinase and peptides derived from it have also been examined as potential substrates (84, 85). A summary of this data is presented in Fig. 2. Peptides 1–5 are based upon the sequence of glycogen synthase, peptides 6–34 on the sequence of phosphorylase. The phosphorylatable residue is Ser-7 in glycogen synthase and Ser-14 in phosphorylase. Peptide 5 is the sequence of the first fifteen residues of glycogen synthase; peptide 6, the first eighteen residues of phosphorylase; and peptide 7, the originally identified tetradecapeptide (80). For comparison the two sequences derived from phosphorylase and glycogen synthase have been aligned for amino acid homology; the sequence of glycogen synthase has five residues deleted which would be approximately a little more than one turn of an α helix.

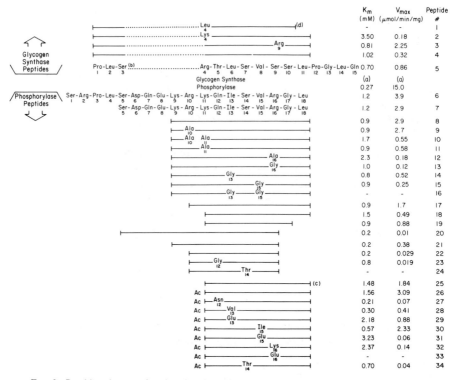

	K_m (mM)	V_{max} (μmol/min/mg)	Peptide #
	-	-	1
	3.50	0.18	2
	0.81	2.25	3
	1.02	0.32	4
	0.70	0.86	5
	0.27	15.0	(a)
	1.2	3.9	6
	1.2	2.9	7
	0.9	2.9	8
	0.9	2.7	9
	1.7	0.55	10
	0.9	0.58	11
	2.3	0.18	12
	1.0	0.12	13
	0.8	0.52	14
	0.9	0.25	15
	-	-	16
	0.9	1.7	17
	1.5	0.49	18
	0.9	0.88	19
	0.2	0.01	20
	0.2	0.38	21
	0.2	0.029	22
	0.8	0.019	23
	-	-	24
	1.48	1.84	25
	1.56	3.09	26
	0.21	0.07	27
	0.30	0.41	28
	2.18	0.88	29
	0.57	2.33	30
	3.23	0.06	31
	2.37	0.14	32
	-	-	33
	0.70	0.04	34

FIG. 2. Peptide substrates for phosphorylase kinase. Data taken from Refs. (*81–84*). Values of K_m and v_{max} denoted by ''-'' indicate that the activity with this substrate was too low to be measured. (a) In separate studies (*86*), the K_m for glycogen synthase has been reported to be approximately the same as for phosphorylase but the V_{max} is about one-half. (b) The sequence of peptides derived from glycogen synthase are aligned with those from phosphorylase with residues 6–10 deleted. (c) The data for peptides 25–34 were from a study separate from that for peptides 7–24. There were minor differences in the values from the two studies, as indicated by data for the peptide designated 18 in the first study and 25 in the second. (d) Data from (*85*).

The primary conclusions that can be derived from these studies are as follows:

1. In the phosphorylase sequence, little change occurs when the first eight N-terminal amino acids are deleted from the native sequence (peptide 8); however, deletion of the two carboxyl-terminal amino acids (Gly-17–Leu-18) dramatically reduces the rate of phosphorylation, albeit that the K_m values are more analogous to that of phosphorylase (peptides 19–22).
2. The simple substitution of the phosphorylatable serine by threonine markedly reduces phosphorylation of the peptide despite the presence of the needed hydroxyl group (peptides 24 and 34).

3. There appears to be a requirement for basic amino acids to be present on both (or either) the N- and C-terminal sides of the phosphorylatable serine, although the results are ambiguous.

Thus, replacement of either Arg-10 or Lys-11 or both (peptides 9–11) in the phosphorylase sequence only modestly affects activity, whereas replacement of Arg-4 in glycogen synthase by Lys (peptide 2) makes it a poor substrate, and by Leu (peptide 1) eliminates activity. Similarly, for the potential requirement of a basic amino acid on the C-terminal side, the native glycogen synthase does not contain a basic amino acid in that position, but replacement of Ser-9 by Arg (peptide 3) markedly improves the peptide as a substrate and substitution of Arg-16 in the phosphorylase sequence by Ala (peptide 12) or Gly (peptide 13) markedly diminishes the V_{max}. Another residue of potential importance in dictating substrate specificity appears to be Gln-12 in phosphorylase, since Asn replacement (peptide 27) markedly decreases the V_{max}; however, in glycogen synthase the equivalent position contains a Thr. A hydrophobic residue on and directly next to the C-terminal side of the phosphorylatable serine also appears essential; in both glycogen synthase and phosphorylase this is Val; substitution by Ile (peptide 15) markedly reduces the rate of phosphorylation. These studies have thus begun to provide information on what dictates substrate specificity of phosphorylase kinase, but in all probability more is involved than the sequence of amino acids. A full elucidation of substrate requirements will most likely require a study, not only of the sequences of the peptides but also of their conformational structure and what conformation they can assume when associated with the enzyme. That phosphorylase is a better substrate than any of the peptides clearly indicates that tertiary structure plays a role in establishing the efficacy of a substrate.

In addition to phosphorylase b and glycogen synthase (*86, 89, 90, 90a*), phosphorylase kinase can, *in vitro,* phosphorylate itself (*18, 61, 62, 87, 88*), and has been reported to phosphorylate troponin I (*91*), troponin T (*92, 93*), the sarcolemmal Na^+, K^+ ATPase (*94*), the Ca^{2+}-dependent (transport) ATPase of sarcoplasmic reticulum (*32, 33,*) casein (*94a*), myosin light chain (*94b*), and several other proteins. It is of interest that the site phosphorylated on glycogen synthase (*95*) and on the β-subunit in autophosphorylation (*88*) are also phosphorylated by the cAMP-dependent protein kinase, whereas the latter enzyme does not phosphorylate phosphorylase. As with glycogen synthase (see Fig. 2), the site phosphorylated on the β-subunit of phosphorylase kinase does not contain an Arg residue on the C-terminal side of the phosphorylated serine. Although peptides containing Thr instead of Ser are not readily phosphorylated, the site phosphorylated in troponin I is threonine.

Whether, in addition to phosphorylase, any of these proteins, or others, are indeed physiological substrates for phosphorylase kinase is not known. The site phosphorylated on glycogen synthase *in vitro* can be phosphorylated within the cell (*96*), but since this site is also phosphorylated by at least three other enzymes, it is not known which one(s) are responsible for the phosphorylation *in vivo*.It will be difficult to prove whether or not glycogen synthase is indeed a cellular substrate for phosphorylase kinase. Currently, it appears unlikely that phosphorylase kinase autophosphorylation occurs physiologically. Were it to occur, it would most probably do so in response to an elevation of Ca^{2+} since autophosphorylation is Ca^{2+}-dependent. However under *in vivo* conditions, where Ca^{2+}-dependent phosphorylase kinase-catalyzed phosphorylation of phosphorylase clearly occurs (*73*), there is no indication of phosphorylase kinase autophosphorylation (and autoactivation). This has been explicitly examined for both subunits in perfused cardiac muscle (*96a*).

Two attributes of phosphorylase kinase suggest that it may well have functions in addition to the regulation of phosphorylase. First, as previously indicated, phosphorylase kinase has a very complex structure and, in comparison to other enzymes, it appears to be more complex than is necessary simply for the regulation of glycogenolysis. Second, phosphorylase kinase is present in skeletal muscle at a very high concentration (0.5% of soluble protein). Since phosphorylase is 2% of the soluble protein, then, using the molar activities for the nonactivated enzyme measured by Chan and Graves (*45*) (99.3 molecules/sec), it can be calculated that the amount of phosphorylase kinase present in skeletal muscle would be sufficient to fully activate all of the phosphorylase in the cell within one-tenth of a second. Activation (by covalent modification) would presumably increase this rate even more. Thus it appears that the amount of phosphorylase kinase present in muscle is far greater than is needed to activate glycogenolysis, even under the most extreme circumstances.

It is of interest that whereas phosphorylase, when isolated, is totally associated with the glycogen particle, a different result is observed with phosphorylase kinase, where 20–30% is glycogen bound and most of the rest is cytosolic (*97*). Jennissen *et al.* (*98*) have shown by cytochemical techniques that a "phosphorylase kinase" has a localization distinct from that of phosphorylase, with most phosphorylase appearing to be glycogen bound but most phosphorylase kinase being present in the region of the sarcolemma. Dombradi *et al.* (*99*) have further shown that purified rabbit muscle T-tubules contain phosphorylase kinase, and Le Peuch *et al.* (*33*) and Horl and Heilmeyer (*32*) reported that a unique form of "phosphorylase kinase" is associated with purified sarcoplasmic reticulum vesicles. All of these data suggest that phosphorylase kinase may well play some role in addition to the regulation of phosphorylase and, in particular, that the sarcolemma Na^+,K^+-ATPase and the sarcoplasmic reticulum Ca^{2+}-dependent ATPase may be target sites for control. If so, then phosphorylase kinase may have an important role in ionic homeostasis as well as metabolite availability.

D. Nucleotide Substrate Specificity

The nucleotide specificity of phosphorylase kinase has been examined by Flockhart et al. (100) by comparing it with the cAMP-dependent and cGMP-dependent protein kinases. As might be expected from the sequence homologies of the catalytic sites (28), the three enzymes are similar with respect to nucleotide affinities. Comparisons between nucleotides were made by competition with ATP. The principal conclusion derived is that the 6-amino and β-phosphoryl groups are primary factors in dictating substrate specificity. No other natural nucleotide triphosphates serve as suitable substrates.

V. Regulation of Phosphorylase Kinase Activity by Allosteric Effectors

Phosphorylase kinase activity can be modulated by a number of effectors that interact in a noncovalent and specific manner and presumably modulate activity by affecting the enzyme's conformation. Included in this group are Ca^{2+}, Mg^{2+}, calmodulin, and glycogen; each is discussed here. Phosphorylase kinase activity is also affected by pH, ADP (see Section IV,A), actin (100a, see also Section V,C), ionic strength, several phosphate-containing compounds, and organic solvents. The latter have been discussed in detail elsewhere (11) and need little additional comment. One comment is pertinent, though, particularly in the area of ions used and ionic strength. The conditions that various investigators have used to investigate phosphorylase kinase have varied widely. Ionic strength, and the type of ions present, have a marked effect on phosphorylase kinase structure, and which of these conditions truly mimics the conformation in which phosphorylase kinase exists in the cell is not known; inappropriate conditions can clearly lead to artifactual observations. Personal bias suggests that phosphorylase kinase is particularly sensitive to such changes. In many of the past studies glycerophosphate has been employed as buffer. In glycerophosphate, however, enzyme activity, the degree of phosphorylation, and the regulation by ADP are each suppressed in comparison to some other buffer system; thus glycerophosphate is clearly not the buffer of choice for several types of studies.

A. Ca^{2+}

The requirement of phosphorylase kinase for Ca^{2+} was first demonstrated by Meyer et al. (70). EGTA, a relatively specific chelator, potently inhibited the enzyme's activity, and this inhibition could be reversed by the addition of excess Ca^{2+} ions (7,70). Besides being required for the phosphorylation of phosphorylase b, Ca^{2+} has since been shown to be necessary for the phosphorylation of glycogen synthase, troponin, and phosphorylase kinase itself (61, 62, 87, 88, 101).

The mechanism by which Ca^{2+} regulates phosphorylase kinase is clearly allosteric. Not only is the δ subunit identified as a Ca^{2+}-binding subunit whose properties can fully account for Ca^{2+} binding by the holoenzyme (see Table I), but several lines of evidence have shown that Ca^{2+} does not participate directly in catalysis. Included in the evidence are the observations that (a) free γ-subunit, obtained by partial dissociation and EGTA treatment plus sucrose gradient centrifugation (46), exhibits catalytic activity that is independent of Ca^{2+}, and (b) in most studies some EGTA-insensitive activity exists (54, 74). Allosteric activation of phosphorylase kinase by Ca^{2+} promotes as much as a 25-fold change in the K_m for phosphorylase (102). Removal of Ca^{2+} by EGTA addition prompts dissociation of the phosphorylase kinase–phosphorylase complex (103).

The activation of phosphorylase kinase by Ca^{2+} requires the binding of Ca^{2+} to at least three of the four Ca^{2+} binding sites per δ-subunit (i.e., at least 12 mol per holoenzyme) (104, 105). Thus, for nonactivated phosphorylase kinase (i.e., dephospho-enzyme), allosteric activation (in the presence of Mg^{2+}) requires Ca^{2+} in the concentration range of 2–25 μM (54, 104, 105). [Some of the variation between reports most probably reflects differences in experimental conditions; another likely cause is varying minor degrees of proteolysis (106)]. This level of Ca^{2+} which is required to activate phosphorylase kinase coincides well with the binding constants determined by direct binding studies (Table I), given the differing conditions needed to study the two parameters. The K_a for Ca^{2+} for phosphorylase kinase activation is reduced about 15- to 30-fold by phosphorylation of the enzyme by the cAMP-dependent protein kinase (106, 107–109) and as much as 300-fold by proteolysis (106). [Here again, there is some discrepancy between the amount of change reported; see Ref. (54) for example.] With cardiac phosphorylase kinase (α' isozyme), in contrast to the skeletal-muscle enzyme, phosphorylation does not appear to modify the requirement for Ca^{2+} (22, 72).

The activation of phosphorylase kinase by Ca^{2+} clearly occurs physiologically. In resting skeletal muscle, the intracellular concentration of Ca^{2+} is in the range of 10–100 nM, which rises to 1–10 μM upon stimulation of muscle contraction (103, 104). These concentration changes are in the appropriate range to allosterically regulate phosphorylase kinase and thus provide a physiological link whereby contractile activity is connected to enhanced glycogenolysis. [Evidence that this indeed occurs in skeletal muscle has been well documented, initially by the work of Drummond et al. (110) and Stull and Mayer (111), and subsequently verified by several investigators.] As discussed in more detail in Section VIII,A, in electrically stimulated muscle, enhanced contraction is clearly associated with the phosphorylation and activation of phosphorylase, without covalent modification of phosphorylase kinase, and without an increase in cAMP or cAMP-dependent protein kinase activity. The cause of this phosphorylase activation appears to be the stimulation of phosphorylase kinase by increased cytosolic Ca^{2+}. Activation of phosphorylase by the Ca^{2+}-dependent stimulation of phosphorylase kinase has also been well documented as the mechanism underlying the α-adrenergic activation of glycogenolysis in both heart (73) and liver (112–115). In skeletal muscle, the primary organelle that sequesters Ca^{2+} is the sarcoplasmic reticulum. Addition of isolated sarcoplasmic reticulum to phosphorylase kinase *in vitro* inhibits its activity, and this inhibition can be reversed by the addition of Ca^{2+} (107). More details on the regulation of phosphorylase kinase by Ca^{2+} are presented in Sections V,C and VII,A.

B. MG^{2+} AND MG^{2+}–CA^{2+} INTERACTIONS

One major role for Mg^{2+} in phosphorylase kinase-catalyzed reactions is to serve as part of the substrate, $MgATP^{2-}$. In addition to this, however, Mg^{2+} modulates phosphorylase kinase activity in a manner that is often manifested as an interaction between Mg^{2+} and Ca^{2+}. These interactions have been studied but have been difficult to unravel. A likely site of Mg^{2+}-binding to phosphorylase kinase is the δ-subunit. Isolated calmodulin binds Mg^{2+} at the Ca^{2+}-binding sites and, although the binding of Mg^{2+} to calmodulin is several orders of magnitude weaker than that of Ca^{2+}, the two interact with calmodulin at physiological concentrations that are the same relative concentrations that affect phosphorylase kinase. Interactions between Mg^{2+} and Ca^{2+} on the δ-subunit could readily explain the variety of effects that have been observed for these two ions on phosphorylase kinase. Whether, in addition to those on the δ-subunit, there are other specific Mg^{2+}-binding sites on phosphorylase kinase is not known. One particular problem in interpreting the results obtained so far is that different investigators have most frequently used different assay conditions. This has made comparisons difficult since ionic strength, pH, and types of ions all seem to affect the responses observed with varying concentrations of Mg^{2+} and Ca^{2+}.

In one of the most complete studies, Kilimann and Heilmeyer (54) described the effects of Ca^{2+} and Mg^{2+} in terms of three types of activities, designated A_0, A_1, and A_2, according to their dependence on Ca^{2+}, Mg^{2+}, ionic strength, and pH. These do not necessarily represent unique catalytic sites, although that has been suggested as one possible explanation. At a minimum, these three activities reflect different conformations of the enzyme brought about by the presence of the various ions and cations.

In the Kilimann and Heilmeyer terminology, the A_0 activity is only a small portion (0.1–1%) of the total activity of nonactivated phosphorylase kinase and is essentially Ca^{2+}-independent. Most likely, this activity is identical to that described by King and Carlson (74, 75) as being "EGTA-insensitive." As discussed in Section IV,B, King and Carlson (74) report that the lag that is characteristic of the phosphorylase kinase-catalyzed phosphorylase conversion progress curve can be removed by preincubation of the enzyme with Mg^{2+} plus Ca^{2+}. This preincubation results in a 2- to 7-fold increase in total (Ca^{2+}-dependent) activity [depending upon conditions (74, 75)], and can account for the removal of the lag in the progress curve. In addition to this effect on total activity, however, this preincubation with Ca^{2+} and Mg^{2+} also results in up to a 30-fold increase in the EGTA-insensitive (A_0) activity (75). Several characteristics described by King and Carlson (75) suggest that the activation of the total (Ca^{2+}-dependent) activity and of the EGTA-insensitive activity are not fully the same process.

In addition to the A_0 activity, Kilimann and Heilmeyer (54) described two other activities, A_1 and A_2, that are Ca^{2+}-dependent. The A_1 activity constitutes most of that seen at pH 6.8 and is a high-affinity Ca^{2+}-dependent activity (K_a, $Ca^{2+} = 1.4 \ \mu M$), which requires free Mg^{2+} ($K_a \simeq 0.25$ mM at pH 6.8). High Mg^{2+} inhibits this activity ($K_i = 3.5$ mM) due to competition with Ca^{2+} ions. The A_2 activity is a low-affinity Ca^{2+}-dependent activity that at pH 6.8 is induced by 20 mM Mg^{2+} and is more prominent at high pH due to an increase in V_{max} and an increased affinity for Mg^{2+}. The A_2 activity is half-maximally stimulated by Ca^{2+} concentrations of 10–70 μM (depending on Mg^{2+} concentration and pH). All three activities (A_0, A_1, and A_2) are inhibited at millimolar concentrations of Ca^{2+} and this inhibition is competitively antagonized by Mg^{2+}.

Kilimann and Heilmeyer (54, 55) have also presented evidence suggesting that the three types of phosphorylase kinase activities have different functions. All three catalyze the phosphorylation of phosphorylase but the characteristics of cation dependencies suggest that autophosphorylation is only catalyzed by A_0 and A_1, troponin I phosphorylation only by A_0 and troponin T phosphorylation only by A_1. The three activities also appear to be differentially regulated. Cyclic AMP-dependent phosphorylation (of only the β subunit or of both α and β subunits) appears to regulate only the A_2 activity, autophosphorylation (of α plus β subunits) appears to stimulate both the A_1 and A_2 activities, and limited trypsin proteolysis increases the V_{max} of all three and the Ca^{2+} affinity of A_1. The exact nature of the A_0, A_1, and A_2 activities of phosphorylase kinase remains to be resolved. Perhaps they reflect different catalytic sites, but it is equally possible that they represent three different conformations of the protein brought about by the relative concentrations of different cations and manifested as different interactions at a single catalytic site. In a further report from this same laboratory (55b), the A_0, A_1, and A_2 activities have been proved using subunit-specific monoclonal antibodies. One anti-α antibody obtained provoked a marked enhancement of Ca^{2+}-independent activity (A_0) without affecting Ca^{2+}-dependent activity; it was suggested that the antibody might in some way uncouple an inhibitory signal between Ca^{2+}-free δ subunit(s) and the catalytic site(s). Analogously, an anti-δ subunit antibody was shown to block Ca^{2+}-dependent activity suggesting that restriction on δ subunit conformation can block δ subunit-catalytic site signal transfer. These studies would appear to be a beginning probe in studies of the major conformational changes and interactions which are clearly integral to the multiple interactions that exist for the regulation of this enzyme.

Other investigations of the effects of Mg^{2+} on phosphorylase kinase (65–69) have provided conclusions that generally concur with those presented by Kilimann and Heilmeyer (54, 55). Singh et al. (67) studied the effects of varying concentrations of Mg^{2+}, either in the presence of maximal Ca^{2+} or with Ca^{2+} absent. With Mg^{2+} in substantial excess of ATP, varying Mg^{2+} resulted in a 5- to 10-fold activation of nonactivated phosphorylase kinase (measured at either pH 6.8 or 8.2, and at pH 6.8 in the presence or absence of Ca^{2+}) and of phosphorylase kinase activated by either the cAMP-dependent protein kinase, autophosphorylation, or limited trypsin proteolysis. This stimulation is accounted for by a decrease in the K_m for both phosphorylase and ATP and an increase in V_{max}, and is apparently due to a direct interaction with phosphorylase kinase since it is also observed with casein as substrate. Hallenbeck and Walsh (18) compared the effects of Mg^{2+} and Mn^{2+} on phosphorylase kinase-catalyzed phosphorylase phosphorylation and autophosphorylation. Varying Mg^{2+} affects both in the same concentration range, whereas varying Mn^{2+} stimulates autophosphorylation but inhibits phosphorylase phosphorylation. This appears to provide a distinction in the role that the metal ions play in these two processes that cannot be attributed simply to allosteric effects modulating a single catalytic site.

As discussed in Section IV,B, King and Carlson (74, 75) have shown that the synergistic presence of both Ca^{2+} and Mg^{2+} causes a slow conformational change in the structure of phosphorylase kinase that leads to its activation. Using cross-linking agents as a probe, the conformational change induced has been shown to be similar to that which occurs upon activation by phosphorylation, proteolysis, and high pH (38). The full relationship between this synergistic effect of the two divalent cations, the three states of phosphorylase kinase termed by Heilmeyer A_0, A_1, A_2, the binding characteristics of Ca^{2+} and Mg^{2+} (Table I), and the effects of phosphorylation and proteolysis on divalent cation binding clearly needs considerable clarification.

Although it is clear that varying Mg^{2+} affects phosphorylase kinase activity, the physiological ramifications of this observation are not apparent. Free Mg^{2+} in muscle is approximately 4 mM (116) and is thought to remain relatively constant. It would certainly be surprising if it varied to an extent that would cause significant differences in phosphorylase kinase activity. The role of Mg^{2+} thus appears to be more that of a constitutive cofactor that is either bound or not, depending upon other factors. Heilmeyer et al. (117), for example, proposed a mechanism wherein cAMP-dependent phosphorylation of phosphorylase kinase influences Mg^{2+}-binding properties of the enzyme, which in turn modifies Ca^{2+} binding and hence Ca^{2+}-dependent regulation. In other words, in the unactivated state, the Ca^{2+}-binding sites on the δ subunit may, as their "counter-ion," be occupied by Mg^{2+} that binds but does not produce an active conformation. Increases in Ca^{2+} promote exchange, or exchange might arise because the relative affinities for Ca^{2+} and Mg^{2+} are modified by other changes occurring on the protein, such as covalent modification. The effects of Ca^{2+} and Mg^{2+} on phosphorylase kinase could all be explained by such a mechanism given the known cooperativity between binding sites (either on the same δ subunit or different subunits). Thus Mg^{2+} may be an important constituent of phosphorylase kinase in the cell and be involved in its regulation, even though it does not itself change in concentration.

C. REGULATION BY EXTRINSIC CALMODULIN AND TROPONIN C

As previously described, phosphorylase kinase activity is regulated by Ca^{2+}, mediated via the δ subunit whose structure is identical to calmodulin except for the presence of trimethyllysine and carboxylation. In addition to this intrinsic calmodulin, however, skeletal muscle phosphorylase kinase can be specifically activated by extrinsic calmodulin (15, 37, 89, 106, 118, 119). This activation also requires Ca^{2+} but, unlike the intrinsic Ca^{2+}-dependent activation occurring via the δ subunit, activation by extrinsic calmodulin is blocked by addition of other calmodulin-binding proteins (which compete for calmodulin) and by phenothiazines. In addition, when Ca^{2+} is removed, the extrinsic calmodulin does not remain bound to the holoenzyme. Cohen (106) has termed this extrinsic calmodulin the "δ'-subunit." Over and above the almost total requirement of holophosphorylase kinase for Ca^{2+}, addition of extrinsic calmodulin activates the enzyme, synergistically, a further 2- to 7-fold. Extrinsic calmodulin has no effect on phosphorylase kinase in the absence of Ca^{2+}. Activation by extrinsic calmodulin is observed not only with phosphorylase b as substrate, but also in the phosphorylation of glycogen synthase (89), phosphorylase kinase itself (12, 89), and troponin I (12). The activation by exogenous calmodulin is very pH dependent.

The activation of phosphorylase kinase by extrinsic calmodulin occurs via a mechanism very similar to that observed for other calmodulin-dependent enzymes, albeit with the added complexity that the enzyme itself, in the absence of extrinsic calmodulin, also binds and is activated by Ca^{2+}. A scheme for the activation of phosphorylase kinase by Ca^{2+} and extrinsic calmodulin, taken in particular from the works of Burger et al. (104) and Cohen (106), is presented in Fig. 3. In the absence of extrinsic calmodulin, phosphorylase kinase activation requires the binding of at least 3 mol of Ca^{2+} per intrinsic δ-subunit. In the

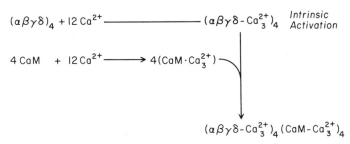

$$(\alpha\beta\gamma\delta)_4 + 12\,Ca^{2+} \longrightarrow (\alpha\beta\gamma\delta\text{-}Ca^{2+}_3)_4 \quad \begin{array}{l} \textit{Intrinsic} \\ \textit{Activation} \end{array}$$

$$4\,CaM \;+\; 12\,Ca^{2+} \longrightarrow 4(CaM\cdot Ca^{2+}_3)$$

$$(\alpha\beta\gamma\delta\text{-}Ca^{2+}_3)_4\,(CaM\text{-}Ca^{2+}_3)_4$$

$$\textit{Intrinsic \& Extrinsic Activation}$$

Fig. 3. Scheme for activation of phosphorylase kinase by calcium and calmodulin. Activation by Ca^{2+} with either intrinsic or extrinsic calmodulin requires a minimum of three of the four Ca^{2+}-binding sites to be occupied. Modified from the data of Burger *et al.* (*104*) and Cohen (*106*).

presence of Ca^{2+}, phosphorylase kinase also binds extrinsic calmodulin stoichiometrically [i.e., one mole of calmodulin per $\alpha\beta\gamma\delta$ (*37*) with a K_d of 2–15 nM (*37, 89, 104, 106*)]. Similar to observations with other calmodulin-requiring enzymes, the binding of extrinsic calmodulin occurs in two steps; namely, initial binding of at least 3 mol of Ca^{2+} per mol of calmodulin, followed by binding of the $(Ca^{2+})_3$–calmodulin complex to the enzyme. The binding of extrinsic calmodulin to phosphorylase kinase does not appear to affect the binding characteristics of Ca^{2+} to the intrinsic δ-subunit, and the binding characteristics of Ca^{2+} to either the intrinsic δ-subunit or extrinsic calmodulin appear to be identical. As a consequence, addition of extrinsic calmodulin does not alter the Ca^{2+} concentration dependence of phosphorylase kinase activity but only the maximum velocity; that is, $(\alpha\beta\gamma\delta\text{-}Ca^{2+}_3)_4$ and $(\alpha\beta\gamma\delta\text{-}Ca^{2+}_3)_4\,(CaM\text{-}Ca^{2+}_3)_4$ have the same Ca^{2+} concentration dependence for formation but the V_{max} of the calmodulin activated form, is 2- to 7-fold greater than V_{max} of the simple holoenzyme. Extrinsic calmodulin has no effect on the affinity of phosphorylase kinase for ATP, phosphorylase, or glycogen (*89*).

Burger *et al.* (*104*) suggested that the inactivation of phosphorylase kinase activated by extrinsic calmodulin does not follow a simple reversal of the activation process. The most likely route of inactivation appears to be first the dissociation of Ca^{2+}, followed by the dissociation of extrinsic calmodulin. The extrinsic calmodulin and the intrinsic δ subunit, however, maintain their separate integrities, and no exchange occurs (within the typical activation–inactivation process) between the two species.

The activation of phosphorylase kinase by extrinsic Ca^{2+}-binding protein occurs not only with calmodulin but also with the homologous protein, skeletal muscle troponin C (*104*). Interestingly, cardiac troponin C, which, though also homologous exhibits distinct Ca^{2+}-binding characteristics (*120–122*), does not serve in this role. This indicates that the activation by the skeletal muscle troponin C has a high degree of specificity. Activation is also not observed with the parvalbumins, though they too are homologous with calmodulin (106). The characteristics of activation of phosphorylase kinase by troponin C are similar to those with calmodulin but with some important differences. Activation by troponin C requires much higher concentrations (K_d = 1–2 μM) than that by calmodulin (K_d = 2–15 nM), but troponin C activates to a greater extent (20- to about 30-fold) and, most importantly, appears to increase the sensitivity to activation by Ca^{2+} by about 5-fold (106). Cohen (106) has argued that troponin C, rather than calmodulin, may well be more physiologically important as the extrinsic activator of phosphorylase kinase. One reason for this proposal is that, while a 100-fold higher concentration of troponin C is required for activation, it is present in the cell at a concentration (100 μM) greater than that required for phosphorylase kinase activation. Potentially, it may be more readily available than the more limited supplies of calmodulin. More importantly, the concentration of Ca^{2+} with which troponin C activates phosphorylase kinase appears to be in a more physiological range than that required of extrinsic calmodulin. Two additional facts support Cohen's proposal. First, he has shown that artificial thin filaments, composed of actin, tropomyosin, and the troponin subunits, are equally as effective as isolated troponin C in the activation of phosphorylase kinase. This observation is particularly pertinent since most (if not all) troponin C in the cell is bound as a component of the myofibrils. Second, Sigel and Pette (123, 124) have shown that the protein-glycogen particle (of which phosphorylase kinase is one component) appears to be associated with the myofibrils. This suggests that the localization of both phosphorylase kinase and troponin C within the cell is highly suitable for one to be regulating the other. An interesting adduction to this is the observation of Livanova et al. (100a), who report that phosphorylase kinase is also separately activated by actin, shown to be through promotion of an increased affinity for phosphorylase and an increased V_{max}. It appears to be the polymerized F-actin form which is the effective activator.

The activation of phosphorylase kinase by extrinsic calmodulin is isozyme-specific. This has been shown by Tam et al. (21) for the α' isozyme isolated from bovine red-skeletal muscle, and by Yoshikawa et al. (125) for this isozyme isolated from rabbit heart, neither of which is activated by addition of calmodulin. Some activation of the heart enzyme has been seen with skeletal muscle troponin C, but since no activation is observed with cardiac troponin C, this suggests that the α' isozyme is not regulated physiologically by extrinsic Ca^{2+}-binding proteins. The α' isozyme is regulated by Ca^{2+} in a manner identical to the α isozyme via the intrinsic δ-subunit (21, 22). Yoshikawa et al. (125) suggested that this difference in regulation between the two isozymes by extrinsic Ca^{2+}-binding proteins may reflect a difference in the needs of the different muscle types for a tighter regulatory linkage between contraction and glycogenolysis. Thus, in fast-twitch muscle, which contains the α isozyme of phosphorylase kinase, contraction relies heavily on glycogenolysis as an energy source. In contrast, in muscles with higher oxidative capacity (cardiac, and FOG and SO skeletal muscle), which contain the α' isozyme, glycogen metabolism is much less important as a source of energy because of a higher aerobic metabolic potential. The difference in the binding of calmodulin to the α and α' isozymes currently serves as the best experimental procedure to separate the two isozymes during purification (15, 102). Of potential note, the skeletal muscle α isozyme, when activated by phosphorylation by the cAMP-dependent protein kinase, is also not significantly activated by extrinsic calmodulin or troponin C but displays an increased sensitivity to Ca^{2+} activation compared to the nonphosphorylated enzyme (106). Thus, epinephrine stimulation enhances glycogenolysis directly, without the need for additional activation by extrinsic Ca^{2+}-binding proteins.

The binding of extrinsic calmodulin to holophosphorylase kinase is quite distinct from that of the intrinsic δ-subunit. Cross-linking studies (37) suggest that its most likely sites of binding are on the α- and δ-subunits (see Fig. 1) and this conclusion is supported by studies of the effects of proteolysis. Apparently the α-subunits must play a distinct role in the interaction with extrinsic calmodulin since no binding occurs with the α' isozyme. Phosphorylase kinase displays an essentially negligible rate of exchange between the intrinsic and extrinsic calmodulin with a $t_{1/2}$ in the neighborhood of several weeks (37).

D. REGULATION BY GLYCOGEN AND IN THE PRESENCE OF THE CONSTITUENTS OF "GLYCOGEN PARTICLE"

Within the cell it appears likely that most, if not all, of the phosphorylase kinase-catalyzed activation of phosphorylase occurs within a complex containing these two enzymes, and others, associated with glycogen. By electron microscopy, glycogen in the cell has been identified as being present in isodiametric particles, termed β-particles, of 150 to 300 Å in diameter, having unique subcellular localization. This is the typical form of glycogen in muscle where glycogen is associated with the sarcoplasmic reticulum and myofibrils. In liver, notably, these β particles form aggregates of a larger size, termed α particles, that are rosette-like in appearance. Such glycogen particles have been isolated from muscle by a variety of methods relying primarily on procedures of differential centrifugation (97, 102, 126–128). The isolated glycogen particles contain all of the enzymes of glycogen metabolism plus sarcoplasmic reticulum vesicles. The high degree of consistency in composition of the glycogen particles prepared by different procedures strongly suggest that they do indeed represent the organization of the enzymes of glycogen metabolism as occurs within the cell. This conclusion is also supported by the observation that the morphology of the isolated glycogen particles is very similar to the structures observed in the intact cell. Wanson and Drochmans (126) have found evidence for specific structures that link the particles to the sarcoplasmic reticulum. The enzymes of glycogen metabolism can be dissociated from the sarcoplasmic reticulum membranes by amylase digestion of the glycogen (97); however, a glycogen particle containing these enzymes but dissociated from sarcoplasmic reticulum vesicles has not been obtained. Using direct precipitation using either ultracentrifugation or acetone typically, glycogen particles contain 70–90% of the glycogen synthase,' 70–100% of the phosphorylase, and 20–40% of the phosphorylase kinase found in skeletal muscle. When an acid precipitation step is also included a lower amount of glycogen is associated with a lower ratio of phosphorylase to phosphorylase kinase, suggesting that there are possibly different species of glycogen particles. Both cAMP-dependent protein kinase and protein phosphatases appear to be specifically associated with glycogen particles, but, fully consistent with the roles they play in other regulatory events within the cell, much of the activity of these enzymes is present in other subcellular locations. It may be of particular significance that only a fraction (20–40%) of phosphorylase kinase is associated with these particles (97); this observation suggests that phosphorylase kinase has other functions within the cell in addition to phosphorylase activation. Depletion of glycogen *in vivo* results in the release of the enzymes of glycogen metabolism from the glycogen particle to the cytosol (127). Of note, even following extensive characterization (128a) there appears to be some residual catecholamine-sensitive adenyl cyclase associated with the SR-glycogen particle.

It appears clear that the phosphorylase kinase-catalyzed activation of phosphorylase occurs in the cell with part or all of these enzymes not being freely soluble in the cytosol, but within a complex with a specific organization of the component parts. Because of this, it becomes particularly pertinent for the reactions involving phosphorylase kinase to be examined with as much of this organization intact as possible. Phosphorylase specifically binds glycogen and studies of its 3-dimensional structure have revealed that, in addition to the substrate site, there is a unique glycogen-binding site that is not involved in the catalytic process (129, 130). Most likely, this is one of the specific interactions that forms the basis of the organization within the glycogen particle. In addition to this, phosphorylase kinase also binds glycogen. Initially, this was recognized from the studies of DeLange *et al.* (8) who observed a Mg^{2+}-dependent glycogen–phosphorylase kinase complex, and also found that phosphorylase kinase autophosphorylation was stimulated by glycogen addition (8). Steiner and Marshall (131) have shown that there is a synergistic effect of both Ca^{2+} and Mg^{2+} upon the interaction of phosphorylase kinase with glycogen, and we have subsequently found that this is also true for the formation of a glycogen phosphorylase–phosphorylase kinase complex (132). Whether this divalent cation synergism is similar to that reported by King and Carlson (53, 74) has not been evaluated.

The effects of glycogen on phosphorylase kinase have been examined by two approaches, one employing purified enzymes, the other the isolated glycogen particle. From both, however, our current knowledge is still only fragmentary. With the isolated enzyme, glycogen activates phosphorylase conversion 3- to 6-fold (8, 22); maximum stimulation occurs with about 0.3% glycogen. Activation is greater at pH 6.8 than at pH 8.2, and arises as a consequence of up to a 12-fold decrease in the K_m for phosphorylase b. Since Tabatabai and Graves (69) found that glycogen has no effect on phosphorylase kinase-catalyzed peptide phosphorylation, it is most likely that glycogen acts via interaction with the substrate, phosphorylase, rather than the catalyst, phosphorylase kinase.

Studies of phosphorylase activation in the glycogen particle were first initiated by Fischer's laboratory (97, 102). As previously discussed, the glycogen particle contains not only the enzymes of glycogen metabolism, but also sarcoplasmic reticulum vesicles. The latter have a maintained and highly active Ca^{2+} accumulation capacity which leads to regulation of Ca^{2+} concentration in the media in which the glycogen particles are suspended. Fischer's laboratory studied what is termed the "flash activation" of phosphorylase in these particles. In the isolated glycogen particle, even if incubated at 30°C with $MgATP^{2-}$, phosphorylase and phosphorylase kinase are inactive and the rate of glycogen degradation minimal. The total lack of activity of phosphorylase kinase reflects the complete removal of any free contaminating Ca^{2+} by the sarcovesicular system. This situation is probably an accurate reflection of what occurs in unstimulated skeletal muscle, where the sarcoplasmic reticulum system dominates the cytosolic Ca^{2+} concentration. Incubation of the glycogen particle with Ca^{2+}-EGTA buffers plus $MgATP^{2-}$ results in flash activation, characterized by a rapid activation of phosphorylase b to a, and a subsequent rapid reversal to basal levels. Typically, with a glycogen particle containing 40 mg/ml protein, the addition of 10 μmol of ATP results in the conversion of 50–80% of phosphorylase b to a within 15 sec, and a return to basal levels within 3 min. The duration of phosphorylase activation is increased by increasing ATP, but under all circumstances inactivation is concomitant with the depletion of ATP, the latter being caused primarily by the Ca^{2+}-dependent ATPase of the sarcoplasmic reticulum Ca^{2+} transport system. All of the properties of flash activation are consonant with the activation being due to allosteric activation of phosphorylase kinase by Ca^{2+}. The Ca^{2+} requirement for this activation appears to be somewhat higher in the glycogen particle than with the isolated enzyme (102). Since α-amylase digestion results in the Ca^{2+}-dependency becoming similar to that for the isolated enzyme, it suggests that glycogen, or the organizational structure of the glycogen particle, modifies the divalent cation dependency of the enzyme (102).

Most of the additional studies on the properties of the glycogen particle report either on the regulation of phosphorylase kinase phosphorylation, which is described in subsequent sections of this review, or on phosphorylase, which is beyond the scope of this review but has been reviewed elsewhere (133, 134).

However, three further observations of phosphorylase kinase activity in the isolated glycogen particle are of potential interest: (a) Srivastava et al. (135) reported a Ca^{2+}-dependent inactivation of glycogen synthase in these particles that, on the basis of antibody studies, could be attributed to phosphorylase kinase-catalyzed phosphorylation. (b) although in their initial report Heilmeyer et al. (102) did not observe Ca^{2+}-dependent autophosphorylation of phosphorylase kinase, we have demonstrated that this can occur (132). It can be masked, however, by concomitant Ca^{2+}-Mg^{2+}-dependent synergistic activation. (c) Heilmeyer et al. (102) reported that, in the glycogen particle, as initially described by Fischer and Krebs (137) with the isolated enzyme, conversion of phosphorylase b to a involves the stepwise addition of phosphate with the formation of intermediary phospho–dephospho hybrids. The potential relevance of these hybrids in metabolic control has, for the most part, been ignored (136). From the fragmentary studies that have been performed with glycogen particles isolated from skeletal muscle or cardiac muscle (138, 139), it seems clear that the organizational structure imposed upon phosphorylase kinase by the glycogen particle is of fundamental importance in dictating the reactions that occur. The ramifications of such controls have yet to be fully realized.

VI. Proteolytic Activation of Phosphorylase Kinase

The activity of phosphorylase kinase towards phosphorylase b can be stimulated by at least two types of covalent modification, phosphorylation and proteolysis. Early observations of Ca^{2+}-dependent activation of partially purified phosphorylase kinase were found to be due in large part to the presence of a contaminating Ca^{2+}-dependent protease, referred to in the early literature as "kinase-activating factor" (KAF) and later as "calcium-activating factor" (CAF) (70). The proteolytic activation was initially distinguished from that due to direct Ca^{2+}-allosteric action on the kinase in that the former required millimolar concentrations of Ca^{2+}, was time dependent, and was irreversible. The Ca^{2+}-concentration dependency is no longer a suitable means for distinguishing the two mechanisms since proteases have since been described that are activated by Ca^{2+} in the more physiological, micromolar range (140). Other proteases, such as trypsin, chymotrypsin, and papain are also capable of activating phosphorylase kinase. The high sensitivity to proteolysis, which can readily be demonstrated at very low protease : protein concentration ratios, in contrast to the lack of specificity as to the type of peptide bond cleaved, suggests that phosphorylase kinase must contain highly exposed regions of peptide chain with little conformational organization, yet which must play a major role in regulating activity. Increases in activity due to proteolysis are on the order of 50- to 100-fold at pH 6.8 and 2- to 3-fold at pH 8.2 (11). The α subunit is the most susceptible to proteolytic degradation; however, both α- and β-subunits are eventually degraded (*14, 44*) whereas the γ-subunit is more resistant. Proteolytic activation eliminates binding by extrinsic calmodulin (*37, 141*) but increases sensitivity to Ca^{2+} activation via the intrinsic δ-subunit (*106*). From cross-linking studies, it can be shown to induce the same conformational rearrangements in the β-subunits, as observed also with activation either by phosphorylation or high pH (*83*). That α-subunit proteolysis results in β-subunit conformational shifts, enhanced activity (presumably of the γ-subunit), and modified binding of Ca^{2+} by the δ-subunits, shows that there must be a high degree of interaction between all of the subunits of this complex molecule.

Most likely, proteolytic activation is of little or no physiological significance because of its irreversibility. However, it does pose a major problem to the maintenance of native enzyme in crude extracts or during purification. Even minor proteolytic ''nicking' gives rise to significant erroneous results as emphasized in particular by the evidence that proteolyzed phosphorylase kinase requires lower Ca^{2+} concentrations for activation (*54, 106*). As noted, however, despite proteolytic nicking, the basic structure and mass of the holoenzyme can remain intact (*35a*). Quite possibly, the results of several of the past studies of phosphorylase kinase may have been compromised because the enzyme used was modified during purification. In general this has been monitored by an assessment of the activity ratio at pH 6.8 to that at pH 8.2, but even this may not be a sufficient criterion (*106*). It appears important that in preparing phosphorylase kinase one should attempt to achieve a stoichiometry of α : β subunits; in most preparations, actual unity of these subunits is rarely reported.

VII. Covalent Regulation: *In Vitro* Studies

A. INTRINSIC PHOSPHATE CONTENT

Nonactivated phosphorylase kinase contains intrinsic alkali-labile phosphate, although the actual amount present is in some dispute. Three reports provide values of 8.5 mol (*141*), 7.18 ± 0.95 mol (*142*), and 7.84 ± 1.32 mol (*143*) per $(\alpha\beta\gamma\delta)_4$; but in an additional study from one of these same laboratories (*27*) the amount was reported as 19.4 ± 0.7 mol per mol, with the α-subunit containing 2.7 ± 0.4 mol/mol subunit, the β-subunit 1.9 ± 0.5 mol/mol subunit and the γ- and δ-subunits less than 0.3 mol/mol subunit. From the latter study, it appears that skeletal muscle phosphorylase kinase can be isolated at one of two levels of phosphorylation; the difference between the two forms or the reason for differences obtained using apparently the same purification procedure are not known.

The function of this intrinsic phosphate is also not known. It is not present in the sites phosphorylated by the cAMP-dependent protein kinase (142), cannot be removed by alkaline phosphatase (143) or protein phosphatase 1 or 2 (143, 144), does not appear to contribute to the activity of the enzyme, and if it does turn over in the cell must do so with a half-life of at least many hours since prolonged incubation (6 h) of intact isolated muscle with ^{32}P-inorganic phosphate does not lead to such levels of intrinsic labeling (145).

B. PHOSPHORYLATION: GENERAL CONSIDERATIONS

Phosphorylation of phosphorylase kinase with accompanying changes in its activity can be catalyzed in vitro by a number of different enzymes. These include the cAMP-dependent protein kinase (13, 14), phosphorylase kinase itself [i.e., Ca^{2+}-dependent autophosphorylation (87)], the cGMP-dependent protein kinase (146, 147), a Ca^{2+}-calmodulin-dependent protein kinase (148), and a Ca^{2+}- and cAMP-independent protein kinase (glycogen synthase kinase-1) (149, 149a). Initial reports that it was also phosphorylated by a Ca^{2+}-phospholipid-dependent protein kinase (or Ca^{2+}-dependent protease-activated kinase) (150, 151) are probably incorrect (152).

The phosphorylations catalyzed by the cAMP-dependent protein kinase and by autophosphorylation have been the most thoroughly studied and are given most attention here. Both β- and α- or α'-subunits of phosphorylase kinase are phosphorylated by these enzymes. The two reactions can be differentiated and thus each can be studied fairly exclusively due to several differences in their catalytic properties. First, the autophosphorylation process requires Ca^{2+} where the cAMP-dependent protein kinase operates independently of this ion (55, 87). Second, autocatalysis, unlike the protein kinase-catalyzed reaction, occurs more rapidly at pH 8.2 than at pH 6.8 (7). One further means of differentiating the two processes is in the use of the inhibitor protein for the cAMP-dependent protein kinase, which, though a potent inhibitor of this enzyme, has no effect on phosphorylase kinase autophosphorylation (87). For some time it was thought that the $MgATP^{2-}$ requirement for autophosphorylation was much greater than that for the protein kinase, so that this was an additional means of separating the two reactions (87). Subsequent studies (discussed in Section VII,D), however, indicate that the two reactions have a similar K_m for the nucleotide (18, 88). Mg^{2+} does stimulate autophosphorylation (67) and this may explain observations that led to the earlier conclusion (87). However, since Mg^{2+} can also affect the kinetic parameters of the protein kinase-catalyzed reaction (153), and because it appears to have an effect on phosphorylase kinase as a substrate (154), it is doubtful whether this ion can be used as a tool in differentiating the two reactions. Equally troublesome, different effects of Mg^{2+} on the cAMP-dependent protein kinase-catalyzed phosphorylation have been reported from different laboratories (152, 154), and divalent cation effects on phosphorylase kinase phosphorylation appear to be different for the different isozyme types (*21, 142, 144, 154*).

C. CYCLIC AMP-DEPENDENT PROTEIN KINASE-CATALYZED REACTIONS

The *in vitro* phosphorylation of skeletal-muscle phosphorylase kinase by the cAMP-dependent protein kinase has been studied by several groups of investigators (*13, 14, 142, 154*). Phosphorylation of both β- and α-subunits has been consistently observed, as has the temporal pattern in which β-subunit phosphorylation is initiated prior to that of the α-subunit and proceeds at a much faster rate. We have presented an extensive characterization of the protein kinase and phosphorylase kinase concentration dependencies of the subunit phosphorylation (*142*). Part of these data are depicted in Fig. 4. These data are compatible with and extend previously published results (*13, 14, 21, 154*), and, taken as a whole, permit the following conclusions:

1. The rates of phosphorylation of either subunit show the expected variation with protein kinase concentration (i.e., they increase in an essentially linear manner with increasing protein kinase concentration (for example, examine the panels vertically in Fig. 4).

2. Increasing the concentration of phosphorylase kinase at a fixed concentration of protein kinase (examine the panels horizontally in Fig. 4) produces the expected increase in the amount of time required to achieve a given stoichiometry of phosphorylation, but a simple Michaelis–Menten relationship is not obeyed.

3. The phosphorylation of the β-subunit always appears to be initiated immediately, and even when rates are low there is no indication of an initial lag.

4. The phosphorylation of the β-subunit always precedes that of the α-subunit and a lag in α-subunit phosphorylation is often apparent. It appears as though α-subunit phosphorylation does not occur until the β-subunit has been phosphorylated to a level of at least 1 mol/mol $(\alpha\beta\gamma\delta)_4$.

5. The phosphorylation of the α-subunit is catalyzed directly by the cAMP-dependent protein kinase. An alternative possibility that could be envisioned from inspection of the progress curves would be that (*a*) the cAMP-dependent protein kinase catalyzed the phosphorylation of the β-subunit, (*b*) this leads to protein kinase catalyzed phosphorylation of the β-subunit, and (*c*) enhanced phosphorylase kinase activity results in an increase in α-subunit phosphorylation via autophosphorylation (i.e., α-subunit phosphorylation catalyzed by phosphorylase kinase). This alternative has been eliminated by studies showing that addition of the inhibitor protein of the cAMP-dependent protein kinase, after β-subunit phosphorylation is maximal, blocks further α-subunit phosphorylation (*142*).

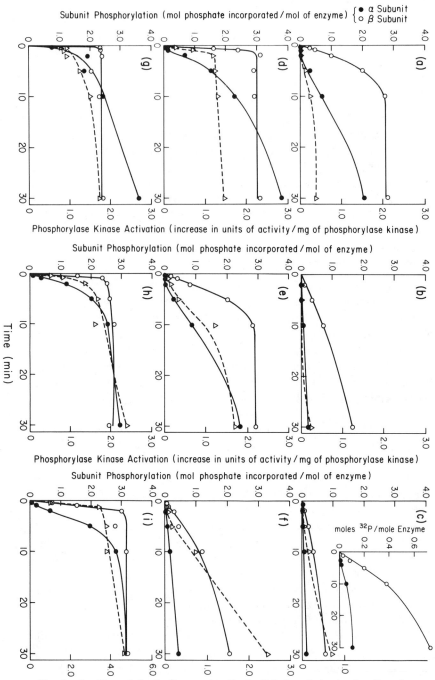

6. The maximal rate of β-subunit phosphorylation is 5- to 10-fold greater than the maximum rate of α-subunit phosphorylation (*142, 155*).

It appears from the various data that the initial phosphorylation of at least one β-subunit is required to produce a conformational change that then permits α-subunit phosphorylation. A speculative possibility from these *in vitro* data is that the nature of the conformational change is similar to that seen upon preincubation with Mg^{2+} and Ca^{2+} (see Section V,B), since both apparently require the same period of time to be manifested. One reason to make this proposal is that there does not appear to be a clear relationship between the time courses of β-subunit and α-subunit phosphorylation. Thus, it might appear that the sequence of events is as shown in Scheme II.

$$\begin{array}{ccccc} \text{β-Subunit} & & \text{conformational} & & \text{α-subunit} \\ \text{phosphorylation} & \rightarrow & \text{shift} & \rightarrow & \text{phosphorylation} \end{array}$$

SCHEME II

where the conformational shift is rate limiting.

Observations on the stoichiometry of subunit phosphorylation by the cAMP-dependent protein kinase have been somewhat discrepant. In early studies with the skeletal muscle enzyme, Hayakawa *et al.* (*13*) observed a maximum incorporation of 1.8 mol of phosphate/mol β_4 and in excess of 4 mol/mol α_4, while Cohen (*14*) reported that phosphorylation plateaued after one phosphate had been incorporated into each α- and β-subunit. These latter values are those most apparently accepted, although there remains some uncertainty since the reports specifically documenting a stoichiometry of 4 mol of phosphate into the β-subunit/mol of $(\alpha\beta\gamma\delta)_4$ have been few. In a subsequent study of ours (*142*), the maximum level of β-subunit phosphorylation ranged from 2.5 to 3.2 mol/mol of $(\alpha\beta\gamma\delta)_4$ with an average of 3.03 ± 0.27 mol/mol of $(\alpha\beta\gamma\delta)_4$ ($n = 25$). In this latter study (*142*) a number of parameters that might have resulted in an aberrantly low value were specifically excluded. The same extent of phosphorylation was obtained whether assayed by $^{32}P_i$ incorporation or colorimetric measurement of total phosphate. An impairment of the conditions of phosphorylation occurring during the reaction was ruled out since addition of more substrate (phosphorylase kinase) resulted in its ready phosphorylation. Since nonactivated phosphorylase kinase contains intrinsic phosphate (see Section VII,A) a less than stoichiometric phosphorylation would be obtained if some of the specific sites in

---

FIG. 4. Time courses of subunit phosphorylation and activation of phosphorylase kinase with varying concentrations of cAMP-dependent protein kinase and phosphorylase kinase. Phosphorylase kinase was phosphorylated under the conditions given in Ref. (*142*) with 3.9 [panels (a), (b), (c)], 39 [panels (d), (e), (f)], or 390 [panels (g), (h), (i)] μunits/ml of protein kinase, at phosphorylase kinase concentrations of 0.12 [panels (a), (d), (g)], 0.72 [panels (b), (e), (h)] or 4.32 [panels (c), (f), (i)] mg/ml. From Pickett-Gies and Walsh (*142*) with permission.

the nonactivated enzyme already contained phosphate. This possibility was eliminated by pretreatment of the enzyme with the appropriate protein phosphatase to dephosphorylate such sites, which did not change the observed degree of phosphorylation.

Accurate determinations of stoichiometry are, however, difficult because the calculated value relies on the accurate measurement of several parameters. Nevertheless, from the available data there is clearly some doubt as to whether fully maximal β-subunit phosphorylation (i.e., 4 mol/$(\alpha\beta\gamma\delta)_4$) of the skeletal-muscle α isozyme readily occurs. In this regard, the data with cardiac phosphorylase kinase are of particular interest. The phosphorylation of bovine cardiac phosphorylase kinase (α′ isozyme) has been studied under a variety of conditions (22, 144, 156). The basic pattern of phosphorylation is consistent with that of the rabbit skeletal-muscle α isozyme; namely, β-subunit phosphorylation precedes that of the α′-subunit and occurs at a faster rate. However, the maximal level of phosphate incorporation into the β-subunit is only 1 mol/mol β_4. Appropriate controls of the type described above for the skeletal-muscle enzyme have shown that this measured level is not aberrant. With the cardiac-muscle enzyme, the magnitude of difference between the observed level of phosphorylation [1 mol/mol of $(\alpha\beta\gamma\delta)_4$] and the phosphorylation of all four β-subunits is such that it can definitely be concluded that the latter does not readily occur. Thus within the cardiac enzyme, interactions must exist of a type akin to negative allosterism whereby, once one of the four β-subunits is phosphorylated, a conformational change must occur that blocks the phosphorylation of the same peptide site in the other β-subunits. Given this conclusion for the cardiac-muscle isozyme, the argument becomes much more tenable that in skeletal-muscle phosphorylase kinase, fully maximal β-subunit phosphorylation [i.e., to 4 mol/mol $(\alpha\beta\gamma\delta)_4$] is significantly inhibited once three of the four sites have been phosphorylated. Of note, the α′ isozyme from bovine skeletal muscle gave different results from those obtained for the cardiac enzyme with the β-subunit being phosphorylated by the cAMP-dependent protein kinase to a level of 3.2 mol/mol $(\alpha\beta\gamma\delta)_4$(21). Possibly this difference is seen because, although both enzymes are designated as α′ isozymes, they are not identical; differences between α isozymes and between α′ isozymes have been noted (64). One reason why phosphorylase kinase may not be stoichiometrically phosphorylated could be that the four β-subunits in a molecule of enzyme may not in fact, be identical. If the differences were minor, like only a few amino acid substitutions, the techniques used would not distinguish differences.

From what is presented here, an additional speculation arises. It appears that the phosphorylation of the first (of the four) β-subunits in phosphorylase kinase markedly affects the enzyme's conformation. With both the cardiac (α′) and skeletal muscle (α) isozymes, the first mole of phosphate incorporated into the β subunits appears essential in order for α subunit phosphorylation to occur. In the cardiac enzyme this also blocks the phosphorylation of the other three β subunits. It is tempting to speculate that the same conformational change directs the subsequent site of phosphorylation; namely, to depress β-subunit phosphorylation but enhance that of the α subunit.

Most generally, cAMP-dependent protein kinase-catalyzed α-subunit phosphorylation achieves a stoichiometry of 4 mol/mol of $(\alpha\beta\gamma\delta)_4$, or somewhat higher. This is observed with either the α isozyme from rabbit skeletal muscle (13, 14, 142, 154) or the α' isozyme from either bovine heart (144, 156) or bovine red-skeletal muscle (21). Singh and Wang (154) studied the effects of Mg^{2+} on the extent of α subunit phosphorylation and showed that by increasing the Mg^{2+} from 1 mM to 10 mM there was an increase in the extent of α-subunit phosphorylation from 4 to 20 mol/mol of $(\alpha\beta\gamma\delta)_4$. Thus Mg^{2+} in this concentration range caused more α sites to become available for phosphorylation, presumably by interacting with Mg^{2+}-specific sites on phosphorylase kinase (see Section V,B). Under one restricted set of conditions we have found evidence that, as previously described for the β-subunit, interactions between α-subunits can limit the degree of their phosphorylation (142). If the free Mg^{2+} concentration is reduced to 20 μM, the phosphorylation of the β-subunit is normal but that of the α-subunit is only to the level of 2 mol/mol of $(\alpha\beta\gamma\delta)_4$. As with β-subunit phosphorylation, this reduced level of α-subunit phosphorylation is not the result of conditions of phosphorylation becoming impaired. It appears again to be indicative of a negative allosteric-type effect whereby the initial phosphorylation of two sites leads to a conformational change in the protein that blocks subsequent phosphorylation of identical sites on other α-subunits. Presumably, Mg^{2+} binding interferes with this conformational change. It is apparent that the phosphorylation of phosphorylase kinase produces a sequence of incremental changes in the conformational organization of the molecule. To what extent these properties of the isolated enzyme pertain to the enzyme in the physiological milieu remains an important question to be addressed. There are also distinctions between the isozyme types; the principal differences documented are in the level of β-subunit phosphorylation and in the effects of Mg^{2+}. Mg^{2+} increases the level of α-subunit phosphorylation of the α isozyme. In contrast, Mg^{2+} inhibits α'-subunit phosphorylation of either the cardiac enzyme (156) or the α' isozyme from red-skeletal muscle (21). Whether these various dissimilarities are indicative of actual regulatory differences in the isozymes remains to be evaluated.

Factors that might modify the cAMP-dependent protein kinase-catalyzed phosphorylation of phosphorylase kinase remain for the large part unexplored. It appears particularly important to examine those factors with which phosphorylase kinase is known to interact within the cell, notably, phosphorylase, calmodulin, glycogen, and the other constituents of the glycogen particle. Cox and Edström (157) reported that extrinsic calmodulin inhibits cAMP-dependent protein kinase-catalyzed phosphorylation of the β-subunit to the extent that, in the presence of calmodulin, the rates of α- and β-subunit phosphorylation were very similar. In the isolated glycogen particle, the cAMP-dependent phosphorylation of phosphorylase kinase can be readily demonstrated (158). Both α- and β-subunits are phosphorylated, reaching stoichiometries similar to those observed with the isolated enzyme, and showing the same temporal pattern, namely, β-subunit phosphorylation is faster and thus appears to precede that of the α-subunit. These subunit phosphorylations are EGTA-insensitive and blocked by addition of the inhibitor protein of the cAMP-dependent protein kinase. They are thus not a consequence of Ca^{2+}-dependent activation as occurs in the flash activation of phosphorylase in these particles (97) (However, autophosphorylation can also be demonstrated). Other parameters of these reactions in the glycogen particle, particularly effects of phosphorylase and glycogen, remain to be explored. Singh and Huang (158) have reported that cAMP-dependent phosphorylation of both the α and β subunits was selectively activated by spermine (10 μM) or spermidine (150 μM), although other substrates for the cAMP-dependent enzyme were not affected.

The sequences of the specific phosphorylation sites of rabbit skeletal-muscle phosphorylase kinase have been identified by Cohen's laboratory (155, 159) as shown below:

β-Subunit:

Val

Ala-Arg-Thr-Lys-Arg-Ser-Gly-Ser(P)-Ile-Tyr-Glu-Pro-Leu-Lys-Ile

α-Subunit:

Phe-Arg-Arg-Leu-Ser(P)-Ile-Ser-Thr-Glu-Ser-Gly

SCHEME III

The Val or Ile of the β-subunit sequence represent two alleles. The sequences for both the α- and β-subunits contain the pair of basic amino acids N-terminal to the target serine that are considered one of the major recognition sequences dictating substrate specificity of the cAMP-dependent protein kinase. It is of interest that for the β-subunit one of the basic amino acids is lysine rather than arginine, and there are two rather than one intervening amino acids. With artificial peptides both of these changes diminish the affinity for the substrate; however, in the inhibitor protein of the cAMP-dependent protein kinase, an arginine located on the N-terminal side of the basic subsite makes a marked difference in inhibitory potency (159a). The sites phosphorylated on the α' and β subunits of the α' isozyme have not yet been characterized.

The phosphorylation of phosphorylase kinase results in its activation but understanding of the relationship between the two is incomplete. Maximum phosphorylation leads to up to a 50-fold increase in activity at pH 6.8 with either the skeletal muscle α isozyme (7) or α' isozyme (21), but only a 2- to 3-fold increase with the cardiac α' isozyme (144). These increases are due primarily to an increase in affinity for phosphorylase, as is also seen in activation by proteolysis or allosterically by Ca^{2+}. As previously described, there are also some changes in the requirements for Ca^{2+} and Mg^{2+}. Phosphorylation by the cAMP-dependent protein kinase results in a conformational change that is similar to that which occurs with activation by proteolysis, increased pH, or Ca^{2+} plus Mg^{2+}; this is evidenced by cross-linking studies that showed a greater proportion of β-β dimers formed with enzyme activated by any of these ways (38).

There has been a lack of concurrence about the molecular events that result in activation. A thesis presented by Cohen (14), which was based on activation and inactivation profiles (under one set of reaction conditions), stated that activation is the result of β-subunit phosphorylation alone. Later, additional support for this was provided by a comparison of activation by the cAMP-dependent and cGMP-dependent protein kinases (147). Each phosphorylated both subunits, but with different ratios of activities towards them. Equal increments of activation, however, were correlated with β-subunit phosphorylation but not α-subunit phosphorylation. Concomitant with the above proposal presented by Cohen (14), Hayakawa et al. (13) reported, in contrast, that a correlation was not observed between phosphorylation of either subunit and enzyme activation; rather it was concluded that the phosphorylation of both contributed to the increase in activity. Most subsequent data supports this latter position. Singh and Wang (154), for example, described a biphasic pattern of phosphorylation and activation at high Mg^{2+} concentrations. Under these conditions an initial rapid rise in activity coincided with either β- or both α- and β-subunit phosphorylation, followed by a slow increase in activity that corresponded to further α-subunit phosphorylation. A similar biphasic pattern was also observed with the α' isozyme from either bovine cardiac (144) or bovine red-skeletal muscle (21) (independent of the Mg^{2+} concentration). With either, the first phase of activation appears to be best correlated with β-subunit phosphorylation, but a second slower phase then occurs after β-subunit phosphorylation is maximal, and this is clearly dependent upon cAMP-dependent α-subunit phosphorylation. Especially for the cardiac enzyme, where β-subunit phosphorylation occurs only to the level of 1 mol/mol of $(\alpha\beta\gamma\delta)_4$, the second α-subunit phosphorylation-dependent phase of activation is marked. Also with the cardiac enzyme, inactivation has also been shown to be correlated with α-subunit dephosphorylation using an enzyme preparation where the β-subunit phosphate was stabilized as the thiophosphate (160). α-Subunit-dependent activation of the skeletal muscle α isozyme is well-exemplified by the studies of Singh et al. (149, 149a). As discussed in more detail later, phosphorylation was examined using two enzymes, the cAMP-dependent protein kinase, and a Ca^{2+} and cAMP-independent enzyme which phosphorylated the β-subunit in the same site as the cAMP-dependent protein kinase, but did not phosphorylate the α-subunit. Phosphorylation with this latter enzyme activated phosphorylase kinase, but not maximally. The activation was clearly due to β-subunit phosphorylation. Following this treatment, addition of the cAMP-dependent protein kinase resulted in no further phosphorylation of the β-subunit, but the α-subunit was phosphorylated and the enzyme further activated. Thus, this latter activation is clearly dependent on α-subunit phosphorylation. Data obtained with the isolated glycogen particle also show that both α- and β-subunit phosphorylation contribute to activation (75a).

Expanding upon this data, we have further examined the relationship between subunit phosphorylation and activation, over a range of protein kinase and phosphorylase kinase concentrations, previously commented on (Fig. 4) in reference to concentration dependency of subunit phosphorylation. The general trends for activity changes are as follows:

1. The rates of activation show the expected variation with protein concentration and at each fixed concentration of phosphorylase kinase increase approximately linearly (examine panels vertically, Fig. 4).
2. The rates of activation with varying substrate concentration do not obey a simple Michaelis-Menten relationship but at the higher phosphorylase kinase concentration increase markedly (examine panels horizontally, Fig. 4).
3. No simple correlation is apparent between enzyme activation and subunit phosphorylation. Some conditions show data similar to that first presented by Cohen (14), namely an apparent correlation with β-subunit phosphorylation alone, whereas others more closely mirror the results of Hayakawa et al. (13) with the phosphorylation of both subunits apparently contributing to the increase in activity.

The data available indicate that the relationship between subunit phosphorylation and enzyme activation is clearly complex. The potential pathways for conversion of $\alpha_4\beta_4\gamma_4\delta_4$ to $(\alpha\text{-P})_4(\beta\text{-P})_4\gamma_4\delta_4$ are illustrated by Fig. 5. Minimally, in the conversion of $\alpha_4\beta_4$ to $(\alpha\text{-P})_4(\beta\text{-P})_4$, seven partially phosphorylated intermediates are generated. Maximally, there is a possibility of 23 intermediates, 40 possible reactions involving unique substrate species, and at least 70 possible routes to go from $\alpha_4\beta_4$ to $(\alpha\text{-P})_4(\beta\text{-P})_4$. Although it is unlikely that all of these are actually utilized, this degree of possible complexity might explain why no simple correlation between subunit phosphorylation and activation is apparent from activation studies. It might be anticipated that the various phospho intermediates will not behave identically as substrates, and this clearly appears to be the case. Thus, as an example, with the α isozyme, phosphorylation of the α-subunits appears to require the prior phosphorylation of at least one β-subunit, but whether subsequent β-subunit phosphorylation further affects the reaction rate of α-subunit phosphorylation is not known. A second question to be considered is whether the incremental phosphorylation of a single subunit produces equal increments in activity. As the reaction from $\alpha_4\beta_4$ to $(\alpha\text{-P})_4(\beta\text{-P})_4$ proceeds

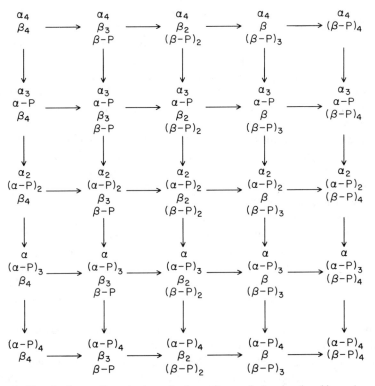

FIG. 5. Phospho intermediates in the activation pathway of phosphorylase kinase (γ- and δ-subunits are not indicated). From Pickett-Gies and Walsh (*142*) with permission.

different intermediates will be generated, each of which may next be phosphorylated in either the α- or β-subunit. Unless phosphorylation of one or another were to be exclusively favored, which does not appear likely from available data, the subsequent phosphorylation event would lead to multiple products. This would lead, as the reaction progressed beyond the phosphorylation of the first β-subunit, to the eventual generation of many of the species depicted in Fig. 5. Quite possibly, which product is formed may depend upon the concentrations of both the enzyme and the substrate. If incremental phosphorylation of a site were not to lead to incremental increases in activity, this could well explain the array of apparent correlations depicted by the type of data presented in Fig. 4.

As previously discussed, there are several lines of evidence that have implicated the phosphorylation of both the α- and β-subunits in the regulation of activity. In what appeared to be in contrast to this, Ganapathi and Lee (*161*) showed with dephosphorylation of skeletal-muscle phosphorylase kinase (phosphorylated in both α- and β-subunits) that a correlation is observed between β-subunit dephosphorylation and enzyme inactivation, and that the enzyme can be fully inactivated while still containing phosphate in the α-subunit. One possibility from this data, however, was that activation required β subunit phosphorylation, but could be further enhanced by phosphorylation of the α subunit, even though α subunit phosphorylation alone does not cause activation. This is what we have now shown to be the case (*161a, 161b*). The dephosphorylation of phosphorylase kinase has been studied using two highly specific protein phosphatase preparations: one (the catalytic subunit of the ATP Mg-dependent protein phosphatase) selective for the β subunit, the other (the catalytic subunit of the polycation dependent protein phosphatase) specific for the site(s) on the α subunit phosphorylated by the cAMP-dependent protein kinase. Using this approach it is seen (*161a, 161b*) that as Ganapathi and Lee (*161*) reported there is a close linear correlation between β subunit dephosphorylation and enzyme inactivation; further, enzyme fully dephosphorylated in the β subunit is fully inactivated. α-Subunit phosphorylation, however, also regulates activity and does so by linearly amplifying the effect of β subunit phosphorylation. Thus, enzyme phosphorylated in both the α and β subunits is inactivated as a consequence of selective α subunit dephosphorylation but the extent of inactivation is dependent on the level of β subunit phosphorylation. Thus, complete α subunit dephosphorylation does not fully inactivate the enzyme. As discussed earlier, with phosphorylation the incorporation of the first mole of phosphate into the β subunit appears to result in a conformational change that permits α subunit phosphorylation. Possibly, it is this same conformational change that permits subsequent α subunit phosphorylation to modify activity, and without this phospho-β-subunit induced conformational change, α subunit phosphorylation is without effect.

The reason for the regulation of phosphorylase kinase by phosphorylation in different sites but by the same enzyme, and presumably in response to the same stimuli, is obscure. One possibility, suggested by several investigators, is that different characteristics or functions might be regulated by phosphorylation of different sites. Phosphorylation of phosphorylase kinase is known to enhance activity toward glycogen synthase (*88, 89*) and phosphorylase kinase itself (*87*); however, the relationship between specific subunit phosphorylation and activity toward these substrates has not been explored. One early suggestion by Cohen and Antoniw (*162*) was that α-subunit phosphorylation regulated the dephosphorylation of the β-subunit. This proposal, termed by the authors ''second site'' regulation, was based upon observations of dephosphorylation catalyzed by a contaminating phosphatase. Later experiments, however, failed to reproduce the initial findings (*163*), and both Ganapathi *et al.* (*164*) and ourselves (*161b*), specifically exploring this possibility with purified enzymes, have found no evidence that the phosphorylation of one subunit modifies the rate of dephosphorylation of the others.

The cAMP-dependent phosphorylation and concomitant activation of phosphorylase kinase has also been explored in the isolated glycogen particle (*75a*). The glycogen particle-associated enzyme can be phosphorylated either by endogenous cAMP-dependent protein kinase (as a consequence of exogenously added or endogenously produced cAMP) or by exogenous addition of protein kinase catalytic subunit. The characteristics of all three are similar with the phosphorylation of β subunit slightly proceeding that of the α subunit and with associated enzyme activation. Examining correlations between subunit phosphorylation and enzyme activation was not practical because of concomitant Ca^{2+}-Mg^{2+}-dependent synergistic activation that occurred as the ATP was consumed. Clearly from these studies, however, phosphorylase kinase in the glycogen particle is equally accessible to intrinsic or extrinsic cAMP-dependent protein kinase. Also evident from these studies were the presence in the glycogen particle of both β subunit and α subunit phosphatases, with the latter being Ca^{2+}-dependent.

D. Autophosphorylation

Phosphorylase kinase can catalyze its own phosphorylation in the presence of $MgATP^{2-}$ and Ca^{2+} (*8, 61, 62, 87*). This is not a novel property of phosphorylase kinase as many protein kinases appear to be capable of autophosphorylation although the physiological significance of this is not known.

Autocatalysis results in the incorporation of phosphate into both α- and β-subunits with an accompanying increase in enzyme activity toward phosphorylase *b*. Activation can be as much as 80- to 100-fold (depending on conditions of the reaction) and thus of greater magnitude than that generally observed upon phosphorylation by the cAMP-dependent protein kinase (*18*). Using the partially dissociated complexes, Chan and Graves (*45*) observed that the γδ complex did not autophosphorylate, but did catalyze EGTA-insensitive phosphorylation and activation of the holoenzyme. The αγδ complex did autophosphorylate, but its activity was unaffected.

The mechanism of autophosphorylation is not yet clearly understood. Studies by King *et al.* (*88*) and Hallenbeck and Walsh (*18*) have reported that the K_m for $MgATP^{2-}$ is in the range of 17–27 μM (a value quite similar to the nucleotide requirement of the cAMP-dependent protein kinase). This $MgATP^{2-}$ requirement is somewhat lower than the reported K_m range for the phosphorylase *b* to *a* reaction (i.e., K_m = 70–240 μM). The K_a for Mg^{2+} in the autocatalytic reaction also differs from that in the phosphorylase *b* to *a* reaction, being approximately 6-fold higher (*18*). Mg^{2+}, in this case, may be affecting not only the catalytic activity of phosphorylase kinase but its ability to serve as a substrate as well. Support for this suggestion can be found in the observation that Mg^{2+} increases the extent of phosphorylation by the cAMP-dependent protein kinase (*154*). It has been suggested by Carlson and Graves (*51*), that autophosphorylation and the phosphorylase b to a reaction may be catalyzed by different catalytic sites on the enzyme. These investigators observed that phosphorylase b and peptide analogs stimulated, rather than inhibited, autophosphorylation. Whether or not the differences in kinetic parameters indeed are a further reflection of different catalytic sites is unclear.

Autophosphorylation could potentially be either an intra- or intermolecular process. Two groups of investigators have reported that the initial velocity of autophosphorylation is linear (i.e., the reaction is first-order) with respect to enzyme concentration (18, 88). This observation is consistent with an intramolecular, but not with an intermolecular, process. Interestingly, in earlier studies, DeLange et al. (8) reported that the activation of nonactivated kinase was stimulated by the addition of activated kinase, and Chan and Graves (45) have shown that the $\gamma\delta$ partial complex can phosphorylate the holoenzyme. These latter observations appear to be more consistent with an intermolecular mechanism of autophosphorylation. Whether or not both mechanisms can occur is not clear.

The pattern and extent of subunit phosphorylation and activation observed during the autocatalytic reaction have, like those of the protein kinase reaction, varied considerably. This appears to be due at least in part to the dependence of this process upon buffers, pH, and divalent cations. At a pH near neutrality there appears to be a lag in both α- and β-subunit phosphorylation as well as in activation (51, 61). This lag is particularly pronounced in glycerophosphate buffer, which appears to inhibit autophosphorylation (as do phosphate and several phosphate-containing compounds) (61). At a pH of 8.2, or at pH 6.8 following preincubation with Mg^{2+} plus Ca^{2+} (88), this lag is not observed. Regardless of pH, the phosphorylation of the α- and β-subunits appears to commence simultaneously (i.e., α-subunit phosphorylation does not lag behind that of the β-subunit as it does in the protein kinase-catalyzed reaction) (18, 61, 88).

The maximal levels of autophosphorylation reported have varied from 1 to 4 mol phosphate/mol β and from 3 to greater than 5 mol phosphate/mol α. While this may suggest that the potential number of sites that can be autophosphorylated is much greater than that which can be phosphorylated by the cAMP-dependent protein kinase, in general autophosphorylation is carried out at higher Mg^{2+} concentrations than the latter reaction. Singh and Wang (154) have reported that the autocatalytic process and the protein kinase-catalyzed reaction at high Mg^{2+} are similar with respect to both the final level of phosphorylation and the activation attained.

Discrepancies also exist as to the correlation between subunit phosphorylation and activation. Wang et al. (61) and Hallenbeck and Walsh (18), who carried out autophosphorylation in buffers near neutrality, observed coincident phosphorylation of both α- and β-subunits. Activation that occurred comcomitantly with phosphorylation could not be correlated with the phosphorylation of a specific subunit. In contrast, King *et al.* (*88*), performing experiments at pH 8.0 reported that activation correlated well with β-subunit phosphorylation. In these experiments, β-subunit phosphorylation clearly reached a maximum prior to the complete phosphorylation of the α-subunits. Activation also plateaued before α-subunit phosphorylation did. Chan and Graves (*45*) have shown that autophosphorylation of the partial complex (αγδ) results in the incorporation of up to 4 mol phosphate per mol without any effect on catalytic activity. This might suggest that autophosphorylation of the α-subunit does not affect activity; however, the αγδ complex has a higher specific activity (at pH 6.8) than the holoenzyme, so possibly it is already in a conformational state equivalent to that promoted by autophosphorylation in the holoenzyme and thus phosphorylation is without further effect. Indeed, a role for α-subunit phosphorylation in autoactivation has been implicated in other experiments. Hallenbeck and Walsh (*18*) found that when MnATP^{2-}, rather than MgATP^{2-}, was used in autocatalytic reactions, α- and β-subunit phosphorylation were no longer coincident, and β phosphorylation plateaued at 1 mol of phosphate/mol β. Under these conditions, both activation and α-subunit phosphorylation continued in the absence of additional β-subunit modification. In another study, Sul *et al.* (*156*) observed an increase in cAMP-independent α′ subunit phosphorylation (most likely autophosphorylation) subsequent to phosphorylation by the cAMP-dependent protein kinase. Concomitant with this increase in α′ phosphorylation was an increase in enzyme activity. Taken together the existing data suggest that autophosphorylation of both α- (or α′) and β-subunits can alter enzyme activity. Clearly, multiple sites can be phosphorylated on both subunits by this process; it is not clear whether all or only some of these sites control activity.

As autophosphorylation and protein kinase-catalyzed phosphorylation have many characteristics in common, the question of whether common sites are phosphorylated by the two enzymes is of interest. This question was approached indirectly in the studies of Wang *et al.* (*61*) who observed that when phosphorylase kinase was phosphorylated to a maximal level by either the cAMP-dependent protein kinase or by autophosphorylation, and subsequently subjected to the alternate process of phosphorylation, additional phosphate was incorporated. In addition, autophosphorylation of enzyme previously phosphorylated by the protein kinase was accompanied by a further increase in enzyme activity. This suggests that the two mechanisms of phosphorylation do not involve common phosphorylation sites. However, in reactions catalyzed by cAMP-dependent protein kinase at high Mg^{2+} concentration, Singh and Wang (*154*) observed a maximal level of phosphate incorporation quite similar to that attained by the autocatalytic process. When both reactions were allowed to proceed simultaneously the same maximal level of incorporation was attained as with either reaction alone. Clearly, conclusive evidence as to the identity of sites phosphorylated by these two mechanisms awaits characterization of all of the phosphorylated sequences. King *et al.* (*88*) have mapped tryptic phosphopeptides derived from phosphorylase kinase that had been autophosphorylated to slightly less than 1 mol/mol β. Two major phosphopeptides were obtained which appeared to be identical to the two allelic phosphopeptides generated from the protein kinase-phosphorylated β-subunit. Therefore, it appears that the β-subunit site phosphorylated by the cAMP-dependent protein kinase can also be phosphorylated by autocatalysis. Information on the other sites phosphorylated by these two processes is still needed.

The physiological significance of phosphorylase kinase autophosphorylation is not known. As discussed in Section VIII,B, the bulk of *in vivo* data suggest that activation by neural or electrically stimulated Ca^{2+} release is not due to covalent modification of the enzyme. The possibility that autophosphorylation is enhanced subsequent to phosphorylation by the cAMP-dependent protein kinase, however, cannot be ruled out, particularly since this appears to occur *in vitro*. The observation that glycogen and phosphorylase appear to stimulate autophosphorylation, coupled with their apparent close association with phosphorylase kinase *in vivo,* suggests that intracellular conditions may permit this type of activation.

E. Phosphorylation and Activation by Other Protein Kinases

Many features of the phosphorylation of phosphorylase kinase by other protein kinases have been presented in previous sections of this chapter. The cGMP-dependent protein kinase catalyzes the phosphorylation of both the α- and β-subunits but in contrast to the cAMP-dependent enzyme, α-subunit phosphorylation is faster than that of the β-subunit and occurs without an initial lag (*146, 147*). Tryptic phosphopeptides obtained with either protein kinase are identical (in size) but, especially for the α-subunit where the peptide is large (40 amino acids), it is not known if the phosphorylation sites are identical. Cohen (*147*) has suggested that activation correlates with β-subunit phosphorylation, but the cGMP-dependent protein kinase maximally phosphorylates the β-subunit to a level of only ~1 mol/mol of $(\alpha\beta\gamma\delta)_4$. The data would fit equally well with the proposal that α-subunit phosphorylation regulates activity only after the β-subunit is phosphorylated.

Phosphorylation of phosphorylase kinase by Ca^{2+} and cyclic nucleotide-independent "casein" protein kinase (that is also a glycogen synthase kinase) has been studied by Singh *et al,* (149, 149a). In the initial report (*149*) using this enzyme and with a glycerophosphate buffer system the β subunit was observed to be phosphorylated to a level of 1 mol/mol of $(\alpha\beta\gamma\delta)_4$, and the α subunit only minimally. The β subunit site phosphorylated appears to be identical to that phosphorylated by the cAMP-dependent protein kinase, and likewise causes activation. Subsequent phosphorylation of the α subunit by the cAMP-dependent protein kinase, however, resulted in further activation. In a follow-up report using this casein protein kinase (149a) it was observed that in a Tris chloride buffer an incorporation of greater than 2 mol into each of the α and β subunits was obtained, but with the prior degree of observed activation. One of the sites on the β subunit that is phosphorylated by the casein kinase is also that phosphorylated by the cAMP-dependent protein kinase and is presumably the site responsible for activation. The sites on the α subunit phosphorylated by the casein kinase and cAMP-dependent protein kinase are distinct, and those phosphorylated by the casein kinase do not appear to affect activity. One of the α sites, however, is only phosphorylated by the casein kinase following prior α subunit phosphorylation by the cAMP-dependent enzyme. The role of casein kinase phosphorylation of phosphorylase kinase needs exploration in vivo. Given that phosphorylase kinase as isolated is a phosphoprotein, it is also important to explore whether any of these "endogenous" phosphates are in sites catalyzed by this casein kinase.

F. PHOSPHORYLASE KINASE DEPHOSPHORYLATION

A detailed characterization is available of the protein phosphatases capable of catalyzing phosphorylase kinase dephosphorylation. These are only briefly reviewed here, as more detailed descriptions are available elsewhere (165–173, 173a, 173b). In general, four protein phosphatases appear to account for most, if not all, of the cellular activity, and are involved in metabolic regulation and phosphorylase kinase dephosphorylation. These have been designated (167) Type 1 (ATP Mg-dependent phosphatase) and Types 2A (Polycation-dependent phosphatase), 2B (Calcineurin), and 2C. This designation is based on primary specificity toward either the β-subunit (type 1) or the α-subunit (types 2) of phosphorylase kinase phosphorylated by the cAMP-dependent protein kinase. The type 1 phosphatase is regulated by an inhibitor protein (inhibitor-1), and both it and an intrinsic modulator protein in the phosphatase are regulated by phosphorylation, catalyzed by the cAMP-dependent protein kinase and a Ca^{2+} and cyclic nucleotide-independent protein kinase, respectively. Type 2B is regulated by Ca^{2+} and calmodulin. There are no known physiological regulators of types 2A or 2C. Both types 1 and 2A phosphatases have a broad substrate specificity and act upon a range of substrates, whereas 2B and 2C have a much narrower specificity. From studies of their relative cellular amounts, Ingebritsen et al. (173) estimated that the β-subunit is primarily dephosphorylated by the type 1 phosphatase, although in liver type 2A may also contribute to a small degree. Dephosphorylation of the α-subunit appears to depend upon the availability of Ca^{2+}. In the presence of Ca^{2+}, dephosphorylation of the α-subunit would be predominantly catalyzed by type 2B. In the absence of Ca^{2+}, dephosphorylation, if it occurs, would probably be a consequence of type 2A activity. The effects of dephosphorylation of phosphorylase kinase on its activity and possible other functions are presented in detail in Section VII,C.

G. ADP-RIBOSYLATION

Recently (173c), the ADP-ribosylation of phosphorylase kinase has been described. Using a hen liver nuclear enzyme, phosphorylase kinase was ribosylated in both the α and β subunits on arginine residues. ADP ribosylation diminished both cAMP-dependent and autocatalytic phosphorylation and, as a result, suppressed phosphorylation-dependent activation. ADP-ribosylation itself did not affect activity.

VIII. Regulation of Phosphorylase Kinase in Intact Cells

A. HORMONAL ACTIVATION

Since the classical paper by Drummond et al. (110), it has been well established that phosphorylase kinase activity in cells is regulated both by Ca^{2+}, allosterically, and in response to cAMP, via cAMP-dependent phosphorylation. The means to distinguish between these two appears straightforward. If changes in the activity state of phosphorylase occur (i.e., the ratio of phosphorylase a to phosphorylase b increases) without increases in either the cellular concentration of cAMP or the covalent activation state of phosphorylase kinase, it can, in most circumstances, be reasonably concluded that the mechanism of regulation is via the allosteric, Ca^{2+}-dependent, activation of phosphorylase kinase. In many such situations (as discussed below), further confirmation of the likelihood that Ca^{2+} is acting as the regulatory agent has come (a) from correlations with what occurs when Ca^{2+} metabolism is manipulated, such as by omission of external Ca^{2+} or drug-promoted influxes of Ca^{2+}; (b) from correlations with changes in other Ca^{2+}-mediated processes, such as muscle contraction; and (c) in some situations by direct measurements of internal Ca^{2+} fluxes. Alternatively, regulation of phosphorylase via cAMP is reasonably concluded as the mechanism of control when there are coordinated changes in cAMP levels, cAMP-dependent protein kinase activity ratios, phosphorylase kinase activation state (i.e., covalent phosphorylation state), and phosphorylase a formation. When such data are obtained, the reasonable conclusion is that the covalent activation of phosphorylase kinase is the main contributory cause for phosphorylase activation. A critical question that remains from such initial data, however, is whether or not this is sufficient to explain phosphorylase activation, and whether or not there is an additional need for cellular Ca^{2+} concentrations to be elevated in order for the enhanced activation state of phosphorylase kinase to be expressed. This is discussed in more detail in Section VIII,B.

Examination of the regulation of phosphorylase kinase in intact cells has occurred primarily in liver, skeletal muscle, and heart. In all three, both mechanisms of regulation are clearly important. With liver, there is excellent evidence that α-adrenergic agonists, vasopressin, and angiotensin can act specifically via Ca^{2+}. Phosphorylase activation by these three hormones occurs without concomitant changes in cAMP and phosphorylase kinase activation state, whereas it is severely depressed if extracellular Ca^{2+} is omitted. In addition, increasing cytosolic Ca^{2+} activates phosphorylase in a similar manner. This system, especially with reference to Ca^{2+}, has been extensively characterized. The details are beyond the scope of this review but have been presented elsewhere [see Ref. (174) for extensive references]. In brief, it is apparent that these hormones promote the cellular translocation of Ca^{2+} leading to an increase in cytosolic Ca^{2+} and, in consequence, phosphorylase activation. The most likely sources of this Ca^{2+} are partly intracellular (from a specific pool in the endoplasmic reticulum) and partly extracellular (by transport into the cell). At least one of the messengers promoting these changes is inositol triphosphate. The activation of hepatic glycogenolysis by glucagon is clearly distinct from that of the three hormones previously listed. Glucagon-stimulated phosphorylase activation is correlated with increases in cAMP and covalent activation of phosphorylase kinase; thus, glucagon is clearly regulating phosphorylase via activation of the cAMP-dependent cascade. The role of Ca^{2+} in this cannot be evaluated with certainty because of differences in reported data. Omission of Ca^{2+} from the medium under conditions that eliminate α-adrenergic phosphorylase activation leads to either no change in glucagon-stimulated activation (175) or a diminished but clearly not abolished effect (113). Most likely the difference in these results is due to differences in the degree to which cellular Ca^{2+} has been depleted. Under both circumstances it is clear that phosphorylase activation mediated by α-adrenergic agonists, vasopressin, or angiotensin (i.e., as catalyzed by nonactivated phosphorylase kinase) has a higher Ca^{2+} concentration requirement than that occurring upon glucagon stimulation (i.e., as catalyzed by covalently activated enzyme). Thus, it appears from the data obtained that covalently activated liver phosphorylase kinase may catalyze a Ca^{2+}-independent activation of phosphorylase, or at least requires a lower concentration of Ca^{2+} for activity than does the nonactivated form. Most likely, the latter explanation is correct.

Intact tissue studies with skeletal muscles show that muscle phosphorylase kinase can likewise be regulated either by Ca^{2+}, allosterically, or by cAMP-dependent covalent modification. This was first shown by the classical work of Drummond et al. (110) and has been well borne out subsequently (111, 176, 177). With skeletal muscle, electrical stimulation, either directly or via an in situ nerve, produces rapid formation of phosphorylase a and brisk glycogenolysis, but neither an increase in cAMP nor a covalent activation of phosphorylase kinase. Electrical stimulation also results in contraction; both the contraction and phosphorylase a formation increase with the frequency of stimulation (*178, 179*), but are diminished by Ca^{2+} depletion. The stimulus for both is clearly an increase in cytosolic Ca^{2+}, arising primarily by release from the sarcoplasmic reticulum. For muscle contraction, the site of Ca^{2+} action is troponin C associated with the myofibrils. For the allosteric activation of phosphorylase kinase, three sites are possible, the intrinsic δ-subunit, extrinsic calmodulin, and extrinsic troponin C. Cohen (*106*) presented cogent arguments in favor of the latter as being the most likely, although this remains to be resolved. The similarity of Ca^{2+}-binding characteristics for the activation of contraction and of glycogenolysis provides for a ready coordination between the two.

In contrast to electrical stimulation, β-adrenergic stimulation of skeletal muscle produces a coordinate rise first in cAMP, then in the activation of phosphorylase kinase and, consequentially, in phosphorylase a formation. This clearly implicates activation of the cAMP cascade as the primary route of regulation. However, EGTA addition to the external medium dampens phosphorylase a formation without diminishing catecholamine-stimulated formation of either cAMP or activated phosphorylase kinase (*176*). Thus, activated phosphorylase kinase still requires Ca^{2+}. Catecholamines, however, do not initiate skeletal muscle contraction; therefore, in contrast to what is observed with electrical stimulation and nonactivated phosphorylase kinase, the concentration of Ca^{2+} required by activated phosphorylase kinase must be lower than the threshold level that would stimulate contraction. This data with intact tissues appears to coincide well with that obtained *in vitro* with the purified enzyme. As described in Section V,A, the phosphorylation of skeletal-muscle phosphorylase kinase increases its Ca^{2+} affinity; presumably this permits Ca^{2+} to bind at a concentration where Ca^{2+} is not bound to troponin C in the myofibrils. Two possible situations could be envisioned with the catecholamine stimulation of muscle glycogenolysis; either phosphorylation of phosphorylase kinase lowers its requirements for Ca^{2+} down to the levels present in ambient muscle, or catecholamines increase cytosolic Ca^{2+} but not to levels sufficient to initiate contraction. Most likely, the latter is correct. Measured levels of ambient Ca^{2+} appear to be substantially lower than that required for activated phosphorylase kinase, and the manipulations showing that activated phosphorylase kinase in the cell still required Ca^{2+} (*176*) would probably not have diminished the ambient free Ca^{2+} concentration. Further, Stull and Mayer (*111*) reported that very low concentrations of isoproterenol stimulate phosphorylase a formation without concomitant increases in cAMP or covalent activation of phosphorylase kinase. Most likely, this concentration of isoproterenol acts indirectly by increasing cytosolic Ca^{2+}. Data of Gross and Johnson (*176*) suggest that isoproterenol stimulates a Na^{2+}-dependent Ca^{2+} channel across from the transverse tubules, which because of its location may be selectively available to phosphorylase kinase rather than the myofibrils. These types of channels, however, have not been observed by other approaches.

In cardiac muscle the regulation of phosphorylase kinase appears similar to that in skeletal muscle, but there are some noted differences. Elevation of external Ca^{2+} (*180, 181*), anoxia (*182, 183*), and α-adrenergic stimulation (*73*) all cause phosphorylase *a* formation without changes in either cAMP or phosphorylase kinase activation state; thus for these the likely regulation is via Ca^{2+} stimulation of unactivated phosphorylase kinase. This situation, however, is different from what occurs in skeletal muscle, where Ca^{2+}-stimulated contraction and Ca^{2+} stimulation of nonactivated phosphorylase kinase (as, for example, with electrical stimulation) appear to be coordinately linked. In contrast, cardiac contraction occurs apparently without concomitant phosphorylase activation and the latter appears to occur either when Ca^{2+} is elevated further or via a different route. This suggests that in cardiac muscle either the two processes are differentially sensitive to Ca^{2+} or are exposed to different pools. An important consideration, however, is not simply whether Ca^{2+} has been increased to a given level, but how long it has been elevated. There are clearly marked differences in the time constants for activation of contraction, and subsequent relaxation, and for phosphorylation of phosphorylase, and subsequent dephosphorylation. Phosphorylase activity is only meaningfully (i.e., measurably) increased if Ca^{2+} is elevated for a considerably longer time than is necessary to evoke a contractile response.

The cAMP-dependent regulation of cardiac phosphorylase kinase has also been well studied. Either β-adrenergic (*184–186*) or glucagon (*187*) stimulation with perfused hearts, or β-adrenergic stimulation with either papillary muscle (*183*) or isolated ventricular strips (*188*) results in coordinated changes in cAMP, the covalent activation of phosphorylase kinase, and phosphorylase activation. Omission of external calcium eliminates phosphorylase activation without altering that of phosphorylase kinase or diminishing the increased levels of cAMP (*186, 187*). Thus, the activated phosphorylase kinase still requires Ca^{2+} for phosphorylase activation. Data with papillary muscle suggests that the level of Ca^{2+} required by activated phosphorylase kinase is less than that needed to stimulate contraction (*183*). Thus, a more stringent removal of Ca^{2+} is required to diminish isoproterenol-stimulated phosphorylase activation than is needed to block contraction, and isoproterenol activates phosphorylase in nonstimulated muscle without promoting contraction. These data, coupled with the observation that in the absence of β-adrenergic stimulation cardiac muscle contraction occurs without phosphorylase activation, suggest that, as with skeletal muscle, nonactivated phosphorylase kinase has a higher Ca^{2+} requirement than does the covalently activated enzyme. These data are in apparent conflict with *in vitro* results since with either purified cardiac enzyme (22), or the enzyme in a cell extract (72), an identical Ca^{2+} dependency has been observed for both nonactivated or activated enzyme. This is in contrast to what was observed for the skeletal-muscle enzyme where phospho-phosphorylase kinase has a higher affinity for Ca^{2+} (106). One important consideration of the requirement of activated phosphorylase kinase for Ca^{2+} during repetitive cardiac contraction comes from the data of Barovsky and Gross (188). These authors, working with isolated mouse right ventricular strips, showed that isoproterenol-stimulated phosphorylase a formation is stimulation-frequency dependent. With stimulation at the higher frequency ($\sim$3.3 Hz) isoproterenol caused phosphorylase activation which was coordinated with the expected changes in both cAMP and phosphorylase kinase. At lower frequencies ($\sim$0.2 Hz), both an increase in cAMP levels and a covalent activation of phosphorylase kinase occurred, as at the higher frequency, but phosphorylase activation was eliminated. In contrast to this, tension development was greater at the lower frequency stimulation that at the higher frequency. The higher tension development at the lower frequency indicates that total calcium during the twitch is most likely higher. The probable explanation for the absence of (measurable) phosphorylase activation at the lower stimulation frequency is that the elevated level of Ca^{2+} is not maintained for a sufficient period of time to permit an accumulation of phosphorylase a. Thus, at the lower frequency (0.2 Hz), Ca^{2+} was elevated above the threshold contractile level for only 3.6 sec out of every minute, whereas at the higher frequency (3.3 Hz) the calcium was above threshold for 40 sec every minute, albeit at a lower total level than at 0.2 Hz. It is apparent that this additional dwell time for elevated Ca^{2+} is essential for the net (overall) formation of phosphorylase a. The response times for Ca^{2+}-dependent phosphorylase formation are thus quite different than those for contraction. At low frequency stimulation, the elevated Ca^{2+} levels quite possibly result in some formation of phosphorylase a which, however, is rapidly dephosphorylated in the long relaxation time between twitches. Observable accumulation of phosphorylase a apparently occurs only if Ca^{2+} levels are repeatedly elevated by more frequent stimulation leaving less time for dephosphorylation to occur. This difference in dwell time accounts for the different time courses of Ca^{2+}-dependent contraction and phosphorylase a formation, despite their having very similar Ca^{2+} concentration dependencies.

B.   CORRELATION BETWEEN PHOSPHORYLASE KINASE
     ACTIVATION AND PHOSPHORYLATION IN INTACT TISSUES

The phosphorylation of phosphorylase kinase in intact tissues was initially examined by Mayer and Krebs (141). Their study utilized skeletal muscle stimulated with epinephrine, and phosphorylation was examined following purification of the enzyme by standard procedures. Despite being able to demonstrate phosphorylase activation and phosphorylation, the results with phosphorylase kinase were negative in that activation was observed but no epinephrine-enhanced phosphorylation detected. Several factors may have obscured the phosphorylation of phosphorylase kinase in these studies, among which was the relatively high amount of phosphate present in the unactivated enzyme and the low yield of phosphorylase kinase obtained following extensive purification. One observation of note was that the phosphate present in the unactivated phosphorylase kinase, presumably arising by exchange, achieved a higher specific activity than the γ-phosphate of intracellular ATP. This suggests that phosphorylase kinase may be exposed to a different pool of ATP than that constituting the major fraction. Such an observation will need to be considered in future studies. The first study to show apparent phosphorylation of phosphorylase kinase in intact tissue was that by Yeaman and Cohen (*189*). For their investigation rabbit skeletal muscle was used as the tissue, animals were injected with a bolus of epinephrine, tissue was excised, the enzyme isolated by standard purification procedures, and tryptic phosphopeptides isolated. By this procedure, from epinephrine-treated animals, two phosphopeptides were identified corresponding to the major tryptic peptides phosphorylated by the cAMP-dependent protein kinase (see Section VII,C). These data appear to demonstrate cAMP-dependent protein kinase-catalyzed phosphorylation of the α- and β-subunits in the intact cell but with two reservations. A potential problem with the described experiment was that tissue was not rapidly frozen, leaving open the possibility of post homogenization events. That such activation might have occurred may be indicated by the higher values of the activity ratio obtained at pH 6.8 to that at 8.2 (*189*) relative to other studies (*110, 111, 176, 177*).

Subsequent studies on the potential correlation between phosphorylase kinase activation and phosphorylation have been pursued in our laboratory with both cardiac and skeletal muscle (*145, 190–192*). For cardiac muscle, the Lagendorff retrograde perfusion was used. For skeletal muscle, the rat *flexor digitorum brevis* was the experimental tissue (*193*). This latter muscle preparation is composed primarily of FOG fibers (>90%) and contains mainly the α' isozyme of phosphorylase kinase (>90%). With each experimental preparation, the tissue was rapidly frozen to permit detailed evaluation of time-course changes. The enzyme was isolated by immunoprecipitation and extensive controls were employed to insure that both the phosphorylation and the activation being measured occurred while the cells were intact [see Ref. (*190*) for example].

In the initial study with cardiac muscle (*190*), enzyme activation and total phosphorylation were examined as a function of time over a range of epinephrine concentrations, and with dephosphorylation following removal of the stimulus. With this approach a linear correlation was observed ($r = 0.94$) between activation and total enzyme phosphorylation.

This study has been expanded (*192*) by an examination of specific subunit phosphorylation, with submaximal and near maximal stimulation by epinephrine, isoproterenol, and glucagon. With each, β-subunit phosphorylation occurs prior to α-subunit phosphorylation, but the lag between the two is dependent upon the stimulus, being greater with isoproterenol. This lag appears to be more pronounced with either the inclusion of verapamil or deletion of external Ca^{2+}. The total level of incorporation is estimated to be $\sim 1.2-1.6$ mol ^{32}P/mol $(\alpha'\beta\gamma\delta)_4$, compared to a basal level of 0.4–0.5 mol/mol. In both basal and maximally stimulated preparations, with most of the conditions tested, $^{32}P_i$ is about equally divided between the α'- and β-subunits of phosphorylase kinase. The time course of activation appears to correlate best with that of β subunit phosphorylation, but a role for α subunit phosphorylation in activation also appears probable. In particular, following removal of the hormonal stimulus there is a brisk dephosphorylation of the β subunit but slower α subunit dephosphorylation and inactivation.

With the isolated skeletal-muscle preparation (*145*) the phosphorylation of both the α'- and β-subunits has also been documented. Preincubation of the tissue with $^{32}P_i$-PO_4 leads to incorporation into both subunits, with that in the α' subunit being 2- to 3-fold higher. Stimulation of the muscle preparation with either epinephrine or isoproterenol leads to enhanced phosphorylation of both subunits, again with a 2- to 3-fold higher level being incorporated into the α'- subunit. The stoichiometry of phosphorylation into the α'-subunit is estimated to be $\sim 0.8-0.9$ mol/mol $(\alpha'\beta\gamma\delta)_4$ compared to 0.2–0.5 mol/mol in the β-subunit. No difference is observed in initial time courses of phosphorylation between the two subunits, although that of the β-subunit plateaued earlier. The best correlation with activation appears to be with total phosphorylation (i.e., α' + β) ($r = 0.82-0.97$).

The data that are being accumulated for *in vivo* phosphorylation and activation of phosphorylase kinase in general appears to be supportive of what has been learned from *in vitro* studies, but with some differences. In both, the cAMP-dependent phosphorylation of both α- and β-subunits appears clearly to be a prime feature of control. If anything, the higher level of skeletal-muscle α'- subunit phosphorylation in the intact muscle preparation points to a greater role for that subunit, rather than the β-subunit, in control. One common feature of both cardiac- and skeletal-muscle intact tissue studies is that the level of subunit phosphorylation appears substantially less than stoichiometric, despite high levels of hormonal stimulation. This observation is consistent with the questions raised concerning the potential importance of the initial phosphorylation events in producing some of the major changes in enzyme conformation and perhaps thus regulation. Clearly, however, many studies remain to be done before a full understanding of the regulation and roles of phosphorylase kinase in muscle function can be appreciated.

Acknowledgments

This chapter is dedicated to Dr. George E. Drummond and Dr. Steven E. Mayer, without whose contributions our knowledge of phosphorylase kinase would be much less.

This work was supported by AM 13613 and AM 21019. The authors greatly appreciate the in-depth comments of K. Angelos, L. Anderson, H. C. Cheng, R. Cooper, L. Garetto, P. Hallenbeck, and C. Ramachandran.

References

1. Sutherland, E. W. (1950). *Recent Prog. Horm. Res.* **5,** 441–463.
2. Cori, C. F., and Cori, G. T. (1945). *JBC* **158,** 341–345.
3. Burnett, G., and Kennedy, E. P. (1954). *JBC* **211,** 969–980.
4. Krebs, E. G., and Fischer, E. H. (1956). *BBA* **20,** 150–157.
5. Lipkin, D., Markham, R., and Cook, W. H. (1959). *J. Am. Chem. Soc.* **81,** 6075.
6. Robison, G. A., Butcher, R. W., and Sutherland, E. W. (1971). "Cyclic AMP." Academic Press, New York.
7. Krebs, E. G., Love, D. S., Bratvold, G. E., Trayser, K. A., Meyer, W. L., and Fischer, E. H. (1964). *Biochemistry* **3,** 1022–1033.
8. DeLange, R. J., Kemp, R. G., Riley, W. D., Cooper, R. A., and Krebs, E. G. (1968). *JBC* **243,** 2200–2208.
9. Walsh, D. A., Perkins, J. P., and Krebs, E. G. (1968). *JBC* **243,** 3763–3774.
10. Chan, K. J., and Graves, D. J. (1984). *In* "Calcium and Cell Function" (W. Y. Cheung, ed.), Vol. 5, pp. 1–31. Academic Press, New York.
11. Carlson, G. M., Bechtel, P. J., and Graves, D. J. (1979). *Adv. Enzymol. Relat. Areas Mol. Biol.* **50,** 41–115.
12. Malencik, D. A., and Fischer, E. H. (1982). *In* "Calcium and Cell Function" (W. Y. Cheung, ed.), Vol. 2, pp. 161–188. Academic Press, New York.
13. Hayakawa, T., Perkins, J. P., Walsh, D. A., and Krebs, E. G. (1973). *Biochemistry* **12,** 567–573.
14. Cohen, P. (1973). *EJB* **34,** 1–14.
15. Sharma, R. K., Tam, S. W., Waisman, D. M., and Wang, J. H. (1980). *JBC* **255,** 11102–11105.
16. Jennissen, H. P., Horl, W. H., and Heilmeyer, L. M. G., Jr. (1973). *Hoppe-Seyler's Z. Physiol. Chem.* **354,** 236–237.
17. Jennissen, H. P., and Heilmeyer, L. M. G., Jr. (1975). *Biochemistry* **14,** 754–760.
18. Hallenbeck, P. C., and Walsh, D. A. (1983). *JBC* **258,** 13493–13501.
19. Gross, S. R., and Bromwell, K. (1977). *ABB* **184,** 1–11.
20. Pocinwong, S., Blum, H., Malencik, D., and Fischer, E. H. (1981). *Biochemistry* **20,** 7219–7226.
21. Tam, S. W., Sharma, R. K., and Wang, J. H. (1982). *JBC* **257,** 14907–14913.
22. Cooper, R. H., Sul, H. S., McCullough, T. E., and Walsh, D. A. (1980). *JBC* **255,** 11794–11801.
22a. Nikolaropoulos, S., and Sotiroudis, T. G. (1985). *EJB* **151,** 467–473.
23. Chrisman, T. D., Jordan, J. E., and Exton, J. H. (1982). *JBC* **257,** 10798–10804.
24. Pohlig, G., Wingender-Drissen, R., and Becker, J. U. (1983). *BBRC* **114,** 331–338.
25. Wingender-Drissen, R., and Becker, J. U. (1983). *FEBS Lett.* **163,** 33–36.
26. Jennissen, H. P., and Heilmeyer, L. M. G., Jr. (1974). *FEBS Lett.* **42,** 77–80.
27. Crabb, J. W., and Heilmeyer, L. M. G., Jr. (1984). *JBC* **259,** 6346–6350, 14314.

28. Reimann, E. M., Titani, K., Ericsson, L. H., Wade, R. D., Fischer, E. H., and Walsh, K. A. (1984). *Biochemistry* **23**, 4185–4192.
29. Grand, R. J., Shenolikar, S., and Cohen, P. (1981). *EJB* **113**, 359–367.
30. Burchell, A., Cohen, P. T. W., and Cohen, P. (1976). *FEBS Lett.* **67**, 17–22.
30a. Pickett-Gies, C. A., Carlsen, R., Angelos, K. L., and Walsh, D. A. (1986). *JBC* (in press).
30b. Lawrence, J. C., Jr., Krsek, J. A., Salsgiver, W. J., Hiken, J. F., Salmons, S., and Smith, R. L. (1986). *Am. J. Physiol.* **250**, C84–C89.
31. Schwartz, A., Entman, M. L., Kaniike, K., Lane, L. K., Van Winkle, B., and Bornet, E. P. (1976). *BBA* **426**, 57–72.
32. Horl, W. H., and Heilmeyer, L. M. G., Jr. (1978). *Biochemistry* **17**, 766–772.
33. Le Peuch, C. J., Le Peuch, D. A., and Demaille,G. (1982). *Ann. N.Y. Acad. Sci.* **402**, 549–557.
34. Hessova, Z., Varsanyi, M., and Heilmeyer, L. M. G., Jr. (1985). *EJB* **146**, 107–115.
35. Cohen, P. (1978). *Curr. Top. Cell. Regul.* **14**, 117–196.
35a. Trempe, M. R., Carlson, G. M., Hainfeld, J. F., Furcinitti, P. S., and Wall, J. S. (1986). *JBC* **261**, 2882–2889.
36. Lambooy, P. K., and Steiner, R. F. (1982). *ABB* **213**, 551–556.
37. Picton, C., Klee, C. B., and Cohen, P. (1980). *EJB* **111**, 553–561.
38. Fitzgerald, T. J., and Carlson, G. M. (1984). *JBC* **259**, 3266–3274.
39. Picton, C., Klee, C. B., and Cohen, P. (1981). *Cell Calcium* **2**, 281–294.
40. Chan, K.-F. J., and Graves, D. J. (1982). *JBC* **257**, 5939–5947.
41. Skuster, J. R., Chan, K.-F. J., and Graves, D. J. (1980). *JBC* **255**, 2203–2210.
42. Kee, S. M., and Graves, D. J. (1985). *FP* **44**, 1074.
43. Cohen, P., Burchell, A., Foulkes, J. G., Cohen, P. T. W., Vanaman, T. C., and Nairn, A. C. (1978). *FEBS Lett.* **92**, 287–293.
43a. Jennissen, H. P., Horl, W. H., Groschel-Stewart, U., Velick, S. F., and Heilmeyer, L. M. G., Jr. (1976). *In* "Metabolic Interconversions of Enzymes" (S. Shaltiel, ed.), pp. 19–26. Springer-Verlag, Berlin and New York.
44. Hayakawa, T., Perkins, J. P., and Krebs, E. G. (1973). *Biochemistry* **12**, 574–580.
45. Chan, K.-F. J., and Graves, D. J. (1982). *JBC* **257**, 5948–5955.
46. Chan, K.-F. J., and Graves, D. J. (1982). *JBC* **257**, 5956–5961.
47. Gulyaeva, N. B., Vul'fson, P. L., and Severin, E. S. (1977). *Biokhimiya* **43**, 373–381.
48. King, M. M., Carlson, G. M., and Haley, B. E. (1982). *JBC* **257**, 14058–14065.
48a. King, M. M., and Carlson, G. M. (1981). *Biochemistry* **20**, 4382–4387.
48b. King, M. M., and Carlson, G. M. (1982). *FEBS Lett.* **140**, 131–134.
49. Fischer, E. H., Alaba, J. O., Brautigan, D. L., Kerrick, W. G. D., Malencik, D. A., Moeschler, H. J., Picton, C., and Pocinwong, S. (1978). *In* "Versatility of Proteins" (C. H. Li, ed.), pp. 133–145. Academic Press, New York.
50. Killilea, S. D., and Ky, N. M. (1983). *ABB* **221**, 333–342.
50a. Sotiroudis, T. G., and Nikolaropoulos, S. (1984). *FEBS Lett.* **176**, 421–425.
50b. Crabb, J. W., Sotiroudis, T. G., and Heilmeyer, L. M. G., Jr. (1986). *JBC* (submitted for publication).
51. Carlson, G. M., and Graves, D. J. (1976). *JBC* **251**, 7480–7486.
52. Dickneite, G., Jennissen, H. P., and Heilmeyer, L. M. G., Jr. (1978). *FEBS Lett.* **87**, 297–302.
53. King, M. M., and Carlson, G. M. (1981). *Biochemistry* **20**, 4387–4393.
54. Kilimann, M. W., and Heilmeyer, L. M. G., Jr. (1982). *Biochemistry* **21**, 1727–1734.
55. Kilimann, M. W., and Heilmeyer, L. M. G., Jr. (1982). *Biochemistry* **21**, 1735–1739.
55a. Georgoussi, Z., and Heilmeyer, L. M. G., Jr. (1986). *Biochemistry* (in press).

55b. Hessova, Z., Thieleczek, R., Varsanyi, M., Falkenberg, F. W., and Heilmeyer, L. M. G., Jr. (1985). *JBC* **260**, 10111–10117.
56. Grand, R. J., Shenolikar, S., and Cohen, P. (1981). *EJB* **113**, 359–367.
57. Heilmeyer, L. M. G., Jr., Groschel-Stewart, U., Jahnke, U., Kilimann, M. W., Kohse, K. P., and Varsanyi, M. (1980). *Adv. Enzyme Regul.* **18**, 121–144.
58. Kilimann, M., and Heilmeyer, L. M. G., Jr. (1977). *EJB* **73**, 191–197.
59. Kohse, K. P., and Heilmeyer, L. M. G., Jr. (1981). *EJB* **117**, 507–513.
60. Stull, J. T., Nunnally, M. H., and Michnoff, C. H. (1986). This volume, Chapter 4.
61. Wang, J. H., Stull, J. T., Huang, T.-S., and Krebs, E. G. (1976). *JBC* **251**, 4521–4527.
62. Carlson, G. M., and Graves, D. J. (1976). *JBC* **251**, 7480–7486.
63. Graves, D. J., Hayakawa, T., Horvitz, R. A., Beckman, E., and Krebs, E. G. (1973). *Biochemistry* **12**, 580–585.
64. Sul, H. S., Dirden, B., Angelos, K. L., Hallenbeck, P., and Walsh, D. A. (1983). "Methods in Enzymology," Vol. 99, pp. 250–259.
64a. Jennissen, H. P., Petersen-Von Gehr, J. K. H., and Botzet, G. (1985). *EJB* **147**, 619–630.
65. Clerch, L. B., and Huijing, F. (1972). *BBA* **268**, 654–662.
66. Kim, G., and Graves, D. J. (1973). *Biochemistry* **12**, 2090–2095.
67. Singh, T. J., Akatsuka, A., and Huang, K.-P. (1982). *ABB* **218**, 360–368.
68. Villar-Palasi, C., and Wei, S. H. (1970). *PNAS* **67**, 345–350.
69. Tabatabai, L. B., and Graves, D. J. (1978). *JBC* **253**, 2196–2202.
70. Meyer, W. L., Fischer, E. H., and Krebs, E. G. (1964). *Biochemistry* **3**, 1033–1039.
71. Cheng, A., Fitzgerald, T. J., and Carlson. G. M. (1985). *JBC* **260**, 2535–2542.
72. Hayes, J. S., and Mayer, S. E. (1981). *Am. J. Physiol.* **240**, E340–E349.
73. Ramachandran, C., Angelos, K. L., and Walsh, D. A. (1982). *JBC* **257**, 1448–1457.
74. King, M. M., and Carlson, G. M. (1981). *ABB* **209**, 517–523.
75. King, M. M., and Carlson, G. M. (1981). *JBC* **256**, 11058–11064.
75a. Hallenbeck, P., and Walsh, D. A. (1986). *JBC* **261**, 5442–5449.
76. Stull, J. T., England, P. J., Brostrom, C. O., and Krebs, E. G. (1972). *Cold Spring Harbor Symp. Quant. Biol.* **37**, 263–265.
77. Tessmar, G., and Graves, D. J. (1973). *BBRC* **50**, 1–7.
78. Krebs, E. G., Kent, A. B., and Fischer, E. H. (1958). *JBC* **231**, 73–83.
79. Shizuta, Y., Khandelwal, R. L., Maller, J. L., Vandenheede, J. R., and Krebs, E. G. (1977). *JBC* **252**, 3408–3413.
80. Nolan, C., Novoa, W. B., Krebs, E. G., and Fischer, E. H. (1964). *Biochemistry* **3**, 542–551.
81. Tessmar, G. W., Skuster, J. R., Tabatabai, L. B., and Graves, D. J. (1977). *JBC* **252**, 5666–5671.
82. Graves, D. J. (1983). "Methods in Enzymology," Vol. 99, pp. 268–278.
83. Viriya, J., and Graves, D. J. (1979). *BBRC* **87**, 17–24.
84. Chan, K.-F. J., Hurst, M. V., and Graves, D. J. (1982). *JBC* **257**, 3655–3659.
85. Kemp, B. E., and John, M. J. (1981). *Cold Spring Harbor Conf. Cell Proliferation* **8**, 331–342.
86. Soderling, T. R., Srivastava, A. K., Bass, M. A., and Khatra, B. (1979). *PNAS* **76**, 2536–2540.
87. Walsh, D. A., Perkins, J. P., Broström, C. O., Ho, E. S., and Krebs, E. G. (1971). *JBC* **246**, 1968–1976.
88. King, M. M., Fitzgerald, T. J., and Carlson, G. M. (1983). *JBC* **258**, 9925–9930.
89. DePaoli-Roach, A. A., Roach, P. J., and Larner, J. (1979). *JBC* **254**, 4212–4219.
90. Walsh, K. X., Millikin, D. M., Schlender, K. K., and Reimann, E. M. (1979). *JBC* **254**, 6611–6616.

90a. Akatsuka, A., Singh, T. J., and Huang, K.-P. (1984). *JBC* **259**, 7878–7883.
91. Stull, J. T., Broström, C. O., and Krebs, E. G. (1972). *JBC* **247**, 5272–5274.
92. Perry, S. V., and Cole, H. A. (1974). *BJ* **141**, 733–743.
93. Moir, A. J. G., Cole, H. A., and Perry, S. V. (1977). *BJ* **161**, 371–382.
94. St. Louis, P. J., and Sulakhe, P. V. (1977). *Eur. J. Pharmacol.* **43**, 277–280.
94a. De Paoli-Roach, A. A., Bingham, E. W., and Roach, P. J. (1981). *ABB* **212**, 229–236.
94b. Singh, T. J., Akatsuka, A., and Huang, K.-P. (1983). *FEBS Lett.* **159**, 217–220.
95. Soderling, T. R., Sheorain, V. S., and Ericsson, L. H. (1979). *FEBS Lett.* **106**, 181–184.
96a. Angelos, K. L., and Walsh, D. A. (unpublished observation).
97. Meyer, F., Heilmeyer, L. M. G., Jr., Haschke, R. H., and Fischer, E. H. (1970). *JBC* **245**, 6642–6648.
98. Jennissen, H. P., Horl, W. H., Groschel-Stewart, U., Velick, S. F., and Heilmeyer, L. M. G., Jr. (1975). *In* "Metabolic Interconversion of Enzymes" (S. Shaltiel, ed.), pp. 19–26. Springer-Verlag, Berlin and New York.
99. Dombradi, V. K., Silberman, S. R., Lee, E. Y. C., Caswell, A. H., and Brandt, N. R. (1984). *ABB* **230**, 615–630.
100. Flockhart, D. A., Freist, W., Hoppe, J., Lincoln, T. M., and Corbin, J. D. (1984). *EJB* **140**, 289–295.
100a. Livanova, N. B., Silonova, G. V., Solovyeva, N. V., Andreeva, I. E., Ostrovskaya, M. V., and Poglazov, B. F. (1983). *Biochem. Int.* **7**, 95–105.
101. Huang, T. S., Bylund, D. B., Stull, J. T., and Krebs, E. G. (1974). *FEBS Lett.* **42**, 249–252.
102. Heilmeyer, L. M. G., Jr., Meyer, F., Haschke, R. H., and Fischer, E. H. (1970). *JBC* **245**, 6649–6656.
103. Gergely, P., Vereb, G., and Bot, G. (1975). *Arch. Biochem. Biophys. Acad. Sci. Hung.* **10**, 153.
104. Burger, D., Cox, J. A., Fischer, E. H., and Stein, E. A. (1982). *BBRC* **105**, 632–637.
105. Burger, D., Stein, E. A., and Cox, J. A. (1983). *JBC* **258**, 14733–14739.
106. Cohen, P. (1980). *EJB* **111**, 563–574.
107. Broström, C. O., Hunkeler, F. L., and Krebs, E. G. (1971). *JBC* **246**, 1961–1967.
108. Ozawa, E., Hosoi, K., and Ebashi, S. (1967). *J. Biochem. (Tokyo)* **61**, 531–533.
109. Ebashi, S., Makoto, E., and Ohtsuki, I. (1969). *Q. Rev. Biophys.* **2**, 351–384.
110. Drummond, G. E., Harwood, J. P., and Powell, C. A. (1969). *JBC* **244**, 4235–4240.
111. Stull, J. T., and Mayer, S. E. (1971). *JBC* **246**, 5716–5723.
112. Vandenheede, J. R., De Wulf, H., and Merlevede, W. (1979). *EJB* **101**, 51–58.
113. Keppens, S., Vandenheede, J. R., and De Wulf, H. (1977). *BBA* **496**, 448–457.
114. Cherrington, A. D., Hundley, R. F., Dolgin, S., and Exton, J. H. (1977). *J. Cyclic Nucleotide Res.* **3**, 263–271.
115. Hutson, N. J., Brumley, F. T., Assimacopoulos, F. D., Harper, S. C., and Exton, J. H. (1976). *JBC* **251**, 5200–5208.
116. Cohen, S. M., and Burt, C. T. (1977). *PNAS* **74**, 4271–4275.
117. Heilmeyer, L. M. G., Jr., Jahnke, U., Kilimann, M. W., Kohse, K. P., and Sperling, J. E. (1981). *Cold Spring Harbor Conf. Cell Proliferation* **8**, 321–329.
118. DePaoli-Roach, A. A., Gibbs, J. B., and Roach, P. J. (1979). *FEBS Lett.* **105**, 321–324.
119. Shenolikar, S., Cohen, P. T. W., Cohen, P., Nairn, A. C., and Perry, S. V. (1979). *EJB* **100**, 329–337.
120. Wilkinson, J. (1980). *EJB* **103**, 179–188.
121. Potter, J. D., Johnson, J. D., Dedman, J. R., Schreiber, W. E., Mandel, F., Jackson, R. L., and Means, A. R. (1977). *In* "Calcium Binding Proteins and Calcium Functions" (R. H. Wasserman, R. A. Corradino, E. Carafoli, R. H. Kretsinger, D. H. MacLennan, and F. L. Siegel, eds.), pp. 239–250. Elsevier/North-Holland, New York.

122. Leavis, P. C., and Kraft, E. L. (1978). *ABB* **186**, 411–415.
123. Sigel, P., and Pette, D. (1969). *J. Histochem. Cytochem.* **17**, 225–237.
124. Pette, D. (1975). *Acta Histochem. Suppl.* **14**, 47–68.
125. Yoshikawa, K., Usui, H., Imazu, M., Takeda, M., and Ebashi, S. (1983). *EJB* **136**, 413–419.
126. Wanson, J.-C., and Drochmans, P. (1972). *J. Cell Biol.* **54**, 206–224.
127. DiMauro, S., Trojaborg, W., Gambetti, P., and Rowland, L. P. (1971). *ABB* **144**, 413–422.
128. Bergamini, C., Buc, H., and Morange, M. (1977). *FEBS Lett.* **81**, 166–172.
128a. Reddy, N. B., Oliver, K. L., Festoff, B. W., and Engel, W. K. (1978). *BBA* **540**, 371–378.
129. Kasvinsky, P. J., Madsen, N. B., Sygusch, J., and Fletterick, R. J. (1978). *JBC* **253**, 3343–3351.
130. Madsen, N. B., Kasvinsky, P. J., and Fletterick, R. J. (1978). *JBC* **253**, 9097–9101.
131. Steiner, R. F., and Marshall, L. (1982). *BBA* **707**, 38–45.
132. Hallenbeck, P., and Walsh, D. A., unpublished observation.
133. Preiss, J., and Walsh, D. A. (1981). *In* "Biology of Carbohydrates" (V. Ginsberg, ed.), Vol. 1, pp. 199–314. Wiley, New York.
134. Busby, S. J. W., Gadian, D. G., Griffiths, J. R., Radda, G. K., and Richards, R. E. (1976). *EJB* **63**, 23–31.
135. Srivastava, A. K., Khatra, B. S., and Soderling, T. R. (1980). *ABB* **205**, 291–296.
136. Gergely, P., Vereb, G., and Bot, G. (1976). *BBA* **429**, 809–816.
137. Fischer, E. H., and Krebs, E. G. (1966). *FP* **25**, 1511–1520.
138. Entman, M. L., Kaniike, K., Goldstein, M. A., Nelson, T. E., Bornet, E. P., Futch, T. W., and Schwartz, A. (1976). *JBC* **251**, 3140–3146.
139. Entman, M. L., Bornet, E. P., Van Winkle, W. B., Goldstein, M. A., and Schwartz, A. (1977). *J. Mol. Cell. Cardiol.* **9**, 515–528.
140. Kayikawa, N., Kishimoto, A., Shiota, M., and Nishizuka, Y. (1983). "Methods in Enzymology," Vol. 102, pp. 279–290.
141. Mayer, S. E., and Krebs, E. G. (1970). *JBC* **245**, 3153–3160.
142. Pickett-Gies, C. A., and Walsh, D. A. (1985). *JBC* **260**, 2046–2056.
143. Kilimann, M. W., Schnackerz, K. D., and Heilmeyer, L. M. G., Jr. (1984). *Biochemistry* **23**, 112–117.
144. Cooper, R. H., Sul, H. S., and Walsh, D. A. (1981). *JBC* **256**, 8030–8038.
145. Pickett-Gies, C. A., Carlsen, R., Anderson, L. J., Angelos, K. L., and Walsh, D. A. (1986). *JBC* **261**, (in press).
146. Lincoln, T. M., and Corbin, J. D. (1977). *PNAS* **74**, 3239–3243.
147. Cohen, P. (1980). *FEBS Lett.* **119**, 301–306.
148. Waisman, D. M., Singh, T. J., and Wang, J. H. (1978). *JBC* **253**, 3387–3390.
149. Singh, T. J., Akatsuka, A., and Huang, K.-P. (1982). *JBC* **257**, 13379–13384.
149a. Singh, T. J., Akatsuka, A., and Huang, K. P. (1984). *JBC* **259**, 12857–12864.
150. Kishimoto, A., Takai, Y., and Nishizuka, Y. (1977). *JBC* **252**, 7449–7452.
151. Takai, Y., Kishimoto, A., Iwasa, Y., Kawahara, Y., Mori, T., and Nishizuka, Y. (1979). *JBC* **254**, 3692–3695.
152. Hallenbeck, P., Ramachandran, C., and Walsh, D. A., unpublished.
153. Cook, P. F., Neville, M. E., Vrana, K. E., Hartl, F. T., and Roskoski, R. R. (1982). *Biochemistry* **21**, 5794–5799.
154. Singh, T. J., and Wang, J. H. (1977). *JBC* **252**, 625–632.
155. Cohen, P., Watson, D. C., and Dixon, G. H. (1975). *EJB* **51**, 79–92.
156. Sul, H. S., Cooper, R. H., Whitehouse, S., and Walsh, D. A. (1982). *JBC* **257**, 3484–3490.
157. Cox, D. E., and Edström, R. D. (1982). *JBC* **257**, 12728–12733.
158. Singh, T. J., and Huang, K.-P. (1985). *BBRC* **130**, 1308–1313.
159. Yeaman, S. J., Cohen, P., Watson, D. C., and Dixon, G. H. (1977). *BJ* **162**, 411–421.

159a. Cheng, H.-C., Kemp, B. E., Pearson, R. B., Smith, A. J., Misconi, L., Van Patten, S. M., and Walsh, D. A. (1986). *JBC* **261**, 989–992.

160. Sul, H. S., and Walsh, D. A. (1982). *JBC* **257**, 10324–10328.

161. Ganapathi, M. K., and Lee, E. Y. C. (1984). *ABB* **233**, 19–31.

161a. Ramachandran, C., Pickett-Gies, C. A., Goris, J., Waelkens, E., Merlevede, W., and Walsh, D. A. (1985). *Adv. Protein Phosphatases* **2**, 355–374.

161b. Ramachandran, C., Goris, J., Waelkens, E., Merlevede, W., and Walsh, D. A. (1986). *JBC* **261**, (in press).

162. Cohen, P., and Antoniw, J. F. (1973). *FEBS Lett.* **34**, 43–47.

163. Stewart, A. A., Hemmings, B. A., Cohen, P., Goris, J., and Merlevede, W. (1981). *EJB* **115**, 197–205.

164. Ganapathi, M. K., Silberman, S. R., Paris, H., and Lee, E. Y. C. (1981). *JBC* **256**, 3213–3217.

165. Jurgensen, S., Shacter, E., Huang, C. Y., Chock, P. B., Yang, S. D., Vandenheede, J. R., and Merlevede, W. (1984). *JBC* **259**, 5864–5870.

166. Vandenheede, J. R., Yang, S. D., and Merlevede, W. (1981). *JBC* **256**, 5894–5900.

167. Ingebritsen, T. S., and Cohen, P. (1983). *EJB* **132**, 255–261.

168. Ingebritsen, T. S. Foulkes, Z. G., and Cohen, P. (1983). *EJB* **132**, 263–274.

169. Ingebritsen, T. S., Blair, J., Guy, P., Witters, L., and Hardie, D. C. (1983). *EJB* **132**, 275–281.

170. Pato, M. D., Adelstein, R. S., Crouch, D., Safer, B., Ingebritsen, T. S., and Cohen, P. (1983). *EJB* **132**, 283–287.

171. Stewart, A., Ingebritsen, T. S., and Cohen, P. (1983). *EJB* **132**, 289–295.

172. Pelech, S., Cohen, P., Fisher, M. Z., Pogson, C., El-Maghrabi, R., and Pilkus, S. J. (1984). *EJB* **145**, 39–45.

173. Ingebritsen, T. S., Stewart, A. A., and Cohen, P. (1983). *EJB* **132**, 297–307.

173a. DiSalvo, J., and Merlevede, W. (1985). *In* "Advances in Protein Phosphatases," Vol. 1. Leuven Press.

173b. DiSalvo, J., and Merlevede, W. (1985). *In* "Advances in Protein Phosphatases," Vol. 2. Leuven Press.

173c. Tsuchiya, M., Tanigawa, Y., Ushiroyama, T., Matsuura, R., and Shimoyama, M. (1985). *EJB* **147**, 33–40.

174. Williamson, J. R., Cooper, R. H., and Hoek, J. B. (1981). *BBA* **639**, 243–295.

175. Assimacopoulos-Jeannet, F. D., Blackmore, P. F., and Exton, J. H. (1977). *JBC* **252**, 2662–2669.

176. Gross, S. R., and Johnson, R. M. (1980). *J. Pharmacol. Exp. Ther.* **214**, 37–44.

177. Posner, J. B., Stern, R. S., and Krebs, E. G. (1985). *JBC* **240**, 982–985.

178. Danforth, W. H., and Helmreich, E. (1964). *JBC* **239**, 3133–3138.

179. Danforth, W. H., and Lyon, J. B. (1964). *JBC* **239**, 4047–4050.

180. Friesen, A. J., Allen, G., and Valadares, J. R. (1967). *Science* **155**, 1108–1109.

181. Friesen, A. J., Oliver, N., and Allen, G. (1969). *Am. J. Physiol.* **217**, 445–450.

182. Dobson, J. G., and Mayer, S. E. (1973). *Circ. Res.* **33**, 412–420.

183. Dobson, J. G., Ross, J., and Mayer, S. E. (1976). *Circ. Res.* **39**, 388–395.

184. Hammermeister, K. E., Yunis, A., and Krebs, E. G. (1965). *JBC* **240**, 986–991.

185. Namm, D. H., and Mayer, S. E. (1968). *Mol. Pharmacol.* **4**, 61–69.

186. Namm, D. H., Mayer, S. E., and Maltbie, M. (1968). *Mol. Pharmacol.* **4**, 522–530.

187. Mayer, S. E., Namm, D. H., and Rice, L. (1970). *Circ. Res.* **26**, 225–233.

188. Barovsky, K., and Gross, S. R. (1981). *J. Pharmacol. Exp. Ther.* **217**, 326–332.

189. Yeaman, S. J., and Cohen, P. (1975). *EJB* **51**, 93–104.

190. McCullough, T. E., and Walsh, D. A. (1979). *JBC* **254,** 7345–7352.
191. Sul, H. S., Cooper, R. H., McCullough, T. E., Pickett-Gies, C. A., Angelos, K. L., and Walsh, D. A. (1981). *Cold Spring Harbor Conf. Cell Proliferation* **8,** 343–355.
192. Angelos, K. L., Ramachandran, C., and Walsh, D. A. (1986). *JBC* (in press.)
193. Carlsen, R. C., Larson, D. B., and Walsh, D. A. (1985). *Can. J. Pharmacol. Physiol.* **63,** 958–965.

11

Muscle Glycogen Synthase

PHILIP COHEN

Department of Biochemistry
The University
Dundee DD1 4HN, United Kingdom

THE ENZYMES, Vol. XVII

I. Introduction

Glycogen synthase was the third enzyme shown to be regulated by a phosphorylation–dephosphorylation mechanism. Following the discovery that glycogen phosphorylase (*1*) and phosphorylase kinase (*2*) were activated by phosphorylation, Larner and co-workers found that glycogen synthase could exist in two forms in mammalian skeletal muscle. One possessed little activity without glucose 6-phosphate (G6P), whereas the other was almost fully active in the absence of this allosteric activator (*3*). The conversion of glycogen synthase from a G6P-independent to a G6P-dependent form was shown to require MgATP and a further protein, and to be stimulated by cyclic adenosine 3′:5′-monophosphate (cAMP) (*3–5*). The basis for these effects became clearer following the purification of glycogen synthase to homogeneity, when the enzyme was shown to be phosphorylated by cAMP-dependent protein kinase (*6, 7*). However, in 1971 Smith *et al.* (*8*) reported that the G6P-dependent form contained ≃6 mol phosphate/86 kDa subunit, and with the subsequent identification of additional glycogen synthase kinases (*9–13*) it became clear that glycogen synthase was regulated by *multisite* phosphorylation. These discoveries provided a major stimulus for research over the subsequent decade. Progress since 1982 has been particularly marked and is reviewed in this chapter.

II. Structure of Glycogen Synthase from Mammalian Skeletal Muscle

The glycogen synthase subunit migrates on SDS-polyacrylamide gel electrophoresis (SDS-PAGE) with an apparent molecular mass of ≃86 kDa [reviewed in Ref. (*12*)]. The primary structure of the first 29 (*13*) and last 124 residues (*14*) of the rabbit skeletal-muscle enzyme have been determined (Fig. 1), establishing that the subunits are identical. The smallest active species is a tetramer [reviewed in Ref. (*12*)], although the tetrameric species has a strong tendency to aggregate. As discussed in Section II, A, all the phosphorylation sites that have been sequenced are contained within the N- and C-terminal regions. Residues in the N-terminal cyanogen bromide peptide (CB-N) are there-

```
                    T  C
         1    P    P    P                    20                      38
CB-N   P L S R T L S V S S L P G L E D W E D E F D L E N S V L F
```

```
                                      C T
       1         10          20       P        P         P         P  40 P       P      50                 60                70
CB-C   A L A K A F P D H F T Y E P H E A D A T Q G Y R Y P R P A S V P P S P S L S R H S S P H Q S E D E E E P R D G L P E E D G E R Y D E D E E A A K
        T       T         80        T       P     90        T T    P                 110                 120      T
       D R R N I R A P Q W P R R A S C T S S S G G S K R S N S V D T S S L S T P S E P L S S A P S L G E E R N
```

FIG. 1. Primary structures of the N-terminal and C-terminal of the cyanogen bromide peptides (CB-N and CB-C) of rabbit skeletal-muscle glycogen synthase. The sequence of residues 108–124 in CB-C are unpublished results from this laboratory, and differ from the sequence published in (15) at a number of positions. Residue 124 appears to be the C-terminus of glycogen synthase. Sites phosphorylated *in vitro* are denoted by P, and the positions at which the native enzyme is cleaved by trypsin and chymotrypsin by T and C, respectively. Data are from Refs. (13–15).

fore prefixed by N (N1, N2, N3, etc.) and those in the C-terminal cyanogen bromide fragment (CB-C) by C (C1, C2, C3, etc.).

Glycogen synthase purified in this laboratory contains a second protein component of apparent molecular mass 44 kDa (16). The molar ratio of the 86-kDa-to the 44-kDa-species is approximately 2 : 1. Whether the 44-kDa protein interacts with glycogen synthase or is merely an impurity is unresolved. The 44-kDa component is not phosphorylated by any glycogen synthase kinase.

The N- and C-terminal regions that have been sequenced contain all the phosphorylation sites (Section III) and are also extremely sensitive to proteinases. Brief incubation of the native enzyme with low concentrations of trypsin initially cleaves the peptide bonds N4–N5, C75–C76, C78–C79, and C84–C85, followed by C39–C40 (14, 17). Cleavage in the region C75–C85 is accompanied by a decrease in apparent molecular mass from 86 kDa to 77 kDa, and cleavage at C39–C40 by a further reduction to 69 kDa (14), as judged by SDS-PAGE. Brief incubation with chymotrypsin cleaves the peptide bond C23–C24 specifically, reducing the apparent molecular mass from 86 kDa to 67 kDa (14). However the actual molecular mass of the peptide C24–C124 is only 11 kDa (Fig. 1). This peptide is extremely hydrophilic and behaves in solution as a random coil (18). Furthermore, it contains only two lysyl residues and is unlikely to bind SDS as well as normal globular proteins. These observations indicate that the unusual C-terminal region of glycogen synthase reduces its mobility on SDS-PAGE. Consequently the subunit may not be as large as 86 kDa.

Brief incubation of glycogen synthase with very low concentrations of subtilisin initially cleaves the peptide bond N6–N7, enhancing activity in the presence of G6P about 5-fold (17). Trypsin also causes a transient (1.5-fold) rise in activity in the presence of G6P (17, 19) presumably resulting from cleavage of N4–N5 (17). In contrast, cleavage at the C-terminal region is accompanied by inactivation, especially in the absence of G6P (19–22).

III. Glycogen Synthase Kinases in Mammalian Skeletal Muscle

It is clear that skeletal-muscle glycogen synthase can be phosphorylated *in vitro* by at least 10 protein kinases. Indeed, of protein kinases that have been tested, only myosin light chain kinase is unable to phosphorylate the enzyme (*23, 24*). In this section, the protein kinases that act on glycogen synthase and the sites that they phosphorylate are identified. A number of these protein kinases are described elsewhere in this volume, and readers are referred to other chapters for more detailed accounts of their structure and properties.

A. Cyclic AMP-Dependent Protein Kinase

The phosphorylation of glycogen synthase by cAMP-dependent protein kinase reaches different plateau values, depending on the concentration of protein kinase in the incubation (*12, 25*). Up to an incorporation of $\simeq 2$ mol/subunit nearly all the phosphate is incorporated into the seryl residues N7 (site-2), C87 (site-1a) and C100 (site-1b). The initial rate of phosphorylation of site-1a is 7- to 10-fold faster than site-2 and 15- to 20-fold faster than site-1b (*26*).

Sheorain *et al.* (*27*) have reported that additional sites can be phosphorylated when extremely high concentrations of cAMP-dependent protein kinase (2.5 μM) are employed. Under these forcing conditons, the total amount of phosphate incorporated exceeded 3 mol/subunit, and following exhaustive digestion with trypsin, ^{32}P-labeled peptides were resolved by reverse-phase high-performance liquid chromatography (HPLC). These experiments revealed the presence of ^{32}P-radioactivity eluting at the positions expected for the tryptic peptides C25–C39 and C40–C53. The maximum levels of phosphorylation of these two regions were estimated to be 0.6 mol/subunit (C25–C39) and 0.2 mol/subunit (C40–C53). Automated Edman degradation of C40–C53 showed a "burst" of ^{32}P-radioactivity at the third cycle, suggesting that C42 was phosphorylated. The location of the phosphate in region C25–C39 was not determined. Although these results may well be correct, primary structure analysis is needed to substantiate the conclusions because the peptides were not obtained in pure form. For example phosphorylation of C42 may have rendered the peptide bond C39–C40 resistant to trypsin (see Section III, G). Consequently, the peptide assigned at C25–C39 could have been C25–C53. If this were true, phosphorylation of C43 might be much more extensive and C25–C39 nonexistent.

B. Phosphorylase Kinase

Roach *et al.* (*28, 29*) were the first to demonstrate that phosphorylase kinase catalyses the phosphorylation of glycogen synthase, and this finding was confirmed by others (23, 30, 31). The phosphorylation occurs at site-2 (23, 32) and the rate of phosphorylation is comparable to that of glycogen phosphorylase (23). The report of an additional phosphorylation site in the C-terminal region after prolonged incubation with high concentrations of phosphorylase kinase (32) can be explained by trace contamination with glycogen synthase kinase-5 (33).

C. Calmodulin-Dependent "Multiprotein" Kinase

The finding that purified preparations of glycogen synthase were contaminated with traces of a protein kinase that was stimulated by Ca^{2+} and calmodulin (24, 34) led to the discovery of a Ca^{2+}–calmodulin-dependent glycogen synthase kinase in liver (35) and skeletal muscle (36). The rabbit skeletal-muscle enzyme has been purified $\approx$5000-fold and shows a major 58-kDa band and a minor 54-kDa species when analyzed by SDS-PAGE. The native enzyme is a dodecamer with a molecular mass of 700 kDa, and the 12 subunits appear to be arranged as two hexagonal rings stacked one upon the other, as judged by electron microscopy (37). The enzyme phosphorylates site-2 and site-1b, the initial rate of phosphorylation of site-2 being 5- to 10-fold faster than site-1b (37).

The calmodulin-dependent glycogen synthase kinase has a broad substrate specificity in vitro, and is capable of phosphorylating a number of proteins at comparable rates to glycogen synthase. These include synapsin I, microtubule-associated protein 2, and tyrosine hydroxylase (38, 39). Glycogen phosphorylase is not a substrate (36, 37). Like many protein kinases the calmodulin-dependent glycogen synthase kinase can phosphorylate itself, and up to 5 mol phosphate/subunit are incorporated via the autophosphorylation reaction. Autophosphorylation does not affect activity measured in the presence of Ca^{2+} and calmodulin (37). However, recent work with the closely related brain enzyme (see below) has demonstrated that autophosphorylation causes the protein kinase to become almost fully active in the absence of Ca^{2+} and calmodulin (39a). This may represent a mechanism for prolonging the Ca^{2+} signal.

A synthetic peptide corresponding to the first 10 residues of glycogen synthase is an excellent substrate for the calmodulin-dependent glycogen synthase kinase, the phosphorylation occurring at N7 (40). If the arginine at N4 is substituted with leucine or alanine the peptide no longer serves as a substrate. Studies with other synthetic peptides have confirmed that the enzyme will only phosphorylate sequences of the type Arg-x-y-Ser-z at significant rates. However, in contrast to cAMP-dependent protein kinase, insertion of a second arginine residue at position x does not improve the kinetics of phosphorylation (40).

A protein kinase with an identical substrate specificity is present in brain, where it has been termed synapsin I kinase-II (38) or calmodulin-dependent protein kinase-II (39). The brain enzyme is composed of two isoenzymes with subunit molecular masses of 50 kDa and 58–60 kDa whose proportions vary from brain region to brain region (41). The 58–60-kDa component from brain is closely related to the skeletal-muscle enzyme, as judged by one-dimensional peptide mapping of phosphopeptides and immunological criteria (38, 39), but does not seem to be identical (41a).

The calmodulin-dependent glycogen synthase kinase has been purified from both rat and rabbit liver (42–44). The enzyme preparations show a protein-staining doublet (57–55 kDa) on SDS-PAGE. However based on a sedimentation constant of 10.6 S and Stokes radius of 70 Å, the native enzyme appears to have a molecular mass of ≈300 kDa (43). This suggests that it may be a hexamer, in contrast to the muscle enzyme which is a dodecamer. The liver enzyme also phosphorylates rabbit skeletal-muscle glycogen synthase (at site-2 and site-1b) (43), synapsin I, microtubule-associated protein 2, and tyrosine hydroxylase (45).

The broad substrate specificity and widespread tissue distribution of this protein kinase suggests that it may mediate many of the actions of Ca^{2+} in vivo. Accordingly, it has been termed the calmodulin-dependent "multiprotein" kinase (38).

D. GLYCOGEN SYNTHASE KINASE-3

Glycogen synthase kinase-3 (GSK-3) has been purified ≈50,000-fold to homogeneity from rabbit skeletal muscle. Its molecular mass estimated by SDS-PAGE (51 kDa) is similar to that obtained by sedimentation equilibrium centrifugation of the native enzyme (47 kDa), demonstrating that GSK-3 is monomeric. However, it is eluted from gel filtration columns slightly earlier than serum albumin (66 kDa), indicating an asymmetric structure (18).

GSK-3 phosphorylates the tryptic peptide comprising residues C28 to C39 specifically (13, 14). Following incubation of glycogen synthase with GSK-3 and MgATP, mono-, di-, and triphosphorylated forms of this tryptic peptide can be resolved, demonstrating that at least three seryl residues are phosphorylated (13). Based on the release of ^{32}P-radioactivity during automated Edman degradation of C28–C39, the residues phosphorylated appear to be C30, C34, and C38 (13); however, the order of phosphorylation is unknown. These serine residues are collectively referred to as sites-3.

The type II regulatory subunit of cAMP-dependent protein kinase is a substrate for GSK-3 and two residues (Ser-44 and Ser-47) are phosphorylated (46). As discussed in Section VI, GSK-3 can also phosphorylate a protein termed inhibitor-2 on a specific threonyl residue (47). GSK-3 phosphorylates itself, and up to 4 phosphates/mol can be incorporated via autophosphorylation, without any apparent effect on activity (46, 48). However, many proteins that are phosphorylated by cAMP-dependent protein kinase are not touched by GSK-3 (49). Further properties of GSK-3 are reviewed in Ref. (49). GSK-3 has also been termed factor F_A by Merlevede and co-workers (48, 50) for reasons discussed in Section VI.

Ahmed *et al.* (*51*) partially purified a protein kinase from skeletal muscle whose activity was stimulated several-fold by heparin ($A_{0.5} = 3\mu g/ml$). Although it was originally suggested that this enzyme was distinguishable from other glycogen synthase kinases (*51*), more recent work suggests that it is GSK3 (*51a*). The heparin-stimulated kinase phosphorylates the tryptic peptide C25-C39 specifically, and has Factor F_A activity (see Section VI). Its apparent molecular mass estimated by gel filtration (70kDa) is identical to GSK3 (*51*). Highly purified GSK3 is not activated by heparin (*33*), but stimulation is lost during exposure to low ionic strength prior to chromatography on DEAE-cellulose. Loss of heparin stimulation is caused by a rise in activity in the absence of the glycosaminoglycan (C. Smythe, unpublished work from this laboratory). It therefore appears that the heparin-stimulated protein kinase may represent the "native" form of GSK3. However, no substances capable of substituting for heparin that might be of physiological importance have been identified so far.

GSK-3 was partially purified from rabbit liver by DePaoli-Roach *et al.* (*52*). Their preparation predominantly phosphorylated the C-terminal cyanogen bromide peptide (CB-C) of rabbit skeletal-muscle glycogen synthase, as expected, but some incorporation of phosphate occurred in CB-N. Ramakrishna *et al.* (*52a, 52b*) also isolated a protein kinase from liver that phosphorylated site-2 in addition to sites-3, and this preparation was capable of phosphorylating other proteins that are not substrates for muscle GSK3, such as ATP-citrate lyase and acetyl-CoA carboxylase (*52a, 52b*). These results suggest that hepatic GSK-3 either has a broader specificity than its muscle counterpart or that the preparations are contaminated with another protein kinase.

E. GLYCOGEN SYNTHASE KINASE-4

Glycogen synthase kinase-4 (GSK-4) has been only partially purified from rabbit skeletal muscle, and its subunit composition is therefore unknown. The apparent molecular mass on gel filtration is ≈ 115 kDa. It phosphorylates glycogen synthase at site-2 and no other protein tested is phosphorylated at a significant rate (*33*). The substrate specificity of GSK-4 demonstrates that it is not a proteolytic fragment of phosphorylase kinase or the calmodulin-dependent multiprotein kinase that has lost its ability to be regulated by Ca^{2+}-calmodulin (*37*). GSK-4 is identical to the enzyme termed $PC_{0.4}$ by Roach and co-workers (*52*) and to certain other glycogen synthase kinases that have been reported [reviewed in Ref. (*33*)]. Further properties of GSK-4 are summarized in Ref. (*49*). No mechanisms for regulating the activity of GSK-4 have been identified.

F. GLYCOGEN SYNTHASE KINASE-5

Glycogen synthase-5 (GSK-5) has been termed variously $PC_{0.7}$ (*52*), casein kinase-II, casein kinase-G, or casein kinase-TS [see discussion in Ref. (*33*)]. The enzyme has been purified to homogeneity from skeletal muscle (54) and other tissues (55–57) and has an $\alpha_2\beta_2$ structure in which the apparent molecular masses of the α- and β-subunits are 43 kDa and 26 kDa, respectively. The 43-kDa component is the catalytic subunit (55, 58). GSK-5 has a number of distinctive properties including up to 40-fold activation by spermine at physiological (1 mM) concentrations of Mg^{2+}, potent inhibition by heparin ($K_i \simeq 0.05$ μg/ml), and the ability to use GTP as a substrate almost as effectively as ATP (33, 55).

GSK-5 phosphorylates glycogen synthase at residue C46, termed site-5 (14, 33). However, the enzyme has a broad substrate specificity, and physiological substrates include the type II regulatory subunit of cAMP-dependent protein kinase (46), troponin T (59, 60), and the β-subunit of protein synthesis initiation factor eIF-2 (60, 61). The enzyme can also phosphorylate its own β-subunit without any effect on activity (53–55). GSK-5 phosphorylates seryl residues that are followed by a number of consecutive acidic residues (46), and this factor is critical for specific substrate recognition (18, 63, 64).

G.   CASEIN KINASE-I

Casein kinase-I (CK-I) has been purified $\simeq 100,000$-fold to near homogeneity from skeletal muscle, and is a monomeric protein of molecular mass $\simeq 35$ kDa (65), like CK-I from other mammalian sources (54, 66). It has also been termed glycogen synthase kinase-1 (67–70) or $PC_{0.6}$ (52, 71).

Incubation of glycogen synthase with high concentrations of muscle or liver CK-I for 2–5 h results in the incorporation of 6 phosphates/subunit (65, 70, 71) and as many as 10 residues may become phosphorylated (65). In CB-N the seryl residues N3, N7, and N10 are major sites of phosphorylation and the threonyl residue N5, a minor site. The electrophoretic mobility of CB-N is slower after phosphorylation by CK-I than after phosphorylation by cAMP-dependent protein kinase. This implies that N7 cannot be the first residue in CB-1 phosphorylated by CK-I; the initial serine phosphorylated must therefore be either N3 or N10.

The C-terminal cyanogen bromide peptide (CB-C) is phosphorylated by CK-I at a similar rate to CB-N, and at least five of the seven serines in the tryptic peptide C28–C53 are phosphorylated. These include the residues phosphorylated by GSK-3 and GSK-5. The exact seryl residues phosphorylated by CK-I are unknown, because phosphorylation renders the Arg–His bond between C39 and C40 completely resistant to trypsin (65). In contrast, the same bond is cleaved readily by trypsin if glycogen synthase is phosphorylated by either GSK-3 or GSK-5 (13, 33). This might suggest that C41 is phosphorylated by CK-I, because the failure of trypsin to cleave sequences of the type Arg-x-Ser(P) is well documented [e.g., see Refs. (72–74)]. Minor phosphorylation by CK-I also occurs in the tryptic peptide C98 to C123, mainly at seryl residues (65).

Casein kinase-I has a very broad substrate specificity *in vitro* and can phosphorylate many proteins in addition to glycogen synthase [see Refs. *(55, 75–77)*]. However no mechanisms for regulating its activity have been identified.

H. Cyclic Guanosine Monophosphate-Dependent Protein Kinase

Cyclic guanosine monophosphate (cGMP) is present in extremely low concentrations in skeletal muscle *(78, 79)*, and phosphorylation of glycogen synthase has only been examined using cGMP-dependent protein kinase from lung. The enzyme phosphorylates the tryptic peptides containing site-2, site-1a, and site-1b *(26)*, the same peptides labeled by cAMP-dependent protein kinase. The order of phosphorylation is also site-1a > site-2 > site-1b, although the difference in initial rate of phosphorylation between site-1a and the other two sites is not as pronounced *(26)*. These observations are not unexpected in view of the known similarity in substrate specificity between these two cyclic nucleotide-dependent protein kinases. However, the rate of phosphorylation of glycogen synthase by cGMP-dependent protein kinase is about 100-fold slower than that of cAMP-dependent protein kinase *(26, 79)*.

I. Diacylglycerol-Dependent Protein Kinase

The activity of diacylglycerol (DG)-dependent protein kinase requires phosphatidylserine and supraphysiological concentrations of Ca^{2+}, but in the presence of DG, the $A_{0.5}$ for Ca^{2+} is decreased over 1000-fold [reviewed in Ref. *(80)*]. The activity of DG-dependent protein kinase is much lower in skeletal muscle than in other mammalian tissues *(81, 82)* and phosphorylation has only been examined using the rat brain enzyme *(83)*. Glycogen synthase can be phosphorylated to >1 mol/subunit and the major tryptic peptides that become labeled are N5–N38 (containing site-2) and C85–C97 (containing site-1a). The C-terminal peptide is phosphorylated at a slightly faster rate. Glycogen synthase is phosphorylated at a comparable rate to mixed histones, and as with most substrates, activity is stimulated $\simeq$10-fold by Ca^{2+} and phospholipid *(83)*.

IV. Effect of Phosphorylation on the Activity of Skeletal-Muscle Glycogen Synthase

For many years it was believed that glycogen synthase existed in just two forms, a phosphorylated species dependent on G6P and a dephosphorylated form that was fully active in the absence of this effector. Subsequently, it was found that activation of the phosphorylated form by G6P could be antagonised by metabolites such as ATP and P_i. The dephosphorylated form was also inhibited strongly by ATP, but this inhibition was reversed by very low concentrations of G6P (84). Thus, it appeared that glycogen synthase could exist in two forms that differed in sensitivity to G6P, ATP, and P_i.

The most detailed study of the effects of phosphorylation on the kinetic properties of glycogen synthase (85, 86) was carried out before the multiplicity of glycogen synthase kinases was appreciated. In these experiments, glycogen synthase was phosphorylated in undefined sites by incubation with MgATP for varying periods of time, at an early stage of purification when the enzyme was still contaminated with protein kinase activities. Glycogen synthase was purified to homogeneity and its phosphate content and kinetic properties examined. It was found that the $S_{0.5}$ for UDP-Glc increased about 1000-fold over the phosphorylation range studied (0.27 to 3.5 mol/subunit). G6P attenuated the effect of phosphorylation on the $S_{0.5}$ for UDP-Glc although the $A_{0.5}$ for G6P also increased about 1000-fold over the same phosphorylation range and G6P-saturation curves became more sigmoidal. Phosphorylation was accompanied by a greater sensitivity to inhibition by substances such as ATP and P_i. However, even with highly phosphorylated enzyme, such inhibition could be counteracted effectively by G6P.

An exhaustive kinetic analysis using glycogen synthase phosphorylated in defined sites by defined protein kinases, alone and in combination, has not been performed. Most studies have simply measured the "activity ratio" of the enzyme, defined as activity in the absence of G6P divided by activity in the presence of saturating G6P (usually measured at $\simeq$ 5 mM UDP-Glc).

Cyclic AMP-dependent protein kinase phosphorylates site-1a much faster than site-1b or site-2 (Section III, A). Conversely, if glycogen synthase labeled in site-1a, site-1b, and site-2 is incubated with protein phosphatase-1 (Section VI), phosphate is removed sequentially from the three sites. The dephosphorylation of site-2 precedes site-1a, and site-1a precedes site-1b (26). These studies demonstrate that site-2 and site-1a are both inactivating sites, although phosphorylation of site-2 depresses the activity ratio to a greater extent than site-1a. In contrast, phosphorylation of site-1b appears to have little or no effect on the activity ratio (26).

There is general agreement that maximal phosphorylation of sites-3 by GSK-3 produces a greater decrease in the activity ratio than the phosphorylation of site-1a + 1b + 2 (33, 52, 70). However, the effects of these phosphorylations are additive, and greater decreases in the activity ratio are observed when all six sites are phosphorylated (87). Since the order of phosphorylation of C30, C34, and C38 is unknown, the relative contributions of these three sites to inactivation is unclear.

Two laboratories have reported that phosphorylation of glycogen synthase by GSK-5 does not decrease the activity ratio (33, 54), whereas another group reported that phosphorylation was accompanied by a small reduction in activity (56, 70).

Phosphorylation of glycogen synthase to 4–6 mol/mol subunit by CK-I is accompanied by a decrease in activity ratio similar to that observed with GSK-3 and cAMP-dependent protein kinase combined (65, 67–71). This is consistent with the finding that CK-I phosphorylates serine residues in the region N3–N10 and C30–C46 (Section III, G).

V. Synergism between Glycogen Synthase Kinase-3 and Glycogen Synthase Kinase-5

Although the phosphorylation of site-5 by GSK-5 does not affect the activity ratio, the presence of phosphoserine at this position is critical for the activity of GSK-3. In this laboratory "dephosphorylated" preparations of glycogen synthase with activity ratios of 0.8 to 0.9 usually contain 0.5–0.6 mol phosphate/subunit, mostly located in the tryptic peptide (C40 to C53) containing site-5 (88). This phosphate is resistant to the action of skeletal-muscle protein phosphatases (Section VI), but can be removed by incubation with potato acid phosphatase. This treatment abolishes phosphorylation of glycogen synthase by GSK-3, without affecting the rate of phosphorylation by cAMP-dependent protein kinase, phosphorylase kinase, the calmodulin-dependent multiprotein kinase, or GSK-4. Rephosphorylation at C46 by GSK-5 restores the ability of GSK-3 to phosphorylate the enzyme (89). The presence of $\simeq 0.5$ mol phosphate/subunit in the tryptic peptide C40–C53 may explain why phosphorylation by GSK-3 usually reaches a plateau near 1.5 mol/subunit in vitro, rather than 3 mol/subunit. This view is supported by the work of DePaoli-Roach et al. (52), who found that phosphorylation by GSK-5 (without prior incubation with potato acid phosphatase) increased the amount of phosphate that could be incorporated by GSK-3 from ~1.3 to >2 mol/subunit.

The effects of dephosphorylating and rephosphorylating the peptide C40–C53 on the phosphorylation of glycogen synthase by GSK-3 can be reproduced using the peptide C24–C124 (18) that can be isolated by brief chymotryptic attack of the native enzyme (Section II). Since peptide C24–C124 is monomeric, this demonstrates that phosphorylation of C40–C53 is essential for phosphorylation of the *same subunit* by GSK-3 (18).

These observations suggest that GSK-5 is a novel protein kinase, whose function is to form the recognition site for another protein kinase. Similar observations have been made for two other proteins. The type II regulatory subunit of cAMP-dependent protein kinase contains 1.5–1.8 phosphates/subunit mostly located in Ser-74 and Ser-76, the sites phosphorylated by GSK-5. Dephosphorylation of these residues by incubation with potato acid phosphatase prevents

GSK-3 from phosphorylating Ser-44 and Ser-47. Rephosphorylation with GSK-5 restores the ability of GSK-3 to phosphorylate the protein (46). The amino acid sequence following Ser-76 (Glu-Asp-Glu-Glu-Asp) is almost identical to that following C46 (Fig. 1).

DePaoli-Roach (90) reported that inhibitor-2 is a substrate for GSK-5, and that phosphorylation by this protein kinase potentiated phosphorylation by GSK-3 (see also Section VI). The residue phosphorylated by GSK-3 is threonine-72 (90a), and the residues phosphorylated by GSK-5 are serines 86, 120, and 121 (90b). These findings suggest that the presence of a C-terminal phosphoserine residue is critical for substrate recognition by GSK-3. However, in the case of glycogen synthase a region C-terminal to residue C64 is also essential for phosphorylation (18).

VI. The Glycogen Synthase Phosphatases in Skeletal Muscle

Relatively few serine- and threonine-specific protein phosphatases are present in the cytoplasm of mammalian cells (91, 92). Two of these enzymes, termed protein phosphatase-1 (PP-1) and protein phosphatase-2A (PP-2A) have broad substrate specificities and account for virtually all detectable glycogen synthase phosphatase and phosphorylase phosphatase activity in skeletal-muscle extracts. Thus the combined addition of inhibitor-2 (I-2), a specific inhibitor of PP-1, and antibody to PP-2A, inhibit glycogen synthase phosphatase and phosphorylase phosphatase activities in rabbit skeletal-muscle extracts by >95% (93). Furthermore, fractionation of the extracts by anion-exchange chromatography and gel filtration fails to detect any other protein phosphatase with significant activity toward glycogen synthase (94). A third protein phosphatase (PP-2C) capable of dephosphorylating glycogen synthase is present in skeletal-muscle extracts, but its contribution to the total activity (1–2%) is negligible (94, 95).

PP-1 and PP-2A dephosphorylate site-1a, site-2, and sites-3 at comparable rates in vitro (91). However, as discussed in Section IV, if glycogen synthase is phosphorylated in site-1a + 1b + 2, PP-1 dephosphorylates site-2 5- to 10-fold faster than site-1a and $\simeq$100-fold faster than site-1b. The dephosphorylation of site-1a occurs more rapidly once site-2 is dephosphorylated, and dephosphorylation of site-1b takes place at a significant rate only after both site-2 and site-1a are dephosphorylated (26). Site-1b, and especially site-5 (91), are dephosphorylated very slowly by PP-1 and PP-2A, and no other protein phosphatases capable of acting on these sites at a significant rate have been detected in skeletal muscle. The extremely weak phosphatase activity toward site-5 may explain the high level of phosphorylation of the region C40–C53 in vivo (Section VII). The large number of acidic residues immediately C-terminal to C46 may act as a negative specificity determinant for protein phosphatases (91).

The order of dephosphorylation of residues C30, C34, and C38 (sites-3), like their order of phosphorylation, is unknown.

A. STRUCTURE OF PROTEIN PHOSPHATASE-1 AND PROTEIN PHOSPHATASE-2A

The catalytic (C)-subunits of PP-1 and PP-2A [termed C-I and C-II by Lee and co-workers (96)] have been purified to homogeneity by procedures that involve precipitation with 80% ethanol at room temperature at an early stage of purification (96, 97). If proteinase inhibitors are included, the C-subunit of PP-1 is recovered as a 37-kDa protein and the C-subunit of PP-2A as a 36-kDa species (97). However, despite their similar molecular mass and substrate specificities, peptide mapping studies have established that the two C-subunits are the products of distinct genes (97).

The C-subunits do not exist as such *in vivo*, but are complexed with other proteins, removed, or denatured during treatment with 80% ethanol. Several of these high-molecular-mass forms have been purified to homogeneity from skeletal muscle and their subunit compositions elucidated.

When skeletal-muscle extracts are centrifuged at 80,000 g, to pellet glycogen and its associated proteins, 50–60% of the PP-1 activity sediments with these glycogen-protein particles (12, 95). This is similar to the proportion of glycogen synthase bound to glycogen (12). The glycogen-bound form of PP-1, termed PP-1$_G$, consists of the 37-kDa C-subunit complexed to a 103-kDa G-subunit, which is the glycogen-binding component (98). It is probable that much of the PP-1 activity that does not sediment with the glycogen-protein particles is also PP-1$_G$, although this remains to be established.

Protein phosphatase-1 can also be isolated in an inactive form, termed PP-1$_I$, which is not associated with glycogen. It consists of the 37-kDa C-subunit complexed to I-2 (99–102), whose molecular mass is 22.8 kDa (90a). PP-1$_I$ has also been termed the MgATP-dependent protein phosphatase, because preincubation with MgATP and another protein (factor F_A) is required to generate catalytic activity (103). Factor F_A has been purified to homogeneity and shown to be identical to GSK-3 (18, 48). Activation of PP-1$_I$ is triggered by the phosphorylation of a threonyl residue on I-2 (18, 47, 90a). The mechanism of activation and deactivation of PP-1$_I$ is discussed in greater detail elsewhere in this volume (Chapter 8).

Protein phosphatase-2A is not associated with the glycogen-protein particles (95) and three forms of this enzyme can be resolved by chromatography on DEAE-cellulose, termed PP-2A$_0$, PP-2A$_1$, and PP-2A$_2$ (91, 105). Each of these species contain a 60-kDa A-subunit and a 36-kDa C-subunit. The A- and C-subunits of PP-2A$_0$, PP-2A$_1$, and PP-2A$_2$ are identical, as judged by peptide mapping, and the C-subunit is identical to the catalytic subunit of PP-2A isolated by treatment with 80% ethanol at room temperature. PP-2A$_0$ contains an additional 54-kDa B'-subunit and PP-2A$_1$ a 55-kDa B-subunit. The B'- and B-subunits display different peptide maps and therefore appear to be distinct gene products. PP-2A$_2$ lacks the B' and B-subunits, and appears to be derived from PP-2A$_0$ and/or PP-2A$_1$ during purification, through dissociation and/or degradation of the B'- and/or B-subunits. Consequently PP-2A$_0$ and PP-2A$_1$ may be the species that are present *in vivo*. PP-2A$_0$ and PP-2A$_1$ have the subunit structures AB'C$_2$ and ABC$_2$, respectively. The structure of PP-2A$_2$ appears to be AC (*105*).

B. REGULATION OF PROTEIN PHOSPHATASE-1 AND PROTEIN PHOSPHATASE-2A

Protein phosphatase-1 is inhibited by nanomolar concentrations of a protein, termed inhibitor-1 (I-1), which functions as an inhibitor only if it is first phosphorylated by cAMP-dependent protein kinase (*106, 107*). The rate of phosphorylation of I-1 *in vitro* is similar to that of glycogen synthase (*108*). The complete primary structure of I-1 has been determined (*109*) and the site of phosphorylation is Thr-35. The concentration of I-1 in muscle is $\simeq 1.8~\mu M$ (*110*), higher than that of PP-1$_I$ + PP-1$_G$, which is about 0.5 μM (*97, 98*).

The state of phosphorylation of I-1 in skeletal muscle is under hormonal control. Epinephrine increases the level of phosphorylation *in vivo* (*111*) or in the perfused hind limb (*112, 113*). The phosphorylation state of I-1 in the perfused hind limb is exquisitely sensitive to β-adrenergic agonists, half-maximal effects being observed at 1 nM isoproterenol (*113*). A concentration of isoproterenol (0.5 nM), which produces a 40% increase in cAMP, causes a 2-fold rise in the phosphorylation state of I-1 (from 15 to 30%). Both effects are prevented by nanomolar concentrations of insulin, added together with isoproterenol. However, at high levels of isoproterenol, where the level of phosphorylation is $\simeq 70\%$, or in the presence of the β-adrenergic antagonist, L-propranolol, where the level of phosphorylation is $<10\%$, insulin has no effect on cAMP levels or I-1 phosphorylation (*113*).

Although inhibition of the C-subunit of PP-1 by I-1 is instantaneous, inhibition of PP-1$_G$ is not. In the standard assay (30°C, 50 mM tris-Cl, pH 7.0), it is necessary to preincubate PP-1$_G$ and I-1 for at least 10 min before initiating the reaction with substrate, to obtain comparable inhibition to that observed with the C-subunit. If assays are initiated with PP-1$_G$, after preincubating I-1 and substrate, and carried out for only a few minutes, little inhibition is obtained (*98*). This phenomenon probably explains the failure of two laboratories to observe significant inhibition of PP-1 by I-1 in skeletal-muscle extracts (*114*) or glycogen-protein particles (*115*).

The time-dependence of inhibition of PP-1_G by I-1 is greatly decreased if assays are performed at physiological ionic strength (0.15 M KCl) and temperature (37°C) (*98*). Furthermore, the G-subunit of PP-1_G is phosphorylated by cAMP-dependent protein kinase at a comparable rate to glycogen synthase (*98*). Phosphorylation of the G-subunit appears to enhance the rate at which PP-1_G can be inhibited by I-1 (*98*). These findings suggest that inhibition of PP-1_G by I-1 is likely to be of physiological importance. The regulation of PP-1 by I-1 can explain how very high levels of phosphorylase *a* (70–80%) are attained in response to epinephrine, even in resting muscle, where the concentration of Ca^{2+} is <0.1 μM and phosphorylase kinase should exhibit no more than a few percentages of its potential activity (*111*).

Activation of PP-1_I requires preincubation with Mg-ATP and GSK-3 and involves the phosphorylation of threonine-72 on I-2 (Sections V and VI, A). The precise mechanism of activation is unclear, but current evidence suggests that phosphorylation converts the catalytic-subunit to an active conformation (*104*). However, in order to express activity toward exogenous substrates, it appears that I-2 must first be dephosphorylated, either via an Mg^{2+}-dependent intramolecular autodephosphorylation catalyzed by PP-1 itself (*116*), or by the action of other enzymes, such as protein phosphatase-2A (*116a*). The dephosphorylated form of I-2 then induces a change in the catalytic subunit, causing it to gradually revert to the inactive conformation.

Inhibitor-2 can also be phosphorylated *in vitro* by GSK-5 on serine residues (*90, 90b*) and by the insulin-receptor kinase on a tyrosine residue(s) (*49*). Neither phosphorylation triggers the activation of PP-1_I, although phosphorylation by GSK-5 increases 5-fold to 10-fold the rate of phosphorylation of I-2 and activation of PP-1_I catalyzed by GSK-3 (*90, 90b*).

PP-1_I is of considerable interest as the first example of a protein phosphatase that can be activated by a protein kinase. However, whether phosphorylation of I-2 by GSK-3 occurs *in vivo* is unknown, and the physiological role of this activation reaction is unclear. A serious problem is that GSK-3 catalyzes two antagonistic reactions, i.e., the phosphorylation of glycogen synthase and the activation of PP-1_I (which can dephosphorylate glycogen synthase). It could be argued that GSK-3 is regulated by a substance not yet identified, which stimulates its glycogen synthase kinase activity but inhibits its I-2 kinase activity (or vice versa). Alternatively, PP-1_I might not be involved in the regulation of glycogen metabolism at all, and its absence from the glycogen-protein particles is consistent with this proposal. Analysis of the *in vivo* phosphorylation state of I-2 is necessary before the physiological role of PP-1_I can be evaluated.

Protein phosphatase-2A is unaffected by I-1 or I-2 (*91*). However, the activity of the C-subunit in PP-$2A_0$ and PP-$2A_1$ is suppressed by interaction with the A-,

B-, and B'-subunits. Thus removal of the A, B, and B' components by freezing and thawing in the presence of 0.25 M 2-mercaptoethanol, or treatment with 80% ethanol at room temperature, enhances the activity of the C-subunit considerably [reviewed in Ref. (92)].

Basic proteins, such as protamine, polylysine, and histone H1 activate PP-2A$_0$, PP-2A$_1$, and PP-2A$_2$ without dissociation to the C-subunit (105, 117). Protamine stimulates the dephosphorylation of glycogen synthase (phosphorylated by GSK-3) about 7-fold. Similar activation is observed with the isolated C-subunit, but the concentration of protamine needed for half-maximal activation (4 μM) is about 10-fold greater than with PP-2A$_1$ (0.4 μM). The physiological significance of these observations is unclear. Protamine and polylysine do not exist in mammalian cells, and histone H1 is located in the nucleus. It is unknown whether PP-2A is present in the nucleus or whether it would be activated by histone H1 if this protein were complexed with DNA in chromatin. In any event, histone H1 could not be involved in regulating glycogen synthase, a cytoplasmic enzyme. Whether a basic protein capable of activating PP-2A is present in the cytoplasm of skeletal muscle is unknown.

The only basic substance capable of activating PP-2A$_0$ and PP-2A$_1$, that may be of physiological relevance, is spermine. This polyamine stimulates the dephosphorylation of sites-3 8- to 15-fold, and the dephosphorylation of site-2 5- to 7-fold. Half-maximal activation occurs at 0.2 mM spermine, with optimal effects at 1–2 mM. At higher concentrations spermine is inhibitory (Fig. 2). The effects of spermine are not mimicked by concentrations of Mg^{2+} thought to exist *in vivo* (1 mM), nor does Mg^{2+} affect the response to spermine. Spermine is a much more effective activator than spermidine, whereas putrescine is ineffective. Spermine stimulates the dephosphorylation of glycogen synthase to a greater extent than seven other protein substrates tested, while the dephosphorylation of glycogen phosphorylase is inhibited (118).

Spermine also stimulates the dephosphorylation of sites-3 by PP-1 2- to 2.5-fold. Similar activation is obtained with either the isolated C-subunit or PP-1$_G$. Half-maximal activation occurs at 0.1 mM spermine, with optimal effects at 1 mM and inhibition at higher concentrations (Fig. 3). Spermidine is much less effective than spermine, and putrescine is ineffective. However, if glycogen synthase is phosphorylated by phosphorylase kinase, the dephosphorylation of site-2 by PP-1 is inhibited by spermine (Fig. 3). Spermine also inhibits the dephosphorylation of glycogen phosphorylase by PP-1 (118).

The effects of spermine on PP-1 and PP-2A appear to result from interaction of the polyamine with both the protein phosphatases and their substrates (118). The amino acid sequence following sites-3 is extremely acidic (16 out of 31 residues, see Fig. 1) and acidic residues C-terminal to phosphorylation sites may act as negative specificity determinants for protein phosphatases, as discussed in Sec-

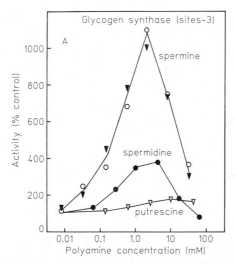

FIG. 2. Effect of different polyamines on the dephosphorylation of glycogen synthase (labeled in sites-3) by PP-2A$_1$. Assays were performed in Tris-Cl pH 7.5 in the presence of spermine (○——○), spermidine (●——●), or putrescine (▽——▽), or in imidazole-Cl pH 7.5 in the presence of spermine (▼——▼). Taken from Ref. (*118*).

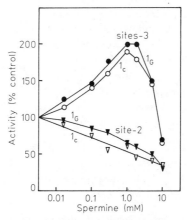

FIG. 3. Effect of spermine in the dephosphorylation of glycogen synthase by PP-1. Open and closed symbols show results with the free catalytic subunit PP-1$_C$ and PP-1$_G$, respectively. Data obtained with glycogen synthase labeled in sites-3 are denoted by circles and with glycogen synthase labeled in site-2 by triangles. Taken from Ref. (*118*).

tion V. Perhaps spermine interacts with this acidic region, thereby neutralizing its inhibitory effects on the dephosphorylation of sites-3.

Protein phosphatase-1 accounts for 65–75% of the glycogen synthase phosphatase activity in skeletal-muscle extracts and PP-2A for the remainder, when assays are performed in the absence of spermine (95). However, in the presence of spermine, PP-2A becomes the major phosphatase acting on glycogen synthase, especially when the enzyme is labeled at sites-3 (118).

VII. Phosphorylation State of Skeletal-Muscle Glycogen Synthase *in Vivo*

It was shown over 20 years ago that the activity ratio of glycogen synthase *in vivo* decreases in response to epinephrine (119) and increases in response to insulin (120). The activity ratio also increases as the glycogen content of the tissue decreases and vice versa (121), an important feedback control by which glycogen regulates the rate and extent to which it is resynthesized. The activity ratio also changes in a complex manner during muscle contraction. When mouse or rat skeletal muscle is tetanized for a few seconds, glycogen phosphorylase is converted to its active phosphorylated state within a second and glycogen levels decline rapidly. Conversely, when electrical stimulation ceases, glycogen phosphorylase is reconverted to its inactive dephosphorylated form within seconds (122, 123). However, the activity ratio of glycogen synthase does not decrease during a muscle tetanus, but rather starts to increase from a resting value of ≈0.25. When contraction ceases, the activity ratio continues to rise for about 5 min reaching a value of 0.8, and then decreases to ≈0.2 over the next 30–60 min (121, 123). The increase in activity ratio during the first 5 min may be a consequence of the depletion of glycogen during a muscle tetanus. Similarly, the progressive decrease in activity ratio after 5 min may reflect the extent to which glycogen has been resynthesized.

In this section, studies of the *in vivo* phosphorylation state of glycogen synthase in resting muscle are reviewed, and the results from different laboratories are compared.

A.   EFFECT OF EPINEPHRINE ON THE PHOSPHATE CONTENT OF GLYCOGEN SYNTHASE *IN VIVO*

In order to determine the *in vivo* phosphorylation state of a protein, it is essential that the tissue be extracted and protein purification performed under conditions that prevent further phosphorylation or dephosphorylation from taking place. This objective is achieved by rapid removal of muscle tissue and homogenization in buffers containing EDTA and sodium fluoride. The EDTA inactivates protein kinases by chelating Mg^{2+}, whereas sodium fluoride inhibits protein phosphatases. The efficacy of this procedure has been validated by measurements of the activity ratio of glycogen synthase and the $A_{0.5}$ for G6P in the muscle extracts, at each step of purification (12), and by alkali-labile phosphate determination (124, 125). In this laboratory, phosphorylation stoichiometry is based on a subunit molecular mass of 86 kDa and an absorbance index, $A^{1\%}_{280nm}$, of 13.4 (16), and corrected for the presence of the 44-kDa protein, which comprises 20% of the material by weight (Section II). This component is not phosphorylated *in vitro* by any glycogen synthase kinase, and it is assumed that it does not contain covalently bound phosphate. Effects of epinephrine are compared to controls performed in an identical manner except that epinephrine is omitted and the β-adrenergic antagonist, L-propranolol, is used to minimize animal-to-animal differences that might be caused by the presence of variable amounts of epinephrine in the circulation. However, omission of propranolol does not alter the kinetic properties significantly (88). In this laboratory epinephrine was observed to decrease the activity ratio from about 0.21 to 0.04, and to increase the $A_{0.5}$ for G6P from about 0.4 to 5.9 mM and the alkali-labile phosphate from about 2.9 to 5.1 mol/subunit (Table I).

Similar results have been obtained by other laboratories. Sheorain *et al.* (126, 127) reported that intravenous injection of epinephrine (66 μg/kg body weight) increases the total phosphate content of the rabbit skeletal-muscle enzyme from 2.35 to 3.85 mol/90-kDa subunit. These values correct to 2.9 and 4.8 mol/86-kDa subunit if allowance is made for the presence of the 44-kDa protein (Section II). In their experiments, glycogen synthase was ashed in the presence of magnesium nitrate, and covalently bound phosphate released from the protein by this treatment measured by reaction with Malachite Green (128). Protein concentrations were determined by the procedure of Lowry *et al.* (129). Chiasson *et al.* (130) perfused rat hind limbs in the presence and absence of 0.1 μM epinephrine. In the absence of hormone, the alkali-labile phosphate content was 3.1 mol/85-kDa subunit, and in the presence of epinephrine, 4.9 mol/subunit. These values correct to 4.0 and 6.5 mol/subunit, respectively, if allowance is made for the 44-kDa protein that comprised 25% of the material in their preparation. Protein was estimated by the method of Bradford (131) and alkali-labile phosphate determined after complexing inorganic phosphate with Malachite Green (128).

B. EFFECT OF INSULIN ON THE PHOSPHATE CONTENT
   OF GLYCOGEN SYNTHASE *IN VIVO*

Effects of insulin on the activity ratio, $A_{0.5}$ for G6P and phosphate content of rabbit skeletal muscle glycogen synthase are summarized in Tables II and III (132). These experiments were carried out using 24-h fasted animals. Similar results have been obtained by Uhing *et al.* in more limited studies using rat

TABLE I

PHOSPHATE CONTENTS OF REGIONS OF GLYCOGEN SYNTHASE ISOLATED FROM SKELETAL MUSCLE OF NORMALLY FED RABBITS[a,b]

| Treatment | Regions of glycogen synthase (mol phosphate/mol peptide) | | | | | | |
	N5–N38	C25–C39	C40–C75	C85–C97	C98–C123	Total[c]	Total[d]
L-Propranolol	0.39 ± 0.08 (14)	1.27 ± 0.10 (17)	0.65 ± 0.06 (5)	0.33 ± 0.07 (11)	0.38 ± 0.05 (13)	3.02	2.92 ± 0.10 (17)
L-Epinephrine	1.03 ± 0.05 (14)	2.43 ± 0.12 (18)	0.65 ± 0.06 (5)	0.59 ± 0.10 (8)	0.65 ± 0.05 (14)	5.35	5.14 ± 0.14 (18)

[a] Data from Parker et al. (88).

[b] The alkali-labile phosphate content of C25–C39 was determined by making use of the differential cleavage by chymotrypsin and trypsin at C23–C24 and C39–C40, respectively. The phosphate contents of other peptides were measured after separation of tryptic peptides by gel-filtration. Values are the means ± SEM; the number of different preparations are given in parentheses.

[c] Sum of phosphate in the individual regions.

[d] Phosphate content of glycogen synthase preparations used for this analysis.

TABLE II

EFFECT OF INSULIN ON ACTIVITY RATIO OF RABBIT SKELETAL MUS-
CLE GLYCOGEN SYNTHASE IN VIVO[a]

Treatment	Activity ratio	$A_{0.5}$ for G6P
Propranolol[b]	0.18 ± 0.02 (11)	1.2 ± 0.1 (11)
Propranolol + insulin[c]	0.35 ± 0.02 (12)	0.6 ± 0.05 (12)
Saline + insulin	0.22 ± 0.03 (3)	1.4 ± 0.1 (3)
Insulin	0.34 ± 0.02 (3)	0.8 ± 0.1 (3)

[a] Data are from Refs. (49) and (132). Values are the means ±
SEM; the number of different preparations are given in parentheses.
[b] 2 mg/kg, 15 min before insulin.
[c] 16.5 μg/kg, 15 and 7.5 min before animals are sacrificed.

skeletal muscle glycogen synthase (133). These results seem definitive, but are
in contrast to Sheorain et al. (127), who reported increases in activity ratios
following insulin administration but did not detect a significant decrease in
phosphate content. The reason for this discrepancy is unclear.

Two laboratories have reported that the activity ratio of glycogen synthase
decreases in rats (135) or rabbits (127, 136) made diabetic with alloxan and that
the total phosphate content of the enzyme increases. These effects are reversed
by twice daily injections of insulin for 4 days (127, 136). In contrast, in this
laboratory, the activity ratio and $A_{0.5}$ for G6P and the alkali-labile phosphate
content of glycogen synthase from alloxan-treated animals were not significantly
different from nondiabetic controls. The reasons for these discrepancies are
unknown, since the animals were clearly diabetic, as judged by their serum
glucose concentrations (132).

C. DISTRIBUTION OF PHOSPHATE WITHIN THE PEPTIDE CHAIN

In order to locate the covalently bound phosphate within the peptide chain,
enzyme preparations phosphorylated in vivo are mixed with traces of glycogen
synthase that have been phosphorylated at defined sites in vitro. Following
digestion of the native enzyme with proteinases, these trace amounts of marker
enzyme enable the purification of particular peptides to be followed and recov-
eries to be calculated, but do not interfere with subsequent analysis of alkali-
labile phosphate. This approach should be valid, provided that in vivo and in
vitro phosphorylated glycogen synthase are cleaved by proteinases at identical
positions and the phosphopeptides do not contain other posttranslational modifications that would alter their elution positions on columns.

TABLE III

PHOSPHATE CONTENTS OF REGIONS OF GLYCOGEN SYNTHASE ISOLATED FROM SKELETAL MUSCLE OF 24-HOUR-STARVED RABBITS[a]

Treatment	Regions of glycogen synthase (mol phosphate/mol peptide)					Total[d]	Total[e]
	N5–N38	C25–C39[b]	C40–C75[c]	C85–C97	C98–C123		
L-Propranolol	0.37 ± 0.05 (7)	1.35 ± 0.05 (13)	0.71 ± 0.06 (9)	0.18 ± 0.02 (5)	0.35 ± 0.05 (7)	2.96	2.74 ± 0.09 (14)
L-Propranolol + insulin	0.31 ± 0.04 (7)	0.90 ± 0.05 (13)	0.82 ± 0.05 (9)	0.12 ± 0.02 (5)	0.29 ± 0.02 (8)	2.44	2.33 ± 0.09 (14)

[a] Data from Parker et al. (132).

[b] The alkali-labile phosphate content of C25–C39 was determined by making use of the differential cleavage by chymotrypsin and trypsin at C23–C24 and C39–C40, respectively. The phosphate contents of other peptides were measured after separation of tryptic peptides by gel-filtration or HPLC. Values are the means ± SEM; the number of different preparations are given in parentheses.

[c] Brief tryptic digestion of native glycogen synthase generates the peptide C40–C75 which can be cleaved to C40–C52 by redigestion with trypsin. The region C52–C75 is devoid of seryl and threonyl residues (Fig. 1).

[d] Sum of phosphate in the individual regions.

[e] Phosphate content of glycogen synthase preparations used for these analyses.

When native glycogen synthase is incubated for a few minutes with low concentrations of trypsin, several peptide bonds in the C-terminal domain are cleaved quantitatively (Fig. 1) and the peptides containing site-1a (C85–C97), site-1b (C98–C123), and site-5 (C40–C75) are released as trichloroacetic acid (TCA)-soluble material. These peptides can be resolved by gel filtration (88, 132) or by HPLC and their content of alkali-labile phosphate determined. The TCA insoluble material can be redigested with trypsin, and the peptide N5–N38 purified to homogeneity by making use of its unusual insolubility in acid, followed by gel filtration (88, 132). If native glycogen synthase is incubated for a few minutes with chymotrypsin, a single peptide bond between C23 and C24 is cleaved quantitatively, releasing the 100-residue peptide C24–C124, which is also soluble in TCA due to its extraordinary hydrophilicity (Fig. 1). The differential effects of trypsin and chymotrypsin can therefore be used as a simple and rapid measurement of the phosphate content of the region C25–C39 containing sites-3 (88, 132).

Table I shows the phosphate contents of various regions of glycogen synthase isolated from normally fed animals after injection of either propranolol or epinephrine. Table III shows the results of a second series of experiments in which 24-hour-starved animals were injected with either propranolol or propranolol plus insulin. In all experiments, the sum of the covalently bound phosphate in the individual regions was in excellent agreement with the total phosphate content of glycogen synthase. These findings suggest that the regions shown in Tables I and III are the only regions phosphorylated to a significant extent in vivo.

Brief tryptic digestion of native glycogen synthase also cleaves the peptide bond between N4 and N5 (Section II), releasing the peptide N1–N4, containing a major site phosphorylated in vitro by CK-I (Section III, G). However, at the time these experiments were performed, the sites phosphorylated by CK-I had not been identified, and hence this peptide was not included in the analyses. The tetrapeptide N1–N4 has since been isolated from glycogen synthase purified after injection of either propranolol or epinephrine, using fast atom bombardment mass spectrometry, and shown to be devoid of phosphate (P. Cohen, B. Gibson, and D. H. Williams, unpublished experiments).

The major conclusions that can be drawn from the results in Tables I and III are the following:

1. Covalently bound phosphate is present in five regions of the protein, N5–N38, C25–C39, C40–C75, C85–C97, and C98–C123. This indicates that at least five residues are phosphorylated in vivo.
2. Twenty-four-hour starvation does not affect either the total phosphate content or the distribution of phosphate between the various regions.
3. Epinephrine increases the phosphate content of N5–N38, C25–C39, C85–C97, and C98–C123, but not C40–C75. The largest increases occur in N5–N38 and C25–C39.

4. Insulin decreases the phosphate content of C25–C39 by 0.4–0.5 mol/sub-
   unit, without a statistically significant change in any other site. However,
   small changes (<0.1 mol phosphate/subunit) at other sites cannot be excluded, because of the limitations of the methodology. A loss of 0.4–0.5
   mol phosphate/subunit from region C25–C39 is consistent with the increase in activity ratio of 0.18 to 0.35 and decrease in $A_{0.5}$ for G6P from
   1.2 to 0.6 mM (*132*).

Sheorain *et al.* (*127, 136*) have also analyzed the distribution of phosphate in
rabbit skeletal-muscle glycogen synthase. In their initial studies (*127*) they measured trypsin-sensitive and trypsin-insensitive phosphate, defined as the proportion released in TCA-soluble form or remaining in the insoluble-TCA precipitate, respectively, after brief tryptic digestion of the native enzyme. However, as
discussed in Ref. (*88*), the problem with this method is that trypsin causes *partial*
cleavage of the peptide bond between C24–C25, the extent of cleavage depending on the concentration of proteinase and time of incubation. Consequently,
region C25–C39 containing sites-3 can be either trypsin-sensitive or trypsin-
insensitive depending on the digestion conditions. The use of both trypsin and
chymotrypsin, and the inclusion of marker enzymes phosphorylated at defined
sites, is necessary to obtain meaningful information (*88*).

In later studies, Sheorain *et al.* (*136*) carried out a total tryptic digestion of
denatured glycogen synthase, and resolved phosphorylated peptides by HPLC
(*137*). Marker enzymes phosphorylated *in vitro* were included to calculate recov-
eries. Covalently bound phosphate was determined by ashing with magnesium
nitrate followed by reaction with Malachite Green. In glycogen synthase isolated
from control animals the phosphate contents (mol/mol) were found to be N5–
N38, 0.43 ± 0.07; C28–C39, 0.46 ± 0.06; C40–C52, 0.62 ± 0.07; C85–C97,
0.29 ± 0.08; C98–C123, 0.23 ± 0.03 (±S.E.M. for 5 preparations). Intravenous injection of epinephrine increased the phosphorylation of N5–N38 2-
fold, C28–C39 1.5-fold, and C98–C123 2-fold, but did not have a significant
effect on the phosphate content of C85–C97 or C40–C52. These results are
similar to those in Table I, except for the much lower phosphate content of C28–
C39, and the failure to observe increased phosphorylation of C85–C97 in response to epinephrine. The reasons for these discrepancies are unclear. However, from the data of Sheorain *et al.* (*136*), the sum of the phosphate in the
different regions can be calculated to be 2.0 mol/subunit in glycogen synthase
from control animals, and about 2.9 mol/subunit in glycogen synthase from
epinephrine-treated animals. These values are much lower than the total phosphate content of glycogen synthase reported in their studies (*127*), which were
2.35 mol/subunit in control animals (2.8 mol/subunit corrected for 20% contamination with the 44-kDa protein) and 3.9 mol/subunit in animals injected with

epinephrine (4.8 mol/subunit corrected for the 44-kDa protein). These differences suggest that the phosphate content of C25–C39 may have been underestimated in their experiments. The covalently bound phosphate in this region is unusually resistant to alkaline hydrolysis (88) due to the large number of prolines in the peptide (Fig. 1). The resistance of phosphoserine and phosphothreonine residues to alkaline hydrolysis when adjacent to proline residues is well documented (138). Whether the phosphoserine residues in region C25–C38 are resistant to ashing in magnesium nitrate as used in their studies is unknown.

Sheorain et al. (136) also reported that in alloxan-diabetic animals the phosphate contents of N5–N38 and C25–C38 were both increased 2-fold compared to controls, without a significant change at any other site.

Hiken and Lawrence (139) incubated rat epitrochlearis skeletal muscles for 5 h at 37°C with ^{32}P-inorganic phosphate. The muscle was extracted with EDTA-fluoride and glycogen synthase isolated by immunoprecipitation. The immunoprecipitates were subjected to SDS-PAGE and the band corresponding to glycogen synthase excised and hydrolyzed in $6M$ HCl for 2 h at 110°C. Electrophoresis at pH 1.9 or 3.5 showed the presence of 8-fold more phosphoserine than phosphothreonine; phosphotyrosine was absent. The detection of a small amount of phosphothreonine is of interest, but the proportion of this phosphoamino acid may have been overestimated because of its greater acid stability. Much of the phosphoserine may have been hydrolyzed to inorganic phosphate; however, the amount of inorganic phosphate generated in these experiments was not reported. The only glycogen synthase kinase capable of phosphorylating threonine residues in rabbit skeletal-muscle glycogen synthase is CK-I. However, the threonines phosphorylated (residue N5 and a further residue(s) in region C98–C123) are only very minor sites of phosphorylation (Section III, G). Furthermore, even the major sites phosphorylated by CK-I in vitro may not be phosphorylated in vivo (Section VIII).

Lawrence et al. (140) incubated rat hemidiaphragms in the presence of ^{32}P-inorganic phosphate. Insulin (25 milliunits/ml) was added to the perfusate for the last 30 min, or epinephrine (10 μM) for the last 10 min. The tissue was frozen and powdered, extracted with EDTA-fluoride, and glycogen synthase immunoprecipitated from the extracts. The material was subjected to SDS-PAGE, and the 90 kDa band corresponding to glycogen synthase cleaved with cyanogen bromide and resubjected to electrophoresis. This analysis revealed two major ^{32}P-labeled peptides, whose mobilities were similar to the N-terminal (CB-N) and C-terminal (CB-C) cyanogen bromide fragments of the rabbit skeletal-muscle enzyme, and were therefore equated with these peptides. Epinephrine decreased the activity ratio of glycogen synthase from 0.24 to 0.16 and increased ^{32}P-radioactivity associated with the enzyme by 50%. Insulin increased the activity ratio from 0.24 to 0.38 and decreased the ^{32}P-radioactivity by 44%. The ratio of ^{32}P-radioactivity CB-C : CB-N was $6.9 \pm 0.7 : 1$ in the absence of added hormones, $3.6 \pm 0.2 : 1$ after incubation with epinephrine and $6.4 \pm 1.0 : 1$ after incubation with insulin. These values are very similar to those calculated for the rabbit skeletal-muscle enzyme from the data in Tables I and III, which are $6.5 : 1$ (control preparations), $4.2 : 1$ (epinephrine), and $7.0 : 1$ (insulin).

From the experiments of Lawrence et al. (140) it can be calculated that 56% of the ^{32}P-radioactivity incorporated in response to epinephrine went into CB-C and 44% into CB-N; ^{32}P-radioactivity in CB-C increased by 35% and that in CB-N by 280%. In response to insulin, 87% of the ^{32}P-radioactivity lost from the enzyme was removed from CB-C and 13% from CB-N. These results are similar to those obtained for the rabbit skeletal-muscle enzyme (Tables I and III).

Lawrence et al. (140) were unable to achieve steady-state labeling of glycogen synthase with ^{32}P$_i$ in their experiments, even though the diaphragms were incubated for 6 h at 37°C and the specific activity of the terminal phosphate of ATP had reached a plateau. Thus the absolute amounts of phosphate and subunit could not be determined in their experiments.

VIII. Interpretation of in Vivo Phosphorylation Experiments

Because of the large number of glycogen synthase kinases, their overlapping site specificities, the presence of two or more potential phosphorylation sites within the same tryptic peptide (Section III), and the limitations of the methodology, the conclusions that can be drawn from the in vivo phosphorylation data in Tables I and III are tentative. The following assumptions have been made to simplify discussion.

1. The levels of cGMP-dependent protein kinase and DG-dependent protein kinase in skeletal muscle are very low compared to most other mammalian tissues (Section III), and in the case of cGMP-dependent protein kinase, the rate of phosphorylation of glycogen synthase is very slow (Section III, I). Furthermore, there are no reports of physiological signals that alter the levels of cGMP or DG in skeletal muscle. It is therefore assumed that these two protein kinases are not involved in the regulation of glycogen synthase in skeletal muscle.

2. The activity of casein kinase-I in skeletal muscle is extremely low. When skeletal-muscle extracts are chromatographed on phosphocellulose and the fractions assayed in the presence of EGTA (to inhibit Ca^{2+}-dependent protein kinases), heparin (to inhibit GSK-5), and the specific protein inhibitor of cAMP-dependent protein kinase, CK-I accounts for no more than 1% of the glycogen synthase kinase activity eluted from the column and GSK-3 (60%) and GSK-4

(40%) account for the remainder [from Ref. (*64*) and unpublished work from this laboratory]. Furthermore, studies have shown that N3, one of the major sites phosphorylated by CK-I *in vitro,* is not phosphorylated *in vivo* (Section VII). In addition, extensive phosphorylation of glycogen synthase by CK-I renders the peptide bond C39–C40 resistant to trypsin (Section III, H), but this bond can be cleaved quantitatively when *in vivo* phosphorylated glycogen synthase is incubated with trypsin (unpublished experiments from this laboratory). For these reasons, it is assumed that CK-I is not a glycogen synthase kinase *in vivo.*

  3. Region C25–C39 and residue C42 are phosphorylated extremely slowly by cAMP-dependent protein kinase. Significant phosphorylation of C25–C39 and C42 requires incubation for 5–15 min with 2.5 μM cAMP-dependent protein kinase [see Ref. (*27*) and Section III, A], about 10-fold higher than the concentration of this protein kinase in skeletal muscle [calculated from the data in Ref. (*141*)]. However, only 30% of the cAMP-dependent protein kinase can be activated *in vivo,* even after injection of high concentrations of epinephrine (*136*); other substrates (e.g., phosphorylase kinase and I-1) compete with glycogen synthase, and glycogen synthase phosphatases capable of dephosphorylating C25–C39 rapidly (i.e., PP-1 and PP-2A) are also present. Furthermore, C25–39 is phosphorylated at least 500-fold more slowly than site-1a (C87) by cAMP-dependent protein kinase *in vitro,* yet is dephosphorylated more rapidly (*132*). If cAMP-dependent protein kinase were a major enzyme acting on C25–C39 *in vivo,* this region should be much less highly phosphorylated than site-1a. Since the reverse is the case (Tables I and III), it is unlikely that the region C25–C39 and residue C42 are phosphorylated by cAMP-dependent protein kinase *in vivo.*

A. Resting Muscle in the Absence of Epinephrine

  If the above assumptions are correct, then the only sites phosphorylated *in vivo* should be site-1a (C87), site-1b (C100), site-2 (N7), sites-3 (C30, C34, C38), and site-5 (C46); and the physiologically relevant protein kinases are cAMP-dependent protein kinase, phosphorylase kinase, the calmodulin-dependent multiprotein kinase, GSK-3, GSK-4, and GSK-5. In this case cAMP-dependent protein kinase would be responsible for the phosphorylation of site-1a, GSK-3 for sites-3, and GSK-5 for site-5. However, site-2 is a target for four protein kinases (Section III) and the enzyme(s) that phosphorylates this residue *in vivo* is difficult to ascertain. Cyclic AMP-dependent protein kinase phosphorylates site-1a 7- to 10-fold faster than site-2 *in vitro* (Section III, B), whereas site-2 is dephosphorylated preferentially by protein phosphatases (Section VI). Therefore if cAMP-dependent protein kinase were the only enzyme acting on site-2 *in vivo,* site-1a should always contain more phosphate than site-2. Since the reverse is the case (Tables I and III), cAMP-dependent protein kinase is unlikely to be the only site-2 kinase *in vivo*. In animals injected with propranolol, the level of phosphorylase *a* is <2% (*88, 132*), indicating that phosphorylase kinase is essentially inactive. The calmodulin-dependent protein kinase should also be inactive in resting muscle, since it requires micromolar concentrations of Ca^{2+} for activity. These considerations suggest that GSK-4 may be responsible for the phosphorylation of site-2 in resting muscle in the absence of epinephrine.

Site-1b is phosphorylated relatively slowly *in vitro* by cAMP-dependent protein kinase and the calmodulin-dependent multiprotein kinase, and dephosphorylated slowly by PP-1 (Section VI). The phosphorylation state of site-1b increases in response to epinephrine (Table I) suggesting that cAMP-dependent protein kinase phosphorylates this site *in vivo*. However, whether the calmodulin-dependent multiprotein kinase also phosphorylates site-1b *in vivo* is unknown (Section VIII, D).

B. RESTING MUSCLE IN THE PRESENCE OF EPINEPHRINE

In our experiments, epinephrine increased the phosphorylation of site-1a and site-1b by 0.25 to 0.3 mol/subunit, site-2 by 0.6 mol/subunit and sites-3 by 1.2 mol/subunit (Table I). Since phosphorylation of site-1a has only a small effect, and site-1b no effect, on the activity ratio (Section IV), the additional phosphorylation at site-2 and sites-3 is largely responsible for inactivation by epinephrine.

Phosphorylation at site-1a and site-1b is presumably catalyzed by cAMP-dependent protein kinase, but the increased phosphorylation at sites-3 is unexpected. The action of epinephrine on glycogen metabolism in skeletal muscle is purely a β-adrenergic effect (*142*), and must therefore be mediated by cAMP. However, the activity of GSK-3 is unaffected by cAMP (*48, 87*). Furthermore, attempts to demonstrate phosphorylation of GSK-3 by cAMP-dependent protein kinase *in vitro,* or activation of GSK-3 following injection of epinephrine *in vivo* have been unsuccessful (unpublished experiments in this laboratory). The possibility that phosphorylation at sites-1a, 1b, and 2 alters the conformation of glycogen synthase in such a way that sites-3 are phosphorylated more rapidly, has also been excluded (*89*).

If the assumption that cAMP-dependent protein kinase does not phosphorylate sites-3 *in vivo* is correct, then there are two possible mechanisms by which epinephrine could increase the phosphorylation of these sites. First, cAMP-dependent protein kinase could inhibit PP-1 through the phosphorylation of I-1 (Section VI, B). Second, the high levels of phosphorylase *a* ($\approx$70%) formed *in vivo* in response to epinephrine [e.g., see Refs. (*88, 142*)] may decrease the rate of dephosphorylation of sites-3 by simple competition for PP-1 and/or PP-2A. In view of the high concentration of phosphorylase *in vivo* (80 μM) relative to glycogen synthase (2.5 μM) this mechanism could be important.

Increased phosphorylation of site-2 in response to epinephrine could occur through (*a*) phosphorylation by cAMP-dependent protein kinase; (*b*) phosphorylation by phosphorylase kinase, which is activated by cAMP-dependent protein kinase; (*c*) inhibition of PP-1 through the phosphorylation of I-1; or (*d*) inhibition of PP-1 and/or PP-2A by phosphorylase *a*. Perhaps all four mechanisms are operative, explaining why site-2 (N5–N38) becomes phosphorylated stoichiometrically in response to epinephrine (Table I).

The high level of phosphorylation of glycogen synthase in the absence of epinephrine ($\simeq$3 mol/mol) makes it extremely difficult to study the effects of low concentrations of epinephrine on glycogen synthase. In the perfused rat hind limb, 0.5 nM isoproterenol doubled the phosphate content of I-1 from 0.15 to 0.3 mol/mol (Section VI, B). However, a similar effect on glycogen synthase might only increase the phosphate content by 5%, which would not register as a significant change in the activity ratio or phosphate content of the enzyme. The ability of insulin to suppress the rise in cAMP produced by epinephrine is observed only in skeletal muscle at low concentrations of β-agonists (*113*). This may explain why antagonism between insulin and epinephrine has not been detected by measuring the activity ratio or phosphate content of glycogen synthase (*143, 144*).

C. The Effect of Insulin

Classical experiments carried out over 20 years ago by Larner established that activation of glycogen synthase by insulin is not a consequence of the increased rate of transport of glucose into muscle, which is also stimulated by the hormone (*119–121*). The results presented in Table III demonstrate that the acute effect of insulin to activate glycogen synthase within minutes is explained by partial dephosphorylation of the region containing sites-3. This indicates that the action of insulin must either involve an inhibition of GSK-3, an activation of PP-1 and/or PP-2A, or a change in the conformation of glycogen synthase, rendering it a poorer substrate for GSK-3 or a more effective substrate for PP-1 and/or PP-2A.

In the experiments shown in Table III, animals were injected with propranolol prior to sacrifice. Under these conditions, the level of phosphorylation of glycogen phosphorylase (<2%) and I-1 (<5%) was negligible in the presence or absence of insulin (*132*). This excludes the possibility that insulin exerts its effect by decreasing the phosphorylation states of these proteins leading to activation of PP-1 and PP-2A.

The results in Table III cannot be explained by a decrease in the concentration of cAMP, or a decrease in the activity of cAMP-dependent protein kinase through some other mechanism. First, insulin does not influence the level of cAMP in skeletal muscle in the absence of epinephrine (*113, 143, 144*). Second, the phosphorylation states of the major sites phosphorylated by cAMP-dependent protein kinase *in vitro* (1a, 1b, and 2) are not decreased significantly by insulin. It could be argued that injection of propranolol selects against detecting an effect of insulin on cAMP-dependent protein kinase, since this β-antagonist decreases cAMP to basal levels, thereby inhibiting the protein kinase even in the absence of insulin. Therefore, if propranolol were omitted, activation of glycogen synthase might occur through two mechanisms, one of which involved inhibition of cAMP-dependent protein kinase. This possibility can be excluded for the following reason. Although propranolol was injected to simplify interpretation of the results, its omission did not affect the activity ratio of glycogen synthase or the $A_{0.5}$ for G6P significantly (Table II).

It is also inconceivable that insulin acts by altering the intracellular concentration of Ca^{2+}. First, the effects of insulin are observed in resting muscle, where the cytoplasmic concentration of Ca^{2+} ($<10^{-7}M$) should be insufficient to activate Ca^{2+}-dependent enzymes. Second, Ca^{2+}-dependent glycogen synthase kinases do not phosphorylate the region affected by insulin. Third, the only known Ca^{2+}-dependent protein phosphatase does not dephosphorylate glycogen synthase (*91, 92*).

The discovery that the insulin receptor is a tyrosine-specific protein kinase (*145*) has led to speculation that the action of insulin might not involve the generation of a "second messenger." The interaction of insulin with the receptor's α-subunit on the outer surface of the plasma membrane activates the tyrosine kinase activity associated with the β-subunit. The tyrosine kinase could then transmit the hormonal signal by phosphorylating serine- or threonine-specific protein kinases and/or protein phosphatases involved in the control of cellular metabolism. However, no evidence to support this hypothesis has been obtained in this laboratory. We have tested the ability of the insulin receptor to phosphorylate GSK-3 and the various forms of PP-1 and PP-2A, and the only protein that can be phosphorylated on a tyrosyl residue is I-2 [Ref. (*49*) and unpublished experiments in this laboratory]. However, unlike phosphorylation by GSK-3, phosphorylation by the insulin receptor does not lead to activation of PP-1$_I$. Furthermore, PP-1$_G$, rather than PP-1$_I$, is likely to be the form of PP-1 involved in the regulation of glycogen synthase *in vivo* (Section VI). To assess whether phosphorylation of I-2 by the insulin receptor has any physiological relevance it will be necessary to determine whether this protein contains significant amounts of phosphotyrosine *in vivo*. Of course, it is possible that the insulin receptor acts more indirectly, first phosphorylating a different protein kinase which then acts on GSK-3, PP-1, or PP-2A. However, evidence for or against such a mechanism is lacking.

The discovery that the insulin receptor is a protein kinase does not exclude the possibility that the rapid metabolic actions of insulin involve the generation of a second messenger. Alternatively, autophosphorylation of the β-subunit of the receptor, which increases its protein kinase activity (146), might allow the receptor to interact with a plasma membrane protein involved in second messenger generation, or to phosphorylate an enzyme that catalyzes the formation of a second messenger.

No mechanism for regulating the activity of GSK-3 has been identified, suggesting that stimulation of glycogen synthase by insulin might be mediated through activation of PP-1 and/or PP-2A. Initially this possibility seemed unlikely because PP-1 and PP-2A dephosphorylate site-2 at a similar rate to sites-3 *in vitro* (91, 132), and yet the effects of insulin are rather specific for sites-3. Furthermore PP-1 and PP-2A are the only phosphorylase phosphatases in skeletal muscle (93, 94), and yet insulin does not decrease the level of phosphorylase *a* in this tissue (119, 132, 143, 144). However, the finding that activation of PP-1 and PP-2A by spermine is targeted relatively specifically toward sites-3 (Figs. 2 and 3) has renewed interest in this possibility. Activation occurs at concentrations of spermine that are present *in vivo*, and is not mimicked by physiological concentrations of Mg^{2+} (1.0 mM).

Several previous observations have suggested a possible role for polyamines in insulin action. For example, when rats are deprived of food for 64 h, the activity of hepatic ornithine decarboxylase, the rate-limiting enzyme in polyamine biosynthesis from ornithine, is decreased over 100-fold (147). The level can be restored by refeeding for 3 h, but is blocked by subcutaneous injection of diaminopropranol. Diaminopropranol also suppresses the 2- to 5-fold increase in hepatic lipogenesis and activation (dephosphorylation) of pyruvate kinase that occurs on refeeding (147). These observations indicate that the level of ornithine decarboxylase activity correlates with some of the longer-term actions of insulin. Similarly, the rapid metabolic effects of insulin on glucose transport and utilization in adipocytes can be reproduced by adding spermine to the incubation medium (148–150). In addition, it has been reported that insulin's activation of glycogen synthase in isolated rat diaphragm muscle can be mimicked by the addition of 2 mM putrescine or spermidine (the precursors of spermine) to the perfusate (151).

The mitochondrial pyruvate dehydrogenase complex, like glycogen synthase, is dephosphorylated and activated in response to insulin [reviewed in Ref. (152)]. Damuni *et al.* (153) have reported that pyruvate dehydrogenase phosphatase, a Mg^{2+}-dependent enzyme distinct from PP-1 and PP-2A (92), can be activated by spermine and spermidine. The extent of activation increases with decreasing concentrations of Mg^{2+}, reaching 20- to 30-fold at 0.3 mM Mg^{2+} (153).

Spermine is present *in vivo* in the millimolar concentration range, but is mostly bound to nucleic acids and proteins (154, 155). If spermine is a second messenger for insulin, then interaction of the hormone with its receptor would have to (a) trigger the activation of ornithine decarboxylase or a later enzyme in the pathway of spermine biosynthesis, or (b) induce the release of spermine from a bound store, or (c) increase the sensitivity of PP-1 and/or PP-2A to activation by spermine. Insulin might also be able to stimulate the transport of polyamines into mitochondria thereby stimulating pyruvate dehydrogenase phosphatase. These suggestions are, of course, extremely speculative, and clearly much more experimental work is required to evaluate this hypothesis. Another possibility is that spermine is merely mimicking the action of another basic substance that is the real second messenger.

D. MUSCLE CONTRACTION

Phosphorylase kinase and the calmodulin-dependent multiprotein kinase require micromolar concentrations of Ca^{2+} for activity. It is therefore assumed that they are involved in suppressing the activity of glycogen synthase during muscle contraction, but the experiments needed to test this hypothesis have not been performed. In particular, it will be necessary to examine whether the phosphorylation of site-2 increases during contraction. To assess which protein kinase is responsible for the phosphorylation of site-2, it will be necessary to perform the experiments using I-strain mice that lack skeletal-muscle phosphorylase kinase activity (121, 122) and a control strain with normal activity. However, due to elevated levels of glycogen, the activity ratio of glycogen synthase in skeletal muscle of I-strain mice is much lower than in control strains (121, 122), and site-2 might already be highly phosphorylated, even in resting muscle. Also brief tetanic stimulation does not decrease the activity ratio of glycogen synthase in vivo (Section VII). Nevertheless, it is possible that phosphorylation of site-2 is counterbalanced by dephosphorylation of the C-terminal region (C25–C39?) resulting from rapid depletion of glycogen during tetanic contraction.

The concentration of glycogen phosphorylase in adult rabbit skeletal muscle (80 μM) is 30-fold higher than the concentration of glycogen synthase (2.5 μM) (12). At these concentrations, glycogen phosphorylase is used preferentially by phosphorylase kinase in vitro (36), suggesting that the calmodulin-dependent multiprotein kinase may be the more important glycogen synthase kinase during the initial phase of contraction.

Oron et al. (156) perfused mouse diaphragms in the presence of EGTA to deplete intracellular Ca^{2+}, and detected a small increase in the activity ratio and $A_{0.5}$ for G6P, as compared to control incubations where Ca^{2+} was present in the perfusate. This could be taken as preliminary evidence for involvement of a Ca^{2+}-dependent protein kinase in the regulation of glycogen synthase in this muscle.

E. Regulation by Glycogen

The activity ratio of glycogen synthase increases as the glycogen content of the tissue decreases, or vice versa (Section VII), suggesting that glycogen either activates a glycogen synthase kinase or inhibits a glycogen synthase phosphatase. Several years ago it was reported that relatively impure preparations of glycogen synthase phosphatase were inhibited by glycogen [reviewed in Ref. (157)]. However, provided glycogen is treated to remove nucleotides and other charged molecules, we have been unable to detect any effect of commercial liver glycogen on the rate of dephosphorylation of glycogen synthase (phosphorylated at site-2 or sites-3) by homogeneous preparations of PP-1 or PP-2A (unpublished experiments from this laboratory); nor does glycogen increase the activity of any glycogen synthase kinase (89). The molecular basis for the effect of glycogen on the activity ratio is therefore unclear. Perhaps the explanation lies in the molecular structure of the glycogen-protein particle itself, and that phosphorylation sites on glycogen synthase become more accessible to protein phosphatases as glycogen is depleted.

The regulation of glycogen synthase exerted by glycogen probably accounts for the lack of effect of 24-hour starvation on the activity ratio or phosphorylation state of glycogen synthase. Thus a decrease in muscle glycogen during starvation (127), would tend to be counterbalanced by the decrease in the serum insulin, and vice versa. The effects of insulin on glycogen synthase may therefore be readily detected only by the acute administration of this hormone.

References

1. Krebs, E. G., and Fischer, E. H. (1956). *BBA* **20**, 150–157.
2. Krebs, E. G., Graves, D. J., and Fischer, E. H. (1959). *JBC* **234**, 2867–2873.
3. Friedman, D. L., and Larner, J. (1963). *Biochemistry* **2**, 669–675.
4. Rosell-Perez, M., and Larner, J. (1964). *Biochemistry* **3**, 81–88.
5. Huijing, F., and Larner, J. (1966). *BBRC* **23**, 259–263.
6. Schlender, K. K., Wei, S. H., and Villar-Palasi, C. (1969). *BBA* **191**, 272–278.
7. Soderling, T. R., Hickenbottom, J. P., Reimann, E. M., Hunkeler, F. L., Walsh, D. A., and Krebs, E. G. (1970). *JBC* **245**, 6317–6328.
8. Smith, C. H., Brown, N. E., and Larner, J. (1971). *BBA* **242**, 81–88.
9. Nimmo, H. G., and Cohen, P. (1974). *FEBS Lett.* **47**, 162–167.
10. Huang, K. P., Huang, F. L., Glinsmann, W. H., and Robinson, J. C. (1975). *BBRC* **65**, 1163–1169.
11. Schlender, K. K., and Reimann, E. M. (1975). *PNAS* **72**, 2197–2201.
12. Cohen, P. (1978). *Curr. Top. Cell. Regul.* **14**, 117–196.
13. Rylatt, D. B., Aitken, A., Bilham, T., Condon, G. D., Embi, N., and Cohen, P. (1980). *EJB* **107**, 529–537.
14. Picton, C., Aitken, A., Bilham, T., and Cohen, P. (1982). *EJB* **124**, 37–45.
15. Huang, T. S., and Krebs, E. G. (1979). *BBRC* **75**, 643–650.

16. Nimmo, H. G., Proud, C. G., and Cohen, P. (1976). *EJB* **68**, 21–30.
17. Huang, T. S., and Krebs, E. G. (1979). *FEBS Lett.* **98**, 66–70.
18. Woodgett, J. R., and Cohen, P. (1984). *BBA* **788**, 339–347.
19. Soderling, T. R. (1976). *JBC* **251**, 4359–4364.
20. Takeda, Y., and Larner, J. (1975). *JBC* **250**, 8951–8956.
21. Huang, K. P., Huang, F. L., Glinsmann, W. H., and Robinson, J. C. (1976). *ABB* **173**, 162–170.
22. Salavert, A., Roach, P. J., and Guinovart, J. J. (1979). *BBA* **570**, 231–238.
23. Embi, N., Rylatt, D. B., and Cohen, P. (1979). *EJB* **100**, 339–347.
24. Srivastava, A. K., Waisman, D. M., Broström, C. O., and Soderling, T. R. (1979). *JBC* **254**, 583–586.
25. Soderling, T. R. (1975). *JBC* **250**, 5407–5412.
26. Embi, N., Parker, P. J., and Cohen, P. (1981). *EJB* **115**, 405–413.
27. Sheorain, V. S., Corbin, J. D., and Soderling, T. R. (1985). *JBC* **260**, 1567–1572.
28. Roach, P. J., DePaoli-Roach, A. A., and Larner, J. (1978). *J. Cyclic Nucleotide Res.* **4**, 245–257.
29. DePaoli-Roach, A. A., Roach, P. J., and Larner, J. (1979). *JBC* **254**, 4212–4219.
30. Soderling, T. R., Srivastava, A. K., Bass, Y. A., and Khatra, B. S. (1979). *PNAS* **76**, 2536–2540.
31. Walsh, K. Y., Millikin, D. M., Schlender, K. K., and Reimann, E. M. (1979). **254**, 6611–6616.
32. Soderling, T. R., Sheorain, V. S., and Ericsson, L. H. (1979). *FEBS Lett.* **106**, 181–184.
33. Cohen, P., Yellowlees, D., Aitken, A., Donella-Deana, A., Hemmings, B. A., and Parker, P. J. (1982). *EJB* **124**, 21–35.
34. Rylatt, D. B., Embi, N., and Cohen, P. (1979). *FEBS Lett.* **98**, 76–80.
35. Payne, M. E., and Soderling, T. R. (1980). *JBC* **255**, 8054–8056.
36. Woodgett, J. R., Tonks, N. K., and Cohen, P. (1982). *FEBS Lett.* **148**, 5–11.
37. Woodgett, J. R., Davison, M. T., and Cohen, P. (1983). *EJB* **136**, 481–487.
38. McGuinness, T. L., Lai, Y., Greengard, P., Woodgett, J. R., and Cohen, P. (1983). *FEBS Lett.* **163**, 329–334.
39. Woodgett, J. R., Cohen, P., Yamauchi, T., and Fujisawa, H. (1984). *FEBS Lett.* **170**, 49–54.
39a. Lai, Y., Nairn, A. C., and Greengard, P. (1986). *PNAS* (in press).
40. Pearson, R. B., Woodgett, J. R., Cohen, P., and Kemp, B. E. (1985). *JBC* **260**, 14471–14476.
41. McGuinness, T. L., Lai, Y., and Greengard, P. (1985). *JBC* **260**, 1696–1704.
41a. Yamauchi, T., and Fujisawa, H. (1986). *BBA* **886**, 57–63.
42. Ahmad, Z., DePaoli-Roach, A. A., and Roach, P. J. (1982). *JBC* **257**, 8348–8355.
43. Payne, M. E., Schworer, C. M., and Soderling, T. R. (1983). *JBC* **258**, 2376–2382.
44. Schworer, C. M., Payne, M. E., Williams, A. T., and Soderling, T. R. (1983). *ABB* **224**, 77–86.
45. Schworer, C. M., and Soderling, T. R. (1983). *BBRC* **116**, 412–416.
46. Hemmings, B. A., Aitken, A., Cohen, P., Rymond, M., and Hofmann, F. (1982). *EJB* **127**, 473–481.
47. Aitken, A., Holmes, C. F. B., Campbell, D. G., Resink, T. J., Cohen, P., Leung, C. T. W., and Williams, D. H. (1984). *BBA* **790**, 288–291.
48. Hemmings, B. A., Yellowlees, D., Kernohan, T. C., and Cohen, P. (1981). *EJB* **119**, 443–451.
49. Cohen, P., Parker, P. J., and Woodgett, J. R. (1985). *In* "Molecular Basis of Insulin Action" (M. Czech, ed.), pp. 213–233. Plenum, New York.
50. Vandenheede, J. R., Yang, S. D., Goris, J., and Merlevede, W. (1980). *JBC* **255**, 11768–11774.

51. Ahmad, Z., DePaoli-Roach, A. A., and Roach, P. J. (1985) *FEBS Lett.* **179**, 96–100.
51a. Ahmad, Z., Lee, F. T., DePaoli-Roach, A. A., and Roach, P. J. (1985). 13th International Congress of Biochemistry, Abstract FR-348 p. 728.
52. DePaoli-Roach, A. A., Ahmad, Z., Camici, M., Lawrence, J. C., and Roach, P. J. (1983). *JBC* **258**, 10702–10709.
52a. Ramakrishna, S., and Benjamin, W. B. (1985). *JBC* **260**, 12280–12286.
52b. Sheorain, V. S., Ramakrishna, S., Benjamin, W. B., and Soderling, T. R. (1985). *JBC* **260**, 12287–12292.
53. DePaoli-Roach, A. A., Roach, P. J., and Larner, J. (1979). *JBC* **254**, 12062–12068.
54. DePaoli-Roach, A. A., Ahmad, Z., and Roach, P. J. (1981). *JBC* **256**, 8955–8962.
55. Hathaway, G., and Traugh, J. A. (1982). *Curr. Top. Cell. Regul.* **21**, 101–127.
56. Huang, K. P., Itarte, E., Singh, T. J., and Akatsuka, A. (1982). *JBC* **257**, 3236–3242.
57. Ahmad, Z., DePaoli-Roach, A. A., and Roach, P. J. (1982). *JBC* **257**, 5873–5876.
58. Feige, J., Cochet, C., Pirollet, F., and Chambaz, E. M. (1983). *Biochemistry* **22**, 1452–1459.
59. Pinna, L. A., Meggio, F., and Deduikina, M. M. (1981). *BBRC* **100**, 449–454.
60. Pearlston, J. R., Carpenter, M. R., Johnson, P., and Smillie, L. B. (1976). *PNAS* **73**, 1902–1906.
61. Traugh, J. A. (1981). *Biochem. Actions Horm.* **8**, 167–207.
62. DePaoli-Roach, A. A., Roach, P. J., Pham, K., Kramer, G., and Hardesty, B. (1981). *J. Biol. Chem.* **256**, 8871–8874.
63. Pinna, L. A., Meggio, F., Marchiori, F., and Borin, G. (1984). *FEBS Lett.* **171**, 211–214.
64. Kuenzel, E. A., and Krebs, E. G. (1985). *PNAS* **82**, 737–741.
65. Kuret, J., Woodgett, J. R., and Cohen, P. (1985). *EJB* **151**, 39–48.
66. Hathaway, G., and Traugh, J. A. (1979). *JBC* **254**, 762–768.
67. Itarte, E., Robinson, J. C., and Huang, K. P. (1977). *JBC* **252**, 1231–1234.
68. Itarte, E., and Huang, K. P. (1979). *JBC* **254**, 4052–4057.
69. Huang, K. P., Lee, S. L., and Huang, F. L. (1979). *JBC* **254**, 9867–9870.
70. Huang, K. P., Singh, T. J., Akatsuke, A., Shapiro, S. G., Vandenheede, J. R., and Merlevede, W. (1984). *ABB* **232**, 111–117.
71. Ahmad, Z., Camici, M., DePaoli-Roach, A. A., and Roach, P. J. (1984). *JBC* **259**, 3420–3428.
72. Cohen, P., Watson, D. C., and Dixon, G. H. (1975). *EJB* **51**, 79–92.
73. Yeaman, S. J., and Cohen, P. (1975). *EJB* **51**, 93–104.
74. Proud, C. G., Rylatt, D. B., Yeaman, S. J., and Cohen, P. (1977). *FEBS Lett.* **80**, 435–442.
75. Singh, T. J., Akatsuka, A., Huang, K. P., Sharma, R. K., Tam, S. W., and Wang, J. H. (1982). *BBRC* **107**, 676–683.
76. Singh, T. J., Akatsuka, A., and Huang, K. P. (1982). *JBC* **257**, 13379–13384.
77. Singh, T. J., Akatsuka, A., Huang, K. P., Murthy, A. S. N., and Flavin, M. (1984). *BBRC* **121**, 19–26.
78. Takai, Y., Kishimoto, A., and Nishizuka, Y. (1982). *In* "Calcium and Cell Function" (W. Y. Cheung, ed.), Vol. 2, pp. 385–412. Academic Press, New York.
79. Lincoln, T. M., and Corbin, J. D. (1977). *PNAS* **74**, 3239–3243.
80. Kaibuchi, K., Kikkawa, U., Takai, Y., and Nishizuka, Y. (1984). *Mol. Aspects Cell. Regul.* **3**, 81–94.
81. Kuo, J. F., Andersson, R. G. G., Wise, B. C., Mackerlova, L., Salomonsson, I., Brackett, N. L., Katoh, N., Shoji, M., and Wrenn, R. W. (1980). *PNAS* **77**, 7039–7043.
82. Nishizuka, Y., and Takai, Y. (1981). *Cold Spring Harbor Conf. Cell Proliferation* **8**, 237–249.
83. Ahmad, Z., Lee, F. T., DePaoli-Roach, A. A., and Roach, P. J. (1984). *JBC* **259**, 8743–8747.
84. Piras, R., Rothman, L. B., and Cabib, E. (1968). *Biochemistry* **7**, 56–66.

85. Roach, P. J., Takeda, Y., and Larner, J. (1976). *JBC* **251,** 1913–1919.
86. Roach, P. J., and Larner, J. (1976). *JBC* **251,** 1920–1925.
87. Embi, N., Rylatt, D. B., and Cohen, P. (1980). *EJB* **107,** 519–527.
88. Parker, P. J., Embi, N., Caudwell, F. B., and Cohen, P. (1982). *EJB* **124,** 47–55.
89. Picton, C., Woodgett, J. R., Hemmings, B. A., and Cohen, P. (1982). *FEBS Lett.* **150,** 191–196.
90. DePaoli-Roach, A. A. (1984). *JBC* **259,** 12144–12152.
90a. Holmes, C. F. B., Campbell, D. G., Caudwell, F. B., Aitken, A., and Cohen, P. (1986). *EJB* **155,** 173–182.
90b. Holmes, C. F. B., Kuret, J., Chisholm, A. A., and Cohen, P. (1986). *BBA* **870,** 408–416.
91. Ingebritsen, T. S., Foulkes, J. G., and Cohen, P. (1983). *EJB* **132,** 255–261.
92. Ingebritsen, T. S., and Cohen, P. (1983). *Science* **221,** 331–338.
93. Alemany, S., Tung, H. Y. L., and Cohen, P. (1984). *EJB* **145,** 51–56.
94. Ingebritsen, T. S., and Cohen, P. (1983). *EJB* **132,** 263–274.
95. Ingebritsen, T. S., Stewart, A. A., and Cohen, P. (1983). *EJB* **132,** 297–307.
96. Silberman, S. R., Speth, M., Nemani, R., Ganapathi, M. K., Dombradi, V., Paris, H., and Lee, E. Y. C. (1984). *JBC* **254,** 4406–4412.
97. Tung, H. Y. L., Resink, T. J., Hemmings, B. A., Shenolikar, S., and Cohen, P. (1984). *EJB* **138,** 635–641.
98. Stralfors, P., Hiraga, A., and Cohen, P. (1985). *EJB* **149,** 295–303.
99. Hemmings, B. A., Resink, T. J., and Cohen, P. (1982). *FEBS Lett.* **150,** 319–324.
100. Resink, T. J., Hemmings, B. A., Tung, H. Y. L., and Cohen, P. (1983). *EJB* **133,** 455–461.
101. Ballou, L. M., Broutigan, D. L., and Fischer, E. H. (1983). *Biochemistry* **22,** 3393–3399.
102. Tung, H. Y. L., and Cohen, P. (1984). *EJB* **145,** 57–64.
103. Goris, J., Defreyn, G., and Merlevede, W. (1979). *FEBS Lett.* **99,** 279–282.
104. Ballou, L. M., and Fischer, E. H. (1986). This volume, Chapter 8.
105. Tung, H. Y. L., Alemany, S., and Cohen, P. (1985). *EJB* **148,** 253–263.
106. Huang, F. L., and Glinsmann, W. H. (1976). *EJB* **70,** 419–426.
107. Nimmo, G. A., and Cohen, P. (1978). *EJB* **87,** 341–351.
108. Cohen, P., Rylatt, D. B., and Nimmo, G. A. (1977). *FEBS Lett.* **76,** 182–186.
109. Aitken, A., Bilham, T., and Cohen, P. (1982). *EJB* **124,** 235–246.
110. Foulkes, J. G., and Cohen, P. (1980). *EJB* **105,** 195–203.
111. Foulkes, J. G., and Cohen, P. (1979). *EJB* **97,** 251–256.
112. Khatra, B. S., Chiasson, J., Shikama, H., Exton, J. H., and Soderling, T. R. (1980). *FEBS Lett.* **114,** 253–256.
113. Foulkes, J. G., Cohen, P., Strada, S. J., Everson, W. V., and Jefferson, L. S. (1982). *JBC* **257,** 12493–12496.
114. Laloux, M., and Hers, H. G. (1979). *FEBS Lett.* **105,** 239–243.
115. Khatra, B. S., and Soderling, T. R. (1983). *ABB* **227,** 39–51.
116. Li, H. C., Price, D. J., and Tabarini, D. (1985). *JBC* **260,** 6416–6426.
116a. Tonks, N. K., and Cohen, P. (1984). *EJB* **145,** 65–70.
117. Pelech, S., and Cohen, P. (1985). *EJB* **148,** 245–251.
118. Tung, H. Y. L., Pelech, S., and Cohen, P. (1985). *EJB* **149,** 305–313.
119. Craig, J. W., and Larner, J. (1964). *Nature (London)* **202,** 971–973.
120. Villar-Palasi, C., and Larner, J. (1960). *BBA* **39,** 171–173.
121. Danforth, W. H. (1965). *JBC* **240,** 588–593.
122. Danforth, W. H., and Lyon, J. B. (1964). *JBC* **239,** 4047–4050.
123. Piras, R., and Staneloni, R. (1969). *Biochemistry* **8,** 2153–2160.
124. Ames, B. N. (1966). "Methods in Enzymology," Vol. 8, pp. 115–118.
125. Guy, P. S., Cohen, P., and Hardie, D. G. (1981). *EJB* **114,** 399–405.

126. Sheorain, V. S., Khatra, B. S., and Soderling, T. R. (1981). *FEBS Lett.* **127**, 94–96.
127. Sheorain, V. S., Khatra, B. S., and Soderling, T. R. (1982). *JBC* **257**, 3462–3470.
128. Stull, J. T., and Buss, J. E. (1977). *JBC* **252**, 851–857.
129. Lowry, O. H., Rosebrough, N. J., Farr, A. L., and Randall, R. J. (1951). *JBC* **193**, 265–275.
130. Chiasson, J. L., Aylward, J. H., Shikama, H., and Exton, J. H. (1981). *FEBS Lett.* **127**, 97–100.
131. Bradford, M. M. (1976). *Anal. Biochem.* **72**, 248–254.
132. Parker, P. J., Caudwell, F. B., and Cohen, P. (1983). *EJB* **130**, 227–234.
133. Uhing, R. J., Shikama, H., and Exton, J. H. (1981). *FEBS Lett.* **134**, 185–188.
134. Roach, P. J., Rosell-Perez, M., and Larner, J. (1977). *FEBS Lett.* **80**, 95–98.
135. Komuniecki, P. R., Kochan, R. G., Schlender, K. K., and Reimann, E. M. (1982). *Mol. Cell. Biochem.* **48**, 129–134.
136. Sheorain, V. S., Juhl, H., Bass, M., and Soderling, T. R. (1984). *JBC* **259**, 7024–7030.
137. Juhl, H., Sheorain, V. S., Schworer, C. M., Jeff, M. F., and Soderling, T. R. (1983). *ABB* **222**, 518–526.
138. Kemp, B. E. (1980). *FEBS Lett.* **110**, 308–312.
139. Hiken, J. F., and Lawrence, J. C. (1984). *FEBS Lett.* **175**, 55–58.
140. Lawrence, J. C., Hiken, J. F., DePaoli-Roach, A. A., and Roach, P. J. (1983). *JBC* **258**, 10710–10719.
141. Beavo, J. A., Bechtel, P. J., and Krebs, E. G. (1974). "Methods in Enzymology," Vol. 39, pp. 299–308.
142. Dietz, M. R., Chiasson, J. L., Soderling, T. R., and Exton, J. H. (1980). *JBC* **255**, 2301–2307.
143. Chiasson, J. L., Dietz, M. R., Shikama, H., Wootten, M., and Exton, J. H. (1980). *Am. J. Physiol.* **239**, E69–E74.
144. Shikama, H., Chiasson, J. L., and Exton, J. H. (1981). *JBC* **256**, 4450–4454.
145. Kasuga, M., Zick, Y., Blithe, D. L., Crettaz, M., and Kahn, C. R. (1982). *Nature (London)* **298**, 667–669.
146. Rosen, O. M., Herrera, R., Olowe, Y., Petruzzelli, L. M., and Cobb, M. H. (1983). *PNAS* **80**, 3237–3240.
147. Brosnan, M. E., and Hu, W. Y. (1983). *Adv. Polyamine Res.* **4**, 563–569.
148. Lockwood, D. H., and East, L. E. (1974). *JBC* **249**, 7717–7722.
149. Livingston, J. N., Gurny, P. A., and Lockwood, D. H. (1977). *JBC* **252**, 560–562.
150. Czech, M. P., Lawrence, J. C., and Lynn, W. S. (1974). *JBC* **249**, 5421–5427.
151. Huang, L. C., and Chang, L. Y. (1980). *BBA* **6?3**, 106–115.
152. Denton, R. M., Brownsey, R. W., and Belsham, G. J. (1981). *Diabetologia* **21**, 347–362.
153. Damuni, Z., Humphreys, J. S., and Reed, L. J. (1984). *BBRC* **124**, 95–99.
154. Tabor, C. W., and Tabor, H. (1976). *Annu. Rev. Biochem.* **45**, 285–306.
155. Tabor, C. W., and Tabor, H. (1984). *Annu. Rev. Biochem.* **53**, 749–790.
156. Oron, Y., Emerson, J. F., Garrison, J. C., and Larner, J. (1984). *Cell Calcium* **5**, 143–158.
157. Curnow, R. T., and Larner, J. (1979). *Biochem. Actions Horm.* **6**, 77–119.

12

Liver Glycogen Synthase

PETER J. ROACH

Department of Biochemistry
Indiana University School of Medicine
Indianapolis, Indiana 46223

THE ENZYMES, Vol. XVII

I. Introduction

Glycogen is widely distributed in nature and functions as a stored reserve of glucose monomers. Eukaryotic glycogen synthase (EC 2.4.1.11), the enzyme responsible for the formation of the α-1,4-glycosidic linkages of glycogen, utilizes UDPglucose as the glucosyl donor and is subject to covalent phosphorylation. In vertebrates, most cells contain glycogen but, in absolute amounts, the liver and skeletal muscle represent the major deposits of the polysaccharide. The physiological role of glycogen in these two tissues differs, however.

Primary utilization of skeletal muscle glycogen is by the muscle cells themselves, either to provide or to supplement, depending on muscle type, the energetic requirements of muscular activity. Only indirectly, via the formation of the gluconeogenic precursors lactate and alanine, does muscle glycogen contribute to the homeostasis of blood glucose. The main signals for glycogen mobilization in skeletal muscle are neural or through stimulation of β-adrenergic receptors by epinephrine. Muscle glycogen synthesis is under the control of insulin.

The predominant fate of hepatic glycogen is not as a metabolic energy source for liver cells but rather to supply glucose to the circulation for utilization by other tissues. Liver glycogen metabolism is tuned to the nutritional status of the organism as reflected in the levels of insulin, glucagon, and glucose in the portal vein. Relevant to this chapter, elevated concentrations of insulin and glucose promote glycogen synthesis and the activation of glycogen synthase (Section VI,D). Considerable interest has addressed the possibility that liver glycogen synthesis during recovery from fasting does not involve direct conversion of glucose to glycogen but passage through 3-carbon intermediates (1, 2). However, irrespective of the past history of the carbons of the UDPglucose precursor, the effect of glucose to stimulate glycogen synthesis is likely to be an important regulatory interaction. Fasting rapidly depletes liver glycogen and the primary positive stimulus is glucagon (Section VI,B). Liver glycogen can be mobilized by other hormones, such as epinephrine, vasopressin, and angiotensin II (Section VI,C). All of these glycogenolytic hormones suppress glycogen formation by inactivating glycogen synthase.

The short-term control of liver glycogen metabolism by extracellular signals has the potential to be more complex than in muscle and, while many common features can be perceived, there are significant differences. These differences in control can reflect specializations at several levels in the signal relay system linking receptor activation to the regulation of intracellular targets such as glycogen synthase. Obviously, the set of hormone receptors expressed in the liver cell provides a primary determinant of how metabolism is controlled. Thus, for example, glucagon controls liver, but not muscle, glycogen metabolism. However, evolved and specific characteristics of intracellular elements of the regulatory chain may also be important. It is now clear that liver and muscle glycogen synthases, though closely related proteins, are distinct isozymes, differing in primary structure and therefore coded by separate genes.

This chapter attempts to review knowledge of mammalian liver glycogen synthase and its short-term regulation by covalent phosphorylation mechanisms. This review is a complement to the excellent accompanying Chapter 11 by Cohen (3) which addresses the muscle isozyme, an enzyme for which more molecular detail is available. The chapter by Cohen should be viewed as a primary resource for information common to the different isozymes but, despite efforts to avoid redundancy, some duplication is inescapable. Any comparative discussion is forced to use rabbit muscle glycogen synthase as the point of reference. This chapter also concentrates on studies that are seeking to advance our understanding of the enzymology and regulation of liver glycogen synthase at a molecular level. Coverage of the earlier development of the field is to be found in previous reviews (4–6).

II. Liver Glycogen Synthase

A. PURIFICATION

Most recent enzymological studies of liver glycogen synthases have been of the enzymes from rabbits or rats (Table I) (7–17). The choice of rabbit liver, besides the enzymologist's predilection for the species, has the advantage of allowing strict comparison, within species, with the well-studied rabbit muscle enzyme. The choice of rat liver has become important in understanding physiological regulation since a large body of work on hormonal control of liver glycogen synthase has used rat hepatocytes. Most often, 1500- to 3000-fold purification was required to obtain enzyme close to homogeneity. Preparations have sought either to maintain endogenous phosphate (D- or b-form) or else to remove it by fostering phosphatase activity at a stage of partial purification (I- or a-form). Isolation of such "dephosphorylated" enzyme from rabbit liver required greater apparent purification (15, 16) than phosphorylated enzyme but this is probably due to the effect of dephosphorylation to activate the enzyme in the standard assay. Specific activities for the purified enzymes were 4–44 μmol/min/mg for rabbit and 2–60 μmol/min/mg for rat (normalizing some rates to 30°C). Judged from gel electrophoretic analyses of the preparations of Table I, this degree of variability cannot be explained by differences in purity. Unappreciated differences in assay conditions may be one factor. The effect of phosphorylation state is another. Another important variable is limited proteolysis.

Purification of liver glycogen synthase has proved more difficult than for the

TABLE I

Preparations of Mammalian Liver Glycogen Synthase

Species	-Fold purification	Specific activities (units/mg)	Activity ratio[a]	Subunit M_r[b]	Native M_r	Phospho- or dephospho-[c]	Phosphates/ subunit[d]	$S_{0.5}$ UDPglucose (mM)	Reference
Rat	2000	3.5[e]	0	85,000	260,000	P	12.4 (17.3)	—	(7)
	1400	22[e]	—	77,000–80,000	310,000	P	—	0.5	(8)
	1300	29	0.98	85,000 ± 5,000	156,000–171,000	deP	—	—	(9)
	700–800	5–8[e]	—	77,000 ± 3,000	—	P	—	0.28–0.56	(10)
	—	50–60	0.95–1.00	85,000, (80,000)	—	deP	< 0.2	—	(11)
	3193	47 ± 4	0.96	87,000 ± 2,000	—	deP	0.2–0.5	0.6	(12)
	8000	15–16	0.83–0.96	85,000, (76,000)	—	deP	—	—	(13)
Rabbit	1800	25	—	85,000	170,000–180,000	P	—	—	(14)
	1473	3–5	0.02–0.05	90,000	—	P	5.2	2.1	(15,16)
	3054	11–17	0.9–1.1	85,000	—	deP	< 0.6	1.3	(16)
	2056	39 ± 2	0.03	78,000	—	P	5.7 (17.3)	31	(17)
	2684	44 ± 4	0.55	85,000	—	deP	2.9 (13.4)	0.08	(17)

[a] Ratio of activity measured without glucose-6-P: activity in presence of glucose-6-P.

[b] Apparent molecular weight estimated from polyacrylamide gel electrophoresis in the presence of SDS.

[c] All preparations sought to maintain covalent phosphate (P) or deliberately to dephosphorylate the enzyme (deP).

[d] Measurements of endogenous alkali-labile phosphate. Values in parentheses are total phosphate measured after ashing.

[e] Reactions at 37°C (otherwise 30°C was used).

muscle enzyme. No single procedure is in general use though most methods have in common attention to speed and the inclusion of protease inhibitors to reduce incidental proteolysis. Almost all preparations have taken advantage of the association of the enzyme with glycogen to effect initial purification. In several instances, the associated glycogen has been removed; methods of removal have included degradation with α-amylase (9, 17) or by phosphorolysis (7, 10); disassociation by reversible chemical modification (8); or a reversible thermal inactivation procedure (18). Others have elected not to attempt removal of the glycogen (11–16) since, under some conditions, the enzyme is unstable in the absence of the polysaccharide (14, 19).

B. SUBUNIT STRUCTURE

1. *Rabbit*

Purified rabbit liver glycogen synthase, containing < 0.6 phosphates/subunit, contained a single polypeptide species as judged by SDS-polyacrylamide gel electrophoresis and had an apparent $M_r = 85,000$ (16). However, enzyme in which ~ 5 phosphates/subunit of endogenous phosphate had been preserved behaved upon electrophoresis as a polypeptide of $M_r = 90,000$ (15). The difference in electrophoretic properties is caused by phosphorylation at certain sites having an effect to reduce mobility out of proportion to the actual mass added. Thus, the 85,000-dalton dephosphorylated subunit form can be converted to one of apparent $M_r = 90,000$ by appropriate phosphorylation (Section IV). Almost identical behavior was observed by DePaoli-Roach et al. (20) in studies of rabbit muscle glycogen synthase (see also Chapter 11).

Another complicating factor in assessing the subunit structure of the enzyme has been its susceptibility to proteolysis during isolation. Camici et al. (15), in attempting to purify enzyme retaining its endogenous phosphate, first obtained several polypeptides, apparent $M_r = 79,000$–90,000, but then found that careful attention to the inclusion of protease inhibitors preserved the subunit as a species of $M_r = 90,000$. Tan and Nuttall (17) purified a highly phosphorylated form of enzyme which was composed of a subunit with $M_r = 78,000$, probably the result of proteolytic breakdown. In the same study, a less phosphorylated form of enzyme (2.9 phosphates/subunit) was obtained with subunit of apparent $M_r = 85,000$. For this preparation, as for the earlier one of Killilea and Whelan (14), it is not simple to know the relative contributions of proteolysis and phosphorylation in determining the observed electrophoretic behavior.

The best hypothesis is that the dephosphorylated, undegraded subunit has apparent $M_r = 85,000$ judged by SDS-polyacrylamide gel electrophoresis, indistinguishable in fact from that of the rabbit muscle enzyme. In its native phosphorylated state or upon appropriate phosphorylation *in vitro*, the subunit behaves electrophoretically as though it had M_r = 90,000. This proposal is consistent with the rapid immunoprecipitation of a species of apparent M_r = 90,000 from isolated rabbit hepatocytes (*21*).

2. Rat

Similar problems of proteolysis have attended the purification of rat liver glycogen synthase. Apparent subunit molecular weights have ranged from 77,000 to 87,000 (Table I), and it now seems possible that the lower values could reflect limited degradation, as has been acknowledged in the studies of Huang *et al.* (*11*) and Ciudad *et al.* (*13*). In both investigations, purified dephosphorylated enzyme contained a major species of M_r = 85,000 as well as significant amounts of polypeptides of M_r = 80,000 or 76,000. Huang and colleagues (*11, 22*) argued that the degraded species are present in the liver and represent enzyme in the process of physiological turnover. Imazu *et al.* (*12*) have described dephosphorylated enzyme that ran as a single band of M_r = 87,000 on Weber-Osborn gel electrophoresis. Refining the methods of Ciudad *et al.* (*13*), Wang *et al.* (*23*) isolated enzyme almost totally composed of the species of M_r = 85,000.

Specific antibodies have been utilized in several studies to examine the phosphorylation of glycogen synthase in isolated rat hepatocytes (*13, 24–26*). There is agreement that the enzyme immunoprecipitated has subunit M_r of ~ 88,000 on SDS-polyacrylamide gel electrophoresis although it is difficult to avoid the presence of lower M_r species, presumed to result from proteolysis. The fact that only the polypeptide of M_r = 88,000 is detected under appropriate conditions makes it likely that the generation of lower M_r forms is an uncontrolled posthomogenization occurrence.

As for the rabbit enzyme, the rat glycogen synthase at its native phosphorylation state has higher apparent M_r by SDS-polyacrylamide gel electrophoresis (M_r ~ 88,000) than the dephosphorylated enzyme (M_r ~ 85,000). Again, the difference can be ascribed to the presence of covalent phosphate and it has been shown that phosphorylation *in vitro* can convert the subunit from apparent M_r = 85,000 to M_r ~ 87,000 (*13*). Most of the preceding results can be rationalized by proposing that the undegraded dephosphorylated rat liver subunit appears as a species of M_r ~ 85,000 that is retarded electrophoretically by naturally occurring phosphorylation.

However, the results of two investigations are not embraced by this proposal. Bahnak and Gold (*27*) who, from conventional purification obtained enzyme with subunit M_r = 77,000–80,000, reported that *in vitro* translation of rat liver poly(A$^+$) RNA followed by immunoprecipitation also yielded a polypeptide of M_r = 77,000–80,000. Rulfs *et al.* (*28*) used antibody raised to rat heart glycogen synthase for Western blotting analysis of both rat heart and liver extracts. In both cases, the antibodies interacted with a species (assigned M_r = 93,000) of lower electrophoretic mobility than phosphorylase. The authors further reported that the antibodies recognized a similar M_r polypeptide formed by *in vitro* translation of rat heart or liver poly(A$^+$) RNA. The antibodies did not, however, recognize a purified rat liver preparation containing a subunit of M_r = 80,000–85,000. An interesting point is that the apparent M_r of the species recognized either by direct Western blotting or by *in vitro* translation was the same. Since the glycogen synthase in rat liver extracts is phosphorylated, either phosphorylation does not influence the electrophoretic mobility of this form of the subunit or else the reticulocyte lysate system used for translation is able to phosphorylate at the sites that influence electrophoretic mobility. The prospect of a still larger form of liver glycogen synthase is interesting but the hypothesis needs confirmation. As suggested by Rulfs *et al.* (*28*), it may prove essential to clone a cDNA corresponding to glycogen synthase so that the length of the coding sequence can be compared with the enzymological data.

3. Oligomeric Structure

Despite residual questions over the true size of the subunit, it is generally believed that liver glycogen synthase is composed of a single subunit type. Estimates of native M_r have indicated that the enzyme exists as a multimer although such measurements, of course, could only be made of enzyme freed from glycogen. Killilea and Whelan (*14*) reported a value of 170,000–180,000 for the rabbit liver enzyme, suggestive of a dimer. Jett and Soderling (*9*) obtained similar results, M_r = 156,000–171,000, with the rat enzyme. Other groups (*7, 8, 29, 30*) observed dimers of the rat enzyme but also larger forms that could correspond to trimers and tetramers. It appears that multiple oligomeric forms or aggregates of glycogen synthases can exist as a poorly understood function of temperature, solvent, phosphorylation state, and effectors [see review in Ref. (*6*)].

C. KINETIC PROPERTIES

The kinetic properties of liver glycogen synthase, like its muscle counterpart, are relatively complex. Much of the kinetic work was derived from earlier studies of partially purified preparations (*4–6*). The apparent K_m or substrate concentration for half-maximal activity ($S_{0.5}$) of the dephosphorylated enzyme is in the range 1–2 mM [Table I, Refs. (*4–6*)]. The dependence of activity on glycogen concentration is a more difficult parameter to assess since a polydisperse substrate is involved. McVerry and Kim (*8*) found nonhyperbolic kinetics, Hill slope 0.55, with estimated K_m 0.03 mg/ml. Steady-state kinetic analysis of purified rat enzyme was consistent with the formation of a ternary complex of the enzyme with its substrates glycogen and UDPglucose, with no obligate order of binding (*8*).

Glucose-6-P is the most important activator of the enzyme, probably through interaction with an allosteric binding site. For dephosphorylated enzyme, values of concentration for half-maximal effect ($M_{0.5}$) have been in the range 0.06–0.1 mM for rat (5) and 0.01–0.5 mM for rabbit (16, 17). Glucose-6-P probably decreases the $S_{0.5}$ for UDPglucose (4–6, 10, 12). There is need for more extensive kinetic characterization of the enzyme now that highly purified preparations are available in several laboratories. Glucose-6-P activation is a function of phosphorylation state and thus, as for muscle glycogen synthase [see Chapter 11 and Refs. (4–6)], the activity in the absence of glucose-6-P divided by activity in the presence of glucose-6-P ("activity ratio") has been used as an index of the phosphorylation state. In general, increased phosphorylation correlates with decreased activity ratio (see Section IV,B).

D. Limited Proteolysis

In studies of the rabbit liver enzyme, the effects of controlled proteolysis of the native enzyme on kinetic properties have been examined. Dephosphorylated enzyme, consisting of a subunit of $M_r = 85,000$, was converted by trypsin to species of 78,000–82,000 and finally 71,000–72,000 (16, 17). The activity ratio $S_{0.5}$ for UDPglucose and $M_{0.5}$ for glucose-6-P were not greatly altered (16, 17) but the V_{max} was significantly reduced (16). Chymotrypsin caused complete conversion to a species of $M_r = 71,000$ that had much lower activity (17). In terms of the proteolytic generation of intermediate fragments of the enzyme subunit, the results resemble those observed with muscle glycogen synthase [see Chapter 11 and Ref. (31)]. However, the effects on activity are quite different since tryptic proteolysis of the muscle enzyme leads ultimately to acquisition of glucose-6-P dependency (i.e., decreased activity ratio).

Exposure of phosphorylated rabbit liver glycogen synthase (5.2 endogenous phosphates/subunit) to trypsin produced a similar degradative pathway except that a species of apparent $M_r = 80,000$ accumulated (15). The presence of phosphate may also affect the electrophoretic mobility of degraded forms. No conversion to a 72,000-dalton species occurred, and this could reflect the influence of phosphorylation to reduce cleavage at certain sites, as has been observed with muscle glycogen synthase [see Chapter 11 and Ref. (20)]. Tryptic proteolysis had profound effects on the kinetic properties of the enzyme. The V_{max} derived from variation of UDPglucose increased some 16-fold and the $S_{0.5}$ for UDPglucose increased 20-fold. Depending on the exact conditions, more than 20-fold activation was seen. These results could explain why Tan and Nuttall (17), whose phosphorylated enzyme had a subunit of $M_r = 78,000$, found no effect of trypsin either on subunit M_r or kinetic properties. Again, the effect of tryptic proteolysis on kinetic properties is different than that observed with the phosphorylated muscle enzyme [see Chapter 11 and Ref. (31)].

III. Converting Enzymes

A. Glycogen Synthase Kinases

With few exceptions, the comparative enzymology of protein kinases is poorly understood with respect either to the existence and function of tissue-specific isozymes or to tissue-specific expression. To the extent studied, the protein kinases that phosphorylate muscle glycogen synthase *in vitro* also act on the liver isozyme. For details of these protein kinases, the reader is referred to Chapter 11 by Cohen and Ref. (*32*). Table II lists the enzymes with the nomenclature used in this review (*33–39*). The enzymes listed have been identified in liver, with the exception of the heparin-activated protein kinase which has not been sought, but their relative importance as glycogen synthase kinases in liver is not established.

Because of the proposed role of Ca^{2+} in mediating several hormones that act on liver (Section VI,C), the Ca^{2+}-activated protein kinases have attracted particular attention. The most definitive structural characterization of liver phosphorylase kinase came with the work of Chrisman *et al.* (*40*) who isolated a

TABLE II

Glycogen Synthase Kinases[a]

Category	Enzyme	Alternate designations	Reference[b]
Cyclic nucleotide-dependent	(1) Cyclic AMP-dependent protein kinase	—	(*33*)
	(2) Cyclic GMP-dependent protein kinase	—	(*33*)
Ca^{2+}-activated	(1) Phosphorylase kinase	—	(*34*)
	(2) Protein kinase C	Ca^{2+}, phospholipid activated protein kinase	(*35*)
	(3) Calmodulin-dependent protein kinase	Calmodulin-dependent "multiprotein kinase"	(*36*)
Other	(1) Casein kinase I	Casein kinase-S, glycogen synthase kinase-1, $PC_{0.6}$	(*37*)
	(2) Casein kinase II	$PC_{0.7}$, casein kinase-TS, casein kinase-G, glycogen synthase kinase-5	(*37*)
	(3) F_A/GSK-3	Glycogen synthase kinase-3, F_A	(*3*)
	(4) Heparin-activated protein kinase	—	(*38*)
	(5) $PC_{0.4}$	Glycogen synthase kinase-4	(*3,39*)

[a] The table lists the established glycogen synthase kinases with the nomenclature used in this chapter, as well as some alternate designations (see also Cohen, Chapter 11).

[b] Where possible, references to general reviews, especially in this volume, are given.

Ca^{2+}-activated form of enzyme with similar native M_r and subunit structure as the muscle enzyme. A distinct calmodulin-dependent glycogen synthase kinase in liver was first described by Payne and Soderling (41). The enzyme was purified (42, 43) and the existence of subunits of $M_r = 51,000$ and 53,000 was proposed by Ahmad et al. (42). The enzyme appears similar to a widely distributed calmodulin-dependent protein kinase (36). Protein kinase C, the Ca^{2+} and phospholipid-dependent protein kinase of Nishizuka (35), requires Ca^{2+} for activity but most interest is directed at the possibility that its activity is modulated physiologically by diacylglycerols.

A protein kinase that is emerging as particularly interesting for glycogen synthase control is glycogen synthase kinase-3 or F_A (F_A/GSK-3). The enzyme also has the property (F_A activity) of activating the ATP-Mg^{2+}-dependent protein phosphatase studied by Merlevede and colleagues (44, 45). F_A/GSK-3 has been purified close to homogeneity only from rabbit muscle by Cohen's group (Chapter 11) but it is present in liver and has been partially purified (20). Others (46–48) have argued that more than one related enzyme might exist, and specifically that an ATP-citrate lyase kinase possessed F_A activity and phosphorylated glycogen synthase. Unequivocal demonstration of the presence of distinct enzymes is lacking but the question needs to be addressed.

Ahmad et al. (38) described a heparin-activated protein kinase activity in rabbit muscle. This enzyme was later found to phosphorylate muscle glycogen synthase in the tryptic peptide containing sites-3, the sites recognized by F_A/GSK-3 (see Chapter 11), and to activate the ATP-Mg-dependent protein phosphatase (49). Exposure of the protein kinase to trypsin caused activation with a significant reduction in the extent of heparin stimulation. Loss of heparin stimulation leaves the enzyme with properties very close to those of F_A/GSK-3 and one hypothesis is that F_A/GSK-3 derives from a heparin-activated form. The interest in this suggestion is that no mechanism for the control of F_A/GSK-3 is known. Though heparin is unlikely to be a physiological regulator, it might be mimicking another effector that interacts with a regulatory domain of F_A/GSK-3 lost during purification.

B. PHOSPHOPROTEIN PHOSPHATASES

Only recently are the protein phosphatases that might act physiologically on glycogen synthase being subjected to rigorous enzymological definition. The reader is directed to Chapter 8 by Fischer (50) and Chapter 11 by Cohen in this volume for reviews. Glycogen synthase phosphatase activity in muscle can be ascribed mainly to two catalytic subunits, type 1 and type 2A, in the nomenclature of Ingebritsen and Cohen (51). Both have been extensively purified and have M_r of $\sim 38,000$. The catalytic components probably function in association with other subunits that may exert regulatory roles.

The comparative enzymology of liver and muscle phosphatases is not understood at a detailed molecular level. Both type 1 and type 2A catalytic subunits are present in liver (51–53) although it is unknown if these are identical to the muscle versions or represent isozymal forms. A significant fraction of the type 1 catalytic subunit in muscle is in association with a regulatory protein, inhibitor-2, as the ATP-Mg-dependent phosphatase (3, 44, 45, 51). Such is probably true in liver in which ATP-Mg-dependent protein phosphatase has been identified (54) and which contains a protein that is very similar to inhibitor-2 (55). Regulation of type 1 phosphatase through inhibitor-2 phosphorylation therefore remains a possibility in liver as in muscle. There exists the same paradoxical role as in muscle for F_A/GSK-3, potentially inactivating glycogen synthase directly and activating it indirectly via the ATP-Mg-dependent phosphatase (see discussion by Cohen, Chapter 11).

A phosphatase analogous to the muscle type 2A catalytic subunit has been purified from liver (52, 56, 57). Also, antibodies to the muscle enzyme inhibit a portion of the phosphatase activity in liver extracts (58). As for muscle, the liver type 2A catalytic subunit is likely to exist as a component of larger multimeric complexes (59, 60). Mechanisms for the control of these enzymes have not been worked out. Several studies describe phosphatases activated by basic compounds such as histone H1, polylysine, and polyamines (61, 62) and, as discussed by Cohen in Chapter 11, these are likely to be type 2A enzymes. Whether this phenomenon provides a clue to physiological control is still uncertain. Isolated rabbit muscle type 2A catalytic subunit, of $M_r = 37,000$, was stimulated no more than 20–30% as a phosphorylase phosphatase by histone H1 or by polylysine. After controlled proteolysis with trypsin, however, to generate subunits of $M_r = 36,000$ then $M_r = 34,000$, the enzyme could be stimulated 2- to 3-fold by polylysine (63). Whether such behavior occurs when the subunit is present in multimeric forms is not known.

Both phosphatases-1 and -2A are active towards glycogen synthase and phosphorylase (51, 64). Several groups, however, have reported that glycogen synthase and phosphorylase phosphatases in liver are distinct enzymes (65–68). The best-defined example of a relatively specific glycogen synthase phosphatase is the enzyme purified by Tsuiki and colleagues (65, 66). This enzyme, of $M_r = 40,000$, appears similar to the Mg^{2+}-stimulated liver enzyme described by McKenzie et al. (69) [designated phosphatase 2C by Ingebritsen and Cohen (51)]. Further molecular definition is required for complete evaluation of glycogen synthase phosphatase(s) associated with the glycogen pellet (67, 68).

Analysis of the relative expression of phosphatases in different tissues is not a trivial undertaking. Alemany et al. (58) used either inhibitor-2 (specific for type 1 phosphatase) or antibodies to muscle type 2A catalytic subunit as specific probes. In assays of glycogen synthase dephosphorylation by liver extracts, they found that 50 to 60% of the activity was suppressed by the antibodies and 50 to

60% by inhibitor-2. The results were not greatly changed whether the muscle glycogen synthase used as substrate was phosphorylated at site-2 or sites-3 (Chapter 11). Virtually all of the glycogen synthase phosphatase activity in the extracts was annulled by the combined presence of the antibody and inhibitor-2. The result is important in suggesting that, at least with the assay conditions used, all the glycogen synthase phosphatase activity can be ascribed to enzymes containing either type 1 or type 2A catalytic subunits. The assays would not of course address any form of phosphatase requiring activation, such as the ATP-Mg-dependent phosphatase. Similar analysis of liver glycogen pellets indicated that most, ~ 80%, of the glycogen synthase phosphatase was associated with the type 1 catalytic subunit. It is not entirely clear, however, whether type 1 or type 2A phosphatase dominates in the regulation of liver glycogen synthase. Also, the previous study of Ingebritsen et al. (53) had argued that the Mg^{2+}-activated enzyme would be a significant glycogen synthase phosphatase in liver.

One question for the dephosphorylation of liver glycogen synthase is completely unexplored, and that concerns the site specificity of protein phosphatases. Since the liver glycogen synthase is an isozyme, its interaction with even identical phosphatases could differ from the muscle isozyme. As described in Section IV,B, differences in the phosphorylation patterns of the liver and muscle enzymes do exist.

IV. Phosphorylation of Liver Glycogen Synthase

A. Endogenous Phosphate

In 1973, Lin and Segal (7) reported that rat liver glycogen synthase, purified to maintain its phosphorylation state, contained approximately 12 mol of alkali-labile phosphate per mol of subunit. Though this value seemed high at the time, it appeared less and less extraordinary as the multiple phosphorylation of the muscle subunit became established (70). No one appears to have repeated the measurement with the rat enzyme. Values of 5–6 phosphates/subunit have been determined for rabbit liver glycogen synthase (15, 17), slightly higher than estimates for the rabbit muscle enzyme [see Chapter 11 and Ref. (31)].

Preparations of dephosphorylated glycogen synthase omit phosphatase inhibitors such as fluoride and include an incubation aimed at allowing phosphatase(s) to act. The alkali-labile phosphate content of liver enzymes purified this way has usually been < 0.6 phosphates/subunit (11, 12, 16). Tan and Nuttall (17) purified what they considered to be a fully activated enzyme (i.e., I-form) but which still contained 2.9 phosphates/subunit. However, the activity ratio was only 0.55 compared with values greater than 0.9 in many studies (Table I) and so it is possible that only partial dephosphorylation occurred.

There is some evidence that not all the endogenous phosphate of liver glycogen synthase is alkali-labile (7, 17). From measurements of total inorganic phosphate after ashing of samples, Lin and Segal (7) found about 5 phosphates/subunit of rat enzyme not attributable to alkali-labile phosphate. Tan and Nuttall (17) found an even greater discrepancy with the rabbit enzyme, to the level of 10–11 phosphates/subunit and, despite further investigation, the chemical nature of his alkali-stable phosphate remains unknown. Where studied, the phosphate introduced into liver glycogen synthase by protein kinases appears to be at serine residues (Section IV,B). The most intriguing possibility is that some novel covalent modification occurs *in vivo*.

B. Phosphorylation by Individual Protein Kinases

1. *Rabbit*

Killilea and Whelan (14) described the inactivation of rabbit liver glycogen synthase with the introduction of ~ 1 phosphate/subunit by cyclic AMP-dependent protein kinase. Phosphorylation of the enzyme by several different protein kinases was addressed in the study of Camici *et al.* (71) (Fig. 1). Phosphate was introduced almost exclusively into one or other of two CNBr-fragments, CB-1 (apparent $M_r = 12,300$) and CB-2 (apparent $M_r = 16,000–17,000$). Cyclic AMP-dependent protein kinase phosphorylated to a stoichiometry of 0.5–0.7 phosphates/subunit, introducing phosphate first into CB-1 and more slowly into CB-2. Most inactivation correlated with the latter phosphorylation (Fig. 2). Phosphorylation also caused a decrease in the electrophoretic mobility of the subunit; in contrast, phosphorylation of the muscle enzyme to a similar stoichiometry by the same protein kinase had no such effect. In work by Wang *et al.* (72), the amino acid sequence of the N-terminal region of a tryptic phosphopeptide corresponding to CB-1 phosphorylation has been determined:

```
                      10              20
                       |               |
      Muscle:  PLSRTLSVSS LPGLEDWEDEFDLENSVLF
      Liver:             SLSVTSLGGLPQWEVEELPVDDLLL
```

A strong homology to the region N5–N20 around site-2 of the muscle enzyme is apparent (see Chapter 11). Analysis (72) of a phosphopeptide corresponding to CB-2 phosphorylation by cyclic AMP-dependent protein kinase indicated significant sequence homology to the region C24–C32 of the muscle enzyme, the region containing site-3a (see Chapter 11).

The three Ca^{2+}-stimulated protein kinases (Table II) all phosphorylated in CB-1. Protein kinase C and phosphorylase kinase also introduced phosphate into CB-2. The enzyme F_A/GSK-3 phosphorylated preferentially in CB-2 with a lesser phosphorylation in CB-1. The action of this protein kinase was the most inactivating and also caused an alteration of the electrophoretic mobility of the enzyme subunit to correspond to apparent $M_r = 90,000$. Casein kinase II introduced $\sim$0.5 phosphates/subunit into CB-2 without effect on the activity ratio. As with muscle glycogen synthase (20), phosphorylation by casein kinase II potentiated the action of F_A/GSK-3 leading to synergistic phosphorylation by the two

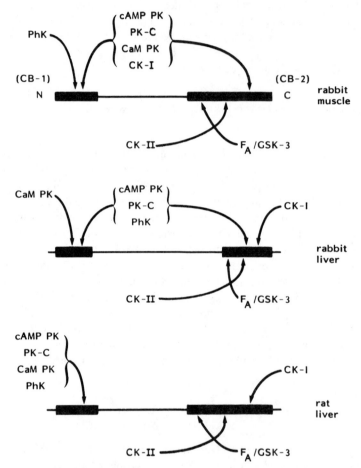

FIG. 1. Comparison of the phosphorylation of liver and muscle glycogen synthase. The figure shows phosphorylation maps for the subunit of glycogen synthase from rabbit muscle, rabbit liver, and rat liver. Indicated are the two phosphorylated CNBr-fragments, CB-1 and CB-2, that are the main regions of phosphorylation in each enzyme. The liver subunits are lined up by analogy with the muscle enzyme but the relative location of the fragments and the N→C directionality have not been formally proved. The dominant specificity of various protein kinases for CB-1 and CB-2 is indicated: PhK, phosphorylase kinase; cAMP PK, cyclic AMP-dependent protein kinase; PK-C, protein kinase C; Cal PK, calmodulin-dependent protein kinase; CK-I, casein kinase I; CK-II, casein kinase II. The "crossed arrows" for CK-II and F_A/GSK-3 indicate synergistic phosphorylation.

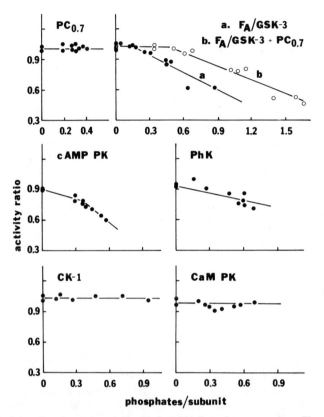

FIG. 2. Activity–phosphorylation relationship for rabbit liver glycogen synthase. The relationship between activity ratio and phosphorylation by several protein kinases is shown. $PC_{0.7}$, casein kinase II; cAMP PK, cyclic AMP-dependent protein kinase; PhK, phosphorylase kinase; CK-1, casein kinase I, CaM PK, calmodulin-dependent protein kinase.

protein kinases. Casein kinase I was relatively ineffective toward rabbit liver glycogen synthase, introducting ~1 phosphate/subunit into CB-2. The phosphorylation was without effect on activity ratio but did cause reduced electrophoretic mobility of the subunit.

The effect of phosphorylation on enzyme kinetic behavior was studied in more detail (Table III). In fact, comparison of enzyme purified to contain < 0.6 or 5 phosphates/subunit indicated that the main influence of the endogenous phosphorylation was to reduce the V_{max} without significant effect on the $S_{0.5}$ for UDPglucose. The extent of activation by glucose-6-P, as reflected in decreased activity ratio, was greater for phosphorylated enzyme though the $M_{0.5}$ for the sugar phosphate was higher. These changes in kinetic properties were partly mimicked by phosphorylation with purified protein kinases. Thus, the only two

TABLE III

Kinetic Properties of Rabbit Liver Glycogen Synthase as a Function of Phosphorylation[a]

Property	Dephosphorylated	cAMP PK	F_A/GSK-3	CK II + F_A/GSK-3	In vivo
		\multicolumn{4}{c}{Phosphorylated by}	
Phosphorylation					
(phosphates/subunit)	<0.6	+0.7[b]	+1.2[b]	+1.7[b]	5.2
Standard assay					
+ glucose-6-P (units/mg)	15–17	12.2	9.6	6.6	3–5
activity ratio	0.95–1.1	0.60	0.46	0.38	0.02–0.05
UDPglucose varied					
V_{max} (units/mg)	21–23	8.0	6.3	4.0	0.39
$S_{0.5}$ (mM)	1–2	1.3	1.9	3.0	2.1
Glucose-6-P varied					
V_{max} (units/mg)	10.5–11.5	8.6	6.2	3.9	0.8
$M_{0.5}$ (mM)[c]	0.1–0.2	0.30	0.38	0.48	7.2

[a] From Camici et al. (15,16,71).

[b] Phosphate added to that of dephosphorylated form by the indicated protein kinase(s): cAMP PK, cyclic AMP-dependent protein kinase; CK II, casein kinase II.

[c] 0.2 mM UDPglucose.

inactivating protein kinases, cyclic AMP-dependent protein kinase and F_A/GSK-3, decreased the V_{max} and increased the $M_{0.5}$ for glucose-6-P. Even greater effects were elicited by the synergistic actions of casein kinase II and F_A/GSK-3. However, the inactivation stopped short of that found for the enzyme phosphorylated in vivo. Possibly the combined actions of several enzymes are required for complete inactivation or, the thought jars, we lack a relevant protein kinase. The extent to which multiple phosphorylation sites combine in determining kinetic properties is not yet known.

The detailed phosphorylation map of the rabbit liver glycogen synthase has still to be worked out. CB-1 contains at least one phosphorylation site, perhaps very similar to site-2 of the muscle enzyme. Phosphorylation in CB-1, however, has little effect on activity. CB-2 contains probably distinct sites for casein kinase I and casein kinase II. The latter interacts positively with the site(s) phosphorylated by F_A/GSK-3. On the basis of similar influence on kinetic properties and electrophoretic mobility, it is likely that F_A/GSK-3 and cyclic AMP-dependent protein kinase modify the same region of CB-2, a region with homology to the muscle site-3. Phosphorylation in this region may be the key to determining the catalytic activity of the enzyme. However, the fact that this is only a secondary site of action for cyclic AMP-dependent protein kinase raises questions about the direct regulatory role of this protein kinase. No analogues of sites-1a and -1b of the muscle enzyme have been identified.

2. *Rat*

Several groups have studied the phosphorylation of rat liver glycogen synthase *in vivo*. Phosphate is distributed between two CNBr-fragments, CB-1 (apparent $M_r = 14,000$) and CB-2 (apparent $M_r = 28,000$). The relative specificity of protein kinases for CB-1 and CB-2, from work in this laboratory, is shown in Fig. 1. Imazu *et al.* (*12*) found rather different electrophoretic mobilities, especially for CB-1 which appeared with apparent $M_r = 17,000$. Virtually all the phosphorylation occurs at serine residues (*23*).

The results of phosphorylating the enzyme with cyclic AMP-dependent protein kinase have been somewhat variable. The earlier study of Jett and Soderling (*9*) reported a stoichiometry of phosphorylation of 3 phosphates/subunit but the values from subsequent work have been much lower (*11–13*). Ciudad *et al.* (*13*), in phosphorylating to 0.5–1.0 phosphate/subunit, found the phosphate almost exclusively in CB-1. All of this phosphate could be recovered in a small phosphopeptide (~ 20 amino acids) generated by proteolysis with trypsin plus chymotrypsin (*23*). Huang *et al.* (*11*) likewise observed a single tryptic phosphopeptide after phosphorylation with cyclic AMP-dependent protein kinase. In contrast, Imazu *et al.* (*12*), observed initial phosphorylation in CB-1 but as the phosphorylation approached and exceeded 1 phosphate/subunit, phosphate was also detected in CB-2. Analysis of tryptic peptides at lower levels of phosphorylation indicated a single main species but at higher levels a complex pattern of at least five phosphopeptides was found on gel filtration (*73*). The results are in accord, however, with the finding that the primary and preferred site of action of cyclic AMP-dependent protein kinase is in a single peptide derived from CB-1. Secondary site(s) in CB-2 may exist. This proposed behavior is similar to that of the rabbit liver enzyme.

The effect of phosphorylation on activity is less clear. Jett and Soderling (*9*) reported that complete inactivation (i.e., activity ratio reduced to zero) required the introduction of 3 phosphates/subunit. Imazu *et al.* (*12*) observed partial inactivation even after phosphorylation to ~ 0.5 phosphates/subunit whereas Huang *et al.* (*11*) saw no inactivation up to 1.0 phosphate/subunit. The explanation for the discrepancies is not apparent but could well correlate with modification of different sites in the various studies. For example, it might be that initial phosphorylation in CB-1 does not influence activity, as for the rabbit enzyme, but that only with phosphorylation in CB-2 is significant inactivation realized. More work is needed.

Phosphorylation by the three Ca^{2+}-activated protein kinases to levels of 0.5–1.0 phosphates/subunit was restricted to CB-1 (*12, 13*). Wang *et al.* (*23*) found the phosphate in the same small proteolytic fragment modified by cyclic AMP-dependent protein kinase. A similar conclusion was reached by Imazu *et al.* (*73*). Akatsuka *et al.* (*74*) reported that phosphorylase kinase modified the same peptide as cyclic AMP-dependent protein kinase. Therefore, the three Ca^{2+}-activated protein kinases recognize the same limited region of CB-1 as cyclic AMP-dependent protein kinase. Imazu *et al.* (*12*) correlated such phosphorylation by phosphorylase kinase or the calmodulin-dependent protein kinase with inactivation. Protein kinase C did not inactivate, suggestive of a distinct phosphorylation site in the same region. Akatsuka *et al.* (*74*), however, found no inactivation due to phosphorylase kinase action.

Casein kinase II phosphorylates in CB-2 to a level of 0.6–1.0 phosphates/subunit (*12, 13*) without effect on the activity ratio (*12*). Phosphorylation by this enzyme is synergistic with F_A/GSK-3, as found for rabbit glycogen synthases (*3, 20, 71*). The site(s) modified by casein kinase II appeared distinct from those of other protein kinases on the basis of peptide mapping (*23, 73*). F_A/GSK-3 phosphorylates to approximately 1.5 phosphates/subunit, predominantly in CB-2, at what appeared as characteristic sites (*12, 13, 23, 73*). The enzyme is the most potent in inactivating rat liver glycogen synthase (*12*). Glycogen synthase kinase-4 ($PC_{0.4}$) is specific for CB-1 and probably modifies the same peptide as the other "CB-1 kinases" (*12, 73*).

The action of casein kinase I is of particular interest. From the studies employing CNBr-fragmentation (*12, 13*), casein kinase I acts preferentially in the CB-2 region of the subunit normally. Huang and colleagues (*11*) correlated phosphorylation to 1.6–2.2 phosphates/subunit with the presence of 4 tryptic peptides and no effect on activity. However, prior phosphorylation of the glycogen synthase with cyclic AMP-dependent protein kinase (also without effect on activity in these studies) directed casein kinase I to modify the same peptide phosphorylated by the cyclic nucleotide-dependent enzyme. Mechanistically, this is similar to the synergistic phosphorylation discussed for casein kinase II and F_A/GSK-3. Of special interest in the present case is the fact that the modification at the novel site correlated with significant inactivation of the glycogen synthase. From the specificity of cyclic AMP-dependent protein kinase, this novel site would be in the CB-1 region of the molecule. The implications of this unique mechanism are not fully explored but this could provide cyclic AMP-mediated control over casein kinase I action.

C. SUMMARY

The overall picture of the phosphorylation and inactivation of liver glycogen synthase is still not completely focused. Both rat and rabbit enzymes contain a region in CB-1 (by conjecture N-terminal) that is the primary site(s) of action of cyclic AMP-dependent protein kinase and the three Ca^{2+}-stimulated enzymes. Some of these enzymes recognize secondarily site(s) in CB-2. F_A/GSK-3 phosphorylates predominantly in CB-2 at site(s) that affect electrophoretic mobility and which are linked, via positive site–site interaction, to a separate site of action of casein kinase II. Casein kinase I preferentially phosphorylates probably multiple sites in CB-2 but may also interact with CB-1. Nothing analogous to the rabbit muscle sites 1a and 1b has been identified thus far.

Less clear is the role of the different phosphorylations in inactivating the enzyme. There is accord that F_A/GSK-3 phosphorylation, presumably in a CB-2 region, potently inactivates both rabbit and rat glycogen synthase. Likewise, casein kinase II action appears without direct influence on kinetic properties. What is most in question is the effectiveness of cyclic AMP-dependent protein kinase and the three Ca^{2+}-activated enzymes in causing inactivation. Based on the work of this laboratory (21, 23, 71) and of Huang and colleagues (11, 74), primary phosphorylation by these protein kinases in CB-1 has relatively little effect on activity ratio and only after secondary action in CB-2, in the region recognized by F_A/GSK-3, does inactivation ensue. This makes some arbitrary assumptions, in particular close functional homology between the rat and rabbit liver enzymes. As discussed in Section VI, many of the hormone-mediated effects on activity would be consistent with altered phosphorylation in CB-2. Exton and colleagues (12) might argue for an effect of CB-1 phosphorylation to inactivate the rat enzyme. Huang and colleagues (11) might also stress the role of the synergistic phosphorylation by casein kinase I, presumably in CB-1, to decrease enzyme activity.

Most of the work has monitored activity ratio, for understandable reasons of practicality. It is certainly safe to infer a significant effect on kinetic properties from a reduced activity ratio, as was documented for the rabbit liver enzyme (71). Over the years, though, there have been repeated questions (75–80), also from this author (6), as to the limitations of the activity ratio as a kinetic parameter. One might argue that lack of observed effects on activity ratio does not rule out significant changes in kinetic properties if measured appropriately. There is some merit to this argument and certainly the various phosphorylation reactions ought to be reevaluated with this in mind. Similar questions can be asked of activity measurements made in crude tissue or cell extracts. However, though activity ratio may not be the ideal or most sensitive measure, major effects on the kinetics for UDPglucose or glucose-6-P probably have not been missed through measurements made in the absence or presence of glucose-6-P, and one is not obliged, of course, to combine such measures to form a ratio.

V. Comparative Enzymology of Glycogen Synthase

A. Species

Though detected by activity in numerous cell types (4, 5), glycogen synthase has been purified to homogeneity from relatively few sources (Table IV). From kinetic measurements, including those of relatively impure preparations, some

TABLE IV

SOME PREPARATIONS OF GLYCOGEN SYNTHASE FROM DIFFERENT SOURCES[a]

Species	Tissue	Subunit M[b]	Reference
Human	Leukocytes	85,000	(81,82)
Rabbit	Skeletal muscle	85,000–86,000	(3,31)
	Liver	85,000	See Table I
Rat	Skeletal muscle	86,000	(83)
	Liver	85,000–87,000	See Table I
Pig	Kidney	92,000	(84)
	Adipose	90,000	(85)
Cow	Heart	88,000	(86,87)
Saccharomyces cerevisiae		77,000	(88)
Neurospora crassa		88,000–90,000	(89)
Escherichia coli B		49,000	(90)

[a] The table seeks to collate enzyme preparations close to homogeneity and for which reasonably clear estimates of subunit molecular weight are presented.

[b] Determined by SDS-polyacrylamide gel electrophoresis. Where possible, values for dephosphorylated enzymes are presented.

generalizations seem safe. In animal cells, glycogen synthase (a) uses UDPglucose as glucosyl donor, (b) is activated by glucose-6-P, and (c) is interconvertible between kinetically distinguishable forms. The last mentioned property is taken to reflect covalent phosphorylation. From Saccharomyces cerevisiae and Neurospora crassa to humans, the enzyme appears to consist of a single subunit of molecular weight 80,000–90,000 (Table IV) (81–90). It seems probable that important catalytic and regulatory properties have been conserved. Only as we pass to bacteria are fundamental differences encountered. For example, the Escherichia coli B glycogen synthase has a subunit of $M_r = 49,000$ (90) and is subject neither to covalent phosphorylation nor activation by glucose-6-P (91). Bacterial glycogen is synthesized, however, by a different pathway in which even the glucosyl donor, ADP-glucose, is different than in eukaryotes. The review of Preiss and Walsh (91) provides an excellent account of comparative aspects of glycogen and starch metabolism.

Besides subunit size, there is relatively little structural information available to compare the eukaryotic enzymes. The amino acid composition of rabbit muscle glycogen synthase (70, 92) bears no particularly strong similarity to that of the enzyme from N. crassa except in a few amino acids (89). However, there is likewise no great similarity between the compositions of the rabbit muscle and pig kidney (84) enzymes. Even less is known at the level of primary structure. Mapping of phosphorylated peptides has been performed with several mammalian enzymes. For example, analyses of phosphopeptides of rabbit (Chapter 11), rat (*83*), and mouse (*93*) muscle glycogen synthases have been made in such a way that meaningful comparisons are possible. Both with respect to the two phosphorylated CNBr-fragments and also smaller tryptic peptides, the three muscle enzymes appeared remarkably similar. Thus, peptides behaving on reverse phase HPLC like the rabbit muscle phosphopeptides N5–N38, C25–C39, C40–C52, C85–C97, and C98–C123 (see Cohen, Chapter 11) were obtained from rat (*83*) or mouse (*93*) muscle glycogen synthase phosphorylated by appropriate protein kinases. The three enzymes also exhibited immunological cross-reactivity (*93, 94*). Obviously, there may still be significant sequence differences not revealed by such analyses. Also, one could argue with justification that highly functional regions of the enzymes, such as phosphorylation sites, are exactly those which are most likely to have been conserved. Rat and rabbit liver glycogen synthase, in contrast, differed even at the level of CNBr-fragmentation (Section IV,B). Such experiments provide only a glimpse of primary structural homologies. Still the impression has to be that at least the mammalian glycogen synthases are closely related enzymes.

B. ISOZYMES

From the preceding section, it follows that the glycogen synthases of different tissues in higher organisms are likely to be quite homologous. Nonetheless, isozymes do exist. The best example is of the liver and muscle isozymes, in particular from rabbit. Comparison of the two isozymes pervades this chapter but, briefly, the main points are as follows. The enzymes share subunit molecular weight, phosphorylation by the same set of protein kinases, and activation by glucose-6-P. The region in CB-1, inferred to be N-terminal for both enzymes, that contains a major cyclic AMP-dependent protein kinase site, is very similar in both enzymes. The actions of casein kinase II and F_A/GSK-3 both in synergism, effects on electrophoretic mobility and in causing inactivation are very similar for both enzymes, suggestive of strongly conserved features.

One of the striking differences is the way in which phosphorylation is translated into inactivation. Whereas phosphorylation of the rabbit muscle enzyme caused mainly increased $S_{0.5}$ for UDPglucose (*31*), for the liver enzyme the primary effect was to reduce the V_{max} (*71*). Though a conspicuous difference, it had to be viewed as circumstantial evidence for isozymes since the procedures for the purification of the enzymes were so different. Preoccupation with proteolysis forced ultimately a very rapid purification of the liver enzyme compared with the muscle, and no attempt was made to remove glycogen. One could argue that differences in kinetic behavior were a function of the purification procedures. However, the observation of distinct CNBr-fragmentation patterns for the two enzymes (*71*) together with the limited amino acid sequence data (Sec-

TABLE V

Effects of Hormones on the Phosphorylation of Glycogen Synthase
in Rat Hepatocytes[a]

Treatment[b]	Activity ratio[c]	Subunit	^{32}P Content of[d]	
			CB-1	CB-2
Control	0.26 ± 0.05	1	0.45	0.55
Glucagon	0.17 ± 0.02	2.21 ± 0.20	0.80	1.41
Epinephrine	0.19 ± 0.01	2.35 ± 0.47	1.04	1.31
Vasopressin	0.18 ± 0.01	2.03 ± 0.34	1.04	0.99

[a] Modified from Ciudad et al. (13).

[b] Hormones were added, for 8 min, to cells incubated with [^{32}P]P_i and the glycogen synthase purified by the use of specific antibodies. Epinephrine 10^{-5} M, glucagon 10^{-7} M, and vasopressin 10^{-5} M.

[c] Activity ratio measured by the low/high glucose-6-P assay of Guinovart et al. (76).

[d] Arbitrary units referred to the control value for the whole subunit.

tion IV,B) provide unequivocal proof of the existence of isozymes. It will be especially interesting to compare the enzymes as more primary structural information becomes available. One extremely interesting point is the apparent lack in the liver enzyme of cyclic AMP-dependent protein kinase sites analogous to site-1a and -1b of the muscle enzyme. Another difference is the lack of effect of phosphorylation of the liver enzyme in CB-1 on activity. In contrast, phosphorylation of the muscle site-2 in CB-1 appears important in determining enzyme activity.

Rat liver and muscle also express distinct glycogen synthase isozymes. The enzymological evidence is less complete but comparison of the phosphorylation of the rat muscle enzyme (83) with the liver enzyme (Section IV,B) reveals some differences. Also, antibodies to the rat muscle enzyme did not recognize the rat liver glycogen synthase in the studies of Kaslow and Lesikar (95). Tan and Nuttall (26) performed the inverse experiment by showing that antibodies to rat liver glycogen synthase did not inactivate the rat muscle or heart enzyme. In some cases, greater immunological cross-reactivity has been observed (96) but this probably depends on the particular antibody preparations. How many different tissue-specific isozymes exist is still unknown. From Western blotting experiments, Kaslow and Lesikar (95) found that their antimuscle antibodies interacted with enzyme from heart, gut, kidney, and brain, and suggested that these tissues expressed a muscle-type (M-type) isozyme.

The overall similarity of the liver and muscle isozymes does raise the question of why distinct enzymes have evolved. It is difficult to rationalize how the properties of the two isozymes fulfill needs specific to their respective tissues.

Obviously, even subtle differences in kinetic properties may be important but no simple rationale comes to mind. Perhaps the possibility that the liver enzyme lacks certain phosphorylation sites present in the muscle isozyme is significant. Even this difference defies teleological explanation but it directs us to consider glycogen synthase in the context of the protein kinases and phosphatases that regulate, in a manner that is undoubtedly tissue-specific, the enzyme activity. Perhaps the interactions with the converting enzymes are what have guided the evolution of separate isozymes. At least in a general way, such tissue-dependent specialization makes some sense in terms of hormonal controls that, even when ostensibly operating by a common intracellular mechanism, elicit responses finely tuned to the needs of each target tissue.

VI. Control of Hepatic Glycogen Synthase

A. SCOPE

The phenomenology of the hormonal control of liver glycogen metabolism covers an extensive literature [see several excellent reviews in Refs. (4, 5, 97–106)]. This section concentrates on the rapid changes in glycogen synthase activity that are considered to be linked to its phosphorylation state. The objective, although elusive, is to correlate the behavior of the enzyme in cells with what is known of its chemical properties. Discussion is restricted to the glycogenic stimuli provided by glucose and insulin, and the glycogenolytic actions of glycagon, epinephrine, angiotensin II, and vasopressin. Much of the most relevant work has been with rat liver.

Though many investigations understandably seek to isolate an individual agent for study, it must be appreciated that the endocrine signals listed, as well as thyroid hormone and the glucocorticoids, are all likely to interact and to modulate each others' actions. Also, though an experiment may focus on short-term consequences of hormonal administration, especially for the liver, the experimental system is nonetheless conditioned by the past hormonal status of the animal (e.g., age, sex, feeding, and fasting). Thus, by default, longer-term adaptive changes, which can alter the levels or functions of proteins from receptors through to intracellular regulatory proteins, may still influence the experimental outcomes.

B. GLUCAGON

1. Effects on Glycogen Synthase

Glucagon has been shown in numerous studies to decrease the activity of liver glycogen synthase in whole animals [see Ref. (106)] and isolated hepatocytes

(*107–110*). Most often, decreased activity ratio was recorded (e.g., Table V). Akatsuka *et al.* (*25*), using rat hepatocytes, found that the main effect of glucagon was on the $M_{0.5}$ for glucose-6-P, with little change in either V_{max} or the $S_{0.5}$ for UDPglucose (measured in the presence of glucose-6-P). Bosch *et al.* (*111*) described a modest increase in $M_{0.5}$ but a more striking change in $S_{0.5}$ (measured in the absence of glucose-6-P). The $S_{0.5}$ values in this study were considerably higher than found with purified rabbit liver enzyme (Section II).

Several groups have demonstrated that the reduced activity correlates with altered phosphorylation of the glycogen synthase subunit. While it is feasible, if not trivial, to purify sufficient muscle enzyme from individual rabbits to make chemical determinations of phosphate as a function of hormonal treatments (see Chapter 11), such is not the case for liver. Thus, almost all the investigations of liver glycogen synthase phosphorylation in cells have employed rat hepatocytes incubated with [^{32}P]P_i to label cellular phosphoproteins (*13, 24–26, 112*). Isolation of glycogen synthase is then effected with specific antibodies. In such experiments it is usually possible to achieve steady-state labelling of the γ-P of the cellular ATP but the ^{32}P-content of glycogen synthase does not reach a plateau value before the metabolic parameters of the cells decline. Thus, the disadvantage of the technique is that absolute phosphate levels are not determined. Theoretically, any hormone-induced increase in the ^{32}P-content of glycogen synthase could reflect an increased level of phosphorylation and/or increased turnover at any site(s) below the specific activity of the γ-P of the ATP. However, altered ^{32}P-content unequivocally identifies a perturbation in the phosphorylation reactions occurring. Also, similar studies of rat muscle enzyme have shown a good correlation between changes in ^{32}P-content and chemically measured covalent phosphate (*94*). An advantage of the technique is that turnover can be studied and, by virtue of using isolated cells, one avoids the endocrine interactions that complicate the administration of hormones to whole animals.

Simply by gel electrophoretic separation of phosphoproteins from ^{32}P-labeled hepatocytes, Garrison (*112*) had tentatively assigned glycogen synthase as an ~83,000-dalton species that was in the set of proteins whose ^{32}P-content was increased by glucagon. This apparent M_r for the phosphorylated form seems somewhat low in light of subsequent information (Section II,B; Table I) and possibly the subunit was proteolytically clipped. However, in later work, Garrison *et al.* (*113*) provided confirmation of the assignment by immunoprecipitation with antibodies to glycogen synthase. Ciudad *et al.* (*13*) found that a 2.2-fold increase in the ^{32}P-content of immunoprecipitated glycogen synthase was correlated with a reduction in activity ratio from 0.26 to 0.17 (Table V). These results are comparable to the 1.6– to 2.2-fold increase observed by Akatsuka *et al.* (*25*). Tan and Nuttall (*26*) describe one experiment in which a smaller 25%

increase was found but this was matched by a modest reduction in activity ratio from 0.1 to 0.08. An interesting observation of Tan and Nuttall (26) was that all the ^{32}P introduced into the glycogen synthase was alkali-labile, consistent with the occurrence of turnover only at P-Ser and/or P-Thr.

Using CNBr-cleavage, Ciudad et al. (13) assessed the distribution of the glucagon-induced phosphorylation between fragments CB-1 and CB-2 (Table V). The ^{32}P-labeled CNBr-fragments obtained from immunoprecipitated glycogen synthase were identical to those derived from purified enzyme as judged by SDS-polyacrylamide gel electrophoresis (13). Most of the increase in ^{32}P-content caused by glucagon was associated with CB-2 with lesser effect on CB-1. Recall that CB-1 is the dominant, perhaps the only, site of action for cyclic AMP-dependent protein kinase (Fig. 1). Akatsuka et al. (25) performed similar experiments and separated multiple tryptic phosphopeptides. The authors concluded that glucagon increased phosphorylation in peptides corresponding to the actions of both cyclic AMP-dependent protein kinase and casein kinase I, but several other peptides, representing a significant amount of the ^{32}P-labeled glycogen synthase, were also affected. Protein kinases specific for these peptides are unassigned. The two investigations are consistent in showing that the increased phosphorylation of glycogen synthase by glucagon cannot be explained solely by the direct action of cyclic AMP-dependent protein kinase. Phosphorylation in CB-2, from the studies in vitro, could well account for the altered activity of the glycogen synthase (Section IV).

2. Mechanism

Since glucagon is known to elevate hepatic cyclic AMP levels, the immediate expectation is for a role of cyclic AMP-dependent protein kinase in the control of liver glycogen synthase. There have, of course, been discussions of cyclic AMP-independent actions of glucagon [see Ref. (106)], but current thinking would favor cyclic AMP as the dominant messenger for glucagon. For the case in question, inactivation of glycogen synthase, elegant and compelling evidence for cyclic AMP involvement comes from the work of Botelho and colleagues (114). These workers have utilized two cyclic AMP analogs, (Sp)- and (Rp)-adenosine 3',5'-phosphorothioates (cyclic AMPs), that are respectively, agonist and antagonist for cyclic AMP-dependent protein kinase (115, 116). The agonist (Sp)-cyclic AMPS was able to inactivate glycogen synthase while (Rp)-cyclic AMPS antagonized glucagon-mediated inactivation of the enzymes. It is difficult to interpret these results as other than an obligate role for cyclic AMP in glycogen synthase control by glucagon.

How does cyclic AMP-dependent protein kinase control glycogen synthase? The situation for rat liver is reminiscent of epinephrine-induced phosphorylation of rabbit muscle glycogen synthase, another cyclic AMP-mediated process in which the observed phosphorylation cannot be ascribed entirely to the direct action of cyclic AMP-dependent protein kinase (see Chapter 11). Three main, but not mutually exclusive, possibilities can be presented.

1. The assessments of the site specificity of cyclic AMP-dependent protein kinase for the phosphorylation of rat liver glycogen synthase *in vitro* misguide us.
2. Cyclic AMP exerts an indirect control over protein kinases able to phosphorylate in CB-2.
3. Cyclic AMP exerts an indirect control over protein phosphatase(s) acting on glycogen synthase.

For the first eventuality listed, there is some controversy. Though the work of this laboratory (*13, 23*) and of Huang *et al.* (*11*) concur in proposing the existence of a single dominant phosphopeptide derived from cyclic AMP-dependent protein kinase action, Exton and colleagues (*12*) argue for multiple sites, and for significant phosphorylation in CB-2. Even so, there is no disagreement that the *preferred* phosphorylation is in CB-1 and so the predominant glucagon stimulation would have to be at a nonpreferred site on glycogen synthase. Could this "reversed" specificity be brought about in the cell by the site specificity and kinetics of glycogen synthase phosphatases? A related question is whether an undiscovered site–site interaction exists such that *in vivo* glycogen synthase does have a preferred cyclic AMP-dependent protein kinase site in CB-2 (cf., the synergistic phosphorylation described in Section IV).

The second option would invoke an indirect control by cyclic AMP-dependent protein kinase of one of the enzymes able to phosphorylate in CB-2; the known candidates are F_A/GSK-3, casein kinase I, and casein kinase II. F_A/GSK-3 is an attractive possibility since phosphorylation by this enzyme has potent effects on glycogen synthase activity. Via site–site interactions, the action of F_A/GSK-3 could equally well be regulated via casein kinase II. Unfortunately, no mechanism for cyclic AMP-mediated or any other physiological control of these protein kinases is known. There is likewise insufficient information available to exclude this eventuality.

The third possibility would be the existence of active control of protein phosphatase(s) via a cyclic AMP-initiated pathway. Obviously, if control of phosphatases occurs, the glucagon-mediated effects on glycogen synthase phosphorylation are no longer dictated simply by the specificity of cyclic AMP-dependent protein kinase. One potential mechanism would be via the phosphatase inhibitory protein, inhibitor-1, that is effective only after phosphorylation by cyclic AMP-dependent protein kinase (see Chapter 11). Analogous proteins are thought to exist in liver (*117, 118*). Another possible mechanism is based on the hypothesis of Hers (*97*) and Stalmans (*98*) in which the dephosphorylation of glycogen synthase would be coupled to the phosphorylation state of phosphorylase. The central mechanistic feature of the proposal is the inhibition of glycogen synthase dephosphorylation by phosphorylase *a*. Thus, cyclic AMP-mediated formation of phosphorylase *a* would reduce dephosphorylation and allow prevailing protein kinases to phosphorylate glycogen synthase. There are reports that glucagon action results in reduced glycogen synthase phosphatase activity measured in extracts (*119–122*). From the use of antibodies to phosphorylase, there is some evidence to correlate an increased level of phosphorylase *a* with the decreased phosphatase activity (*119, 122*). Others have challenged this interpretation (*120, 121*). Further discussion of the mechanism of Hers and Stalmans is presented in Section VI,D. Finally, we cannot yet exclude some unknown mechanism of cyclic AMP-mediated control of glycogen synthase phosphatase activity.

C. CATECHOLAMINES, VASOPRESSIN, AND ANGIOTENSIN II

1. *Effects on Glycogen Synthase*

Catecholamines (*77, 80, 107, 109, 123*), vasopressin (*77, 110*), and angiotensin II (*110*) have all been shown to cause inactivation of glycogen synthase in rat hepatocytes (see also Table V). Most often, only decreases in activity ratio were measured, but Bosch et al. (*111*) reported that the primary effect of epinephrine was to increase the $M_{0.5}$ for glucose-6-P. Thomas et al. (*80*), in contrast, reported that the main effect of phenylephrine was to decrease by 74% the V_{max}, with little change in $S_{0.5}$, consistent with the effects of CB-2 phosphorylation on the behavior of the rabbit liver enzyme (Section IV).

From electrophoretic separation of hepatocyte phosphoproteins, Garrison and colleagues (*110, 112*) had found that the same ~83,000-dalton species, whose ^{32}P-content was increased by glucagon, responded also to vasopressin, norepinephrine, and angiotensin II. Garrison et al. (*113*) later used antibodies to substantiate identification of this species as glycogen synthase. Using methods identical to those described for glucagon in Section IV,B, Ciudad et al. (*13*) analyzed the phosphorylation of glycogen synthase in hepatocytes exposed to epinephrine or vasopressin (Table V). In these cells, epinephrine action on glycogen synthase is predominantly α-adrenergic in nature. Reduction of the activity ratio from 0.26 to 0.18–9.19 correlated with 2- to 3-fold increase in the ^{32}P-content of the glycogen synthase subunit. From analysis of CNBr-fragments, the increase was associated with both CB-1 and CB-2. In absolute terms, a greater increase in the ^{32}P-content of CB-2 was observed. The responses to epinephrine and vasopressin were qualitatively similar. Ariño et al. (*24*) described similar effects of the two hormones in terms both of whole subunit phosphorylation and the ^{32}P-contents of CB-1 and CB-2. It is to be recalled that the dominant site of action of the three Ca^{2+}-activated protein kinases in phosphorylating rat liver glycogen synthase is in CB-1.

In these studies, the effects of vasopressin and epinephrine on glycogen synthase phosphorylation differed slightly from the results of glucagon administration. Specifically, glucagon elicited a less decided increase in CB-1 phosphorylation (13). This could imply differential control of phosphorylation sites by glucagon as compared with the other two hormones. Bosch et al. (111) had made this argument on the basis of different alterations in the kinetic properties of glycogen synthase caused by epinephrine and glucagon. Garrison et al. (113) made the same suggestion from partial V8 protease digestions of glycogen synthase immunoprecipitated from vasopressin- or glucagon-treated cells. However, none of the evidence for differential control of rat liver glycogen synthase phosphorylation sites can be considered conclusive and this exciting possibility awaits more incisive experimentation.

2. Intracellular Messengers

Over several years, work on the control of hepatic function by epinephrine underlined the importance of α-adrenergic receptors, particularly in rat liver (100, 102–105). It also became apparent, however, that the relative contributions of α- and β-adrenergic receptors in rat liver are not fixed, but are a function of a variety of factors including age (124–126), sex (127), status of other endocrine systems (128, 129), and, for isolated cells, time and conditions of culture (130). Thus, for example, phosphorylase activation by catecholamines in mature male rats is completely α-adrenergic whereas in young males there is a significant β-adrenergic component (126, 131). The α-adrenergic control of phosphorylase has been further characterized as due to α_1-adrenergic rather than α_2-adrenergic receptors (132). Fewer investigations of catecholamine action have addressed glycogen synthase but the results clearly indicate a predominance of α-adrenergic mechanisms (77, 107, 109, 110). In fact, glycogen synthase inactivation is mainly α-adrenergic even when the control of phosphorylase involves significant β-adrenergic contributions, as in young male rats (131). The inactivation of glycogen synthase by epinephrine can be further categorized as α_1-adrenergic (133). Although catecholamine control of glycogen synthase, with appropriate species and/or conditions, may involve β-adrenergic mechanisms, the present discussion is limited to α_1-adrenergic regulation in rat.

Angiotensin II, vasopressin, and α_1-adrenergic agonists, though operating through distinct receptors, exhibit certain mechanistic features in common; the first to draw attention was the modulation of cellular Ca^{2+} distribution (100, 102–105). Subsequent interest also focused on their ability to stimulate phosphatidylinositol breakdown leading to the production of polyphosphoinositols and diacylglycerols (134–136). The concept of multiple second messengers is thus evolving [e.g., Ref. (137)]. Inositol trisphosphate has been invoked as the messenger (138, 139) to elicit Ca^{2+} release from intracellular deposits such as the endoplasmic reticulum. Diacylglycerols have been proposed as endogenous activators of protein kinase C (*35*). Details of the mechanism of signal transmission are actively being sought but the idea of two separate regulatory pathways, one via elevation of cytosolic Ca^{2+} and the other via activation of protein kinase C, seems eminently reasonable. With regard to the phosphorylation of hepatic proteins, significant support for the scheme comes from a study of Garrison and colleagues (*140*). They showed that angiotensin II, vasopressin, and α_1-adrenergic agonists stimulated the phosphorylation of a set of 10 hepatocyte proteins. Exposure of cells to the Ca^{2+} ionophore A23187, which should bypass receptor control of cytosolic Ca^{2+}, led to increased phosphorylation of a subset of 7 of the hormonally sensitive phosphoproteins, in which phosphorylase was included. Treatment of cells with tetradecanoyl phorbol acetate (TPA), a tumor-promoting phorbol ester which can directly activate protein kinase C (*141*), stimulated phosphorylation of the other three hormonally controlled proteins. Unfortunately, technical reasons excluded glycogen synthase from the analysis.

The role of Ca^{2+} in mediating the control of phosphorylase appears well established (*99–105*). Strong support for phosphorylase kinase as the Ca^{2+}-sensitive control element has come from experiments using *gsd/gsd* rats, a genetic variant defective in hepatic phosphorylase kinase activity (*142*). Blackmore and Exton (*143*) found that phenylephrine, vasopressin, and ionophore A23187 could not activate phosphorylase in cells from these animals. Garrison *et al.* (*140*) reported no change in the phosphorylation of phosphorylase after exposure of cells from *gsd/gsd* rats to angiotensin II or vasopressin.

The second messenger system by which angiotensin II, vasopressin, and α_1-adrenergic agonists regulate hepatic glycogen synthase is not certain. Consideration must be given to both Ca^{2+}-mediated and Ca^{2+}-independent pathways. For several years the dominant idea has been for a Ca^{2+}-dependent mechanism, encouraged by the discovery of Ca^{2+}-activated glycogen synthase kinase (see Section III,A). The role of Ca^{2+}, however, is somewhat enigmatic. Garrison *et al.* (*110*) reported that the inactivation of rat hepatocyte glycogen synthase by vasopressin or angiotensin II was abolished in cells depleted of Ca^{2+} by incubation with EGTA, and similar results were obtained by Strickland *et al.* (*77*) using vasopressin. Experiments with ionophore A23187 provided conflicting results. Strickland *et al.* (*77*) showed that A23187 could antagonize the activation of hepatocyte glycogen synthase caused by addition of glucose (i.e., A23187 and glucose added together). Garrison (*112*), though, had reported that the hepatocyte phosphoprotein of $M_r = \sim 83,000$, later confirmed as glycogen synthase (*113*), did not increase in phosphorylation in cells treated with A23187.

Other studies have also indicated that glycogen synthase inactivation is not a simple parallel to phosphorylase activation, as might be predicted if Ca^{2+} concentration was the central regulator of both target enzymes. For example, the glycogen synthase response is more sensitive to epinephrine than phosphorylase activation (*77, 131*), at least under some circumstances, so that glycogen synthase inactivation can occur without phosphorylase activation. More striking uncoupling of the control of the two enzymes came from the study of Roach and Goldman (*144*) who treated cells with tumor-promoting phorbol esters. While low concentrations of TPA caused a reduction in the glycogen synthase activity ratio comparable to that elicited by epinephrine, no effect on phosphorylase activity was detected. The latter result is consistent with the results of Garrison and colleagues (*140*). Therefore, the pathway stimulated by TPA is sufficient to cause glycogen synthase inactivation but insufficient to provoke phosphorylase activation (Fig. 3). The obvious extrapolation is that TPA activates the protein kinase C, Ca^{2+}-independent pathway discussed previously, and that this does not lead, at least on its own, to phosphorylase activation. The fact that TPA does not activate phosphorylase argues against stimulation of either cyclic AMP- or Ca^{2+}-mediated controls. It is not completely clear as to the relative importance of the Ca^{2+} and protein kinase C pathways in controlling liver glycogen synthase. My opinion is that the Ca^{2+}-independent mechanism is of particular significance. The operation of separate regulatory links to glycogen synthase and phosphorylase might then correlate with the different sensitivities of the enzymes to epinephrine stimulation.

3. *Mechanisms of Controlling Phosphorylation*

Analysis of the phosphorylation of glycogen synthase from hepatocytes stimulated with either vasopressin or epinephrine has not enabled conclusive identifi-

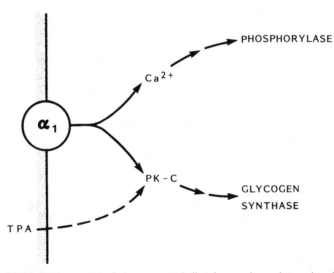

FIG. 3. Model for the control of glycogen metabolism by α_1-adrenergic agonists. The model proposes separable Ca^{2+}-dependent and Ca^{2+}-independent controls of glycogen phosphorylase and glycogen synthase, respectively. PK-C, protein kinase C; TPA, tetradecanoyl phorbol acetate. Based on the results of Roach and Goldman (*144*) and formulations such as by Nishizuka (*137*).

cation of the second messenger system involved. Increased phosphate was associated with both CB-1 and CB-2. Phosphorylase kinase, calmodulin-dependent protein kinase, and protein kinase-C are all specific for CB-1 (Fig. 1). Therefore, whether a Ca^{2+}- or protein kinase C-dependent pathway is involved, the phosphorylation of glycogen synthase cannot be explained fully by the direct action of any of the three protein kinases listed. The situation is analogous to that obtained with glucagon (Section VI,B) and similar arguments can be advanced. If a Ca^{2+}- or protein kinase C-mediated control is operative, there must be either an indirect activation of protein kinases acting on CB-2 or else inhibition of glycogen synthase phosphatase(s).

No phosphatase or phosphatase regulatory protein with appropriate properties is known. The one Ca^{2+}-sensitive phosphatase that has been characterized is *activated* by Ca^{2+} (*51, 145*). Miller *et al.* (*122*), in studies of perfused rat liver, reported that vasopressin caused a reduction in glycogen synthase phosphatase activity that they attributed to the production of phosphorylase *a*. Strickland *et al.* (*146*), using rat liver extracts, reached the conclusion that Ca^{2+}-stimulated inactivation of glycogen synthase was mediated primarily through formation of phosphorylase *a* and consequent inhibition of glycogen synthase phosphatase. Both studies invoke, therefore, the mechanism of Stalmans (*98*) and Hers (*97*), whereby phosphorylase *a* levels determine glycogen synthase phosphatase activity. Since phosphorylase activation is Ca^{2+}-mediated, the above proposals could only account for a Ca^{2+}-dependent control of glycogen synthase and cannot explain the inactivation of glycogen synthase by TPA (*144*).

The major alternative to explain hormone-induced phosphorylation of CB-2 is by the activation of appropriate protein kinases. For Ca^{2+}-mediated control, a link is required between phosphorylase kinase or the calmodulin-dependent protein kinase and one or more of the CB-2 kinases, F_A/GSK-3, casein kinase I, or casein kinase II. For the diacylglycerol pathway, protein kinase C would need to regulate one or more CB-2 kinase(s). No such regulatory connections have been described.

The mechanism of control of glycogen synthase by α_1-adrenergic agonists, vasopressin, and angiotensin II is not resolved. Compared with the analogous discussion for glucagon, an important difference is that identification of the intracellular messenger is less secure. While Ca^{2+} and especially diacylglycerol are the obvious candidates, it is certainly not impossible that some other of the compounds generated by receptor stimulation might function in protein kinase or phosphatase control.

D. GLUCOSE AND INSULIN

1. *Effects on Glycogen Synthase*

Elevated circulating levels of insulin and glucose promote hepatic glucogen synthesis (*97, 98*). There is reasonable accord that exposure of liver cells to glucose alone results in decreased phosphorylase activity and increased glycogen synthase activity (77, 80, 147–150). Van de Werve and Jeanrenaud (151), however, reported that feeding of starved rats permitted glycogen synthesis without activation (i.e., interconversion) of glycogen synthase. Most studies have been able to document activation, usually from measurements of glycogen synthase activity ratio. Thomas et al. (80) correlated the altered glycogen synthase activity caused by exposure of rat hepatocytes to glucose with a 3-fold increase in V_{max} and only small effects on $S_{0.5}$ and $M_{0.5}$ for glucose-6-P. The changes in V_{max} are consistent with the effects of phosphorylation in vitro of the purified rabbit liver enzyme (Section IV,B). Direct correlation of the activation of rat hepatocyte glycogen synthase by glucose with decreases in its phosphorylation state are limited to the work of Tan and Nuttall (26). These authors describe an experiment in which exposure of ^{32}P-labeled hepatocytes to glucose increased the activity ratio from 0.10 to 0.53 with a concomitant reduction in the glycogen synthase ^{32}P-content by 60%.

Direct and independent effects of insulin to activate liver glycogen synthase, though documented in several studies (77, 80, 108, 148, 152–154), have tended to be small. Though insulin alone can influence glycogen synthase, maximal activation is seen in the presence of glucose (77, 80, 148). Again, most experiments have monitored some form of activity ratio. Ciudad et al. (155) measured decreased $S_{0.5}$ and $M_{0.5}$ for glucose-6-P for glycogen synthase from hepatocytes exposed to insulin in the absence of glucose. The $S_{0.5}$ values were rather higher than those of purified rabbit liver enzyme (see also Section II) and those determined by Thomas et al. (80) in hepatocyte extracts. Thomas et al. (80) found that insulin treatment caused some 64% increase in V_{max} and a small decrease in $S_{0.5}$. No effect on the $M_{0.5}$ for glucose-6-P was observed. No study of insulin action on liver glycogen synthase has directly monitored changes in enzyme phosphorylation, obviously an important parameter to measure in the future.

2. Mechanism

How glucose and insulin regulate liver glycogen synthase activity is of considerable importance to hepatic metabolism. Most investigators would agree that dephosphorylation of glycogen synthase is important in such regulation. There is, in my view, no unequivocal mechanism to explain how such dephosphorylation comes about.

An important hypothesis for the mechanism of glucose control of hepatic glycogen synthesis was formulated by Hers, Stalmans and colleagues (97, 98, 119, 156, 157). Mechanistically, the proposal is based on two sets of molecular interactions: (a) that glucose binds to phosphorylase a rendering the protein a better substrate for its phosphatase; and (b) that phosphorylase a (but not phosphorylase b) is a potent inhibitor of glycogen synthase phosphatase. These two aspects must be clearly distinguished. The proposed regulatory sequence is that glucose binding to phosphorylase would promote its dephosphorylation (and inactivation) so that ultimately the decreased level of phosphorylase *a* would, by relieving inhibition of glycogen synthase phosphatase, permit dephosphorylation (and activation) of glycogen synthase. In essence, the model involves a coupling between the phosphorylation states of glycogen synthase and phosphorylase. Note also that the second mechanistic postulate (i.e., phosphorylase *a* inhibition of glycogen synthase phosphatase) can be invoked to explain glycogen synthase inactivation by glycogenolytic hormones (Sections VI,B and VI,C).

An important part of the evidence cited in support of the hypothesis (*97, 98, 158*) derives from experiments in which glucose causes sequential inactivation of phosphorylase and activation of glyocgen synthase, whether in whole animals (*156*), isolated hepatocytes (*147, 149*), or cell-free systems (*119*). In other words, glycogen synthase activation occurs only after a lag which is interpreted as the period during which the phosphorylase *a* concentration, though falling, remains high enough to suppress glycogen synthase phosphatase. Hers (*97, 157*) and Stalmans (*98*), in fact conceive of a specific threshold value of phosphorylase *a* and only below this level would glycogen synthase dephosphorylation be enabled. The existence of a precise threshold is not, in my opinion, necessarily required by the mechanistic postulates, at least as stated above.

The interpretation of the lag in glycogen synthase activation has been much discussed and sometimes questioned (*101, 158*). Insofar as phosphorylase and glycogen synthase activities both vary in response to a single experimental perturbation (addition of glucose) it is not possible to infer from the temporal sequence alone a strict cause and effect relationship between the behaviors of the two target enzymes. It is also important to address situations in which the predicted correlations are not observed [see also Refs. (*101, 159*)]. One example, already cited, is the ability of the phorbol ester, TPA, to inactivate glycogen synthase in hepatocytes from starved rats without affecting phosphorylase (*144*). Another important study was that of Watts *et al.* (*159*) who examined hepatocytes from *gsd/gsd* rats which, due to defective liver phosphorylase kinase, contain very low levels of phosphorylase *a*. After exposure of the cells to glucose, the kinetics of glycogen synthase activation were not greatly altered as compared to control cells in which 7–8 times more phosphorylase *a* was present. The temporal lag in glycogen synthase activation was observed in both cases. There are certainly many examples in which the kinetics of the changes in glycogen synthase and phosphorylase activities are consistent with the Hers–Stalmans hypothesis but there are also instances where the predicted behavior is not observed. This does not disprove the hypothesis but implies that, if true, its operation is not universal.

Other experiments have addressed the molecular interactions required of the model. An effect of glucose to stimulate the conversion of phosphorylase *a* to phosphorylase *b* has been known for some time and, furthermore, the site of interaction was proposed to be the protein substrate rather than the phosphatase (*156, 160–163*). A molecular rationalization of these results came from X-ray diffraction studies of the muscle isozyme of phosphorylase (*164, 165*). A glucose binding site was identified and its occupancy caused an alteration in the conformation of the N-terminal region of the subunit such that the phosphorylated serine, residue 14, became more exposed. Though one has to extrapolate to the *liver* isozyme of phosphorylase and its interaction with the appropriate *liver* phosphatase, there is a reasonable basis to propose that a substrate-directed interaction of glucose can account for phosphorylase activation in liver cells.

The second molecular interaction of the model, inhibition of glycogen synthase phosphatase by phosphorylase *a,* is less well defined. First, there is no clear consensus (a) as to whether in liver glycogen synthase and phosphorylase are dephosphorylated by the same or different phosphatases and (b) as to the native structure and properties of the glycogen synthase phosphatase(s) (Section III,B). Much of the direct evidence for the proposed inhibition, therefore, comes from experiments with crude extracts (*119, 121, 146*). Added phosphorylase *a* inhibited glycogen synthase activation whereas treatment of extracts with antibodies to phosphorylase led to increased glycogen synthase activation. Using a highly purified low M_r phosphatase, Killilea *et al.* (*166*) have shown competition between phosphorylase and glycogen synthase as substrates for the phosphatase. If, in the cell, both glycogen synthase and phosphorylase are recognized by a common phosphatase then, depending on reactant concentrations and enzyme kinetic properties, a degree of competition would be anticipated. Whether such an occurrence would be sufficient to dictate the strict coupling between glycogen synthase and phosphorylase dephosphorylation implied in the Hers–Stalmans model is in my opinion hard to say. An alternative, that would provide compelling support for the spirit of the hypothesis of Hers and Stalmans, would be if a glycogen synthase phosphatase (whether the same as or different from phosphorylase phosphatase) contained a specific *regulatory* site recognized by phosphorylase *a.* Ultimately these questions should be answered using well-defined enzymes purified from liver. The main obstacle is the identification and characterization of native hepatic glycogen synthase phosphatase(s). Many aspects of the model proposed by Hers and Stalmans are attractive as a control mechanism, and the model has provoked much useful attention to the problem. In my opinion, unequivocal proof will have to await molecular definition of the glycogen synthase phosphatase that is central to the concept of coupling between glycogen synthase and phosphorylase.

If the dephosphorylation of glycogen synthase caused by glucose action is not mediated by phosphorylase *a,* then an alternate control mechanism would have to be invoked, and another interaction of glucose or a metabolite identified. No effects of glucose to cause significant inhibition of any glycogen synthase kinase has been found. Some other possibilities for phosphatase activation have been presented. Nuttall and colleagues (*101, 167, 168*) have reported that ATP inhibits glycogen synthase phosphatase in glycogen pellets and that glucose can counteract this inhibition. Other work, though not primarily of liver systems, has shown that glucose-6-P can stimulate glycogen synthase dephosphorylation, presumably through interaction with the glycogen synthase substrate (*158, 169–172*). The effect was in fact also found with a highly purified low M_r phosphatase preparation by Killilea *et al.* (*166*). However, as noted in regard to the Hers–Stalmans mechanism, these observations will need to be substantiated with defined native forms of hepatic glycogen synthase phosphatases, but a role for glucose-mediated or glucose-6-P mediated control of phosphatase, independent of phosphorylase *a,* cannot be excluded.

The detailed mechanism of the acute effects of insulin is as obscure for liver as for other tissues. In fact, though glycogen synthase has long been an important intracellular marker of insulin action (*173*), liver glycogen synthase has not, overall, been especially illuminating of short-term regulation by insulin. Part of the problem has been the relatively small changes in glycogen synthase obtained in isolated systems and the difficulty in disentangling insulin- and glucose-dependent controls. For direct effects of insulin on glycogen synthase activity, and by inference phosphorylation state, regulation of protein kinases and/or phosphatases must be sought just as in evaluating other hormonal controls. Direct measures of enzyme phosphorylation are not, however, available as a guide.

Numerous proposals have been made to explain the mechanism of insulin action [see Ref. (*174*)]. Much attention has been directed to two models for the events immediately following insulin binding to its receptor. One, deriving originally from the work of Larner and colleagues (*175*) proposes the generation of specific intracellular mediators that would then interact with specific cellular targets such as protein kinases and phosphatases. Activation of phosphatase and inhibition of cyclic AMP-dependent protein kinase have been reported (*176*). The second model follows from the recognition that the insulin receptor has associated a tyrosine protein kinase whose activity is stimulated by insulin binding [see Chapter 7 (*177*)]. Functional intracellular targets for the tyrosine kinase have not been identified but, for the present discussion, one could envision the phosphorylation and control of glycogen synthase kinase or phosphatase. No such interactions have been proved in liver or nonliver systems (see also Cohen, Chapter 11). Unequivocal identification of the protein kinases and/or phosphatases that mediate the short-term activation of liver glycogen synthase by insulin remains to be made.

VII. Conclusion

The role of covalent phosphorylation in mediating the regulation of liver glycogen synthase is firmly established. Exploration of the molecular mechanisms whereby glucose and hormones control that phosphorylation is, in my opinion, only just beginning. Some may disagree with this appraisal but I do not believe that we know enough to write a complete and incontrovertible regulatory sequence for any of the controls discussed in this chapter. It is sufficient to examine the case of glucagon for which the "classical" cyclic AMP → cyclic AMP-dependent protein kinase → glycogen synthase scheme was expected. Instead, as for muscle, a more complex mechanism must be operating. Part of the overall problem is uncertainty as to how some of the hormones, such as insulin and the Ca^{2+}-mobilizing glycogenolytic agents, function, obviously not a problem specific to glycogen synthase regulation. Much of the difficulty, however, is imperfect definition of various intracellular enzymes including some of the protein kinases, the protein phosphatases, and liver glycogen synthase itself. Progress is being made and, I think, significant advances are to be expected in the coming few years. Only with much more sophisticated knowledge of glycogen synthase properties and control are we likely to understand the evolutionary rationale for the existence of tissue-specific isozymes of this complex regulatory enzyme.

Acknowledgments

Work from the author's laboratory described in this chapter has been supported by research grants from the National Institutes of Health, AM27221 and AM27240, from the Indiana Affiliate of the American Heart Association, and from the Grace M. Showalter Foundation, for which extreme gratitude is expressed in these uncertain times. Likewise, receipt of Research Career Development Award AM01089 from the National Institutes of Health is gratefully acknowledged. I thank Mary Jo Coffin for typing the manuscript.

References

1. Foster, D. W. (1984). *Diabetes* **33,** 1188–1199.
2. Katz, J., and McGarry, J. D. (1984). *J. Clin. Invest.* **74,** 1901–1909.
3. Cohen, P. (1986). This volume, Chapter 11.
4. Larner, J., and Villar-Palasi, C. (1971). *Curr. Top Cell. Regul.* **3,** 195–236.
5. Stalmans, W., and Hers, H. G. (1973). "The Enzymes," 3rd ed., Vol. 9, pp. 309–361.
6. Roach, P. J., and Larner, J. (1977). *Mol. Cell Biochem.* **15,** 179–200.
7. Lin, D. C., and Segal, H. L. (1973). *JBC* **248,** 7007–7011.
8. McVerry, P. H., and Kim, K.-H. (1974). *Biochemistry* **13,** 3505–3511.
9. Jett, M. F., and Soderling, T. R. (1979). *JBC* **254,** 6739–6745.
10. Bahnak, B. R., and Gold, A. H. (1982). *ABB* **213,** 492–503.
11. Huang, K.-P., Akatsuka, A., Singh, T. J., and Blake, K. R. (1983). *JBC* **258,** 7094–7101.
12. Imazu, M., Strickland, W. G., Chrisman, T. D., and Exton, J. H. (1984). *JBC* **259,** 1813–1821.
13. Ciudad, C., Camici, M., Ahmad, Z., Wang, Y., DePaoli-Roach, A. A., and Roach, P. J. (1984). *EJB* **142,** 511–520.

14. Killilea, D., and Whelan, W. J. (1976). *Biochemistry* **15,** 1349–1356.
15. Camici, M., DePaoli-Roach, A. A., and Roach, P. J. (1982). *JBC* **257,** 9898–9901.
16. Camici, M., DePaoli-Roach, A. A., and Roach, P. J. (1984). *JBC* **259,** 3429–3434.
17. Tan, A. W. H., and Nuttall, F. Q. (1983). *JBC* **258,** 9624–9630.
18. Steiner, D. F., Younger, L., and King, J. (1965). *Biochemistry* **4,** 740–751.
19. Camici, M., this laboratory, unpublished results.
20. DePaoli-Roach, A. A., Ahmad, Z., Camici, M., Lawrence, J. C., Jr., and Roach, P. J. (1983). *JBC* **258,** 10702–10709.
21. Ciudad, C., DePaoli-Roach, A. A., and Roach, P. J. (1984). *BBA* **804,** 261–263.
22. Akatsuka, A., Singh, T. J., and Huang, K.-P. (1984). *ABB* **235,** 186–195.
23. Wang, Y., Camici, M., Lee, F.-T., Ahmad, Z., DePaoli-Roach, A. A., and Roach, P. J. (1985). *FP* **44,** 1073.
24. Ariño, J., Mor, A., Bosch, F., Baanante, I. V., and Guinovart, J. J. (1984). *FEBS Lett.* **170,** 310–314.
25. Akatsuka, A., Singh, T. J., Nakabayashi, H., Lin, M. C., and Huang, K.-P. (1985). *JBC* **260,** 3239–3242.
26. Tan, A. W. H., and Nuttall, F. Q. (1985). *JBC* **260,** 4751–4757.
27. Bahnak, B. R., and Gold, A. H. (1983). *BBRC* **117,** 332–339.
28. Rulfs, J., Wolleben, C. D., Miller, T. B., and Johnson, G. L. (1985). *JBC* **260,** 1203–1207.
29. McVerry, P., and Kim, K.-H. (1972). *BBRC* **48,** 1636–1640.
30. Sanada, Y., and Segal, H. L. (1971). *BBRC* **45,** 1159–1168.
31. Roach, P. J. (1981). *Curr. Top. Cell Regul.* **20,** 45–105.
32. Roach, P. J. (1984). "Methods in Enzymology," Vol. 107, pp. 81–101.
33. Beebe, S. J., and Corbin, J. D. (1986). This volume, Chapter 3.
34. Pickett-Gies, C. A., and Walsh, D. A. (1986). This volume, Chapter 10.
35. Kikkawa, U., and Nishizuka, Y. (1986). This volume, Chapter 5.
36. Stull, J., Nunnaly, M. H., and Micknoff, C. H. (1986). This volume, Chapter 4.
37. Hathaway, G. M., and Traugh, J. A. (1982). *Curr. Top. Cell. Regul.* **21,** 101–127.
38. Ahmad, Z., DePaoli-Roach, A. A., and Roach, P. J. (1985). *FEBS Lett.* **179,** 96–100.
39. DePaoli-Roach, A. A., Roach, P. J., and Larner, J. (1979). *JBC* **254,** 12062–12068.
40. Chrisman, T. D., Jordan, J. E., and Exton, J. H. (1982). *JBC* **257,** 10798–10804.
41. Payne, M. E., and Soderling, T. R. (1980). *JBC* **259,** 8054–8056.
42. Ahmad, Z., DePaoli-Roach, A. A., and Roach, P. J. (1982). *JBC* **257,** 8348–8355.
43. Payne, E. M., Schworer, C. M., and Soderling, T. R. (1983). *JBC* **258,** 2376–2382.
44. Vanderheede, J. R., Yang, S.-D., Goris, J., and Merlevede, W. (1980). *JBC* **255,** 11768–11774.
45. Yang, S.-D., Vandenheede, J. R., Goris, J., and Merlevede, W. (1980). *JBC* **255,** 11759–11767.
46. Sheorain, V. S., Ramakrishna, S., Benjamin, W., and Soderling, T. R. (1985). *JBC* **260,** 12287–12292.
47. Hegazy, M., Brown, D., Schlender, K. K., and Reimann, E. (1985). *FP* **44,** 1073.
48. Ramakrishna, S., and Benjamin, W. B. (1985). *FP* **44,** 704.
49. Ahmad, Z., Lee, F.-T., DePaoli-Roach, A. A., and Roach, P. J. (1985). *13th IUB Meet.,* p. 728.
50. Fischer, E. H. (1986). This volume, Chapter 8.
51. Ingebritsen, T. S., and Cohen, P. (1983). *EJB* **132,** 255–261.
52. Ingebritsen, T. S., Foulkes, J. G., and Cohen, P. (1980). *FEBS Lett.* **119,** 9–15.
53. Ingebritsen, T. S., Stewart, A. A., and Cohen, P. (1983). *EJB* **132,** 297–307.
54. Goris, J., Doperé, F., Vandenheede, J. R., and Merlevede, W. (1980). *FEBS Lett.* **117,** 117–121.

536   PETER J. ROACH

55. Roach, P., Roach, P. J., and DePaoli-Roach, A. A. (1985). *JBC* **260**, 6314–6317.
56. Brandt, H., Capulong, Z. L., and Lee, E. Y. C. (1975). *JBC* **250**, 8038–8044.
57. Lee, E. Y. C., Silberman, S. R., Ganapathi, M. K., Petrovic, S., and Paris, H. (1980). *Adv. Cyclic Nucleotide Res.* **13**, 95–131.
58. Alemany, S., Tung, H. Y. L., Shenolikar, S., Pilkis, S. J., and Cohen, P. (1984). *EJB* **145**, 51–56.
59. Tamura, S., Kikuchi, H., Kikuchi, K., Hiraga, A., and Tsuiki, S. (1980). *EJB* **104**, 347–355.
60. Tamura, S., and Tsuiki, S. (1980). *EJB* **114**, 216–224.
61. Mellgren, R., and Schlender, K. K. (1983). *BBRC* **117**, 501–508.
62. Di Salvo, J., Wallkens, E., Gifford, D., Goris, J., and Merlevede, W. (1983). *BBRC* **117**, 493–500.
63. DePaoli-Roach, A. A., and Klingberg, D. (1985). In "Advances in Protein Phosphatases" (W. Merlevede and J. DiSalvo, eds.), Vol. 2, p. 396. Leuven Univ. Press, Belgium.
64. Ingebritsen, T. S., Foulkes, J. G., and Cohen, P. (1983). *EJB* **132**, 263–764.
65. Kikuchi, K., Tamura, S., Hiraga, A., and Tsuiki, S. (1977). *BBRC* **75**, 29–37.
66. Hiraga, A., Kikuchi, K., Tamura, S., and Tsuiki, S. (1981). *EJB* **119**, 503–510.
67. Mvumbi, L., Doperé, F., and Stalmans, W. (1983). *BJ* **212**, 407–416.
68. Gilboe, D. P., and Nuttall, F. Q. (1984). *ABB* **228**, 587–591.
69. MacKenzie, C. W., Bulbulian, G. J., and Bishop, J. S. (1980). *BBA* **614**, 413–424.
70. Takeda, Y., Brewer, H. B., Jr., and Larner, J. (1975). *JBC* **250**, 8943–8950.
71. Camici, M., Ahmad, Z., DePaoli-Roach, A. A., and Roach, P. J. (1984). *JBC* **259**, 2466–2473.
72. Wang, Y., Bell, A. W., Hermodson, M. A., and Roach, P. J. (1986). ASBC Meeting, Washington, D.C.
73. Imazu, M., Strickland, W. G., and Exton, J. H. (1984). *BBA* **789**, 285–293.
74. Akatsuka, A., Singh, T. J., and Huang, K.-P. (1984). *JBC* **259**, 7878–7883.
75. Kaslow, H. R., Eichner, R. D., and Mayer, S. E. (1979). *JBC* **254**, 4674–4677.
76. Guinovart, J. J., Salavert, A., Massague, J., Ciudad, C., Salsas, E., and Itarte, E. (1979). *FEBS Lett.* **106**, 284–288.
77. Strickland, W. G., Blackmore, P. F., and Exton, J. H. (1980). *Diabetes* **29**, 617–622.
78. Van Patten, S. M., and Walsh, D. A. (1980). *Anal. Biochem.* **109**, 432–436.
79. Kochan, R. G., Lamb, D. R., Reimann, E. M., and Schlender, K. K. (1981). *Am. J. Physiol.* **240**, E197–202.
80. Thomas, A. P., Martin-Reguero, A., and Williamson, J. R. (1985). *JBC* **260**, 5963–5973.
81. Juhl, H. (1981). *Mol. Cell. Biochem.* **35**, 77–92.
82. Sølling, H., and Esmann, V. (1977). *EJB* **81**, 119–128.
83. Hiken, J. F., and Lawrence, J. C., Jr. (1985). *ABB* **236**, 59–71.
84. Issa, H. A., and Mendicino, J. (1973). *JBC* **248**, 685–696.
85. Miller, R. E., Miller, E. A., Fredholm, B., Yellin, J. B., Eichner, R. D., Mayer, S. E., and Steinberg, D. (1975). *Biochemistry* **14**, 2481–2488.
86. Mitchell, J. W., Mellgren, R. L., and Thomas, J. A. (1980). *JBC* **255**, 10368–10374.
87. Sivaramakrishnan, S., High, C. W., and Walsh, D. A. (1982). *ABB* **214**, 311–325.
88. Huang, K.-P., and Cabib, E. (1974). *JBC* **249**, 3851–3857.
89. Takahara, H., and Matsuda, K. (1979). *J. Biochem. (Tokyo)* **85**, 907–914.
90. Fox, J., Kawaguchi, K., Greenberg, E., and Preiss, J. (1976). *Biochemistry* **15**, 849–857.
91. Preiss, J., and Walsh, D. A. (1981). In "Biology of Carbohydrates" (V. Ginsburg, and P. Robbins, eds.), Vol. 1, pp. 199–314. Wiley, New York.
92. Nimmo, H. G., Proud, C. G., and Cohen, P. (1976). *EJB* **68**, 21–30.
93. Lee, F.-T., Ahmad, Z., DePaoli-Roach, A. A., and Roach, P. J. (1985). *FP* **44**, 682.
94. Lawrence, J. C., Jr., Hiken, J. F., DePaoli-Roach, A. A., and Roach, P. J. (1983). *JBC* **258**, 10710–10719.

95. Kaslow, H. R., and Lesikar, D. D. (1984). *FEBS Lett.* **172**, 294–298.
96. Wititsuwannakul, D., and Kim, K.-H. (1979). *JBC* **254**, 3562–3569.
97. Hers, H. G. (1976). *Annu. Rev. Biochem.* **42**, 167–189.
98. Stalmans, W. (1976). *Curr. Top. Cell. Regul.* **11**, 51–97.
99. Hems, D. A., and Whitton, P. D. (1980). *Physiol. Rev.* **60**, 1–50.
100. Exton, J. H. (1980). *Am. J. Physiol.* **238**, E3–E12.
101. Nuttall, F. Q., Gilboe, D. P., Tan, A. W. H., Doorneweerd, D. D., Theen, J. W., Gannon, M. C., and Chou, B. B. (1981). *In* "The Regulation of Carbohydrate Formation and Utilization in Mammals" (C. M. Veneziale, ed.) pp. 315–343. University Park Press, Baltimore, Maryland.
102. Williamson, J. R., Cooper, R. H., and Hoek, J. B. (1981). *BBA* **639**, 243–295.
103. Hue, L., and Van de Werve, G. (1981). "Short-term Regulation of Liver Metabolism." Elsevier/North-Holland Biomedical Press, Amsterdam.
104. Exton, J. H., Chrisman, T. D., Strickland, W. G., Prpic, V., and Blackmore, P. F. (1983). *In* "Isolation and Use of Hepatocytes" (R. A. Harris and N. W. Cornell, eds.), pp. 401–410. Elsevier-Biomedical, New York.
105. Williamson, J. R., Joseph, S. K., Thomas, A. P., Coll, K. E., and Marks, J. S. (1983). *In* "Isolation and Use of Hepatocytes" (R. A. Harris and N. W. Cornell, eds.), pp. 419–432. Elsevier-Biomedical, New York.
106. Stalmans, W. (1983). *Hand. Exp. Pharmak.* **66**, Part 1, 291–313.
107. Hutsen, N. J., Brumley, F. T., Assimacopoulos, F. D., Harper, S. C., and Exton, J. H. (1976). *JBC* **251**, 5200–5208.
108. Massagué, J., and Guinovart, J. J. (1977). *FEBS Lett.* **82**, 317–320.
109. Hue, L., Feliu, J. E., and Hers, H. G. (1978). *BJ* **176**, 791–797.
110. Garrison, J. C., Borland, M. K., Florio, V. A., and Twible, D. A. (1979). *JBC* **254**, 7147–7156.
111. Bosch, F., Ciudad, C. J., and Guinovart, J. J. (1983). *FEBS Lett.* **151**, 76–78.
112. Garrison, J. C. (1978). *JBC* **253**, 7091–7100.
113. Garrison, J. C., Borland, M. K., Moylan, R. D., and Ballard, B. J. (1981). *In* "Protein Phosphorylation" (O. M. Rosen and E. G. Krebs, eds.), pp. 529–545. Cold Spring Harbor Lab., Cold Spring Harbor, New York.
114. Rothermal, J. D., Perillo, N. L., Marks, J. S., and Botelho, L. H. P. (1984). *JBC* **259**, 15294–15300.
115. deWit, R. J., Hoppe, J., Stec, W. J., Baraniak, J., and Jastorff, B. (1982). *EJB* **112**, 95–99.
116. O'Brian, C. A., Roczniak, S. O., Bramson, H. N., Baraniak, J., Stec, W. J., and Kaiser, E. T. (1982). *Biochemistry* **21**, 4371–4376.
117. Goris, J., DeFreyn, G., Vandenheede, J. R., and Merlevede, W. (1978). *EJB* **91**, 457–464.
118. Khandelwal, R. L., and Zinman, S. M. (1978). *JBC* **253**, 560–565.
119. Stalmans, W., DeWulf, H., and Hers, H. G. (1971). *EJB* **18**, 582–587.
120. Gilboe, D. P., and Nuttall, F. Q. (1978). *JBC* **2253**, 4078–4081.
121. Nuttall, F. Q., and Gilboe, D. P. (1980). *ABB* **203**, 483–486.
122. Miller, T. B., Jr., Garnache, A., and Vicalvi, J. J., Jr. (1981). *JBC* **256**, 2851–2855.
123. Massagué, J., and Guinovart, J. J. (1978). *BBA* **543**, 269–272.
124. Blair, J. B., James, M. E., and Foster, J. L. (1979). *JBC* **254**, 7579–7584.
125. Blair, J. B., James, M. E., and Foster, J. L. (1979). *JBC* **254**, 7585–7590.
126. Morgan, N. G., Blackmore, P. F., and Exton, J. H. (1983). *JBC* **258**, 5103–5109.
127. Studer, R. K., and Borle, A. B. (1982). *JBC* **257**, 7987–7993.
128. Chan, T. M., Blackmore, P. F., Steiner, K. E., and Exton, J. H. (1979). *JBC* **254**, 2428–2433.
129. Malbon, C. C., Li, S., and Fain, J. N. (1978). *JBC* **253**, 8820–8825.
130. Itoh, H., Okajima, F., and Ui, M. (1984). *JBC* **259**, 15464–15473.

131. Ananth, S., DePaoli-Roach, A. A., and Roach, P. J. (1983). *In* "Isolation and Use of Hepatocytes" (R. A. Harris and N. W. Cornell, eds.), pp. 591–595. Elsevier-Biomedical, New York.

132. Aggerbeck, M., Guellan, G., and Hanoune, J. (1980). *Biochem. Pharmacol.* **29**, 643–645.

133. Roach, P. J., and Goldman, M., unpublished results.

134. Michell, R. H., Kirk, C. J., Jones, L. M., Downes, C. P., and Creba, J. A. (1981). *Philos. Trans. R. Soc. London, Ser. B* **296**, 123–138.

135. Creba, J. A., Downes, C. P., Hawkins, P. T., Brewster, G., Michell, R. H., and Kirk, C. J. (1983). *BJ* **212**, 733–747.

136. Williamson, J. R., Cooper, R. H., Joseph, S. K., and Thomas, A. P. (1985). *Am. J. Physiol.* **248**, C203–C216.

137. Nishizuka, Y. (1983). *Trends Biochem. Sci.* **8**, 13–16.

138. Joseph, S. K., Thomas, A. P., Williams, R. J., Irvine, R. F., Berridge, M. J., and Shultz, I. (1983). *Nature (London)* **306**, 67–69.

140. Garrison, J. C., Johnson, D. E., and Campanile, C. P. (1984). *JBC* **259**, 3283–3292.

141. Castagna, M., Takai, Y., Kaibuchi, K., Sano, K., Kikkawa, U., and Nishizuka, Y. (1982). *JBC* **257**, 7847–7851.

142. Malthus, R., Clark, D. G., Watts, C., and Sneyd, J. G. T. (1980). *BJ* **188**, 99–106.

143. Blackmore, P. F., and Exton, J. H. (1981). *BJ* **198**, 379–383.

144. Roach, P. J., and Goldman, M. (1983). *PNAS* **80**, 7170–7172.

145. Stewart, A. A., Ingebritsen, T. S., Manalan, A., Klee, C. B., and Cohen, P. (1982). *FEBS Lett.* **137**, 80–84.

146. Strickland, W. G., Imazu, M., Chrisman, T. D., and Exton, J. H. (1983). *JBC* **258**, 5490–5497.

147. Hue, L., Bontemps, F., and Hers, H. G. (1975). *BJ* **152**, 105–1144.

148. Witters, L. A., and Avruch, J. (1978). *Biochemistry* **17**, 406–410.

149. Kasvinsky, P. J., Fletterick, R. J., and Madsen, N. B. (1981). *Can. J. Biochem.* **59**, 387–395.

150. Nuttall, F. Q., Theen, J. W., Niewoehner, C., and Gilboe, D. P. (1983). *Am. J. Physiol.* **245**, E521–E527.

151. Van de Werve, G., and Jeanrenaud, B. (1984). *Am. J. Physiol.* **247**, E271–E275.

152. Miller, T. B., Jr., and Larner, J. (1973). *JBC* **248**, 3483–3488.

153. Van de Werve, G., Stalmans, W., and Hers, H. G. (1977). *BJ* **162**, 143–146.

154. Massagué, J., and Guinovart, J. J. (1978). *BBA* **543**, 269–272.

155. Ciudad, C. J., Bosch, F., and Guinovart, J. J. (1981). *FEBS Lett.* **129**, 123–126.

156. Stalmans, N., DeWulf, H., Hue, L., and Hers, H. G. (1974). *EJB* **41**, 127–134.

157. Hers, H. G. (1981). *In* "Short-Term Regulation of Liver Metabolism" (L. Hue and G. Van de Werve, eds.), pp. 105–117. Elsevier/North-Holland Biomedical Press, Amsterdam.

158. Curnow, R. T., and Larner, J. (1979). *Biochem. Actions Horm.* **6**, 77–119.

159. Watts, C., Redshaw, J. R., and Gain, K. R. (1982). *FEBS Lett.* **144**, 231–234.

160. Holmes, P. A., and Mansour, T. E. (1968). *BBA* **156**, 275–284.

161. Bailey, J. M., and Whelan, W. J. (1972). *BBRC* **46**, 191–197.

162. Martensen, T. M., Brotherton, J. E., and Graves, D. J. (1973). *JBC* **248**, 8329–8336.

163. Stalmans, W., Laloux, M., and Hers, H. G. (1974). *EJB* **49**, 415–427.

164. Madsen, N. B., Kasvinsky, P. J., and Fletterick, A. J. (1978). *JBC* **253**, 9097–9101.

165. Fletterick, R. J., and Madsen, N. B. (1980). *Annu. Rev. Biochem.* **49**, 31–61.

166. Killilea, S. D., Brandt, H., Lee, E. Y. C., and Whelan, W. J. (1976). *JBC* **251**, 2363–2368.

167. Gilboe, D. P., and Nuttall, F. Q. (1973). *BBRC* **53**, 164–171.

168. Gilboe, D. P., and Nuttall, F. Q. (1974). *BBA* **338**, 57–67.

169. Hizukuri, S., and Takeda, Y. (1970). *BBA* **211**, 179–181.

170. Kato, K., and Bishop, J. (1972). *JBC* **247**, 7420–7429.

171. Nakai, C., and Thomas, J. A. (1974). *JBC* **249,** 6459–6467.
172. Lawrence, J. C., Jr., and Larner, J. (1978). *JBC* **253,** 2104–2113.
173. Villar-Palasi, C., and Larner, J. (1960). *BBA* **39,** 171–173.
174. Czech, M. P. (1977). *Annu. Rev. Biochem.* **46,** 359–384.
175. Larner, J., Huang, L. C., Brooker, G., Murad, F., and Miller, T. B. (1974). *FP* **33,** 261.
176. Larner, J., Galasko, G., Cheng, K., DePaoli-Roach, A. A., Huang, L., Daggy, P., and Kellogg, J. (1979). *Science* **206,** 1408–1410.
177. Kahn, R. (1985). This volume, Chapter 7.

Author Index

Numbers in parentheses are reference numbers and indicate that an author's work is referred to although the name is not cited in the text.

A

Aaronson, S. A., 214(185, 186, 187, 191), 215(191, 198), *242*
Abel-Ghany, M., 13(51), *19*
Abell, C. W., 86(308), *108*
Abel-Liatif, A. A., 171(33), *184*
Ackerman, P., 22(48), 23(48), *38*
Adachi, A.-M., 345(203), *360*
Adachi, K., 60(142), 93(385), 94(385), *104, 110, 120, 161*
Adelstein, R. S., 22(98), 24(98), 26(98), *40,* 73(254), *107,* 118(51), 119(73), 120(51, 78, 80, 87), 124, 125(51, 77, 111), 126(51, 80), 130(80), 131(51, 111, 148, 151), 132(51, 80, 151), 133(51, 151), 136(80), 137(87, 108, 164, 165), 138(94, 165, 166), 139(168), 154(51, 80), *160, 161, 162, 163,* 172(48), 179(134), 180(134), *185, 187,* 331(119, 125), 332(125, 127), 333(119, 125, 127), 334(125, 127), 335(125), 336(125), 337(125), 338(125), 346(119), 347(119, 207), 348(207), 349(207), *358, 360,* 445(170), *458*
Adkins, B., 200(72), 203(72), 232(72), *238*
Aggerbeck, M., 526(132), *537*
Agronoff, B. W., 171(35, 37), 172(37), 176(86), *184, 186*
Aguilera, G., 176(99), 177(99), *186*
Ahmad, Z., 142(183), 147(183), 148, 149(183), 150(183), 151(183, 218), 152(183), 153(183), 157(183), *164, 165,* 180(162), *188,* 266(216), 280(216),

283(216), *308,* 466(42), 467(51a, 52), 468(54, 57), 469(54, 83), *494, 495, 497,* 501(13), 502(13), 503(13, 20), 504(13, 23), 506(20), 507(38), 508(20, 42, 49), 511(71), 512(20), 514(71), 515(13, 23), 516(13, 20, 23, 71), 517(23, 71), 519(71, 93), 520(13), 522(13), 523(13), 524(13, 23), 525(13), 526(13), *534, 535, 536*
Aitken, A., 22(33), 23(33), *38,* 53(101), 59(134), *103, 104,* 156(261), *166,* 319(57), 322(81, 89), 323(95), 339(170), 341(170, 186), *356, 357, 359, 360,* 462(13, 14), 463(13, 14), 465(33), 466(13, 14, 46, 47), 467(33), 468(13, 14, 33, 47, 71), 470(33, 46), 471(46), 472(47, 90a), 473(90a), 474(109), *493, 494, 496*
Akanuma, Y., 265(228), 282(228), *309*
Akatsuka, A., 22(70, 112), 23(70), 24(112), *39, 40,* 133(153), *163,* 409(67), 415(90a, 94b), 421(67), 430(67, 149, 149a), 437(149, 149a), 444(149, 149a), 445(149a), *455, 456, 457,* 468(56, 70), 469(75–77), 471(70), *495,* 501(11), 502(11), 503(11), 504(11, 22, 25), 510(11), 515(11), 516(11), 517(11, 74), 522, 523, 524(11), *534, 535, 536*
Akers, R. F., 181(173, 178), *188*
Akhtar, R. A., 171(33), *184*
Akiyama, T., 22(60), 23(60), *39,* 209(150), 216(210), *240, 242,* 265(228), 282(228), *309*
Aksoy, M. O., 141(175), *163*
Alaba, J. O., 404(49), 409(49), *454*

541

D

Subject Index

A